Biochemistry, Cell and Molecular Biology, and Genetics

An Integrated Textbook

Zeynep Gromley, PhD
Associate Professor of Biochemistry and Chair of Molecular Sciences
DeBusk College of Osteopathic Medicine
Lincoln Memorial University
Harrogate, Tennessee, USA

Adam Gromley, PhD
Director of Research and Associate Professor of Cell/Molecular Biology
DeBusk College of Osteopathic Medicine
Lincoln Memorial University
Harrogate, Tennessee, USA

437 illustrations

Thieme
New York • Stuttgart • Delhi • Rio de Janeiro

Library of Congress Cataloging-in-Publication Data is available with the publisher.

© 2021. Thieme. All rights reserved.

Thieme Medical Publishers, Inc.
333 Seventh Avenue, 18th Floor
New York, NY 10001, USA
www.thieme.com
+1 800 782 3488, customerservice@thieme.com

Cover design: Thieme Publishing Group
Typesetting by DiTech Process Solutions, India

Printed in USA by King Printing Company, Inc

ISBN 978-1-62623-535-9

Also available as an e-book:
eISBN 978-1-62623-536-6

FSC
www.fsc.org
100%
Paper from well-managed forests
FSC® C103101

We dedicate this book to our son, Eren Gromley, and our daughter, Evangelina Gromley.

Zeynep and Adam Gromley

Contents

Contents

Contents

Contents

Contents

Preface

When the opportunity was presented to write a new textbook for students in the health care professions, we began by reviewing current offerings to identify any specific needs that have not yet been fulfilled. It quickly became clear to us that a textbook integrating biochemistry, genetics, and molecular and cell biology would be most beneficial, as these topics are commonly taught together in various curricula, but no textbook is currently available that integrates all of these subjects.

As we began to formulate our plans on how to develop this text, we consulted with colleagues regarding specific topics and areas of focus. In addition, we took into account students' suggestions with regards to the organization and presentation of material. The end result is a textbook that not only covers the most important aspects of biochemistry, genetics, and molecular and cellular biology, but also aligns these topics with other preclinical disciplines, such as pathology, pharmacology, microbiology, immunology, histology, and physiology, and attempts to bridge the gap between preclinical medical sciences and clinical medicine.

In addition to an integrative approach, our goal was to make the text as clear and concise as possible. Not only does this help the reader to grasp cellular processes, molecular mechanisms, and biochemical pathways more readily, but it also provides a quick reference for examinations. It also allows the textbook to be more portable for those who choose the printed format. We believe that we have created a balanced textbook by providing clear and to-the-point explanations for underlying molecular mechanisms of diseases.

Zeynep Gromley, PhD
Adam Gromley, PhD

Acknowledgment

We thank our former and current students and colleagues for providing constructive feedback about the content of our textbook. We also thank our contributors for their commitment to this project. Last but not least, we want to thank the Thieme staff for their assistance, continued encouragements, and support.

Zeynep Gromley, PhD
Adam Gromley, PhD

Contributors

Emine Ercikan Abali, PhD
Assistant Dean for Basic Science Curriculum
CUNY School of Medicine
New York, New York, USA

Somkenechi Ajene, MS, DO
General Surgery Resident
University of Texas Medical Branch
Galveston, Texas, USA

Alesia Cloutier, MS, DO
Harvard South Shore Psychiatry Resident
Department of Psychiatry
Harvard Medical School
Veterans Affairs Boston Healthcare System
Boston, Massachusetts, USA

Louis Fanucci, DO
Radiology Resident
Department of Radiology
University of California
Irvine, California, USA

Justin M Harrell, MS, DO
Internal Medicine Resident
University of Kentucky
Lexington, Kentucky, USA

Alexander D. Lake, DO
Internal Medicine Resident
Hospital Corporation of America Healthcare
University of South Florida
Hudson, Florida, USA

Corey Mayer, MBA, DO
Internal Medicine Resident
Beaumont Hospital
Royal Oak, Michigan, USA

Amanda G. Noyes, MS, LCGC
Senior Genetic Counselor
GeneDx Clinical Genomics Program
Gaithersburg, Maryland, USA

Abbey M. Putnam, MS, CGC
Licensed Genetic Counselor
East Tennessee Children's Hospital Genetics Center
Knoxville, Tennessee, USA

Hira Rashid, MS, MBA, DO
Transition year resident
DeBusk College of Osteopathic Medicine
Lincoln Memorial University
Harrogate, Tennessee, USA

Eric Seronick, DO
Rheumatology Fellow
Department of Medicine
Rheumatology Division
University of California
Irvine, California, USA

Anastassiia Vertii, PhD
Instructor
Department of Molecular, Cell and Cancer Biology
University of Massachusetts Medical School
Worcester, Massachusetts, USA

Bei Zhang, MD, PhD, CLS (ASCP)
Associate Professor
Director of Innovative Curriculum Design and Instruction
Department of Pathology and Laboratory Medicine
The Robert Larner, College of Medicine
University of Vermont
Burlington, Vermont, USA

Part I

1 Anatomy of the Cell and Organelles

At the conclusion of this chapter, students will be able to:
- Identify the lipid, protein, and carbohydrate components of the plasma membrane
- Describe the basic functions of the plasma membrane, including selective permeability, cellular communication, and physical barrier functions
- Identify the major ions found inside the cell and in the extracellular environment as well as explain the variations of ion concentrations across cell membranes
- Identify the six major membrane-bound organelles of the human cell
- Describe the structure and major functions of each organelle
- Identify genetic diseases associated with dysfunction of the nuclear lamina, mitochondria, lysosomes, peroxisomes, and golgi apparatus and explain the molecular process affected in each

All living organisms are composed of building blocks called cells. These individual units consist of an aqueous solution enclosed by a lipid-rich membrane. Inside the cell exist all of the chemicals and molecules to sustain itself as well as those necessary to perform its requisite functions including cellular metabolism, chemical signaling, and, in some cases, the synthesis of products that will ultimately be expelled from the cell. In addition, the cell contains the machinery necessary to make an exact copy of itself, preserving the hereditary information that defines the individual organism.

Living organisms can be categorized into two main groups: eukaryotes, and prokaryotes (▶ Fig. 1.1**a**, **b**). The prokaryotes are unicellular organisms that can be further classified as eubacteria and archaebacteria. Prokaryotes typically possess a flagellum to allow for movement. In addition, they do not have membrane-bound organelles, nor do they have a nucleus. Instead, all of the genetic material is concentrated in an irregularly shaped region of the cytoplasm termed the nucleoid. The cell membrane of most prokaryotes is encased by a cell wall, which is itself surrounded by a cross-linked, mesh-like structure, called the peptidoglycan layer, which is composed of short amino acid chains linked to sugar molecules. The thickness of the peptidoglycan layer is used to distinguish between two main groups of bacteria: the gram positive and gram negative bacteria. Gram positive bacteria have a thick peptidoglycan layer and, hence, retain gram stain, whereas this layer in gram negative bacteria is thin and readily gives up the stain. Some classes of antibiotics work by inhibiting the synthesis of this peptidoglycan layer, which is essential for the bacteria's survival. These include the beta lactam penicillin and the glycopeptide antibiotics vancomycin and bleomycin. (▶ Fig. 1.1)

In multicellular organisms, such as humans, organ systems are comprised of individual organs with specific functions. In

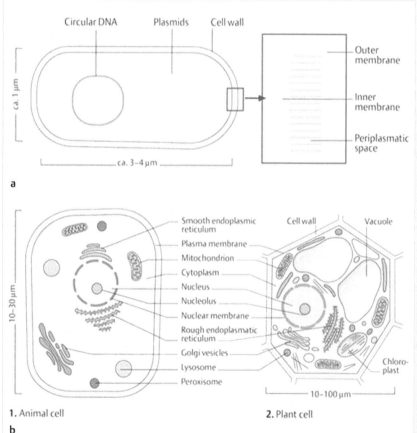

Fig. 1.1 (a) Anatomy of a typical prokaryotic cell. **(b)** Anatomy of (1) an animal cell and (2) a plant cell. (Source: Passarge E, ed. Color Atlas of Genetics. 4th Edition. New York, NY. Thieme; 2012.)

turn, each organ is comprised of tissues with physiological roles tailored to that particular organ. These tissues are themselves made up of individual cells with specialized abilities that allow them to contribute to the overall function of the tissue. And so, cell function contributes to tissue function which contributes to organ function that ultimately leads to the proper functioning of the organ system. The importance of the individual cell cannot be overstated, as the ability of the organ system to function properly is determined by the health of the cells that make up the individual organs.

In humans, the body is made up of trillions of individual cells representing hundreds of different cells types. Despite this tremendous variation, nearly all vertebrate cells share the same general anatomy (▶ Fig. 1.2). Inside the cell are membrane-enclosed organelles such as the nucleus, mitochondria, lysosomes, peroxisomes, endoplasmic reticulum, and golgi apparatus. In addition

to these organelles, the cells also contain a cytoskeleton that serves multiple functions in cellular integrity, cell motility, intracellular transport, and cellular division. Details of the cytoskeleton will be discussed in Chapter 5. All of these components are surrounded by a lipid-rich structure called the plasma membrane, which separates the inside of the cell from its exterior environment. (▶ Fig. 1.3)

1.1 The Plasma Membrane

The plasma membrane is formed from a lipid bilayer, serving as a selectively permeable barrier that controls the passage of ions and organic molecules into and out of the cell. This bilayer is asymmetric, having different components in the inner lipid layer versus the outer layer lipid layer. The layer facing the

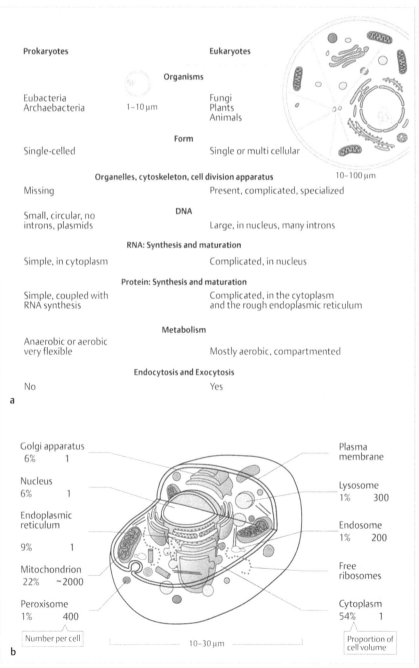

Fig. 1.2 (a) Comparison of prokaryotes and eukaryotes. **(b)** Components of a typical animal cell. (Source: Koolman J, Röhm K, ed. Color Atlas of Biochemistry. 3rd Edition. New York, NY. Thieme; 2012.)

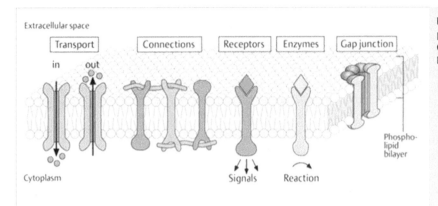

Fig. 1.3 Classes of membrane-embedded proteins and complexes. (Source: Passarge E, ed. Color Atlas of Genetics. 4th Edition. New York, NY. Thieme; 2012.)

extracellular environment is known as the outer leaflet, and the layer facing the cytoplasm is termed the inner leaflet. This asymmetric structure is important for many cellular processes, as we will see later. The most abundant lipids within the plasma membrane are the phospholipids, which have a hydrophilic head connected to two hydrophobic tails via a glycerol backbone. The presence of both hydrophobic and hydrophilic parts means that these molecules are amphipathic, a characteristic that favors their formation into bilayers in an aqueous environment. There are four major phospholipids found in the plasma membrane. These are phosphotidylethanolamine, phosphatidylserine, phosphatidylcholine, and sphingomyelin. In addition to their role as a structural component of the plasma membrane, these phospholipids also serve as important mediators of other processes in the cell, including signal transduction and programmed cell death (Chapters 10 and 13). Another type of phospholipid, the phosphoinositides, are less abundant in the plasma membrane but perform important functions in transport vesicle targeting within the cell (Chapter 6) as well as cell signaling (Chapter 10). Besides phospholipids, the plasma membrane also contains cholesterol, glycolipids, and proteins (▶ Fig. 1.4**a, b**). Cholesterol represents approximately 20% of the lipid content by weight of the plasma membrane. The cholesterol molecule is short and rigid, allowing it to fill the spaces created by kinked unsaturated hydrocarbon rings of adjacent phospholipid molecules in the membrane. Like phospholipids, glycolipids are also amphipathic. However, these molecules differ from phospholipids in three fundamental ways: their hydrophilic head contains an oligosaccharide chain, the backbone is a sphingosine molecule instead of glycerol, and it lacks a phosphate group. The glycolipid molecules have important roles in cell recognition. For example, glycolipids are the antigenic determinant of the ABO blood groups. Nerve cell membranes have high concentrations of two specific kinds of glycolipids, the cerebrosides and the gangliosides. A defect in the normal turnover of these glycolipids is responsible for the devastating effects seen in individuals with lysosomal storage diseases, as we will see later in this chapter. (▶ Fig. 1.4**a, b**)

Membrane phospholipids are capable of moving and changing places with one another within the plane of the bilayer, a characteristic known as membrane fluidity. In contrast, they rarely exchange between the layers. When this does occur, it is mediated by specialized enzymes called flippases. This fluid nature of the plasma membrane is important for many of the cell's functions. It enables membrane proteins to diffuse rapidly in the plane of the bilayer and to interact with one another, which is important for many aspects of cell signaling. It also permits membrane lipids and proteins to diffuse from sites where they are inserted into the bilayer after their synthesis to other regions of the cell. It ensures that membrane molecules are distributed evenly between daughter cells when a cell divides, and it allows membranes to fuse with one another and mix their molecules.

Two key factors determine the degree of fluidity of the membrane. The first is based on the length of the hydrocarbon tails of the phospholipids. Short hydrocarbon chains in adjacent lipid molecules have less tendency to form bonds with one another and, hence, allow for a more fluid membrane. In contrast, lipids with long chains would have more interactions with one another and would therefore decrease the fluidity. The second factor that determines the fluidity of the membrane is the amount of cholesterol. Since cholesterol packs within the spaces between adjacent phospholipid molecules, a higher percentage of cholesterol would create a more rigid, and therefore less fluid, membrane.

In addition to lipids and cholesterol, the plasma membrane is home to many different types of proteins (▶ Fig. 1.3). Some of these proteins are embedded in the plasma membrane and are called integral proteins, while others, known as peripheral proteins, are merely associated with the plasma membrane through attachment to integral membrane proteins. Other proteins are anchored to the plasma membrane through a covalently-linked lipid molecule that is embedded in the inner or outer leaflet of the membrane. There are four functional classes of transmembrane proteins, each of which will be discussed in later chapters. In Chapter 7 we will discuss the transporters and how these structures allow the movement of molecules into and out of the cell. We will also see the importance of anchors in holding the cells of a tissue together, as well as their role in connecting cells to the extracellular matrix. Chapter 10 will cover the integral membrane proteins known as receptors and how these structures can lead to the activation of downstream effector molecules within the cell.

The plasma membrane of eukaryotic cells serves several important functions. It creates a physical barrier, separating the inner components of the cell from the outside environment and, in this capacity, it gives the cell the ability to regulate the entry and exit of ions, nutrients, and other molecules across the membrane. It also contains receptors that receive external signals from the extracellular environment and that elicit a response within

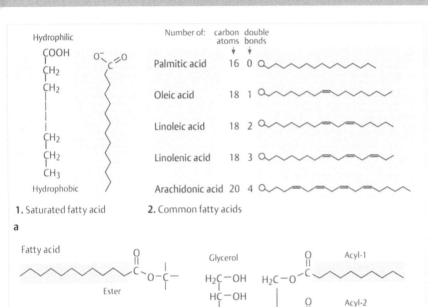

Fig. 1.4 **(a)** Structure of a saturated fatty acid molecule (1) and common fatty acids in the cell that have no double bonds (saturated) and varying numbers of double bonds (unsaturated) (2). **(b)** Comparison of the structures of fatty acids, triglycerides, glycolipids, and phospholipids. (Source: Passarge E, ed. Color Atlas of Genetics. 4th Edition. New York, NY. Thieme; 2012.)

the cell. Finally, the plasma membrane contains proteins and complexes that allow for connections between adjacent cells, as well as between the cell and the extracellular matrix.

Although small, nonpolar molecules such as gases (O_2, CO_2, N_2) and small uncharged polar molecules (H_2O and ethanol) can diffuse readily across the lipid bilayer, the plasma membrane retains tight control over the movement of large, uncharged polar molecules and ions. Since these molecules cannot diffuse through the membrane, the cell must use integral transmembrane complexes, such as transporters and channels, to uptake or expel these molecules. This tight control allows for the maintenance of precise concentrations of ions inside and outside of the cell. In vertebrate cells, the concentration of K^+ is higher inside the cell than outside the cell. In contrast, the concentrations of Cl^-, Na^+, and Ca_2^+ are kept low inside the cell and high outside. The importance of these concentration differences will be discussed in Chapter 7.

1.2 The Nucleus

The plasma membrane is not the only lipid membrane of the cell. Inside the cell there are multiple organelles that are surrounded by a lipid bilayer. The nucleus is one such structure. When cells are observed under a light microscope, the nucleus is often the most prominent organelle. This structure is the site of storage for a majority of the DNA in the cell. The nucleus is composed of DNA surrounded by a structure known as the nuclear envelope, which consists of two separate lipid bilayers. One of the lipid bilayers, the inner nuclear membrane or INM, is associated with a matrix of proteins known as the nuclear lamina, formed from members of the lamin family of proteins. This structure contributes to the shape and rigidity of the nucleus. The other lipid bilayer, the outer nuclear membrane or ONM, is continuous with the membrane of another organelle, the endoplasmic reticulum. Within the nucleus are nucleoli, dense structures composed of ribosomal RNA (rRNA) and proteins (▶ Fig. 1.5).

Dysfunction of nuclear lamina components has been shown to be responsible for some human diseases. Progeria, an extremely rare, fatal childhood disease characterized by premature aging of the musculoskeletal and cardiovascular systems is one such disease. Individuals with this condition have a mutation in the gene that encodes lamin A, which is an integral component of the nuclear lamina. This mutation creates a cryptic splice site in the gene, resulting in the production of a truncated form of the protein, known as progerin. Progerin is not properly integrated into the nuclear lamina, resulting in a disfigured nucleus and altered

I

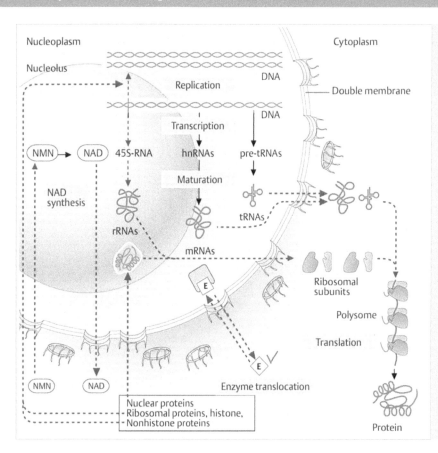

Fig. 1.5 Anatomy and functions of the eukaryotic nucleus. (Source: Koolman J, Röhm K, ed. Color Atlas of Biochemistry. 3rd Edition. New York, NY. Thieme; 2012.)

genomic integrity, which is thought to contribute to the symptoms associated with this disease.

1.3 The Endoplasmic Reticulum

The endoplasmic reticulum (ER) is a membrane-bound system of interconnected sacs and tubes that is continuous with the outer nuclear membrane. There are two types of ER: smooth ER and rough ER (▶ Fig. 1.6). Smooth ER is the major site of synthesis of new membrane within the cell, whereas rough ER is associated with ribosomes and is the site of synthesis for proteins that will ultimately be secreted or embedded in either the plasma membrane or in the membrane of another organelle in the cell. The details of both protein synthesis in the ER as well as intracellular transport of these proteins will be described later in this chapter and in Chapter 6. In addition to protein and new membrane synthesis, the ER serves as a major site of storage for intracellular calcium ions (Ca_2^+). As we will see in Chapter 10, tight control over the cytoplasmic levels of this molecule are very important for many signaling events within the cell. The ER also contains the components of a system that monitors for the proper folding of proteins within the lumen of the ER, called the Unfolded Protein Response (UPR). If the cell is not able to monitor this process appropriately, an accumulation of misfolded proteins would occur, leading to cell death. (▶ Fig. 1.6)

1.4 The Golgi Apparatus

The Golgi apparatus is situated between the ER and the plasma membrane (▶ Fig. 1.6**a**). It is formed from a series of stacks of flattened, membrane-enclosed sacs and has polarity due to the presence of two different faces: the cis face and the trans face. The cis face lies near the ER and receives transport vesicles from this organelle that contain newly synthesized proteins. Within the lumen of the Golgi, these proteins are modified by the addition of tagging molecules such as carbohydrates that will allow them to later be packaged into the appropriate transport vesicle. The trans face of the Golgi is oriented toward the plasma membrane and sorts these tagged proteins into transport vesicles so that their contents can be delivered to their proper destination. Depending on the contents, this may involve shuttling to the lysosomes, incorporation into the plasma membrane, or secretion from the cell. Details regarding the intracellular trafficking of proteins and transport vesicles will be discussed in Chapter 6.

1.4.1 Link to Pathology

I-cell disease is a rare inherited condition clinically characterized by defective physical growth and mental retardation. It is caused by a deficiency in a phosphorylating enzyme, GlcNac-1-phosphotransferase, normally present in the Golgi apparatus. The function of this enzyme is to allow the tagging of lysosomal enzymes with mannose-6-phosphate, a molecule that marks the enzyme as destined for the lysosome. In the absence of this tag, the lysosomal enzyme is not sent to the lysosome but is, rather, sent to the plasma membrane, where it is secreted. Since the lysosome lacks this enzyme, it is non-functional, resulting in the accumulation of un-degraded cellular components and cell death. Because the lysosomal enzymes are expelled from the cell, the presence of the enzyme in blood can be used to diagnose individuals with I-cell disease.

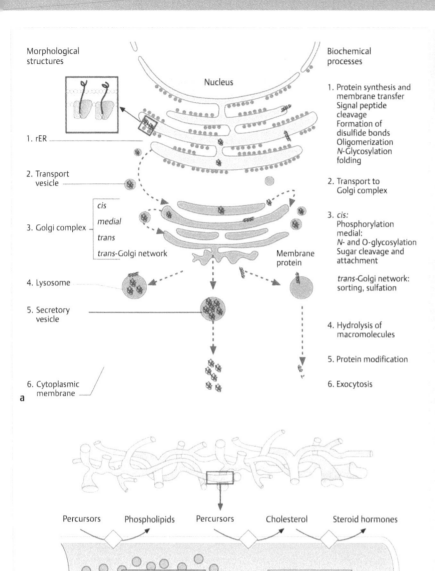

Morphological
structures

Nucleus

1. rER

2. Transport
vesicle

3. Golgi complex
- cis
- medial
- trans
- trans-Golgi network

4. Lysosome

5. Secretory
vesicle

6. Cytoplasmic
membrane

a

Membrane
protein

Biochemical
processes

1. Protein synthesis and
membrane transfer
Signal peptide
cleavage
Formation of
disulfide bonds
Oligomerization
N-Glycosylation
folding

2. Transport to
Golgi complex

3. cis:
Phosphorylation
medial:
N- and O-glycosylation
Sugar cleavage and
attachment

trans-Golgi network:
sorting, sulfation

4. Hydrolysis of
macromolecules

5. Protein modification

6. Exocytosis

Fig. 1.6 (a) Structure and functions of the rough endoplasmic reticulum and Golgi apparatus. **(b)** Structure and functions of smooth endoplasmic reticulum. (Source: Koolman J, Röhm K, ed. Color Atlas of Biochemistry. 3rd Edition. New York, NY. Thieme; 2012.)

Percursors   Phospholipids   Percursors   Cholesterol   Steroid hormones

Calcium store

Biotransformation

ATP   ADP

Ca^{2+}

Xeno-
biotics

Metabolites

Lipid metabolism
enzyme systems

b

1.5 Mitochondria

Within the cell are hundreds to thousands of small, oblong structures spread throughout the cytoplasm and are known as mitochondria (▶ Fig. 1.7). Each of these organelles is enclosed by two separate membranes: the outer membrane and the inner membrane. The outer membrane contains pores and channels that allow for the movement of molecules inside and out of the mitochondria. Between the inner and outer membrane is a space known as the intermembrane space. This compartment contains a molecule called cytochrome c, which has an integral role in apoptosis, as we will see in Chapter 13. Inside the inner membrane is the matrix, which contains much of the machinery needed to perform the mitochondrion's main function, which is to produce energy for the cell in the form of ATP. (▶ Fig. 1.8)

The number of mitochondria in a cell depends on the cell's energy needs. Those cells with higher energy needs, such as the hepatocytes of the liver, contain many thousands of mitochondria. On the other hand, relatively inactive cells, such as the epithelial cells of the skin, only have 200 to 300 mitochondria.

There are two ways in which the mitochondria produces ATP: through oxidative phosphorylation, which occurs on the inner membrane, and through the TCA cycle, which occurs in the matrix. Besides ATP production and the aforementioned role in apoptosis, the mitochondria also carry out β oxidation of fatty acids, as wells as several steps in the synthesis of both heme and steroids.

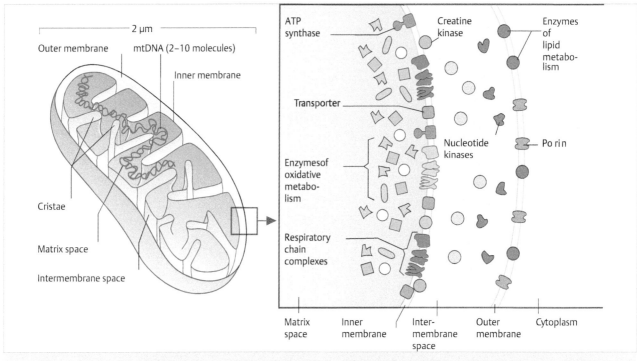

Fig. 1.7 Mitochondrial structure. (Source: Koolman J, Röhm K, ed. Color Atlas of Biochemistry. 3rd Edition. New York, NY. Thieme; 2012.)

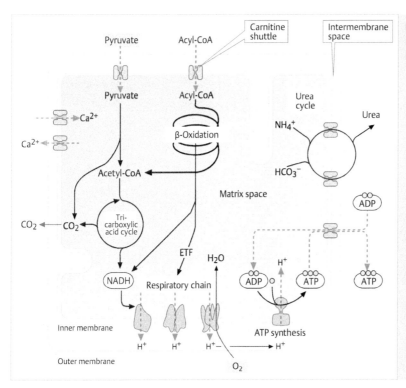

Fig. 1.8 Metabolic functions of mitochondria. (Source: Koolman J, Röhm K, ed. Color Atlas of Biochemistry. 3rd Edition. New York, NY. Thieme; 2012.)

Mitochondria have several unique properties that set them apart from the other organelles in the cell. Since mature sperm do not have any mitochondria of their own, the egg cell is the only source of mitochondria for the developing embryo and, therefore, all of an individual's mitochondria are inherited from the mother. Mitochondria also have their own genome, in the form of DNA molecules called mDNA or mtDNA. Unlike nuclear DNA, which is linear, the mtDNA is in a circular form. Since mitochondria are only inherited from the mother, this means that all of the mtDNA is of maternal origin and there is no contribution of genetic diversity from the father. This mtDNA, however, only encodes for some, but not all, of the proteins that are required for proper function of the mitochondria. The rest are encoded in the nuclear DNA. Another property that sets mitochondria apart from most of the other organelles of the cell is that, similar to the nucleus, mitochondria self-replicate by fission, rather than forming de novo.

There are a number of mitochondrial diseases that fall under the category of mitochondrial myopathies. Due to the high energy requirements of muscle and neurons, defects in mitochondrial function often lead to neuromuscular symptoms. Since components of the energy-generating machinery are encoded in both nuclear DNA and mitochondrial DNA, mutations in either compartment can lead to disease. However, mutations in mtDNA tend to occur at a higher frequency than do in nuclear DNA due to a much lower fidelity of the DNA polymerase used to replicate mtDNA. Because the number of affected mitochondria can vary greatly, the symptoms of these diseases differs among afflicted individuals. The diagnosis of these diseases is difficult since severity of the disease depends on which organ is affected and the number of organs affected, as well as the ratio of mutation-harboring mitochondria to normal mitochondria within the individual cells. (▶ Fig. 1.9)

1.6 Lysosomes

Like mitochondria, the lysosome is another small organelle in the cell existing in multiple copies. These structures are often called the "stomach of the cell" as they are responsible for digesting a large variety of cellular components (▶ Fig. 1.10a, b). Lysosomes form an acidic compartment filled with many different types of hydrolytic enzymes that are capable of breaking down nearly all macromolecules in the cell. The main function of the lysosome is to degrade worn out or damaged organelles and structural components within the cell. In addition, specialized cells such as the phagocytes of the immune system contain lysosome-related organelles called phagolysosomes that are responsible for breaking down pathogens that have been engulfed by the cell. (▶ Fig. 1.10)

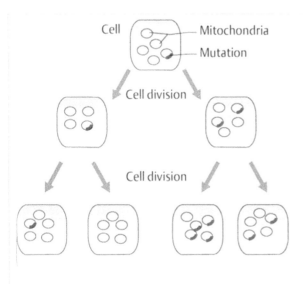

Fig. 1.9 Unequal distribution of mutant mitochondria during cell division (Heteroplasmy). (Source: Passarge E, ed. Color Atlas of Genetics. 4th Edition. New York, NY. Thieme; 2012.)

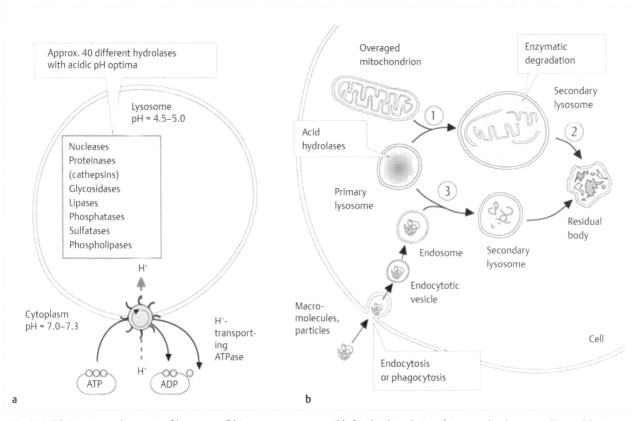

Fig. 1.10 (a) Structure and contents of lysosomes. **(b)** Lysosomes are responsible for the degradation of macromolecules, organelles, and foreign organisms. (Source: Koolman J, Röhm K, ed. Color Atlas of Biochemistry. 3rd Edition. New York, NY. Thieme; 2012.)

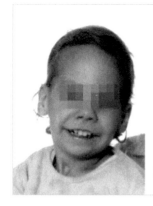
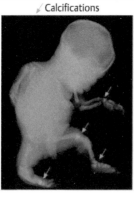
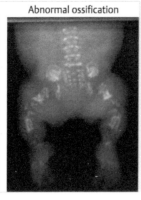

Calcifications   Abnormal ossification

Fig. 1.11 Facial features of individuals with Hurler syndrome. (Source: Passarge E, ed. Color Atlas of Genetics. 4th Edition. New York, NY. Thieme; 2012.)

There are several autosomal recessive diseases, called lysosomal storage diseases, which result from the loss of one of the lysosomal enzymes. In individuals afflicted by these diseases, the inability of their lysosomes to breakdown cellular components causes an abnormal accumulation of substances inside the cell, resulting in the increase in mass of the affected tissue or organ, and ultimately leading to death of the cell. For most lysosomal storage diseases, symptoms are caused by the neurodegeneration that results from the accumulation of old and damaged cellular components in the neurons. These conditions are often fatal within the first few years of life. Enzymatic assays that test for reduced enzyme levels, loss of enzymatic activity, or accumulation of enzyme in the blood can be used to diagnose these lysosomal storage diseases.

1.6.1 Link to Pathology

Tay-Sachs is a lysosomal storage disease due to a deficiency in the enzyme hexosaminidase A, brought about by mutations that prevent the synthesis of an active form of this enzyme. In the brain, this enzyme is responsible for the breakdown of gangliosides in the lysosome. Therefore, deficiency of this enzyme results in an accumulation of gangliosides in the neurons, leading to neuronal cell death. A common clinical finding in these individuals is a cherry-red spot on the retina of the eye.

Another type of lysosomal storage disease is Hurler syndrome. In this case, the defective enzyme is α-L-iduronidase,

which normally breaks down glycosaminoglycans (GAGs). Without this enzyme, GAGs accumulate in the cell, leading to death of the cell, manifesting in skeletal abnormalities and mental retardation. This condition can be detected by the presence of non-degraded GAGs in the urine. (▶ Fig. 1.11)

1.7 Peroxisomes

Peroxisomes, also known as microbodies, are small, membrane-bound vesicles, similar in size to lysosomes. As is true of lysosomes and mitochondria, the cell has multiple copies of peroxisomes. These organelles serve as a contained environment for oxidative reactions, such as those that produce H_2O_2, as well as β oxidation of very long chain fatty acids. The presence of an enzyme called catalase within the peroxisomes allows for the conversion of the toxic H_2O_2 to inert H_2O and O_2. In addition, peroxisomes are the site of the first reactions that ultimately lead to the formation of plasmalogen, the most abundant phospholipid in myelin. Cells within the liver and kidneys have a large number of peroxisomes, which help to metabolize ethanol and detoxify various substances, including some prescription drugs. (▶ Fig. 1.12)

Dysfunction of the peroxisome can lead to peroxisomal biogenesis disorders. These are autosomal recessive diseases caused by mutations in peroxins, proteins that are structural components of the peroxisomes. Without functional peroxins, an accumulation of very long chain fatty acids occurs, ultimately resulting in death of the cell. Zellweger syndrome is one example of a peroxisomal biogenesis disorder. (▶ Fig. 1.13)

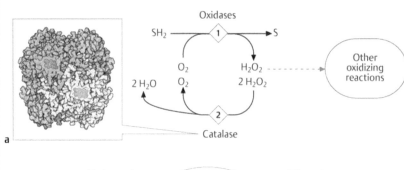

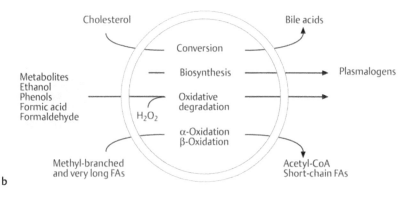

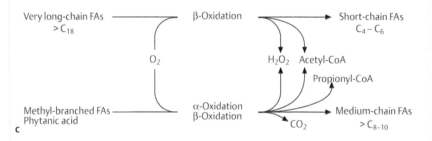

Fig. 1.12 **(a)** Catalase enzymes in the peroxisome convert H_2O_2 to O_2 and H_2O. **(b)** Functions of peroxisomes. **(c)** Beta oxidation reactions of very long chain fatty acids in the peroxisome. **(d)** Perixisomal diseases. (Source: Koolman J, Röhm K, ed. Color Atlas of Biochemistry. 3rd Edition. New York, NY. Thieme; 2012.)

Disease	Cause	Defect
Zellweger syndrome	Disturbed peroxisome biosynthesis due to enzyme transport defect	Organ damage especially in liver and brain, conspicuous facial changes
Adreno-leukodystrophy	Deficient ABC transporter for very long-chain fatty acids	Neurological disorders, weakness, dizziness, adrenal medullary atrophy, demyelination in the brain
Refsum syndrome	Defect of phytanic acid degradation	Neuropathies, vision, hearing and cardiac disorders

d

I

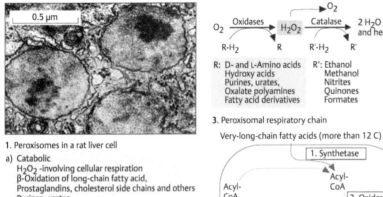

0.5 µm

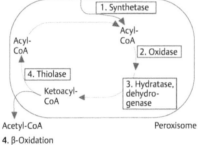

$$O_2 \xrightarrow[R-H_2]{\text{Oxidases}} H_2O_2 \xrightarrow[R'-H_2]{\text{Catalase}} \begin{array}{c} 2\,H_2O \\ \text{and heat} \end{array}$$

R: D- and L-Amino acids   R': Ethanol
Hydroxy acids   Methanol
Purines, urates,   Nitrites
Oxalate polyamines   Quinones
Fatty acid derivatives   Formates

3. Peroxisomal respiratory chain

Very-long-chain fatty acids (more than 12 C)

1. Synthetase

Acyl-CoA

Acyl-CoA

2. Oxidase

4. Thiolase

3. Hydratase, dehydrogenase

Ketoacyl-CoA

Acetyl-CoA   Peroxisome

4. β-Oxidation

1. Peroxisomes in a rat liver cell

a) Catabolic
H_2O_2 -involving cellular respiration
β-Oxidation of long-chain fatty acid,
Prostaglandins, cholesterol side chains and others
Purines, urates
Pipecolic acid, dicarboxy acids
Ethanol, methanol

b) Anabolic
Phospholipids (plasmalogen)
Cholesterol, bile acids
Gluconeogenesis
Glyoxalate transamination

2. Function of peroxisomes

a

Fig. 1.13 **(a)** Biochemical reactions in perioxisomes **(b)** Examples of perioxismal diseases **(c)** Zellweger cerebrohepatorenal syndrome. (Source: Passarge E, ed. Color Atlas of Genetics. 4th Edition. New York, NY. Thieme; 2012.)

214100 Zellweger cerebrohepatorenal syndrome
202370 Neonatal adrenoleukodystrophy
266510 Infantile Refsum disease

239400 Hyperpipecolic acidemia
215100 Rhizomelic chondrodysplasia punctata
259900 Primary hyperoxaluria type I
and others

b

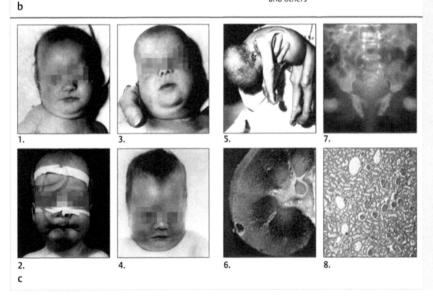

1.

2.

3.

4.

5.

6.

7.

8.

c

Review Questions

1. A 3-year-old girl presents to her pediatrician with developmental delay and neurological deterioration. Urine tests reveal an abnormally high level of glycosaminoglycans. What cellular function is most likely disrupted in this child?
   A) Expression of Hydrolases
   B) Production of ATP
   C) Release of secretory vesicles
   D) Breakdown of very long chain fatty acids
   E) Formation of plasmalogen
2. Progressive decline in motor and sensory functions is noted in a 3-year-old girl. Imaging reveals bilaterally symmetric demyelination, and laboratory studies indicate high levels of saturated very long-chain fatty acids in her body fluids. What is a characteristic of the organelle that is defective in this patient?
   A) It is responsible for sorting secretory proteins
   B) It is inherited in a non-Mendelian fashion
   C) It is composed of a matrix formed by lamins
   D) It contains an enzyme that converts H_2O_2 to H_2O and O_2
   E) It separates intracellular components from the outside environment

Answers

1. **The correct answer is A.** An elevated level of glycosaminoglycans and neurological deterioration early in life are indicative of the lysosomal storage disease Hurler syndrome. This disease is caused by mutations in α-L-iduronidase, which affect the expression of this hydrolase enzyme and lead to the absence of this enzyme in the lysosomes.
2. **The correct answer is D.** The symptoms described for this individual are consistent with a peroxisome biogenesis disorder. Recall that β oxidation of very long chain fatty acids occurs in the peroxisome. This child most likely has a peroxisome defect due to the accumulation of very long chain fatty acids in her body fluids. In addition to the oxidation of very long chain fatty acids, another normal function of the peroxisome is the conversion of H_2O_2 to H_2O and O_2, through the activity of an enzyme called catalase

13

2 DNA Replication, Gene Mutations, and Repair

At the conclusion of this chapter, students should be able to:
- Describe the composition and structure of a DNA molecule
- Apply their knowledge of nucleotide pairings to produce the complimentary sequence of a given nucleic acid sequence
- Describe the mechanism of action of nucleoside analogs and how they can be used to inhibit retroviral infections
- Identify the enzymes involved in DNA replication and describe the activities of each of these enzymes
- Explain the processes of leading strand synthesis and lagging strand synthesis
- Explain the function of telomerase
- Describe the causes of DNA mutations and the major types of mutations
- Compare and contrast the five major types of DNA repair processes
- Identify specific molecules involved with each type of DNA repair
- Associate types of DNA mutations with the DNA repair process that cells would use to correct them

2.1 The Structure of DNA

The heritable information in the cell exists in the form of DNA, a polymer of nucleotide molecules. Each nucleotide has three general components: a nitrogenous base, a sugar molecule, and a phosphate group. The nitrogenous bases are heterocyclic aromatic compounds that contain nitrogen atoms and are weakly basic. These nitrogenous bases are divided into two families: the pyrimidines and purines. The pyrimidines include cytosine and thymine, and the purines include adenine and guanine (▶ Fig. 2.1). The purines are physically bigger than pyrimidines due to their heterocyclic rings consisting of pyrimidine fused to an imidazole ring. Each nucleotide in the chain is connected to the next by a phosphodiester backbone created through the linkage of the 3′-carbon atom of one sugar molecule to the 5′ carbon atom of the next sugar molecule via the phosphate group. This linkage results in a free O⁻ at physiological pH, which gives the DNA molecule an overall negative charge. The chemical structure of the nitrogenous bases in the nucleotides favor a pairing scheme such that adenines of one strand bind to thymines of the opposite strand through two hydrogen bonds, and cytosines bind to guanines through three hydrogen bonds. The pairing of two polynucleotide strands to form a DNA molecule creates a three dimensional structure in which the strands are wound into a double helix (▶ Fig. 2.2).

A defining characteristic of a DNA molecule is that each strand of the double helix is oriented in the opposite direction to the other. One strand has the 5′carbon of the sugar above the 3′ carbon and is designated as being the 5′ to 3′ direction, and the other strand has the 3′carbon above the 5′ carbon and is designated as the 3′ to 5′ strand. The importance of this antiparallel property will be discussed in the context of DNA replication later on.

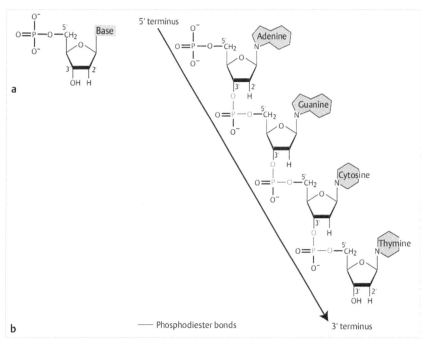

Fig. 2.1 DNA composition and structure (a) An example of the general structure of a 5′-phospho-2′-deoxyribose unit. (b) DNA is typically represented in the 5′ to 3′ direction. The 3′-OH group of a nucleotide is connected in a phosphodiester linkage to the 5′-phosphate of the next nucleotide. Thus, every DNA strand has a phosphate residue at the 5′ terminus and a free hydroxyl group at the 3′ terminus. (Source: Panini S, ed. Medical Biochemistry—An Illustrated Review. New York, NY. Thieme; 2013.)

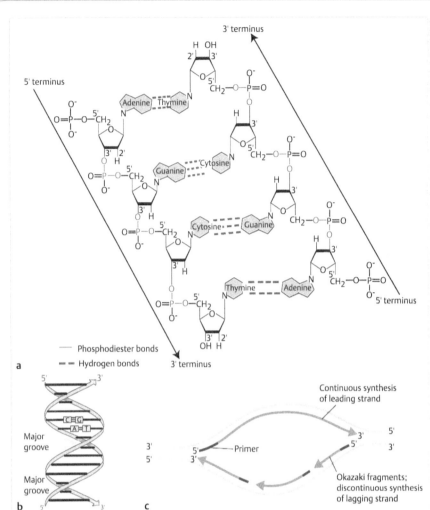

Fig. 2.2 Complementary base pairing in DNA and DNA forms. **(a)** The two strands of the DNA double helix are arranged antiparallel to each other. Base pairing occurs between a purine (double-ring) and a pyrimidine (single-ring) by the formation of noncovalent hydrogen bonds. Adenine and thymine are linked by two hydrogen bonds, whereas guanine and cytosine are linked by three such bonds. **(b)** As a result, the two antiparallel strands typically create a B-form, right-handed double helix with the hydrophobic bases in the interior and the hydrophilic sugar phosphate backbone exposed to the aqueous exterior. The major and minor grooves of one complete turn of the helix are indicated. **(c)** Structure of double helix.

2.1.1 Link to Pharmacology

Nucleoside analogues are drugs that are similar in structure to the naturally occurring nucleotides and are, thus, capable of being incorporated into newly synthesized DNA strands. These molecules have proven to be effective treatments for several types of retroviral infections. This is due to the absence of a 3′ OH group on the sugar group in these molecules, which is normally required for the addition of the next nucleotide in the growing polynucleotide chain. Therefore, when these analogues are incorporated by the DNA polymerase, chain termination occurs. Because the analogues are preferentially used by retroviral enzymes over the host cell's enzymes, the retroviral infection is treated effectively. Examples of these nucleoside analogues include azidothymidine (AZT) and acyclovir (acycloguanosine).

2.2 DNA Replication

Accurate duplication of the cell's genomic material is essential for its own survival, and can ultimately determine the survival of the whole organism, if multicellular. This process must occur with minimal changes to the parent DNA molecule, and if changes do occur, the cell must be able to reliably repair them. The duplication of the genomic information of a cell is called

DNA replication and it occurs in a specific phase of the cell cycle, the S phase. The S phase lies between the two gap phases of the cell cycle, G1 and G2, and in animal cells takes on average between 6 and 8 hours to complete. The DNA replication process results in two complete double helices derived from the original DNA molecule, with each new DNA molecule being identical in nucleotide sequence to the parental DNA. The exact mechanism of how this occurs was first proposed by Watson and Crick and was referred to as the semi-conservative model. Their hypothesis was subsequently proven experimentally by Meselson and Stahl. In the semi-conservative model, the DNA serves as a template for its own duplication. The two strands must first separate, and each strand serves as a template for the formation of the newly synthesized strand, resulting in a complete DNA molecule that is a hybrid of one old strand and one new strand (▶ Fig. 2.3).

The process of DNA replication can be split into three main steps: initiation, elongation, and termination.

Initiation occurs at specific nucleotide sequences within the genome known as the origins of replication (Ori). Because only two hydrogen bonds exist between adenine and thymidine pairings, less energy is needed to break these bonds compared to the three hydrogen bonds present in cytosine-guanine pairs. Therefore these Ori's tend to be rich in A-T base pairs. Prokaryotes only have one origin of replication, which is sufficient to duplicate

I

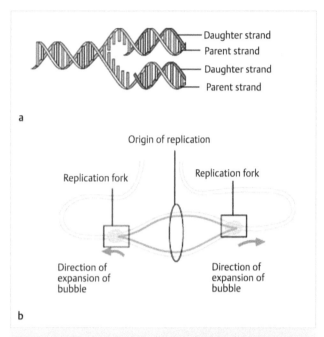

a

b

Fig. 2.3 Semiconservative, bidirectional, semidiscontinuous 5′ → 3′ synthesis of DNA. **(a)** Each of the parental strands of DNA serves as a template for the synthesis of a complementary daughter strand. Thus, each of the replicated DNA molecules contains one parental strand and one newly synthesized daughter strand. **(b)** Bidirectional replication of DNA in a eukaryotic chromosome. (Source: Panini S, ed. Medical Biochemistry—An Illustrated Review. New York, NY. Thieme; 2013.)

their circular genome. However, in eukaryotes there are multiple origins of replication spread throughout each chromosome in order to allow for efficient and timely replication of their relatively large genome (▶ Fig. 2.4).

In order for these origins of replication to initiate DNA synthesis, several factors must bind. The first to bind to the Ori is the ORC, or origin recognition complex. Consisting of six separate components, the function of this complex is to recruit another complex called the minichromosome maintenance complex, or MCM. In turn, the MCM recruits a DNA helicase that is required to separate the two DNA strands, forming a small bubble. To keep the separated strands from binding back together, single strand binding proteins, such as the Replication Protein A, or RPA, associate with the single stranded DNA and hinder complementary nucleotide associations. These components form Y-shaped junctions, called replication forks, on each end of the Ori (▶ Fig. 2.5). As DNA synthesis continues, these replication forks progress in opposite directions, allowing for the duplication of large amounts of DNA in a relatively short period of time.

Following the formation of the replication fork, a series of proteins are recruited to form a complex called the replisome, with each component having a specific function needed for successful replication. In addition to the aforementioned DNA helicase, the replisome includes topoisomerase, DNA polymerase, and RNA primase. As the helicase separates and unwinds the strands of DNA, the DNA upstream of the replication fork becomes more tightly wound or supercoiled, creating torsional stresses on the DNA molecule. Topoisomerase's activity is to break phosphodiester bonds in the backbone of the DNA

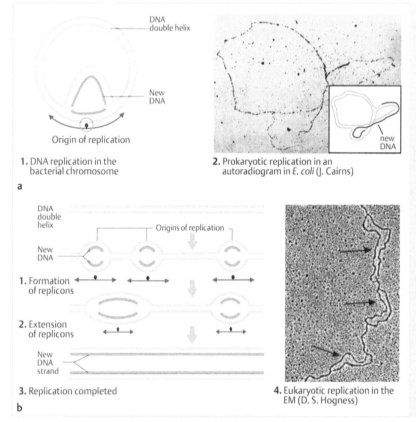

a

b

Fig. 2.4 (a) Prokaryotic replication begins at one site. **(b)** Eukaryotic replication begins at several sites. (Source: Passarge E, ed. Color Atlas of Genetics. 4th Edition. New York, NY. Thieme; 2012.)

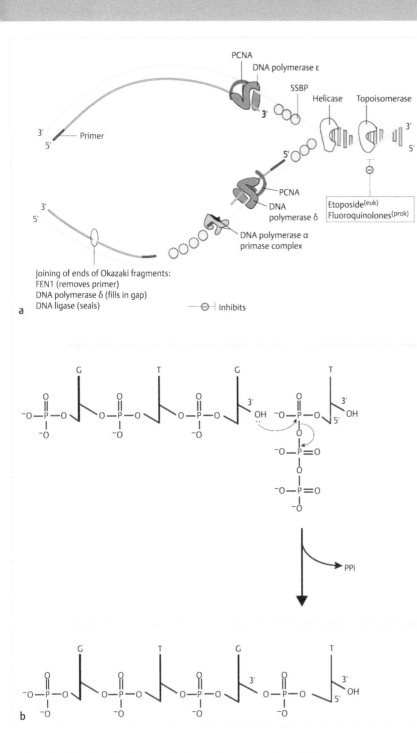

Joining of ends of Okazaki fragments:
FEN1 (removes primer)
DNA polymerase δ (fills in gap)
DNA ligase (seals)

⊖─| Inhibits

a

b

Fig. 2.5 DNA synthesis requires a template and a primer. **(a)** DNA polymerase cannot initiate DNA synthesis without a primer. It extends a preexisting oligonucleotide primer by adding new nucleotides to the 3' end. The template strand is read by the polymerase in the 3' → 5' direction, but the complementary nucleotides are added to the primer in the 5' → 3' direction. DNA polymerases are kept bound to the template with the aid of the sliding clamp proliferating cell nuclear antigen (PCNA). DNA helicase melts the hydrogen bonds and unwinds the DNA. Topoisomerases relieve the supercoils ahead of the replication fork. DNA polymerase α synthesizes the short RNA primers. Single-stranded DNA-binding proteins keep the template strands separated. SSBP, single-stranded DNA-binding proteins. **(b)** The polymerase uses free nucleotide triphosphates (NTPs) as substrates. The free 3'-OH group of the primer attacks the α-phosphate of the incoming NTP, thereby releasing the terminal two phosphates as pyrophosphate (PPi). Subsequent hydrolysis of the pyrophosphate drives the reaction forward. (Source: Panini S, ed. Medical Biochemistry—An Illustrated Review. New York, NY. Thieme; 2013.)

molecule, allowing for the revolution on its long axis, which relieves this supercoiling and prevents mechanical damage to the DNA. After the supercoiling is resolved, topoisomerase then rejoins the backbone.

The next step in replication is the **elongation** phase, in which the synthesis of the new DNA strand occurs. However, the anti-parallel nature of the individual strands of the DNA molecule, and the fact that DNA polymerase can only add nucleotides in the 5' to 3' direction, means that the two strands cannot be replicated in the same way. Recall that the strands of the DNA molecule are oriented antiparallel to each other. Since DNA polymerase, the enzyme that produces the new strand, can

only add nucleotides to free 3' ends of the DNA, they must move in a 5' to 3' direction. However, both strands must be replicated at the same time and in the same direction. To overcome this problem, each strand is synthesized in a different way. For the strand that is running in the 3' to 5' direction, designated the leading strand, there is always a free 3' OH group on the newly synthesized strand so the polymerase can move in a continuous manner along this strand. In the other strand, which is running 5' to 3', the new strand must be made in a discontinuous fashion in which the polymerase synthesizes short fragments, known as Okazaki fragments, that must be later joined together to produce one continuous DNA molecule. This

I

lagging strand is the strand of template DNA that is oriented so that the replication fork moves along it in a 5′ to 3′ manner. The lagging strand is replicated in a discontinuous fashion, producing small DNA fragments that are later joined by DNA ligase to form a continuous new strand (▶ Fig. 2.3). A key initial step in this process is the addition of a short RNA primer to provide the free 3′OH needed to start the DNA polymerase. This addition is accomplished through the action of a specialized RNA polymerase called RNA primase.

Several other enzymes are also necessary for the lagging strand to be synthesized properly. After the DNA polymerase is initiated, the RNA primer is no longer needed and must be removed to prevent the formation of a DNA-RNA hybrid. The enzyme responsible for removing the RNA primer is RNAseH. Also, each Okazaki fragment possesses a 5′ end that must be processed to allow for the joining to the 3′ end of the fragment lying upstream. FEN1 is the enzyme responsible for the processing of these 5′ ends. After the ends have been processed, another enzyme called DNA ligase joins the 3′ end to the 5′ end, creating a single DNA strand (▶ Table 2.1).

In eukaryotes, there are multiple versions of DNA polymerase, each with a specific function. Pol α exists in a complex with RNA primase and assists in the placing of the RNA primer. Pol δ synthesizes the lagging strand, and Pol ε synthesizes the leading strand. As we saw earlier, the mitochondria have their own DNA, mtDNA, and must therefore have their own machinery for replicating their DNA. In this case, Pol γ, encoded by a nuclear gene, is responsible for this process. The cell also contains several other polymerases that are responsible for DNA repair.

The binding of DNA polymerase to the DNA molecule is enhanced by a protein complex composed of the proliferating cell nuclear antigen (PCNA) protein. This complex forms a sliding clamp which increases the processivity of the polymerase, keeping it attached to the DNA molecule for a longer period of time. A protein called replication factor C (RCF), or clamp loader, is responsible for the initial recruitment of the sliding clamp to the DNA (▶ Fig. 2.5)

Termination of elongation occurs when adjacent replication forks meet and resolve each other, or when the replication fork encounters the end of the chromosome.

Table 2.1 Enzymes involved in eukaryotic DNA replication

Enzyme	Function
Helicase	Unwinds DNA helix
Topoisomerase	Relieves overwound supercoils (called DNA gyrase in bacteria)
Single-stranded DNA-binding protein	Binds the single-stranded DNA that has been separated
DNA polymerase αα (in complex with primase)	Synthesizes RNA-DNA primer
DNA polymerases δ and ε	Synthesize new DNA chain in the 5′ → 3′ direction (DNA Pol III in prokaryotes)
Flap endonuclease 1 (FEN1)	Removes RNA primers (DNA Pol I in prokaryotes)
DNA polymerase δ	Fills in gaps (DNA Pol I in prokaryotes)
DNA ligase	Seals nicks

2.3 Replication of Telomeres Requires Telomerase

The ends of linear chromosomes in eukaryotes, called the telomeres, contain thousands of tandem repeats of the sequence TTAGGG (▶ Fig. 2.6). The telomeres allow the cell to identify the ends of the chromosomes and to distinguish them from the free ends that may occur through double strand DNA breaks, which require repair. Due to the lagging strand's requirement for an upstream sequence to lay down the primer, the very end of the telomere cannot be replicated appropriately in this molecule. In order to replicate these telomeres, the activity of an enzyme called telomerase is required. This enzyme contains both a RNA template complimentary to the repeat sequences found in the telomere and a reverse transcriptase activity which allows for the synthesis of DNA from this RNA template. At the very end of the telomere, the telomerase binds to the template strand via its RNA template component and extends the template strand. This results in a longer sequence to allow for the incorporation of an RNA primer and continued synthesis of the lagging strand at the end of the telomere. The ultimate effect of this process is to prevent the shortening of the telomere in this new DNA molecule.

As cells age, the activity of telomerase falls progressively. This loss of telomerase activity necessarily leads to the shortening of telomeres, which is considered an indication of cellular aging. After many rounds of shortening, the telomere is reduced to a threshold length which the cell identifies as DNA damage, resulting in cessation of cell division.

2.4 DNA Mutations and Repair

The DNA of a cell is bombarded on a daily basis with factors that can induce changes in the nucleotide composition or structure of the DNA molecules. If these changes are permanent, they are called mutations. Somatic cell mutations only affect the host in which they occur, but mutations in germ cells can result in inherited conditions that manifest in the individual's progeny. In cells that have experienced DNA mutations, the potential consequences include loss of a gene's normal function, a disruption in the synthesis of the protein encoded by the affected gene, or in some cases the mutation may have no effect.

External insults are the most common source of DNA mutations. However, there are also internally-derived spontaneous mutations that occur at the rate of thousands of times per day. These spontaneous mutations are naturally occurring and come in the form of depurinations and deaminations (▶ Fig. 2.7**a, b**). As the name suggests, depurinations involve the specific loss of a purine nitrogen base, while the sugar molecule and the phosphate group of the nucleotide remain unchanged. These mutations also do not involve the breakage of the phosphodiester backbone. When the replication machinery encounters the missing purine on the template strand, it skips to the next complete nucleotide, producing a nucleotide deletion in the newly synthesized strand.

In contrast to depurinations, deaminations do not lead to a loss of a nitrogenous base but, rather, the conversion of one nitrogenous base to another. In this case, the loss of an amine group from cytosine converts the nitrogenous base to uracil.

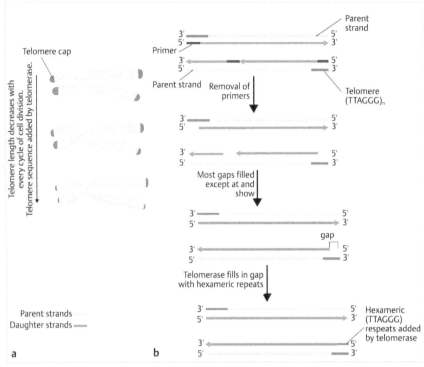

Fig. 2.6 (a) Chromosome ends are capped by telomeres. **(b)** A class of enzymes called telomerases add hexameric repeat sequences (TTAGGG) to the ends of chromosomes to form the telomere caps. Telomerases are ribonucleoproteins that contain a built-in RNA template. (Source: Panini S, ed. Medical Biochemistry—An Illustrated Review. New York, NY. Thieme; 2013.)

When the DNA replication machinery encounters this uracil, it will insert an adenine instead of the cytosine partner guanine, resulting in a G to A nucleotide change at this position in the newly synthesized DNA molecule.

The deletions and nucleotide substitutions that result from depurinations and deaminations are examples of point mutations. Point mutations are mutations that cause a change in a single nucleotide in the DNA sequence. If these mutations occur in the coding region of a gene, they can have one of several different consequences. If the mutation changes one amino acid within the encoded protein, then it is called a missense mutation. When the mutation produces a premature stop codon, resulting in the synthesis of a truncated protein, then it is called a nonsense mutation. Insertions or deletions of nucleotides can change the reading frame of the encoded protein, resulting in a frame shift mutation.

As mentioned earlier, induced mutations are the most common type of mutation that results in detrimental effects on the cell. Examples are x rays, which produce double strand breaks, ultraviolet light, creating thymine dimers (covalent bonds between carbons in adjacent thymines), and a multitude of carcinogenic chemicals such as benzo[a]pyrene, a chemical found in cigarette smoke that, when it is metabolized by cytochrome p450 enzymes, combines with guanine to make a mutagenic bulky adduct.

Trinucleotide repeat expansions are increases in the number of tandem repeats of a specific three-nucleotide sequence. This type of mutation is due to a defect during DNA replication, in which the newly synthesized strand dissociates from the template strand, shifts a short distance along the template strand and re-anneals, creating a kink. During this process, the polymerase replicates a number of the trinucleotides a second time (▶ Fig. 2.8).

In addition to mutations that affect the DNA molecule on the nucleotide level, gross chromosomal rearrangements that result from breaking and re-annealing of regions of two different chromosomes can occur. These rearrangements are called chromosomal translocations. These genomic alterations are the result of double stranded DNA breaks that create free ends that reseal with the break point of another broken chromosome. Chromosomal translocations are frequently observed in cancer cells, and some hereditary diseases are associated with chromosomal translocations.

2.5 Nucleotide Excision Repair and Base Excision Repair

In normal healthy cells, there are several ways in which DNA mutations can be repaired, depending on what type of mutation has occurred. Single nucleotide defects are repaired by either excision repair or mismatch repair. Excision repair includes two separate types of repair: nucleotide excision repair and base excision repair. The type of nucleotide defect present will determine which excision repair mechanism is used. For example, thymine dimers and bulky adducts are repaired by nucleotide excision repair and deaminations and depurinations are repaired by base excision repair. The basic processes for nucleotide excision repair and base excision repair are similar, although each uses a different set of enzymes to carry out its function (▶ Fig. 2.9**a**).

2.5.1 Link to Pathology

Xeroderma pigmentosum is a genetic disease in humans associated with mutations in one of the excision repair proteins that is required for the removal of UV-induced thymine dimers. The inherited mutation occurs in a member of the XP

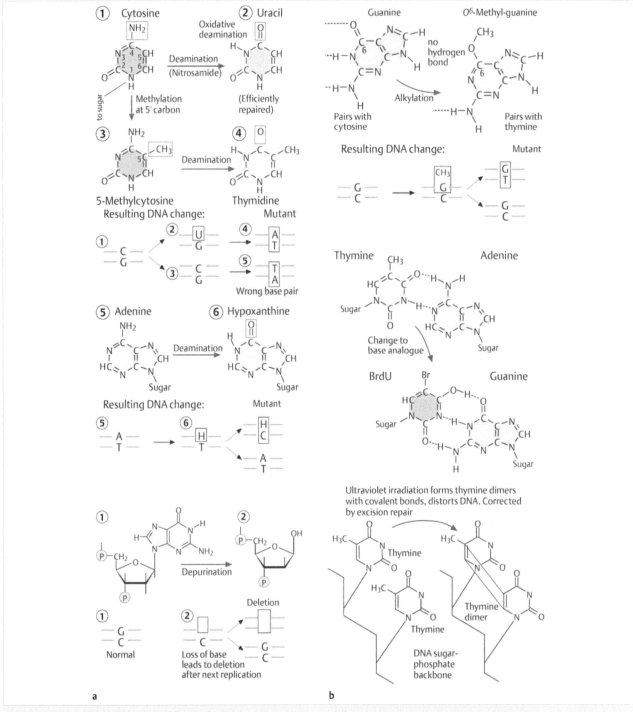

Fig. 2.7 (a) Deamination and methylation, Depurination. **(b)** Alkylation of guanine. Base analogues. UV-light-induced thymine dimers. (Source: Panini S, ed. Medical Biochemistry—An Illustrated Review. New York, NY. Thieme; 2013.)

family of genes, such as XPA, XPB, XPC, etc. Due to impaired excision repair, individuals with this disease are sensitive to light and readily accumulate UV-induced DNA damage, including thymine dimers, upon exposure to sunlight. The increased rates of mutation predispose them to developing multiple pigmented growths on the skin and place them at a high risk of skin cancer.

2.6 Mismatch Repair

During the DNA replication process, DNA polymerase occasionally incorporates the incorrect base, creating a mismatch with the base present on the template strand. The repair mechanism used to correct this mistake is called mismatch repair. Because the mismatched bases do not form a normal Watson-Crick base

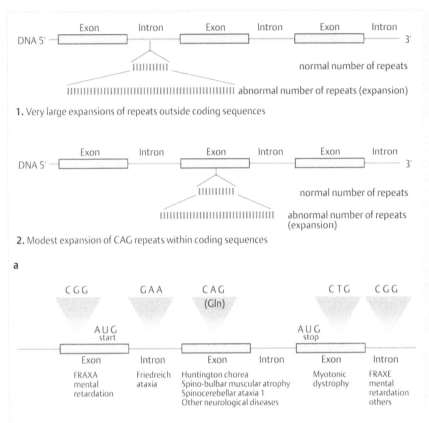

Fig. 2.8 (a) Different types of trinucleotide repeat expansion (b) Unstable trinucleotide repeats in different diseases. (Source: Panini S, ed. Medical Biochemistry—An Illustrated Review. New York, NY. Thieme; 2013.)

pairing, this area is bulged out, producing a bubble in the DNA. This bubble is recognized by the mismatch repair enzyme complex, which removes a segment of the newly synthesized DNA that includes the mismatched bases. DNA polymerase then fills in the gap that is produced, incorporating the correct base this time. DNA ligase is then used to connect the 3′ end of the newly synthesized fragment to the 5′ end of the unaffected downstream strand (▸ Fig. 2.9**b**).

2.6.1 Link to Pathology

Hereditary NonPolyposis Colorectal Cancer (HNPCC), also known as **Lynch syndrome**, is a form of hereditary colon cancer caused by mutations in members of the mismatch repair machinery, most commonly MSH2 or MLH1. When these mutations are present, the cell is unable to repair nucleotide mismatches. They also exhibit variations in the length of tandem repeat regions in the genome, known as microsatellite regions, producing a condition known as microsatellite instability. In addition to colon cancer, females with Lynch syndrome are at higher risk for endometrial cancer.

2.7 Non-Homologous End Joining and Homologous Recombination

In response to a mutagenic challenge that produces double strand breaks, the cell has two choices for repair: non-homologous end joining and homologous recombination (▸ Fig. 2.9**c**). In somatic cells, **non-homologous end joining** is the most common mechanism for repairing double strand breaks due to the fact that it does not require a homologous chromosome for a template. Following a double strand break, nucleases process the broken ends to form blunt ends, resulting in the loss of nucleotides in the overhang region. The ends are then brought together by a specialized group of enzymes and rejoined by DNA ligase. Although this is a relatively quick method for repairing double strand breaks, it is error-prone since the loss of nucleotides could result in a mutation if it occurs in an expressed gene. Consequently, cells that have undergone DNA replication (post-S phase) preferentially use another type of double-strand break repair known as homologous recombination.

In contrast to non-homologous end joining, **homologous recombination** is an error-free method for repairing double strand breaks in DNA. This is due to the use of a homologous chromosome as a template to re-synthesize the damaged area. Homologous recombination is commonly used for the repair of newly replicated DNA. The process is as follows: 1) Two homologous chromosomes become aligned, 2) a nuclease generates single-stranded ends at the break by chewing back one of the complementary strands 3) One of the single strands then invades the homologous DNA duplex by forming base pairs with its complementary strand. A significant number of bases must pair to produce a branch point where one strand from each duplex crosses 4) The invading strand is elongated by DNA polymerase, using the complementary strand as a template 5) The branch point migrates as the base pairs holding together the duplexes break, and new ones

I

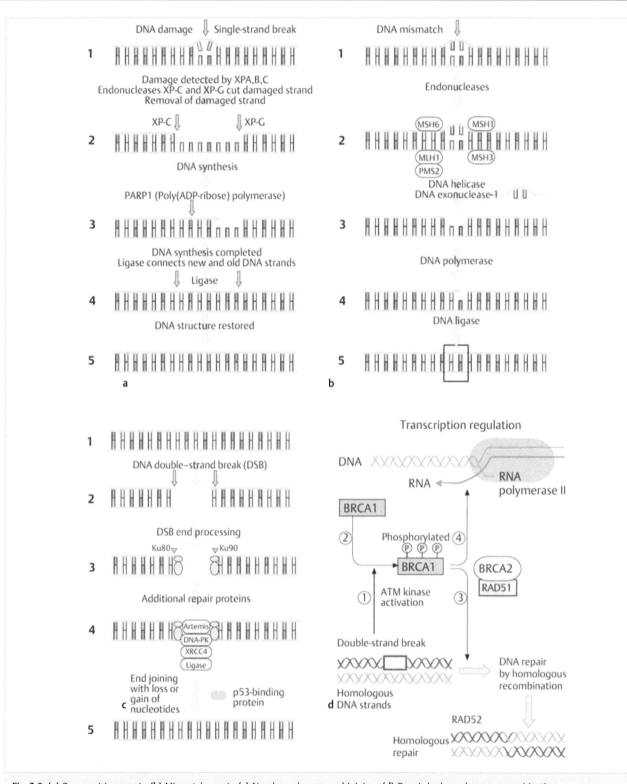

Fig. 2.9 (a) Base-excision repair. **(b)** Mismatch repair. **(c)** Non-homologous end joining. **(d)** Repair by homologous recombination.

form 6) Additional DNA synthesis and ligation completes the repair. Since sister chromatids are required for this type of repair, this process must occur after replication, in either G2 or M. Therefore, cells that have not transitioned through S phase (i.e.,- G1 cells) must use the error-prone non-homologous end joining to repair double-strand breaks, whereas G2 or M phase cells will preferentially use the error-free homologous recombination.

2.7.1 Link to Pathology

Ataxia telangiectasia is a disease caused by mutations in the gene ATM, which is an important component of both non-homologous end joining and homologous recombination. In normal cells, ATM is recruited to the double strand breaks by binding to the proteins that recognize the break points and becomes activated. The activated ATM, in turn, activates additional downstream molecules that are necessary for repair of the DNA. Individuals with this disease have an impaired DNA damage response in both NHEJ and homologous recombination repair and, as such, are susceptible to agents that cause double stranded breaks in DNA, such as X-rays, radiation therapy, and radiomimetic agents (▶ Fig. 2.10).

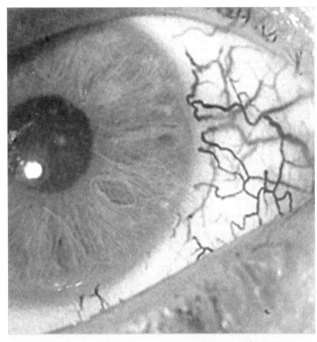

Fig. 2.10 Telangiectasias of the conjuctiva. Ataxia-telangiectasia (AT).

Review Questions

1. Which of the following represents a DNA sequence that is complimentary to the sequence
   5'-ATGGCCACCTTGGTTAACCGC-3'?
   A) 5'-ATGGCCACCTTGGTTAACCGC-3'
   B) 5'-GCGGTTAACCAAGGTGGCCAT-3'
   C) 5'-TACCGGTGGAACCAATTGGCG-3'
   D) 5'-CGCCAATTGGTTCCACCGGTA-3'
   E) 5'-AUGGCCACCUUGGUUAACCGC-3'

2. The structure of which of the following differs between DNA molecules and RNA molecules?
   A) Phosphodiester backbone
   B) Cytosine nucleotides
   C) Ribose sugar
   D) Phosphate group

3. The replication of nuclear DNA occurs in which phase of the cell cycle?
   A) G1
   B) S
   C) G2
   D) M

4. Which of the following complexes contains DNA helicase, DNA topoisomerase, DNA polymerase, and RNA primase?
   A) Minichromosome maintenance complex (MCM)
   B) Mismatch repair complex
   C) Origin of replication complex (ORC)
   D) Replisome complex

5. Mutations in the ATM gene are most likely to affect the repair of which type of DNA damage?
   A) Deamination
   B) Depurination
   C) Double strand breaks
   D) Nucleotide mismatches
   E) Thymine dimers

Answers

1. **The correct answer is B**. Watson-Crick base pairing rules dictate that A's always pair with T's and C's always pair with G's. In addition, recall that the two strands of a double-stranded DNA molecule are oriented in an antiparallel fashion. Therefore, answer choice B shows the complimentary sequence to the given sequence and it is in the correct orientation.

2. **The correct answer is C**. The nucleotides of DNA molecules contain a ribose sugar that lacks an OH group on the 2' carbon, whereas RNA molecules contain ribose that possesses an OH group on the carbons in both the 2' and 3' positions. Both DNA and RNA molecules contain nucleotides that possess phosphate groups (answer choice D) that form a phosphodiester backbone (answer choice A) that connects neighboring nucleotides within the polynucleotide chain. Both DNA and RNA molecules can contain cytosine (answer choice B), although RNA has uracil in the place of thymidine.

3. **The correct answer is B**. During the S-phase of the cell cycle, the DNA is replicated. In addition, the centrosome is duplicated during this phase. G1 (answer choice A) and G2 (answer choice C) are the two gap phases, and M phase (answer choice D), or mitosis, is the period in which the cell physically divides, producing two new daughter cells.

4. **The correct answer is D**. The replisome is a complex containing helicase, topoisomerase, DNA polymerase, and RNA primase that is required for the replication of the genomic DNA. The minichromosome maintenance complex (answer choice A), or MCM, is a complex of proteins that recruits DNA helicase to the origin of replication so that separation of the two DNA strands can occur. The origin of replication complex (answer choice C) binds to sequences within the genome known as origins of replication and recruits the MCM. Mismatch repair complexes (answer choice B) correct nucleotide mismatches that occur during the process of DNA replication.

5. **The correct answer is C**. ATM which is an important component of both non-homologous end joining and homologous recombination. Since double strand breaks are corrected by these repair mechanisms, mutations in ATM would affect double strand break repair. The disease Ataxia telangiectasia is caused by mutations in ATM. Deaminations (answer choice A) and depurinations (answer choice B) are internally-derived spontaneous mutations that occur at the rate of thousands of times per day. These types of mutations are fixed by an excision repair mechanism. During the DNA replication process, DNA polymerase occasionally incorporates the incorrect base, creating a nucleotide mismatch (answer choice D) with the base present on the template strand. The repair mechanism used to correct this mistake is called mismatch repair. Thymine dimers (answer choice E) are induced by ultraviolet light exposure and are repaired by excision repair mechanisms.

3 Transcription and Regulation of Transcription

At the conclusion of this chapter, students should be able to:
- Describe the different types of RNA molecules and their roles
- Compare the transcription of prokaryotic and eukaryotic genes
- Describe the synthesis of mRNA in prokaryotes and in eukaryotes
- Describe the regulatory mechanisms of transcription and RNA processing
- Identify compounds, mutations, and errors that disrupt the process of transcription and/or regulation of transcription

The synthesis of RNA from a DNA template is carried out by RNA polymerases. Cells contain three types of RNA molecules: ribosomal RNA, messenger RNA, and transfer RNA (▶ Fig. 3.1). Ribosomal RNA (rRNA) and transfer RNA (tRNA) play important roles in translation. rRNA molecules function as structural components of ribosomes as well as catalyze peptide bond formation during protein synthesis. tRNAs are the carriers of activated amino acids to ribosomes during protein synthesis. mRNAs encode the amino acid sequences of proteins and mediate the transmission of genetic information from the DNA form to the protein form.

RNA polymerases synthesize RNA molecules from the DNA template. In prokaryotes, all three major classes of RNA are transcribed by one multi-subunit RNA polymerase. However, in eukaryotes the rRNA is synthesized by RNA polymerase I, mRNA is synthesized by RNA polymerase II, and tRNA is synthesized by RNA polymerase III (▶ Fig. 3.2).

RNA polymerase II reads the template DNA strand in the 3'-to-5' direction and synthesizes the mRNA molecule in the 5'-to-3' direction. Therefore, the mRNA sequence is antiparallel and complimentary to the template or "antisense" DNA strand. However, the mRNA sequence is parallel and corresponds to the coding or "sense" strand of the DNA. Unlike DNA polymerase, RNA polymerase cannot proof read its

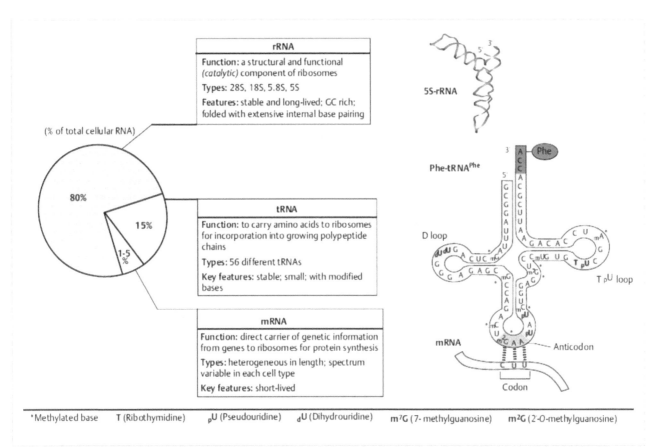

Fig. 3.1 RNA types and their secondary/tertiary structures. The three major types of RNA in eukaryotic cells are rRNA (80% of total), tRNA (15% of total), and mRNA (1 to 5% of total). All but mRNA are long-lived species. rRNA (e.g., 5S-rRNA) and tRNA (e.g., phenylalanyl-tRNAPhe) exhibit extensive secondary structures, whereas mRNA is generally a linear molecule (although some mRNAs do contain stem-loop or hairpin structures in their untranslated regions). tRNA not only assumes a cloverleaf-like secondary structure but also contains a tertiary structure (not shown) that facilitates its function as an adapter in protein synthesis. tRNA is "charged" at its 3' end (red) when attached to its cognate amino acid. The anticodon loop (pink) contains a triplet nucleotide sequence that is complementary to the codon sequence on the mRNA. These complementary sequences form hydrogen bonds with each other on the assembled ribosome. tRNA also contains several unusual nucleotides, such as ribothymidine (T), dihydrouridine (dU), pseudouridine (pU), 7-methylguanosine (m7G), and 2'-O-methylguanosine (m2G). (Source: Panini S, ed. Medical Biochemistry—An Illustrated Review. New York, NY. Thieme; 2013.)

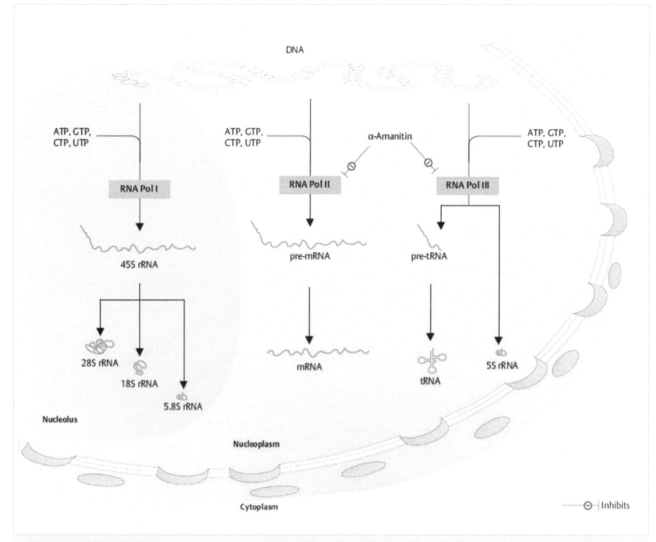

Fig. 3.2 RNA polymerases and the RNAs they transcribe. In the nucleolus, RNA polymerase I synthesizes a 45S precursor rRNA, which is cleaved into the 28S, 18S, and 5.8S rRNAs. In the nucleoplasm, RNA polymerase II synthesizes a precursor mRNA, which is then processed into mature mRNA. Also in the nucleoplasm, RNA polymerase III synthesizes small RNAs such as tRNAs and the 5S rRNA. α-Amanitin is a more potent inhibitor of RNA polymerase II than of RNA polymerase III. (Source: Panini S, ed. Medical Biochemistry—An Illustrated Review. New York, NY. Thieme; 2013.)

errors and cannot repair the errors during transcription. This inability to repair errors is due to the absence of 3′-to-5′ exonuclease activity.

3.1 Link to Microbiology

Most of the RNA viruses, including influenza and HIV viruses, develop resistance to anti-viral drugs because of the increased rate of mutations in their genome caused by the lack of proofreading activity in the RNA polymerase.

3.2 Structure and Transcription of Prokaryotic Genes

Both the process of mRNA transcription and the organization of gene structure are less complex in prokaryotes than in eukaryotes (▶ Fig. 3.3).

In prokaryotes, the sigma (σ) factor of RNA polymerase binds the promoter region and allows initiation of transcription. Two consensus sequences, located at positions -10 and -35 in the DNA template strand, are recognized by the RNA polymerase. Unlike DNA polymerase, RNA polymerase does not require a primer to initiate polynucleotide synthesis. By convention, the first nucleotide that is transcribed is designated as the +1 nucleotide. During the elongation phase of transcription, the RNA polymerase synthesizes the nascent RNA molecule in the 5′-to-3′ direction. In prokaryotes, transcription and translation occur at the same time. As the RNA molecule is being transcribed, the ribosomes bind to a Shine-Dalgarno sequence at the 5′UTR region of the RNA. This allows for the concurrent synthesis of both the RNA and the protein for which it encodes.

Due to the coupling of transcription and translation in prokaryotes, most of the regulation of gene expression occurs during transcription, and the rate of transcription is regulated by the interactions of proteins with specific DNA sequences.

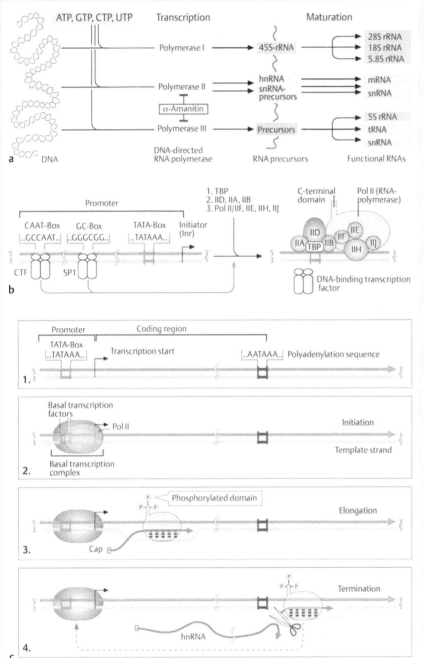

Fig. 3.3 (a) transcription and maturation of RNA: overview. **(b)** Basal transcription complex. **(c)** Process of transcription. (Source: Koolman J, Röhm K, ed. Color Atlas of Biochemistry. 3rd Edition. New York, NY. Thieme; 2012.)

RNA synthesis is terminated by either a rho-dependent process or a rho-independent process.

In the rho-independent termination process, 6–8 uracil nucleotides form weak bonds with the complementary 6–8 adenine nucleotides in the DNA template. This is followed by a hairpin loop structure created by intramolecular complementary sequence within the RNA molecule. The hairpin loop and the weak hydrogen bonding between AU pairs stimulates the dissociation of the RNA from the template DNA strand.

The rho-dependent termination process requires Rho factor, which is a protein with RNA helicase activity that binds to the newly formed RNA molecule and promotes the displacement of RNA polymerase, terminating transcription.

3.2.1 Link to Pharmacology

Many antibiotics interfere with prokaryotic RNA transcription or protein synthesis, without an effect on eukaryotes. **Rifampin**, which is used for the treatment of tuberculosis, inhibits the initiation of RNA synthesis by binding to the β subunit of the prokaryotic RNA polymerase. **Actinomycin D** inhibits movement of RNA polymerase on the DNA template. **Fidaxomicin** binds to the sigma factor of the RNA polymerase and inhibits the transcription. This drug is used for the treatment of *C. difficile* infection. Symptoms of *C. difficile* infection include abdominal cramps, watery diarrhea, blood or pus in stool, dehydration, and fever.

3.3 Structure and Transcription of Eukaryotic Genes

There are two major differences between prokaryotic and eukaryotic gene structures. While the prokaryotic coding region is a continuous sequence, eukaryotic genes contain exons (coding regions) and introns (non-coding regions). The gene regions in prokaryotes may be polycistronic meaning they encode for several different proteins, whereas gene regions in eukaryotes are always monocistronic.

The transcription of eukaryotic DNA will be reviewed in three phases: initiation, elongation, and termination.

3.4 Assembly of the initiation complex

RNA polymerase II synthesizes mRNA and is dependent on transcription factors (TFIIs), proteins that recognize and bind to specific DNA sequences in the promoter region of the DNA to initiate transcription (▶ Fig. 3.4). There are short sequences within eukaryotic gene promoters that are recognized by transcription factors. These are known as the "*TATA box*", "*CAAT box*" and "*GC box*". The transcription factor TFIID binds the TATA box with the help of another protein called TBP (TATA-Binding Protein) and TFIIA. TFIIB serves as a bridge between TFIID and the RNA polymerase II, while the ATP-dependent helicase TFIIH is opens up the double helix of the DNA to give the RNA polymerase access to the template strand (▶ Fig. 3.5**a**). Unlike DNA polymerase, RNA polymerase II can initiate synthesis of the RNA chain without a primer.

3.4.1 Link to Pharmacology and Toxicology

The α-amanitin compound, found in the death cap mushroom *Amanita phalloides*, inhibits transcription in eukaryotes by binding to RNA polymerase II, thus ceasing protein production and subsequent cellular metabolism and leading to cell death.

The liver and kidneys are most susceptible to the toxic effects of α-amanitin, with the initial symptoms being abdominal pain, watery diarrhea, nausea, and vomiting. In severe cases, hypotension, tachycardia, hypoglycemia, and acid-base disturbances arise. Due to liver and kidney failure, life threatening complications occur within two weeks.

3.5 Elongation and Termination

The processing of pre-mRNA including 5'-capping, polyadenylation, and RNA splicingis carried out by factors bound to the tail of RNA polymerase II, and these processes occur as the mRNA molecule is synthesized (Fig. 3.5b). The 5'-cap is added early in the elongation process, whereas the poly-A tail is synthesized at the end of elongation by an enzyme called poly-A polymerase, following cleavage at the 3' near the specific sequence AAUAAA. Both the 5'-cap and the poly-A tail increase the stability of newly synthesized RNA by protecting it from nucleases. In addition, the 5'cap and poly-A tail also function as transport signals from the nucleus to the cytoplasm. Splicing consists of the removal of the introns from the pre-mRNA molecule and joining of the exons together to form the mature mRNA. This process of splicing occurs in the spliceosomes, which are composed of the small nuclear ribonucleo proteins (snRNPs) U1, U2, U4, U5, and U6. The GU sequence at the beginning and the AG sequence at the end of each intron are recognized by the snRNPs in the spliceosome (▶ Fig. 3.6).

Alternative splicing of the introns yields several different mRNAs from the same primary transcript. For example, a whole family of different tissue-specific tropomyosins are the result of alternative splicing of the tropomyosin primary transcript.

3.5.1 Link to Pathology

β-Thalassemia, cystic fibrosis, Gaucher, and Tay-Sachs are examples of diseases that are caused by mutations in splice sites, which result in abnormal or truncated protein products. In some cases of β-Thalassemia, mutations within the splice site of the first intron cause the production of an abnormally long mRNA. This is because the mutation prevents the intron from

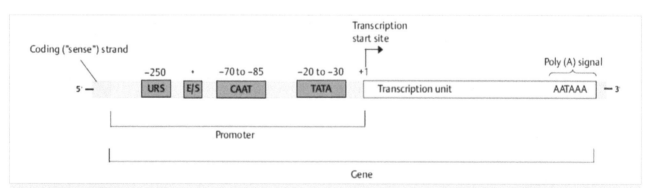

Fig. 3.4 Functional organization of genes. A transcription unit is the linear DNA sequence from the transcription start site (+1) to the polyadenylation (poly A) signal sequence (AATAAA). Eukaryotic RNA polymerases are re cruited to the core promoter by basal (general) transcription factors that form a pre-initiation complex on the promoter. Formation of the initiation complex is complete when it includes RNA polymerase II. The TATA box is usually found 20 to 30 base pairs (bp) upstream of the transcription start site, and the CAAT box may be found 70 to 85 bp upstream of the start site. The upstream regulatory sequence (URS), to which specific transcription factors bind, may extend up to or beyond 250 bp upstream of the start site. Enhancer (E) and silencer (S) sequences can be anywhere (*) upstream, downstream, or within the transcription unit and on either the coding or the noncoding strand. (Source: Panini S, ed. Medical Biochemistry—An Illustrated Review. New York, NY. Thieme; 2013.)

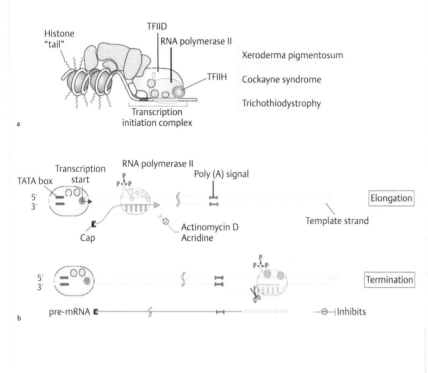

Xeroderma pigmentosum

Cockayne syndrome

Trichothiodystrophy

Elongation

Template strand

Termination

Inhibits

Fig. 3.5 Process of transcription, **(a)** Key components of the transcription initiation complex of eukaryotes are TFIID (which recognizes and binds to the TATA box), TFIIH (which separates the DNA strands), and RNA polymerase II (which catalyzes the polymerization of ribonucleotides into an RNA chain). **(b)** The elongation of mRNA chains proceeds in the 5′ → 3′ direction, 500 to 2000 nucleotides past the polyadenylation signal until RNA polymerase II encounters nonspecific termination signals. Poly(A) polymerase-associated enzymes then cleave the primary transcript at a site 10 to 35 nucleotides downstream from the polyadenylation signal. Actinomycin D and acridine inhibit elongation by intercalating between the bases of the DNA double helix. (C-1) DNA becomes accessible via the acetylation of lysine residues on histone tails. (C-2) RNA polymerase II initiates mRNA synthesis by catalyzing the formation of a dinucleotide from two ribonucleoside triphosphates. This (and subsequent) phosphodiester bond formation is driven by the hydrolysis of pyrophosphate (PPi) into two inorganic phosphates (2 Pi). (C-3) Shortly after synthesis begins, a 7-methylguanosine cap is added to the 5′ end of the mRNA transcript. HAT, histone acetyltransferase; HDAC, histone deacetylase. (Source: Panini S, ed. Medical Biochemistry—An Illustrated Review. New York, NY. Thieme; 2013.)

being recognized by spliceosome, resulting in the inclusion of the intron in the mature mRNA molecule that is produced.

3.6 Regulation Mechanisms of Eukaryotic Transcription

3.6.1 Chromatin Modifications such as Methylation of Cytosines on DNA or Acetylation of Histones

The highly methylated DNA found in CpG "islands" forms condensed chromatin known as heterochromatin. Because transcription factors have little or no physical accessibility to bind to promoter regions of the genes in these heterochromatin areas, the genes are held in an inactive state. In contrast, euchromatin regions have highly acetylated histones, which cause loose packing of nucleosomes, and unwinding of DNA. Since the promoter regions are accessible to transcription factors, the rate of transcription is higher in euchromatin regions.

Link to Pathology

In a normal individual, the DNA sequence of the Fragile X syndrome gene (*FMR1*) contains around 30 CGG repeats in the promoter region. In the case of fragile X syndrome, a trinucleotide-expansion of this region occurs, with the CGG repeat number increasing to > 200. This leads to an increase in methylation of the promoter of the *FMR1* gene and a loss of transcription of this gene and subsequent reduction in production of the protein product. The symptoms of Fragile X syndrome are mild to severe intellectual disability, a delay in talking, hyperactive behavior, and anxiety. Physical symptoms are long face, large ears, and prominent jaw and forehead.

Binding of gene-specific transcription factors to enhancer and silencer elements on DNA Although enhancer and silencer regions in DNA can be very far (~1000 bases) upstream or downstream from the promoter, they are capable of interacting with transcriptional activators or repressors and influencing the activity of the general transcription factors bound to the promoter regions.

Transcription factors and transcriptional activators share common secondary structural motifs. The most common secondary structural motifs are found in the homeodomain proteins, the helix-loop-helix family, the leucine zipper family, and the zinc-finger family.

Homeodomain Proteins: The homeodomain is a conserved amino acid sequence that is found in some transcription factors. The genes that encode these proteins are called **Hox genes** and are involved in regulation of anatomic development during early embryogenesis. Mutations in these genes causes congenital malformations such as synpolydactyly, which is reviewed in Chapter 55.

Helix-loop-helix Proteins: The helix-loop-helix proteins comprise a are large class of transcriptional regulators that share a structural motif of two α-helices connected by a short loop. Helix-loop-helix proteins regulate expression of genes that are involved in sex determination, as well as the development of muscle and the nervous system.

I

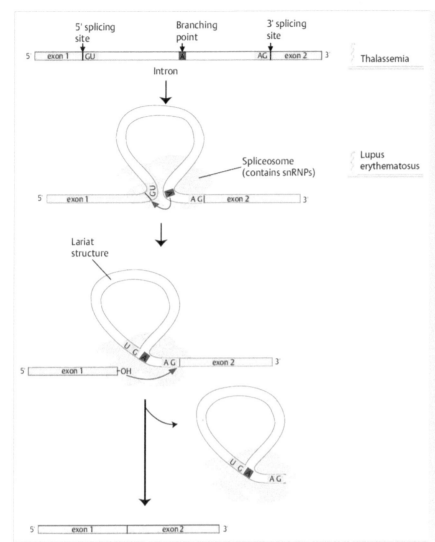

5' splicing site

Branching point

3' splicing site

Thalassemia

Lupus erythematosus

Fig. 3.6 Mechanism for splicing pre-mRNA, Intron sequences usually have GU at their 5' end and AG at their 3' end. An adenosine (A) is usually found at the branching point within the intron sequence. The snRNPs of the spliceosome recognize intron–exon junctions and splice out the intron as a "lariat" structure. Thalassemia is associated with a mutation (G to A) in the exon–intron junction. In systemic lupus erythematosus, the body produces antibodies against its own snRNPs. (Source: Panini S, ed. Medical Biochemistry—An Illustrated Review. New York, NY. Thieme; 2013.)

Leucine Zipper Proteins: Some transcription factors have amino acids sequences that contain leucine residues in every seventh amino acid position. This results in a secondary structure called the leucine zipper dimerization domain, which assists in the dimerization, and thus activation, of these transcription factors. Leucine zipper proteins interact with other proteins that are involved in cell proliferation and apoptosis. Some examples of leucine zipper proteins are c-jun and myc family members.

Zinc Finger Proteins: These transcription factors contain a conserved domain, called the zinc finger motif, made up of two cysteine and two histidine residues that interact with Zn^{2+} ions. The zinc finger proteins bind to DNA through this zinc finger motif. In addition to the regulation of transcription, these proteins have diverse functions including apoptosis, development, differentiation, metabolism, and autophagy.

Review Questions

1. Which of the following inhibits transcription by binding to RNA polymerase II?
   A) Actinomycin D
   B) α-amanitin
   C) Fidaxomicin
   D) Rifampin

2. Given the DNA template sequence 5'-GCTACT-3', RNA polymerase II would synthesize which of the following products?
   A) 5'-ATCGGC-3'
   B) 5'-GCTAUA-3'
   C) 5'-AGUAGC-3'
   D) 5'-AUCGTG-3'
   E) 5'-GCUAAT-3'

3. Which of the following molecules binds to the TATA box in the promoter region of a gene?
   A) Rho factor
   B) Sigma factor
   C) TFIID
   D) TFIIH

4. Which of the following mRNA processing steps requires the activity of snRNPs?
   A) 5'-capping
   B) Methylation
   C) Polyadenylation
   D) Splicing

Answers

1. **The correct answer is B**. α-amanitin is a toxin found in the death cap mushroom *Amanita phalloides* and inhibits transcription in eukaryotes by binding to RNA polymerase II. Actinomycin D (answer choice A) inhibits movement of RNA polymerase on the DNA template, while fidaxomicin (answer choice C) binds to the sigma factor of the RNA polymerase and inhibits transcription, and rifampin (answer choice D) inhibits initiation of RNA synthesis by binding to the β subunit of the RNA polymerase.

2. **The correct answer is C**. The DNA template is 5'-GCTACT-3'. RNA polymerase reads the template DNA strand 3'to 5' direction and synthesizes the RNA molecule in the 5' to 3' direction. So the newly synthesized RNA sequence will be in the 5' to 3' direction AGUAGC. Remember that T is replaced with U in the RNA molecule. Also, note that if the given DNA sequences' ends aren't specified, it is always written in the 5' to 3' direction.

3. **The correct answer is C**. Short sequences in the promoters of genes, known as TATA boxes, are recognized by the transcription factor TFIID. This protein binds to the TATA box through interactions with TBP (TATA-Binding Protein) and TFIIA. TFIIH (answer choice D) is an ATP-dependent helicase, which separates the two strands of the double helix DNA. In prokaryotic cells, the Rho factor (answer choice A) is a protein that binds to the newly formed RNA during transcription and causes displacement of RNA polymerase and termination of transcription. Also in prokaryotes, the sigma factor (answer choice B) of RNA polymerase binds the promoter region and allows initiation of transcription.

4. **The correct answer is D**. The splicing of a pre-mRNA molecule involves the removal of the introns within the sequence and joining of the remaining exons together. In this process, the spliceosome, which is made up of snRNP molecules (U1, U2, U4, U5, and U6) brings together a GU sequence at the beginning of the intron and an AG sequence at the end of the intron to form a lariat structure that allows for the intron's removal. 5'-capping (answer choice A) and addition of the poly-A tail at the 3' end, also called polyadenylation (answer choice C), increase the stability of newly synthesized RNA by protecting it from nucleases. Methylation (answer choice B) is not a step in mRNA processing, but is an important mechanism for silencing of promoters in DNA.

4 Translation and Regulation of Translation

At the conclusion of this chapter, students should be able to:
- Describe the genetic code, reading frames and how mutations affect the genetic code
- Outline the roles of ribosomes, rRNA, and tRNA in the process of translation
- Compare translation in prokaryotic and eukaryotic cells
- Explain the mechanisms of action of antibiotics that interfere with prokaryotic protein synthesis
- Describe the regulatory mechanisms of translation
- Identify compounds, mutations, and errors that disrupt the process of translation

4.1 Genetic Code and Mutations

The messenger RNA molecule (mRNA) is read in a series of three-nucleotide units, or codons, which correspond to either a specific amino acid or a signal that terminates translation (▶ Fig. 4.1). Nearly all of the twenty amino acids that make up the proteins of the cell are encoded by more than one codon. The only two exceptions are methionine and tryptophan, each being encoded by only one unique codon. In addition, there are three codons that do not encode an amino acid but, rather, signal the termination of translation. These terminating codons are called the stop codons. Due to the redundant nature of the genetic code, it is often described as being degenerate.

The manner in which the codons are read by the translational machinery in the cell is referred to as the reading frame. Each mRNA molecule can have three separate reading frames depending on when the grouping of the trinucleotide units starts. The position of the AUG start codon, which encodes methionine, is used to determine the reading frame of the mRNA molecule.

There are multiple types of mutations that can affect the genetic code. These include missense, nonsense, silent, and frameshift mutations. A **missense mutation** is a change in the nucleotide sequence that leads to a new codon encoding a different amino acid, whereas **nonsense mutations** are changes in the nucleotide sequence that create a stop codon and lead to termination of the peptide chain. **Silent mutations** are changes in the nucleotide sequence that lead to a change in the codon, but it does not change the amino acid encoded due to the degenerate nature of the genetic code. In contrast, **frameshift mutations** are deletions or insertions of a nucleotide that change the reading frame of the downstream codons and amino acid sequence. However, not all insertions or deletions are frameshift mutations. If a unit of three nucleotides, or multiples thereof, are inserted or deleted, then a frameshift will not occur, but there will be an addition or loss of some amino acids.

4.1.1 Link to Pathology

A classic example of a missense mutation is the Glu to Val substitution in the β-globin gene that causes structural changes in the β-globin subunit of hemoglobin. This mutation, if found on both alleles, can give rise to sickle cell anemia.

4.2 Basic Components of Translation

In order for the cell to translate an mRNA molecule into a protein, several key components are required. These include ribosomes, aminoacyl-tRNAs, mRNAs, and translation factors. Ribosomes are created by the pairing of one large subunit with one small subunit, with each subunit being composed of both proteins and ribosomal RNA (rRNA) (▶ Fig. 4.2). Although this fundamental configuration of the ribosome is similar in prokaryotes and eukaryotes, there is a distinct difference in the size, as well as the protein and rRNA content, of these structures.

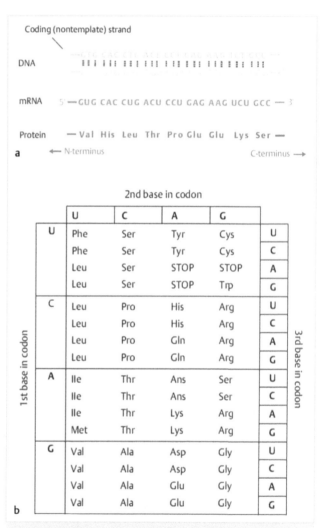

Fig. 4.1 Colinearity of nucleotide and amino acid sequences and the genetic code. **(a)** The sequence of triplet deoxyribonucleotides in the coding (sense or nontemplate) strand of DNA determines the corresponding sequence of ribonucleotide codons in mRNA (except that U replaces T), which in turn determines the sequence of amino acids in a protein. **(b)** A standard genetic code. (Source: Panini S, ed. Medical Biochemistry—An Illustrated Review. New York, NY. Thieme; 2013.)

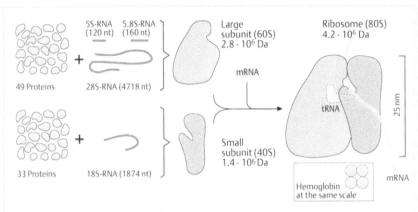

Fig. 4.2 Structure of eukaryotic ribosomes. (Source: Koolman J, Röhm K, ed. Color Atlas of Biochemistry. 3rd Edition. New York, NY. Thieme; 2012.)

In prokaryotes, the ribosomes are composed of a 50S large subunit and a 30S small subunit, which come together to form the 70S ribosome. The prokaryotic large subunit contains 5S and 23S rRNA components and 31 proteins whereas the small subunit contains a 16S rRNA and 21 proteins. The eukaryotic ribosome is a larger complex of 80S, consisting of a 60S subunit, made of 5S, 5.8S, and 28S rRNAs and 46 proteins, and a 40S subunit containing an 18S rRNA and 33 proteins (▶ Fig. 4.2).

Aminoacyl-tRNAs are transfer RNA (tRNA) molecules that are associated with a specific amino acid. Their function is to deliver the correct amino acid to the ribosome to be incorporated into the growing polypeptide chain during translation. The tRNA portion is a single stranded RNA molecule with self-complementary sequence which causes intra-molecular binding interactions that generate a secondary structure that resembles a cloverleaf (▶ Fig. 4.3). Within the tRNA molecule are two functional domains: an anticodon domain and an amino acid attachment site. The anticodon domain is a hairpin loop that is found on one end of the tRNA and has complementary sequence to the codon that encodes the amino acid that is bound to the amino acid attachment site found on the opposite end of the molecule, which is the 3′ end of the tRNA strand.

The enzyme that links an amino acid to its specific tRNA is called the aminoacyl-tRNA synthetase, and this process is called tRNA charging. There is a separate aminoacyl-tRNA synthetase for each of the amino acids, with the name of the enzyme identifying the amino acid involved. For example, the aminoacyl-tRNA synthetase that is responsible for the charging of isoleucine-containing tRNAs is called isoleucyl-tRNA synthetase.

4.3 Translation in Prokaryotes

Although the end result of translation, the synthesis of protein, is the same in eukaryotes and prokaryotes, the way in which this occurs differs. In prokaryotes, the mRNA is translated as soon as it is synthesized by the RNA polymerase, as there is no nuclear membrane to separate the transcriptional machinery from the translational machinery. In eukaryotes however, the mRNA must first be transferred from the nucleus to the cytoplasm for translation. In addition, the mRNAs of prokaryotes are polycistronic, meaning that each mRNA molecule can encode several different proteins. However, the mRNA molecule is not translated into one long protein but, instead, each of the protein-encoding sequences in the mRNA must undergo the process of translation separately (▶ Fig. 4.4).

In the initiation phase of translation in prokaryotes, the small 30S ribosomal subunit binds the mRNA at a specific six-base consensus sequence located 8 to 10 bases upstream from the AUG start codon, known as the Shine-Dalgarno sequence. During this time, the translation factors IF-1 and IF-3 bind to the 30S subunit and prevent it from associating with the 50S subunit prematurely. The translation factor IF-2 then binds the initiator-tRNA, which is attached to formyl-methionine, and recruits it to the AUG start codon in the mRNA in a GTP-dependent manner. The 50S subunit then binds to the 30S subunit and dissociates the translation factors IF-1, IF-2, and IF-3, producing a functional ribosome and creating three tRNA-binding pockets known as the A-site, P-site, and E-site. The A-site, or aminoacyl site, binds the appropriate aminoacyl-tRNA, depending on the codon present at this site. The P-site, or peptidyl site, is the pocket that contains the tRNA that is bound to the growing polypeptide chain. Once the tRNA is no longer associated with the peptide chain, it transitions to the E-site, or ejection site, where it is expelled from the ribosome.

During elongation, the translation factor EF-Tu escorts the next aminoacyl-tRNA to the A-site of the ribosome in a GTP-dependent fashion. The ribosome catalyzes peptide bond formation between the nascent polypeptide chain and the incoming amino acid through its peptidyl-transferase activity. As translation occurs, the ribosome slides along the mRNA molecule. With each shift in location, the ribosome binds the elongation factor EF-G, which mediates the translocation and allows for the freeing of the A-site for the next aminoacyl-tRNA to bind.

Termination of translation occurs once a stop codon is reached. At this point, a release factor, or RF is recruited to the A-site to bind the stop codon. Because there is no amino acid associated with the RF, peptidyl transferase adds a water molecule to the peptidyl tRNA in the P-site. This frees the carboxyl end of the polypeptide chain, releasing the completed protein. Once the protein is released, ribosomal recycling factor, or RRF, binds the ribosome, separating the small and large subunits, and releasing the mRNA (▶ Fig. 4.4).

I

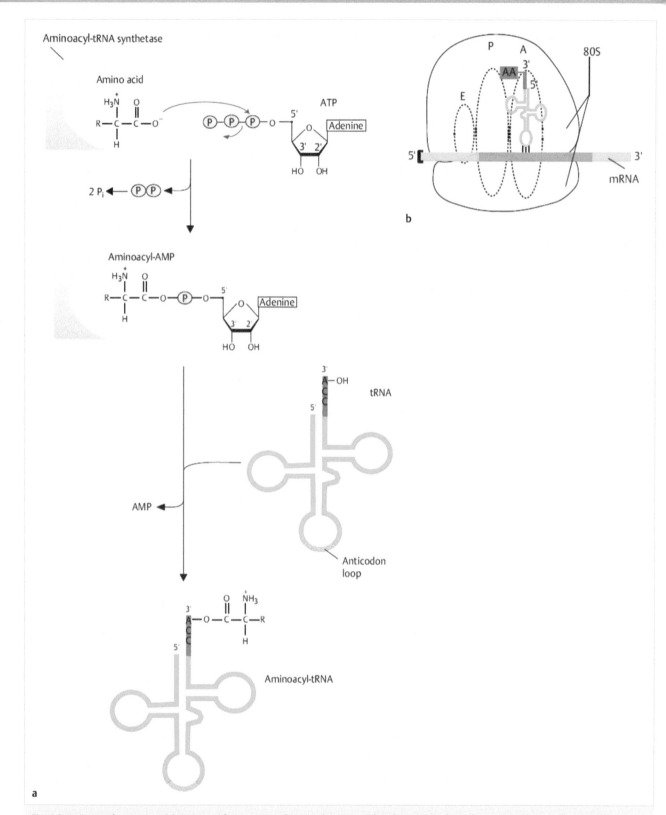

a

b

Fig. 4.3 Amino acid activation. **(a)** Amino acids are activated in an adenosine triphosphate (ATP)–dependent manner by specific cytoplasmic aminoacyl-tRNA synthetases that catalyze the transfer of an adenosine monophosphate (AMP) residue onto the carboxyl residue of the amino acid (forming aminoacyl-AMP). This reaction is driven by the energy obtained from the hydrolysis of pyrophosphate (PPi) into two phosphates. tRNA synthetase then catalyzes the transfer of the amino acid from aminoacyl-AMP onto tRNA. This releases AMP and creates an aminoacyl-tRNA molecule, which brings the amino acid to the ribo. **(b)** Aminoacyl-tRNAs associate in a manner that is antiparallel to the mRNA template. Cyt, cytosine. (Source: Panini S, ed. Medical Biochemistry—An Illustrated Review. New York, NY. Thieme; 2013.)

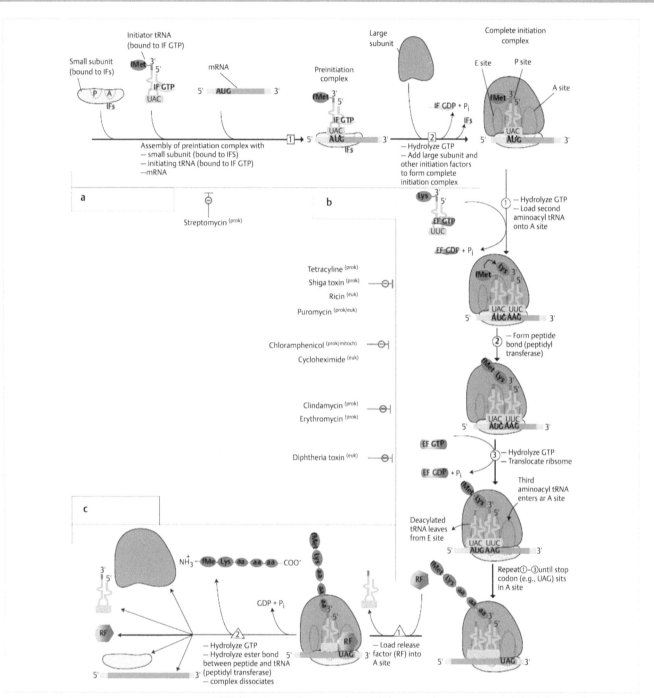

Fig. 4.4 General mechanism of translation (in prokaryotes). The mechanisms of translation for prokaryotes and eukaryotes are very similar. This figure summarizes prokaryotic translation, and ▶ Table 4.1 offers a comparison of the specific factors and sources of energy for translation in both systems. **(a)** Initiation of eukaryotic translation involves the assembly of the ribosomes of the complete initiation complex (70S in prokaryotes and 80S in eukaryotes) from small and large subunits, mRNA, and the initiator tRNA. This assembly is achieved with the help of proteins known as initiation factors (IFs in prokaryotes). **(b)** Elongation of the nascent polypeptide chain follows the initiation of translation. Amino acids are added sequentially to the initiating amino acid via multiple elongation cycles. Elongation consists of the binding of aminoacyl-tRNAs to guanosine triphosphate (GTP)– bound elongation factors (step 1). Binding of aminoacyl-tRNAs to the acceptor (A) site is driven forward by the energetically favorable hydrolysis of GTP (step 2). The formation of peptide bonds is catalyzed by peptidyl transferase, a ribozyme on the rRNA portion of the large ribosomal subunit (step 3). The C terminus of the amino acid occupying the P site is transferred onto the amino group of the residue in the A site. GTP hydrolysis of the GTP-bound elongation factor (EF) powers the translocation of the ribosome along the mRNA. This moves the peptidyl-tRNA to the P site and exposes a new mRNA codon in the A site. The next aminoacyl-tRNA can bind in the A site. **(c)** Termination of translation. A stop codon in the A site of the ribosome is recognized by release factors (RFs). Their binding to the A site causes cleavage of the peptide–tRNA linkage by peptidyl transferase and the release of the polypeptide chain. Following GTP hydrolysis, the ribosomal subunits dissociate. (Source: Panini S, ed. Medical Biochemistry—An Illustrated Review. New York, NY. Thieme; 2013.)

4.3.1 Inhibitors of Translation in Prokaryotes

The translational machinery in prokaryotes is also targeted by a variety of substances that are toxic to the organism (▶ Fig. 4.4). Both chloramphenicol and the macrolide class of antibiotics bind the 50S ribosomal subunit. Chloramphenicol inhibits peptidyl-transferase activity, whereas the macrolides prevent the translocation of tRNAs from the A site to the P site. Linezolid also binds the 50S ribosomal subunit, specifically on the 23S rRNA component, and prevents the formation of the 70S complex.

Other antibiotics inhibit ribosome function by binding to the 30S subunit. The tetracyclines and glycylcyclines bind the 30S ribosomal subunit and block aminoacyl-tRNAs from entering the A site. In addition to blocking the ribosome directly, some antibiotics, like mupirocin, bind to specific aminoacyl-tRNA transferases and prevent the charging of the tRNA. This prevents the amino acid from being incorporated into the growing polypeptide chain and results in premature chain termination.

4.3.2 Link to Pathology and Pharmacology

Chloramphenicol is used to treat bacterial meningitis. Chloramphenicol treatment may cause a serious side effect known as "gray baby syndrome," especially if it is used in premature infants. The metabolism and excretion of chloramphenicol requires UDP-glucuronyl transferase activity in the liver. Since the activity of this enzyme is low in premature infants, treatment with chloramphenicol may result in an accumulation of the drug to toxic levels in the blood, which can lead to low blood pressure and cyanosis, giving the infant's skin a grayish hue.

4.4 Translation in Eukaryotes

Due to the physical separation of the transcription and translation machinery, the translation process in eukaryotes involves many more molecules and is much more complex (▶ Table 4.1). In order for translation to occur on the ribosomes, which are either free in the cytoplasm or bound to the endoplasmic reticulum, the mRNA needs to be transported from the nucleus to the cytoplasm prior to translation. This nuclear export is mediated by the exon junction complex, or EJC, which allows for the mRNA to escape the nucleus via nuclear pore complexes. Once in the cytoplasm, the 5′ cap of the mRNA binds to eIF4E, a component of the eIF4F complex. Other components of this complex include eIF4A and eIF4G. eIF4G binds to the poly-A binding protein, or PABP, on the 3′ poly-A tail of the mRNA, and eIF4A acts as a helicase to unwind any secondary structures that may have formed in the mRNA, important for creating a linear molecule that can associate with the ribosomes.

During the eukaryotic initiation of translation, eIF3 binds to the 40S ribosomal subunit and forms the initiation complex by recruiting the mRNA-associated eIF4F complex, as well as the ternary complex, consisting of eIF2, the methionine initiator tRNA, and GTP. eIF5B then binds to the 60S subunit and

Table 4.1 Components involved in prokaryotic and eukaryotic translation

Components	Prokaryotes	Eukaryotes	Key points to remember
Ribosome subunits	30S subunit 50S subunit	40S subunit 60S subunit	• Streptomycin binds to the 30S subunit to disrupt initiation of translation • Shiga toxin binds to the 60S subunit to disrupt elongation • Clindamycin and erythromycin bind to the 50S subunit to disrupt translocation of the ribosome • Tetracyclines bind to the 30S subunit to disrupt elongation • Peptidyl transferase activity is housed in the large subunits
Initiation: requires hydrolysis of one GTP (equivalent to one ATP)			
Initiation factor(s) bound to small ribosomal subunit	IF-1, IF-3	eIF1, eIF1A, eIF3, eIF5	• The initiation factors facilitate binding of the small ribosomal subunit to the initiator tRNA and base pairing between the anticodon and codon
Initiation factors bound to initiator tRNA (fMet-tRNA$_i$^{Met}/Met-tRNA$_i$^{Met})	IF-2-GTP	eIF2-GTP	• Hydrolysis of GTP to GDP + P$_i$ provides the energy for assembly of the initiation complex
Features of mRNA important for assembly of initiation complex	• AUG start codon • Shine-Delgarno sequence	• AUG start codon • 5′ cap and poly(A) tail	• Additional initiation factors such as the eIF4 complex and eIF5B-GTP are required for the assembly of the final 80S initiation complex
Termination: requires hydrolysis of one GTP (equivalent to one ATP)			
Elongation factors attached to aminoacyl tRNAs	Tu-GTP	EF1α-GTP	
Elongation factors that power the ribosome's translocation	EF-GTP	EF2-GTP	• Diphtheria toxin inactivates EF2-GTP and inhibits elongation
Termination: requires hydrolysis of one GTP (equivalent to one ATP)			
Release factor that recognizes the stop codon	RF-1 or RF-2	eRF-1	
Factor (s) that power dissoxiation of the ribosomal complex	EF-GTP	eRF-3-GTP	

facilitates the formation of a functioning ribosome through the association with the 40S subunit.

The elongation phase of translation in eukaryotes is similar to that in prokaryotes with the main difference being the names of the factors involved. In the case of eukaryotes, the translation factor that escorts the next aminoacyl-tRNA to the A-site of the ribosome is called eEF-1, and the factor that mediates the translocation of the ribosome along the mRNA is called eEF-2. eRF is the eukaryotic version of the releasing factor that binds to the stop codon and causes release of the newly synthesized protein, and, like prokaryotes, RRF binds to the ribosome and separates the large subunit from the small subunit and releases the mRNA.

In both prokaryotes and eukaryotes, multiple ribosomes can be recruited to the same mRNA molecule and move along the mRNA translating it simultaneously. This allows for the production of a large amount of protein from a single mRNA molecule. The structure that is formed from the association of many ribosomes with a single mRNA molecule is called a polysome.

4.5 Inhibitors of Translation in Eukaryotes

Many antibiotics and some bacterial toxins are inhibitors of translation that directly bind to and interfere with the ribosome or other components of the translational machinery. The A

subunit of diphtheria toxin inhibits translation in eukaryotic cells by catalyzing the ADP-ribosylation of eEF2, which renders it inactive and prevents the elongation phase of translation. This ultimately leads to cell death. Symptoms of exposure to diphtheria toxin include sore throat, fever, and the development of a thick coating on the tongue and throat that impedes respiration.

Another eukaryotic toxin that affects translation is ricin, which is extracted from castor beans and has been used as a biological weapon. Ricin binds to the 60S subunit of the ribosome and removes an adenine nucleotide from the 28S rRNA component. This creates an inactive ribosome and, therefore, prevents translation. Ricin is toxic by ingestion, inhalation, or injection, and death occurs 36–72 hours after exposure.

In addition to diphtheria toxin and ricin, shiga toxin from *Shigella dysenteria* and Verotoxin (also known as shiga-like toxin) from enterohemorrhagic *E. coli* have RNA glycosylase enzymatic activity and remove the adenine nucleotide from the 28S rRNA in eukaryotes, effectively shutting down translation.

4.5.1 Link to Pharmacology

As reviewed above, most of the antibiotics are targeted to inhibit translation in prokaryotes, however, there are many other therapeutically important antibiotics that act on bacterial metabolism in different pathways (▶ Fig. 4.5). Intercalators, such as actinomycin D and daunomycin, insert between the bases in the DNA, thereby interfering with DNA replication and

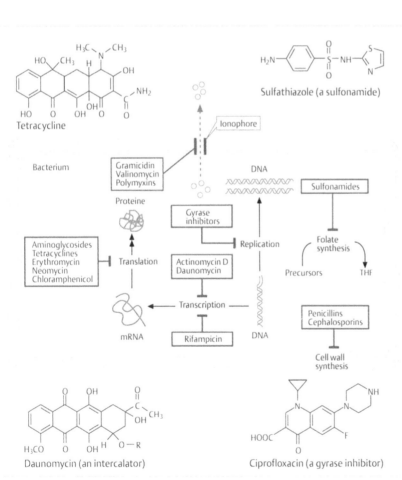

Fig. 4.5 Antibiotics overview. (Source: Koolman J, Röhm K, ed. Color Atlas of Biochemistry. 3rd Edition. New York, NY. Thieme; 2012.)

the transcription of genes. Intercalators are not only toxic to prokaryotes, they are also toxic to eukaryotes. Another family of antibiotics, known as gyrase inhibitors, are capable of inhibiting replication of bacteria, without effecting eukaryotes. Penicillin and cephalosporin are both inhibiters of bacterial wall biosynthesis. Therefore, these compounds are effective treatments for infections caused by gram-positive bacteria. Sulfonamides are anti-metabolites that prevent the synthesis of folic acid.

4.6 Regulation of Translation in Eukaryotes

The process of translation can be regulated either globally or at the level of a specific mRNA. In the case of global regulation, the synthesis of all proteins in the cell is affected. This allows for a sudden increase or decrease in protein production in response to specific conditions and is independent of transcription. In most cases, global regulation involves the modulation of translation factor activity, thereby influencing the translation of all mRNAs.

One way in which the global translation of proteins is controlled is through the phosphorylation state of eIF2. Under times of stress, such as low nutrients or growth factors, infection, or an increase in temperature, eIF2 is phosphorylated. The phosphorylated eIF2 sequesters eIF2B, a protein required for the activity of eIF2. This sequestration reduces the available pool of eIF2B and slows the rate of translation, which is the normal cellular response to stress. Leukoencephalopathy with vanishing white matter (VWM disease) is caused by a mutation in the gene encoding eIF2B, which results in a loss of eIF2 activity and diminishing overall protein levels. Symptoms include delayed motor development that progresses to spasticity and ataxia, mainly due to loss of myelin production in the brain and spinal cord.

The regulatory elements that govern the translation of specific mRNAs can be found in their untranslated regions (UTRs). Both the 5′ UTR and 3′ UTR can serve as regulatory units for the translation of the mRNA. Within these regions, specific sequences can bind proteins or regulatory RNAs that can either prevent the translation of the mRNA, through sequestration and loss of its interactions with the ribosome, or induction of protein synthesis by stabilizing the mRNA. An example of these regulatory mechanisms can be found in the 5′ UTR of the ferritin mRNA and the 3′ UTR of the transferrin receptor mRNA. In both the ferritin mRNA's 5′ UTR and the transferrin receptor mRNA's 3′ UTR, there is a sequence called the iron response element (IRE). The iron response element-binding protein (IRP) binds to this sequence when iron levels are low in the cell. The IRP stabilizes the transferrin receptor mRNA, leading to increased production of this receptor, and, at the same time, it blocks the translation of the ferritin mRNA. Since the transferrin receptor is responsible for the uptake of iron and ferritin is responsible for the storage of iron, the effect is an increase in available iron in the cell. When iron in the cell is high, the opposite occurs. IRP is released from both of these mRNAs, resulting in destabilization of the transferrin receptor mRNA and reduction in iron uptake, as well as stabilization of the ferritin mRNA and increased iron storage.

Another means by which translation can be regulated is through the activity of microRNAs, or miRNAs. miRNAs are short noncoding RNAs with sequence that is complementary to a sequence found in the 3′UTR of a mRNA. Each miRNA binds to and regulates a specific mRNA. When the miRNA binds the 3′ UTRs of the mRNA, the mRNA is either targeted for degradation or is translationally repressed, with eventual degradation. A higher level of sequence complementarity will favor direct degradation, whereas a lower level of sequence complementarity is more likely to cause translational repression.

Review Questions

1. Which of the following types of mutations creates a stop codon and leads to the termination of the growing peptide chain during translation?
   A) Frameshift
   B) Missense
   C) Nonsense
   D) Silent

2. The translation initiation factor IF-1 has which of the following functions?
   A) Binds to the formyl-methionine tRNA
   B) Binds to the ribosome 30S subunit
   C) Recruits the initiator tRNA to the AUG start codon in the mRNA
   D) Unwinds secondary structures in the mRNA

3. Chloramphenicol inhibits which of the following?
   A) 30S ribosomal subunit
   B) 60S ribosomal subunit
   C) eEF2
   D) Peptidyl transferase

4. Which of the following factors mediates the translocation of the ribosome along the mRNA molecule during translation in eukaryotes?
   A) eEF-1
   B) eEF-2
   C) eIF-3
   D) eRF

Answers

1. **The correct answer is C.** A mutation in the mRNA sequence that produces a stop codon and leads to termination of the peptide chain is known as a nonsense mutation. Missense mutations (answer choice B) change the nucleotide sequence in a way that leads to a new codon encoding a different amino acid. Silent mutations (answer choice D) are changes in the nucleotide sequence that lead to a change in the codon, but it does not change the amino acid encoded due to the degenerate nature of the genetic code. Insertions or deletions of nucleotides within the mRNA sequence that changes the reading frame of the downstream codons and subsequent amino acid sequence are known as frame shift mutations (answer choice A).

2. **The correct answer is B.** In prokaryotes, the translation factors IF-1 and IF-3 bind to the 30S subunit and prevent it from associating with the 50S subunit prematurely. The translation factor IF-2 binds the formyl-methionine tRNA (answer choice A), the initiator-tRNA, and recruits it to the AUG start codon in the mRNA (answer choice C). During eukaryotic translation, eIF4A acts as a helicase to unwind any secondary structures that may have formed in the mRNA (answer choice D).

3. **The correct answer is D.** Chloramphenicol is an antibiotic that inhibits the activity of the prokaryotic peptidyl transferase. The tetracyclines and glycylcyclines bind the 30S ribosomal subunit (answer choice A) and block aminoacyl-tRNAs from entering the A site, while the eukaryotic toxin ricin inhibits the 60S ribosomal subunit (answer choice B). Diphtheria toxin blocks eukaryotic translation by inhibiting the activity of eEF2 (answer choice C) through ADP-ribosylation.

4. **The correct answer is B.** In eukaryotes, the translation factor that escorts the next aminoacyl-tRNA to the A-site of the ribosome is called eEF-1 (answer choice A), and the factor that mediates the translocation of the ribosome along the mRNA is called eEF-2. During the initiation of translation, eIF-3 (answer choice C) binds to the 40S ribosomal subunit and forms the initiation complex. eRF (answer choice D) is the releasing factor that binds to the stop codon and causes release of the newly synthesized protein.

Review Questions

1. A 3-year-old girl presents to her pediatrician with developmental delay and neurological deterioration. Urine tests reveal an abnormally high level of glycosaminoglycans. What cellular function is most likely disrupted in this child?
   A) Expression of hydrolases
   B) Production of ATP
   C) Release of secretory vesicles
   D) Breakdown of very long chain fatty acids
   E) Formation of plasmalogen

2. A 4-year-old girl presents with impaired motor and sensory functions. Brain imaging reveals extensive demyelination, and laboratory studies indicate high levels of saturated very long-chain fatty acids in her blood. What is a characteristic of the organelle that is defective in this patient?
   A) It is responsible for sorting secretory proteins.
   B) It is inherited in a non-Mendelian fashion.
   C) It is composed of a matrix formed by lamins.
   D) It contains an enzyme that converts H_2O_2 to H_2O and O_2.
   E) It separates intracellular components from the outside environment.

3. A 7-year-old boy with multiple hyperpigmented lesions on his face and upper arms develops a skin nodule on the back of his right hand. Biopsy of this skin nodule discloses a squamous cell carcinoma. Genotyping reveals that this patient has germline mutations in the gene encoding a protein involved in nucleotide excision repair. What gene product is most likely defective?
   A) ATM
   B) MSH2
   C) p53
   D) XPA

4. A 36-year-old female presents with bloody stools and persistent abdominal cramps. Her family history is positive for endometrial cancer and colon cancer. Colonoscopy reveals multiple adenomas, which are removed. The gene that is most likely mutated in this patient is a component of which of the following DNA repair mechanisms?
   A) Homologous recombination
   B) Nucleotide excision repair
   C) Non-homologous end joining
   D) Base excision repair
   E) Mismatch repair

5. A 47-year-old mine worker is diagnosed with Non-Hodgkin lymphoma. DNA analysis of one of his excised lymph nodes identifies mutations in a component of non-homologous end joining DNA repair. In what stage of the cell cycle is DNA repair most affected?
   A) G1
   B) S
   C) G2
   D) M

6. A 69-year-old woman presents with a painful rash across the left side of her trunk. Her physician prescribes acyclovir for her infection. What is the mechanism of action of this drug?

   A) It causes chain termination due to absence of a 3' OH group.
   B) It is a nucleoside analog of adenosine.
   C) It is efficiently converted to the nucleotide form by cellular thymidine kinase.
   D) It directly inhibits translation of viral proteins.

7. DNA isolated from an individual with Crigler-Najjar syndrome is sequenced. The analysis of the UDP-glucuronyl transferase gene reveals that two additional nucleotides are found within the TATA box region of the gene. What effect would this change have?
   A) It wouldn't have any effect on gene expression
   B) It would affect the RNA processing mechanisms
   C) It would affect the initiation of transcription
   D) It would affect the translation of the mRNA in the cytoplasm
   E) It would affect the alternative splicing of the pre-mRNA

8. In the death cap mushroom, which toxic substance causes symptoms of diarrhea and abdominal cramps after a few hours of ingestion, and what enzyme does it inhibit?
   A) α-amanitin/Aminoacyl tRNA synthetase
   B) α-amanitin/Topoisomerase
   C) α-amanitin/RNA polymerase II
   D) muscarine/DNA polymerase
   E) muscarine/Topoisomerase

9. A 7-year-old girl presents with fever, cough, and sore throat. Physical examination reveals lethargy and breathing difficulties. A throat culture is positive for diphtheria infection. The toxin produced by this infection most likely inhibits which step of translation?
   A) Charging of aminoacyl-tRNA synthetases
   B) Translocation of mRNA from nucleus to cytoplasm
   C) Movement of the ribosome along the mRNA molecule
   D) Release of the polypeptide chain

10. A researcher is studying a newly-discoveredpathogenic strain of *E. coli*. Biochemical analysis identifies a defect in global translation within these cells, and sequencing of the genome reveals a mutation in the gene that encodes IF-2. What is the most likely cause of the translation defects in these cells?
    A) Failure to initiate translation with formyl-methionine
    B) Premature association of the 30S and 50S ribosomal subunits
    C) Failure of release factor (RF) to bind to the stop codon
    D) Inability of the ribosome to move along the mRNA

Answers

1. **The correct answer is A.** An elevated level of glycosaminoglycans and neurological deterioration early in life are indicative of the lysosomal storage disease Hurler syndrome. This disease is caused by mutations in the gene encoding ?-L-iduronidase. These mutations prevent the cell from efficiently degrading glycosaminoglycans, allowing for their accumulation in the cell and subsequent cell death, especially in the neurons. This results in the marked neurological deterioration seen in these patients. Production of ATP (answer choice B) would be compromised in the

presence of mitochondrial defects, such as those seen in mitochondrial myopathies. The defect in the release of secretory vesicles (answer choice C) would occur in cells that have defective ER or Golgi components. The breakdown of very long chain fatty acids (answer choice D) and the formation of plasmalogen (answer choice E) are affected in peroxisomal storage diseases, since both of these processes are important functions of the peroxisome.

2. **The correct answer is D**. The symptoms described for this individual are consistent with a peroxisome biogenesis disorder. Recall that α oxidation of very long chain fatty acids occurs in the peroxisome, which explains why individuals with defective peroxisomes, such as this patient, have an accumulation of very long chain fatty acids in the bodily fluids. This condition also damages the myelin sheaths of neurons, thus explaining the extensive demyelination. Besides the oxidation of very long chain fatty acids, another normal function of the peroxisome is the conversion of H_2O_2 to H_2O and O_2 due to the presence of the enzyme catalase. The sorting of secretory proteins (answer choice A) is the function of the Golgi, whereas the separation of intracellular components from the outside environment (answer choice E) is provided by the plasma membrane. The organelles that are inherited in a non-Mendelian fashion (answer choice B) are the mitochondria. The nucleus is composed of a matrix formed by lamins (answer choice C).

3. **The correct answer is D**. Multiple pigmented lesions and the presence of squamous cell carcinoma in a child of this age is consistent with an inherited mutation in a DNA damage repair process. One of the most common inherited diseases associated with these symptoms is xeroderma pigmentosum, which is caused by mutations in XPA, a protein that functions in the nucleotide excision repair of the DNA damage caused by UV light exposure. Exposure to UV light produces a structural change in neighboring thymidines that results in thymidine dimers. These thymidine dimers are normally removed by nucleotide excision repair, but in individuals with xeroderma pigmentosum, this process is defective and they develop multiple skin neoplasms. ATM (answer choice A) is an important component of the homologous recombination DNA repair mechanism, whereas MSH2 (answer choice B) plays a role in the repair of mismatched DNA bases. p53 (answer choice C) is a transcription factor that has multiple functions within the cell, including DNA damage response to double strand breaks in the DNA molecule.

4. **The correct answer is E**. Because this patient has multiple adenomas in her colon at a relatively young age and she has a family history of colon cancer, she likely has an inherited mutation associated with colorectal cancer. In addition, her family history is also positive for endometrial cancer. These findings are consistent with the inherited disease known as Hereditary Non-Polyposis Colorectal Cancer (HNPCC), or Lynch Syndrome. HNPCC is caused by a mutation in one of the components of the DNA mismatch repair machinery. Due to the inability to repair mismatches, DNA mutations accumulate in these individuals, ultimately leading to cancer. HNPCC is not caused by mutations in the DNA repair mechanisms in answer choices A, B, C, and D.

5. **The correct answer is A**. Non-homologous end joining is a DNA repair mechanism in which a double strand DNA break is repaired by trimming the broken ends to form blunt ends, which are then mended together to produce a continuous DNA molecule. This type of repair is error prone, as the nucleotides within the trimmed region are lost. Therefore, this type of DNA repair is not the preferred method in the cell. However, the other type of double strand break repair that the cell can undergo, homologous recombination, requires the presence of a homologous chromosome to serve as a template for the damaged DNA molecule. This repair mechanism is favored by the cell due to the fact that the use of an unaffected DNA molecule as a template means that it is error free. However, homologous recombination repair is only possible in a cell that has already completed DNA replication. As a result, non-homologous end joining is the only repair process that can correct double strand breaks in G1 (before S phase), and homologous recombination is the preferred method of double strand break repair in cells that have undergone DNA replication: S (answer choice B), G2 (answer choice C), and M (answer choice D). Therefore, the cell cycle most affected by a defect in non-homologous end joining DNA repair is G1.

6. **The correct answer is A**. Nucleoside analogues such as acyclovir and azidothymidine lack a 3' OH group in their structure. This 3' OH is required for the replicating enzyme to add the next nucleotide in the sequence. Therefore, when these analogues are incorporated into a growing DNA strand, the next nucleotide cannot be added and chain termination occurs. Acyclovir is a nucleoside analogue of guanosine, not adenosine (answer choice B), and it is efficiently converted to the nucleotide form inside the cell by the retroviral thymidine kinase, not the cellular thymidine kinase (answer choice C). Acyclovir does not directly inhibit translation of viral proteins (answer choice D).

7. **The correct answer is C**. Since the mutation is not within the coding region, the translation of the mRNA in the cytoplasm would not be affected (answer choice D). Similarly, the alternative splicing of the pre-mRNA (answer choice E) would not be affected because this would require a mutation in either the donor or the acceptor splice sites. Under normal conditions, the TATA box is recognized by TFIID and recruits RNA polymerase II to initiate transcription. Therefore, the initiation of the UDP-glucuronyl transferase gene transcription would be affected by a mutation (addition of two nucleotides, in this case). Individuals with Crigler-Najjar syndrome remain asymptomatic. However, due to decreased levels of the UDP-gluconyl transferase enzyme, they have chronic hyperbilirubinemia with slightly higher levels of unconjugated bilirubin, although their liver enzyme levels are found to be normal.

8. **The correct answer is C**. The death cap mushroom produces the toxin α-amanitin. This toxin is known to bind to and specifically inhibit the eukaryotic RNA polymerase II. Therefore, exposure to this toxin would result in an inhibition of eukaryotic transcription within the cell.

9. **The correct answer is C.** Diphtheria toxin, which is an exotoxin, has ADP-ribosylation activity. This exotoxin can enter the cells and catalyzes the ADP-ribosylation of eEF-2, inhibiting its activity. Since eEF2 is necessary for the translocation of the ribosomes along the mRNA molecule during translation, this toxin prevents translation. As discussed in this chapter, ricin, another eukaryotic toxin, also causes ADP-ribosylation of eEF2 and inhibition of translation. Charging of aminoacyl-tRNA synthetases (answer choice A) is inhibited by the antibiotic mupirocin, but is not affected by diphtheria toxin. Some viruses have mechanisms that disrupt the translocation of mRNA from the nucleus to the cytoplasm (choice B), but they are unrelated to diphtheria infection. Diphtheria toxin does not inhibit the release of the polypeptide chain (choice D). However, the antibiotic chloramphenicol can have this effect since it inhibits the activity of peptidyl transferase.

10. **The correct answer is A.** In prokaryotes, the translation factor IF-2 is responsible for binding to the initiator tRNA, containing formyl-methionine, and recruits it to the AUG start codon in the mRNA molecule. A mutation in IF-2 would inhibit this process and prevent initiation of translation with formyl-methionine. Premature association of the 30S and 50S ribosomal subunits (answer choice B) is normally inhibited by the translation factors IF-1 and IF-3. A mutation in the release factor (RF) would result in the failure of RF to bind to the stop codon (answer choice C) and result in a defect in the termination of translation. Movement of the ribosome along the mRNA is mediated by the elongation factor EF-G. A mutation in the gene that encodes EF-G would result in the inability of the ribosome to move along the mRNA (answer choice D).

Part II

5 Cytoskeleton

At the conclusion of this chapter, students should be able to:

- Identify the three major constituents of the cytoskeleton and describe the function of each
- Identify the two classes of motor proteins that bind to microtubules, as well as the family of motor proteins that associates with actin and describe the functions of each
- Describe how myosin thick filaments and actin thin filaments contribute to the process of muscle contraction

During the course of their lives, mammalian cells undergo dramatic changes in their metabolism, number and shape. The remodeling of cellular shapes, organization of cytoplasm, and cell-cell attachments largely rely on the cytoskeleton, a dynamic network of intracellular protein filaments (▶ Fig. 5.1). There are three types of cytoskeleton filaments: actin, microtubules, and intermediate filaments. Besides specific structural proteins that build these filaments, each type of cytoskeleton structure employs additional accessory proteins to regulate the dynamics of the filaments, traffic organelles throughout the cytoplasm, and communicate with other types of cytoskeletal components. This allows for a well-orchestrated response to certain stimuli, like migration during wound healing.

This chapter describes each of the three types of filaments that comprise the cytoskeleton and discusses examples of medical conditions that are associated with defects in cytoskeleton function.

5.1 Actin Cytoskeleton

The actin cytoskeleton is involved in many essential cellular activities including cell motility, cellular shape, cytokinesis, muscle contraction and inflammation. For example, during inflammation, the recruitment of immune cells to the sites of infection involves well-coordinated actin remodeling events. These diverse functions occur through the spatial and temporal regulation of the actin cytoskeleton and are mediated by a wide repertoire of accessory proteins that nucleate, elongate and link the actin cytoskeleton to the plasma membrane or other cytoskeletal components.

The actin cytoskeleton consists of actin filaments that are built from monomeric globular actin (G-actin) into long, two-stranded polymers (F-actin). In addition, filament-interacting proteins associate with these polymers and assist in their formation and stabilization. These accessory proteins can be divided into several groups based on how they affect actin dynamics. Among these groups are proteins that nucleate actin, proteins that affect the stability of the filaments, proteins that move along the filaments, and actin cross-linkers.

5.2 Actin Filaments

Actin filaments, also called microfilaments, are thin, flexible cytosolic filaments 7–9 nm in diameter that consist of polymerized actin molecules. Actin is present in all eukaryotic cells and, like other filament proteins, is one of the most abundant proteins in the cell. Cytoplasmic actin can be found not only in polymerized form but also as soluble monomers, which possess ATP hydrolyzing activity.

The actin protein exists in different states in the cell (▶ Fig. 5.2). In the monomeric globular form, referred to as G-actin, the molecule can be bound to either ATP (ATP-G-actin) or ADP (ADP-G-actin). Actin can also be polymerized into filaments, which is referred to as F-actin. F-actin, similar to monomeric G-actin, can have either ATP (ATP-F-actin) or ADP (ADP-F-actin) bound forms.

The major feature of G-actin as a structural protein of microfilaments is its ability to polymerize into F-actin filaments. Once F-actin is formed, the bound ATP is slowly hydrolyzed, and therefore the majority of F-actin is in the ADP-bound form.

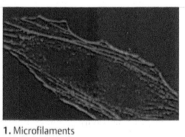

1. Microfilaments

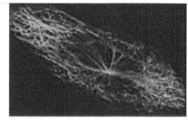

2. Microtubules

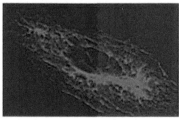

3. Intermediate filaments

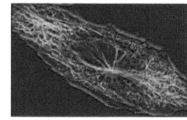

4. 1–3 superimposed

Fig. 5.1 Major components of the cytoskeleton. (1) Microfilaments, (2) Microtubules, (3) Intermediate filaments, (4) combination of all four cytoskeleton components. (Source: Koolman J, Röhm K, ed. Color Atlas of Biochemistry. 3rd Edition. New York, NY. Thieme; 2012.)

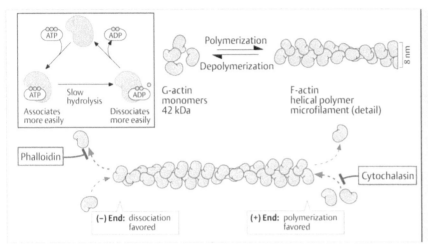

Fig. 5.2 Actin microfilament structure and dynamics. ATP-associated G-actin monomers assemble to form F-actin microfilament polymers. Polymerization occurs at the + end and dissociation occurs at the – end of the microfilament. (Source: Koolman J, Röhm K, ed. Color Atlas of Biochemistry. 3rd Edition. New York, NY. Thieme; 2012.)

ATP hydrolysis is not required for the polymerization to occur, instead, it regulates the kinetics of polymerization. Indeed, the polymerization and depolymerization of actin is one of the critical features of actin that defines its temporal and spatial function within the cell.

F-actin filaments have polarity, possessing a pointed, or "-", end and a barbed, or "+" end (▸ Fig. 5.2). On the pointed end, actin's ATP-binding clefts are exposed to the environment. The opposite is true of the barbed ends, where the ATP-binding clefts of actin are turned inward towards the adjacent actin subunit. Such structural polarization of the filament affects the orientation of actin-binding proteins. For example, myosin, a motor protein that binds to actin filaments, helps to define the "–" and "+" ends of the filament due to the direction in which it binds. The polarity of actin filaments also influence its dynamics. The addition of actin monomers occurs five to ten fold faster at the "+" or Barbed end, whereas actin depolymerizes at its "–" or pointed end.

5.3 Actin Filament Organization and Dynamics

The main functional sites for actin in the cell include the cell cortex, stress fibers, nucleus, and the lamellopodia and filopodia in the leading edge of migrating cells.

Rapid changes in actin filament organization provide the cell with the ability to adapt to environmental challenges. For example, cell migration, endocytosis, and cell division all require actin cytoskeleton remodeling. Actin remodeling includes polymerization-depolymerization events and their regulation by cell signaling.

Actin assembly can be promoted, inhibited, or stabilized by different classes of actin-binding proteins. Inhibition of actin polymerization is mediated through the sequestration of monomeric actin by the cytosolic protein thymosin β, and this interaction between G-actin and thymosin β prevents G-actin incorporation into the growing F-actin polymer. Thus the amount of thymosin β regulates available G-actin monomers that can be used by the cell to build the actin cytoskeleton. In contrast, the binding of another cytosolic protein, profilin, promotes polymerization of actin, specifically at the barbed, or

"+", end. In addition, profilin plays a role in the anchoring of actin filaments to the plasma membrane.

The presence of actin-binding proteins and their expression in the cell does not only regulate available G-actin monomers but also can affect already established F-actin filaments. For example, the length of actin filaments can be altered by the actin severing proteins gelsolin and cofilin. These proteins bind to actin and change the conformation of the actin subunits, breaking the filament.

5.4 Actin Bundles and Networks

Within the cell, F-actin is organized into two major higher-order structures: bundles and networks. Although both types of actin filament organization function to support the cellular membrane, thus determining cellular shape, the way in which F-actin filaments are positioned within each differs. Parallel tightly positioned F-actin filaments are characteristic of bundles, whereas mesh-like organization of filaments is a feature of networks. Networks can be either gel-like 3D structures, or a web-like planar type that localize beneath plasma membrane.

Stress fibers are bundles of actin filaments that also exhibit contractile properties in non-muscle cells. They appear as a result of mechanical tension, and were proposed to exert tension on matrix collagen surrounding the cells, which is important for wound healing. Stress fibers are positioned in a specific way inside the cell, with one end connected to focal contacts on the plasma membrane, while the other end is either connected to focal contacts as well, or associated with the intermediate filaments around the nucleus. These stress fibers are temporary structures, as they disappear during each cell division when the cell loses its tight attachment to its substrate.

5.5 Cell Cortex

The cell cortex is one of the most actin-rich areas within the cell. A web-like actin network is formed just underneath the plasma membrane. Actin organization at the cellular cortex involves two distinct actin nucleators: Arp2/3 and formin.

5.6 Actin Cytoskeleton Motors Myosins

Although very important, actin function is not restricted to structural support for the cell's shape. Actin also mediates transport of proteins and vesicles to different sites within the cell, as well as muscle contraction through its interactions with molecular motor proteins. The membrane-bound cellular compartments known as vesicles carry cargo proteins to different cellular locations. These vesicles move along the actin filaments using actin motor proteins called myosins (▶ Fig. 5.3). All myosins have the common structural features of a head, neck and tail domain, each with a distinct function. The head of myosin molecules binds to actin and ATP and is responsible for the actual movement along the actin filament, using the hydrolysis of the bound ATP as the source of energy. The neck domain is responsible for the regulation of the head domain's function. The tails bind to the cargo and, therefore, specify the function of each type of myosin. There are more than 10 members of the myosin family, each with a specific function. Myosins I and II are the best-studied of the myosins and can be found in almost all cells. Myosin I, a monomer that has a calmodulin light chain bound to its neck, is important for vesicle transport, whereas myosin II, a dimer with two different light chains, is essential for muscle contraction and a final step during cell division, cytokinesis. Myosins are regulated through their light chains by calcium ions, second messengers involved in many signaling events in the cell (discussed in theme 10 Signal Transduction). The tails of myosin I interact with either vesicle membranes or with the plasma membrane inside the cell. Tails of multiple myosin II dimers are combined together to form a thick filament with myosin molecule heads on both sides and a bare zone in the middle. These thick filaments are components of actin –myosin contractile machinery in sarcomeres, functional units of muscle fibers, or myofibrils.

Myosins are, therefore, powerful molecular motors that slide along the actin filaments with their head domains and mediate essential functions of actin, namely contraction and transport.

5.6.1 Link to Physiology

Cell Migration and Actin Remodeling

Cell motility has an indispensable role in many vital processes including embryonic development, vascularization, and the immune response. Another aspect of cell migration in human health is the ability of cancer cells to migrate and metastasize in different organs. Cell movement largely relies on actin remodeling with the assistance of actin assembly factors.

In a migrating cell, one part of the cell polarizes in the direction of the movement and becomes a feature called the leading edge. This polarization includes the re-positioning of some of the organelles, including the Golgi and the centrosome, along with many of the microtubules. The leading edge of the migrating cell has characteristic formations termed lamellipodia and filopodia. During migration, the lamellipodia extend outward, leading to the formation of new integrin-based adhesions. At

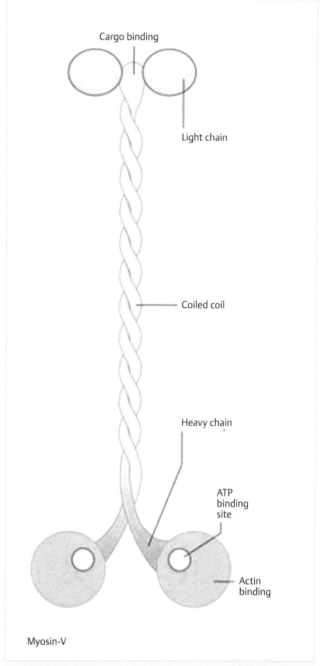

Fig. 5.3 Myosin structure. (Source: Koolman J, Röhm K, ed. Color Atlas of Biochemistry. 3rd Edition. New York, NY. Thieme; 2012.)

the same time, the old mature adhesions at the rear end of the cell are disassembled.

5.6.2 Microvilli

Microvilli are finger-like extensions that are found on the apical surface of cells and whose primary function is to increase the surface area and, therefore, absorptive capacity of the cell (▶ Fig. 5.4). One location in the body in which microvilli play an important physiological role is in the small intestine. Here, epithelial cells known as enterocytes are responsible for the

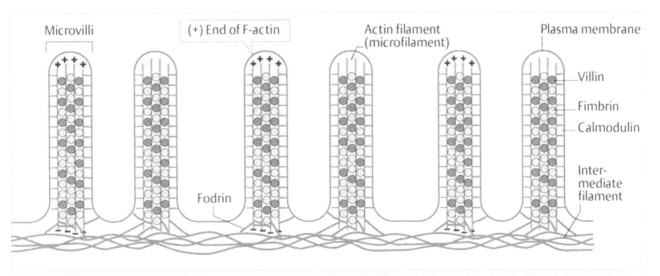

Fig. 5.4 Microvilli formed by the assembly of actin microfilaments and intermediate filaments. (Source: Koolman J, Röhm K, ed. Color Atlas of Biochemistry. 3rd Edition. New York, NY. Thieme; 2012.)

absorption of nutrients from the lumen of the intestine. The large number of microvilli on their surface, which could number into the thousands, allows them to perform this function. These specialized protrusions are covered with plasma membrane that is enriched with enzymes and polysaccharides. Inside, microvilli have a highly ordered structure consisting of actin bundles. Microvilli are very different from cilia, another type of cellular protrusions, which we will cover later in this chapter.

5.7 Cytokinesis

Cell division is a part of the cell cycle of proliferating cells and culminates in cytokinesis- a final step that separates the two newly-formed daughter cells. At this stage of cell division, the DNA is segregated into daughter cells, and the nuclear envelope begins to reform, while the two cells are still interconnected by a common cytoplasm. Actin filaments and the contractile complexes composed of actin and myosin mediate the thinning of the cytoplasmic bridge that connects the daughter cells. The contraction of contractile ring generates pull forces that help the formation of thin cytokinetic bridge between the two separating cells. During the final step of cytokinesis, the abscission of this bridge as a result of trafficking and signaling events separates two cells.

5.8 Muscle Contraction

Skeletal muscles form during development by fusion of multiple cells together. The resulting gigantic cell is called myofiber. Most of the cell space in myofibers is occupied by myofibrils, bundles of filaments, which in turn can be separated into multiple sarcomeres, structural and functional units of muscle contraction machinery (▶ Fig. 5.5). Each sarcomere consists of alternating light and dark bands that represent the I band and the A band, along with the H zone and a Z disc. On the molecular level, the I band consists of similar length thin filaments that are organized in a bundle. These thin filaments are composed

of actin and two specialized proteins, tropomyosin and troponin. Tropomyosin and troponin are the proteins that prevent continuous muscle contraction in the absence of a cellular signal. Tropomyosin covers myosin-binding sites along the thin filament, thus preventing myosin binding to the actin filaments. When calcium in the cell is increased, it interacts with troponin, and this interaction leads to dissociation of troponins and tropomyosin molecules from thin actin filaments, allowing myosin to slide along the filaments.

The thin actin filaments in the I band are attached, via their plus ends, to a Z disc region. The minus ends of the thin filaments extend towards the M line of the sarcomere. The M line is located in the middle of the A band and in the middle of the myosin thick filaments.

The activated nerve that contacts the muscle cell transmits the signal into the muscle cell and this results in a massive release of Ca^{2+} ions from sarcoplasmic reticulum (the name for the endoplasmic reticulum in muscle cells) where Ca^{2+} ions are typically stored in resting cell. The rapid increase in cytosolic Ca^{2+} levels causes release of troponin and tropomyosin from the thin actin filaments and allows muscle contraction by sliding of myosin thick filaments onto the actin thin filaments. High levels of Ca^{2+} also act on the light chains of myosin molecules. The heads of myosin molecules that are located at both sides of the thick filament "walk" onto actin filaments, causing muscle contraction.

Some similarity between the muscle contractile mechanisms can be observed in cardiac and smooth muscle cells.

5.9 Microtubules and Their Organization

The microtubule network is organized around microtubule nucleating centers and serves to ensure cell division, immune response, cell migration, and trafficking (▶ Fig. 5.6). In most vertebrate cells, the microtubule cytoskeleton is mainly organized by the centrosome, also called the microtubule organizing

II

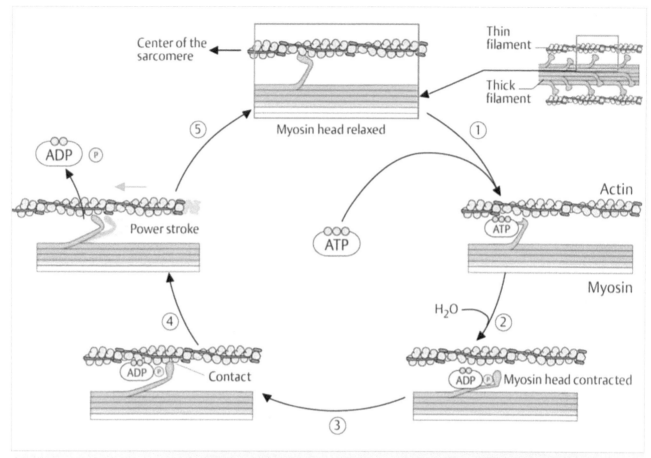

Fig. 5.5 Mechanism of muscle contraction. (Source: Koolman J, Röhm K, ed. Color Atlas of Biochemistry. 3rd Edition. New York, NY. Thieme; 2012.)

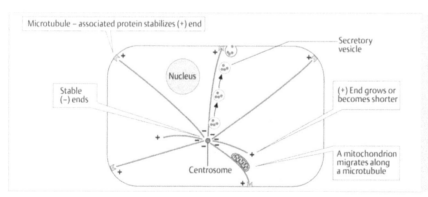

Fig. 5.6 Organization of the microtubule network within the cell. (Source: Koolman J, Röhm K, ed. Color Atlas of Biochemistry. 3rd Edition. New York, NY. Thieme; 2012.)

center (MTOC). The centrosome is a non-membranous organelle that consists of hundreds of proteins. The centrosome duplicates once per cell cycle in a semi-conservative manner. Structurally, the centrosome consists of a pair of microtubule-based centrioles and a dynamic but well organized pericentriolar material. During mitosis, the centrosome forms spindle poles. Microtubules that emanate from the centrosome attach to the kinetochore proteins on the condensed chromosomes and pull sister chromatids to their respective daughter cells. A subset of microtubules, called aster microtubules, is attached to the plasma membrane to ensure the proper mitotic spindle orientation during cell division. Disruption in spindle orientation results in the severe disorder microcephaly (discussed later in this chapter). In terminally differentiated cells, the centrosome moves towards the plasma membrane and serves as a template for microtubules that form cilium, a protrusion from cell that is critical for cell signaling. Defects in cilium underlie a whole set of diseases collectively termed ciliopathies. In immune cells, the centrosome and microtubules polarize towards the immunological synapse, a place of contact between T lymphocytes or natural killer cells with target cells. This polarization is important in mediation killing function of immune cells. Another function of microtubules is the transport of cargoes, proteins or vesicles, to different cellular locations, like cilium or spindle poles.

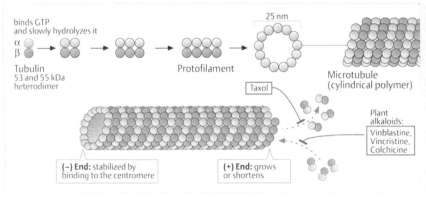

Fig. 5.7 Tubulin heterodimers assemble to form microtubule polymers. This assembly occurs specifically at the plus end of the microtubule. (Source: Koolman J, Röhm K, ed. Color Atlas of Biochemistry. 3rd Edition. New York, NY. Thieme; 2012.)

5.10 Microtubule Structure and Dynamics

Microtubules consist of heterodimers of two separate proteins, α and β tubulin, which are arranged in a hollow tube that is about 25 nm in diameter. The α/β tubulin pairs assemble longitudinally to form a linear polymer known as a protofilament. Thirteen of these protofilaments come together to form the walls of the microtubule (▶ Fig. 5.7). Other tubulins such as γ, δ, and ε, are not a part of the microtubule, but are essential components of the centrosome, such as the microtubule-nucleating γ tubulin ring complexes (γTURC). The β tubulin unit of the α/β tubulin dimer binds to GTP and possesses GTP hydrolysis activity. Upon incorporation of the GTP-bound tubulin dimer into the growing microtubule polymer, the β subunit begins to hydrolyze the GTP to form GDP. Thus, newly incorporated tubulin dimers at the end of the microtubule have GTP-bound β subunits, whereas β tubulin found further down the microtubule are GDP-bound. Microtubules are polar structures due to the polar arrangement of the tubulin dimers within the microtubule. The microtubule end that contains exposed α subunits is termed the "– end" or "minus end", whereas the end that has exposed β subunits is known as the "+end" or "plus end." The minus end is often embedded in pericentriolar material of the centrosome, while the plus end of the microtubule is rapidly growing and is characterized by GDP-bound β tubulins. This polarity has functional consequences in microtubule motor movements and trafficking events (discussed later).

The highly dynamic nature of microtubules, referred to as dynamic instability, is due to their constant turnover and is characterized by continuous polymerization and shrinkage events. The shrinkage of microtubules occurs in response to the hydrolysis of GTP by the β tubulin subunits. The presence of GTP-bound β tubulins at the ends of microtubules, also called the GTP-cap, inhibits the depolymerization of the microtubule, thus contributing to its stabilization.

5.11 Kinesins and Dynein Are Microtubule-based Motors

Cells utilize microtubules as highways to deliver cargo (proteins, vesicles) across the cell. For example, microtubules mediate the recruitment of proteins and vesicular components to the spindle poles during mitosis, the transport of vesicles within neuron axon, or the secretion of signaling molecules during an immune response. Microtubule-based motors execute this function by mediating the physical connection between the cargoes and the microtubules and undergoing unidirectional movements along the microtubules. There are two types of microtubule motors: plus end motors called kinesins and minus end motors called dyneins (▶ Fig. 5.8). Since many of the microtubules in the cell are oriented such that their plus ends are near the plasma membrane or cell cortex, kinesins tend to transport cargo in this direction as well. On the other hand, the minus end of microtubules typically face the centrosome in the middle of the cell and, therefore, dynein mediates transport in this direction. The transport requires energy generated by ATP hydrolysis.

Both types of microtubule-based motors, kinesins and dyneins, consist of two heavy chains and two lights chains. The heavy chains of the motors have globular heads that bind to microtubules and tails that bind to cargoes. The globular heads "walk" along the microtubule, with each step requiring the hydrolysis of one molecule of ATP (▶ Fig. 5.9).

The transport function of microtubules relies on these motor proteins and mediates proper localization of vesicles and proteins components during cell division, cilia assembly, neuronal function, and immune response. Moreover, transport of whole organelles, like mitochondria is also possible.

5.12 Spindle Poles in Mitosis

During mitosis, the microtubule cytoskeleton orients the plane of cell division and functions to physically separate the two sister chromatids into the newly-forming daughter cells. This process begins with the movement of the centrosomes toward the opposite sides of the cell. At this stage, the microtubules that are nucleated by the centrosome/spindle poles begin to acquire different functions at different locations. One pool of microtubules interacts with the cellular cortex and connects the spindle pole with the cellular membrane. These microtubules are known as astral microtubules. Another pool of microtubules reaches out to the condensed chromosomes and binds to kinetochores of each sister chromatid. These microtubules are called kinetochore microtubules. The third type of microtubules, interpolar microtubules, also grows in the direction of condensed chromosomes, but does not attach to the chromosomes. These microtubules, together with microtubule motors, are critical to align and position spindle poles.

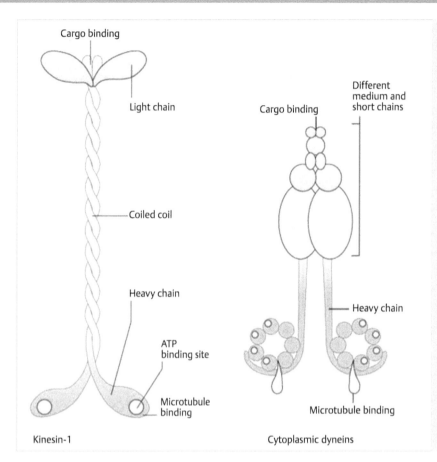

Fig. 5.8 Structure of kinesins and dyneins. (Source: Koolman J, Röhm K, ed. Color Atlas of Biochemistry. 3rd Edition. New York, NY. Thieme; 2012.)

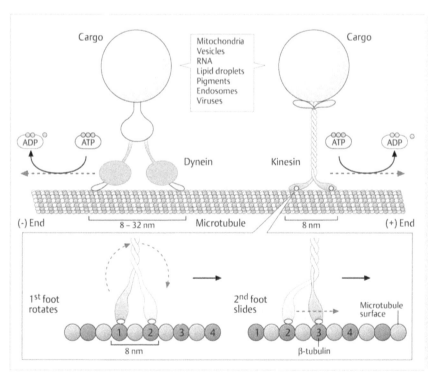

Fig. 5.9 Kinesin and dynein movement on microtubules. (Source: Koolman J, Röhm K, ed. Color Atlas of Biochemistry. 3rd Edition. New York, NY. Thieme; 2012.)

5.13 Cilia and Flagella

Some cells of the body possess multiple motile membrane extensions on their apical surface called cilia. One such cell type is the polarized epithelial cells of the respiratory tract. The movement of these cilia are important for clearing dirt, debris, and mucus out from the lungs. Another motile structure that is similar to cilia is the flagella. In humans, these structures are found only in sperm cells and, in contrast to cilia, there is only one per cell.

In these ciliated cells, the centrosome serves as the base, or basal body, from which the cilia are formed. Microtubules are the main structural base for the cilium, arranged as a pair of individual microtubules surrounded by a collection of nine microtubule doublets around the periphery of the structure. This 9 + 2 structure is known as the axoneme.

5.14 Intermediate Filaments

The third type of cytoskeletal structure is the intermediate filament (▶ Fig. 5.10). Intermediate filaments are structural elements of the cell that enable cellular resistance against mechanical stresses. As we will see later in this chapter, epithelial cells use intermediate filaments, along with the membrane junction complexes desmosomes and hemidesmosomes, to connect to neighboring cells and join their cytoskeletons together. This allows the epithelial sheet to distribute the forces over many cells and prevents rupture of the individual cells, thus retaining the integrity of the tissue.

The name "intermediate filament" describes its relative size (~10 nm diameter), being intermediate between actin microfilaments (7 nm) and microtubules (24 nm). Intermediate filaments are composed of several α -helical protein subunits. These helical segments pair together to form a dimer. These dimers are further assembled into tetramers to form a protofilament, and finally, into twisted-together filaments.

Four different classes of intermediate filaments exist, three of which are specific for certain types of cells: keratins in epithelial cells, vimentins in connective tissue, and neurofilaments in neurons. The fourth class is the nuclear lamins. These proteins are unique among intermediate filaments in that instead of forming rope-like structures, they assemble into a two dimensional mesh just inside the nuclear membrane, where they provide structure and strength to this organelle. This mesh-like network of nuclear lamins is found in all nucleated cells and is referred to as the nuclear lamina.

5.14.1 Link to Pathology

Myopathy

Dystrophin is a very larger cytoplasmic protein (more than 400kDa) that connects the cytoskeleton inside a myofibril to the plasma membrane, or sarcolemma, of the cell. Through its interactions with the actin, intermediate filament, and microtubule components of the cytoskeleton, dystrophin provides stability to the muscle fibers during physical tension. In addition, this protein participates in signaling events that mediate muscle function and Ca²⁺ levels. Mutations in the dystrophin gene (*DMD*) result in several different forms of muscular dystrophy, including **Duchenne muscular dystrophy**, characterized by severe defects in muscular function which progressively affects the quality of life of the affected individual.

Erythrocyte Cytoskeleton and Spherocytosis

In red blood cells, the cytoskeleton must be flexible in order to allow the deformations necessary to squeeze through very narrow capillaries. Because of this, red blood cells do not have the typical cytoskeleton that is found in other cell types. Instead, their cytoskeleton is a highly flexible structure consisting of the proteins ankyrin, glycophorin, Band 3, and Band 4.1, as well as spectrin and actin filaments. This spectrin –actin cytoskeleton is located just underneath the plasma membrane inside the cell. The spectrin and actin filaments are connected to one another through Band 4.1, whereas the entire structure is anchored to the membrane by the integral membrane protein glycophoryn and Band 3, via associations with ankyrin and Band 4.1.

Genetic mutations in the proteins that compose the spectrin-actin cytoskeleton of erythrocytes result in a loss of their typical biconcave disc shape. This gives rise to a spherically shaped cell that cannot navigate the tight passages of the capillaries and the splenic sinuses, leading to their destruction within the spleen. Individuals that have inherited mutations in these genes have the disease **hereditary spherocytosis** and often undergo a splenectomy in order to prevent hemolytic anemia.

Ciliopathies

Dysfunction of cilia has been linked to numerous human disorders, collectively termed ciliopathies, and can be associated with a multitude clinical symptoms including cystic kidneys, polydactyly, situs inversus, obesity, respiratory defects, infertility, and encephalocele. One example of such a disease is **Kartagener syndrome**, a genetic disease caused by mutations in the gene that encodes for ciliary dynein. This disorder is characterized by a clinical triad of symptoms that includes bronchiectasis, chronic sinusitis, and situs inversus. In addition to the recurrent lung

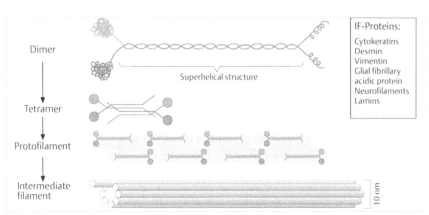

IF-Proteins:

Cytokeratins
Desmin
Vimentin
Glial fibrillary acidic protein
Neurofilaments
Lamins

Fig. 5.10 Structure and assembly of intermediate filaments. (Source: Koolman J, Röhm K, ed. Color Atlas of Biochemistry. 3rd Edition. New York, NY. Thieme; 2012.)

infections caused by defective cilia in the respiratory tract, infertility in males is common due to loss of motility in the sperm flagella. Female fertility can also be compromised since the transport of the ovum down the fallopian tubes to the uterus requires synchronized ciliary movements.

5.14.2 Link to Toxicology and Pharmacology

There are several drugs that modulate the dynamics of both actin filaments and microtubules. **Phalloidin** is a toxin found in the death cap mushroom. It binds to actin filaments inside the cell and stabilizes them by preventing depolymerization. In contrast, **cytochalasin**, a fungal toxin, blocks the polymerization of actin filaments.

Due to the importance of microtubule dynamics during mitosis, drugs that disrupt this process have proven to be effective treatments for the uncontrolled cell division which characterizes cancer. These include two families of chemotherapeutic drugs: the **taxanes** and the **vinca alkaloids**. The taxanes, which include **paclitaxel**, are microtubule stabilizers whereas the vinca alkaloids (**vinblastine, vincristine**) promote depolymerization of microtubules. Another type of microtubule inhibitor, **nocodazole**, also promotes microtubule depolymerization.

Review Questions

1. Which of the following cytoskeleton components is responsible for the proper segregation of sister chromatids during mitosis?
   A) Actin filaments
   B) Intermediate filaments
   C) Microtubules
   D) Nuclear lamin

2. Which of the following motor proteins uses the energy from ATP hydrolysis to travel along microtubules toward its plus end?
   A) Dynein
   B) Kinesin
   C) Myosin
   D) Tubulin

3. Which of the following molecular processes results in the contraction of a sarcomere subunit of skeletal muscle?
   A) Binding of calcium to troponin
   B) Movement of thick filaments away from the Z disc
   C) Growth of actin thin filaments at their plus end
   D) Movement of kinesins along microtubules

4. Which of the following is a function of intermediate filaments?
   A) Cell motility
   B) Cytokinesis
   C) Intracellular trafficking
   D) Mechanical strength

5. A defect in which of the following cytoskeleton components causes Kartagener syndrome?
   A) Actin
   B) Dynein
   C) Intermediate filaments
   D) Kinesin

Answers

1. **The correct answer is C.** The dynamic instability of microtubules, created by rapid assembly and disassembly of the polymers at their plus ends, is responsible for the movement of sister chromatids towards opposite spindle poles during anaphase of mitosis. Although actin filaments (answer choice A) are also highly dynamic, they are not involved in the movement of chromatids in mitosis. Instead, actin filaments function a variety of other cellular functions including cell motility, cytokinesis, and in muscle contractions. Intermediate filaments (answer choice B) give mechanical strength to the cell and prevent damage by external forces. Nuclear lamins (answer choice D) are a type of intermediate filament.

2. **The correct answer is B.** The kinesins are a family of motor proteins that use the energy gained by hydrolysis of ATP to move cargo along microtubules within the cell, specifically towards the plus end of the microtubule. Likewise, dyneins (answer choice A) also move cargo along microtubules, but in this case the motor protein is directed toward the minus end of the microtubule. Myosins (answer choice C) do not associate with microtubule, but they do move cargo along actin filaments. Tubulin (answer choice D) is a structural component of microtubules and does not travel along these cytoskeletal structures, although it has the ability to hydrolyze GTP.

3. **The correct answer is A.** The contraction of skeletal muscle occurs in response to an increase in cytosolic Ca^{2+} levels. The Ca^{2+} binds to troponin, a protein associated with a filamentous molecule called tropomyosin. In the absence of Ca^{2+}, tropomyosin blocks myosin binding sites on the actin thin filaments. When Ca^{2+} binds troponin, the tropomyosin shifts and reveals the myosin binding sites. The myosin is then able to walk along the actin thin filaments. Since myosin moves toward the plus end of actin filaments, and the plus end of the actin thin filaments is embedded in the Z disc, this results in contraction of the sarcomere. In contrast, if the thick filaments move away from the Z disc (answer choice B), this would cause relaxation of the muscle. Growth of actin thin filaments at their plus end (answer choice C) does not occur during muscle contraction. And movement of kinesins along microtubules (answer choice D) is not involved with contraction of the sarcomere.

4. **The correct answer is D.** Intermediate filaments are structural elements of the cytoskeleton composed of several α -helical protein subunits. The main function of the intermediate filaments is to provide the cell with resistance to mechanical stresses. Cell motility (answer choice A) and cytokinesis (answer choice B) are functions of the actin cytoskeleton. The microtubule cytoskeleton serves as lanes of transport for the intracellular trafficking (answer choice C) of cargo such as vesicles and organelles.

5. **The correct answer is B.** Dynein is a molecular motor that associates with microtubules. It travels in a minus end directed manner along the microtubules using the hydrolysis of ATP as an energy source. One dynein family member, known as ciliary dynein, is an important component of the cilia that are found on the surface of many different cell types in the body, including those in the respiratory tract, and the flagella of sperm cells. Mutations in the gene that encodes this protein result in the genetic disease Kartagener syndrome, which is characterized by a defect in ciliary function that leads to a clinical triad of bronchiectasis, chronic sinusitis, and situs inversus. Actin (answer choice A) and intermediate filaments (answer choice C) do not contribute to ciliary function and therefore are not defective in this disease. Although kinesin (answer choice D) is also a molecular motor, it does not have a role in ciliary function.

6 Protein Sorting, Modifications, and Intracellular Traffic

At the conclusion of this chapter, students should be able to:

- Describe the process of vesicle budding, transport, and fusion from the endoplasmic reticulum (ER) to the cis Golgi, and identify the type of coated vesicle involved as well as the molecules that make up the coat
- Describe the process of retrograde vesicle transport from the Golgi back to the ER, what molecules are recycled back to the ER, and identify the type of coated vesicle involved.
- Describe the process of endocytosis and what type of membrane-bound vesicles are involved, as well as identify the type of coated vesicle used and what molecule is needed for the budding of these vesicles
- Explain the function of the recycling endosome and how this structure is important in increasing the number of glucose transporters on the plasma membrane in response to insulin
- Explain the mechanisms that allow for the specificity in targeting transport vesicles to their proper target membrane, and identify the molecules involved in this process

The inside of the cell is a highly dynamic environment. Proteins, lipids, and other molecules are synthesized in one part of the cell and must move to another location in order to serve their function (▶ Fig. 6.1). Likewise, membrane bound organelles and other structures must also be taxied to and from various locations within the cell. In order to achieve this movement, organelles are carried along cytoskeletal structures via molecular motors, while the soluble molecules are shuttled in membrane-bound structures called transport vesicles.

Intracellular trafficking begins when a transport vesicle buds off from a membrane-bound compartment, or organelle. After it is formed, this vesicle is then ferried to its ultimate destination, known as the target compartment or organelle, where it fuses with that structure's membrane in order to deliver its contents. Typically, these transport vesicles carry material as cargo from the lumen and membrane of the donor compartment to the lumen and membrane of the target compartment. The process of producing a vesicle, called budding, involves a much different mechanism than that of vesicle fusion. Budding requires extroversion of membrane from the donor compartment, which results in the formation of a membrane bud neck. This bud neck ultimately undergoes fusion, causing release of the vesicle. In contrast, fusion of a transport vesicle to the target compartment only requires a membrane fusion event initiated from the cytoplasmic side of both the donor and target membranes (▶ Fig. 6.2).

The budding of transport vesicles involves the concerted effort of a family of proteins that form a coat around the nascent vesicle. This coat not only initiates the budding process, but is also important for recruiting the cargo of the transport vesicle from the lumen of the donor compartment. This occurs through the concentration of membrane-bound receptors that bind to the cargo. There are three types of vesicle coats that are responsible for the majority of intracellular transport: clathrin coats, COP-I coats, and COP-II coats. Vesicles bearing clathrin are transported from the plasma membrane to the Golgi apparatus, from Golgi to late endosomes, and from secretory vesicles to Golgi. COP-I coated vesicles mediate retrograde transport from Golgi to ER, as well as movement from one compartment to another within the Golgi. COP-II coated vesicles transport cargo from ER to the cis face of the Golgi.

Following the successful translation and folding of a secretory protein in the ER, a transport vesicle must move this cargo to the Golgi, where it will be processed before it is directed to either an intracellular destination or to the plasma membrane for integration or secretion (▶ Fig. 6.3). This ER to Golgi transport requires the formation of a COP-II coat and is initiated by a molecule called Sar1, a coat recruitment GTPase responsible for the earliest stages of COP-II coat assembly. Sar1 is found at high concentrations in the cytoplasm in an inactive, GDP-bound state. When Sar1-GEFs (guanine nucleotide exchange factors) embedded in the ER membrane come in contact with the soluble Sar1, the associated GDP is released and replaced by GTP. This GTP-bound Sar1 undergoes a conformational change, exposing an amphipathic helix that inserts into the ER membrane, anchoring the Sar1 to the ER. This ER membrane-bound Sar1 then directs the recruitment of two structural components of the COP-II coat: Sec23 and Sec24. The Sec24 component contains a binding site for the cytoplasm-facing tail of cargo receptors that are embedded in the ER membrane, allowing for the sequestration of these receptors and their cargo to the developing transport vesicle. The binding of this cargo to the receptor is based on the presence of a specific amino acid sequence, called the exit signal, in the cargo protein. Membrane-bound cargo, which are not dependent on binding to the cargo receptors, also contain exit signals that allow them to be recruited to COP-II coated vesicles.

Next, a complex of two additional proteins, Sec13 and Sec31, associates with Sec23 and Sec24 and forms the outer shell of the coat. As the COP-II coat assembles, it begins to bud off the membrane. Once the bud narrows to a point when the membranes of the bud neck are close enough together, the membranes fuse, pinching off and releasing the coated vesicle. Shortly after its release, the COP-II coat falls off, leaving a naked vesicle that can then fuse with the cis face of the Golgi.

6.1 Vesicular Tubular Clusters

Each of the ER-derived transport vesicles possesses surface-bound molecules called SNAREs that, in a homotypic fashion, mediate the fusion of multiple vesicles to form a larger structure called the vesicular tubular cluster (▶ Fig. 6.4). Homotypic membrane fusion, or fusion of vesicles that originated from the same organelle, occurs. These vesicular tubular clusters use the activity of molecular motors to move along microtubules toward the cis Golgi. Because all of the vesicles originating from the ER have the Golgi as their destination, the formation of these vesicular tubular clusters enhances the efficiency of this process.

In addition to the movement from the ER to the Golgi, transport vesicle use components of the cytoskeleton to translocate to other intracellular locations. As mentioned, the microtubules,

Morphological structures

Biochemical processes

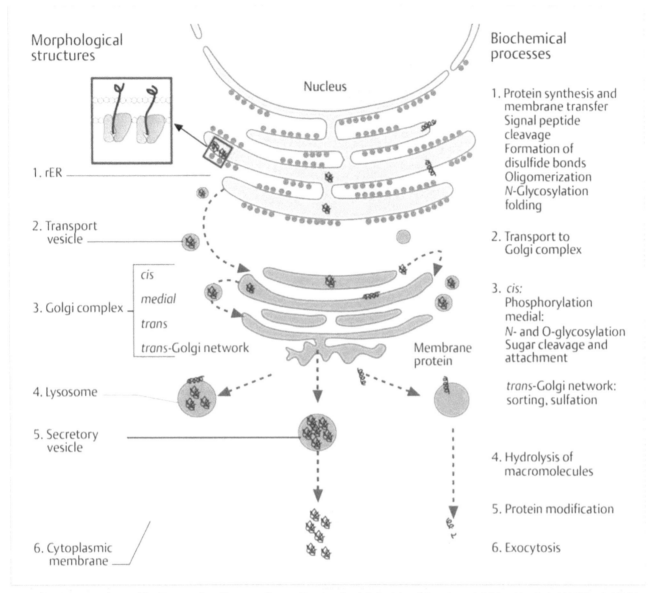

Nucleus

1. rER

2. Transport vesicle

3. Golgi complex
 - cis
 - medial
 - trans
 - trans-Golgi network

Membrane protein

4. Lysosome

5. Secretory vesicle

6. Cytoplasmic membrane

1. Protein synthesis and membrane transfer Signal peptide cleavage Formation of disulfide bonds Oligomerization N-Glycosylation folding

2. Transport to Golgi complex

3. cis: Phosphorylation medial: N- and O-glycosylation Sugar cleavage and attachment

trans-Golgi network: sorting, sulfation

4. Hydrolysis of macromolecules

5. Protein modification

6. Exocytosis

Fig. 6.1 Structure and assembly of intermediate filaments. (Source: Passarge E, ed. Color Atlas of Genetics. 4th Edition. New York, NY. Thieme; 2012.)

along with their associated molecular motors kinesin and dynein, are required for some of these processes. In addition, actin filaments, along with some myosin family members, play a role in several intracellular trafficking events.

6.2 Retrieval of ER Cargo Receptors and Soluble ER Proteins

After fusion of the vesicular tubular cluster with the cis Golgi, the ER cargo receptors have to be recycled back to the ER membrane. During the process of forming the COP-II coated vesicle, some soluble ER-resident proteins are occasionally packaged by mistake. These resident ER proteins contain a specific amino acid sequence (Lysine-Aspartate-Glutamate-Leucine, or KDEL) that is recognized by membrane-bound receptors in the ER called KDEL receptors (▶ Fig. 6.3). As the KDEL receptors are

constantly cycled between the ER and Golgi, they pick up soluble ER resident proteins in the Golgi and are then packaged into COP-I coated vesicles via interactions between the cytoplasmic tails of the KDEL receptor and components of the COP-I coat. The COP-I coated vesicles are then shuttled back to the ER, where they fuse with the ER membrane and return the ER resident proteins back to the lumen of the ER.

6.3 Clathrin Coated Vesicles

Clathrin coated vesicles transport material from the plasma membrane into the cell and between endosomal and Golgi compartments within the cell (▶ Fig. 6.5). These coats are made up of subunits called triskelions, three legged structures composed of three large and three small polypeptide chains. The triskelions form a basket-like structure on the cytosolic surface of membranes. Just as COP-II coats require the GTPase Sar1 for

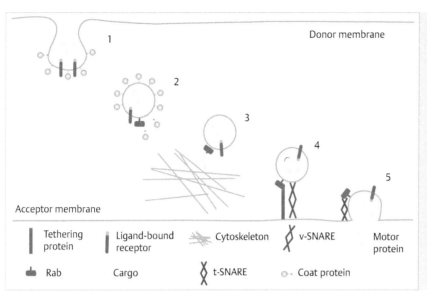

Fig. 6.2 Vesicle formation on donor membrane and delivery of cargo to acceptor membrane.

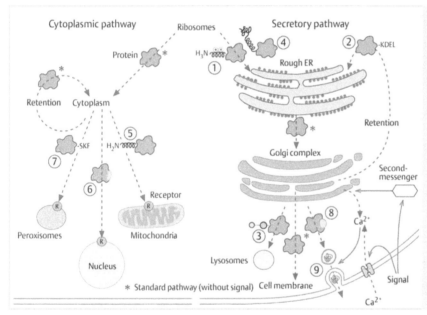

Fig. 6.3 Cytoplasmic pathway versus secretory pathway of protein sorting. (Source: Passarge E, ed. Color Atlas of Genetics. 4th Edition. New York, NY. Thieme; 2012.)

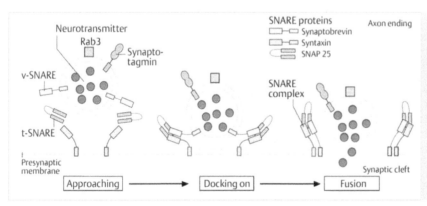

Fig. 6.4 SNARE complexes mediate the fusion of transport vesicles to their acceptor membranes to deliver cargo. (Source: Passarge E, ed. Color Atlas of Genetics. 4th Edition. New York, NY. Thieme; 2012.)

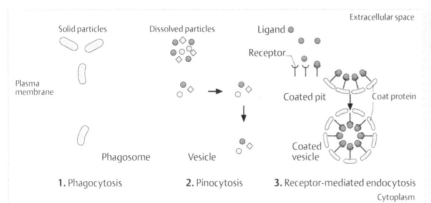

Fig. 6.5 Types of endocytosis. (Source: Passarge E, ed. Color Atlas of Genetics. 4th Edition. New York, NY. Thieme; 2012.)

II

assembly, both COP-I and clathrin coats rely on the GTPase Arf (ADP Ribosylation Factor) for their assembly. And like the COP-I and COP-II coats, clathrin coat proteins bind to cargo receptors in the membrane via adaptor proteins. The adaptor proteins form a discrete second layer of the coat, positioned between the clathrin cage and the membrane, and trap transmembrane proteins that capture soluble cargo molecules inside the vesicle. There are several different types of adaptor proteins, each packaging a specific type of cargo. Clathrin coated vesicles that bud from different membranes use different adaptor proteins and, therefore, package different cargo molecules.

In contrast to the COP-I and COP-II coats, whose assembly ultimately results in the pinching off of the vesicle, the release of budding clathrin coated vesicles requires the activity of an enzyme called dynamin. Dynamin is a GTPase that forms a spiral around the neck of the developing bud by binding to the phosphatidylinositol PI(4,5)P2 in the membrane of this region. Through the energy released by GTP hydrolysis, dynamin is able to bend this patch of membrane by directly distorting the bilayer structure, resulting in fusion of the adjacent membranes of the neck and release of the vesicle.

6.4 Specificity of Vesicular Targeting and Membrane Fusion

Each organelle in the secretory and endocytic pathways expresses specific types of Rab proteins. When the vesicle buds off, these Rab proteins are displayed on its cytoplasmic surface and specifies the target location of the vesicle. On the membranes of the target compartment, proteins called Rab effectors are exposed to the cytoplasmic environment. Just like each transport vesicle has a specific type of Rab molecule, each target membrane has its own type of Rab effector that specifically binds one version of the Rab protein. The function of the Rab effectors is to bring the vesicle close enough to the target membrane to allow for interactions between v-SNARE and t-SNARE molecules that will mediate fusion of the two membranes, culminating in the delivery of the vesicle's cargo to the lumen of the target compartment.

The v-SNAREs are found on vesicle membranes and are single polypeptide chains. t-SNAREs are found on target membranes and may be composed of two or three proteins. Helical domains of the v-SNAREs and t-SNAREs wrap around each other to form the trans-SNARE complex that holds the vesicle membrane close to the target membrane. In regulated exocytosis, inhibitory proteins prevent the v-SNAREs and t-SNAREs from forming fully functional complexes. A localized influx of Ca2 + is therefore needed to release the inhibitory proteins, allowing for completion of trans-SNARE complex and fusion with the plasma membrane to release the vesicle's contents (▶ Fig. 6.3).

6.5 Endosomes

Endocytic vesicles originating from the plasma membrane bud off using clathrin coats and fuse with membrane-bound structures called early endosomes, which ultimately fuse with late endosomes. Early endosomes and late endosomes differ in their protein compositions, pH, and locations within the cell. Early endosomes are found just inside the plasma membrane and have a pH of 6, whereas late endosomes are closer to the Golgi and nucleus and have a pH around 5. In addition, only late endosomes can fuse with lysosomes, forming endolysosomes.

6.6 Transport of Hydrolases to the Lysosome

Recall that lysosomes contain many different enzymes that are capable of degrading most of the macromolecules of the cell (▶ Fig. 6.6a, b). Initial synthesis of these enzymes occurs in the ER, with a majority of their post-translational modifications taking place in the Golgi. As these enzymes are modified, a tag in the form of mannose-6-phosphate (M6P) is added (▶ Fig. 6.7). This M6P serves as a signal to the cell that the enzyme must be delivered to the lysosome. The process of lysosomal targeting involves the binding of this M6P tag to receptors found in the membrane of the trans Golgi network. Adaptor proteins bound to the cytoplasmic tail of these receptors recruits components of the clathrin coat, leading to budding off of the enzyme-containing vesicle and transport to an early endosome. These early endosomes eventually become a lysosome. During this process, the M6P is removed from the lysosomal enzymes in the endosome so that the enzymes are not brought back to the Golgi when the M6P receptors are recycled. Since this process is required for all lysosomal enzymes to reach the lysosome, defects in the addition of the

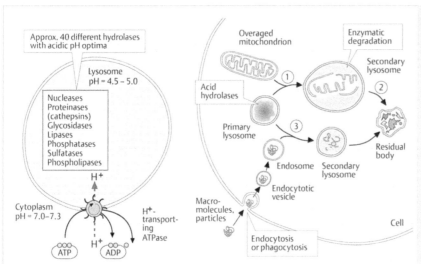

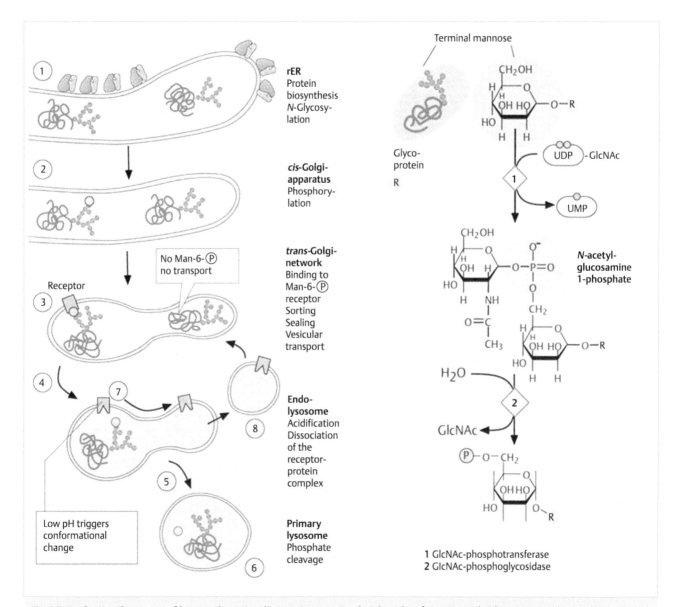

Fig. 6.6 (a) Structure and contents of lysosomes **(b)** functions of the lysosome. (Source: Koolman J, Röhm K, ed. Color Atlas of Biochemistry. 3rd Edition. New York, NY. Thieme; 2012.)

Fig. 6.7 Synthesis and transport of lysosomal proteins. (Source: Passarge E, ed. Color Atlas of Genetics. 4th Edition. New York, NY. Thieme; 2012.)

M6P tag lead to a complete loss of lysosomal function, as is seen in **I-Cell disease**.

6.6.1 Link to Pathology

I-Cell disease. The I-cell disease is caused by mutations that result in a loss in activity of GlcNac-1-phosphotransferase. **GlcNac-1-phosphotransferase** is the enzyme responsible for adding **mannose-6-phosphate** to hydrolytic enzymes destined for the lysosome (▶ Fig. 6.7). A defect in this enzyme would, therefore, prevent these hydrolytic enzymes from reaching the lysosome, which would lead to accumulation of multiple substrates for these enzymes in the blood. Individuals with I-cell disease commonly display coarse facial features such as depressed nasal bridge, long and narrow head, and epicanthal folds. In addition, developmental delay, hypotonia, skeletal abnormalities, and hepatosplenomegaly are also common in I-cell disease.

There are three pathways to degradation in lysosomes. Phagocytosis is the engulfment of bacteria by phagocytic cells, such as macrophages. Endocytosis is the internalizing of extracellular molecules. Autophagy is the degradation of cellular components that are old, damaged, or nonfunctional (▶ Fig. 6.6b).

6.7 Receptor Mediated Endocytosis

Some extracellular ligands must be internalized following binding to receptors on the surface of the cell. This occurs through a process known as receptor-mediated endocytosis (▶ Fig. 6.8).

Examples of this type of uptake are endocytosis of blood lipoprotein LDL via LDL receptor and of iron-bound transferrin (ferrotransferrin) from the blood via the transferrin receptor. Upon binding of ligand (LDL or ferrotransferrin) to its receptor on the cell surface, a clathrin coat is formed on the cytoplasmic surface of the plasma membrane that encapsulates the bound receptor. Once the clathrin coat is fully assembled, the vesicle buds off and the coat dissociates. This vesicle then fuses with an early endosome, which ultimately becomes a late endosome. Since the internal environment of the late endosome is more acidic than the early endosome, the ferric ions are released from the transferrin. This naked transferrin, now called apotransferrin, remains bound to the receptor and is shuttled back to the plasma membrane via a recycling endosome, while the ferric ions are stored by the cell.

Besides being recycled back to the plasma membrane, there are several other possible fates for endocytosed transmembrane receptor proteins. Transcytosis, in which the receptor is delivered to a different membrane than the one from which it came, is another possibility. The receptor can also be degraded if it is no longer needed.

Recycling endosomes can also be used to store transmembrane proteins or transporters that can be quickly shuttled to the plasma membrane in response to external stimuli. In muscle and adipose cells, the glucose transporter GLUT4 is stored in a steady state inside recycling endosomes. Upon binding of insulin to the insulin receptor on the surface of these cells, a signaling cascade is initiated that directs the budding off of the GLUT4 transporters from the recycling endosomes and migration of these transporters to the plasma membrane, where they increase the uptake of glucose into the cells (reviewed in theme 21).

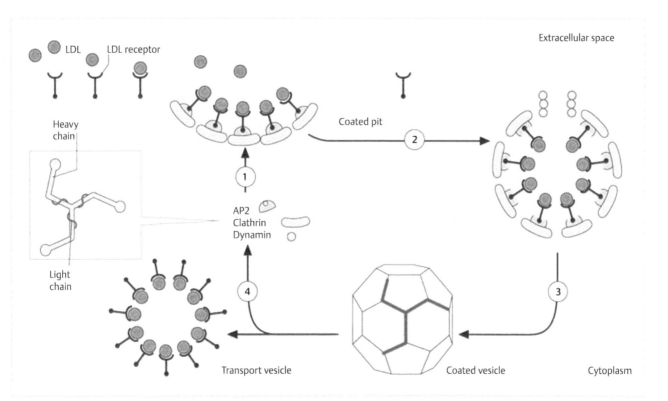

Fig. 6.8 Receptor-mediated endocytosis of LDL. (Source: Passarge E, ed. Color Atlas of Genetics. 4th Edition. New York, NY. Thieme; 2012.)

6.8 Secretory Vesicle Formation

Cells that are specialized for secreting molecules rapidly on demand concentrate and store these products in secretory vesicles, also known as secretory granules. Proteins that will be secreted are packaged in the trans Golgi by a mechanism that involves the selective aggregation of the secretory proteins. The way in which the secretory proteins are selected and packaged into vesicles, as well as the method of secretory vesicle budding from the Golgi, is unknown.

6.9 Insulin Secretion in the Pancreas

The beta cells in the pancreas produce secretory granules packed with insulin. When blood glucose levels are high, glucose enters the beta cells by facilitated diffusion through passive GLUT2 transporters (▶ Fig. 6.9). As the glucose is metabolized in the mitochondria, ATP is produced. The increase in ATP causes the closing of ATP-sensitive potassium channels in the plasma membrane, resulting in the accumulation of K+ in the cell and subsequent depolarization of the membrane. The depolarization of the plasma membrane opens up voltage-gated calcium channels, creating a calcium influx and release of the insulin secretory granules.

6.9.1 Link to Pharmacology

Like ATP, the sulfonylurea class of drugs inhibits the ATP-sensitive K+ channels, causing membrane depolarization and an increase in insulin release from the beta cells of pancreas.

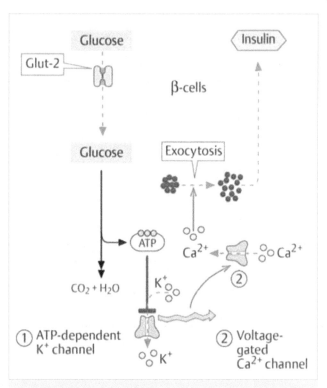

Fig. 6.9 Regulation of insulation secretion. (Source: Passarge E, ed. Color Atlas of Genetics. 4th Edition. New York, NY. Thieme; 2012.)

Review Questions

1. Which of the following molecules is responsible for initiating the formation of the coated vesicle that transports cargo from the ER to the Golgi?
   A) Clathrin
   B) KDEL receptor
   C) Rab
   D) Sar1

2. Which of the following molecules coordinates the pinching-off of endocytic vesicles from the plasma membrane?
   A) Dynamin
   B) Mannose-6-phosphate
   C) Sar1
   D) Sec23

3. Which of the following molecules is found on the surface of target membranes and mediates membrane fusion?
   A) Arf GTPase
   B) Rab effector
   C) t-SNARE
   D) v-SNARE

4. Which of the following molecules allows for specificity in directing transport vesicles to the appropriate target membrane?
   A) Clathrin
   B) Dynamin
   C) Rab
   D) Sar1

Answers

1. **The correct answer is D.** COPII coats form on the surface of the ER to allow for the formation of transport vesicles that shuttle cargo from the ER to the Golgi. Sar1 is the molecule that initiates the formation of the COPII coat. Sar1 first embeds in the ER membrane and recruits the components of the COPII coat, such as Sec23/24, which ultimately leads to the formation of the complete coat and pinching off of the vesicle. Clathrin (answer choice A) is a component of the coat that surrounds vesicles that are involved with endocytosis. KDEL receptors (answer B) bind to ER-resident proteins in the Golgi lumen and help to transport them back to the ER from the Golgi. The specificity of a transport vesicle for its target membrane is determined by the type of Rab protein on its surface.

2. **The correct answer is A.** Endocytic vesicles are formed by the assembly of a clathrin coat around a specific area on the plasma membrane. However, in contrast to COPII coated vesicles, the clathrin coat is not capable of finishing the final step of vesicle pinching-off. This process is mediated by dynamin, which uses the hydrolysis of GTP for energy to bring the membranes of the vesicle neck close enough together to allow for fusion. Mannose-6-phosphate (answer choice B) is a molecular tag placed on enzymes that are synthesized in the ER and destined for lysosomes. Sar1 (answer choice C) and Sec23 (answer choice D) are components of transport vesicles that transit from the ER to Golgi.

3. **The correct answer is C.** When a transport vesicle comes in close contact with its specific target membrane, the t-SNAREs on the surface of target membranes associate with v-SNAREs (answer choice D) present on the surface of transport vesicles. Arf GTPases (answer choice A) are involved in the initial formation of COPI-coated vesicles, but do not have a function in vesicle fusion. Rab effectors (answer choice B) are molecules present on the surface of target membranes that recognize the Rab identifiers on the surface of transport vesicles and bring them close enough to the target membrane to allow for interactions between the t-SNAREs and v-SNAREs.

4. **The correct answer is C.** Members of the Rab family of proteins are present on the surface of transport vesicles. The type of Rab protein is determined by the membrane-bound organelle from which the transport vesicle originated. In turn, the Rab protein determines which membrane-bound organelle receives the vesicle, as each target membrane within the cell displays a different Rab effector protein on its surface. Dynamin (answer choice B) binds to the neck of a budding clathrin (answer choice A) coated vesicle and allows for the fusion of the membrane and the release of the vesicle. Sar1 is a GTPase that is required for assembly of the COPII coated vesicles that transport cargo from the ER to the Golgi.

7 Membrane Transport

II

At the conclusion of this chapter, students should be able to:
- Identify the major ions found inside the cell and in the extracellular environment as well as explain the variations of ion concentrations across cell membranes
- Distinguish between passive and active modes of membrane transport
- Describe the three methods of active transport
- Identify the three types of ion-gated channels and describe the mechanism of action of each
- Give specific examples of membrane transporters and channels

As mentioned in chapter 1, the plasma membrane of the cell functions as a selectively permeable barrier, controlling the movement of molecules between the extracellular and intracellular environments (▶ Fig. 7.1). In order to maintain tight control over these processes, the cell makes use of various transmembrane transporters and channels in the plasma membrane. These structures are capable of discriminating among a large variety of molecules based on characteristics such as charge, size, shape, and abundance.

Transporters are complexes that have moving parts and are capable of transferring molecules from one side of the membrane to the other through changes in its shape. Channels, on the other hand, form hydrophilic pores in the membrane that allow the passage of small molecules and ions. The movement of molecules through channels and transporters can be described as being either passive transport or active transport (▶ Fig. 7.2**a, b**). In passive transport, also known as facilitated diffusion, molecules move down their concentration gradient from a site of highest concentration to a site of lowest concentration. This type of transport requires no expenditure of energy. On the other hand, active transport does require energy as the molecules are transferred against their concentration gradient, from a site of lowest concentration to a site of highest concentration.

7.1 Passive Transporters

Passive transporters (▶ Fig. 7.2**a**) transition between two states. These transitions occur randomly and independent of whether solute is bound. In state A, the binding sites for the solute are exposed to the outside of the membrane, whereas in state B, the binding sites are exposed to the inside of the membrane. The net effect is that solute will move from the side of highest concentration to the side of lowest concentration. This movement does not require energy. An example of a passive transporter is the family of GLUT transporters. These are characterized as uniporters because they move only one type of molecule at a time.

7.2 Active Transporters

In order to understand the driving forces that influence the direction of movement of charged particles, electrochemical gradients must be considered. An electrochemical gradient is the net force responsible for driving a charged solute, such as an ion, across a membrane. There are two components to these electrochemical gradients: the concentration gradient of the solute, and the voltage across the membrane.

There are three methods of active transport (▶ Fig. 7.2**b**): coupled transport, ATP-driven pumps, and light-driven pumps. In coupled transport, the movement of one molecule down its concentration gradient accompanies the movement of another molecule against its concentration gradient. In contrast, ATP-driven pumps use the energy generated through ATP hydrolysis to transport molecules against their concentration gradient. Light-driven pumps, which are predominantly found in prokaryotes, use light energy to transport molecules against their concentration gradient.

Coupled transporters can be separated into two functional groups based on the direction in which the transported molecules move. In one of these groups, called symporters, two different molecules are transported in the same direction across the membrane. Antiporters, on the other hand, transport the two different molecules in opposite directions.

An example of a symporter is the sodium glucose family of transporters, or SGLTs. These symporters use the energy generated from the movement of Na + ions down their electrochemical gradient into the cell to transport glucose against its concentration gradient, also into the cell. Different organs within the body express different types of SGLT, although they all have the same basic function. One member of this family, SGLT1, is expressed in the enterocytes of the small intestine, where it is needed for the uptake of glucose from the intestinal lumen. Another member, SGLT2, is found in the proximal

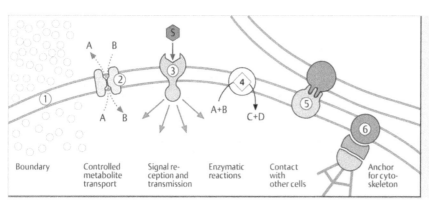

Fig. 7.1 Functions of the plasma membrane. (Source: Passarge E, ed. Color Atlas of Genetics. 4th Edition. New York, NY. Thieme; 2012.)

| Boundary | Controlled metabolite transport | Signal reception and transmission | Enzymatic reactions | Contact with other cells | Anchor for cytoskeleton |

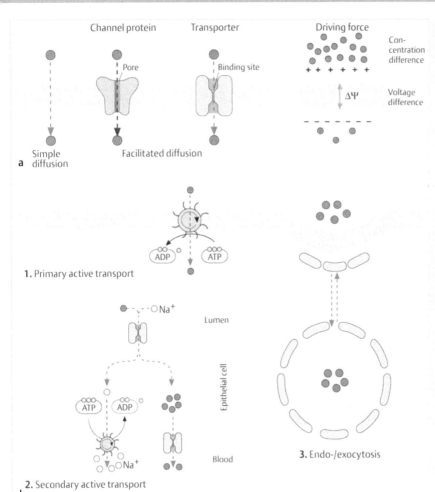

Channel protein   Transporter   Driving force

Pore   Binding site   Concentration difference

$\Delta\Psi$ Voltage difference

Simple diffusion   Facilitated diffusion

a

Fig. 7.2 (a) Types of passive transport across the plasma membrane and **(b)** types of active transport. (Source: Passarge E, ed. Color Atlas of Genetics. 4th Edition. New York, NY. Thieme; 2012.)

1. Primary active transport

ADP   ATP

Na^+ Lumen

ATP   ADP

Epithelial cell

Na^+ Blood

2. Secondary active transport

3. Endo-/exocytosis

b

tubule cells of the kidney and is responsible for glucose reabsorption in this organ.

Along with the SGLT transporter, these cells also have glucose uniporters (GLUT). In the intestine, the lumen-facing side, known as the apical surface, of the enterocytes contains SGLT transporters that move glucose across the membrane from a site of low concentration to a site of high concentration, using the energy produced from movement of sodium ions down its electrochemical gradient into the cell. On the basal surface, GLUT transporters move glucose in a passive manner from the inside of the cell, where glucose levels are high, to the extracellular fluid, where glucose levels are low.

Likewise in the kidney, SGLT and GLUT transporters mediate the absorption of glucose from the urine.

7.2.1 Link to Pharmacology

A family of drugs have been developed that take advantage of this process to reduce glucose uptake in diabetic individuals, effectively decreasing their blood glucose concentrations. This family of drugs is known as the **gliflozins**, and specific examples are canagliflozin and dapagliflozin. Their mechanism of action is to inhibit the SGLT transporters on the surface of the proximal tubules cells, thereby preventing uptake of glucose and resulting in the expulsion of glucose in the urine.

In contrast to symporters, antiporters move molecules in opposite directions across the membrane, with the movement of an ion down its electrochemical gradient providing the force necessary to push another molecule or ion against its concentration or electrochemical gradient. An example of an antiporter is the Na^+-Ca^{2+} exchanger, a plasma membrane resident transporter responsible for the export of Ca^{2+} from the cell. This antiporter moves Na^+ ions down their electrochemical gradient into the cell and, at the same time, Ca^{2+} against its electrochemical gradient to the outside of the cell.

Not only is Ca^{2+} kept low inside the cell through the action of the $Na+$-$Ca2+$ exchanger, but there are also ATP-dependent pumps, described later, that pump $Ca2+$ out of the cell, against its concentration gradient. As we will see in Chapter 10, this precise control over cytoplasmic $Ca2+$ levels is very important, as $Ca2+$ is an intracellular second messenger in many signal transduction events.

As we have seen, $Na+$ gradients allow the cells, via symporters and antiporters, to import and export nutrients and other molecules required by the cell. In addition, the intracellular and extracellular concentrations of other ions, such as $Ca2+$, $Cl-$, and $K+$, must be tightly controlled in order to ensure a healthy cell with normal physiologic processes. It is through the actions of transporters, channels, and pumps that $Na+$, $Ca2+$, and $Cl-$

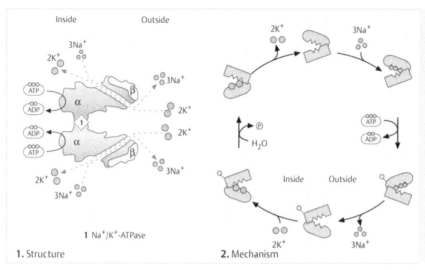

Fig. 7.3 Active transport of Na+ and K+ ions across the plasma membrane by the Na+/K+ ATPase. (Source: Passarge E, ed. Color Atlas of Genetics. 4th Edition. New York, NY. Thieme; 2012.)

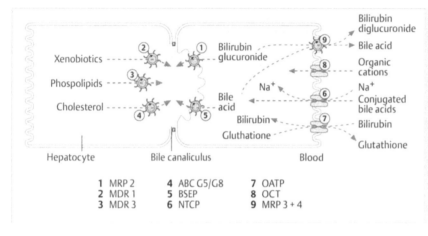

Fig. 7.4 Transport systems for bile components. (Source: Passarge E, ed. Color Atlas of Genetics. 4th Edition. New York, NY. Thieme; 2012.)

1 MRP 2	4 ABC G5/G8	7 OATP
2 MDR 1	5 BSEP	8 OCT
3 MDR 3	6 NTCP	9 MRP 3 + 4

is kept at low concentrations inside the cell, while K+ concentrations are kept high.

The high concentrations of Na+ outside the cell and K+ inside the cell are maintained through the action of an ATP-dependent pump known as the Na+-K+ ATPase (▶ Fig. 7.3). These pumps are found in the plasma membrane of all animals cells, require the hydrolysis of ATP, and are responsible for one third of the cell's energy expenditure (two thirds of energy expenditure in neurons).

The Na+-K+ pump transports ions in a cyclic manner. In one cycle, three Na+ ions are exported from the cell, against the Na+ electrochemical gradient. At the same time, two K+ ions are imported, also against its electrochemical gradient. The entire cycle results in a net negative charge (3+'s out and 2+'s in), creating a negative charge near the inner side of the plasma membrane known as the resting potential. In the case of the enterocytes of the small intestines, this activity is necessary to maintain the high extracellular concentrations of Na+ that is needed for the SGLT transporter to move glucose from the lumen into the cell.

Another type of transporter is the ABC (ATP-Binding Cassette) transporters. As the name suggests, they contain ATP-binding domains and are active transporters, requiring the hydrolysis of ATP for their energy to move substances against their electrochemical or concentration gradients. Some members of this family are ion channels, whereas others transport metabolic products, lipids, or sterols across the plasma membrane. A majority are effluxers, meaning that they remove the transported substances from the cell.

7.2.2 Link to Anatomy and Physiology

Various ABC transporters in between the hepatocytes and the bile canaliculi assist bile formation. Some of these important ABC transporters are shown in ▶ Fig. 7.4.

7.2.3 Link to Pathology

Adrenoleukodystrophy is a peroxisomal disease that is due to deficiency of ABC transporter for very long chain fatty acids in the membrane of peroxisomes. The disease is characterized by increased levels of very long chain fatty acids in the plasma. ▶ Table 7.1 summarizes the other peroxisomal diseases.

Another example of an ABC transporter is the **Cystic Fibrosis Transmembrane Conductance Regulator (CFTR)**. CFTR is a chloride channel responsible for moving Cl- against its electrochemical gradient to the outside of the cell (▶ Fig. 7.5). Mutations in the gene encoding the CFTR (CFTR gene) are responsible for one of the most common autosomal recessive inherited diseases, cystic fibrosis (▶ Fig. 7.6). The activity of this

Table 7.1 Peroxisomal diseases

Disease	Cause	Defect
Zellweger syndrome	Disturbed peroxisome biosynthesis due to enzyme transport defect	Organ damage, especially in liver and brain, conspicuous facial changes
Adreno-leukodystrophy	Deficient ANV transporter for very long-chain fatty acids	Neurological disorders, weakness, dizziness, adrenal medullary atrophy, demyelination in the brain
Refsum syndrome	Defect of phytanic acid degradation	Neuropathies, vision, hearing and cardiac disorders

Source: Koolman J, Röhm K, ed. Color Atlas of Biochemistry. 3rd Edition. New York, NY. Thieme; 2012.

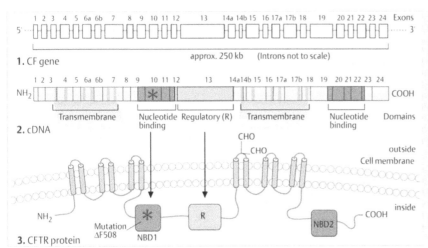

Fig. 7.5 Organization of the *CFTR* gene and the structure of the CFTR protein. (Source: Passarge E, ed. Color Atlas of Genetics. 4th Edition. New York, NY. Thieme; 2012.)

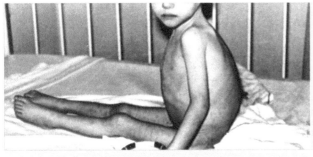

Fig. 7.6 A child with cystic fibrosis. (Source: Passarge E, ed. Color Atlas of Genetics. 4th Edition. New York, NY. Thieme; 2012.)

transporter is regulated by both ATP binding and phosphorylation by protein kinase A (PKA). In addition to Cl-, the CFTR can also transport bicarbonate (HCO3-), which is an important component of the physiological pH buffering system.

7.3 Channels

Channels allow the passage of specific molecules and are capable of discriminating based on size and electrical charge of the solute. In the case of a non-gated channel, the solute will move from the side of highest concentration to the side of lowest concentration, without any type of regulation. An example of a non-gated channel is the aquaporin channel. This channel can be found in the plasma membrane of all mammalian cells. It selectively transports water molecules while preventing the passage of ions and other solutes. The function of the aquaporin channels is to regulate the cell's volume and internal osmotic pressure.

In addition to non-gated channels, multiple types of gated channels exist. These channels can be grouped into three major categories: voltage-gated, ligand-gated, and stress-gated.

Voltage-gated channels are important for propagating electrical signals in nerve cells. They are also found on other types of electrically stimulated cells, such as muscle cells and egg cells. These channels contain voltage sensors, which are specialized charged protein domains that open or close based on the electrical charge or membrane potential of the membrane. In order for the status to change, the channel must encounter an electrical signal that changes the membrane potential above a certain threshold to produce enough force to cause a conformational change in the channel, resulting in its opening or closing. Examples of voltage-gated channels are Na + channels (responsible for the creation and propagation of action potentials), Ca2 + channels (muscle excitation and contraction), and K + channels (repolarization following an action potential).

Another type of gated channel is the **ligand-gated ion channel**. These channels act as a receptor, with their opening being controlled by the binding of a ligand to a pocket on the channel. This ligand binding site can face either the outside of the cell (to interact with extracellular ligands) or the inside of the cell (to interact with intracellular ligands). The opened channel unveils a selective pore that allows for the flow of specific ions down their electrochemical gradient, either into or out of the cell. This effectively converts a chemical signal into an electrical signal.

An example of a ligand-gated ion channel is the nicotinic acetylcholine receptor. These receptors are found on the surface of

II

some neurons and at the postsynaptic side of neuromuscular junctions. They bind to the neurotransmitter acetylcholine. When acetylcholine binds, a channel is opened that allows for the influx of Na+ down its electrochemical gradient. This flood of sodium into the cell results in a flip in the polarity of the plasma membrane's charge, initiating an action potential in the postsynaptic cell.

Many gated ion channels function in the activation of a synapse and act to transform chemical signals into electrical signals and vice versa. At synapses, voltage-gated ion channels open in response to action potentials, allowing for the influx of Ca2+. This induces the release of neurotransmitter into the synaptic cleft. In the synaptic cleft, the neurotransmitter binds to ligand-gated ion channels on the postsynaptic cell, resulting in a change in the membrane potential in the postsynaptic cell.

The resting membrane potential of a cell is the charge difference between the inside and outside of the cell near the plasma membrane. For all cells, including unstimulated nerve cells, the resting membrane potential hovers around -70 mV. This negative charge difference is generated in part by the outward movement of K+ ions through open non-gated K+ channels, known as leak channels. This movement is driven by the high K+ concentration inside the cell and low concentration outside the cell.

The sequential opening and closing of voltage-gated Na+ and K+ channels in the plasma membrane of excitable cells results in sudden membrane depolarizations followed by rapid repolarizations. If the depolarization reaches a certain threshold value, then a propagating wave of electrical excitation is created along the membrane of a cell, also known as an action potential.

The final type of gated channel is the **stress-gated ion channel**. In the auditory hair cells of the inner ear, there are projections called stereocilia on the cell's surface. Stress-gated ion channels on one stereocilia are physically connected to adjacent cilia through a linking filament. Sound vibrations cause the basilar membrane to vibrate, resulting in the tilting of the stereocilia to one side. This movement opens the stress-gated ion channels and allows the flow of ions into the hair cell. This induces an electrical signal that is transmitted to the auditory nerve.

Review Questions

1. The concentration of which of the following ions is higher in the cytoplasm of the cell than outside the cell?
   A) Ca^{2+}
   B) Cl^-
   C) K^+
   D) Na^+
2. The function of which of the following transporters is independent of an energy source?
   A) GLUT
   B) Na^+-Ca^{2+} exchanger
   C) Na^+-K^+ ATPase
   D) SGLT
3. Which of the following is an example of a non-gated channel?
   A) ABC transporters
   B) Aquaporin
   C) CFTR
   D) Nicotinic acetylcholine receptor
4. Which of the following is an example of passive transport?
   A) Coupled transporters
   B) ATP-driven pumps
   C) Light-driven pumps
   D) Uniporters
5. The opening of which of the following channels is controlled by the binding of a ligand?
   A) Acetylcholine
   B) Aquaporin
   C) Stress-gated
   D) Voltage-gated

Answers

1. **The correct answer is C.** The concentrations of K^+ are higher in the cytoplasm of the cell than in the extracellular environment. All of the other listed ions (answer choices A, B, and D) are found at higher concentrations outside the cell than inside the cell.
2. **The correct answer is A.** The glucose transporter GLUT is a passive transporter that moves glucose across the plasma membrane down its concentration gradient without the need for an expenditure of energy. The Na^+-Ca^{2+} exchanger (answer choice B) uses the energy of the Na^+ electrochemical gradient to move Ca^{2+} out of the cell against its electrochemical gradient. Similarly, SGLT (answer choice D) transfers glucose across the membrane against its electrochemical gradient through the energy generated by the Na^+ electrochemical gradient. The Na^+-K^+ ATPase (answer choice C) uses energy released from the hydrolysis of ATP to move Na^+ outside the cell and K^+ inside the cell.

3. **The correct answer is B.** Aquaporin is a passive, non-gated channel that allows the selective transfer of water molecules into and out of the cell. This channel plays an important function in the regulation of a cell's volume and internal osmotic pressure. The CFTR (answer choice C) is a type of ABC transporter (answer choice A). These transporters use the energy of ATP hydrolysis to move ions against their electrochemical gradient through an ion channel. The nicotinic acetylcholine receptor (answer choice D) is a ligand-gated ion channel that opens in response to binding of the ligand acetylcholine and allows the flow of Na^+ ions into the cell.
4. **The correct answer is D.** Passive transport is a method of transporting molecules into or out of the cell that does not require the expenditure of energy. Uniporters, such as the GLUT family of glucose transporters, are passive transporters that move glucose down its concentration gradient across the plasma membrane. All of the other answer choices are active transporters that require a source of energy for their function. Coupled transporters (answer choice A) use the movement of ions down their electrochemical gradient to transport other molecules across the membrane, as is seen with the SGLT family of Na^+-Glucose symporters. ATP-driven pumps (answer choice B) uses the energy of ATP hydrolysis for its function, whereas light-driven pumps (answer choice C) uses the energy of light to transport molecules.
5. **The correct answer is A.** The nicotinic acetylcholine channel is characterized as a receptor because the opening of its Na^+ channel occurs upon the binding of its ligand, acetylcholine. Aquaporin (answer choice B) is a passive channel that allows for the movement of water molecules. The opening of stress-gated channels (answer choice C) occurs through physical connections with adjacent structures. Voltage-gated (answer choice D) channels open in response to a change in the electric membrane potential across the membrane.

8 Extracellular Matrix

At the conclusion of this chapter, students should be able to:
- Identify the five major extracellular matrix components that make up the basal lamina and explain how they contribute to the organization of this structure
- Describe the structure and function of collagen, fibronectin, glycosaminoglycans (GAGs), proteoglycans, and elastin fibers
- Identify the mutations associated with the diseases Osteogenesis Imperfecta, Ehlers-Danlos Syndrome, α1-Antitrypsin Deficiency, and Marfan Syndrome, and describe how defective extracellular matrix components give rise to these diseases
- Identify the major family of enzymes that are responsible for degradation of extracellular matrix, and describe the three ways that the activity of these enzymes can be regulated
- Explain how the normal activity of these enzymes contributes to the cell's ability to migrate, and describe how cancer cells can use these enzymes to help them become metastatic

The extracellular matrix, or ECM, is a network of proteins and glycosaminoglycan molecules outside the cells. This structure provides support to the cells and tissues by serving as a scaffold to which the cells of the body are connected. Although the ECM is located outside of the cell, its components are made inside the cell, secreted, and assembled into higher order structures in the extracellular space.

In connective tissue, a major function of the extracellular matrix is to carry the mechanical load to which these tissues are subjected. In contrast, the extracellular matrix of epithelial tissues functions to hold the epithelial cells together. The mechanical load in these tissues is mitigated by the complexes that connect the individual cells together (discussed later in this chapter).

8.1 Basal lamina

The basal lamina, a thin and flexible sheet of extracellular matrix, is the major ECM structure in epithelial tissues (▶ Fig. 8.1). Not only is it an essential component of epithelial sheets, but it also surrounds individual muscle cells, fat cells, and Schwann cells. In the kidney glomerulus, the basal lamina lies between two cell sheets and functions as a selective filter. The basal lamina is made up of five distinct molecules. The glycoprotein component includes laminin, type IV collagen, nidogen, and fibronectin. Another component, perlecan, is a member of the proteoglycan family.

In order to form a functional basal lamina, these molecules must interact in specific ways with one another. For example, nidogen and perlecan serve as linkers to connect the laminin and the type IV collagen networks, and laminin molecules are anchored at the cell surface through binding of their tail regions to receptors on the surface of the associated cell. Their head regions are then free to interact with other molecules, allowing laminin to organize the rest of the ECM sheet structure.

The primary organizer of the sheet structure of the basal lamina is laminin. Laminin is formed by a heterotrimeric complex composed of alpha, beta, and gamma subunits. This structure results in a large, flexible protein that can bind to multiple components of the basal lamina, as well as provides structural support.

8.2 Collagen

Not only is collagen a major component of the basal lamina in epithelial sheets, but it is also the most abundant protein in connective tissues such as bone, tendon, and muscle (▶ Fig. 8.2). In fact, collagen constitutes approximately 25% of the total protein mass in the body. In these tissues, the main function of collagen is to provide resistance to tensile forces.

Collagen is composed of three polypeptides, known as α chains, that are wound around one another to form a super helix structure called tropocollagen (▶ Fig. 8.3). Once the tropocollagen molecules are made, they are secreted from the cell. Outside the cell, multiple tropocollagen complexes are then assembled into polymers called collagen fibrils. The collagen fibrils further assemble into large, fibrous structures called collagen fibers by packing tightly together.

Collagen undergoes extensive post-translational modifications. First, it is hydroxylated on prolines and lysines in the ER. Then it is glycosylated on the hydroxylated lysines

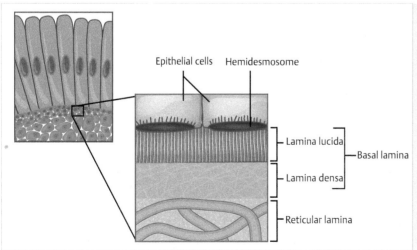

Fig. 8.1 Components of the basal lamina of epithelial tissue.

Epithelial cells   Hemidesmosome

Lamina lucida
Basal lamina
Lamina densa

Reticular lamina

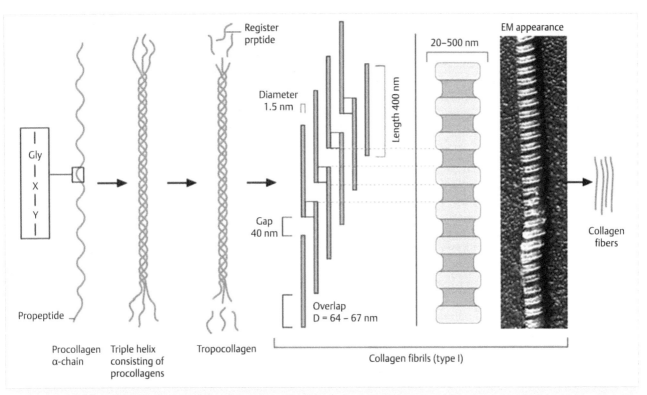

Fig. 8.2 Collagen Type I structure and organization of collagen fibers. (Source: Koolman J, Röhm K, ed. Color Atlas of Biochemistry. 3rd Edition. New York, NY. Thieme; 2012.)

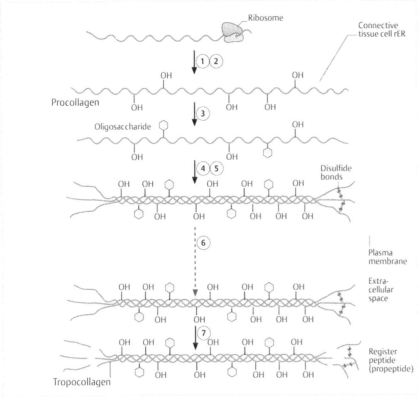

Fig. 8.3 Biosynthesis of collagen. (Source: Koolman J, Röhm K, ed. Color Atlas of Biochemistry. 3rd Edition. New York, NY. Thieme; 2012.)

(hydroxylysines) in the Golgi. It is then secreted from the cell and undergoes cleavage of the pro domain by proteases outside the cell. The mature collagen molecules then self-assemble to form fibril collagen, which aggregates to form collagen fibers.

Many different types of collagen can be found in the extracellular matrix. The different collagens can be organized into three major families based on their structural characteristics. These families include fibrillary collagen, found in bone, sheet-forming collagen, found in basal lamina, and anchoring/linking collagen, found in skeletal muscle. Fibrillar collagen provides the tensile strength necessary for the proper functioning of the tendons, ligaments, bones, and dense connective tissue. They also form the scaffolding for cartilage, where they trap glycosaminoglycans and proteoglycans, which attract water. This results in the formation of the gel-like substance that is characteristic of cartilage and that contributes to its resistance to compressive forces.

Some collagens polymerize into sheets rather than linear fibrils and are referred to as sheet-forming collagens. These sheets, predominantly composed of type IV collagen, assemble into net-like polymers that surround organs and muscle, as well as form the basal lamina beneath epithelial sheets (▶ Fig. 8.4).

As discussed earlier, collagen is made inside the cell in a precursor form and is subsequently processed to a mature, functional molecule outside the cell. In many tissues, the specialized cell that synthesizes collagen, as well as other ECM components, is the fibroblast. As the fibroblast crawls through the tissue, it extrudes collagen precursors that then form collagen fibers. The directional movement of the fibroblasts determines the arrangement of the collagen fibers in the ECM. In addition, fibroblasts can also reorganize already deposited collagen by crawling along them and pulling on them as they move.

8.2.1 Link to Pathology

Diseases of the Extracellular Matrix are summarized in ▶ Table 8.1.

Osteogenesis imperfecta is a genetic disorder characterized by brittle bones, short stature, and spinal curvature. There are several different types of osteogenesis imperfecta, with varying degrees of severity. Most types are caused by mutations in one or more collagen genes, resulting in either poor quality collagen or insufficient quantities of collagen. Of all of the existing types of this disease, Type I is the most prevalent form.

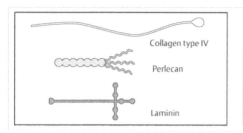

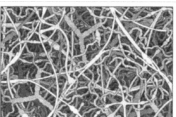

Fig. 8.4 Components of the basement membrane of epithelial tissue. (Source: Koolman J, Röhm K, ed. Color Atlas of Biochemistry. 3rd Edition. New York, NY. Thieme; 2012.)

Table 8.1 Diseases of the extracellular matrix

Congenital disorders (examples)		
Disease	Molecular cause	Clinical picture
Osteogenesis imperfecta	Various defects in the gene for collagen 1	Disorder of bone formation, bone fragility
Ehlers-Danlos syndrome	Various disorders of collagen biosynthesis due to defects of modifying enzymes	Elastic skin, over-extensible joints, deformation of the spine, defective blood vessels
Mucopolysaccharidoses type A-VII	Disorders of proteoglycans degradation: many different enzymes can be affected	Growth disorders, skeletal deformation, disorder of brain development
Epidermolysis bullosa	Various defects of ECM proteins involved in anchoring	Blistering of the skin after mechanical stress
Marfan syndrome	Genetic defects in the fibrillin-1 gene	Tall stature, spider fingers, lens changes, aneurysms, aortic rupture
Glanzmann disease	Genetic defect of the fibrinogen receptor integrin on platelets	Clotting disorder, bleeding tendency
Congenital muscular dystrophy	Defective laminin 2, disorder of the connection between the cytoskeleton of muscle cells with the ECM	Weakened muscles
Acquired disorders (examples)		
Osteoporosis	Reduction of bone mass, deterioration of microarchitecture	Skeletal shrinkage, bone pain, fractures
Scurvy	Ascorbic acid deficiency, disorder of collagen hydroxylation	Defects of connective tissue, bleeding, tooth loss
Rheumatoid arthritis	Joint inflammation (immune reaction), increase in cytokines	Inflammatory swelling of the joints

Source: Koolman J, Röhm K, ed. Color Atlas of Biochemistry. 3rd Edition. New York, NY. Thieme; 2012.

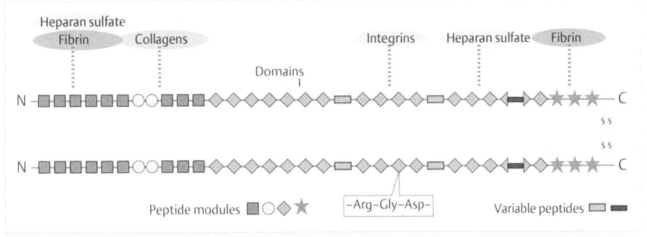

Fig. 8.5 Structure of fibronectin with binding sites for other extracellular matrix components indicated. (Source: Koolman J, Röhm K, ed. Color Atlas of Biochemistry. 3rd Edition. New York, NY. Thieme; 2012.)

Another genetic disorder resulting from the loss of functional collagen molecules is **Ehlers-Danlos syndrome**. It is characterized by hyper-flexible joints, fragile skin, and blood vessels that easily rupture. This disease can also be categorized into several different types, based on the gene affected. Types 1–4 are due to mutations in the collagen genes, resulting in defective collagen synthesis, whereas Type 6 is due to a mutation in the gene encoding lysyl hydroxylase. This enzyme is responsible for addition of the hydroxyl groups to lysine during collagen synthesis. As a result of this mutation, collagen molecules are not processed properly and are therefore nonfunctional.

8.3 Fibronectin

Fibronectin is a large glycoprotein composed of two monomeric subunits joined by disulfide bonds at their carboxyl end (▶ Fig. 8.5). It can exist in either a soluble form, as in blood and other body fluids, or as insoluble fibronectin fibrils, as is found in ECM. Through its free amino ends, fibronectin can clamp onto components of the ECM, such as collagen fibers. On the other end of the molecule, the disulfide-bound carboxyl end is capable of binding to integrins on the surface of cells. It is through these types of interactions that fibronectin provides a linkage between the ECM and cells.

8.4 Glycosaminoglycans and Proteoglycans

Glycosaminoglycans (GAGs) are unbranched polysaccharide chains composed of repeating disaccharide units (▶ Fig. 8.6). Most glycosaminoglycans are covalently linked to core proteins, in which case they are known as proteoglycans. These molecules are highly negatively charged due to their polysaccharide composition. The major GAGs can be separated into four groups based on their structure. These four groups are hyaluronan, chondroitin sulfate/dermatan sulfate, heparan sulfate, and keratin sulfate. GAGs have a highly extended conformation that occupies a large volume relative to their mass. In fact, most of the

extracellular space is filled by GAGs. In addition, their negative charge makes them strongly hydrophilic, allowing them to readily form gels at very low concentrations in the presence of water.

Hyaluronan, also known as hyaluronic acid, is the simplest GAG, consisting of repeats of up to 25,000 disaccharide units (▶ Fig. 8.6). It is the only GAG not linked to a core protein and, therefore, is not synthesized in the ER. Instead, it is synthesized by an enzyme complex, called hyaluronan synthase, embedded in the plasma membrane. The newly synthesized hyaluronan molecules are directly released into the extracellular space as they are made.

Proteoglycan molecules exist in a wide variety of sizes. One of the smallest proteoglycans is decorin, a molecule that binds to collagen fibrils and regulates fibril assembly. In contrast, the cartilage component aggrecan is a very large proteoglycan. Aggrecans can also combine with hyaluronan to produce even larger complexes called aggrecan aggregates. These molecules are found in the cartilage matrix and help to resist compressive forces. In addition, aggrecans perform important roles in cell signaling. For example, they bind to fibroblast growth factor (FGF), allowing them to cross-link and activate their cell surface receptor. They also immobilize chemokines at sites of inflammation and inhibit transforming growth factor β by binding and sequestering it.

The synthesis of aggrecan aggregates begins inside the cell. A core protein is first made in the rough ER, where addition of polysaccharide chains occurs in the lumen. Additional modifications occur in the Golgi, creating a proteoglycan that is released from the cell via exocytosis. At the same time, hyaluronan is made by the hyaluronan synthase at the plasma membrane and released directly into the extracellular environment. Outside the cell, the proteoglycans and hyaluronan combine to form aggrecans, which then combine to form large aggrecan aggregates.

Not all proteoglycans are secreted components of the extracellular matrix. Some are integral components of the plasma membrane and serve as co-receptors. Syndecans are membrane spanning proteoglycans found on the surface of fibroblasts. Their intracellular domains interact with the actin cytoskeleton and signaling molecules within the cell, whereas their extracellular domain interacts with fibronectin in the extracellular matrix. Another example of membrane bound proteoglycans is

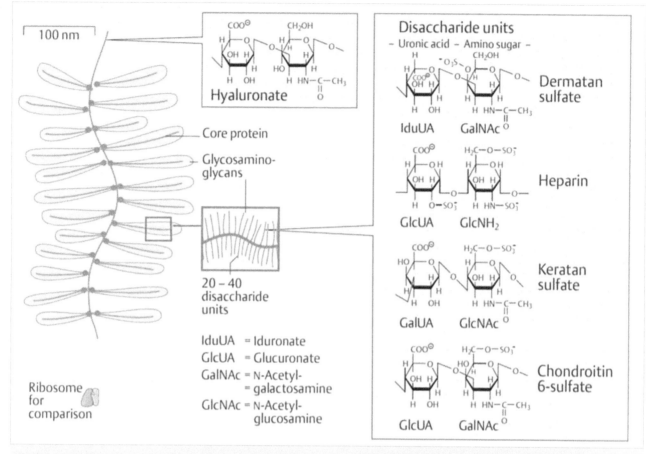

Fig. 8.6 Proteoglycan structure. (Source: Koolman J, Röhm K, ed. Color Atlas of Biochemistry. 3rd Edition. New York, NY. Thieme; 2012.)

betaglycan. These molecules can bind to extracellular TGFβ and present the ligand to its neighboring TGFβ receptor.

8.5 Elastic Fibers

Elastic fibers are loose and unstructured polypeptide chains that are covalently cross-linked into a rubber-like elastic meshwork. They are found in ECM of connective tissue of smooth muscle, blood vessels, skin, and lungs. Elastic fibers allow these tissues to stretch and then relax to their original form. The two major components of elastic fibers are elastin and microfibrils, which are polymers composed of the glycoprotein fibrillin. In addition, collagen fibers are often interwoven with elastic fibers to limit the extent of stretching and prevent the tissue from tearing.

8.5.1 Link to Pathology

α-1 antitrypsin deficiency is an inherited disorder in which the gene encoding α-1 antitrypsin, *SERPINA1*, is mutated, resulting in an absence of this enzyme's activity. This enzyme is normally produced by the liver and acts to inhibit the activity of a neutrophil specific enzyme called neutrophil elastase, which cleaves elastin. In individuals with this condition, the lack of α-1 antitrypsin means that neutrophil elastase activity will be high, leading to increased destruction of elastin. This is especially true in the lungs of afflicted individuals who smoke or are exposed to

cigarette smoke. In this case, the smoke-induced inflammation in the lungs recruits neutrophils, but there is no α-1 antitrypsin available to quench the resulting elastase activity. This causes destruction of the lung tissue and could ultimately give rise to emphysema or COPD (reviewed in chapter 18).

Marfan syndrome is a genetic disorder of connective tissue resulting from a mutation in the gene encoding fibrillin-1. Individuals with this condition have an above-average height with long, slender limbs. Since approximately half of normal aortic tissue contains elastin fibers, Marfan syndrome is associated with a high risk of aortal aneurysm and aortal dissection.

8.6 Extracellular Matrix Degrading Enzymes

Matrix metalloproteinases, or MMPs, are a major class of enzymes that are capable of degrading components of the extracellular matrix (▶ Fig. 8.7). Their activity requires binding of Ca^{2+} or Zn^{2+}.

There are several members of the matrix metalloproteinase family, each having a specific target substrate in the ECM. For example, MMP-1 specifically degrades fibrillary collagen, whereas the target of MMP-12 is elastin.

Some members of the serine protease family can also degrade ECM components. Plasmin is a member of this family and functions to break up blood clots by degrading the ECM

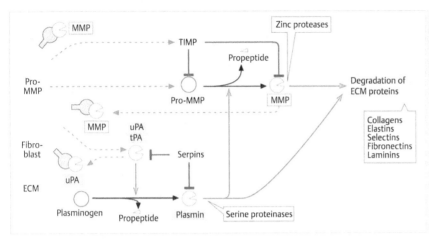

Fig. 8.7 Degradation of extracellular matrix components. (Source: Koolman J, Röhm K, ed. Color Atlas of Biochemistry. 3rd Edition. New York, NY. Thieme; 2012.)

molecule fibrin. Another serine protease is urokinase-type plasminogen activator (uPA), which is responsible for cleaving plasminogen, thus producing active plasmin.

Degradation of the ECM is necessary for both cell migration and cell division. For migration, cells need to create a path through the ECM to allow for their movement, and in cell division, the constraints of the ECM around the cell must be relieved in order to provide the space necessary for division. In addition, some growth factors are deposited in the ECM and must be freed by degradation of the ECM components surrounding them so that they have access to their receptors on the surface of target cells.

Review Questions

1. Which of the following components of the basal lamina is composed of three separate subunits and links all of the other basal lamina components together?
   A) Collagen IV
   B) Laminin
   C) Nidogen
   D) Perlecan

2. A patient with above average height and long, slender limbs presents to the emergency department with an aortic aneurysm. Which of the following extracellular matrix components is most likely defective?
   A) Basal lamina
   B) Collagen
   C) Elastic fibers
   D) Glycosaminoglycans

3. Which of the following extracellular matrix components is composed of two subunits connected by disulfide bonds?
   A) Collagen
   B) Elastin
   C) Fibronectin
   D) Glycosaminoglycan

4. Which of the following molecules can degrade components of the extracellular matrix?
   A) α-1 antitrypsin
   B) Hyaluronan synthase
   C) Lysyl hydroxylase
   D) Matrix metalloproteinase

Answers

1. **The correct answer is B**. Laminin is a multi-subunit complex consisting of an α, β, and γ subunits. The structure of laminin creates multiple binding sites that allow it to link together the other components of the basal lamina (answer choices A, C, and D).

2. **The correct answer is C**. Above average height and long, slender limbs are common symptoms associated with Marfan syndrome. In addition, these individuals are at high risk for aortic aneurysms and aortic dissections. The gene that is mutated in this disease is fibrillin-1, which encodes a component of elastic fibers. Mutations in collagen (answer choice B) can give rise to osteogenesis imperfecta or Ehlers-Danlos syndrome. The basal lamina (answer choice A) is the thin extracellular matrix underlying epithelial tissues. Glycosaminoglycans (answer choice D) are long polysaccharide chains that form a major component of joints.

3. **The correct answer is C**. Fibronectin is an extracellular matrix protein composed of two monomeric subunits joined by disulfide bonds at their carboxyl end. Collagen (answer choice A) is composed of three polypeptides that are wound around one another to form a super helix structure called tropocollagen. Elastin (answer choice B) is a component of the elastic fibers that provide tissue with the ability to stretch. Glycosaminoglycans (answer choice D) are unbranched polysaccharide chains composed of repeating disaccharide units.

4. **The correct answer is D**. The matrix metalloproteinase family includes enzymes that can degrade many different types of extracellular matrix components in the presence of either $Zn2+$ or $Ca2+$. α-1 antitrypsin (answer choice A) is a protein encoded by the *SERPINA1*. It is produced by the liver and functions to inhibit elastase, preventing the breakdown of the extracellular matrix component elastin. The glycosaminoglycan hyaluronan is made by the enzyme hyaluronan synthase (answer choice B). Lysyl hydroxylase (answer choice C) is an enzyme that adds hydroxyl groups to lysines in procollagen molecules.

9 Cell Adhesion and Membrane Junctions

At the conclusion of this chapter, students should be able to:
- Identify the four superfamilies of cell adhesion molecules and explain how they function in extracellular interactions
- Describe the process of leukocyte extravasation, identify the cell adhesion molecules involved, and explain the interactions of the adhesion molecules necessary for each step in the process
- Identify the five major types of intercellular junctions and the proteins that make up each
- Describe the function of each intercellular junction and explain how disruption of these structures could affect the epithelial tissue

As discussed in Chapter 8, the extracellular matrix plays important roles in both the resistance to mechanical forces as well as in the joining of cells and tissues together within the body. These characteristics of the extracellular matrix require the association of cells with each other within tissues, as well as the connection of the individual cells with components of the extracellular matrix. The molecules that allow for these connections are known as cell adhesion molecules.

There are four super-families of cell adhesion molecules (▶ Fig. 9.1). They can be functionally categorized into those that require calcium for their function and those that do not. Among the calcium dependent cell adhesion molecules are the cahderins, integrins, and selectins. Those that are calcium-independent are the immunoglobulin-like cell adhesion molecules (I-CAMs).

9.1 I-CAMs

The immunoglobulin-like cell adhesion molecules (I-CAMs) can interact with both integrins, which would be a heterophilic interaction, or with other I-CAMs, which represent a homophilic interaction. The name I-CAM is derived from its general structure, which consists of several loop structures produced by intramolecular disulfide bonds. Due to this organization, the I-CAM molecule resembles an immunoglobulin. Members of the I-CAM family include N-CAM (Neural CAM), VCAM-1 (Vascular CAM), and PECAM-1 (Platelet Endothelial CAM).

9.2 Cadherins

The cadherins are a family of calcium-dependent cell-cell adhesion molecules containing calcium binding domains, called cadherin domains, which are repeated multiple times. The binding of calcium to the hinge regions between the cadherin repeat domains prevents the molecule from flexing, resulting in a rigid molecule that can associate with other cadherins. In the absence of calcium, the molecule becomes floppy and adhesion fails. Cadherin binding is generally homophilic, meaning that cadherin molecules of a specific subtype on one cell bind to cadherin molecules of the same subtype on the other cell.

Cadherins are integral components of both adherens junctions and desmosomes, providing the extracellular linkages that hold two adjacent cells together. Examples of specific members of the cadherin family include E-cadherin (epithelial), N-cadherin (neural), and P-cadherin (Placenta). All cadherins have similar extracellular regions with multiple copies of the cadherin domain motif. In contrast, their intracellular domains are more varied, with some forms interacting with components of the cell's cytoskeleton, such as actin filaments.

9.2.1 Link to Physiology

Both I-CAMs and cadherins are needed at synapses to physically hold the membrane of the presynaptic cell close to the membrane of the postsynaptic cell, creating an extracellular

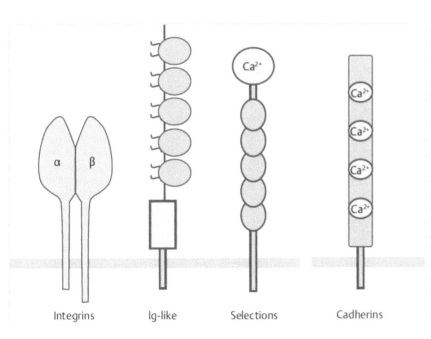

Fig. 9.1 Four Superfamilies of Cell Adhesion Molecules.

Integrins   Ig-like   Selections   Cadherins

pocket known as the synaptic cleft. Cadherins in each membrane bind to one another in a homotypic fashion. Similarly, I-CAMs in each membrane create a homotypic interaction, providing a tight connection to hold the two cells together. Each of these adhesion molecules is connected to the actin cytoskeleton in their respective cells, resulting in a sturdy physical connection of the cytoskeletons between the two cells.

9.3 Integrins

Integrins are transmembrane adhesion molecules that form functional heterodimeric complexes made up of different combinations of α and β subunits (▶ Fig. 9.2). The affinity of the integrin complex for specific extracellular molecules is determined by its α and β subtype combination. For example, the extracellular domains of some integrins bind to extracellular matrix components such as laminin and fibronectin, while other integrins bind to I-CAMs on the surface of other cells. Inside the cell, the intracellular domains of integrins are often connected to the cell's actin cytoskeleton via adaptor proteins.

Integrins have two main functions. They can redirect extracellular stresses from the fragile plasma membrane to the sturdier cytoskeleton inside the cell, and they can act as intermediates for signaling cascades within the cell. In this case, binding of activation molecules on the extracellular domain of the integrin heterodimers can induce signaling events inside the cell that ultimately lead to a cellular response. This type of signaling is called outside-in activation. Signaling can also work in an inside-out fashion. In this scenario, signaling cascades inside the cell cause the tight association of integrin with molecules outside the cell, such as extracellular matrix proteins or cell adhesion molecules on other cells.

An example of both inside-out and outside-in signaling can be seen in the interactions between a T-cell and an antigen presenting cell. When the T-cell encounters an antigen-presenting cell, the two cells first adhere weakly through integrin-ICAM interactions. This brings the cells close enough together to allow for binding of the T-cell receptor to the MHC-antigen complex. An outside-in activation then occurs in which the intracellular tail of the T-cell receptor is activated. This signal is transmitted to the intracellular tail of the neighboring integrin, resulting in an inside-out activation that causes the extracellular domains of the integrin molecule to bind more tightly to its associated I-CAM on the antigen-presenting cell's surface.

9.4 Fibronectin

Fibronectin, an ECM component discussed in Chapter 8, can also mediate cell adhesion interactions. Its structure, consisting of two large subunits arranged in a pincher-like shape and connected at one end by disulfide bonds, allows for the grasping of collagen fibers in the extracellular matrix (▶ Fig. 9.2). At the same time, the other end of fibronectin is connected to integrin heterodimers in the plasma membrane of a nearby cell, creating a ECM-cell bridge.

9.5 Selectins

Selectins are molecules embedded in the plasma membrane that have extracellular domains, called lectin domains, that are capable of binding to carbohydrates-containing molecules, known as glycans, on the surface of other cells. There are three subsets of selectins, each binding to a specific type of glycan. However, this selectin-glycan interaction is weak, which results in a weak adhesion between the two cells. E-selectin is found on endothelial cells and is induced by inflammation. L-selectin is found on leukocytes, and P-selectin is found on platelets and endothelial cells. One of the main roles of selectins is in directing white blood cells, or leukocytes, to the sites of inflammation, a process that includes extravasation of the leukocyte.

9.5.1 Link to Immunology

Leukocytes extravasation involves the movement of a leukocyte out of the circulation and into a site of tissue damage or infection (▶ Fig. 9.3). In the first step of the extravasation process, known as the chemo-attraction step, macrophages in the affected tissue release cytokines, such as IL-1 and TNFα that attract the leukocytes by inducing the expression of selectins on the surface of the endothelial cells lining the blood vessel. During the next step, rolling adhesion occurs as the carbohydrates in the leukocyte plasma membrane bind to selectins on the plasma membrane of endothelial cells lining the inner wall of blood vessels. This is a weak interaction, resulting in the binding and releasing of the cell, causing the rolling of the cell along the endothelial wall. In the tight, or firm, adhesion step, integrins on leukocytes bind to

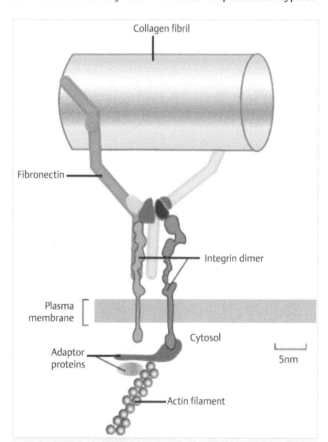

Fig. 9.2 Structure and function of integrins. (Source: Koolman J, Röhm K, ed. Color Atlas of Biochemistry. 3rd Edition. New York, NY. Thieme; 2012.)

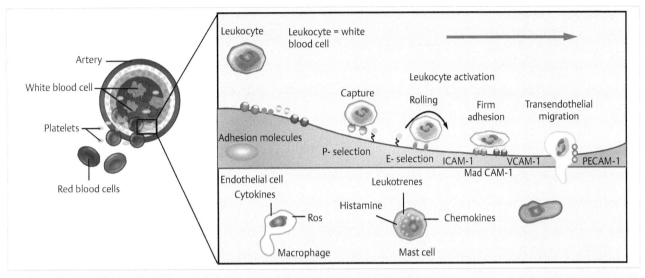

Fig. 9.3 Multiple types of cell adhesion molecules are important for leukocyte extravasation.

I-CAMs on endothelial cell surfaces with high affinity, immobilizing the leukocyte. Transmigration refers to the process wherein the leukocyte passes through gaps between endothelial cells and enters the affected tissue.

Leukocyte adhesion deficiency is an immunodeficiency disease characterized by recurrent infections and is caused by a mutation in the gene that encodes a component of the integrin complex. In these individuals, leukocytes do not make functional integrins, therefore their leukocytes cannot bind to the I-CAMs on the surface of endothelial cells lining the lumen of the blood vessels and the tight adhesion step in extravasation does not occur. This prevents the leukocytes from migrating out of the blood stream through the blood vessel walls. The resulting lack of leukocyte infiltration in the affected tissue gives rise to the diminished ability to fight infections that is seen in afflicted individuals.

9.6 Membrane Junctions in Epithelia

The epithelial tissues of the body are formed from a sheet of cells, each attached to its neighbor laterally. A majority of the tissues of the body are epithelial and, therefore, a majority of the body's cells are epithelial cells. These epithelial sheets cover all of the external surfaces of the body as well as line all internal cavities. Within the sheets, the cells are classified based on their morphology (columnar, cuboidal, or squamous) and the number of cell layers involved (simplified or stratified) (▶ Fig. 9.4).

Epithelial sheets are polarized, with functionally distinct apical and basal sides. In most epithelial tissues, the apical side is exposed to the lumen or cavity of an organ while the basal side rests on and is attached to a structure called the basal lamina. As discussed in Chapter 8, the basal lamina is a type of connective tissue that provides sites of adhesion for the integrin molecules in the plasma membranes of associated epithelial cells. This structure consists of collagen, laminin, and other extracellular matrix proteins.

Protein complexes found on the lateral and basal surfaces of the cells within the epithelial sheet connect the cells to one another and to the underlying basal lamina. These protein complexes are called membrane junctions and are responsible for keeping the epithelial tissue intact. Membrane junctions can be grouped into the following five families based on their structure and location: tight junctions, adherens junctions, gap junctions, desmosomes, and hemidesmosomes (▶ Fig. 9.5).

9.7 Tight Junctions

Tight junctions are formed from transmembrane proteins called claudins and occludins that are arranged in strands along the lines of the junction to create a tight seal and form a ribbon around the entire cell near the apical end. These tight junctions not only help to physically hold cells together, but they also serve as a protective barrier by preventing the passage of molecules and ions through the spaces between the cells. This ensures that materials must pass through the cells themselves and allows for filtering by the cell. In addition, tight junctions also form a functional barrier which helps cells maintain polarity by preventing lateral diffusion of membrane proteins between the apical and basal/lateral surfaces of the cell.

9.8 Adherens Junctions

Adherens junctions contain cadherin molecules that are exposed to the outside of the cell and linker proteins inside the cell that associate with the actin cytoskeleton. These cadherin molecules on the cell surface form homophilic bonds with like molecules in the plasma membrane of the neighboring cell. The adherens junctions are found on the lateral surface of the cells and are commonly located next to tight junctions. Like tight junctions, adherens junctions form a ribbon around the entire cell, near the apical end.

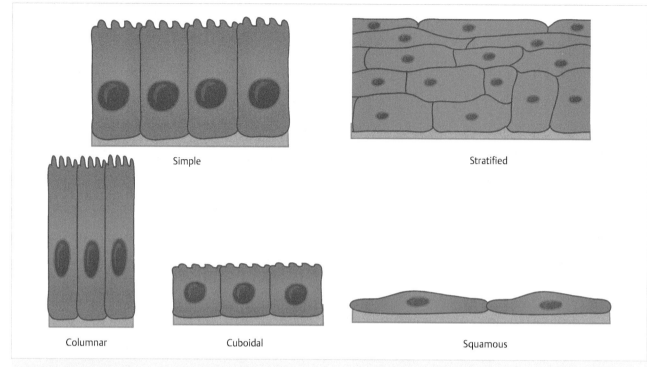

Fig. 9.4 Types of epithelial cells.

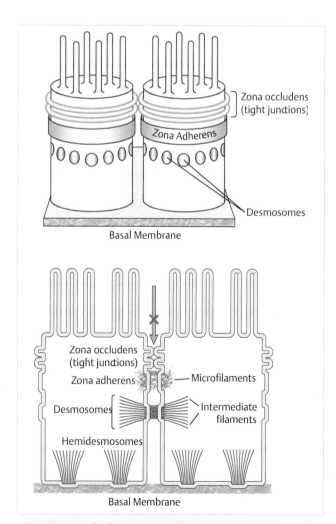

Fig. 9.5 Types of cell junctions in epithelial tissue.

9.9 Gap Junctions

Gap junctions are channels that span the plasma membranes of two adjacent cells, physically connecting the cytoplasms of the cells. This allows the movement of small water soluble molecules and ions to pass from cell to cell. In the heart, for example, gap junctions allow ions to pass from one cardiac muscle cell to the next, providing the electrical coupling necessary to allow for the synchronized contractions of a heartbeat.

Gap junctions are formed by complexes called connexins found in the plasma membrane of the cell. Each connexin consists of six subunits of connexin arranged radially to form a central channel. Most cell types express more than one type of connexin, so that the connexin complexes are heteromeric. The connexins in adjacent cells are aligned end-to-end to create the functional gap junction.

9.10 Desmosomes

By linking the keratin filaments inside two adjacent cells, desmosomes increase the mechanical strength of the cells to allow for resistance to shear forces. These membrane junctions are also found on the lateral surface of the cell and contain a cytoplasmic plaque that is made up of plakoglobin/plakophilin and desmoplakin. To this plaque are connected the keratin filaments of each cell. On the outside of the cell, the cadherin family members desmoglein and desmocollin bind to like molecules on adjacent cells and link the cells together.

9.11 Hemidesmosomes

Hemidesmosomes attach epithelial cells to the underlying basal lamina and are composed of a plaque containing dystonin and plectin, which create a bridge between the collagen and laminin of the basal lamina and the keratin fibers inside the cell. This results in an asymmetrical structure in which one side of the junction consists of the hemidesmosome and the other side is extracellular matrix. This is in stark contrast to the other types of membrane junctions, which involve the linkage of identical complexes on adjacent cells and are, therefore, symmetrical.

9.11.1 Link to Immunology

There are several **blistering skin diseases** that are due to dysfunction of hemidesmosomes and desmosomes. These are autoimmune diseases in which the body makes antibodies against components of these structures. **Bullous pemphigoid** is due to antibodies against dystonin, a component of hemidesmosomes. In this condition, epithelial sheets lose their connections to the underlying basal lamina. In **pemphigus foliaceus**, antibodies against the cadherin protein desmoglein-1 are produced. Since this protein is a component of desmosomes, the adhesion between the adjacent epithelial cells is lost. The same is true for **pemphigus vulgaris**, although in this case the antibodies recognize another component of the desmosome, desmoglein-3. In all of these conditions, blisters are readily formed in response to any friction against the skin.

II

Review Questions

1. Which of the following cell adhesion molecules binds to carbohydrates on the surface of nearby cells?
   A) Cadherins
   B) ICAMs
   C) Integrins
   D) Selectins

2. Which of the following molecular interactions is required for the firm adhesion step of the leukocyte extravasation process?
   A) Cadherin-Cadherin
   B) ICAM-ICAM
   C) ICAM-Integrin
   D) Selectin-Glycan

3. A defect in dystonin would affect the function of which membrane junction?
   A) Adherens junction
   B) Desmosome
   C) Gap junction
   D) Hemidesmosome
   E) Tight junction

4. Which of the following membrane junctions creates a seal between adjacent cells and prevents the movement of molecules through the space between the two cells?
   A) Adherens junction
   B) Desmosome
   C) Hemidesmosome
   D) Tight junction

5. Adherens junctions of neighboring cells are connected by which type of molecule?
   A) Cadherin
   B) Connexin
   C) Integrin
   D) Occludin

Answers

1. **The correct answer is D**. Selectins are embedded in the plasma membrane of cells and contain lectin domains that bind to carbohydrate-containing molecules, known as glycans, on the surface of other cells. Cadherins (answer choice A) bind to other cadherins, while ICAMs (answer choice B) and integrins (answer choice C) can bind to one another.

2. **The correct answer is C**. The firm adhesion step of extravasation requires a tight interaction between two cell adhesion molecules, which is true for ICAM-integrin interactions. Selectin-Glycan (answer choice D) binding is a weak interaction that is involved in the rolling phase of extravasation. Cadherin-cadherin (answer choice A) binding is found in adherens junctions and desmosomes.

3. **The correct answer is D**. Dystonin, along with plectin, form the hemidesmosome plaque that creates a bridge between the collagen and laminin of the basal lamina and the keratin fibers inside the cell. Therefore, a defect in this plaque would affect the normal function of the hemidesmosome. None of the other membrane junctions (answer choices A, B, C, or E) contain the protein dystonin.

4. **The correct answer is D**. Tight junctions are found on the lateral surface of adjacent epithelial cells in an epithelial sheet, near their apical surface. They are composed of claudins and occludins, which form tight seals between the cells that prevent the movement of molecules in between the cells. Although adherens junctions (answer choice A) are also on the lateral side of cells near the apical surface, these junctions are composed of cadherin molecules and connect the actin cytoskeletons of the adjacent cells. Desmosomes (answer choice B) are also found on the lateral side of cells. These complexes connect the intermediate filaments of the cells, providing the epithelial sheet with resistance to mechanical stresses. Hemidesmosomes (answer choice C) reside on the basal surface of epithelial cells and connect the cell to the underlying basal lamina.

5. **The correct answer is A**. Adherens junctions connect the actin cytoskeletons of neighboring cells within an epithelial sheet. These connections occur through extracellular interactions between cadherin molecules embedded in the lateral surfaces of the plasma membranes of the cells. Connexins (answer choice B) are proteins that form gap junctions. Gap junctions create pores that connect the cytoplasm of two cells together and allow for the transfer of soluble molecules. Integrins (answer choice C) on the surface of leukocytes interact with ICAMs on the surface of endothelial cells in blood vessels to allow for leukocyte extravasation. Tight junctions are produced by the binding of occludins (answer choice D) and claudins.

10 Signal Transduction

At the conclusion of this chapter, students should be able to:
- Identify ligand gated ion channel receptors and G protein coupled receptors and give examples of each
- Explain how ligand gated ion channel receptors and G protein coupled receptors initiate a signal transduction pathway within the cell and give specific examples
- Associate each of the three major subclasses of Gα proteins that are activated by G protein coupled receptors with their corresponding second messenger
- Describe the function of each of the second messengers
- Identify the two major types of enzyme coupled receptors
- Explain how enzyme coupled receptors initiate a signal transduction cascade within the cell and identify downstream components of those signaling pathways
- Describe the functions of cytokine receptors and their associated signaling cascades

In multicellular organisms, cells have specialized functions that are influenced by external cues that originate from either their environment or from other cells. These external cues can cause a cell to divide, differentiate, exit the cell cycle, or die. Such events are only possible if the external signals are received and "understood" by the cell so that the cells can respond to the need of the organism. This process of a cell receiving an external stimulus and responding to it is called "signal transduction". Signal transduction mediates all activities of cellular life, from development till apoptotic cell death. Perturbations in signal transduction can pave the way towards disease because they result in loss of proper responses of the cells to the stimuli.

10.1 General Principles of Signal Transduction

Most signal cascades begin with receptors found on the cell surface of the cell. These receptors are capable of receiving the specific external signal by binding to it. The receptor then transmits the signal through the plasma membrane to other proteins within the cell that relay the signal, ultimately leading to a response by the cell. This response is often mediated by the activation of transcription factors or expression of target genes (▶ Fig. 10.1).

10.2 Extracellular Signaling

Extracellular (external) signals can be generated by neighbor cell or cells from another tissue, or even come from the outside of the body. The nature of the signal varies: it can be a temperature change (mechanical), proteins or small peptide (Wnt proteins, cytokines), RNA, lipids or lipopolysaccharides (bacterial LPS). Some extracellular signals need to be transported to the site of action and recognized by the target cell. Usually this recognition is mediated by specific receptors on the target cell's surface. Not all cells express the same repertoire of receptors, and therefore not all external signals have the same effect on all tissues in the body.

There are three types of signaling that are mediated by secreted molecules: endocrine, paracrine, and autocrine. In endocrine signaling, the signaling molecule is produced by a cell and enters the bloodstream and travels to a distant location within the body before it binds to the target cell. Paracrine signaling is produced when a signaling molecule is released from one cell and binds to adjacent target cells. In autocrine signaling, the cell secretes signaling molecules that bind to surface receptors on the signal-producing cell, thus stimulating itself. This type of signaling is important for keeping the cell active for longer periods and to coordinate the behavior of multiple cells of the same type.

10.3 Hormones

Hormones are the signaling molecules of endocrine system and include such molecules as the eicosanoids, steroid hormones, and peptide hormones. Steroid hormones are lipophilic, they can pass through the plasma membrane and interact with intracellular receptors. Eicosanoids are a group of hormones derived from a fatty acid, arachidonic acid, which is reviewed in chapter 41. Examples of steroid hormones include glucocorticoids, androgens, and estrogens. Insulin, adrenocorticotropic hormone, and prolactin are all categorized as peptide hormones.

10.4 Growth Factors

Growth factors are polypeptide molecules that can either promote or inhibit cellular growth and division, depending on the cell type and type of receptor encountered. For example, VEGF promotes proliferation and growth of cells needed for vascularization, while myostatin (GDF-8) inhibits myogenesis (muscle growth). Other examples of growth factors include epidermal growth factor (EGF) and fibroblast growth factor (FGF).

10.5 Cytokines

Cytokines are small proteins that act in endocrine, paracrine and autocrine manner usually as a part of immune response. There are several types of cytokines: chemokines, interferons, interleukins, and tumor necrosis factors. These signaling molecules are secreted by immune cells and are important for the activation and proliferation of various types of cells of the immune system.

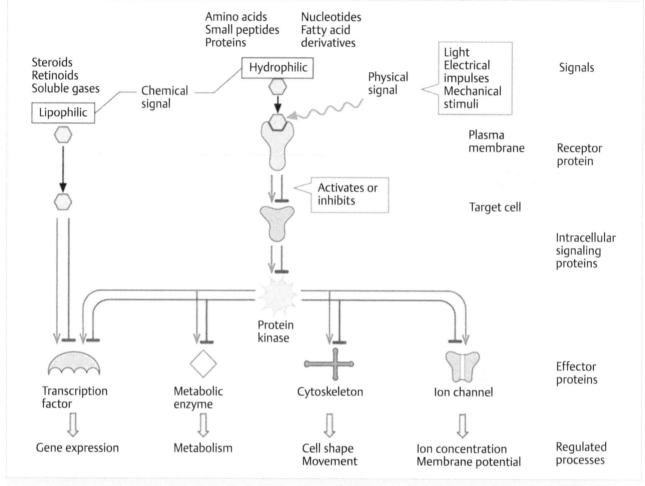

Fig. 10.1 Signal transduction. (Source: Koolman J, Röhm K, ed. Color Atlas of Biochemistry. 3rd Edition. New York, NY. Thieme; 2012.)

10.6 Membrane Diffusible Molecules

Recall that the plasma membrane surrounds the cell and provides a physical barrier that separates the intracellular compartment from its external environment. The lipid bilayer of the plasma membrane is impermeable to solutes such as glucose, amino acids, nucleosides, and ions like sodium, potassium, calcium, chloride, and many others. However, small uncharged polar molecules like water and ethanol and small nonpolar molecules like oxygen, carbon dioxide (CO_2), and nitric oxide (NO) readily diffuse across the lipid bilayers. In addition, steroid hormones are capable of crossing the plasma membrane. Once inside the cell, these molecules can bind to intracellular receptors to elicit a response.

10.7 Receptors

Different cell types can react differently to the same extracellular stimulus. This is because cells recognize signals through signal-specific receptors (▶ Fig. 10.2). Most of the receptors are located at the cell membrane surface, while some like steroid hormone receptors interact with intracellular receptors. The cell surface receptors can be further categorized into four major classes: 1) ligand-gated ion channel receptors, 2) enzyme-linked receptors, 3) cytokine receptors and 4) G-protein-coupled receptors.

10.8 Ligand-gated Ion Channel Receptors

These receptors are found on electrically excitable cells like neurons and muscle cells. Ligand-gated ion channel receptors are able to transduce a chemical signal into an electrical signal and mediate a very fast response to stimuli. In non-stimulated cells the ion channel closed, but upon binding of the signal molecule, the channel opens and the ions enter or exit the cell along their electrochemical gradients (▶ Fig. 10.3).

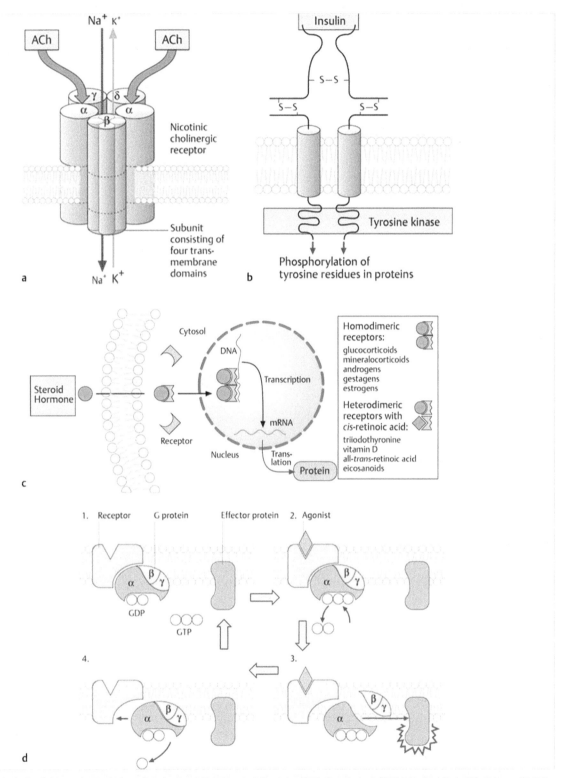

Fig. 10.2 (a–d) Types of cellular receptors. (Source: Koolman J, Röhm K, ed. Color Atlas of Biochemistry. 3rd Edition. New York, NY. Thieme; 2012.)

II

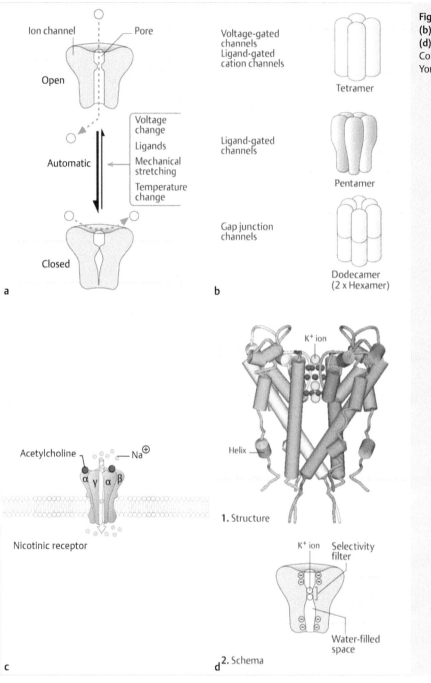

Fig. 10.3 (a) General description, **(b)** composition, **(c)** nicotinic ACh receptor, and **(d)** K$^+$ channel. (Source: Koolman J, Röhm K, ed. Color Atlas of Biochemistry. 3rd Edition. New York, NY. Thieme; 2012.)

10.9 Nicotinic Acetylcholine Receptors

One example of ligand-gated ion channel receptors is the nicotinic acetylcholine receptor (▶ Fig. 10.3b). This receptor changes its conformation upon interaction with its ligand, the neurotransmitter acetylcholine. As the name suggests, it can also respond to nicotine. This change in conformation results in the opening of a pore that allows sodium ions to enter the cell and initiate an action potential.

Termination of a nicotinic acetylcholine signaling event requires the degradation of the neurotransmitter in the synaptic cleft. This is accomplished by an enzyme called acetylcholinesterase. This enzyme eliminates the acetylcholine through enzymatic digestion, shutting down the signal to the post synaptic cell and terminating the response.

10.9.1 Link to Pathology

Myasthenia gravis

Myasthenia gravis is a neuromuscular disorder that is characterized by muscle weakness. It is an autoimmune disease in which the body produces autoantibodies against the nicotinic acetylcholine receptor, thus preventing signaling at neuromuscular junctions.

GABA receptors

Another example of a ligand-gated ion channel is the GABA receptor, which responds to the neurotransmitter Gamma-Amino Butyric Acid (GABA). There are two types of GABA receptors: $GABA_A$ and $GABA_B$. $GABA_A$ is a ligand-gated ion channel, while $GABA_B$ belongs to a different receptor family, the G-protein coupled receptors, which will be discussed later. Binding of GABA to $GABA_A$ receptor opens up a Cl⁻ channel that allows the influx of Cl- ions. Cl⁻ ion influx increases the membrane potential, leading to membrane hyperpolarization and preventing the neuron from firing. Thus, GABA has a sedative and overall inhibitory effect. This sedative effect can also be achieved by using GABA agonists and GABA positive allosteric modulators that interact with the GABA receptors. Examples include barbiturates, benzodiazepines, and alcohol (ethanol).

10.10 Enzyme-coupled Receptors

Enzyme-coupled receptors are receptors that possess intrinsic enzyme activity. Upon binding of ligand to these receptors, a kinase activity is induced, leading to the phosphorylation of downstream targets and propagation of the signal. The largest class of enzyme-coupled receptors is Receptor Tyrosine Kinases (RTKs).

10.11 Receptor Tyrosine Kinases

The receptor tyrosine kinase (RTK) class of receptors interacts with many extracellular growth factors that mediate cell proliferation and growth. A few of these include the receptors for epidermal growth factor (EGF), nerve growth factor (NGF), platelet-derived growth factor (PDGF), and insulin. RTKs share common structural features: they extend their N-terminus domain outside of the cell, and this domain interacts with extracellular signal. In the middle of the molecule, RTKs have only one transmembrane α helix domain, while the tyrosine kinase activity is located on their C-terminal cytosolic domain.

In the absence of ligand, most RTKs exist as inactive monomeric subunits in the plasma membrane. Binding of ligand to the RTKs initiates receptor dimerization and activation of the tyrosine kinase in its intracellular tail. This activation results in the cross phosphorylation of the tails, creating phosphorylated tyrosine docking sites for adaptor proteins downstream in the signaling cascade.

The adaptor proteins possess a specific domain called the SH2 domain, which allows them to bind to the phosphorylated tyrosines. Following the association with the receptor, the adaptor proteins recruit other molecules, such as other enzymes, that further relay the intracellular signal.

10.12 Insulin Signaling

Insulin signaling is one of the key pathways that mediates energy metabolism. Defects in this pathway underlie metabolic syndrome, a complex disorder that includes insulin resistance, a pre-requisite for type II diabetes, reviewed in nutrition and metabolism chapter 45.

Insulin's central function is initiated by binding to the insulin receptor, which is a member of the RTK family (▶ Fig. 10.4). Upon binding, the intracellular domains of the insulin receptor dimer phosphorylate each other and create docking sites for an adaptor protein called insulin receptor substrate (IRS) protein. The IRS then becomes phosphorylated by the intracellular domains of the receptor. Phosphorylated IRS, in turn, recruits and activates PI3 kinase. Active PI3 kinase phosphorylates its substrate, PIP_2 and the result of this phosphorylation is the conversion of PIP_2 into PIP_3. Next, PIP_3 recruits Phosphoinositide-Dependent kinase 1 (PKD1), which binds to PIP_3. PIP_3 also recruits protein kinase B (PKB1, also called Akt1), a kinase similar to PKD1. Close proximity of active PKD1 and Akt/PKB results in phosphorylation of Akt/PKB by PDK1. Activated Akt/PKB dissociates from the membrane-bound PIP_3 into the cytosol, where it enhances glucose uptake by increasing GLUT4 concentration on the cell membrane and glycogen synthesis by activating glycogen synthase.

10.13 Ras-MAP Kinase Signaling

Ras is a small GTP-binding protein that is activated by multiple RTKs in response to growth factor binding. For example, binding of epidermal growth factor (EGF) to its specific RTK, epidermal growth factor receptor (EGFR), leads to autophosphorylation and activation of cytosolic domains of RTKs. Activation of RTKs creates binding sites on RTKs for the SH2 domain of the adaptor protein GRB2 that, in turn, recruits the Ras-activating protein SOS. In its GDP-bound state, Ras is in its inactive form. Ras-activating protein SOS is a guanine nucleotide exchanging factor (GEF) that removes the GDP from Ras and replaces it with GTP. The GTP-bound form of Ras is now in an active form and can activate the downstream component of this pathway Raf, a MAP kinase kinase kinase. The activated Raf phosphorylates Mek, a MAP kinase kinase. Mek then phosphorylates the MAP kinase Erk, which can phosphorylate and activate transcription factors and other enzymes to induce cell cycle progression and cell division (▶ Fig. 10.4).

10.14 Receptor Serine/Threonine Kinases

Similar to RTKs, receptor Serine/Threonine kinases also possess enzymatic activity. However, unlike RTKs, these receptors phosphorylate serine and threonine amino acid residues on the substrate protein molecules, rather than tyrosines. An example of a serine/threonine kinase receptor is the TGF-β receptor that binds to transforming growth factor beta (TGF-β), a multifunctional protein that depending on the cell type it acts on, can lead to programmed cell death, or alterations in the cell cycle. Upon binding of TGF-β to its receptor, the intracellular tails of the receptor are phosphorylated. This recruits a protein called SMAD. The receptor then phosphorylates SMAD, which causes it to dissociate from the receptor, dimerize with other SMAD molecules, and migrate to the nucleus, where this dimer influences gene expression.

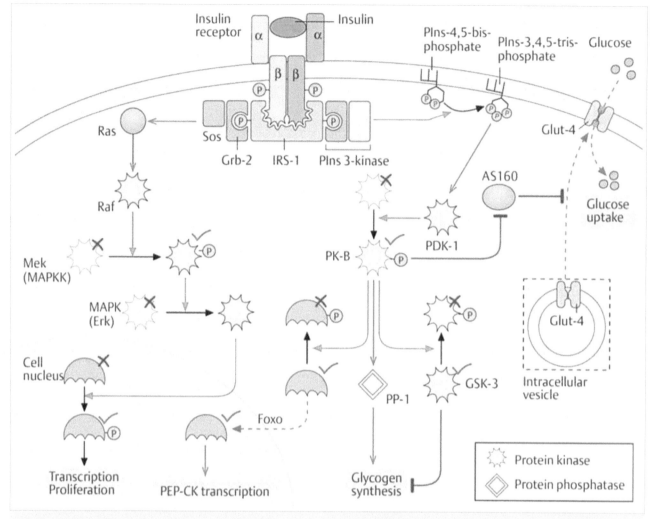

Fig. 10.4 Insulin signaling pathways. (Source: Koolman J, Röhm K, ed. Color Atlas of Biochemistry. 3rd Edition. New York, NY. Thieme; 2012.)

10.15 Cytokine Receptors

Unlike the previously described enzyme-linked receptors, cytokine receptors do not have intrinsic enzyme activity. Instead, their intracellular tails interact with other molecules that provide this function. In this case, a tyrosine kinase called JAK (Janus kinase) is the molecule that provides the enzyme activity (▶ Fig. 10.5). Binding of ligand leads to activation of the receptor-associated JAK. Once activated, JAK phosphorylates the intracellular tail of the receptor at specific tyrosine sites. These phosphorylated sites serve as docking sites for SH2-domain containing protein STAT. Phosphorylation of STATs leads to their dimerization and translocation to the nucleus. In the nucleus, STATs regulate transcription of specific target genes. Due to the molecules involved, the cytokine receptor pathway is commonly referred to as the "JAK-STAT" pathway.

10.16 G-protein–coupled Receptors (GPCRs)

A wide variety of extracellular signals are transmitted through G-protein-coupled receptors. These receptors are the largest family of cell surface receptors and mediate the signaling from a number of signaling molecules. Some examples of GPCRs are the receptors for adrenaline, glucagon, somatostatin, dopamine, histamine, serotonin, prostaglandin, olfactory receptors, and some taste receptors. Nearly half of known drugs work by modulating the activation of GPCRs.

GPCRs are proteins that are characterized by seven alpha-helical transmembrane domains. When ligand binds to its extracellular domain, this causes the intracellular domain to interact with G-proteins (▶ Fig. 10.6). G proteins are associated with the cytosolic side of the plasma membrane and consist of three subunits: α, β, and γ. Due to this arrangement, this complex is often called the heterotrimeric G protein complex. In the absence of ligand, the α subunit of the G protein complex is bound to GDP and exists as part of heterotrimeric complex, together with β and γ subunits. Upon stimulation with extracellular ligands of GPCRs, the G protein complex associates with the GPCR. This stimulates the release of GDP from the α subunit, which is exchanged for GTP. When bound to GTP, the α subunit dissociates from heterotrimeric complex. Although both the GTP-bound α subunit and the β/γ subunit can affect downstream targets, the α subunit is involved in more signaling events than the β/γ subunit. In most cases, the α subunit activates an enzyme that

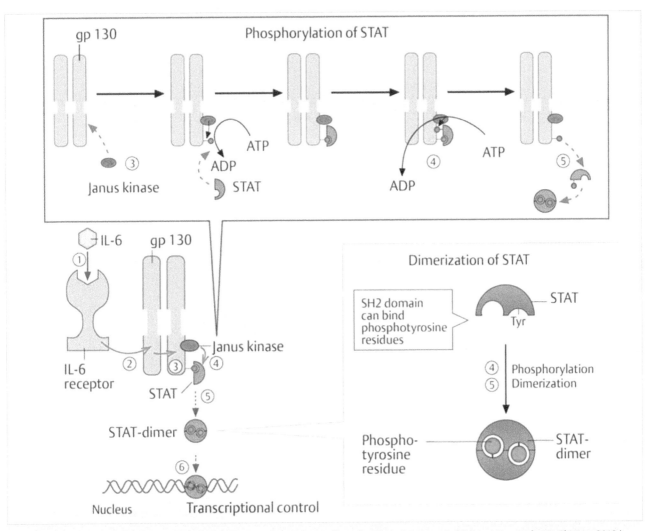

Fig. 10.5 Cytokine signaling pathways. (Source: Koolman J, Röhm K, ed. Color Atlas of Biochemistry. 3rd Edition. New York, NY. Thieme; 2012.)

produces a second messenger molecule that is important for the propagation of the intracellular signaling cascade.

The de-activation of the GTP-bound α subunit happens when the hydrolysis of GTP into GDP occurs, and the inactive, now GDP-bound α subunit, re-associates with β/γ subunits at the plasma membrane.

Gα subunit proteins can be divided into several subclasses including $G_{\alpha S}$, $G_{\alpha i}$, and $G_{\alpha q}$. $G_{\alpha S}$ and $G_{\alpha i}$ have the opposite effects on the plasma membrane-bound enzyme adenylyl cyclase. The $G_{\alpha S}$ stimulates the activity of adenylyl cyclase, whereas the $G_{\alpha i}$ inhibits this activity (▶ Table 10.1). Adenylyl cyclase is an enzyme responsible for converting adenosine triphosphate (ATP) to cyclic adenosine monophosphate (cAMP). Thus, activation of $G_{\alpha S}$ leads to an increase in the production of cAMP, and activated $G_{\alpha i}$ results in decreased amounts of cAMP (▶ Fig. 10.7). cAMP is as second messenger that is critical for many physiological events within the cell, like lipid and sugar metabolism. $G_{\alpha q}$ stimulates a different enzyme, phospholipase C, leading to the production of the second messengers inositol triphosphate (IP_3) and diacylglycerol (DAG) through the phospholipase C-mediated cleavage of the phospholipid PIP2 (▶ Fig. 10.8).

10.17 Second Messengers

Often cell signaling requires simultaneous activation of several signaling cascades within the cell in response to the binding of an extracellular signaling molecule to its cell surface receptor. This can be achieved by the release of molecules within the cell, known as second messengers, that can simultaneously activate multiple targets.

Second messengers can be grouped into two general classes: the hydrophobic, or plasma membrane associated, second messengers, and the hydrophilic, or cytoplasmic, second messengers. Diacylglycerol (DAG) is an example of a hydrophobic second messenger, whereas cAMP, IP3, and Ca2 + are all examples of hydrophilic second messengers.

10.18 cAMP

Cyclic adenosine monophosphate (cAMP) was the first second messenger to be discovered. It acts downstream of G protein coupled receptors in response to hormones like adrenaline or epinephrine, as well as many others that are summarized in

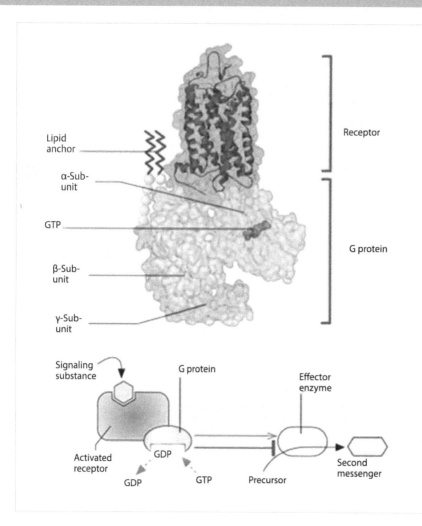

Fig. 10.6 G-protein-coupled receptor. (Source: Koolman J, Röhm K, ed. Color Atlas of Biochemistry. 3rd Edition. New York, NY. Thieme; 2012.)

Table 10.1 Effects of trimeric G-proteins

G class	Direct effects	Intracellular effect
G_s	Activates adenylate cyclases and Ca^{2+} channels	cAMP↑, Ca^{2+}↑
G_{olf}	Activates adenylate cyclases in the olfactory system	cAMP↑
G_i	Inhibits some adenylate cyclases and calcium channels, activates cGMP-specific phosphodiesterase, activates inward K^+ channels	cAMP↑·, Ca^{2+} ↑· cGMP↑·, K^+↑
G_o	Activates inward K^+ channels, inactivates Ca^{2+} channels, activates phosphorylase C	K^+↑·, Ca^{2+}↑ ·InsP³ and DAG ↑·
G_t	Activates cGMP-specific phosphodiesterase	cGMP ↑
G_q	Activates phospholipase C	InsP₃ and DAG↑
$G_{12/13}$	Activates phospholipase A_2 and other effectors	Complex effect

▶ Table 10.2. As discussed earlier, activated GPCRs lead to the association of the G_α subunit of G protein heterotrimeric complex with GTP, creating an active G_α subunit. In the case of G_αs, this GTP-bound form stimulates the membrane-bound enzyme adenylyl cyclase. Adenylyl cyclase then converts ATP into the second messenger cAMP. Once formed, cAMP can activate downstream enzymes such as cAMP-dependent protein kinase A (PKA). In the example of adrenaline signaling, the activated PKA in turn phosphorylates and thus activates phosphorylase kinase, an important step in glycogen breakdown. PKA can also activate a number of transcription factors like cAMP-response element-binding protein (CREB), thereby influencing gene expression.

The action of cAMP is terminated by the conversion of cAMP into 5`-AMP by the plasma membrane-bound enzyme cAMP phosphodiesterase (▶ Fig. 10.7). Inhibition of phosphodiesterase by caffeine results in keeping the intracellular cAMP levels high for a longer period of time. This creates a prolonged

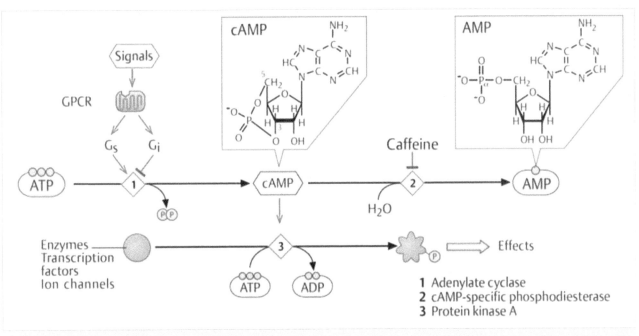

Fig. 10.7 GPCR signaling leading to production of cAMP. (Source: Koolman J, Röhm K, ed. Color Atlas of Biochemistry. 3rd Edition. New York, NY. Thieme; 2012.)

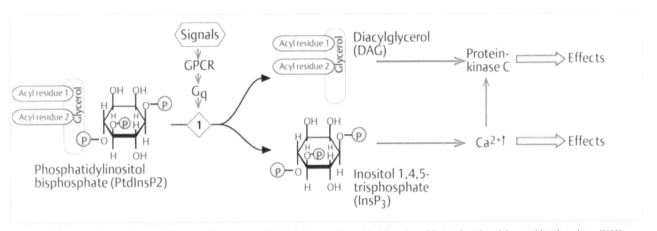

Fig. 10.8 Second messengers inositol 1,4,5-triphosphate (IP3) and diacylglycerol (DAG) produced from phosphatidylinositol bisphosphate (PIP2) cleavage in response to GPCR signaling. (Source: Koolman J, Röhm K, ed. Color Atlas of Biochemistry. 3rd Edition. New York, NY. Thieme; 2012.)

Table 10.2 Examples of c-AMP-mediated hormone effects

Hormone	Target tissue	Response
Thyroid stimulating hormone, TSH	Thyroid	Thyroxine synthesis and secretion
	Adipose tissue	Hydrolysis of triglycerides
Adrenocorticotropic hormone, ACTH	Adrenal cortex	Cortisol synthesis and secretion
Luteinizing hormone, LH	Ovary	Progesterone synthesis and secretion
Epinephrine	Muscle	Glycogenolysis
	Heart	Heart rate ↑, contractility ↑
	Adipose tissue	Hydrolysis of triglycerides
Parathyroid hormone, PTH	Bone	Bone resorption
Glucagon	Liver	Glycogenolysis
	Adipose tissue	Hydrolysis of triglycerides
Vasopressin	Kidney	Water absorption

II

response to hormone glucagon in the liver. The prolonged response to glucagon hormone in the liver may result in increased glucose output from liver glycogenolysis and gluconeogenesis pathway. This is why type-2 diabetic patients are advised to limit their caffeine intake.

10.19 IP3 and DAG

As mentioned previously, G αq stimulates the activity of phospholipase C, which cleaves the plasma membrane component PIP2 into the second messengers inositol 1,4,5- triphosphate (IP3) and diacylglycerol (DAG). DAG can then serve as a docking site for protein kinase C (PKC) at the plasma membrane (▶ Fig. 10.8).

Meanwhile, IP3 migrates to the endoplasmic reticulum (ER) and binds to a ligand-gated calcium channel receptor in the ER membrane to release another second messenger, Ca^{2+} ions, into the cytosol. Both, DAG and Ca^{2+} ions activate PKC, a multifunctional kinase that has several isoforms and that regulates various signaling cascades in the cell that affect cell growth and differentiation.

10.20 Ca^{2+}

Due to its importance in the activation of a number of cytosolic enzymes, the concentration of Ca^{2+} ions within the cytoplasm must be tightly controlled. This is achieved through Ca^{2+} ATPase pumps and transporters that either export these ions outside of the cell or shuttle them into the endoplasmic reticulum (ER), also known as the sarcoplasmic reticulum in

muscle cells, for storage. GPCRs, when activated, lead to production of two second messengers, IP3 and DAG, discussed in more details earlier.

Upon stimulation of the Gαq-type of GPCRs, Ca^{2+} is released from its ER storage site in response to the production of the second messenger IP3. The effects of Ca^{2+} signaling in the cytoplasm are mediated chiefly by the binding of Ca^{2+} to calcium-responsive proteins such as calmodulin (CaM) (▶ Fig. 10.9). Conformational changes in the CaM molecule that result from its interaction with Ca^{2+} ions allow CaM interactions with other signaling proteins such as Ca^{2+}/Calmodulin-dependent protein kinases (CaMKII).

CaM kinases belong to a family of serine/threonine protein kinases that exhibit a large variety of activities in the cell. Depending on the cell type, the activated CaM kinases can activate transcription factors that affect the expression of various genes, participate in the synthesis of neurotransmitters, and regulate the activity of some channels and metabolic enzymes.

Inactivation of G-protein-couples receptor occurs when G_α subunit of G protein heterotrimeric complex inactivates itself by hydrolyzing GTP and thus, returning to GDP-bound form that allows its interaction with other G proteins and restoration of heteroteimeric G protein complex. The process of self-inactivation of G_α serves as a timer that allows for the intracellular signal to only last for a specified amount of time.

10.21 Other GPCRs

Not all GPCRs require association with a heterotrimeric G protein complex for their activity. Some signaling pathways, such

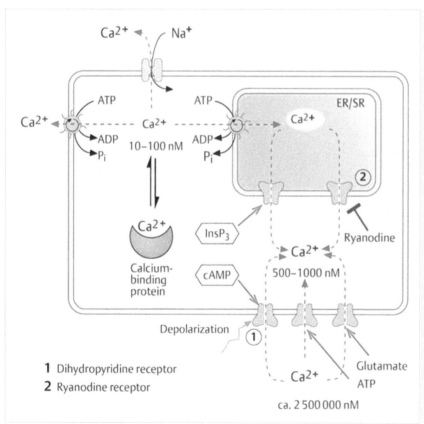

Fig. 10.9 Calcium transport within the cell. (Source: Koolman J, Röhm K, ed. Color Atlas of Biochemistry. 3rd Edition. New York, NY. Thieme; 2012.)

1 Dihydropyridine receptor
2 Ryanodine receptor

as the Wnt pathway and the Shh pathway, work through other molecules to induce a response in the cell.

10.21.1 Wnt

Wnt signaling is implicated in many processes including embryonic development, asymmetric cell division and differentiation, and cancer. The signaling molecule Wnt mediates a paracrine type of signaling and belongs to a family of secreted growth factors that appears to be well conserved among species. Upon binding of Wnt to its GPCR-like receptor, Frizzled, phosphorylation of the cytoplasmic molecule Dishevelled occurs.

Active Dishevelled inhibits a complex that contains glycogen synthase kinase-3 (GSK-3) and the adenomatous polyposis coli (APC) protein. In the absence of Wnt, this complex phosphorylates and degrades a protein called β-catenin. This phosphorylated β-catenin is then targeted for degradation. When Wnt signal is initiated, active Dishevelled inhibits the complex, so that β-catenin is no longer phosphorylated. This allows for the accumulation of stable β-catenin. Free cytoplasmic β-catenin then translocates into the nucleus where it binds to other regulatory proteins and directs gene expression. Target genes of this complex include those that encode for cell cycle promoting proteins, such as myc and cyclin D, as well as proteins involved in stem cell fate.

10.21.2 Shh

Sonic Hedgehog signaling (Shh) is another important pathway that regulates development, cell fate and cell polarity. Hedgehog proteins are secreted molecules that interact with cells through the Patched transmembrane receptor. Patched negatively regulates another molecule in the plasma membrane, Smoothened. Upon binding of Shh to its receptor Patched, the inhibitory signal on Smoothened is relived. This leads to the activation of the transcription factor Gli. Gli then translocates to the nucleus to regulate gene expression. Gli target genes include those that contribute to stem cell identity, cell proliferation, growth factors, and genes that code for proteins that contribute to cell migration and apoptosis. Target genes of the Shh pathway also include Wnt signaling genes, and therefore, link the two pathways, as they both contribute to organism development. Deregulation of both the Shh and Wnt signaling pathways have been shown to contribute to cancer onset and progression.

10.22 Receptor Desensitization

Cells that are continuously exposed to a stimulus over a long period of time often cease to respond to this signal. The mechanism underlying this loss of receptor activation is known as receptor desensitization. Receptor desensitization, sometimes referred to as adaptation, can occur in three different manners: 1) inactivation of receptor; 2) internalization of receptor; and 3) down-regulation of receptor. Inactivation of the receptor can result from a structural modification of the receptor. For example, the receptor might be blocked by binding with inhibitory proteins, uncoupling its interactions with the rest of signaling cascade. Alternatively, receptors/ligand complexes can be incorporated into recycling endosomes, a trafficking vesicle that can either fuse with lysosomes and degrade their content, or recycle back to the cell surface and bring the receptor back. Such recycling decreases the number of the receptors at the cell surface and prevents extracellular ligands from activating the pathway. If the endosomes fuse with lysosomes and degrade their content, the receptor is destroyed, so the cell will no longer be able to respond to the signal unless new receptors are made.

10.23 Other Types of Cell-Cell Communication

In addition to the aforementioned endocrine, paracrine, and autocrine modes of cell signaling, there are also types of cell-cell communication that involve direct physical contact between cells (juxtacrine) and signaling through a compartmentalized intercellular space (synaptic). Juxtacrine type of communication is characterized by membrane-bound signaling molecules on one cell binding to the membrane-bound receptor on another cell. During synaptic communication, the signaling molecules are released into the confined space between two cells, known as the synaptic cleft, and the signal is directly received by one specific target cell. An example of synaptic signaling is signal transmission between neurons (▶ Fig. 10.10).

10.24 Delta-notch Signaling

An important example of juxtacrine signaling is seen in the Delta-Notch pathway, which is crucial for many events during development. Notch is the cell surface receptor that is activated by the interaction with it`s ligand, Delta, a protein that is expressed on the receiving cell. After Delta binds to Notch, the intracellular domain of Notch is cleaved and translocated into the nucleus, where it mediates the expression of target genes. During neurogenesis, the cell that expresses Delta on its surface will become a neuron. This cell interacts with neighboring Notch-expressing cells, and the Notch signaling inhibits each of these cells from differentiating into neuron. The process of one cell preventing its neighbors from taking on a specific cellular fate is called lateral inhibition.

10.25 Death Signals and Apoptosis

The death signal, mediated by the Fas receptor, is another example of juxtacrine signaling. Death signals that the cell receives from the outside trigger signaling events that in turn, mediate programmed cell death, or apoptosis. In the extrinsic pathway of apoptosis, these signals are sensed through cell surface receptors called the Fas receptor, also known as the death receptor. The extrinsic pathway is initiated by the death ligand (Fas ligand) expressed on the surface of one cell making physical contact to the death receptor (Fas receptor) on another cell. This interaction leads to the activation of caspases in the receptor-presenting cell, ultimately leading to apoptotic cell death.

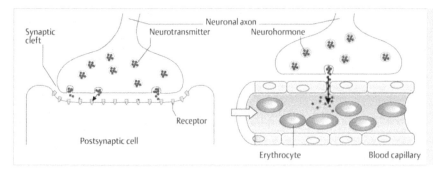

Fig. 10.10 Neurotransmitter and neurohormone signaling. (Source: Koolman J, Röhm K, ed. Color Atlas of Biochemistry. 3rd Edition. New York, NY. Thieme; 2012.)

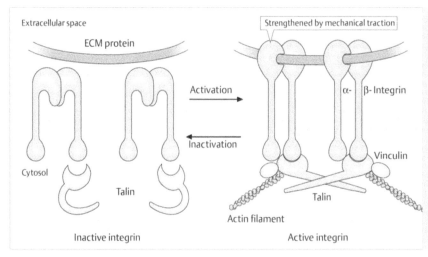

Fig. 10.11 Connections between integrins and the extracellular matrix. (Source: Koolman J, Röhm K, ed. Color Atlas of Biochemistry. 3rd Edition. New York, NY. Thieme; 2012.)

10.26 Synaptic Signaling

Neurons communicate with their target cells, including other neurons, via synaptic type of signal transmission. The neuron transmits this signal through a type of extracellular signaling molecule known as a neurotransmitter. These neurotransmitters are released into the synapse, a confined space between neuron and target cell, where the neurotransmitters find their corresponding receptors on the surface of the post synaptic cell (▶ Fig. 10.10). Most neurotransmitters are amino acids and their derivatives with the exception of acetylcholine and neuropeptides, reviewed in amino acid derivatives chapter 35.

Neurotransmitters are stored in synaptic vesicles, small membranous organelles about 50 nm size in diameter. Depolarization of plasma membrane triggers release of cytosolic Ca^{2+} which, in turn, results in fusion of synaptic vesicles with the plasma membrane and their release into the synapse through exocytosis.

10.27 Integrin Signaling

Besides signaling from cell to cell, communication can also occur between a cell and the extracellular matrix. This type of signaling is often mediated by molecules of integrin, transmembrane proteins that form dimers and that transmit the signal between extracellular matrix components, such as collagen, fibronectin, and laminin, and the actin cytoskeleton inside the cell (▶ Fig. 10.11).

Integrins are transmembrane proteins that interact with intracellular actin filaments via adaptor proteins. Integrin domains that are located outside of the cell act as pinchers to grab hold of extracellular matrix components. Tension between the cell and the extracellular matrix or presence of signaling molecules in the extracellular matrix then transmits the signal inside the cell.

Not only are signals received by integrins outside the cell, but signals can also originate at the intracellular domain of the integrin molecules. When the signal comes from inside the cell and elicits a response on the extracellular domains of the integrin molecules, it is called inside-out activation. If the signal originates on the outside, it is known as outside-in activation. An outside-in activation can affect changes in the actin cytoskeleton inside the cell, whereas inside-out activation can cause the integrin molecules to bind more tightly to an extracellular matrix component.

10.28 Nitric Oxide

Nitric oxide signaling is extremely important for the regulation of cardiovascular system. Activated neurons release

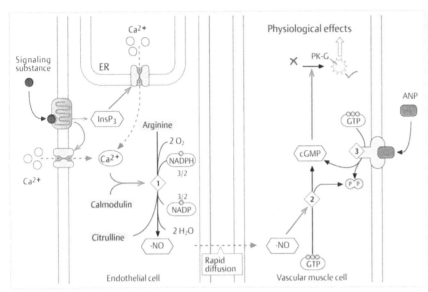

Fig. 10.12 Nitric oxide signaling causes dilation of vascular tissue. (Source: Koolman J, Röhm K, ed. Color Atlas of Biochemistry. 3rd Edition. New York, NY. Thieme; 2012.)

acetylcholine that acts on endothelial cells, leading to release of Ca^{2+} from the endoplasmic reticulum into the cytoplasm. In turn, cytosolic Ca^{2+} activates nitric oxide synthase (NOS), an enzyme that converts arginine into nitric oxide (NO). NO easily diffuses through plasma membrane of the endothelial cell and enters the neighboring smooth muscle cell. Inside the smooth muscle cells, NO binds to and activates guanylyl cyclase, which converts GTP into the second messenger cGMP and ultimately leading to relaxation of the smooth muscle and vasodilation (▶ Fig. 10.12).

Review Questions

1. An inhibitor of RAF would prevent the activation of which of the following molecules?
   A) EGFR
   B) GRB2
   C) JAK
   D) MEK
   E) PKB

2. The insulin receptor is a member of which family of cell surface receptors?
   A) Cytokine
   B) G-protein coupled
   C) Ligand-gated ion channel
   D) Receptor tyrosine kinase

3. The disease myasthenia gravis affects which of the following receptors?
   A) Acetylcholine
   B) EGF
   C) GABA
   D) Insulin

4. Which of the following second messengers activates protein kinase A (PKA)?
   A) cAMP
   B) Ca2 +
   C) DAG
   D) IP3

5. Which of the following molecules is a component of a juxtacrine signaling pathway?
   A) Notch
   B) Ras
   C) SMAD
   D) Wnt

Answers

1. **The correct answer is D**. RAF is a member of the RAS-MAPK pathway. In this pathway, RAS is activated by the binding of a signaling molecule to its cell surface receptor. The activated RAS then activates RAF. RAF phosphorylates MEK to activate it, and MEK phosphorylates ERK to cause its activation. Therefore, inhibition of RAF would prevent the activation of MEK. EGFR (answer choice A) is the receptor for epidermal growth factor (EGF) and is the first component of a RAS-MAPK type of pathway. Immediately downstream of the EGFR is GRB2 (answer choice B), an adaptor protein that recruits the enzyme responsible for the activation of RAS. Since both EGFR and GRB2 are upstream of RAF, inhibition of RAF would not have an effect on the activity of these molecules. JAK (answer choice C) and PKB (answer choice E) are not components of the same pathway as RAF and, therefore, RAF inhibition would not affect these proteins. JAK is a member of the cytokine signaling pathway, and PKB is a component of one type of tyrosine kinase pathway.

2. **The correct answer is D**. Upon binding of insulin to the insulin receptor, the intracellular tails of the receptor become activated and phosphorylate each other on tyrosines. Therefore, the insulin receptor is a receptor tyrosine kinase. The phosphorylated tyrosines recruit an adaptor protein that binds to the enzyme PI-3 kinase. This enzyme phosphorylates the phospholipid PIP2 to create PIP3, which serves as a docking site at the plasma membrane for PDK1 and PKB. PDK1 phosphorylates and activates PKB, resulting in increased uptake of glucose and glycogen formation. Although the activation of cytokine receptors (answer choice A) also results in tyrosine phosphorylations, the receptor itself does not have kinase activity. The insulin receptor is not a member of the G-protein coupled receptors (answer choice B) or ligand-gated ion channels (answer choice C).

3. **The correct answer is A**. Myasthenia gravis is an autoimmune disease in which the body produces antibodies that bind to the acetylcholine receptors found at the neuromuscular junctions. This binding prevents the neurotransmitter acetylcholine from interacting with its receptor, thus causing a loss of muscle response. The EGF (answer choice B), GABA (answer choice C), and insulin (answer choice D) receptors are not affected in myasthenia gravis.

4. **The correct answer is A**. The second messenger cAMP is produced in response to the activation of G-protein coupled receptors. Upon binding of ligand to the G-protein coupled receptor, activation of the Gαs component of the heterotrimeric G-protein complex occurs. Activated Gαs then activates adenylyl cyclase, which converts ATP into cAMP. One of the enzymes that is activated by cAMP is protein kinase A. Ca2 + (answer choice B) can activate many different types of enzymes in the cell, including protein kinase C (PKC) and, along with calmodulin, the Ca2 + / Calmodulin dependent protein kinases (CaM kinases). DAG (answer choice C) is a second messenger derived from the cleavage of the phospholipid PIP2 that collaborates with Ca2 + to activate PKC. The second messenger IP3 is also produced from the cleavage of PIP2. This molecule binds to ligand-gated calcium channels in the membrane of the endoplamic reticulum and causes the release of Ca2 + into the cytoplasm.

5. **The correct answer is A**. Juxtacrine signaling occurs when a signaling molecule found on the surface of one cell interacts with a receptor molecule on the surface of another cell. This interaction requires that the two cells physically contact one another. The molecule notch, along with its interacting partner delta, is part of a juxtacrine type of signaling cascade. Notch is embedded in the membrane of one cell and serves as the receptor of the signaling molecule delta, which is embedded in a separate cell. When the two cells come in physical contact with each other, the signal is transmitted in the notch-bearing cell. Ras (answer choice B) is a component of the Ras-MAPK pathway inside the cell. SMAD (answer choice C) is part of the serine/threonine receptor kinase signaling pathway and is activated in response to TGFβ signaling. Wnt (answer choice D) is a soluble extracellular signaling molecule that induces β catenin-mediated signaling within the cell.

11 Cell Cycle and Control of the Cell Cycle

At the conclusion of this chapter, students should be able to:

- Identify the four phases of the cell cycle and describe their basic characteristics
- Briefly explain each stage of mitosis and give some general characteristics of each
- Describe how cyclins and cyclin dependent kinases control the progression of the cell cycle, giving examples of specific molecules that these complexes target
- Identify the cyclin-Cdk complexes and the cell cycle phase in which each is active
- Describe the three main cell cycle checkpoints and how they function
- Explain how the understanding of cell cycle controls can be used in the treatment of cancer, giving specific examples of different classes of chemotherapeutics and describing their mode of action

Actively cycling cells transition through a sequence of growth states known as the cell cycle. The cell cycle consists of four distinct stages: G_1, S, G_2, and M phase (or mitosis) (▸ Fig. 11.1). The first three stages (G_1, S, and G2) are collectively known as interphase. During both G_1 and G_2, cellular growth occurs, whereas in the interlaying S phase the cell's DNA and centrosomes are duplicated. The final stage of cell division, mitosis, involves the separation of the duplicated DNA and the physical splitting of the cell into two new daughter cells.

11.1 G1 Phase

The amount of time that a cell spends in the G_1 phase varies by cell type. For example, some somatic cells have a G_1 phase that lasts more than 10 hrs, while G_1 in most embryonic cells is nearly non-existent. In addition to cellular growth, the G_1 phase is also the stage in which the cell becomes committed to either continue division or exit from the cell cycle, at which point the cell is considered to be in G_0 or quiescence. If they choose to exit the cell cycle, then they are able to re-enter the cycle at some point in the future in response to external cues such as growth factors or available nutrients. Alternatively, some

external clues may direct them to become one of the terminally differentiated, non-cycling cells of the body, such as a neuron.

Once the cell has committed to cycle, it must transition from G_1 to S phase. The G_1 to S transition is indicated by the presence of an extracellular mitogenic stimulus, which is mediated by a signaling molecule such as a growth factor. Once the mitogenic signal binds to a receptor on the cell's surface, a signal transduction cascade is initiated within the cell that ultimately results in the expression of genes that are needed to proceed to S-phase.

11.2 S phase

The S-phase of the cell cycle requires approximately eight hours to complete in a typical animal cell. It is during this stage that the genomic material is replicated in a semi-conservative fashion, meaning that each strand of the double helix is used as a template to synthesize a new strand. Once this process is complete, each DNA molecule is a hybrid consisting of one old strand and one newly synthesized strand.

Before S phase, each chromosome is composed of one molecule of double-stranded DNA called the chromatid. Following replication of the genome during S phase, each chromosome now consists of two identical double stranded DNA molecules. These structures are called sister chromatids (▸ Fig. 11.2). The sister chromatids are joined along their lengths by proteins complexes called cohesin rings, consisting of the two core subunits SMC1 and SMC3.

Along with the replication of DNA, S-phase is also the time in which the cell duplicates its centrosome. Recall from chapter 5 that the centrosome is the microtubule organizing center of the cell and that it forms the poles of the bipolar spindle during mitosis. Since a pre S-phase cell only has one centrosome, this centrosome has to be duplicated in order for the mitotic spindle to form.

11.3 G2 Phase

Once the cell has successfully replicated its DNA and duplicated its centrosome, it exits S-phase and enters the G_2 phase. Like G_1, the timing of G_2 varies depending on cell type. The G_2 phase is marked by a period of rapid cell growth and protein synthesis, as the cell doubles its content in preparation for dividing in

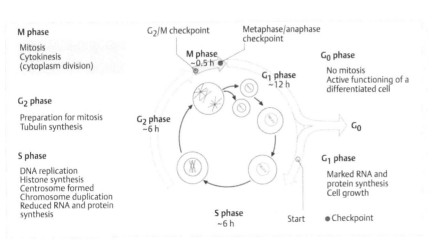

M phase

Mitosis
Cytokinesis
(cytoplasm division)

G_2 phase

Preparation for mitosis
Tubulin synthesis

S phase

DNA replication
Histone synthesis
Centrosome formed
Chromosome duplication
Reduced RNA and protein
synthesis

G_2/M checkpoint

Metaphase/anaphase
checkpoint

M phase
~0.5 h

G_1 phase
~12 h

G_2 phase
~6 h

S phase
~6 h

Start

G_0 phase

No mitosis
Active functioning of a
differentiated cell

G_0

G_1 phase

Marked RNA and
protein synthesis
Cell growth

● Checkpoint

Fig. 11.1 Eukaryotic cell cycle. (Source: Koolman J, Röhm K, ed. Color Atlas of Biochemistry. 3rd Edition. New York, NY. Thieme; 2012.)

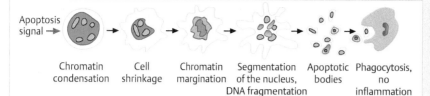

Fig. 11.2 Sister chromatids. (Source: Passarge E, ed. Color Atlas of Genetics. 4th Edition. New York, NY. Thieme; 2012.)

Apoptosis signal → Chromatin condensation → Cell shrinkage → Chromatin margination → Segmentation of the nucleus, DNA fragmentation → Apoptotic bodies → Phagocytosis, no inflammation

two. In addition, an important G$_2$/M DNA damage checkpoint is active in this stage which prevents the cell from continuing in the cell cycle if DNA damage is detected. If no DNA damage is detected, then the cell is ready to divide and enters mitosis, also known as M-phase.

11.4 Mitosis

Mitosis is the shortest phase of the cell cycle, lasting approximately 90 minutes in most cells. Due to the complexity associated with evenly dividing the contents of the cell, a well-regulated progression through mitosis is essential for ensuring the generation of healthy daughter cells. Because of this, mitosis is separated into five distinct, tightly controlled phases: prophase, prometaphase, metaphase, anaphase, and telophase (▶ Fig. 11.3).

11.5 Prophase

During prophase, the chromatin begins to condense due to the activity of protein complexes known as condensins (▶ Fig. 11.4). Condensins function as packaging units, condensing the chromosomes into small physical packets that can be more easily segregated in the later stages of mitosis. Like cohesins, the condensins form rings around the DNA molecule. These rings then assemble into higher-order structures that are responsible for the bundling of the chromatin. The condensing rings consist of two core subunits: SMC2 and SMC4.

In addition to chromatin condensation, prophase also marks the beginning stages of separation of the duplicated centrosomes, which must ultimately move to opposite ends of the nucleus. As they migrate, the centrosomes start to form a framework of microtubules, called the mitotic spindle, that will be used to separate the two sister chromatids.

11.6 Prometaphase

In prometaphase, more condensation of the chromatin occurs and the nuclear envelope breaks down, giving the spindle microtubules access to the sister chromatids. The dissolution of the nuclear envelope is caused by the phosphorylation of nuclear pore proteins and lamins, and requires mitotic cyclin-Cdk activity. At the centromere of each chromatid, a protein structure called the kinetochore is formed. All of the kinetochores will ultimately be bound by spindle microtubules (▶ Fig. 11.5).

11.7 Metaphase

During metaphase, each kinetochore must be attached by a spindle microtubule, which results in the alignment of the

sister chromatids along an axis called the metaphase plate. This process ensures that each daughter cell will receive only one copy of each chromosome. The mitotic spindle checkpoint, one of the major regulatory points of the cell cycle, is also active during metaphase.

Three classes of microtubules make up the mitotic spindle. The aster microtubules hold the centrosome in place at the poles of the mitotic spindle, while the kinetochore microtubules attach to the kinetochores on each side of the sister chromatids and link the chromosomes to the spindle poles. At the same time, microtubules from opposite poles interact with each other through various microtubule-associated proteins and motor proteins in the middle of the spindle pole. These microtubules are not associated with kinetochores and are called interpolar microtubules.

11.8 Anaphase

From S-phase until metaphase, sister chromatids remain connected by the cohesin complexes. In order for these sister chromatids to be properly partitioned into the future daughter cells during the transition from metaphase to anaphase, the cohesion complexes have to be removed to allow for this separation. This is accomplished through the activity of an enzyme called separase. Until this stage, separase is kept in an inhibited state by the protein securin. Once anaphase begins, a mutiprotein complex called the APC complex, or cyclosome, becomes activated. The APC complex catalyzes the ubiquitination of the securin, thereby mediating its degradation. Once securin is degraded, separase is relieved from its inhibition and cleaves the cohesin complexes. This leads to the spindle microtubule-mediated separation of the sister chromatids. As the sister chromatids move toward their respective poles, motor proteins in the middle of the mitotic spindle travel along the interpolar microtubules, elongating the cell.

The movement of the chromatids during anaphase, as well as the elongation of the cell, requires an increase in the dynamic instability of the microtubules. In the early stages of anaphase, known as anaphase A, the progression of the chromatids toward the spindle poles is made possible by depolymerization of kinetochore associated microtubules at their plus ends. In anaphase B, the elongation and sliding of interpolar microtubules past one another pushes the two poles apart. At the same time, the forces exerted by outward-pointing astral microtubules at each spindle pole pull the poles away from each other.

11.9 Telophase

In telophase, the two sets of chromosomes arrive at the poles of the spindle and reformation of nuclear membrane and nucleolus occurs. During prometaphase, the nuclear

II

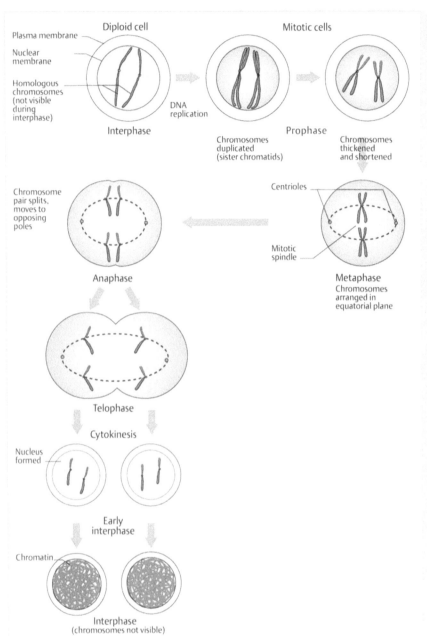

Fig. 11.3 Stages of mitosis. (Source: Passarge E, ed. Color Atlas of Genetics. 4th Edition. New York, NY. Thieme; 2012.)

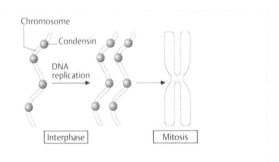

Fig. 11.4 Role of condensins. (Source: Passarge E, ed. Color Atlas of Genetics. 4th Edition. New York, NY. Thieme; 2012.)

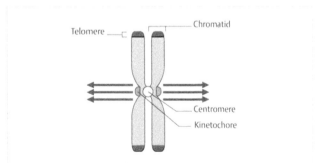

Fig. 11.5 Metaphase chromosome. (Source: Passarge E, ed. Color Atlas of Genetics. 4th Edition. New York, NY. Thieme; 2012.)

pore proteins and lamins were phosphorylated to allow for nuclear envelope breakdown. Now these components are dephosphorylated by protein phosphatases to allow the nuclear envelope to reform.

Meanwhile, the short and thick chromosomes begin to elongate to form long, thin chromatin. It is also during this time that the division of the cytoplasm begins with the assembly of the contractile ring at the cleavage furrow, a shallow groove in the cell near the old metaphase plate. This contractile ring consists of myosin motor proteins associated with actin microfilaments. As the myosin moves along the actin polymers, constriction of the membrane occurs, similar to the tightening of a purse string. This leads to invagination and subsequent narrowing of the cleavage furrow and creating a thin intracellular bridge that connects the two newly developed cells.

At the end of telophase, cytokinesis occurs (▶ Fig. 11.3). During this process, the cytoplasm is divided in two by a final constriction of the contractile ring and a tightly orchestrated fusion of membrane, leading to a pinching off of the connecting intracellular bridge and creating two individual daughter cells, each with a single nucleus.

11.10 Partitioning of Other Organelles

In addition to the partitioning of the genetic material found in the nucleus, the other membrane bound organelles of the cell must also be divided into the two daughter cells during mitosis in order to ensure that each of the daughter cells that will be generated will inherit an equal amount/number of each organelle. Multiple-copy organelles, such as mitochondria and lysosomes, double in number during interphase and are distributed in equal quantities between the two cells during mitosis. Single copy organelles, such as the endoplasmic reticulum and Golgi, double in size before they are divided between the two cells. However, recall from Chapter 1 that the Golgi is a polar organelle

that is only located on one side of the nucleus. Therefore a slightly different mechanism is used for its division during mitosis. The Golgi must first break up into small, uniform vesicles. These vesicles can then be partitioned equally between the two subsequent daughter cells. Following mitosis, they will assemble into functional Golgi.

11.11 Cell Cycle Regulation

The complex processes that occur during the cell cycle must be controlled in a very precise way so that the cell does not move on to the next stage until the appropriate time. So what allows for these tight controls over the cell cycle? The master regulators of cell cycle progression belong to two distinct families of proteins: the Cdk's and the cyclins (▶ Fig. 11.6). The name Cdk, or **c**yclin-**d**ependent **k**inase, is derived by this protein's requirement for association with a member of the cyclin family for its function. The Cdk's themselves are serine/threonine kinases. However, this kinase activity is only present when the Cdk is bound to a cyclin molecule.

There are three major Cdk's in humans: cdk1, cdk2, and cdk4/6. In addition, there are multiple cyclins that constitute a family of cell cycle proteins that are expressed only at specific times during the cell cycle. Each Cdk partners with a specific cyclin during different phases of the cell cycle, resulting in the kinase activity of the Cdk at that time (▶ Table 11.1). Although the activity of the different Cdk's oscillates during different stages of the cell cycle, the absolute concentrations of Cdk remain constant throughout the entire cell cycle. The control of the transition between stages is therefore determined by the stage-specific concentrations of cyclins.

11.12 Cyclin-Cdk Activation

Although the association of a Cdk molecule with its partner cyclin is necessary for the formation of an active complex, it is

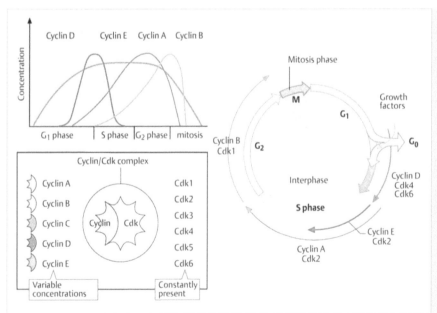

Fig. 11.6 Control of the cell cycle. (Source: Koolman J, Röhm K, ed. Color Atlas of Biochemistry. 3rd Edition. New York, NY. Thieme; 2012.)

not sufficient for a functional kinase due to the presence of additional molecular controls. When cyclin-cdk complexes are initially formed, they are in an inactive state that is maintained through inhibitory phosphorylations. While in this inactive state, an activating phosphorylation is acquired, but this activating phosphorylation cannot override the initial inhibitory phosphorylation. This results in a cyclin-Cdk complex that is poised to become active. At the appropriate time, cell cycle cues activate a phosphatase which is able to remove the inhibitory phosphorylations, allowing for cyclin-CDK activity and transition through the cell cycle.

11.13 Termination of Cyclin Signals

The cell stage-specific changes in concentration of the cyclins is due to rapid increases in expression of the cyclin genes followed by degradation of the resulting protein through ubiquitination. As discussed in Chapter 17, ubiquitination is a posttranslational modification involving the covalent linkage of a small molecule, ubiquitin, to the lysines of a protein. The ubiquitin tag directs the protein to a large multi-subunit complex called the proteasome, whose function is to degrade the ubiquitin-linked proteins. At the appropriate time, cell cycle cues direct the ubiquitination of the specific cyclins, resulting in their destruction.

Table 11.1

Cell cycle phase	Cdk	Cyclin family	Cyclin/Cdk complex substrates
G1	Cdk4 Cdk6	Cyclin D	Rb proteins
G1/S	Cdk2	Cyclin E	Rb proteins, enzymes for modifying histones, DNA replication, DNA repair, centrosome maturation
S	Cdk2 Cdk1	cyclin A	Transcription factors, enzymes for modifying histones, DNA replication, DNA repair, centrosome maturation
M	Cdk1	Cyclin B	Histones, laminins, proteins of the nuclear pore complex, Golgi complex, DNA replication and translation, microtubulebinding proteins

Source: Panini S, ed. Medical Biochemistry- An Illustrated Review. New York, NY. Thieme; 2013.

11.14 Cyclin D-Cdk4/6 and Transition to S-phase

When receptors on the surface of a cell encounter mitogenic signals, a sequence of events inside the cell is initiated that ultimately leads to exit from a G1 state and transition into S-phase. The major cyclin-cdk complexes involved with this process are cyclin D-cdk4/6 and cyclin E-cdk2.

In the absence of a proliferative signal, a protein called Rb binds to the transcription factor E2F and prevents it from inducing the expression of its target genes, holding the cell in the G1 phase. When the cell receives a mitogenic stimulus, such as a growth factor, a series of signaling events occurs inside the cell that ultimately leads to activation of cyclin D-cdk4/6. The active cyclin D-cdk 4/6 then phosphorylates Rb, causing liberation of the transcription factor E2F. E2F then induces the expression of genes required for DNA synthesis (S-phase), such as PCNA and DNA polymerases (▶ Fig. 11.7). Another gene that is expressed at this time encodes for cyclin E, which will pair with Cdk2 and cause additional phosphorylations of Rb protein, further enhancing S-phase transition.

11.15 Cyclin E-Cdk2 and S-phase Initiation

As discussed in Chapter 2, the origin recognition complex (ORC) binds to A-T rich regions of the genome known as origins of replication during the G_1 phase of the cell cycle. The ORC is then bound by MCM. Throughout G_1, the ORC/MCM complex is kept inactive through association with Cdc6. During the transition to S-phase, cyclinE-cdk2 complexes phosphorylate cdc6, marking it for degradation. The ORC/MCM complex is then free to initiate assembly of the replication fork.

11.16 Activation of Mitotic Cyclin-cdk

Accumulation of cyclin B starts at S-phase and rises gradually through the G_2 phase and peaks in mitosis (▶ Fig. 11.8). This increase in cyclin B leads to a corresponding increase in cyclinB-cdk1 complexes, also known as MPF, or maturation promoting factor. However, when the cyclinB-cdk1 complexes are initially formed, they are in an inactive state. This inactivity is maintained through inhibitory phosphorylations by the

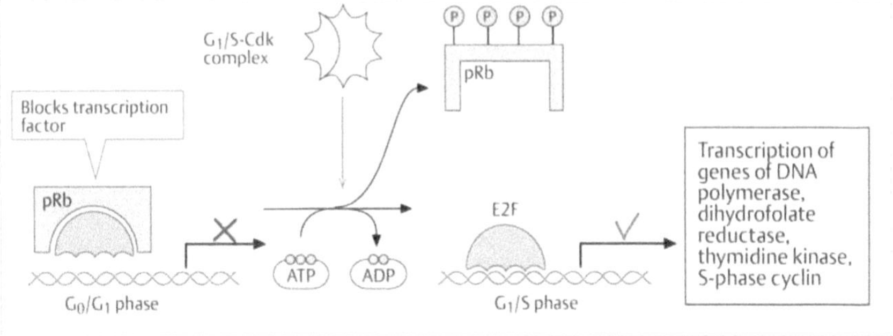

Fig. 11.7 Functions of the retinoblastoma protein (pRb). (Source: Koolman J, Röhm K, ed. Color Atlas of Biochemistry. 3rd Edition. New York, NY. Thieme; 2012.)

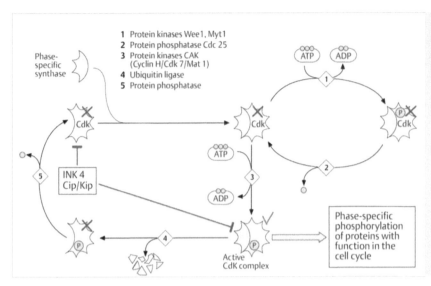

Fig. 11.8 Regulation of cyclin-dependent kinases. (Source: Koolman J, Röhm K, ed. Color Atlas of Biochemistry. 3rd Edition. New York, NY. Thieme; 2012.)

Wee1 kinase. While in this inactive state, the complex also gains an activating phosphorylation through the action of cdk-activating kinase (Cak), but this activating phosphorylation is not capable of overriding the inhibitory phosphorylation. The result is an accumulation of cyclinB-cdk1 complexes that are poised to become active. Near the end of G$_2$, cell cycle cues activate cdc25 phosphatase, which is able to remove the inhibitory phosphorylations, allowing for cyclinB-cdk1 activity.

One of the functions of the activated cyclinB-cdk1 complex is to phosphorylated, and therefore activate, more cdc25 phosphatases. This positive feedback results in a rapid amplification of cyclinB-cdk1 activity. Functional cyclinB-cdk1 has many targets throughout the different phases of mitosis. In early prophase, this complex phosphorylates the condensin subunits, resulting in their assembly on chromosomes. In prometaphase, cyclinB-cdk1 is responsible for the phosphorylation and resulting breakdown of nuclear envelope and lamin components. And in metaphase, cyclinB-cdk1 phosphorylates microtubule-associated proteins, resulting in the formation of the interpolar microtubule connections and formation of the mitotic spindle.

11.17 Checkpoints in the Cell Cycle

There are multiple checkpoints that monitor cellular health and environmental conditions and prevent the cell from transitioning to the next stage of the cell cycle until any abnormalities are corrected or until environmental conditions are favorable for growth. This inhibition is accomplished by the inactivation of the cyclin/cdk complexes that are normally responsible for the transition to the next cell cycle phase (▶ Fig. 11.9). There are three major checkpoints that monitor

these conditions: the restriction point, the G$_1$/S phase DNA damage checkpoint, and the G$_2$/M phase DNA damage checkpoint. In addition, there is a mitotic spindle checkpoint that prevents the cell from transitioning from metaphase to anaphase until all of the sister chromatid pairs are aligned properly on the metaphase plate.

The restriction point is a point in the G$_1$ phase in which a cell becomes committed to enter S-phase and complete a division cycle. This decision is based on the environmental conditions surrounding the cell and is commonly initiated by the binding of a mitogenic signal, such as a growth factor, to a receptor on the cells surface. This leads to the propagation of an internal signal ultimately leading to activation of cyclinD-cdk4/6, Rb phosphorylation, and entry into S-phase.

DNA damage or incomplete DNA replication can lead to the activation of kinases that phosphorylate p53 (▶ Fig. 11.10). The phosphorylated p53 is then stabilized and accumulates in the nucleus, where it functions as a transcription factor to induce the expression of the cdk inhibitor p21. p21 binds to and inhibits cyclinD-cdk4/6, as well as cylcinE-cdk2. The loss of activity of these two complexes results in a cell cycle arrest in either G1 or S, depending on when the DNA damage occurred. This same process is activated in the G2/M DNA damage checkpoint. Of course, in this case cyclinB-cdk1 is inhibited, preventing the cell from entering mitosis until the damage is corrected.

One of the latest cell cycle checkpoints, the mitotic spindle checkpoint, is active during metaphase. This checkpoint monitors the attachment of the spindle microtubules to the kinetochores during metaphase. If the microtubules are not attached to both kinetochores of the sister chromatid pair, the checkpoint is activated and the cell cannot proceed to anaphase until this problem is corrected. As we will see later in this chapter, defects in these cell cycle checkpoints can give rise to abnormal, highly proliferative cells, characteristics often associated with cancer.

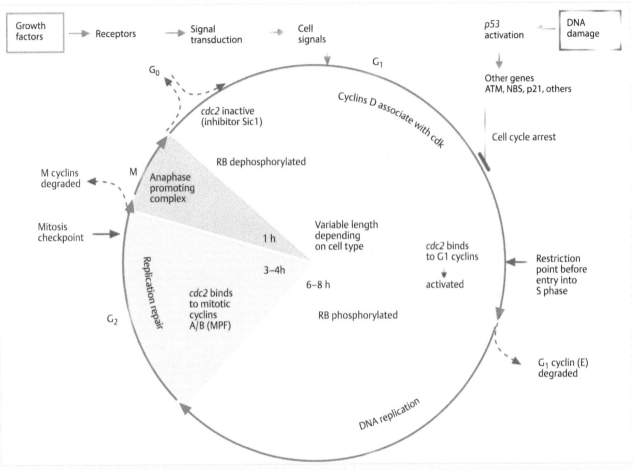

Fig. 11.9 Cell cycle control systems. (Source: Passarge E, ed. Color Atlas of Genetics. 4th Edition. New York, NY. Thieme; 2012.)

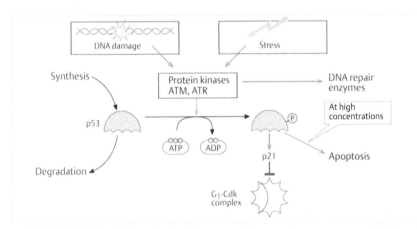

Fig. 11.10 Functions of the p53 protein. (Source: Passarge E, ed. Color Atlas of Genetics. 4th Edition. New York, NY. Thieme; 2012.)

Review Questions

1. A defect in the function of cyclin D would directly affect which phase of the cell cycle?
   A) G1
   B) S
   C) G2
   D) M

2. When G1 phase cells are exposed to ionizing radiation, the expression of which of the following molecules is increased?
   A) Wee1
   B) Cdc25
   C) p21
   D) Securin

3. A mutation in cyclin E would affect the activity of which of the following kinases?
   A) Cdk1
   B) Cdk2
   C) Cdk4
   D) Cdk6

4. A loss of function of which of the following molecules would prevent sister chromatids from separating during anaphase of mitosis?
   A) Condensin
   B) Securin
   C) Separase
   D) SMC2

Answers

1. **The correct answer is A**. In G1 cells that have received a mitogenic extracellular signal, a complex of cyclin D and Cdk4/6 is activated. This complex has kinase activity and phosphorylates the protein Rb. This phosphorylation event releases a transcription factor, E2F, that induces the cell to transition from G_1 to the S phase of the cell cycle. Therefore, a defect in cyclin D would directly affect a cell in G_1. The cyclin that is chiefly responsible for many of the activities in S phase (answer choice B) is cyclin E. Cyclin A is major cyclin in cells in G_2 (answer choice C). In M phase (answer choice D), cyclin B is responsible for transitioning the cell through the phases of mitosis.

2. **The correct answer is C**. Exposure to ionizing radiation causes double strand breaks in the DNA of cells. These double strand breaks induce a DNA damage response that includes the activation of p53. In turn, active p53 acts as a transcription factor and induces the expression of the Cdk inhibitor p21. p21 inhibits the cyclinD-Cdk4/6 complex, which arrests the cell in G_1 and allows for repair of the DNA damage before the cell can proceed through the cell cycle. Wee1 (answer choice A) is a kinase that adds an inhibitory phosphorylation to the cyclinB-Cdk1 complex, which is removed by Cdc25 (answer choice B) at the initiation of mitosis, thus producing an active complex. Securin (answer choice D) is an inhibitory molecule that binds to the enzyme separase. As a mitotic cell transitions from metaphase to anaphase, securin is degraded. Separase is then active and cleaves the cohesin complexes that hold sister chromatids together.

3. **The correct answer is B**. During S phase, cyclin E is synthesized and combines with Cdk2 to create an active complex. Cdk1 (answer choice A) partners with cyclin B during mitosis, whereas Cdk4 (answer choice C) and Cdk6 (answer choice D) combine with cyclin D during G1 to form an active complex.

4. **The correct answer is C**. Beginning in S phase and continuing until anaphase of mitosis, the replicated sister chromatids are held together by a series of ring complexes called cohesins. These cohesins must be cleaved in anaphase to allow for the separation of the sister chromatids into the future daughter cells. These cohesin rings cleaved by the enzyme separase. Therefore, loss of separase function would prevent sister chromatids from separating. SMC2 (answer choice D) is a component of the condensin (answer choice A) complex that allows for the tight packaging of the chromatids during early mitosis. Loss of function of the condensin complex would not prevent sister chromatids from separating in anaphase. Since securin (answer choice B) normally inhibits separase and, therefore, prevents sister chromatids from separating, loss of securin function would have the opposite effect and promote separation of sister chromatids.

12 Stem Cells and Hematopoiesis

At the conclusion of this chapter, students should be able to:
- Describe the general characteristics of stem cells
- Explain the different levels of differentiation potential of stem cells (potency)
- List the common molecular markers of stem cells
- Describe the stem cell niches found in adult skin, gut epithelium, brain, and skeletal muscle
- Describe the entire hematopoietic lineage from stem cell to terminally differentiated cell
- Explain the two methods for generating pluripotent stem cells from adult somatic cells

12.1 What is a Stem Cell?

There are several general characteristics that define stem cells. The first of which is that they remain in an undifferentiated state. Secondly, when they divide, they can make another stem cell, but they can also produce a new cell that is committed to differentiation. They can also undergo an unlimited number of cell divisions. And finally, stem cells have potency, which refers to a stem cell's potential to differentiate into different cell types.

12.2 Stem Cell Potency

There are many levels of stem cell potency. Totipotent cells have the ability to produce all of the differentiated cells in an organism, including extra-embryonic tissues (ie-placenta). Pluripotent cells can give rise to any fetal or adult cell type except those of extra-embryonic tissues. Multipotent cell can give rise to multiple lineages within a tissue. And unipotent cells can only give rise to one type of differentiated cell.

12.3 Asymmetric Divisions

As mentioned previously, when stem cells divide they can produce two new stem cells, but they can also give rise to one daughter cell that is a stem cell and one daughter cell that is committed to a particular type of cell lineage. This type of division, in which each newly formed daughter cell has a different fate, is known as asymmetric division (▶ Fig. 12.1).

There are two types of asymmetrical divisions that stem cells can undergo. In environmental asymmetry, daughter cells produced by the division of a stem cell are initially the same, but environmental influences direct one daughter cell to commit to a specific cell lineage. In divisional asymmetry, internal factors direct one of the daughter cells to follow a path of differentiation.

12.4 Cellular Differentiation

The process of cellular differentiation occurs in a step-wise manner. When a stem cell divides asymmetrically, one of the resulting daughter cells remains a stem cell, while the other

daughter cell is committed to differentiate. In many somatic tissues, this committed cell is called a transit amplifying cell, meaning that it is in transit from a stem cell to a differentiated state. These cells divide frequently, but only for a limited number of divisions, resulting in the amplification of the subsequent differentiated cell population (▶ Fig. 12.2). Extracellular signaling molecules mediate the environmental cues that cause the progeny of these transit amplifying cells to differentiate.

12.5 Extracellular Matrix

In addition to soluble extracellular signaling molecules, the extracellular matrix can also influence the differentiation path of a stem or transit amplifying cell. Bone marrow stromal cells are found in the bone marrow and they have the ability to differentiate into fat cells, cartilage cells, or bone cells. The specific fate of these bone marrow stromal cells is known to be influenced by the physical properties of the extracellular matrix associated with these cells. On stiff matrix, there is strong

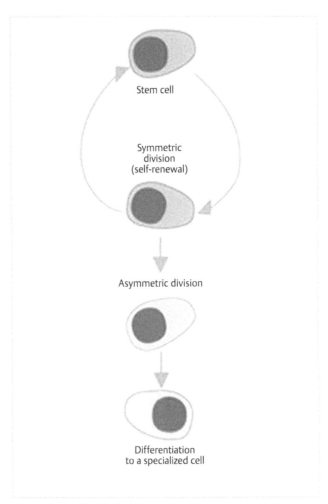

Stem cell

Symmetric division (self-renewal)

Asymmetric division

Differentiation to a specialized cell

Fig. 12.1 Characteristics of stem cells. (Source: Passarge E, ed. Color Atlas of Genetics. 4th Edition. New York, NY. Thieme; 2012.)

II

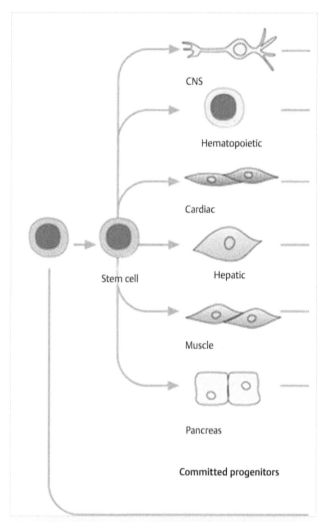

CNS

Hematopoietic

Cardiac

Stem cell   Hepatic

Muscle

Pancreas

Committed progenitors

Fig. 12.2 Stem cells can give rise to many different cell types of the body. (Source: Passarge E, ed. Color Atlas of Genetics. 4th Edition. New York, NY. Thieme; 2012.)

adhesion between the cell and the ECM, which activates the transcription factors YAP and TAZ. These transcription factors then induce the expression of downstream genes that induce the cell to differentiate into a bone cell. On soft matrix however, there is a weak adhesion with the bone marrow stromal cell, preventing the activation of YAP and TAZ, and causing the cell to differentiate into a fat cell.

12.6 Stem Cell Genes

There are four distinct genes known to be highly expressed in pluripotent stem cells: Oct3/4, Sox2, Nanog, and Lin28. These genes encode for transcription factors whose function is to maintain the stemness of the cell. As the cells commit to a more differentiated lineage, these genes are shut off, and other genes are expressed in a tightly controlled sequence that allows for the gradual transition to a terminally differentiated state. Hence, each stage of this differentiation process is characterized by a distinct pattern of gene expression.

12.7 Stem Cell Niches

In the adult, stem cells can be found in many different tissue types. Within these tissues, the stem cells reside in specific compartments known as the stem cell niche. These niches represent microenvironment within the tissue that protects the stem cells and also provides the signals necessary for them to maintain their stemness. Some of the best studied stem cell niches are found in the epidermis, gut epithelium, neural tissue, skeletal muscle, and hematopoietic tissue. Since the stem cells in adult tissues are capable of producing many different cell types, but not all of the cell types in the body, they are considered multipotent stem cells.

12.8 Stem Cell Niches of the Skin

In skin that is devoid of hair follicles, stem cells are found in the peaks of the dermal papillae in the epidermis. These stem cells give rise to transit amplifying cells that migrate to the valleys of the dermal papillae, where they produce the differentiated keratinocytes that will form the many layers of the epidermis. In skin that contains hair follicles, the stem cells are found in a region of the hair follicle that lies just below the epidermis. This region "bulges" out from the side of the follicle and is, therefore called the hair follicle bulge. The bulge stem cells found here are capable of giving rise, via committed transit amplifying cells, to multipole cell types including keratinocytes, sebaceous gland cells, and hair cells.

12.9 Gut Epithelium Stem Cells

In the gut epithelium, the stem cell niche is found at the bottom of structures called crypts. These crypts are protrusions of the epithelium into the underlying connective tissue that occur along the luminal side of the intestinal wall. These gut epithelium stem cells are capable of differentiating into all of the different cell types that make up the epithelial layer of the intestine, including absorptive cells (enterocytes), goblet cells, enteroendocrine cells, and Paneth cells. This differentiation process in the crypts begins with the stem cells at the bottom of the crypt producing transit amplifying cells that migrate upward along the epithelial layer, eventually differentiation into one of the four cell types.

12.10 Notch Signaling in Gut Epithelium

In the intestinal crypts, high levels of Wnt ligand lead to the expression of Notch in the stem cells and Delta in the adjacent Paneth cells. Through juxtacrine signaling, the Delta ligand transmits a signal to the Notch receptor on the stem cell which maintains the stemness of that cell. When the crypt stem cell divides, a transit amplifying cell is produced that migrates upward out of the crypt. The Notch-Delta interactions are also important to keep the transit amplifying cells from differentiating. Eventually, the transit amplifying cell will reach an area that is devoid of Wnt signal, resulting in the loss of Notch expression, and leading to differentiation into one of the terminally differentiated cell types.

12.11 Neural Stem Cells

In the brain, neural stem cells can be found in two distinct niches: the subventricular zone of the lateral ventricle and the hippocampus. These neural stem cells are multipotent, as they can differentiate into neurons and, glial cells such as astrocytes and oligodendrocytes.

In mice, the neural stem cells continuously produce progenitor cells that migrate to the olfactory bulb, where they differentiate into neurons that are needed to replace the ones that are continually lost in this structure. In humans, the neural stem cells in the hippocampus are responsible for providing the new neurons needed for the constant turnover of the cells in this region.

12.12 Skeletal Muscle Stem Cells

The stem cells of skeletal muscle are located between the sarcolemma of skeletal muscle fibers and the surrounding basal lamina, and are known as satellite cells or myosatellite cells. These satellite cells lay in a quiescent state until a stimulus, such as muscle damage, induces them to re-enter the cell cycle and produce progenitor cells. Upon activation, they express high levels of the transcription factor MyoD that results in differentiation into myoblasts and ultimately formation of new myofibers.

12.13 Hematopoietic Stem Cells

The stem cell niche of the hematopoietic system lies within the bone marrow. Here, the hematopoietic stem cells (HSCs) are kept in a state of stemness through interactions with stromal cells. Of all of the stem cell lineages within the body, the hematopoietic lineage is the most diverse and best characterized. The start of this lineage is marked by the hematopoietic stem cell, which is capable of indefinite self-renewal and gives rise to a cell type termed the MultiPotent Progenitor cell (MPP). As the name suggests, these MPPs can give rise to cells from multiple, but limited, lineages. Based on external signals, the MPP can commit to the development of either the myeloid or lymphoid lineages. If the myeloid lineage is induced, the MPP will give rise to the common myeloid progenitor or CMP. The CMP will produce granulocyte/macrophage progenitor (GMP) cells and megakaryocyte/erythrocyte progenitor (MEP) cells. Downstream of these progenitors are types of cells known as blast cells. These cells have a proliferative rate and can, therefore, quickly amplify their numbers. Each type of blast has the ability to differentiate into a specific type of terminally differentiated blood cell. For example, megakaryoblasts produce megakaryocytes and monoblasts produce monocytes.

Besides the myeloid lineage, MPP can produce the components of the lymphoid lineage. The first cell along this path of differentiation is the common lymphoid progenitor (CLP). This cell gives rise to Pro-B cells and to the NK/T cell precursor. The Pro-B cells will give rise to Pre-B cells, which ultimately produce B-cells. The NK/T cell precursor produces both T cells and Natural Killer (NK) cells.

12.14 Hematopoietic Cell Surface Markers

The different cell types of the hematopoietic system can be identified based on the presence of specific proteins found on the outer surface of their plasma membranes. For example, HSCs express KIT, SCA1, and low levels of CD34 and FLK2 on their surface, while more differentiated cells, like MPPs, express different levels of these markers. Flow cytometry can be used to distinguish among these different types of expression patterns and, therefore, can separate each type of cell.

12.15 Hematopoietic Stem Cell Maintenance

As mentioned earlier, the hematopoietic stem cells within the bone marrow must remain physically connected to stromal cells in order to maintain their stem cell characteristics. In the niche, the stem cells express a tyrosine kinase receptor called Kit on their cell surface. This receptor is embedded in the plasma membrane and binds to the Kit ligand, that is embedded in the plasma membrane of neighboring stromal cells. The Kit receptor/ligand interaction initiates a signaling cascade within the stem cell that allows the cell to remain a stem cell. When the stem cell divides, one of the daughter cells will lose its connection with the stromal cell and will, therefore, no longer have the Kit signaling. This cell is then committed to differentiate into one of the blood cells, while the other daughter cell retains the Kit ligand/receptor connection and, therefore, remains as a stem cell.

Although many tissues in mammals have their own source of stem cells, some tissues are thought to be devoid of stem cells. These include the auditory epithelium and retinal epithelium. Because there are no stem cells to replace them, damage to cells in these tissues can result in permanent loss of that tissue's function.

12.16 Exogenous Sources of Stem Cells

In recent years, it has been discovered that a terminally differentiated cell can be forced to convert back to a state of pluripotency, under certain conditions. There are now two established methods that can be used experimentally to accomplish this: somatic cell nuclear transfer (SCNT) and induced pluripotent stem cells (iPS).

12.17 Somatic Cell Nuclear Transfer

Somatic cell nuclear transfer (SCNT) is a cloning technique in which the nucleus of a terminally differentiated somatic cell, or donor nucleus, is transferred to the cytoplasm of an enucleated egg. Since the genomic DNA originates from the donor somatic cell, the resulting zygote is a genetic clone of the donor source.

There are two possible approaches for using this technology: reproductive cloning and therapeutic cloning. In reproductive cloning, the embryo that is produce from this technique is introduced into a pseudopregnant female, and the resulting offspring that is produced will be a clone of the organism that originally donated the nucleus. This method has proven to be valuable for agricultural purposes, as animals with desirable characteristics can be produced in large numbers.

The other approach for using SCNT is in therapeutic cloning, in which personalized embryonic stem (ES) cells can be generated for the purposes of treating a specific disease or condition in an individual. In this case, pluripotent stem cells from the inner cell mass of the embryo are collected and grown in culture. These cells can then be induced to differentiate into whatever cell type is needed to treat the individual. Since these resulting cells are clones of the individual that donated the nucleus, the potential for immunological rejection is minimized. However, there are ethical concerns with this procedure because it requires the destruction of a human embryo in order to collect the pluripotent stem cells.

12.18 iPS Cells

Another method for creating pluripotent stem cells from a terminally differentiated somatic cell involves the forced expression of the stem cell genes Oct3/4, Sox2, and Klf4, and in some cases c-myc. Since this method induces a differentiated cell to revert back to a pluripotent state, it is called induced pluripotent stem cells (iPS cells). A common source of these differentiated cells is skin fibroblasts from a skin biopsy. Under culture conditions, these iPS cells can then be forced to differentiate into any cell type that is needed. There are fewer ethical concerns with this process since it does not require the creation of a human embryo.

There are many potential uses for iPS cells. iPS cells generated from a patient with a genetic disease can be used to analyze the disease mechanism or for discovery of therapeutic drugs. iPS cells can also be used to repair a genetic defect, by being induced to differentiate in vitro and grafted back into the patient without initiating an immune response.

Review Questions

1. Which of the following is an example of a pluripotent stem cell?
   A) Bulge stem cell
   B) Crypt stem cell
   C) Embryonic stem cell
   D) Neural stem cell
2. The signaling pathway that maintains hematopoietic cells in a stem cell state involves which of the following molecules?
   A) KIT
   B) TAZ
   C) WNT
   D) YAP
3. The genes Oct3/4, Lin28, Sox2, and Nanog are highly expressed in which of the following types of stem cell?
   A) Multipotent
   B) Pluripotent
   C) Totipotent
   D) Unipotent
4. Which of the following characteristics differentiates a totipotent stem cell from a multipotent stem cell?
   A) They can undergo an unlimited number of cell divisions.
   B) They divide asymmetrically.
   C) They have the ability to give rise to all cell types of the body.
   D) They remain in an undifferentiated state.

Answers

1. **The correct answer is C.** Embryonic stem cells are capable of giving rise to any cell type of the body, with the exception of extra-embryonic tissue. They are therefore pluripotent stem cells. All of the other answer choices (A, B, and D) are adult somatic stem cells that are multipotent since they can give rise to more than one type of somatic cell found within their tissue of residence. However, they cannot give rise to any cell type of the body.

2. **The correct answer is A.** The KIT receptor is embedded in the plasma membrane of hematopoietic stem cells. In the bone marrow, these cells are found next to stromal cells. The stromal cells possess KIT ligands on their surface, which bind to the KIT receptors to induce a signal inside the stem cell that maintains its stemness. When this interaction is lost, the stem cell becomes committed to differentiate into one of the hematopoietic cell lineages. TAZ (answer choice B) and YAP (answer choice D) are transcription factors that determine the cell fate of bone marrow stromal cells. When the bone marrow stromal cell is in contact with a hard surface, TAZ and YAP are turned off, and the cell becomes a bone cell. On the other hand, when the bone marrow stromal cell is in contact with a soft surface, it becomes a fat cell. In the crypt cells of the small intestine, WNT (answer choice C) signaling maintains the crypt cells in a stem cell state.

3. **The correct answer is B.** Oct3/4, Lin28, Sox2, and Nanog are genes that have been identified as being highly expressed in embryonic stem cells, which are pluripotent stem cells. This combination of genes is not expressed at high levels in multipotent (answer choice A), totipotent (answer choice C), or unipotent (answer choice D) stem cells.

4. **The correct answer is C.** Totipotent stem cells are capable of producing all cell types of the body, as well as extra embryonic tissue. Multipotent stem cells can give rise to several different cell types, but they cannot produce all of the cell types of the body. The other answer choices (A, B, and D) are general characteristics of all stem cells and would therefore be true of both totipotent and multipotent stem cells.

13 Cell Injury, Apoptosis, and Necrosis

At the conclusion of this chapter, students should be able to:

- Explain the four major types of cellular adaptations (hypertrophy, hyperplasia, atrophy, and metaplasia) and give specific examples of each
- Compare and contrast the cellular characteristics associated with necrosis and apoptosis
- Describe the events of the intrinsic apoptotic pathway as well as identify the key molecules involved and their function
- Describe the events of the extrinsic apoptotic pathway as well as identify they key molecules involved and their function
- Identify pro-apoptotic and anti-apoptotic family members and describe their function
- Distinguish between the members of the initiator and executioner caspase families

Under normal physiological conditions, a cell is said to be under a state of homeostasis. In this state, the cell is able to maintain a condition of equilibrium, or stability, within its internal environment when dealing with external changes. Any disruptions to homeostasis, such as cold, heat, infection, poison, radiation, physical injury, or even normal physiologic activities like hormone signaling will induce a cellular response to cope with the encountered stress.

13.1 Cellular Responses to Stress

When stresses are encountered, one of the ways that a cell can respond is through an adaptation. Adaptations are reversible functional and structural responses to more severe physiologic stresses and some pathologic stimuli. The ability of the cell to undergo these adaptations allows for the survival of the cell. Once the insult is removed, the cell can return to the original homeostatic state without consequence. The types of cellular adaptations include hypertrophy, hyperplasia, atrophy, or metaplasia.

13.2 Hypertrophy

Hypertrophy refers to an increase in the size of a particular tissue that is due to an increase in the size of the individual cells that make up that tissue, rather than an increase in the number of cells. This type of increase in tissue mass is typically found in tissues that are composed of non-dividing cells. Hypertrophy can be in response to either a physiologic or pathologic condition. The most common stimulus for hypertrophy of muscle is increased workload. A pathologic example of this is the cardiac hypertrophy seen in the heart as a response to hypertension. Physiologic hypertrophy can be seen in skeletal muscle after the increased workload associated with exercise.

13.3 Hyperplasia

In contrast to hypertrophy, hyperplasia is the increase in size of a tissue due to an increase in the number of cells within the tissue. This occurs only if the cellular population is capable of dividing. There are also both physiologic and pathologic causes of hyperplasia. Physiologic hyperplasia can include changes that are induced in response to changes in the production of certain hormones. Typically, these changes increase the functional capacity of the tissue. Once the hormonal stimulation is eliminated, the affected tissues regress. Specific examples of this type of hyperplasia include the endometrial hyperplasia seen during pregnancy and breast hyperplasia during puberty. In response to damage or loss of tissue, compensatory hyperplasia can occur, as seen in the regeneration of the liver following resection. Pathologic hyperplasia can be caused by abnormal excesses of hormones or growth factors, or due to viral infections. Although hyperplasia cannot occur in tissues composed of non-dividing cells, both hypertrophy and hyperplasia can occur in the same tissue if the cell population is capable of dividing.

13.4 Atrophy

Atrophy is the reduction in size of an organ or tissue resulting from a decrease in cell size and number. Decrease in protein synthesis along with an increase in protein degradation can cause atrophy. Atrophy also can be associated with autophagy, in which the cell digests its own contents for a source of nutrients. Atrophy can be either physiologic or pathologic. Causes of physiologic atrophy include loss of endocrine stimulation such as endometrial atrophy following menopause. Causes of pathologic atrophy include decreased workload such as skeletal muscle atrophy following a broken bone, loss of innervation due to damage to nerves that lead to muscle atrophy, diminished blood supply, inadequate nutrition such as cachexia or muscle wasting seen in cancer patients, or pressure exerted by a growing tumor on surrounding tissue.

13.5 Metaplasia

Metaplasia is a reversible change in which one differentiated cell type is replaced by another cell type. This type of adaptation is commonly seen in tissues that are exposed to chronic irritation. During the metastatic process, cells that are sensitive to the encountered stress are replaced with another cell type that is able to withstand the adverse environment. In epithelial tissue, metaplasia involving a change from columnar to squamous epithelial cells can be found in the respiratory tract of smokers. In this case, the chronic irritation caused by exposure to the smoke changes the architecture of the tissue from columnar to squamous since the squamous cells are better adapted to withstand these stresses than the columnar cells that are normally found in this tissue. Another example of epithelial metaplasia is Barrett esophagus. This condition develops in individuals that suffer from acid reflux. Because columnar cells withstand acid conditions more readily than squamous cells, the normal squamous epithelium in the esophagus of these individuals is replaced with a columnar form.

The aforementioned cellular adaptive responses only occur if the cell experiences a mild injury that is reversible and allows cell to revert back to a homeostatic state. Characteristics of cells undergoing reversible injury include cellular swelling, plasma membrane alterations (blebbing), mitochondrial changes, dilation of the endoplasmic reticulum, and minor nuclear alterations.

If the injury is severe, then it is considered an irreversible injury. This results in cell death in one of two ways, either necrosis or apoptosis. Characteristics of irreversible injury include severe membrane damage, efflux of intracellular enzymes and proteins into the circulation (a clinical marker of cell injury and death), mitochondrial swelling and their complete loss of function, rupture of lysosomes, autolysis, and major nuclear changes such as condensation, fragmentation, and dissolution.

13.6 Necrosis

Necrosis is a passive, pathological process induced by acute injury or disease. Generally a group of cells in a localized region of the tissue undergoes necrosis at the same time due to some type of external injury. The cellular characteristics of necrosis include denaturation of intracellular proteins and enzymatic digestion of injured cells caused by release of lysosome contents, the compromise of plasma membrane integrity leading to cell lysis and cellular contents leaking into extracellular space, and the formation of large, whorled phospholipid masses, called myelin figures, derived from damaged cell membranes. This process is not tightly regulated and leads to considerable destruction of the tissue. A key consequence of necrosis is the induction of an inflammatory response due to the release of cellular debris in the extracellular space and the recruitment of phagocytes to eliminate this debris.

13.7 Apoptosis

In contrast to necrosis, apoptosis is a tightly regulated, programmed cell death (▶ Fig. 13.1). Characteristics include a plasma membrane that remains intact, activation of specific self-degrading enzymes, shrinkage of cells, and the pinching off of small pieces of the cell, forming structures known as apoptotic bodies. Apoptotic cells are engulfed and eliminated by phagocytes in the absence of an inflammatory response.

Apoptosis is an important response to both pathologic and normal developmental conditions. Following a pathologic insult, apoptosis can be used to limit collateral damage by eliminating irreversibly-damaged cells without a host response. Some causes of pathologic apoptosis are DNA damage, misfolded proteins, and infections. Apoptosis is also essential for many normal developmental processes, such as the formation of digits during embryogenesis, establishment of neuronal connections, and lymphocyte maturation in the thymus.

13.7.1 Biochemical Features of Apoptosis

There are several biochemical features that are characteristic of apoptosis (▶ Fig. 13.2) One of the central events in apoptosis is the activation of a family of enzymes called caspases. Caspases are cysteine proteases, having a cysteine residue in their catalytic active site. In the absence of an apoptotic signal, caspases exist in an inactive form and contain a prodomain that must be cleaved off for their activation. Due to the presence of this prodomain, inactive caspases are known as procaspases. The initial activation of caspases results in an amplification cascade, resulting in the activation of many other caspase molecules downstream. This creates a massive reaction in the cell in response to a small stimulus.

Another event in the apoptotic process is the breakdown of DNA and protein in the cell following the caspase-dependent activation of specific nucleases and proteases. These nucleases are capable of cleaving the genomic DNA at specific sites between nucleosomal units, resulting in the liberation of pieces of DNA that are multiples of 180 base pairs in length. In addition, membrane alterations are also induced during apoptosis. Normally, the phospholipid phosphotidylserine is only found in the inner leaflet of the plasma membrane. When a cell is going through apoptosis, these phosphotidylserine molecules flip from the inner leaflet to the outer leaflet, where they are exposed to the extracellular environment. These phospholipidserines then serve as ligands for receptors found on the surface of phagocytes, which allow for the identification of the apoptotic cell and engulfment by the phagocyte.

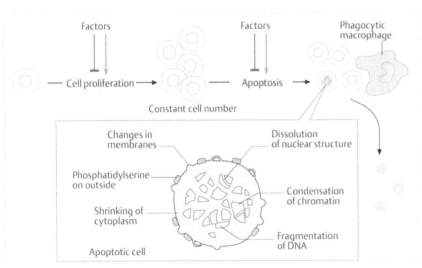

Fig. 13.1 Cell proliferation and apoptosis. (Source: Koolman J, Röhm K, ed. Color Atlas of Biochemistry. 3rd Edition. New York, NY. Thieme; 2012.)

II

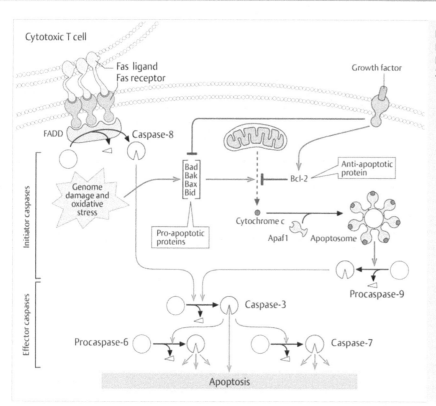

Fig. 13.2 Regulation of apoptosis. (Source: Koolman J, Röhm K, ed. Color Atlas of Biochemistry. 3rd Edition. New York, NY. Thieme; 2012.)

13.8 Caspases

There are two families of caspases: the initiator caspases and the executioner caspases. Initiator caspases are the first type of caspase activated in apoptosis. Their function is to cleave and activate executioner caspases. Members of this family include caspase 8, 9, and 10. Once activated, the executioner caspases induce the activity of downstream targets that are responsible for the molecular events that ultimately lead to the death of the cell, such as nuclease cleavage of the DNA and protease activity. Members of the executioner family include caspase 3, 6, and 7.

13.9 Intrinsic Pathway

There are two distinct pathways for apoptosis: the intrinsic pathway and the extrinsic pathway. The intrinsic pathway results from an increase in mitochondrial permeability and subsequent release of pro-apoptotic molecules. This pathway is activated when the cell is stressed by such things as deprivation of nutrients, DNA damage, or misfolding of proteins.

The intrinsic pathway is controlled by members of the Bcl family of proteins, which includes both pro-apoptotic and anti-apoptotic members. Among the pro-apoptotics is the BH3-only family consisting of Bim, Bid, and Bad. In addition, there is a channel forming family that includes the proteins Bax and Bak. Among the anti-apoptotics are Bcl-2, Bcl-X, and Mcl-1.

Bax and Bak are transmembrane proteins located in the outer membrane of the mitochondrial. In the presence of an apoptotic stimulus, Bax and Bak oligomerize to form pores. This allows for the release of proteins that are normally sequestered in the intermembrane space, including cytochrome c, which is directly responsible for initiating the apoptotic signaling cascade.

In the presence of survival factors, the anti-apoptotic proteins Bcl-2, Bcl-X, and Mcl-1 are produced. They bind to Bax and Bak and prevent their oligomerization. This maintains mitochondrial membrane integrity and prevents leakage of cytochrome c, thus inhibiting apoptosis.

BH3-only proteins Bim, Bad, or Bid are activated in response to cellular stress/apoptotic stimulus. They bind to and inhibit the anti-apoptotics Bcl-2/Bcl-X/Mcl-1. This relieves the inhibition placed on Bax and Bak and allows them to oligomerize to form the channels that allow leakage into the cytoplasm of cytochrome c, resulting in activation of the caspase cascade.

In the cytoplasm, the released cytochrome c combines with the adaptor protein Apaf-1 to form a large complex called the apoptosome. The apoptosome recruits procaspase-9, an initiator caspase. Procaspase-9 molecules are then cleaved to produce the active form of caspase-9. The caspase-9 within the apoptosome mediates the activation of downstream executioner caspases.

13.10 Extrinsic Pathway

The extrinsic apoptotic pathway is also known as the death receptor pathway as it involves the FasR, or death receptor. In some cells that have received an apoptotic signal, expression of Fas receptor (FasR) is induced. When these cells come in physical contact to a cell expressing the Fas ligand (FasL) on their surface, engagement of FasL with FasR occurs, forcing the FasR-bearing cell to undergo apoptosis.

The molecular mechanisms involved in the extrinsic pathway are distinctly different than those required for the intrinsic pathway. Instead of mitochondrial involvement, the extrinsic pathway leads directly to activation of executioner caspases, without the need for cytochrome c release. After binding of FasL to FasR on the cell surface, there is a clustering of three FasR molecules and binding of the adaptor protein FADD (Fas associated death domain) to the intracellular tails of FasR through a region of their intracellular tail that is termed the death domain. FADD then recruits two pro-caspase 8 molecules. The pro-caspase 8 molecules cleave each other, creating active caspase 8. Caspase 8 then cleaves the procaspases 3, 6, 7 to produce the active executioner caspases 3, 6, 7, leading to subsequent cell death.

Although the extrinsic pathway is separate from the intrinsic pathway and does not require mitochondrial involvement, under some circumstances activation of the extrinsic pathway can also lead to activation of the intrinsic pathway. This is due to activation of the BH3-only proteins in response to caspase-8 activation and can result in an amplification of the cellular response.

II

Review Questions

1. What histologic feature will most likely be seen in an esophageal biopsy from an individual with chronic acid reflux disease?
   A) A larger number of columnar epithelial cells than squamous epithelial cells
   B) Decreased size of overall tissue due to a decrease in number of squamous epithelial cells
   C) Increased numbers of squamous epithelial cells
   D) Normal numbers of squamous epithelial cells with an increase in size of individual cells

2. Which of the following terms refers to an increase in the size of a tissue due to an increase in the size of its individual cells?
   A) Atrophy
   B) Hyperplasia
   C) Hypertrophy
   D) Metaplasia

3. The anti-apoptotic family of molecules includes which of the following?
   A) Apaf1
   B) Bad
   C) Bim
   D) Bcl-2

4. Which of the following caspases binds to the FAS receptor in the extrinsic apoptosis pathway?
   A) Caspase 3
   B) Caspase 6
   C) Caspase 8
   D) Caspase 9

Answers

1. **The correct answer is A**. The squamous epithelial cells that normally line the esophagus are sensitive to acidic conditions. Because these squamous cells are continuously exposed to stomach acids in individuals with chronic acid reflux, the cells are destroyed. To counteract this loss of cells, the esophagus undergoes metaplasia, where the squamous cells are replaced by columnar cells. These columnar cells are inherently more resistant to the stomach acids than the squamous cells. Therefore, this condition would result in a larger number of columnar epithelial cells than squamous epithelial cells due to metaplasia. A decrease in size of the overall tissue due to a decrease in the number of cells (answer choice B) is an example of atrophy. Answer choice C, increased numbers of squamous epithelial cells, would result in hyperplasia, whereas an increase in the size of individual cells while maintaining normal numbers of cells is called hypertrophy.

2. **The correct answer is C**. An increase in the size of a tissue due to an increase in the size of individual cells is called hypertrophy. This adaptation is common in tissues in which the individual cells are not capable of cell division. Hyperplasia (answer choice B), on the other hand, is an increase in tissue size due to an increase in the number of cells within the tissue. This can only occur in tissues that contain cells that are capable of dividing. It is also possible to have both hypertrophy and hyperplasia in these types of tissues. Atrophy (answer choice A) is a decrease in the size of a tissue due to a decrease in the number of cells and the size of the individual cells. Metaplasia (answer choice D) is a change from the normal type of cell found in a tissue to another type that is better adept at handling a specific form of stress.

3. **The correct answer is D**. The Bcl family of proteins includes both pro-apoptotic and anti-apoptotic molecules that are mediators of the intrinsic apoptotic pathway. The anti-apoptotic members include Bcl-2, Bcl-X, and Mcl-1. The function of these anti-apoptotic proteins is to inhibit the release of cytochrome c from the mitochondria by preventing the assembly of Bax/Bak pores in the mitochondrial outer membrane. The pro-apoptotic proteins Bad (answer choice B), Bid, and Bim (answer choice C) block the action of the anti-apoptotics and, thus, allow for the formation of Bax/Bak pores and release of cytochrome c. After cytochrome c is released into the cytoplasm, it combines with Apaf1 (answer choice A) and caspase 9 to form the apoptosome, which will activate executioner caspases and leads to cell death.

4. **The correct answer is C**. In the extrinsic apoptosis pathway, a juxtacrine type of mechanism is used to transmit a signal from one cell to another cell through a physical interaction. In this case, the cell that is providing the signal expresses the FAS ligand on its surface, while the cell that receives this signal displays FAS receptor molecules. Once the FAS ligand binds to the FAS receptor, the adaptor protein FADD associates with the intracellular tail of the FAS receptor. FADD then recruits caspase 8, which becomes active and, in turn, activates the downstream executioner caspases which ultimately leads to cell death.

14 Hallmarks of Cancer and Cancer Biology

II

At the conclusion of this chapter, students should be able to:
- Identify the factors contributing to cancer cell immortality
- Describe how oncogenes produce a constant proliferative signal that promotes cancer development
- Explain how tumor suppressors inhibit cancer development and how tumor suppressors can become inactivated
- Explain how cancer cells bypass apoptosis
- Describe the process of angiogenesis
- Describe the steps associated with metastasis
- Identify both hereditary and environmental factors that contribute to tumor development
- Describe the techniques used to test the carcinogenic potential of compounds
- Describe the features associated with dysplastic and anaplastic growths
- Explain the characteristics that distinguish a benign tumor from a malignant tumor

There are six characteristics that are generally applicable to cancer. These are considered the hallmarks of cancer and include immortality, persistent proliferative signal, inactivation of anti-proliferative signals, resistance to cell death, angiogenesis, and metastasis.

14.1 Replicative Capacity

Somatic cells have a predetermined number of divisions, which is termed the replicative capacity of the cells. In normal cells, there are two regulatory mechanisms that govern replicative capacity: senescence and telomere length. Senescence is the process by which a cell irreversibly exits the cell cycle. The molecular events leading to senescence begin with the cumulative buildup of reactive oxygen species (ROS) that occurs as a byproduct of normal cellular metabolism. Over an extended period of time, this ROS produces a sufficient amount of DNA damage to activate p53, resulting in expression of the cell cycle inhibitor p21, which in turn causes the cells to stop cycling, irreversibly exit the cell cycle, and become senescent. Cancer cells must bypass this senescence mechanism in order to divide an unlimited number of times.

As discussed in Chapter 2, telomeres are sections of a repeating sequence of bases (TTAGG) found at the ends of linear chromosomes (▶ Fig. 14.1). These chromosome ends cannot be replicated by the DNA polymerase in the same way as the rest of the chromosome and must, therefore, be synthesized using the enzyme telomerase. As cells age, their telomerase activity diminishes, resulting in the gradual shrinking of the telomeres (▶ Fig. 14.2). When the telomeres reach a certain threshold length, the cell's DNA damage response machinery is activated. This results in an arrest of the cell cycle and eventual senescence of the cell. In a majority of human cancers, telomerase activity is abnormally high, contributing to the immortality of the cancer cells.

14.2 Persistent Proliferative Signal

Proto-oncogenes are genes that encode proteins that function to receive or transmit cellular growth-promoting signals. The proto-oncogenes can become oncogenes if they undergo a mutation that causes them to be constitutively active, or always "on". Most oncogenes are dominant, meaning that only one allele has to be affected to have an effect on the cell. Oncogenic mutations are also considered "gain-of-function" mutations as the mutations enable the encoded protein to function in a normal, albeit unregulated, fashion (▶ Fig. 14.3).

The mitogenic pathway of greatest importance to human cancer is the Ras-MAP kinase pathway, which was reviewed earlier. Within this pathway, several components are known to be mutated in many forms of cancer (▶ Fig. 14.4). These components include the receptor tyrosine kinase, Ras, and Raf. There are several different mechanisms that can result in the transformation of a proto-oncogene into an oncogene (▶ Fig. 14.5). First, a point mutation can occur in the coding region, resulting in the production of a protein with activity that cannot be regulated appropriately. Second, the gene can be duplicated, causing the production of an abnormal amount of the protein. In this case, the gene/protein itself is not changed, but large amounts of the normal protein overwhelm the control mechanisms of the cell. And lastly, a chromosomal translocation can occur. If the gene is moved to a locus with high transcriptional activity, then an abnormal amount of the normal protein will be made. This would be similar to the duplicated gene scenario, where too much normal protein has an abnormal effect on the cells. If the chromosomal translocation creates a fusion gene, ultimately being made into a fusion protein, this hybrid protein may possess an increase in activity/effect on the cell with the fused portion preventing the protein from being regulated properly.

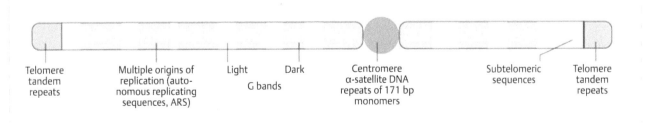

Fig. 14.1 Basic features of a eukaryotic chromosome. (Source: Passarge E, ed. Color Atlas of Genetics. 4th Edition. New York, NY. Thieme; 2012.)

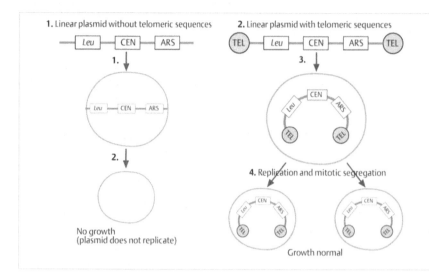

Fig. 14.2 Requirement for telomeric sequences. (Source: Passarge E, ed. Color Atlas of Genetics. 4th Edition. New York, NY. Thieme; 2012.)

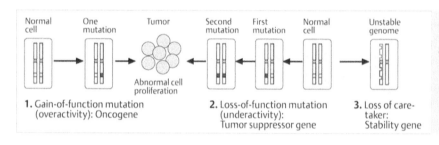

Fig. 14.3 Three categories of cancer genes. (Source: Passarge E, ed. Color Atlas of Genetics. 4th Edition. New York, NY. Thieme; 2012.)

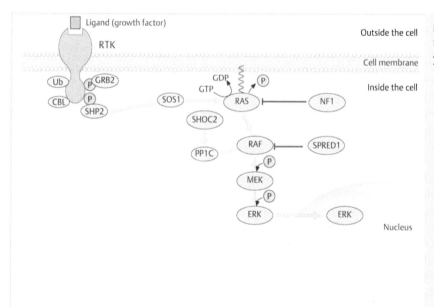

Fig. 14.4 RAS mutations, the RAS-MAPK signaling cascade. (Source: Passarge E, ed. Color Atlas of Genetics. 4th Edition. New York, NY. Thieme; 2012.)

14.3 Inactivation of Antiproliferative Signals

In addition to the induction of a persistent proliferative signal, cancer cells must also inactivate inherent anti-proliferative signals. These anti-proliferative signals are collectively known as tumor suppressors. Therefore, cells must mutate tumor suppressors in order to eliminate their normal function, allowing the cells to progress to a cancerous state. Many tumor suppressor genes encode proteins whose normal function is to control progression through the cell cycle. Some examples of tumor suppressors that are commonly mutated in cancers include p53, Rb, and the Cdk inhibitors p21 and p16^{Ink4a}.

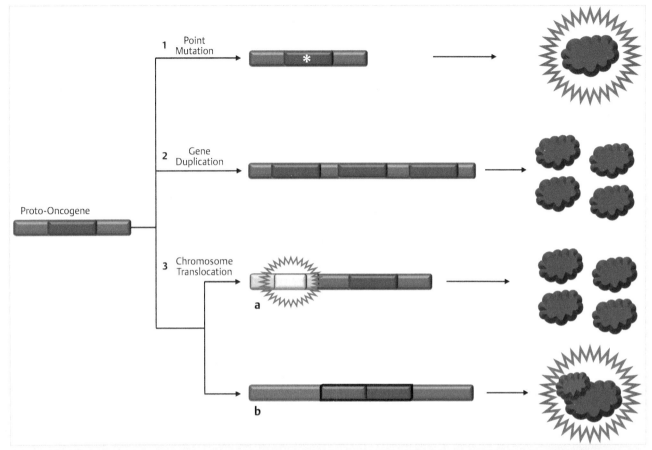

Fig. 14.5 Mechanisms that transform a proto-oncogene into an oncogene. (1) A point mutation in the coding region can cause production of hyperactive protein. (2) Gene duplication can result in over production of the protein. (3) A chromosomal translocation can occur. (3a) If the gene is moved to a locus with high transcriptional activity, then an abnormal amount of the normal protein will be synthesized. (3b) If the chromosomal translocation creates a fusion gene, this can result in synthesis of a hybrid protein that cannot be regulated properly.

p53 is a master sensor of cellular stress and is normally found at very low levels in the cell. This low level of expression is attributed to a tight post-translational regulation involving the ubiquitination and subsequent degradation of p53 by Mdm2, a type of E3 ligase enzyme (▶ Fig. 14.6). Under normal physiological conditions, p53 is continuously made and quickly degraded by this system. When the cell encounters stresses such as DNA damage, hyperproliferative signals, or hypoxia, Mdm2 is inactivated. This allows p53 to accumulate and become active. Active p53 then functions as a transcription factor to induce the expression of genes necessary for cell cycle arrest, senescence, or apoptosis.

As discussed in the cell cycle section of this chapter, Rb is responsible for holding a cell in a G1 phase state until a mitogenic signal is encountered, and is thus categorized as a tumor suppressor. If mutations occur in both alleles of the Rb gene, as is seen in several different types of cancer, there is no longer a requirement for growth factor signaling, and the cell will continue to divide even in the absence of such signals.

Also recall from the cell cycle chapter that p21 is an important Cdk inhibitor that is induced by several different cellular stresses, including DNA damage. If a cell undergoes mutations that eliminate both alleles of the p21 gene, it can no longer arrest in the presence of DNA damage. This could allow for the accumulation of mutations due to unrepaired DNA damage and lead to uncontrolled cell division. Therefore, p21 is another important tumor suppressor.

Genetically, tumor suppressor genes are recessive, meaning that both alleles have to be affected in order to have an effect on the cell. Because the normal function of these genes has to be disrupted, mutations in tumor suppressors are known as "loss-of-function" mutations (▶ Fig. 14.7). In addition, since both alleles of the tumor suppressor gene must be inactivated, the loss of tumor suppressor function is said to follow a "two hit" model. In other words, since a mutation event that inactivates one allele of a tumor suppressor would leave the second allele unaffected and functional, a second mutation in the other allele must occur for the complete functional elimination of the tumor suppressor gene.

The most common first mutation, or first hit, in tumor suppressor genes is a point mutation. These mutations occur in the coding region of the gene and can have one of several possible effects on the way the messenger RNA that is produced is read during translation. If the mutation changes a nucleotide in the sequence but does not change the codon, it would be a silent mutation and would have no effect on the protein that is subsequently made. On the other hand, if the mutation generates a new stop codon, it would be considered a nonsense mutation and would produce a nonfunctional, shortened version of the

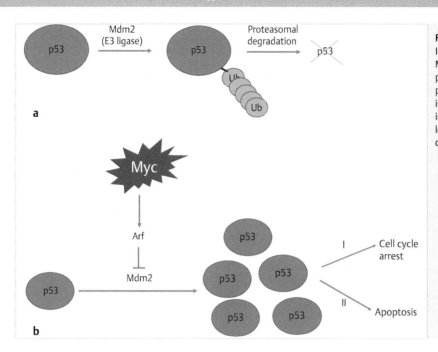

Fig. 14.6 Regulation of p53 levels by Mdm2. **(a)** In the absence of oncogenic Myc, the E3 ligase Mdm2 ubiquitinates the p53. The ubiquitinated p53 is degraded by the proteasomes. **(b)** In the presence of oncogenic Myc, the E3 ligase Mdm2 is inhibited by Arf tumor suppressor protein. The inhibition of the Mdm2 results in increased p53 levels. The high p53 levels can either cause cell cycle arrest (I) or induce apoptosis (II).

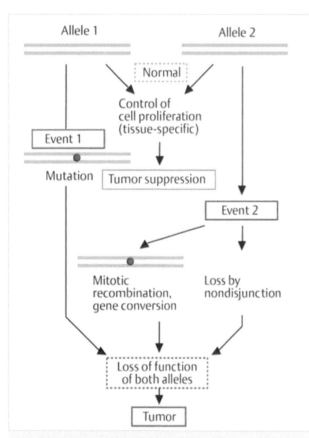

Fig. 14.7 The two hit model of tumor suppressor genes. (Source: Passarge E, ed. Color Atlas of Genetics. 4th Edition. New York, NY. Thieme; 2012.)

protein. A mutation that changes a codon to encode a different amino acid than the original codon, known as a missense mutation, could produce a protein that lacks the normal function. Also, if the mutation is an insertion or a deletion of a nucleotide, then a frameshift mutation is present that would change the way the messenger RNA is read during translation and result in a protein with a completely different amino acid sequence and, likely, a completely different function.

When one allele of a tumor suppressor gene has undergone a mutation (first hit), the alleles are said to be in a heterozygous state, meaning that there is one normal allele and one mutant allele. Mutation in the other allele, or "second hit", would therefore represent a loss of heterozygosity (LOH). Thus, LOH is a process that must occur in order for a cell to lose the normal function of a tumor suppressor gene and advance to a cancerous state. There are several possible ways in which LOH can occur (▶ Fig. 14.7). One is through a point mutation in the second allele that generates a nonfunctional protein. Another is through loss of either the entire chromosome containing the second allele, or through deletion of the region of the chromosome containing the allele. And finally, LOH can occur by promoter methylation on the second allele. In this case, although the allele itself is not mutated, the methylation silences the gene by inactivating the promoter and preventing the expression of the allele.

The process of promoter methylation requires the activity of an enzyme known as a DNA methylase. These enzymes are capable of methylating cytosine bases in the promoters of genes when the cytosine is located directly 5′ to a guanosine. Thus, DNA methylase activity results in "CpG methylation". Once the cytosines are methylated, they recruit proteins known as methylcytosine binding proteins that bind to the promoter and prevent the association of transcription factors that would

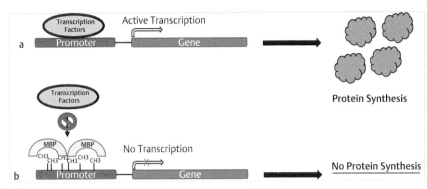

Fig. 14.8 Promoter methylation of tumor suppressor genes. **(a)** Transcription factors bind to the promoter region of genes and induce the expression of the gene. **(b)** Methylated cytosines (CpG repeats) on the promoter region recruits methylcytosine binding proteins (MBP). Binding of MBP prevents transcription factors from associating with the promoter. This silences (turns off) the gene expression. Cancer cells commonly turn off the expression of tumor suppressor genes by promoter methylation.

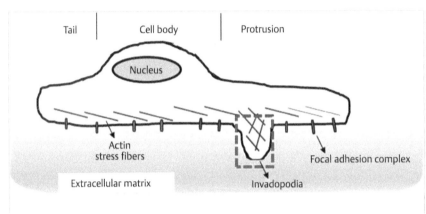

Fig. 14.9 Process of metastasis. Cancer cell invasion through the extracellular matrix starts with down-regulation of the cell adhesion molecules such as cadherin. Next cancer cells synthesize and secrete collagenases such as membrane type I matrix metalloproteinase (MT1-MMP), which cleaves the components of the ECM and allows cancer cells to create membrane protrusions and invadopodia. After invadopodia formation, cancer cells can migrate to a blood vessel, intravasate, and travel to a secondary site in the body. (Source: Brusilovskaya K, Königshofer P, Schwabl P, et al. Vascular Targets for the Treatment of Portal Hypertension. Seminars in Liver Disease. 2019; 39(04): 483-501.)

normally induce the expression of the gene (▶ Fig. 14.8). This results in the silencing of the gene in the absence of any mutations within the promoter or coding region of the gene. In cancer cells, promoter methylation is a common method for turning off tumor suppressor genes.

14.4 Resistance to Cell Death

Another hallmark of a cancer cell is its ability to bypass programmed cell death, also known as apoptosis. In the previous chapter, the process of apoptosis, as well as the molecules involved, were introduced. Under normal conditions, the persistent proliferative signal caused by oncogenic mutations would induce an apoptotic response in the cell. Therefore, loss-of-function of the pro-apoptotic molecules, or gain-of-function of anti-apoptotic molecules is necessary to allow a cancer cell to continue to proliferate. One of the important mediators of apoptosis is the p53 protein. There are several ways in which p53 activity evokes an apoptotic response in the cell. First, p53 can induce the expression of the Fas receptor, allowing for the activation of the extrinsic apoptotic pathway. It can also induce the intrinsic pathway by upregulating the expression of the pro-apoptotic Bax, or by increasing the expression of the IGF-binding protein-3. Following its synthesis, IGFBP-3 is released

into the extracellular space, where it sequesters IGF-1 and IGF-2 and prevents them from binding to survival receptors. This ultimately leads to activation of Bax inside the cell and subsequent apoptosis.

14.5 Angiogenesis

Angiogenesis, the process by which new blood vessels are formed, is an important step in the evolution of many types of solid tumors. As these tumors grow, the cancer cells in the center of the tumor mass are oxygen deprived and exposed to low nutrient levels, high levels of carbon dioxide, and metabolic wastes. In order to continue growth, the tumor has to find a source of oxygen and nutrients, as well as a way to eliminate the carbon dioxide and metabolic waste buildup. As a result, the tumor initiates an "angiogenic switch," due to the acquired ability to synthesize angiogenic factors such as the vascular endothelial growth factor (VEGF). Once expressed, the VEGF induces the growth of new blood vessels, which infiltrate the tumor and provide the nourishment that they need to continue growing, as well as eliminate the toxic byproducts of metabolism (▶ Fig. 14.9). In addition, these new blood vessels also create a path for the cancer cells to metastasize to other sites in the body by traveling through the blood stream.

14.6 Metastasis

The spreading of cancer cells throughout the body from the initial site of the tumor is known as metastasis and generally occurs through interactions with either blood vessels or the lymphatic system. The process of metastasis can be separated into two general phases: invasion of the surrounding tissue, and vascular/lymph dissemination to distant sites. In order for the cancer cell to invade surrounding tissue, it must first escape its local environment (▶ Fig. 14.9). One of the earliest events is the down-regulation of cell adhesion molecules such as cadherin, which allows the cells to release their tight association with neighboring cells. Next, the cells begin to synthesize and secrete enzymes, such as collagenases, that are capable of cleaving the components of the ECM. This creates an unobstructed route of escape into the underlying tissue, and also produces novel binding sites for cell adhesion molecules on the surface of the cell that stimulates migration. At this point, the cell is able to migrate to a blood vessel, intravasate, and travel to a secondary site in the body.

14.7 Causes of Cancer

Approximately 90% of all cancers can be attributed to environmental exposures, diet and lifestyle factors, or infectious agents. The remaining 10% are due to known inherited gene mutations. These hereditary cancer syndromes are discussed in more detail in chapter 56.

14.8 Tests for Mutagenic Potential of Compounds

Due to the substantial influence that environmental exposures and diet/lifestyle factors have on the manifestation of cancer, several tests have been developed to test compounds for their mutagenic potential. One of the most commonly used tests is the Ames test. In this test, a compound is first "activated" by liver homogenate. This step is meant to replicate the normal metabolism of the compound in the body. The activated compound is then added to a bacterial strain that requires histidine for its growth. If the bacteria are able to grow in the absence of histidine, this indicates a random mutation has occurred in the gene

responsible for histidine independence and that compound is therefore a mutagen. To date, this method has been used to identify the mutagenic properties of 1000's of organic compounds.

Another test for mutagenic potential is the rodent carcinogenicity test. In these tests, mice or rats are used as model systems for testing the carcinogenicity of specific chemicals or compounds. The rodents are exposed to very high doses of the test compound for extended periods of time and monitored for development of tumors. There are a few caveats to these tests that must be taken into consideration when analyzing the results. First, the doses used are often thousands of times higher than what a human would be exposed to under normal conditions. Also, the delivery method may not correspond to the way in which the exposure would likely occur.

14.9 Infectious Agents

Several viruses have been implicated in promoting the formation of tumors in humans. For example, the hepatitis virus can induce hepatocarcinoma, and infections with the human papilloma virus (HPV) are known to cause cervical cancer. In the case of HPV, the molecular mechanisms leading to carcinogenesis are well known (▶ Fig. 14.10). Two proteins encoded by the viral genome are capable of turning off two important tumor suppressors inside the infected cell. The E6 viral protein binds to and inhibits p53, while the E7 viral protein binds to and inhibits Rb. Therefore, two key tumor suppressor pathways are eliminated by the virus, leading to the development of cancer.

14.10 Tumor Growth Rate

It has been estimated that the smallest clinically detectable tumor mass contains ~ 10^9 cells and weighs ~1 gram. In order for a single cancer cell to produce a tumor of this size, it must go through roughly 30 population doublings. Only 10 additional doublings would produce a tumor of 1 kilogram in weight, the maximum size compatible with life. Depending on the type of cancer, it can take anywhere from several months to many years for the tumor to reach this detectable size. This tumor growth rate is largely determined by the fraction of cells within the tumor that are actively replicating. Since many chemotherapeutic agents target components of the cell cycle, tumors with a

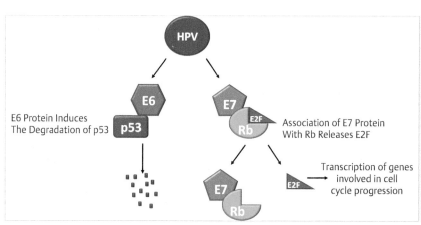

Fig. 14.10 Mechanism of HPV in promoting cancer development. Inside the host cell, two proteins, E6 and E7, are produced by the virus. E6 associates with p53 and causes its degradation, while E7 binds to Rb and prevents it from inhibiting E2F. Liberated E2F. Results in transcription of genes that cause over proliferation of the cells.

large percentage of dividing cells are more sensitive to chemo-therapeutics than tumors with a slow growth rate.

14.11 Multistep Carcinogenesis

Cancer results from the accumulation of multiple mutations over time. For some cancers, such as sporadic colorectal cancer, a specific sequence of mutational events ultimately leads to a malignant tumor. The first mutation that arises in this disease occurs in one allele of the *APC* gene, encoding the adenomatous polyposis coli protein. Next, LOH occurs on the other allele of this gene, resulting in complete loss of APC protein function in the cells. Later mutations involve the conversion of the KRAS gene to an oncogene, loss of p53 function, and increased activity of telomerase.

14.12 Histologic Changes Leading to Cancer

Distinct histological changes can be an early indication of a developing cancer. One of the first changes that the tissue may undergo is an accumulation of excessive numbers of morpho-logically normal cells within the tissue. This condition is referred to as hyperplasia. Further progression may lead to dysplasia, a disordered growth resulting in disordered architec-ture of the tissue. Anaplasia is a more severe histological abnor-mality and is seen in later stages of progression. In anaplastic tissue, both the tissue and cellular architecture lacks the differ-entiated characteristics of an identifiable tissue-of-origin and are highly disorganized. Pleomorphism, or the variation in size and shape of both cells and nuclei of cells, is also seen. Atypical mitosis, including tripolar and multipolar mitotic spindles are abundant, and tumor giant cells are often present.

Tumors can be grouped into one of two categories based on their general characteristics. Benign tumors are an abnormal mass of cells that lacks the ability to invade neighboring tissue or metastasize. The cells within this mass, however, are not abnormal in their morphology and resemble their parent tis-sue. Benign tumors are often surrounded by a fibrous capsule that separates the tumor from the surrounding normal tissue and is easier to excise surgically. Due to their relatively less aggressive characteristics, benign tumors are less likely to be life threatening. Malignant tumors, on the other hand, contain cells that have diverged significantly from the parent tissue in morphology. These tumors readily invade surrounding tissues and are capable of producing metastases, may recur after attempted removal, and are likely to cause death unless adequately treated.

Review Questions

1. What factor is most likely responsible for the motility of a metastatic cancer cell?
   A) Diminished apoptosis of tumor cells
   B) Gain of function mutation in p53
   C) Increased matrix metalloproteinase activity
   D) Loss of heterozygosity in the Ras gene
   E) Tight binding of keratin to hemidesmosomes

2. During a human papilloma virus (HPV) infection, the E6 protein that is produced by the virus directly inhibits which of the following cellular proteins?
   A) p21
   B) p53
   C) Ras
   D) Rb

3. A mutation in the gene encoding which of the following proteins is the most common first mutation in sporadic colorectal adenocarcinoma?
   A) APC
   B) BRAF
   C) BCR-ABL
   D) HER2

4. Increased expression of which of the following molecules is responsible for inducing cellular senescence?
   A) Cyclin D
   B) p21
   C) Ras
   D) Telomerase

Answers

1. **The correct answer is C.** The process through which a cancer cell leaves its local environment and travels to a distant site is known as metastasis. In order for this to occur, the cell must remove any extracellular matrix (ECM) that is encapsulating it in the tissue of origin. There are several families of enzymes that have the ability to degrade these ECM components, including the matrix metalloproteinases. Each member of this family can degrade a specific type of ECM substance. Once the ECM is removed, the cell can migrate to a blood vessel and take up residence in another part of the body. Diminished apoptosis (answer choice A) is a common characteristic of cancer cells and is due to either inactivating mutations in pro-apoptotic genes or activating mutations in anti-apoptotic genes. Since p53 is a tumor suppressor and normally functions to prevent a cancer cell from forming, a gain of function mutation in p53 (answer choice B) would not be expected in a cancer cell. Since Ras activity normally promotes cell growth and division, loss of heterozygosity in this gene (answer choice D) would not be expected in a cancer cell. As mentioned earlier, cancer cells must eliminate tight binding to the ECM through hemidesmosomes in order to escape their local environment and metastasize. Therefore, this type of tight binding (answer choice E) would not be a factor responsible for metastasis.

2. **The correct answer is B.** When a cell is infected with the human papilloma virus, two proteins are made in the cell from the viral genome that are capable of shutting down different tumor suppressor pathways. These proteins are E6 and E7. E6 binds to and inhibits p53, whereas E7 inhibits another major tumor suppressor, Rb (answer choice D). p21 (answer choice A) and Ras (answer choice C) are not bound to or directly inhibited by the viral E6 protein.

3. **The correct answer is A.** A mutation in the *APC* gene is the first event that occurs in many cases of sporadic colorectal adenocarcinoma, as well as some forms of inherited colorectal cancer such as familial adenomatous polyposis (FAP). These mutations allow for a persistent proliferative signal, although subsequent acquired mutations in other genes ultimately leads to the development of the cancer. BRAF (answer choice B) mutations are common in the sporadic form of melanoma, while HER2 (answer choice D) mutations are associated with sporadic breast cancer. The BCR-ABL (answer choice C) fusion protein is a product of the fusion between chromosome 9 and 22 that results in the creation of the Philadelphia chromosome. The unregulated activity of this protein is responsible for the development of chronic myeloid leukemia (CML).

4. **The correct answer is B.** As cells age, the levels of p21 accumulate. p21 inhibits the activity of the cyclin D-Cdk4/6 complex in G1 cells. As p21 levels increase, the cell becomes arrested in G1 and eventually exits the cell cycle irreversibly and enters a senescent state. In contrast, an increase in cyclin D (answer choice A) would promote cell division, not senescence. Since Ras (answer choice C) is a component of growth factor signaling, an increase in its activity would also promote cell division. Telomerase activity (answer choice D) directly correlates with the lifespan of a cell. Therefore, an increase in telomerase activity would not cause senescence, but would instead increase the life of the cell.

15 Molecular Principles of Cancer Treatments and Therapies

At the conclusion of this chapter, students should be able to:
- Describe the four major types of cancer therapies
- Identify the five ways in which surgery can be used in the treatment of cancer
- Describe the mechanism of action of the three families of chemotherapeutics
- Identify the proteins affected by different targeted therapies and the specific type of cancer treated by each
- Explain the advantages of using targeted therapeutics over chemotherapeutics
- Describe the three different types of radiation therapies

The number of available cancer treatments is ever growing. Currently, options include surgical therapy, chemotherapy, targeted therapy, radiation therapy, and cancer immunotherapy.

15.1 Surgical Therapy

There are several ways in which surgery can be used in the treatment of cancer. Surgery can be used to diagnose the cancer, such as when a biopsy is taken and analyzed to identify the specific type of cancer present. There are two main types of surgical biopsies: incisional biopsy and excisional biopsy. In an incisional biopsy, a piece of the suspicious area is removed for examination whereas in excisional biopsy, the entire suspicious area is removed.

Surgery can also be used for staging of the tumor and to determine if or where it has spread. In this case, lymph nodes near the cancer are removed and examined for the presence of cancer cells. This type of surgery is important for deciding what treatment is best and for predicting the patient's prognosis.

Curative surgery is the most common type of cancer surgery. It involves the complete removal of the tumor, as well as some of the surrounding normal tissue. This may be combined with chemotherapy or radiation therapy.

Mohs surgery is a precise surgical technique used to treat skin cancer. In this procedure, the cancer is removed one layer at a time. After each layer is removed, it is microscopically examined to determine if all of the cancerous cells have been excised. If not, another layer is removed, and the process continues until all of the cancer is gone. This process allows for the preservation of healthy tissue surrounding the cancer.

Debulking is a surgical method in which, although as much of the tumor is removed as possible, a portion of the tumor remains. This procedure is performed when the complete removal of a tumor is not possible or might cause excessive damage. Other treatments, such as radiation therapy or chemotherapy, can then be used to eliminate the remaining cancer cells.

In palliative surgery, side effects caused by the tumor are relieved. This can improve the quality of life for patients with advanced cancer or widespread disease. Examples include relief of pain or restoration of physical function if a tumor presses on a nerve or the spinal cord, blocks the bowel or intestines, or creates pressure or blockage elsewhere in the body. It can also be used to stop bleeding, such as suture ligation, where blood vessels are tied closed. In addition, palliative surgery includes procedures that help prevent broken bones.

15.2 Chemotherapy

Because cancer cells have a high rate of division, chemotherapeutic drugs are designed to target various aspects of the cell cycle. Many of them affect mitosis, but some target S phase processes, while others directly damage DNA. Since these drugs do not specifically recognize neoplastic cells, both normal and abnormal cells can be affected. However, the majority of cells in many tissues are non-cycling, so cancer cells in these tissues will be preferentially targeted due to their high rate of mitosis. Unfortunately, tissues with a high turnover, such as those of the GI tract, hair follicles, and germ cells, will also be severely affected by these drugs.

15.3 Mitotic Inhibitors

The mitotic inhibitors are a group of chemotherapeutics that freeze the cancer cells in mitosis. Examples of these types of chemotherapeutics include the vinca alkaloid and taxane families of drugs. Vinca alkaloids are chemicals originally derived from the periwinkle plant that were found to destabilize microtubules by binding to free tubulin dimers and preventing them from being incorporated into microtubule polymers, thus inhibiting miccrotubule growth. Vincristine and vinblastine are two vinca alkaloids that are commonly used as chemotherapy.

Taxanes are chemicals derived from plants of the Taxus genus. In contrast to the vinca alkaloids, taxanes are microtubule stabilizers. Their function is to bind to the cap on the plus end of the microtubule and prevent GTP to GDP conversion, inhibiting the shrinkage of the microtubule. Commonly used chemotherapeutics of the taxane family include paclitaxel and docetaxel. Whether it's destabilizing the microtubules or enhancing their stabilization, the end result of treating with vinca alkaloids and taxanes is the same: loss of microtubule dynamics and arrest in mitosis.

15.4 Antimetabolites

The antimetabolites are another family of chemotherapeutics. These chemicals inhibit S phase by targeting DNA replication. Examples of antimetabolites include methotrexate and fluorouracil (▶ Fig. 15.1) Methotrexate is an analogue of folic acid and competitively inhibits dihydrofolate reductase (DHFR). DHFR is an enzyme that converts dihydrofolate to tetrahydrofolate, a reaction that is essential for the synthesis of thymidine. Fluorouracil (5-FU) is a pyrimidine analogue that inhibits thymidylate synthase, an enzyme that methylates dUMP to form dTMP. This also blocks thymidine synthesis. As both of these drugs cause death of the cell due to loss of thymidine synthesis, their effect on the cells has been termed "thymineless death."

15.5 DNA Damaging Agents

There are several chemotherapeutic agents, including some antibiotics that have anti-tumor activity due to their ability to produce a large amount of DNA damage, primarily mediated by topoisomerase inhibition or DNA intercalation. Because many cancers lack a DNA damage checkpoint, the cells continue through the cell cycle even though they have significant DNA damage. Eventually, the cells are overwhelmed with an increasing amount of DNA damage and die. Examples of DNA damaging agents include doxorubicin, dactinomyocin, and alkylating agents such as streptozocin and busulfan (▶ Fig. 15.2). Since these agents cause DNA damage to normal cells as well, secondary cancers can arise. In fact, for many of these chemotherapeutic drugs there is an increased risk of leukemia years after the treatment.

15.6 Targeted Therapy

Targeted cancer therapy refers to drugs that prevent the progression of cancer by interfering with specific molecules that are involved with the growth, progression, or spread of the cancer. Because they target specific proteins that are known to be mutated in specific cancers, there are fewer side effects as compared to the aforementioned general chemotherapeutics. Most targeted therapies are cytostatic, not cytotoxic, meaning that they block tumor cell proliferation rather than directly induce death of the tumor cells.

Targeted therapeutics can be divided into two main groups: small molecule inhibitors and therapeutic monoclonal antibodies. Small molecule inhibitors are small compounds that bind to and prevent the activity of specific target proteins within the cell. Therapeutic monoclonal antibodies are antibodies that have been modified to effectively bind and inhibit the activity of cancer promoting molecules. Due to the natural characteristics of antibodies, these tend to be more specific for their target than small molecules.

15.7 Small Molecule Inhibitors

Many small molecule inhibitors have been designed to target the Ras-MAPK pathway since this pathway is one of the most commonly mutated in human cancers (▶ Fig. 15.3). Inhibitors of the receptor tyrosine kinases in these pathways

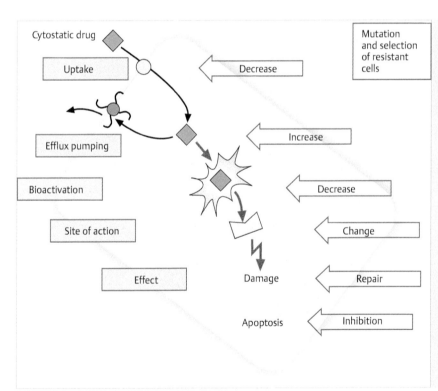

Fig. 15.1 Anti-metabolites used for chemotherapy. (Source: Koolman J, Röhm K, ed. Color Atlas of Biochemistry. 3rd Edition. New York, NY. Thieme; 2012.)

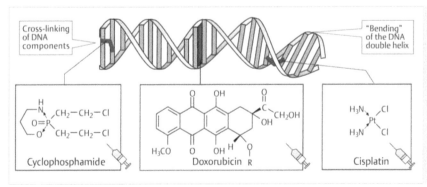

Fig. 15.2 Alkylating agents, anthracyclines. (Source: Koolman J, Röhm K, ed. Color Atlas of Biochemistry. 3rd Edition. New York, NY. Thieme; 2012.)

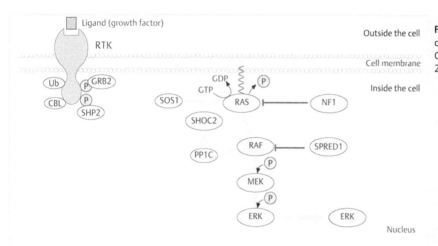

Fig. 15.3 RAS mutations, RAS-MAPK signaling cascade. (Source: Passarge E, ed. Color Atlas of Genetics. 4th Edition. New York, NY. Thieme; 2012.)

have proven to be effective in treating several different types of cancers.

Another small molecule inhibitor has been developed to target the oncogenic Ras member of this pathway. This inhibitor acts by blocking the activity of the enzyme farnesyltransferase, preventing the farnesylation of Ras. Since the farnesylation of Ras is necessary for its localization to the plasma membrane, this inhibitor prevents Ras from getting to its site of activity within the cell. In addition to tyrosine kinase receptor and Ras inhibitors, small molecule inhibitors of Raf and MEK have also been developed. Because mutations in BRAF are common in many forms of melanoma, Raf inhibitors are commonly used to treat these types of cancers.

Other signaling pathways, such as the cytokine pathway and sonic hedgehog (SHH) pathway, can also be targeted with Jak kinase and smoothened inhibitors, respectively. In addition, small molecule inhibitor of cell cycle, such as CDK4/6 inhibitors and inhibitors of tyrosine kinase fusion proteins, such as BCR-ABL inhibitors, have also proven to be effective in treating some forms of cancer.

15.8 Therapeutic Monoclonal Antibodies

Several monoclonal antibodies have been recently developed that target molecules known to be inappropriately activated in certain cancers. For example, antibody inhibitors of the mutated HER2 receptor have been developed to treat HER2 positive breast cancer. Similarly, several other types of cancers may be treated with monoclonal antibodies that bind to and inhibit EGF receptor dimerization, thus preventing the proliferative signal that this would normal induce in the cell.

In addition to targeting the receptor, some antibody inhibitors can also sequester the extracellular signaling ligand itself, preventing its association and, thus, activation of the mitogenic receptor. For example, therapeutic monoclonal antibodies are available that bind to VEGF outside the cell and block its activation of the VEGF receptor. These agents have proven to be particularly effective in treating certain forms of metastatic cancer. Antibodies that induce cell death can also be used to treat some types of cancers. One such therapeutic binds to CD20 on the surface of B cells and mediates their destruction by natural killer (NK) cells. These drugs have been used successfully to treat some types of B-cell non-Hodgkin lymphoma.

15.9 Radiation Therapy

Some forms of cancer are not effectively treated with small molecule inhibitors or therapeutic antibodies, but do respond to another type of treatment: radiation therapy. There are many ways in which a patient can be treated with radiation including systemic radiation therapy, internal radiation therapy, and machine radiation therapy. Systemic radiation therapy involves the administration of a radioactive isotope either by IV or orally. Thyroid cancer is commonly treated in this way. Internal radiation therapy, such as brachytherapy, uses radioactive seeds that are placed in or near the tumor, giving a high radiation dose to the tumor while reducing the radiation exposure in the surrounding healthy tissues. In machine radiation therapy, also known as external radiation therapy, the radiation is delivered by a machine outside of the body. This usually occurs in the form of high energy rays that are directed into the tumor. An advantage to this method is that the patient's body does not become radioactive. There are two common types of machine radiation therapy: gamma knife and proton therapy.

Gamma knife is typically used for individuals with small to medium size brain tumors. In this procedure, several gamma rays are focused on the tumor at the same time, creating a very intense dose of radiation that produces DNA damage that kills the cancer cells. Cyber knife is another technology that works in a similar way, but uses lower doses of gamma rays and involves multiple treatments, whereas gamma knife is usually a single treatment.

Proton therapy is a type of particle beam therapy that uses a beam of protons, precisely focused on the tumor, to induce

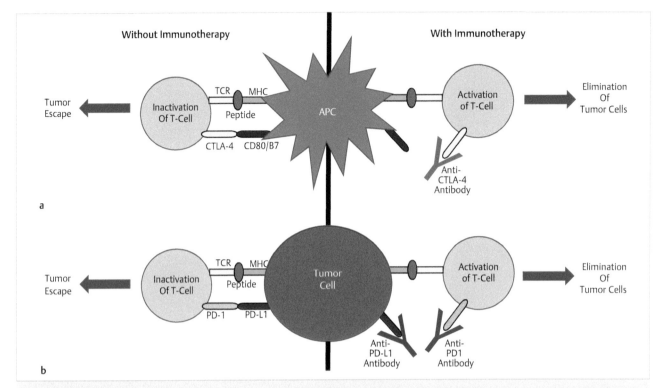

Fig. 15.4 Immune checkpoint inhibitors. **(a)** The checkpoint proteins CD80/B7 on antigen-presenting cells (APC) and CTLA-4 on T cells help to regulate the activation of the immune system. When the T-cell receptor (TCR) binds to antigen and major histocompatibility complex (MHC) proteins on the APC, the T cell can be activated. However, the binding of CD80/B7 to CTLA-4 keeps the T cells in the inactive state (left side of figure). Blocking the binding of CD80/B7 to CTLA-4 with an immune checkpoint inhibitor (anti-CTLA-4 antibody) allows the T cells to be active and to kill tumor cells (right side of figure). **(b)** The checkpoint proteins PD-L1 on tumor cells and PD-1 on T cells also help regulate the immune system's ability to recognize cancer cells. The binding of PD-L1 to PD-1 keeps T cells from killing tumor cells (left side of figure). Blocking the binding of PD-L1 to PD-1 with an immune checkpoint inhibitor (anti-PD-L1 or anti-PD-1) allows the T cells to recognize and kill tumor cells (right side of figure).

DNA damage in the cancer cells. This method is used to treat cancers in which surrounding tissue needs to be preserved, such as prostate cancer and some pediatric brain cancers.

15.10 Cancer Immunotherapy

Immunotherapy is a type of cancer treatment that uses the body's own immunological defenses to eliminate the cancer. This class of drugs is also known as the "checkpoint inhibitors." These drugs target molecules that normally work to suppress immune responses and limit autoimmunity. One family of checkpoint inhibitors affects the activation of T-cells (▶ Fig. 15.4a). When CTLA-4 molecules on the surface of T-cells interacts with CD80

molecules on the surface of the antigen presenting cell, a signal is induced inside the T-cell that inhibits the activation of that cell. Some checkpoint inhibitors bind to the CTLA-4 molecules, thus inhibiting the inhibitory effects of this protein and allowing the activation of T-cells (▶ Fig. 15.4b). This can lead to an increase in the number of T-cells that recognize cancer cells.

Another family of checkpoint inhibitors target the effector phase of the immune response. It is known that some tumor cells can express a molecule on their surface known as PD-L1, which binds to PD-1 on the surface of T-cells and prevents the T-cell from killing the tumor cell. These checkpoint inhibitors bind to PD-1, preventing its interaction with the PD-L1 on the tumor cell, allowing for the T-cell to kill the tumor cell.

Review Questions

1. The drug methotrexate has which of the following effects on cells?
   A) Destabilization of microtubules
   B) DNA damage
   C) Inhibition of thymidine synthesis
   D) Stabilization of microtubules

2. Therapeutic monoclonal antibodies that bind to the CD20 antigen on the surface of B-cells are capable of killing the cell through which mechanism?
   A) Direct DNA damage
   B) Microtubule instability
   C) Natural killer cell activity
   D) Loss of MAP kinase signaling

3. Which of the following drugs is capable of arresting cells in the mitotic phase of the cell cycle?
   A) 5-Fluorouracil
   B) Dactinomycin
   C) Doxorubicin
   D) Taxanes

4. Which of the following is a target of checkpoint inhibitors for cancer immunotherapy?
   A) CD20
   B) HER2
   C) PD-1
   D) VE

Answers

1. **The correct answer is C.** Methotrexate is an enzyme of folic acid that inhibits the enzyme dihydrofolate reductase, which is responsible for converting dihydrofolate to tetrahydrofolate, an essential step in thymidine synthesis. Destabilization of microtubules (answer choice A) is caused by vinca alkaloids, while stabilization of microtubules (answer choice D) is the mechanism of action of taxanes. DNA damage (answer choice B) can be induced by several chemical agents, including doxorubicin and dactinomycin.

2. **The correct answer is C.** Binding of monoclonal antibodies to the CD20 antigen on the surface of B-cells causes the recruitment of natural killer cells, which induces death of the B-cell. These therapeutic monoclonal antibodies are effective in treating some forms of B-cell lymphoma. Direct DNA damage (answer choice A) would be induced by doxorubicin and dactinomycin. The instability of microtubules (answer choice B) is increased by a class of drugs known as vinca alkaloids, whereas loss of MAP kinase signaling (answer choice D) would occur following treatment with an inhibitor of an upstream component of the RAS-MAP kinase pathway, such as an inhibitor of RAS, RAF, or MEK.

3. **The correct answer is D.** The taxanes are a family of drugs whose mechanism of action is to bind to the cap on the plus end of microtubules and prevent the hydrolysis of GTP to GDP. This stabilizes the microtubules and prevents the dynamic instability that is necessary for the transition of the cell through mitosis. This results in a mitotic arrest of the cells. Dactinomycin (answer choice B) and doxorubicin (answer choice C) are DNA damaging agents. 5-fluorouracil is a pyrimidine analogue that inhibits thymidylate synthase, preventing the cell from synthesizing thymidine for DNA replication processes.

4. **The correct answer is C.** PD-1 is a cell surface molecule embedded in the plasma membrane of T-cells. When PD-1 comes in contact with PD-L1 molecules on the surface of tumor cells, this interaction prevents the T-cell from killing the tumor cell. Monoclonal antibodies that bind to PD-1 prevent this interaction and, thus, allow the T-cell to kill the tumor cell. These therapeutic monoclonal antibodies belong to a drug class known as the checkpoint inhibitors. CD20 (answer choice A) is found on the surface of B-cells and is a target of therapeutic monoclonal antibodies that recruit natural killer cells. HER2 (answer choice B) is a receptor tyrosine kinase belonging to the epidermal growth factor receptor (EGFR) family and is often mutated in sporadic breast cancer. The extracellular signaling molecule VEGF (answer choice D) is expressed during the angiogenic switch phase of tumorigenesis, and induces the formation of new blood vessels.

Review Questions

1. A 9-year-old boy with chronic sinusitis is brought to his pediatrician with severe nasal congestion. Crackles and wheezing are heard upon chest auscultation. An endoscopic exam shows excessive mucus in the lungs and abnormal, non-motile cilia of the respiratory epithelial cells. The family of molecules most likely disrupted in this child has what normal cellular function?
   A) They move in an ATP-dependent manner toward the plus end of microtubules.
   B) They are responsible for movement of intracellular cargo along microtubules toward the centrosome.
   C) They walk along thin filaments in the sarcomere of muscle cells and allow for muscle contraction.
   D) They move intracellular cargo along actin filaments toward the plus end.
   E) They are responsible for providing mechanical strength to the cell.

2. When cultured human cells are exposed to a newly discovered compound, disruption of the cytoskeleton is observed. Upon examination of the treated cells under an electron microscope, it is observed that the organization of the long, hollow tube structures composed of heterodimer subunits is disrupted. What is the most likely explanation for the disruption in the cytoskeletal organization?
   A) The ATPase activity of the heterodimer subunits has been inhibited.
   B) The pericentriolar material surrounding the centrioles has been destroyed.
   C) The growth of actin filaments at the plus end has been affected.
   D) Hydrolysis of GTP has been inhibited.

3. A 22-year-old female with chronic fatigue and pallor visits her primary care physician. Numerous sphere-shaped erythrocytes and erythrocyte fragments are found in a sample of her peripheral blood. A mutation in the gene encoding what protein is most likely responsible for her erythrocyte abnormality?
   A) Alpha tubulin
   B) Troponin
   C) Myosin
   D) Ankyrin
   E) Lamin

4. A researcher is studying vesicle transport in cultured enterocytes. Upon examination of these cells with an electron microscope, it is noted that the endoplasmic reticulum contains many nascent transport vesicles that have not fully released from the organelle's membrane. Biochemical analysis reveals a defect in one of the molecules required for the proper assembly of the coat structure associated with these vesicles. What molecule is most likely defective in these cells?
   A) Sar1
   B) Dynamin
   C) Clathrin
   D) Arf
   E) KDEL receptor

5. The insulin sensitivity of muscle cells in culture is measured following exposure to insulin. It is noted that this treatment results in an increase in facilitated transport of glucose at the plasma membrane. The glucose transporters at the plasma membrane most likely originated from which of the following membrane-bound intracellular structures?
   A) Autophagosome
   B) Endolysosome
   C) Lysosome
   D) Recycling endosome

6. A 56-year-old female with atrial fibrillation is prescribed a Na+/K+ ATPase inhibitor. What is the most likely effect that this drug will have onher cardiomyocytes?
   A) Increased intracellular K+
   B) Increased intracellular Na+
   C) Decreased extracellular K+
   D) Decreased intracellular Na+
   E) No effect on intracellular K+ or Na+ levels

7. A3-year-old boy is brought to his pediatrician with parental concerns about chronic bruising. Upon examination, it is found that he has extremely pliable skin and hyperextensible joints. Biochemical tests reveal a deficiency of lysyl hydroxylase. The extracellular matrix component most likely affected in this child has which of the following characteristics?
   A) They are composed of ?,?, and ? subunits.
   B) They are glycosylated prior to secretion.
   C) They are made of repeating disaccharide units.
   D) They are defective in Marfan Syndrome.
   E) They are covalently linked to hyaluronan.

8. A 36-year-old male presents with shortness of breath and wheezing. During consultation with his physician, it is revealed that he moved in with his girlfriend, who is a smoker, two months ago. A distended abdomen is discovered upon physical exam. Microscopic examination of a liver biopsy shows evidence of cirrhosis. What is the normal function of the molecule most likely defective in this patient?
   A) Increases elastin production
   B) Decreases neutrophil elastase activity
   C) Increases cleavage of elastin
   D) Increases ?-1 antitrypsin levels
   E) Decreases collagen synthesis

9. A3-day-old neonate is found to have numerous blisters on the trunk and extremities. A skin biopsy discloses separation of the basal layer of the epidermis from its basal lamina and an absence of inflammatory cells. Which of the following cellular structures is most likely disrupted?
   A) Adherens junctions
   B) Gap junctions
   C) Desmosomes
   D) Hemidesmosomes
   E) Tight junctions

10. An 8-month-old boy with recurrent skin infections is found to have a white blood cell differential with an abnormally high number of circulating neutrophils. Flow cytometry indicates the absence of an adhesion molecule on the surface of these neutrophils. The adhesion molecule most likely absent in this child normally interacts with which molecule on the surface of endothelial cells in the blood vessels?

A) Connects the actin cytoskeleton of adjacent cells

B) Associates with immunoglobulin-like cell adhesion molecules

C) Forms a connection between the lateral surface of two epithelial cells

D) Prevents the passage of molecules between cells

E) Interacts with molecules of the same subtype in the presence of Ca2+

Answers

1. **The correct answer is B**. The symptoms described for this child are consistent with Kartagener syndrome. This is a disease characterized by reduced or absent clearance of mucus from the lungs and chronic recurrent respiratory infections. It is caused by a mutation in the gene encoding ciliary dynein, a motor protein that allows for the movement of cilia on the surface of epithelial cells in the respiratory tract. The loss of ciliary movement prevents the clearance of mucus from the lungs, resulting in the crackles and wheezing, and from the sinuses, resulting in chronic sinus infections. The normal function of dynein is to walk along microtubules in a directed fashion toward the minus end of the microtubule. Since the minus end of the microtubules are located at the centrosome, then answer B correctly describes the normal function of dynein, the molecule most likely disrupted in this child.

Choice A describes the molecular motor protein kinesin, which moves toward the plus end of microtubules. Choices C and D describe myosin, a molecular motor that travels along actin filaments. Choice E describes the function of intermediate filaments.

2. **The correct answer is B**. The cytoskeletal component that is composed of heterodimer subunits and forms a hollow tube is the microtubules. Within the cell, microtubules are organized by a structure called the centrosome, which is composed of pericentriolar material. Since this compound is affecting the organization of the microtubules, it is likely causing destruction of the pericentriolar material and, therefore, eliminating the function of the centrosome. Inhibition of the ATPase activity of the heterodimer subunits (Choice A) would result in stabilization of actin filaments, but it would not cause disorganization of the microtubule cytoskeleton within the cell. Since actin filaments are not involved with microtubule organization, then an effect on the growth of actin filaments at the plus end (Choice C) would not be expected. Since the hydrolysis of GTP causes microtubule disassembly, an inhibition (Choice D) of this process would result in stabilization of microtubules. However, this would not affect the organization of the microtubule cytoskeleton.

3. **The correct answer is D**. In erythrocytes, a cytoskeleton structure made up of several proteins including actin, spectrin, and ankryrin lies just below the plasma membrane. This cytoskeleton is responsible for giving these cells their distinctive concave shape. If any of these components is disrupted, as in the genetic disease hereditary spherocytosis, the cells become sphere-shaped and are unable to traverse the narrow passages of the capillaries and the sinuses of the spleen. This results in the fragmentation of the erythrocytes, as seen in this patient's peripheral blood. Alpha tubulin (answer choice A) is a component of microtubules, which are not affected in this disease. Troponin (answer choice B) and myosin (answer choice C) are components of the sarcomere, the contractile unit of skeletal muscle. Lamins (answer choice E) are intermediate filaments that form a meshwork inside the nucleus of the cell. Since erythrocytes do not have a nucleus, lamins are not affected in hereditary spherocytosis.

4. **The answer is choice A**. Sar1 is a small GTP binding protein that is involved in the early stages of the formation of the COPII-coated vesicles that ultimately bud off from the endoplasmic reticulum and travel to the cis face of the golgi apparatus. Since vesicles are accumulating at the endoplasmic reticulum without releasing, these cells most likely have a defect in the Sar1 protein. Arf (answer choice D) is a component of the clathrin (answer choice C) coats. These coats assemble on endocytic vesicles that pinch off from the plasma membrane. Dynamin (answer choice B) is a GTPase that forms around the bud neck of clathrin-coated vesicles and uses the energy generated by GTP hydrolysis to pinch the membranes together, causing them to fuse and release the endocytic vesicle. KDEL receptors (answer choice E) are receptors that specifically bind to ER resident proteins and allow for the return of these proteins from the Golgi to the ER during retrograde transport.

5. **The correct answer is D**. GLUT4 is the passive glucose transporter responsible for the uptake of glucose in skeletal muscle and adipose tissue. In these cells, GLUT4 is stored in recycling endosomes found in the cytoplasm near the plasma membrane. Upon insulin binding to receptors on the surface of these cells, these transporters are shuttled to the plasma membrane, where they increase the uptake of glucose. The lysosome (answer choice C) is a membrane-bound organelle filled with hydrolytic enzymes that is the chief site of breakdown of cellular components. Endolysosomes (answer choice B) are formed by the fusion of a late endosome with a lysosome. The autophagosome (answer choice A) is a structure that forms around cellular components, such as organelles, that are old or damaged and causes the breakdown of their individual components so that they can be recycled by the cell.

6. **The correct answer is B**. The Na+/K+ ATPase is an ion transporter that maintains the internal concentrations of Na+ and K+ by using the energy generated from ATP hydrolysis to move three Na+ ions out of the cell while bringing two K+ ions into the cell. Inhibition of this pump would result in an increase in Na+ concentrations inside the cell. In cardiomyocytes, this would have the effect of enhancing action potentials and, therefore, increasing contraction of the heart. Inhibition of the Na+/K+ ATPase would not cause an increase in intracellular K+ (answer choice A) or decreased intracellular Na+ (answer choice D), nor would it decrease the extracellular K+ concentrations (answer choice C).

7. **The correct answer is B**. This child's symptoms are consistent with Ehlers-Danlos syndrome. Characteristics of Ehlers-Danlos syndrome include hyperextensible joints,

abnormally flexible skin, and blood vessels that rupture easily (causing bruising). This disease can be caused by mutations in one of the collagen genes, or in the gene in coding lysylhydoxylase, as seen in this patient. This enzyme adds a hydroxyl group to lysine residues in the procollagen molecule that are synthesized in the ER/Golgi of the cell. The hydroxylated lysines are subsequently glycosylated to produce a secreted protein. The extracellular matrix component that consists of ?,?, and ? subunits (answer choice A) is laminin. Proteoglycans are composed of repeating disaccharide units (answer choice C) and may be covalently linked to hyaluronan (answer choice E), but defects in these ECM components do not produce the symptoms seen in this patient. Marfan syndrome (answer choice D) is a genetic disease of elastic fibers caused by mutations in the fibrillin-1 gene and is characterized by an above average height with long, slender limbs, and an increased risk of aortic aneurysms and dissections.

8. **The correct answer is B**. The symptoms seen in this patient are consistent with ?-1 antitrypsin deficiency. This disease is caused by a mutation in the SERPINA1 gene that encodes the protein ?-1 antitrypsin, which is normally made in the hepatocytes of the liver. This mutation results in the production of a misfolded protein that accumulates in the ER of the cells and ultimately causes hepatocyte death, leading to liver damage (cirrhosis). The normal function of this protein is to inhibit the enzyme elastase. Elastase is secreted from neutrophils when inflammation is present in the lungs, which is common when the lungs are exposed to smoke. The elastase enzyme cleaves the protein elastin in the lungs, causing destruction of the lung tissue. Therefore, deficiency in ?-1 antitrypsin can lead to both lung and liver disease. An increase in elastin production (answer choice A), decrease in neutrophil elastase activity (answer choice B), and increase in ?-1 antitrypsin (answer choice D) would all have the opposite effect of what is seen in this patient, namely preservation of the elastic tissue of the lungs. A decrease in collagen synthesis (answer choice E) would not affect the liver.

9. **The correct answer is D**. This child displays the symptoms associated with the blistering skin disease bullous pemphigoid, an autoimmune disease. The immune system of individuals with this disease produce antibodies against components of the hemidesmosomes. The hemidesmosomes are cell adhesion complexes that connect the epithelial sheets of the skin with the underlying basal lamina. When these complexes are disrupted by the antibodies, separation of the skin from the basal lamina occurs, which allows for the accumulation of fluid in this space and produces blisters. Adherens junctions (answer choice A) are composed of cadherins and connect the lateral surfaces of adjacent epithelial cells to the actin cytoskeletons inside the cells. Gap junctions (answer choice B) are pores that are produced on the lateral surfaces of epithelial cells through the assembly of connexins. These pores allow the cells to share small cytoplasmic molecules, such as ions. Tight junctions (answer choice E) hold cells tightly together to form a seal between them and prevent the movement of molecules in between the cells. Dysfunction of desmosomes (answer choice C) is also caused by autoimmune antibodies in the blistering skin diseases pemphigus foliaceus and pemphigus vulgaris. However, since the desmosome functions to hold adjacent cells together in the epithelial sheet by connecting their intermediate filament cytoskeletons, separation between epithelial cells is seen, rather than separation of the epithelial sheet from the basal lamina.

10. **The correct answer is B**. The process of leukocyte extravasation can be broken down into three distinct steps. In the first step, glycan molecules on the surface of the leukocytes interact with selectins lining the endothelial cells of the blood vessel wall. This interaction is relatively weak, producing a rolling effect on the leukocyte. The next step is characterized by a firm adhesion between the leukocyte and the endothelial cell. This is caused by interactions between integrins on the surface of the leukocyte and immunoglobulin-like cell adhesion molecules (ICAMs) on the surface of the endothelial cells. The final step involves the migration of the leukocyte between the endothelial cells to the site of infection. This child exhibits the symptoms associated with the genetic disease leukocyte adhesion deficiency (LAD). This disease is caused by mutations in one of the integrin genes, resulting in a loss of the firm adhesion step of leukocyte extravasation. This leads to an absence of leukocyte homing to sites of infection, resulting in multiple bouts of infection, and an accumulation of leukocytes (neutrophils) in the blood. The other answer choices (A, C, D, and E) describe components of membrane junction complexes that are found on the lateral surfaces of epithelial cells and connect adjacent cells together to produce an epithelial sheet. Answer choices A and E describe adherens junctions. Choice D refers to tight junctions, and choice C is true for all membrane junctions except the hemidesmosomes.

Review Questions

1. A 32-year-old male presents with a 7cm lump on his right upper arm. This mass is found to harbor mutations in the gene encoding Wee1 that result in an inactive protein. This mutation is most likely to lead to high levels of activity of which of the following proteins?
   A) Cdk1
   B) Cdc25
   C) CyclinD
   D) Cyclin E

2. A 4-year-old child complains of vision in herleft eye. Ophthalmoscopic exam reveals a mass in the left eye that nearly fills the entire globe. Following enucleation of the eye, molecular analysis is performed which indicates the absence of both copies of a tumor suppressor gene that controls the transition from the G1 to the S phase of the cell cycle. A mutation in the gene encoding which of the following proteins is most likely responsible for the formation of this neoplasm?
   A) MAP kinase
   B) PI-3 kinase
   C) p53
   D) RAS
   E) Rb

3. A researcher is studying mitosis in cultured fibroblast cells. Treatment of these cells with an inhibitor of the anaphase promoting complex (APC) results in arrest of the cells in mitosis. This compound is most likely to affect what mitotic event?
   A) Breakdown of the nuclear envelope
   B) Condensation of chromosomes
   C) Formation of aster microtubules
   D) Formation of the cohesin ring
   E) Segregation of sister chromatids

4. Dermal fibroblasts that have been harvested from a skin biopsy specimen are transduced with the genes encoding for Sox2, Oct3/4, and Klf4. The resulting cells can be induced to differentiate into cells from the endoderm, mesoderm, and ectoderm germ layers. What type of cell was generated from the transduction of these genes?
   A) Multipotent cell
   B) Pluripotent cell
   C) Totipotent cell
   D) Transit amplifying cell
   E) Unipotent cell

5. Three months after treatment with a chemotherapeutic agent, a sample from a breast cancer metastasis is assessed for the presence of apoptotic cell death. Which of the following findings would most likely be true of the cells within this sample that are resistant to cell death?
   A) Bax and Bak have oligomerized
   B) Bcl-2 has been inactivated
   C) Bid cannot inhibit Bcl-X
   D) The Bim gene is being overexpressed
   E) There are very low levels of Mcl-1

6. A 27-year-old patient presents with intense pain in his upper leg following a fracture of his femur. Dead tissue due to blood vessel obstruction is found upon bone scan. What is a likely characteristic of the dying cells in this tissue?

   A) Cytochrome c is released from the mitochondria in a regulated manner
   B) Myelin figures are formed in the cytoplasm
   C) They do not elicit an inflammatory response
   D) They undergo cellular shrinkage

7. A large, irregularly-shaped mole from the upper arm of a 27-year-old female is analyzed for molecular defects in the Ras-MAPK pathway. It is found that there is a high level of serine and threonine phosphorylations on MEK, although Ras activity is not enhanced. Inhibition of which of the following molecules would most likely stop her tumor cells from growing?
   A) EGFR
   B) B-Raf
   C) HER2
   D) Ras

8. A 67-year-old female heavy smoker with metastatic lung cancer presents with a round face and excess fat deposits around her torso and neck. What is most likely true about the cells within this patient's metastasis?
   A) They are producing angiogenesis inhibitors
   B) They express low levels of matrix metalloproteinases
   C) They have a low level of cytosine methylation
   D) They have lost telomerase activity
   E) They have undergone loss of heterozygosity

9. A 43-year-old male with Philadelphia chromosome positive (Ph+) CML is treated with a small molecule drug that inhibits the mutated enzyme that is responsible for his disease. What effect will this drug have on his cancer cells?
   A) Decreased phosphorylations of tyrosines
   B) Loss of EGFR signaling
   C) Loss of Raf activity
   D) Reduction in myc activity

10. Cells isolated from a human malignant neoplasm are found to continuously replicate in culture, while similar cells isolated from normal tissue do not. A change in the activity of which of the following molecules is most likely responsible for the immortality of these cancer cells?
    A) Cdk2
    B) Cyclin D
    C) Telomerase
    D) VEGF
    E) Wee1

Answers

1. **The correct answer is A.** Just before mitosis, cyclin B-Cdk1 complexes accumulate in the cell. These complexes are activated through phosphorylation by Cdk Activating Kinase (CAK), but subsequently held in an inactive state through phosphorylation by Wee1 kinase. In order for the cell to enter mitosis, the Wee1 phosphorylation must be removed by the phosphatase Cdc25 (answer choice B). In this case, Wee1 is inactive. This means that the inhibitory phosphorylations do not occur and the cell is able to enter mitosis immediately in an unregulated manner, which is likely contributing to the growth of this patient's tumor. Cyclin D (answer choice C) activity is high in the G1 phase,

whereas cyclin E (answer choice D) activity occurs in early S phase. Inactivation of Wee1 would not have any effect on the activity of Cdc25, cyclin D, or cyclin E.

2. **The correct answer is E.** The tumor described here is consistent with retinoblastoma, a cancer that forms on the retina of the eye at an early age. Retinoblastoma arises from mutations in the gene that encodes Rb, a protein that normally inhibits the G1 to S phase transition in the absence of mitogenic stimulus. Mutation in this gene eliminates this inhibition and allows for the continued cell cycle stimulation that gives rise to this tumor. MAP kinase (answer choice A) and RAS (answer choice D) are components of the mitogenic signaling cascade upstream of Rb and can be mutated in some forms of cancer, but mutations in these genes are not generally associated with the development of retinoblastoma. PI-3 kinase (answer choice B) is a component of receptor tyrosine kinase signaling, such as the insulin signaling pathway. P53 (answer choice C) is a tumor suppressor that is mutated in many forms of cancer, but not generally associated with the development of retinoblastoma.

3. **The correct answer is E.** Before anaphase, the enzyme separase is held in an inactive state by a protein called securin. This prevents separase from cleaving the cohesin rings connecting sister chromatids. Once the cell enters anaphase, the anaphase promoting complex (APC) is activated . This complex ubiquitinates securin and targets it for degradation, releasing separase and allowing it to remove the cohesin connections between the sister chromatids. This results in the segregation of a sister chromatid into each of the daughter cells. The breakdown of the nuclear envelope (answer choice A) and formation of aster microtubules (answer choice C) occurs in prometaphase, whereas chromosomes condense (answer choice B) in prophase, and formation of the cohesion ring (answer choice D) occurs during S phase. Since all of these mitotic events occur before the APC is activated, none of them would be affected by the inhibition of APC.

4. **The correct answer is B.** Transduction of somatic cells, such as dermal fibroblasts, with the transcription factors Sox2, Oct3/4, and Klf4 causes the cells to become pluripotent stem cells. The term iPS (induced Pluripotent Stem) cell is used to describe this type of cell. Because these cells can be forced to differentiate into any adult cell type of the body under culture conditions, they are categorized as pluripotent stem cells. None of the other answer choices can be produced from the expression of the indicated genes in somatic cells. Multipotent cells (answer choice A) are stem cells that can differentiate into many different types of cells, but not all of the cell types of the body. Totipotent stem cells (answer choice C) are capable of giving rise to any cell type of the body, in addition to extraembryonic tissue (placenta). Transit amplifying cells (answer choice D) can differentiate into more than one type of cell, depending on the tissue it resides in, but they do not have the limitless replicative capacity that is characteristic of stem cells. Unipotent stem cells (answer choice E) can only give rise to one type of cell.

5. **The correct answer is C.** Bcl-X is a member of the anti-apoptotic protein family in the intrinsic apoptotic pathway. In the presence of an apoptotic signal, a member of the BH3-only pro-apoptotic family (Bim, Bid, or Bad) binds to and inhibits anti-apoptotic members like Bcl-X. This inhibition allows for the formation of Bax/Bak-containing pores in the outer membrane of the mitochondria and the subsequent release of cytochrome c. In the cytoplasm, cytochrome c forms a complex with Apaf-1 and caspase 9 to produce the apoptosome, which activates downstream enzymes necessary for programmed cell death. Therefore, if Bid cannot inhibit Bcl-X, the anti-apoptotic activity of Bcl-X will continue, even in the presence of a strong death signal, and the cell would be resistant to apoptosis. All of the other choices would result in an increase in cell death, rather than a resistance to cell death. Oligomerization of Bax and Bak (answer choice A) would cause release of cytochrome c from the mitochondria and, therefore, more apoptotic cell death. Anti-apoptotic proteins like Bcl-2 and Mcl-1 normally inhibit cell death. So an inactivation of Bcl-2 (answer choice B) or low levels of Mcl-1 (answer choice E) would enhance cell death, as would overexpression of the pro-apoptotic gene Bim (answer choice D).

6. **The correct answer is B.** The blood vessel obstruction, or ischemia, seen in this patient is preventing the delivery of oxygen and nutrients to this area of the bone. These elements are essential to the viability of the cells within this region and, therefore, necrotic cell death is occurring. This type of necrotic death in the bone is known as osteonecrosis. A cellular feature of cells undergoing necrosis is the formation of myelin figures, which are derived from damaged membrane lipids, in the cytoplasm of the cell. All of the other answer choices (A, C, and D) are characteristics of apoptotic cell death, which is not induced by ischemia.

7. **The correct answer is B.** This patient most likely has a melanoma, which is commonly associated with mutations in the gene encoding B-Raf. In the Ras-MAPK pathway, B-Raf lies downstream of Ras, but upstream of MEK. The fact that there is an increase in phosphorylation events on MEK, although Ras activity is not changed, suggests that this patient has mutations in B-Raf. Therefore, inhibition of B-Raf would have the greatest effect on the growth of this tumor. Since EGFR (answer choice A) and Ras (answer choice D) are upstream of the mutated component, B-Raf, inhibition of these molecules would not have an effect on this tumor. In addition, it is stated that Ras activity is not changed. HER2 (answer choice C) is a receptor tyrosine kinase that belongs to the EGFR family. HER2 is mutated in some forms of breast cancer, but it is not associated with melanoma.

8. **The correct answer is E.** In order for most cancers to become metastatic, they must eliminate the normal function of one or more tumor suppressor genes. Tumor suppressors are considered recessive genes, as they must undergo a loss-of-function on both alleles to have an effect on the cell. In many cases, cells lose the function of one allele of a tumor suppressor early on in the tumorigenic process. However, in order to move further down the path to tumorigenesis, the other allele must be eliminated. This loss of the second allele is known as loss of heterozygosity

(LOH). All of the other answer choices (A, B, C, D) would prevent metastasis, not promote it. Inhibitors of angiogenesis (answer choice A) would block the formation of new blood vessels, which is one of the events that allows for metastasis. Since expression of matrix metalloproteinases is essential for degrading the extracellular matrix around a cancer cell and allowing for its mobilization, low levels of these enzymes (answer choice B) would prevent metastasis. Cytosine methylation is one method by which cancer cells can induce a loss of heterozygosity in a tumor suppressor gene. Therefore, low levels of cytosine methylation (answer choice C) would not be expected in a metastatic cancer. The immortality necessary for the survival of a cancer cell is derived by a high level of telomerase activity. Low levels of telomerase activity (answer choice D) would therefore not be expected in this patient.

9. **The correct answer is A**. Philadelphia chromosome positive (Ph+) CML is caused by a chromosomal translocation that produces a fusion between the BCR gene and ABL gene, resulting in the synthesis of the fusion protein Bcr-Abl. Since Bcr-Abl is a tyrosine kinase, tyrosine kinase inhibitors, such as imatinib, are effective in treating CML. Inhibition of Bcr-Abl would cause a decrease in tyrosine phosphorylations. The activity of EGFR (answer choice B), Raf (answer choice C) and myc (answer choice D) are not affected by in CML.

10. **The correct answer is C**. The immortality of cancer cells is due to high levels of telomerase activity. Telomerase is an enzyme that is responsible for the replication of the ends of chromosomes. As cells age, telomerase activity is reduced, leading to shorter chromosomes. Eventually, the chromosomes reach a threshold length, and the cells stop dividing. Many cancer cells become mutated to allow for a high level of telomerase expression. This prevents the shortening of the chromosomes that is associated with cellular senescence and results in a cell that is immortal. Although increased levels of Cdk2 (answer choice A) or cyclin D (answer choice B) would promote cell division, they would not cause a cell to be immortal. Similarly, a change in Wee1 (answer choice E) activity could also affect cell division, but not the lifespan of the cell. VEGF (answer choice D) is expressed during the angiogenic switch and induces new blood vessel formation, which promotes tumor growth. However, this would not affect the mortality of the cell either.

Part III

16 Enzymes and Enzyme Kinetics

At the conclusion of this chapter, students should be able to:
- Describe the molecular basis of enzymes and enzyme catalysis including substrate-binding sites, active sites, regulatory sites, specificity, and catalytic efficiency
- Describe the role of coenzymes/cofactors in catalysis
- Discuss catalytic strategies that are commonly employed by enzymes
- Explain what factors affect reaction velocity
- Define and compare zero-order and first-order kinetics
- Describe enzyme kinetics based on the Michaelis-Menten equation and the meanings of the kinetic parameters K_m, V_{max} and k_{cat}.
- Differentiate among the various mechanisms of enzyme inhibition (competitive reversible, noncompetitive reversible, irreversible)
- Discuss and analyze the therapeutic use of enzyme inhibitors

16.1 Enzymes: Definition and Fundamental Properties

Enzymes catalyze biological reactions such as the breakdown of nutrients in the digestive system, synthesis of proteins and membranes in the cells, replication of DNA, and production of energy for neural functions and muscle contractions. All enzymes are proteins with the exception of catalytic RNA molecules called ribozymes. Enzymes are **catalysts** that are capable of accelerating reaction rates up to ~10^6 to 10^{11} times. By simply lowering the activation energy required for the reaction to proceed (▶ Fig. 16.1), enzymes are able to speed up reactions

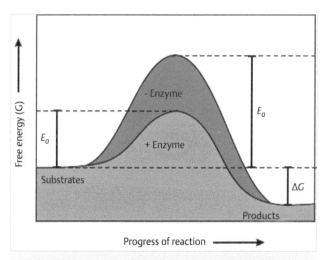

Fig. 16.1 Energetics of enzyme catalysis. An enzyme speeds up a reaction by lowering the activation energy (E a). The E a of a reaction without enzyme (purple) is higher than that of the same reaction with enzyme (orange). The presence of an enzyme does not change the concentration of the substrates or products, or the reaction's net free energy change (ΔG) at equilibrium. (Source: Panini S, ed. Medical Biochemistry—An Illustrated Review. New York, NY. Thieme; 2013.)

without being degraded in the process. These reactions are highly specific and usually involve the conversion of only one substrate to form a single product.

16.2 Substrate Binding Site is Highly Specific to Substrate

Two main hypotheses exist to explain how substrates bind to specific enzymes. The first hypothesis is referred to as the lock and key hypothesis, in which the substrate simply fits into the binding site to form a reaction intermediate. This process is similar to how a key fits into a lock. The second is the induced-fit hypothesis, which says that as the substrate molecule gets closer to the enzyme's binding site, the shape of the enzyme changes or is "induced" to accommodate the substrate for binding.

16.3 The Active Site is the Catalytic Site

In most cases, the substrate binding site and the active site are located in the same region on the enzyme. Thus, when the substrate binds to the active site, the catalytic reaction is able to occur instantly. The shape (proximity and orientation) and the chemical environment (electrostatic, hydrophobic interactions) inside the active site permit a chemical reaction to proceed more easily. The active site is prepared to facilitate the reaction because it also contains the necessary functional groups and cofactors/coenzymes needed for the reaction.

16.4 Role of Coenzymes and Cofactors in Enzyme Catalysis

Coenzymes and cofactors are non-protein, organic or inorganic molecules that participate in enzyme catalysis. Most of the cofactors are metal ions such as Fe^{+2}, Mg^{+2}, Zn^{+2} that contribute to the catalytic process by acting as electrophiles. Coenzymes that are non-protein organic molecules are most often B and C vitamins that facilitate the reaction. There is a large array of coenzymes within the body.) ▶ Table 16.1 summarizes the coenzymes and their roles in human metabolism.

16.5 Principles of Enzyme Catalysis

Enzymes function to speed up reaction rates by reducing the activation energy of the reaction. The enzyme binds the substrate, sequestering it within the active site so the reaction can occur in a confined space with all of the necessary components. This enzyme-substrate interaction stabilizes the transition state and makes the substrate more likely to react with the enzyme. The transition state refers to the enzyme-substrate complex, in which the bond(s) undergoes transformations. In such transformations, the bonds may be distorted, strained or forced to have unlikely electrostatic arrangements. At this point (the

Table 16.1 Coenzymes: sources and some enzymes that require them for activity

Coenzyme	Vitamin Source	Enzyme/Protein/Function
Adenosylcobalamin, methylcobalamin	B_{12}(cobalamin)	• Methylmalonyl coenzyme A (CoA) mutase • Methionine synthase
Ascorbate	C (ascorbic acid)	• Hydroxylases (e.g., lysyl hydroxylase)
Biotin	Biotin	Carboxylases, such as • Pyruvate carboxylase • Acetyl CoA carboxylase
Flavin nucleotides: flavin mononucleotide (FMN), flavin adenine dinucleotide (FAD)	B_2(riboflavin)	Redox enzymes, such as • NADH dehydrogenase/complex I • Succinate dehydrogenase/complex II • Acyl CoA dehydrogenase • Retinal dehydrogenase • Vitamin-activating enzymes • α -Ketoglutaratedehydrogenase complex • Branched-chain α-keto acid dehydrogenase • Pyruvate dehydrogenase complex
Heme	–	• Hemoglobin, myoglobin • Cytochromes • Catalases and peroxidases
Lipoic acid	–	• α -Ketoglutaratedehydrogenase complex • Branched-chain α-keto acid dehydrogenase • Pyruvate dehydrogenase complex
Nicotinamide adenine dinucleotides (NAD^+, $NADP^+$)	B_3 (niacin)	• α -Ketoglutaratedehydrogenase complex • Branched-chain α-keto acid dehydrogenase • Pyruvate dehydrogenase complex • Redox enzymes, such as those found in the tricarboxylic acid (TCA) cycle, and enzymes involved in the biosynthesis of lipids
Pantothenic acid, coenzyme A (CoA)	B_5(pantothenic acid)	• α -Ketoglutarate dehydrogenase complex • Branched-chain α-keto acid dehydrogenase • Pyruvate dehydrogenase complex • Fatty acid synthase complex
Pyridoxal phosphate	B_6(pyridoxine)	• Transaminase • Decarboxylases • Glycogen phosphorylase • Aminolevulinic acid (ALA) synthase
Tetrahydrofolate (THF)	Folate	• A source of one-carbon groups for enzymes that transfer them between molecules (e.g., thymidylate synthetase)
Thiamine pyrophosphate (TPP)	B_1(thiamine)	• α-Ketoglutarate dehydrogenase complex • Branched-chain α-keto acid dehydrogenase • Pyruvate dehydrogenase complex • Transketolase

Note: The following five coenzymes—coenzyme A, $NAD+$, FAD, lipoic acid, and thiamine pyrophosphate—are all found in the following three enzyme complexes— α-ketoglutarate dehydrogenase, branched-chain α-keto acid dehydrogenase, and pyruvate dehydrogenase.
(Source: Panini S, ed. Medical Biochemistry—An Illustrated Review. New York, NY. Thieme; 2013.)

transformation) the substrate-enzyme complex resembles neither the starting material nor the final product. The energy that must be supplied in order for this process to proceed and the reaction to occur is called the activation energy. Ultimately, the role of the enzyme is to lower the activation energy requirement for the transition barrier, thereby increasing the rate of the reaction.

16.6 Factors Affecting Enzyme's Activity

Enzymes catalyze reactions at the optimum rate only if the conditions are at the optimum. There are several factors that affect the rate of enzyme-catalyzed reactions. These factors including the concentrations of substrate, temperature and pH of the environment, and presence of effectors such as inhibitors and activators.

1. **The concentration of enzyme and substrate:** The velocity for non-enzymatic reactions is proportional to the substrate concentration. However, enzyme-catalyzed reactions exhibit saturation kinetics. If the enzyme concentration remains constant, then low concentrations of the substrate will result in a reaction velocity that is proportional to the substrate concentrations. However, when all of the enzyme molecules are occupied with substrate (100% saturation), the reaction velocity will plateau. This is called the saturation point, or the Vmax. At Vmax, the reaction velocity is constant and

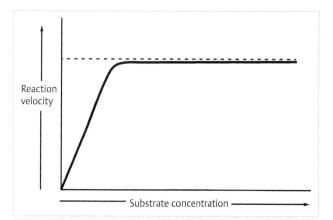

Fig. 16.2 Saturation point (Vmax). When plotting the reaction velocity against the substrate concentration, the enzyme saturation point is illustrated. The Vmax illustrated by the dotted line represents the point at which 100% of the enzymes are bound to substrate molecules. Once the enzyme reaches the maximum saturation, Vmax, then the velocity of the reaction plateaus and is independent of the substrate concentration.

functions independent of the substrate concentration (▶ Fig. 16.2).

2. **The effect of pH:** Each enzyme-catalyzed reaction has an optimal pH. At this optimum pH, the enzyme and substrate exhibit the most efficient interactions, and the velocity of the reaction is at its maximum rate. When the pH levels are in the extremes (either too high or too low), the enzyme will denature, disrupting the active site, and preventing the substrate from interacting with the substrate binding site. Therefore, the pH level is an extremely important determinant in enzyme function. Even slight increases or decreases in pH levels will cause the enzyme-substrate interactions to alter and the reaction will proceed at a slower rate.

3. **The effect of temperature:** The velocity of a reaction is at maximum at 37 °C in humans. Deviations from the optimal temperature have an effect on enzyme function. Because at both high and e low temperatures enzymes are denatured, the reaction velocity is decreased.

4. **Activators (agonists) and Inhibitors (antagonists):** To be able to understand the effects of inhibitors and activators on the reaction velocity, we first need to review enzyme kinetics.

16.7 Enzyme Kinetics

Enzyme kinetics is the study of the rate of enzyme-catalyzed reactions and provides information on enzyme specificities and mechanisms. Understanding enzyme mechanisms (e.g., the number of kinetic steps, detailed chemistry of enzymes) and inhibition mechanisms is paramount to the development of medication therapies. The most common way to express the kinetic properties of enzymes is the Michaelis-Menten (MM) model.

The Michaelis-Menten equation states that the enzyme (E) and substrate (S) combine at the rate of complex formation (k_1), to produce the enzyme-substrate complex (ES). The ES

then proceeds at the rate of product formation (k_2) to produce the product (E + P). The k_{-1} represents the rate of dissociation of the enzyme-substrate complex to the reactants.

There are several assumptions that are used to derive the Michaelis-Menten model.
- An enzyme-substrate complex is formed.
- The concentration of the substrate is much greater than the concentration of enzyme.
- The ES does not change with time (the **steady state** assumption). The rate of formation of the ES is equal to the rate of breakdown of the ES.
- Initial reaction velocities are used in the analysis of enzyme reactions. The M-M model assumes that the ES complex is in rapid equilibrium with free enzymes.

Therefore, from these assumptions about the kinetic scheme, a relationship can be derived for the rate or velocity of product formation: $v_i = k_2 \times [ES]$

v_i = The rate of the enzymatic reaction

k_2 = The rate of the formation of the product from the ES complex

[ES] = Concentration of the enzyme-substrate complex

From this equation, Michaelis-Menten equation is derived. Michaelis-Menten equation describes how velocity of reaction varies with substrate concentration.

16.8 Michaelis-Menten Graph

The Michaelis-Menten graph visually represents the above mentioned Michaelis-Menten equation. The Michaelis-Menten graph is formed when the velocity, **v**, is plotted against the substrate concentration [**S**]. The result is the formation of a hyperbolic curve that describes how the reaction velocity varies with the substrate concentration (▶ Fig. 16.3).

The Michaelis-Menten graph displays a combination of zero-order and first-order kinetics. When the substrate concentration is low, the rate behaves according to "first order kinetics," because the velocity of the reaction is **proportional** to the substrate concentration [S]. On the other hand, when the substrate concentration is high and the enzyme is saturated with respect to the substrate, then the rate is said to behave according to "zero order kinetics," because the velocity of the reaction is **independent** of the [S] (▶ Fig. 16.4).

16.8.1 Link to Pharmacology: First-order and Zero-order Kinetics in Pharmacokinetics

Most drugs exhibit first-order elimination kinetics (linear kinetics): Metabolism and/or excretion of drug from plasma is known as drug elimination or termination of drug action. Most drugs are eliminated from plasma with first-order elimination kinetics. With first-order elimination kinetics, the rate of drug elimination is proportional to the plasma drug concentration at any given time, meaning that a constant *fraction* of the drug is eliminated per unit time. When the drug concentration is high, the rate of disappearance is high too. The elimination process is, therefore, not saturated (▶ Fig. 16.5).

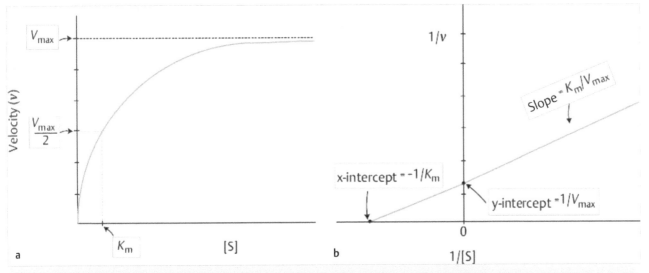

Fig. 16.3 **Michaelis-Menten and Lineweaver-Burk graphs. (a)** The Michaelis-Menten graph is obtained by plotting an enzyme's reaction velocity (v) against the concentration of its substrate [S]. The V max is the velocity the curve approaches (but never really reaches), and K_m is the substrate concentration at which the reaction velocity is at half of V_{max}. **(b)** The Lineweaver-Burk graph is obtained by plotting the inverse of velocity (1/v) against the inverse of substrate concentration (1/[S]) to generate a straight-line graph. The point at which the line intercepts the x-axis is identified as $-1/K_m$, and that at which it intercepts the y-axis is 1/V max, and the slope of the line is K m/V max. (Source: Panini S, ed. Medical Biochemistry- An Illustrated Review. New York, NY. Thieme; 2013.)

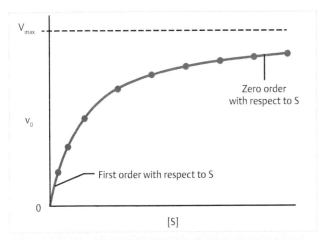

Fig. 16.4 **Michaelis-Menton Model.** The Michaelis-Menten model is the combination of zero-order and first-order kinetics. When the substrate concentration is low, the rate is first order because the velocity of the reaction is proportional to the substrate concentration [S]. However, when the substrate concentration is high and the enzyme is saturated with respect to the enzyme ability to bind to the substrate, then the rate is zero-order. The velocity of the reaction is independent of the substrate concentration [S].

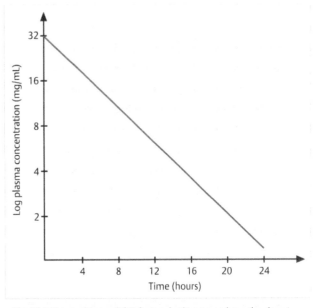

Fig. 16.5 Some drugs exhibit first order kinetics, where the drug is eliminated at a constant fraction. While other drugs exhibit zero order kinetics, they are eliminated at a constant amount.

Some drugs exhibit zero-order kinetics (saturation kinetics): In zero-order kinetics, a constant or fixed *amount* of a drug is eliminated per unit time, but without a fixed half-life. In other words, the drug elimination mechanisms (metabolism of drug and/or excretion) have become saturated. For example, phenytoin, ethanol, and salicylates exhibit zero-order kinetics. If a high dose of any of these drugs is administered, or a hepatic or renal disease has impaired the drug elimination, then the elimination process becomes saturated and exhibits zero order kinetics.

16.9 Significance of K_m and V_{max} values

The Michaelis-Menten graph can be used for estimating the K_m and the V_{max}. Determining the K_m and V_{max} values are useful in understanding the mechanisms of enzymes, which helps in determining the true substrate for the enzyme. Effects of specific drugs or inhibitors on the rates of the enzymes can also be analyzed by comparing the K_m and the V_{max} values. Thus, the study of enzyme kinetics contributes to the development of better drugs and medicines.

Understanding the K_m: The K_m is the substrate concentration at which half of the V_{max} is achieved. K_m measures the affinity of an enzyme for its substrate. For example, if an enzyme has a low K_m value for a specific substrate, then the enzyme achieves V_{max} at a low substrate concentration. Therefore, a low K_m indicates that an enzyme has a high affinity for a specific substrate. Conversely, a high K_m means that an enzyme has a low affinity for a specific substrate, and the enzyme is only able to weakly bind to its substrate.

Understanding the V_{max}: The V_{max} is the maximal velocity that can be achieved with an infinite amount of substrate. As enzyme concentration changes, the V_{max} value also changes. Because of this, the rate of product formation or k_2, also known as catalytic rate (k_{cat}), is a more useful kinetic parameter than V_{max}. We can calculate the k_{cat} with following equation:

$$k_{cat} = V_{max}/[E]_0$$

In this equation, the k_{cat} becomes independent of enzyme concentration, thus it is conventionally used in place of V_{max}. The unit of k_{cat}, or turnover number, is the reciprocal of time (e.g. sec-1) and it is a measure of the maximum number of substrate molecules converted to product per enzyme molecule per unit of time.

16.10 Determining V_{max} and K_m by a Lineweaver-Burk or "Double Reciprocal" Plot

The Michaelis-Menten equation can be re-arranged to produce a linear plot, in which the reciprocal of the velocity, (1/v), can be plotted against the reciprocal of the substrate concentration, (1/[S]). This linear graph can be utilized more easily to determine the V_{max} and K_m, and is referred to as a Lineweaver-Burk, or the "double reciprocal", plot. The x-intercept of the line represents the reciprocal of the K_m, (-1/K_m), while the y-intercept represents the reciprocal of the V_{max}, (1/V_{max}) (▶ Fig. 16.3**b**).

16.11 Enzyme Inhibition

Enzyme inhibitors are molecules such as pharmaceutical agents, metals, or toxins that interact with the enzyme to decrease the rate of the enzymatic reaction. There are two main classifications of enzyme inhibitors: irreversible inhibitors and reversible inhibitors.

Irreversible inhibitors are molecules that irreversibly inactivate enzymes by destroying or covalently modifying key amino acid residues. Irreversible inhibitors are also called suicide or mechanism-based inhibitors because the effect of the irreversible inhibitor can only be overcome by the synthesis of new enzyme.

16.11.1 Link to Pharmacology: Irreversible Inhibitors

Penicillin: Is used as an antibiotic because it irreversibly inhibits the glycopeptidyl transferase required for bacterial cell wall synthesis.

Aspirin: Irreversibly inactivates both COX-1 and COX-2 to decrease the production of prostaglandins and thromboxane, thereby reducing inflammation and preventing the formation of clots.

Organophosphate compounds: Organophosphate compounds such as malathion, parathion, sarin nerve gas, and diisopropylphosphofluoridate (DFP) irreversibly inactivate acetylcholinesterase by forming a covalent complex with the active-site. This can result in symptoms such as salivation, lacrimation, urination, diarrhea, and vomiting.

Reversible inhibitors: There are two main types of reversible inhibitors: competitive and noncompetitive.

Competitive inhibitors are structural analogs that resemble the substrate and *compete* with the substrate for binding to the enzyme. Competitive inhibitors decrease the affinity of the enzyme for its substrate, and change the K_m value, while the V_{max} remains unchanged. In the presence of competitive inhibitors, the K_m value increases, and the affinity decreases (▶ Fig. 16.6**a**). The effects of a competitive inhibitor can be overcome by increasing the concentration of the substrate [S], because it increases the likelihood that the substrate will bind to the active site.

16.11.2 Link to Pharmacology: Competitive Inhibitors

Statin Drugs are the structural analogs of HMG CoA. Statin drugs, such as lovastatin, compete for binding to HMG CoA reductase, which is the rate limiting step of de novo cholesterol biosynthesis.

Methanol and ethylene glycol poisoning can be treated with ethanol: Methanol and ethylene glycol poisoning share similar clinical characteristics. Methanol is oxidized to formaldehyde, and ethylene glycol is oxidized to glycoaldehyde by alcohol dehydrogenase (ADH). The enzyme aldehyde dehydrogenase (ALDH) metabolizes formaldehyde to formic acid, and glycoaldehyde to glycolic acid. These toxic metabolites can attack the optic nerve and cause blindness. Methanol and ethylene glycol poisoning can be treated with ethanol as an antidote because ethanol has ~20 times more affinity for ADH and competitively inhibits this enzyme. This results in a slowdown in the oxidation of methanol and ethylene glycol to its toxic metabolites.

Methotrexate: Methotrexate is an antineoplastic drug used in chemotherapy because it decreases thymidine biosynthesis. Methotrexate is a structural analog of folic acid that inhibits dihydrofolate reductase, which blocks the progression of DNA replication and subsequent cell division.

Noncompetitive inhibitors are molecules or ions that bind to an enzyme in a location separate from the active site. When the noncompetitive inhibitor binds to the enzyme, it causes a conformational change in the enzyme that renders it inactive. Thus, when the enzyme is bound to the noncompetitive inhibitor, it is still able to bind to the substrate but it is unable to perform the necessary catalytic reaction. Noncompetitive inhibitors change the V_{max} value of the enzyme, because the enzyme can no longer function at maximal capacity. On the other hand, the K_m value, or the affinity of the enzyme for its substrate, remains unchanged (▶ Fig. 16.6**b).** The effects of the

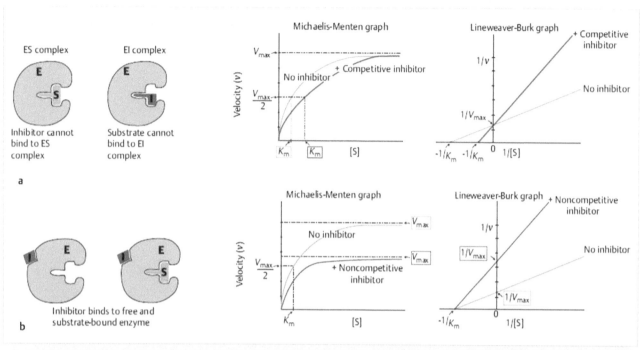

Fig. 16.6 Enzyme inhibition (reversible). Agents that reversibly inhibit enzymes do so by forming noncovalent bonds with the enzyme. **(a)** Some competitive inhibitors compete with the substrate for binding to an enzyme's active site. Others (not shown) bind to regions outside of the active site and cause a conformational change in the enzyme that prevents substrate binding. The competitive inhibitor and substrate cannot bind the enzyme at the same time. **(b)** Noncompetitive inhibitors do not resemble the substrate and bind to either the free or the substrate-bound enzyme at a region other than the active site. (Source: Panini S, ed. Medical Biochemistry- An Illustrated Review. New York, NY. Thieme; 2013.)

noncompetitive inhibitor cannot be overcome by increasing the concentration of the substrate because their effects are independent of the enzyme-substrate binding capacity.

16.11.3 Link to Pharmacology: Noncompetitive Inhibitors

Physostigmine: Physostigmine is used for treatment of myasthenia gravis. Physostigmine binds reversibly to a site on acetylcholine esterase that is not the active site and decreases its catalytic activity, or V_{max}. The therapeutic effects of physostigmine result from an increased amount of acetylcholine available at neuromuscular junctions by decreasing the catalytic rate of acetylcholine esterase, which breaks down acetylcholine.

Captopril: Captopril is used in the treatment of hypertension because it can reversibly bind to the angiotensin-converting enzyme (ACE). This drug induces a conformational change in ACE that inactivates it, making the enzyme unable to proteolytically cleave angiotensin I to form angiotensin II. Decreased angiotensin II levels limit the body's ability to increase renal fluid and electrolyte retention, therefore resulting in a decrease in blood pressure.

Review Questions

1. The elimination rate of drug A shows that 10% of this drug is eliminated from plasma per unit of time. Based on this information, which of the following statements is correct?
   A) A constant amount of drug A is eliminated from plasma
   B) Drug A exhibits first-order elimination kinetics
   C) Drug A exhibits zero-order elimination kinetics
   D) The elimination processes are saturated
   E) The elimination of drug A is slow compared to other drugs

2. Based on the following data, what is the K_m of the enzyme for its substrate?

   [S] (mM) v_i (mmol/sec)
   0.010 2.0
   0.050 9.1
   0.100 17
   0.500 50
   1.00 67
   5.00 91
   10.0 95
   50.0 99
   100.0 100
   A) 0.010 mM
   B) 0.050 mM
   C) 0.500 mM
   D) 5.00 mM
   E) 50 mM

3. The affinity of hexokinase for glucose is reported to be 0.05 mM, versus an affinity for fructose of 1.5 mM. Another in-vitro study reports that the affinity of hexokinase for galactose is 15 mM. Based on the given data, which one is a better substrate for hexokinase?
   A) Fructose
   B) Galactose
   C) Glucose

4. An investigator is studying the effects of compound A on the synthesis of the bacterial wall. When compound A is added to the bacterial culture, the growth and the survival of the bacteria was affected. Which of the following best describes the inhibition mechanism of compound A?
   A) It is a competitive inhibitor
   B) It is an irreversible inhibitor
   C) It is a noncompetitive inhibitor
   D) It decreases the K_m of the enzyme and its substrate
   E) It decreases the V_{max} of the enzyme and its substrate

5. Which of the following best describes an effect of noncompetitive inhibitors?
   A) They competes with substrate for binding to the enzyme
   B) They covalently modify the active site residues
   C) They decrease the affinity of the enzyme for its substrate
   D) They decrease the rate of catalysis
   E) They increase the V_{max} of the reaction

Answers

1. **The correct answer is B.** If the rate of drug elimination is proportional to the plasma drug concentration at any given time and a constant *fraction* of a drug is eliminated per unit time, then the elimination kinetics is first-order. When the drug concentration is high, the rate of disappearance is high too. The elimination process is, therefore, not saturated (answer choice D is incorrect). If elimination processes are saturated, which is zero order, then a constant amount of drug is eliminated. Based on the given information, this drug eliminated with first-order kinetics, therefore answer choices A, C, and D are incorrect. Answer choice E is incorrect because there is no comparison of this drug with other drugs.

2. **The correct answer is C.** The K_m is equal to half of the V_{max} (or 50% saturation of the enzyme). Therefore, if the V_{max} is 100 mmol/sec, then 50% saturation would be 50 mmol/sec, which correlates to a 0.500 mM.

3. **The correct answer is C.** The affinity of an enzyme and for its substrate is measured by the kinetic parameter K_m. A low K_m indicates that an enzyme has high affinity for a specific substrate, and is able to form a strong bond with its substrate. The lowest K_m value among these three substrates is for glucose. Therefore, glucose is the best substrate. Since fructose and galactose have higher K_m values compared to glucose, hexokinase is only able to weakly bind to these substrates.

4. **The correct answer is B.** The compound most likely is inhibiting bacterial wall biosynthesis by irreversibly inhibiting an enzyme that is involved in the synthesis of the bacterial wall. The inhibition mechanism is most likely irreversible since bacterial growth and survival is affected by the compound. Irreversible inhibitors inactivate the enzyme by covalently modifying the key amino acid residues of the active site. Therefore, the effects are irreversible, which destroys the enzyme. Answer choices A and D, which describe the effects of competitive inhibitors, are incorrect because if the compound would be a reversible inhibitor, then only the growth of the bacteria would be affected. For the same reason, answer choices C and E are incorrect because noncompetitive inhibitors are reversible inhibitors as well.

5. **The correct answer is D.** In the presence of noncompetitive inhibitors, the rate of catalysis decreases. Noncompetitive inhibitors bind to the enzyme in a site that is distant from the substrate binding site. The K_m value, and therefore the affinity of the enzyme for its substrate, remains unchanged. Hence answer choice A and C are incorrect. The interaction between the noncompetitive inhibitor and the enzyme often causes conformational changes in the enzyme such that the enzyme can no longer function at maximum capacity. Thus, noncompetitive inhibitors decrease the V_{max} value of the enzyme. Therefore, answer choice E is incorrect. Answer choice B is also incorrect because covalent modifications are irreversible. Therefore both competitive and noncompetitive inhibitors are reversible inhibitors.

17 Regulation of Enzyme Activity

At the conclusion of this chapter, students should be able to:
- Describe and differentiate among the mechanisms involved in the regulation of enzyme activity including allosteric regulation and covalent modifications such as phosphorylations, protein-protein interactions, and proteolytic cleavage
- Discuss examples of each regulatory mechanism affecting enzymatic reactions

17.1 Reasons for the Regulation of Enzymes

The purpose of regulation is to maintain homeostasis, a balanced metabolic state, without wasting resources. To optimize metabolic functions and to be able to rapidly respond to environmental changes, the human body relies on biological processes to occur at specific times, places and speeds. Therefore, effective regulation of metabolic pathways is essential for us to survive. The regulation of metabolic pathways generally occurs through the regulation of one or two key enzymes in each metabolic pathway. These key enzymatic steps are called regulatory steps or rate-limiting steps.

There are two main approaches in the regulation of enzyme activity. First is a change in the activity level of the regulatory enzymes and the second is a change in the quantity of these enzymes. With the quantity change, cells are able to increase or decrease the V_{max} of a specific reaction by increasing or decreasing the amount of enzyme available to perform that reaction. Cells are able to regulate the amount of enzymes by regulating either enzyme synthesis or degradation, or in some cases both. Synthesis is generally regulated during transcription, which is reviewed in chapter 3. The regulation of degradation occurs through either the ubiquitin- or lysosomal-dependent pathways. These pathways are described in chapter 33. In this chapter, we will focus on the regulation of enzymes through changes in their activity.

There are several mechanisms for the alteration of the catalytic efficiency of enzymes. The most common include:
1. Proteolytic cleavage of proenzymes (zymogen activation)
2. Allosteric regulation
3. Covalent modifications (e.g. phosphorylation)
4. Protein-protein interactions such as association and dissociation
5. Feed-back regulation of metabolic pathways
6. Compartmentalization

17.2 Zymogen Activation

Zymogens, or proenzymes, are the inactive precursors of enzymes. To indicate an inactive form of the enzyme, the zymogen is given the suffix of "-ogen" or the prefix "pro." Zymogens are activated through proteolytic cleavage or hydrolysis of one or more peptide bonds, which converts it to its enzymatically active form. Proteolytic cleavage of the zymogen to its active form is an irreversible process.

The synthesis of zymogens allows the body to be in a "ready state" so that the enzymes can be quickly activated to catalyze reactions once the proteolytic cleavage occurs. Zymogens are also compartmentalized and only activated by precise cleavage reactions, which ensure that their activity only occurs in specific locations.

Loss of zymogen synthesis would be disastrous because the enzymes involved in digestion, blood clotting, and bone and tissue remodeling would be able to freely catalyze reactions at inappropriate times and locations. For instance, if blood clotting factors are activated at times when they are not necessary, they would initiate clotting throughout the body. Other examples of zymogens include insulin, which is synthesized as proinsulin, digestive proteins such as trypsinogen, which is activated to trypsin, blood clotting factors such as fibrinogen and prothrombin, and connective tissue collagen, which is initially synthesized as procollagen.

17.3 Conversion of Proinsulin to Insulin

Insulin is synthesized by the β-cells of the pancreas as preproinsulin. Proinsulin is generated from preproinsulin by the cleavage of the N-terminal signal peptide in the endoplasmic reticulum. Prohormone convertase then cleaves proinsulin at two sites, generating 21- and 30- amino acid long peptides and a 33 residue-long C-peptide. The 21- and 30- amino acid long peptides form disulfide bonds to generate a mature insulin molecule. Because the insulin and the C-peptide are formed in the secretory granules of the β-cells at the same time, they are also secreted into blood together. Therefore, determination of C-peptide levels can be used as a diagnostic test to determine endogenous insulin levels in type 1 and type 2 diabetic patients.

17.4 Allosteric Regulation

The activity of allosteric enzymes can be adjusted by reversible, non-covalent binding of specific modulators at regulatory sites within the enzymes. Allosteric enzymes are usually composed of multiple subunits and these subunits can act as regulatory or catalytic entities. These enzymes are unique because they do not follow the Michaelis-Menten relationship between substrate concentration [S] and reaction velocity, v. Instead, allosteric enzymes display a sigmoidal curve, rather than the Michaelis-Menten associated hyperbolic curve. The formation of the sigmoidal curve is due to the general features of allosteric enzymes that dictate their response to positive and negative modulators.

Positive effectors activate allosteric enzymes, while negative effectors are able to inhibit them. When positive effectors bind to the regulatory subunit of the allosteric enzyme, it induces a conformational change in the enzyme that increases the

III

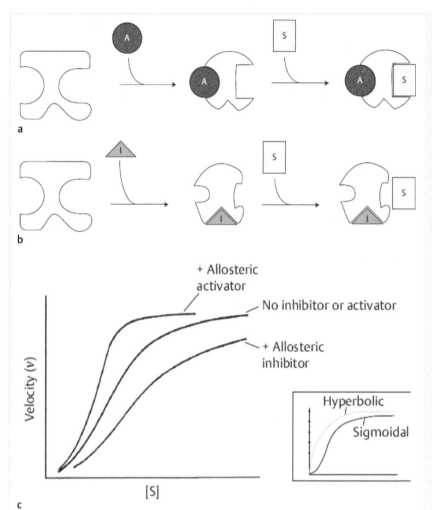

Fig. 17.1 **(a)** Binding of positive allosteric effector (allosteric activator). **(b)** Binding of negative allosteric effector (allosteric inhibitor). **(c)** Allosteric enzyme kinetics. (Source: Panini S, ed. Medical Biochemistry- An Illustrated Review. New York, NY. Thieme; 2013.)

adjacent subunit's ability to bind to the substrate (▶ Fig. 17.1**a**). This unique method of communication between subunits in response to a positive effector is referred to as positive cooperativity. On the other hand, if a negative effector binds to a regulatory subunit, the substrate binding potential of the adjacent subunit decreases, which is referred to as negative cooperativity (▶ Fig. 17.1**b**). The sigmoidal kinetic behavior reflects the cooperative interaction between multiple subunits within the allosteric enzyme (▶ Fig. 17.1**c**). Positive and negative effectors change the K_m value of the enzyme for its substrate by increasing or decreasing the enzymes ability to bind to its substrate. Therefore, positive effectors shift the sigmoidal curve to the left, and the negative effectors will shift the curve to the right.

An example of an allosteric enzyme is phosphofructokinase-1 (PFK-1) which is involved in the rate limiting step of glycolysis (reviewed in chapter 22). ATP is a negative effector that decreases the rate of PFK-1, while AMP is a positive effector that increases the rate of the enzyme. This modulation allows for the rate of glycolysis to increase during low energy (high AMP) states when the body needs energy to function.

Another example of an allosteric regulation is that of hemoglobin and oxygen in the presence of 2,3 bisphosphoglyceric acid (2,3 BPG). The 2,3 BPG is a negative effector of hemoglobin and decreases hemoglobin's affinity for oxygen, which allows bound oxygen to be released in the extra-pulmonary tissues. High levels of 2,3 BPG are present in individuals who live in high altitudes where the oxygen availability is low, and survival depends on the body's ability to deliver oxygen to the extra-pulmonary tissues.

17.5 Covalent Modifications

A variety of chemical compounds can modify enzymes by reversible covalent reactions. There are many covalent modifications of proteins and enzymes including phosphorylation, glycosylation, adenylation, methylation, uridylation, ribosylation, and acetylation (▶ Table 17.1). These modifications allow the cell to promptly respond to meet physiological demands. Of all the covalent modifications, phosphorylation is the most common. Usually a phosphate from an active donor, typically ATP, is transferred to a specific amino acid on the regulatory enzyme, such as serine/threonine, or tyrosine.

Phosphorylation of an enzyme induces a conformational change that can either increase or decrease the enzyme's activity depending on the enzyme. For example, glycogen synthase

Table 17.1 Posttranslational protein modification

Modification		Functional group	Residue affected
Acylation	N-terminus bonded to acetyl or long-chain acyl residue Thioesterification with a long-chain acyl group	Amine () Sulfhydryl (-SH)	N-terminus Cys
Acetylation	Covalent linkage to amine	Amine ()	Lys
Glycosylation	O-glycosylation	Hydroxyl (–OH)	Ser, Thr
	N-terminus bonded to acetyl or long-chain acyl residue Thioesterification with a long-chain acyl group	Amine () Sulfhydryl (-SH)	N-terminus Cys
Phosphorylation	Phosphate linked via esterification	Hydroxyl (–OH)	Ser, Tyr, Thr; also Asp and His
γ-Carboxylation	Addition of –COOH group	γ-carbon	Glu
Ubiquitination	Covalent modification with ubiquitin	Amine ()	Lys
Hydroxylation	Addition of –OH group	C-4	Pro and Lys
Disulfide bonds	Oxidation to achieve covalent linkage of cysteine residues	Sulfhydryl (-SH)	Cys
renylation	An isoprenoid group such as farnesyl or geranylgeranyl group added to cysteines in C-terminal CAAX or CC motifs		Cys

Source: Panini S, ed. Medical Biochemistry- An Illustrated Review. New York, NY. Thieme; 2013.

is deactivated when phosphorylated, while phosphorylation activates the enzyme glycogen phosphorylase. The enzymes that add phosphate groups to proteins are referred to as protein kinases, whereas protein phosphatases are those that remove phosphate groups.

The benefit of being able to regulate enzymes via phosphorylation is that it can be quickly reversed, switching between active and inactive states. Phosphorylation is also an energetically efficient, or rather an energetically inexpensive, process because modifications do not require the synthesis of a new enzyme for a reaction to proceed. Another benefit of phosphorylation is that its affects can be rapidly transmitted to downstream pathways via an amplification cascade. For example, in response to epinephrine, cAMP is increased, which activates protein kinase A (PKA). The active PKA then phosphorylates and activates phosphorylase kinase, which phosphorylates glycogen phosphorylase. This sequence of activations is referred to as a phosphorylation cascade because the act of phosphorylating one enzyme results in the activation and phosphorylation of many other downstream enzymes. This enables the initial signal to be greatly amplified.

Some enzymes are regulated by both allosteric and covalent modifications. Glycogen phosphorylase is one such enzyme that responds to both allosteric effectors and covalent modifications. This enzyme can be allosterically modulated by AMP and also activated by phosphorylation.

17.6 Regulation by Protein-Protein Interactions

Enzymes can also be regulated by association and dissociation of monomers. For example, in the presence of fatty acyl-CoA, acetyl-CoA carboxylase disassociates into dimers and becomes inactive. While on the other hand, in the presence of citrate, acetyl-CoA carboxylase polymerizes into its active configuration.

Protein kinase A (PKA) plays an important role in the body's ability to control hormone-induced metabolic pathways. For example, the binding of glucagon or epinephrine to their receptors results in upregulation of the adenylyl cyclase enzyme, which converts ATP to cAMP. Increased concentrations of cAMP activate PKA by binding to the two regulatory (R) subunits of the PKA, which causes a conformational change in the enzyme that releases the two catalytic (C) subunits from the enzyme. Upon their release from PKA, the protein C monomers become enzymatically active.

Enzymes can be directly regulated through interactions with so called modulator proteins, which can bind to and induce a conformational change in the active site of an enzyme. An example of this regulation involves the Ca^{2+}-calmodulin family of modulator proteins. Ca^{2+}-calmodulin proteins are able to directly activate or deactivate enzymes by inducing a conformational change in their active site. One such protein that Ca^{2+}-calmodulin is able to directly activate is CaM kinase. The binding of Ca^{2+} to the calmodulin protein induces a conformation change that causes it to bend. This Ca^{2+}-calmodulin molecule hinges around CaM kinase, which induces a conformational change to activate the enzyme. Activated CaM kinases are responsible for multiple activities including phosphorylation of ion channels, metabolic enzymes, and transcription factors within the cell.

17.7 Feedback Regulation

Metabolic pathways are a series of enzymatic reactions where the product from one reaction serves as the substrate for another reaction. The reactions and pathways of metabolism

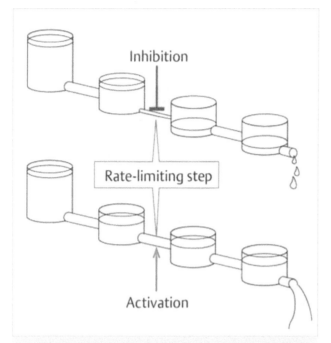

Fig. 17.2 Regulation of metabolic pathways occurs at the rate-limiting step. Inhibition of the rate-limiting enzyme slows down the rate of the pathway. Activation of the rate-limiting enzyme speeds up the rate of the pathway. (Source: Koolman J, Röhm K, ed. Color Atlas of Biochemistry. 3rd Edition. New York, NY. Thieme; 2012.)

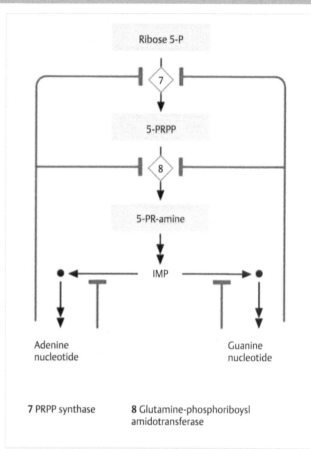

7 PRPP synthase   8 Glutamine-phosphoriboysl amidotransferase

Fig. 17.3 Metabolic pathways are regulated by negative-feedback mechanism. Generally the end-product of the pathway binds and reduces the activity of the rate-limiting enzyme of the pathway. (Source: Koolman J, Röhm K, ed. Color Atlas of Biochemistry. 3rd Edition. New York, NY. Thieme; 2012.)

are tightly regulated in order to ensure that the metabolic demands of the human body are met quickly and in an energetically efficient manner. Regulation of metabolic pathways occurs primarily at the rate-limiting step (▶ Fig. 17.2), which refers to a key enzyme that catalyzes a reaction early on in the pathway. The rate-limiting step is the slowest reaction within the metabolic pathway. This reaction cannot be reversed, and therefore serves as the first committed step within the pathway. Metabolic pathways are intricately intertwined to create a complex tapestry of reactions that rely on the regulation of the enzymes to work.

Feedback regulation is one method by which enzymes within the metabolic pathways are regulated. This refers to the process by which the end product of one reaction feeds back to an enzyme earlier on in the pathway to regulate its own synthesis. Typically, the product will allosterically modulate the enzyme that catalyzes the rate-limiting step to control its own synthesis. There are two types of feedback regulation: feedback inhibition or feed-forward regulation. Feedback inhibition occurs when one of the first enzymes of a metabolic pathway is inhibited by the end product. An example of feedback inhibition is HMG-CoA reductase. HMG-CoA reductase is an enzyme that catalyzes the rate-limiting step of cholesterol biosynthesis. Cholesterol feeds backward in the metabolic pathway to inhibit HMG-CoA activity. Another example for negative feedback inhibition is in the regulation of purine nucleotide biosynthesis (▶ Fig. 17.3). The two key regulatory enzymes, phosphoribosyl pyrophosphate (PRPP) synthase and phosphoribosyl amino-transferase, are inhibited by the final products adenine and guanine nucleotides.

In feed-forward regulation, an intermediate or a regulatory molecule, also called a metabolite, increases the rate of the pathway. One example for feed-forward regulation is the regulation of pyruvate kinase, the third regulatory step of glycolysis, by the glycolytic intermediate fructose 1,6-bis phosphate. Another example is seen in the urea cycle. An increased concentration of NH_4, a toxic product of amino acid degradation, increases the rate of the urea cycle pathway. In fact, this feed-forward regulation of the urea cycle protects the body from the toxic effects of NH_4.

17.8 Compartmentalization

Compartmentalization of enzymes into separate organelles also plays an important role in enzyme regulation. For example, the enzymes of fatty acid metabolism are divided into two separate compartments (▶ Fig. 17.4). The enzymes involved in fatty acid oxidation, or the degradation of activated fatty acids, are sequestered into the mitochondria. Meanwhile, the enzymes involved in the biosynthesis of fatty acids are in the cytoplasm. This compartmentation is important because it ensures that metabolic products are not degraded immediately after they are produced.

III

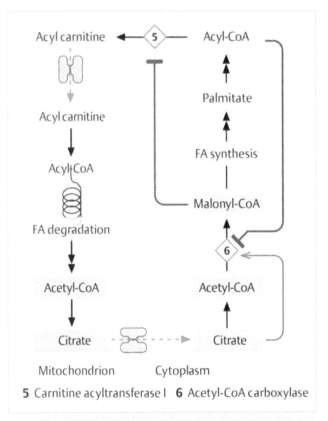

5 Carnitine acyltransferase I   6 Acetyl-CoA carboxylase

Fig. 17.4 Metabolic pathways are regulated by compartmentalization. For example, while the fatty acid synthesis occurs in the cytoplasm, the fatty acid oxidation takes place in the matrix of the mitochondria. (Source: Koolman J, Röhm K, ed. Color Atlas of Biochemistry. 3rd Edition. New York, NY. Thieme; 2012.)

Review Questions

1. Which of the following enzymes is NOT synthesized as a zymogen?
   A) Carboxypeptidase
   B) Chymotrypsin
   C) Elastase
   D) Enteropeptidase
   E) Trypsin

2. . Which of the following regulatory mechanisms best describes the type of regulation mediated by Ca^{2+}-calmodulin proteins?
   A) Allosteric regulation
   B) Covalent modifications
   C) Compartmentalization
   D) Protein-protein interactions
   E) Zymogen activation

3. When oxygen binds to one of the four subunits of hemoglobin, it triggers a conformational change in the adjacent hemoglobin subunits which allows them to bind oxygen more easily. The interaction between the subunits in response to oxygen is an example of what type of regulation?
   A) Allosteric regulation
   B) Covalent modification
   C) Protein-Protein interaction
   D) Zymogen activation

Answers

1. **The correct answer is D.** Unlike many pancreatic digestive enzymes, enteropeptidase is synthesized in the intestinal epithelial cells as an active enzyme. Answer choices A, B, C, and E are incorrect because all of these enzymes are synthesized as zymogens in the pancreas and secreted into intestine, where they become active enzymes.

Enteropeptidase or enterokinase is involved in the activation of trypsinogen to trypsin. After conversion of trypsinogen to trypsin by enteropeptidase, the active trypsin activates other zymogens, such as procarboxypeptidase, chymotrypsinogen, and proelastase, to their active forms.

2. **The correct answer is D.** Ca^{2+}-calmodulin proteins are able to directly activate or deactivate other proteins by inducing a conformational change in their active site. This is an example of enzyme regulation that involves protein-protein interactions. Glycogen phosphorylase kinase contains a calmodulin subunit. Muscle contraction during exercise causes release of Ca^{2+} from smooth endoplasmic reticulum. This liberated Ca^{2+} binds to the calmodulin subunit of glycogen phosphorylase kinase and induces a conformational change, which activates the enzyme. Active glycogen phosphorylase kinase phosphorylates and activates glycogen phosphorylase, which releases glucose from muscle glycogen stores to be oxidized for ATP production for the muscle contraction.

3. **The correct answer is A.** Activity of allosteric enzymes are regulated by reversible, non-covalent binding of specific modulators/effectors at the regulatory sites of the enzymes. Allosteric enzymes are usually composed of either multiple identical or non-identical subunits that act as regulatory or catalytic subunits. When positive effectors bind to the regulatory subunits of the allosteric enzyme, a conformational change is induced in the enzyme that increases the adjacent subunits ability to bind to the substrate. Therefore, oxygen binding to hemoglobin is an example of allosteric regulation by oxygen. In this example oxygen is both a substrate and the positive effector. Answer choice B is incorrect because oxygen binding is not a covalent interaction. The binding of oxygen does change the conformation of hemoglobin. However, this is not through protein-protein interactions and, therefore, C is incorrect. Answer choice D is incorrect because there is no hydrolysis or removal of peptide involved in this regulation.

18 Structure and Function of Proteins

III

At the conclusion of this chapter, students should be able to:
- Analyze the amino acid structure and classification of amino acids based on their side chains
- Describe how histidine's imidazole group can be useful for buffering in physiological systems
- Explain what forces stabilize the primary, secondary, tertiary, and quaternary structure of proteins
- Describe the structure of hemoglobin and myoglobin
- Explain the mechanisms of O_2 binding to myoglobin and hemoglobin
- Discuss how the structures of myoglobin and hemoglobin determine their function
- Explain the effects of 2,3 BPG, pH and pCO_2 on hemoglobin's affinity for oxygen
- Explain the affinity of HbF for oxygen in comparison to HbA and state its significance
- Describe the hemoglobinopathies and the molecular basis of sickle cell and thalassemia
- Explain how protein folding errors cause disease
- Analyze the protein folding errors in cystic fibrosis, α1-anti trypsin deficiency, and amyloidosis

Proteins can be described as molecular machines of the cell, with their structures determining their function. In order for a protein to be functional, it must maintain its stable native conformation. A good example of this concept is the structure and function of hemoglobin, which is also commonly compared to the structure and function of myoglobin. In this chapter, we will first review the amino acid building blocks of proteins and how these components contribute to the overall structure of the protein, with a focus on the structure and function of hemoglobin and myoglobin. Because structure determines function, a protein that is misfolded will not function properly. Therefore, folding errors often lead to disease. In this chapter, we will also review human diseases that are caused by the misfolding and/or aggregation of proteins.

18.1 Proteogenic Amino Acids

Proteins are polymers of amino acids linked by peptide (amide) bonds. Each monomer of amino acid in a protein is also known as a "residue". There are twenty proteogenic amino acids for the building blocks of proteins. Based on their side chain, amino acids can be classified as hydrophobic (nonpolar) and hydrophilic (polar) amino acids.

18.2 Hydrophobic Amino Acids

Nonpolar amino acids are water insoluble and are therefore mostly found in the interior part of the protein, or they interact with hydrophobic membranes. At physiological pH, the nonpolar amino acids have a net charge of zero. The hydrophobic amino acids include alanine, glycine, isoleucine, leucine, methionine, phenylalanine, tryptophan, and proline.

Alanine and glycine are the smallest amino acids. The function of alanine in the transport of nitrogen from muscle to liver is an important concept and reviewed in chapter 33. Valine, leucine and isoleucine are branched chain amino acids. Deficiency of the branched chain α-keto acid dehydrogenase enzyme causes maple syrup urine disease, which is reviewed in chapter 34. Many proline residues are found in the amino acid sequence that makes up collagen, and some of these prolines are hydroxylated by the enzyme proline hydroxylase. This enzyme requires vitamin C for its activity, therefore deficiency of vitamin C affects collagen synthesis and can lead to scurvy. Deficiency of the phenylalanine hydroxylase enzyme in the phenylalanine metabolism pathway causes PKU, which is reviewed in chapter 34. Tryptophan is a precursor for serotonin and melatonin, reviewed in chapter 35.

18.3 Hydrophilic Amino Acids

Side chains of polar amino acids can form hydrogen bonds with water. Therefore, these amino acids are classified as water soluble. The water soluble amino acids are cysteine, serine, threonine, asparagine, glutamine, aspartic acid, glutamic acid, tyrosine, arginine, histidine, and lysine.

Cysteine and threonine are sulfur-containing amino acids. Cysteine plays an important role in protein structure, as it is responsible for the formation of intramolecular and/or intermolecular disulfide bonds. Serine and threonine can be phosphorylated by serine-threonine kinases. While aspartic acid and glutamic acid are negatively charged amino acids, arginine and lysine are positively charged. Histidine is also a positively charged amino acid at physiological pH. Histidine is also important for the buffering capacity of physiological solutions, since the pKa of its side chain is about 6.0, the closest to physiological pH. Tyrosine is precursor for melanin and catecholamines (epinephrine, norepinephrine, and dopamine). Tyrosine residues in proteins can be phosphorylated by tyrosine kinases. Glutamine is also a nitrogen carrier in the blood, which is reviewed in chapter 33. Asparagine residues in proteins can be glycosylated on its amide group.

18.4 Structure of Proteins

The primary structure of proteins is simply the amino acid sequence. The local folding of residues into a regular pattern compose the secondary structure of the proteins. Secondary structures consist of common elements such as alpha helices, beta sheets and irregular loops or coils. (▶ Fig. 18.1). The global folding of a protein forms the tertiary structure, which is the further folding of the secondary structure. Some proteins are composed of more than one polypeptide. Such higher-order assembly of a protein complex is the quaternary structure. While the molecular force that determines the primary structure is the peptide bond, which is a covalent bond, the secondary, tertiary, and quaternary structures of proteins are stabilized by weak forces. Hydrogen bonds stabilize the alpha helices and beta sheets of the secondary structure. Although hydrogen bonds are also involved in stabilizing tertiary and quaternary structures, the most important forces for

III

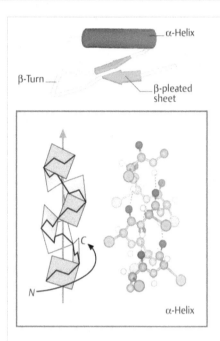
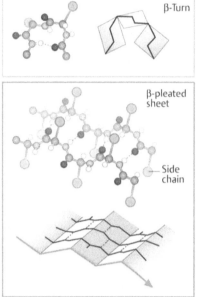

Fig. 18.1 Secondary structures of proteins. (Source: Koolman J, Röhm K, ed. Color Atlas of Biochemistry. 3rd Edition. New York, NY. Thieme; 2012.)

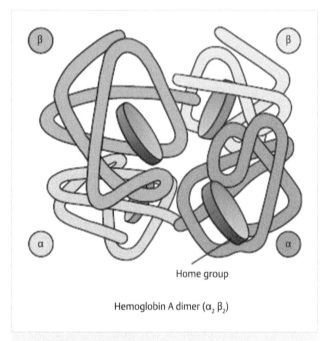

Hemoglobin A dimer ($\alpha_2 \beta_2$)

Fig. 18.2 Quarternary structure of hemoglobin contains 2 alpha and 2 beta globin chains and 4 heme molecules. (Source: Panini S, ed. Medical Biochemistry- An Illustrated Review. New York, NY. Thieme; 2013.)

18.5 The Structure of Hemoglobin

The main function of red blood cells is to transport oxygen from the lungs to peripheral tissues. This is accomplished by 640 million molecules of hemoglobin in each red blood cell. Hemoglobin is a tetrameric protein composed of four globin chains: two α subunits and two β subunits. Each of these subunits contains a heme group, which is responsible for hemoglobin's ability to bind to oxygen (▶ Fig. 18.2). The two α and β subunits join together to form αβ dimers.

The heme group consists of porphyrin (protoporphyrin IX) with Fe^{2+} chelated in its center. The iron found in heme can present in two different forms, ferrous iron (Fe^{2+}) or ferric iron (Fe^{3+}). The ferrous iron (Fe^{2+}) is able to reversibly bind oxygen, while the oxidized ferric iron (Fe^{3+}) is unable to bind to oxygen. The hemoglobin with ferric iron (Fe^{3+}) is called methemoglobin that is unable to bind oxygen causing, deoxygenation of hemoglobin. While the color of oxygenated hemoglobin is red, deoxygenated hemoglobin is blue. Therefore, patients with methemoglobinemias present with bluish/brown color of the skin, known as cyanosis, due to a reduced amount of oxygenated hemoglobin.

18.5.1 Link to Pathology: Congenital Methemoglobinemia

Normally, small amounts of methemoglobin are produced due to oxidation of Fe^{2+} to Fe^{3+}. Such oxidation is corrected by cytochrome b_5 methemoglobin reductase, also known as diaphorase, which reduces Fe^{3+} to Fe^{2+}. Congenital methemoglobinemia is caused by a genetic mutation in methemoglobin reductase that renders it non-functional and leads to accumulation of methemoglobin in the blood. Children with congenital methemoglobinemia present with cyanosis.

determining and stabilizing these higher order structures are ionic, van der Waals, and hydrophobic interactions.

Based on their structure, proteins can be classified into fibrous and globular proteins. Fibrous proteins are thread-like and water insoluble. Some examples of fibrous proteins are keratin, fibrin and collagen. More details of the structure and post-translational modifications of collagen can be found in chaptero 8. In contrast, globular proteins are ball-like and water soluble, both myoglobin and hemoglobin are examples of globular proteins.

18.5.2 Link to Pharmacology: Acquired Methemoglobinemia

Acquired methemoglobinemia can be caused by oxidizing drugs such as nitrites, dapsone, and benzocaine. In these cases, too much methemoglobin quickly accumulates in the blood, with potentially life-threatening consequences.

18.6 Structure of Hemoglobin Facilitates Cooperative Binding to Oxygen

Hemoglobin exists in two forms, the tense (T) form and the relaxed (R) form (▶ Fig. 18.3). The T-form of hemoglobin refers to the form of hemoglobin when oxygen is absent. In the T-form the αβ dimers are more restricted in their movements relative to one another. When oxygen is bound to hemoglobin, it is in the R-form, in which the αβ dimers are able to freely move.

The tetrameric structure of hemoglobin facilitates the loading of oxygen in the lungs and the release of oxygen in the tissues. This reversible binding of hemoglobin to oxygen is due to the presence of the heme group within hemoglobin. Hemoglobin has four separate heme sites for oxygen to bind, although these four heme molecules do not bind oxygen simultaneously. The first O_2 molecule binds to a subunit in the T form, which initiates a conformational change and converts this subunit to the R form. The binding of the first O_2 to the hemoglobin and conformational change of the subunit facilitates the binding of the second O_2 molecule. Likewise, the binding of the second O_2 molecule facilitates the binding of the third O_2 molecule and so on. This is referred to as cooperative binding. In the lungs, where the partial pressure of oxygen is high, hemoglobin is able to rapidly load O_2 for transport since the initial O_2 is able to easily bind the first subunit and facilitate quick loading of the other three subunits with O_2 molecules. When hemoglobin arrives at the extra-pulmonary tissues, the partial pressure of oxygen is low and the first O_2 molecule is able to dissociate. The dissociation of O_2 is also cooperative, as the release of the first O_2 to the extra-pulmonary tissue facilitates the unloading of the rest of the O_2 molecules. This allows for O_2 to be successfully transported from the lungs to the rest of the body.

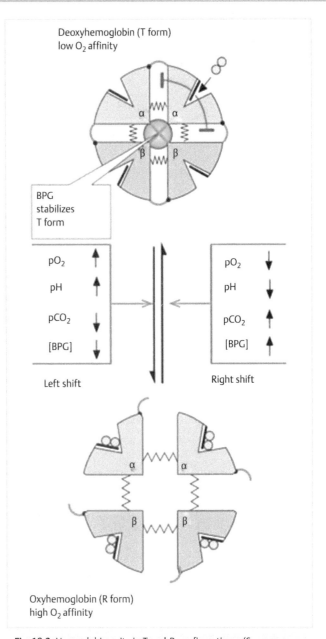

Fig. 18.3 Hemoglobin exits in T and R configurations. (Source: Koolman J, Röhm K, ed. Color Atlas of Biochemistry. 3rd Edition. New York, NY. Thieme; 2012.)

18.7 Oxygen Saturation Curve

The oxygen saturation curve is generated by plotting the saturation of hemoglobin with oxygen against the partial pressure of oxygen (▶ Fig. 18.4). The resulting curve is sigmoidal (S-shape) because of the cooperativity of hemoglobin's subunits for binding to O_2. The oxygen saturation curve for myoglobin is different, however. It forms a hyperbolic shape, which is due to lack of cooperative binding of oxygen by myoglobin. The myoglobin is made of a single peptide, which is structurally homologous to the β globin of hemoglobin. Myoglobin is responsible for storing oxygen in the heart and skeletal muscle for extreme deoxygenation states. While the tetrameric structure of hemoglobin facilitates loading and unloading of oxygen in response to changes in the oxygen level, myoglobin does not release its oxygen until the tissue is in an extreme deoxygenated state.

Hemoglobin and myoglobin also differ in their affinity for oxygen. Hemoglobin is 100% saturated with oxygen when all four hemes are bound to oxygen, which occurs when the partial pressure of oxygen is 100%. Therefore, in the high partial pressure environment of the lungs, hemoglobin saturation is 100%. Hemoglobin is 50% saturated (P_{50}) when the partial pressure of oxygen is 26 mmHg. The O_2-myoglobin dissociation curve is shifted to the left with 50% oxygen saturation at only 2.8 mmHg partial pressure of oxygen, indicating myoglobin has a higher affinity for oxygen than hemoglobin and the oxygen saturation curve for myoglobin is hyperbolic. Therefore, it is the differences in the structures and oxygen affinities of hemoglobin and myoglobin that allow hemoglobin to act as an oxygen transporter, and myoglobin to act as an oxygen storage molecule.

18.8 Agents That Affect O_2 Binding to Hemoglobin

In response to changes in tissue oxygen demands and environmental conditions, molecules within the blood can impact hemoglobin's affinity for oxygen. The most important agents that affect hemoglobin's ability to bind oxygen include: 2,3-Bisphosphoglycerate (2,3-BPG), the hydrogen ion concentration (blood pH), temperature, and CO_2 pressure.

Increased concentrations of 2,3 BPG, H^+ ions, and CO_2 decreases the affinity of hemoglobin for oxygen. Therefore, binding of 2,3 BPG to hemoglobin favors release of oxygen.

Bohr Effect: In the peripheral tissues, delivery (release) of oxygen is facilitated by increased $[H^+]$ and CO_2 (► Fig. 18.5). The oxidation of carbohydrates, lipids, and amino acids produce CO2, which is taken up into erythrocytes. The carbonic anhydrase enzyme within the erythrocytes catalyzes conversion of CO_2 to HCO^{3-}. This reaction also produces H^+ ions, therefore the pH decreases as CO2 produced from catabolism increases.

$$CO_2 + H_2O \leftrightharpoons H^+ + HCO_3^-$$

This acidic environment reduces the affinity of hemoglobin for oxygen, therefore increased CO_2 and H^+ concentration or decreased pH facilitates the release of oxygen from hemoglobin. This phenomenon is known as the Bohr Effect.

A decrease in hemoglobin's affinity for oxygen is referred to as a right shift of the oxygen saturation curve. The oxygen saturation curve shifts to the right when the concentrations of 2,3-BPG, H^+, and CO_2 increase. The left shift of the oxygen saturation curve, on the other hand, is favored by decreased concentrations of 2,3 BPG, H^+, and PCO_2 (► Fig. 18.6).

18.9 Fetal Hemoglobin

During pregnancy, the human fetus produces its own hemoglobin, referred to as hemoglobin F (HbF). Fetal hemoglobin has a higher affinity for oxygen compared to the adult hemoglobin, which is mostly hemoglobin A (HbA). This ensures that the fetus is able to obtain oxygen from the mother's circulation during development. The oxygen dissociation curve for HbF is sigmoidal and shifted to the left, illustrating its higher affinity for oxygen compared to HbA (► Fig. 18.4).

18.10 Protein Folding Errors

In order to function properly, proteins have to be folded into their native three-dimensional structure, either in the tertiary or quaternary conformations. Such higher order structures not only create necessary binding sites, but also impart flexibility, rigidity and stability for proteins to function properly. Misfolding or perturbations in the protein structure may trigger disease. Most of these structural perturbations are caused by mutations within the genes that encode the proteins. For example, individuals with sickle cell disease have a point mutation within the beta chain of hemoglobin that replaces glutamic acid with valine. This substitution results in a conformational change of hemoglobin, causing red blood cells to take on a sickle-shaped appearance.

18.10.1 Link to Pathology: Hemoglobinopathies: Sickle Cell Anemia and β-Thalassemia

Patients with sickle cell anemia or thalassemia do not produce normal hemoglobin. Sickle cell anemia is an autosomal recessive disorder, meaning that homozygotes develop sickle cell disease, whereas heterozygotes are generally asymptomatic

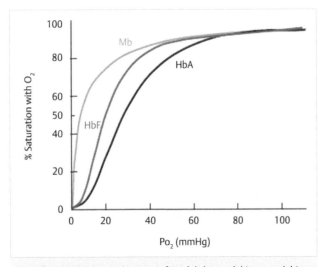

Fig. 18.4 Oxygen saturation curve for adult hemoglobin, myoglobin, and fetal hemoglobin. (Source: Panini S, ed. Medical Biochemistry- An Illustrated Review. New York, NY. Thieme; 2013.)

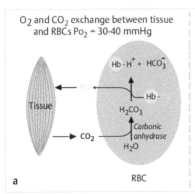

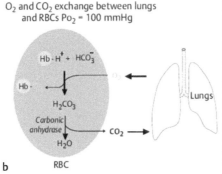

Fig. 18.5 (a, b) Hemoglobin-mediated O_2 and CO_2 transport/exchange. (Source: Panini S, ed. Medical Biochemistry- An Illustrated Review. New York, NY. Thieme; 2013.)

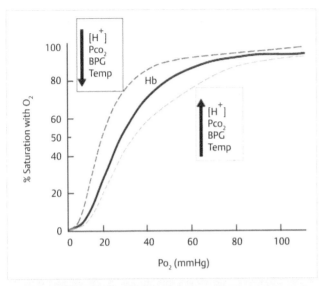

Fig. 18.6 Right and left shifts of the O2-hemoglibin dissociation curve. The right shift represents decreased affinity for oxygen. The left shift represents increased affinity for oxygen.

Fig. 18.7 Sickle cell erythrocytes.

carriers. Carriers do not develop symptoms unless there are extreme circumstances, such as severe hypoxia, deoxygenation of hemoglobin, high temperatures, or a low pH environment. The amino acid substitution from glutamic acid to valine in the β-globin gene of these individuals causes the formation of long polymers between the deoxyhemoglobin molecules. The valine residue in sickle hemoglobin (HbS) creates a hydrophobic patch that increases the hydrophobic interactions between deoxy-HbS molecules. This leads to formation of rod-shape HbS polymers within the erythrocytes, which causes the sickle shape of the erythrocytes (▶ Fig. 18.7). These sickle cells can easily block the small capillary blood vessels, especially when HbS releases its oxygen. Dehydration and low pH also trigger sickling of HbS and result in "sickling" crisis. Patients often present with pain in their legs. The frequency of sickle cell hemoglobin is high in the Mediterranean region, North Africa, and South East Asia.

Deficiency of β-globin synthesis results in β-thalassemia and deficiency of α-globin synthesis results in α-thalassemia. An imbalance in the ratio of alpha and beta chains of hemoglobin causes hemolysis in both thalassemias. In α-thalassemia, excess β-globin chains aggregate, while in β-thalassemia, excess α-globin chains aggregate. In either disease heterozygote are generally asymptomatic, although they may develop anemia in varying degrees due to both ineffective erythropoiesis and hemolysis. The insufficient hemoglobin production presents with microcytic hypochromic anemia. Rapid cell turnover of the erythrocytes increases erythropoetic activity and the absorption of iron, which results in an iron overload over time.

Improper degradation: The proper conformation of proteins is also important for their degradation. Proteins can be degraded by the lysosome or through the ubiquitin-proteasome pathway. These pathways play a critical role in preventing misfolded proteins from accumulating and wreaking havoc within the cell. When these processes do not function properly, whether overactive or underactive, improper degradation of proteins can lead to disease states.

18.10.2 Link to Pathology: Cystic Fibrosis

(CF) is an example of improper protein degradation resulting in a disease state. It is an autosomal recessive disorder that is caused by a defect in the cystic fibrosis transmembrane conductance regulator gene (CFTR) on chromosome 7, most commonly by a deletion of phenylalanine at position 508 (ΔF508) (▶ Fig. 18.8). The ΔF508 interferes with the proper folding of the CFTR protein, resulting in a defective or partially functioning chloride channel. Under normal circumstances, the mutant protein would be tagged for proteasome degradation, rather than being translocated to the cell membrane. In CF patients, however, the chaperone proteins involved in tagging the protein for degradation are unable to properly interact with the mutant CFTR protein and this results in an accumulation of the misfolded proteins within the endoplasmic reticulum.

The CFTR gene encodes an ATP-gated chloride channel that secretes Cl^- within the gastrointestinal tract and the lungs, but reabsorbs Cl^- in the sweat glands. Without properly functioning chloride channels within the lungs, the mucus layer becomes dehydrated due to a decrease in chloride secretion and an increased in reabsorption of Na^+ and H_2O. Because of this dehydration, the secreted mucus in the lungs becomes thicker and more viscous. The viscous mucous forms plugs within the airways and creates the perfect breeding ground for bacteria such as Staphylococcus aureus and Pseudomonas aeruginosa. Therefore, patients with CF present with recurrent pulmonary infections, as well as chronic bronchitis and bronchiectasis (▶ Fig. 18.9). Thickened mucosal secretions also occur in the pancreas, causing decreased secretion of a variety of digestive enzymes from the pancreas. Such pancreatic insufficiency in CF patients results in fat soluble vitamin malabsorption (vitamins A, D, E and K) and steatorrheas (defecation of an abnormal amount of fat in the feces). CF patients are treated with pancreas lipase drugs, which contain varying levels of amylase, lipase and proteases to help with carbohydrate, fat and protein breakdown.

In contrast to the lungs and gastrointestinal tract, the mutant CFTR causes an increased secretion of chloride in the sweat. Therefore, CF can be diagnosed by measuring the amount of chloride present in the sweat. A sweat level of chloride > 60mEq/L is diagnostic for CF.

Improper Localization: Besides gene mutations, other molecular or cellular errors during protein synthesis can also cause protein misfolding. These cellular errors can happen during translation or sorting of the proteins to specific organelles after translation. If the protein is misfolded, the protein cannot be trafficked into its target organelle. This can result in a disease state with dual toxicities due to both the lack of functional protein, and an abnormal accumulation of misfolded proteins

18.10.3 Link to Pathology: α1-Antitrypsin Deficiency

α1-Antitrypsin deficiency is an example of improper localization of a protein contributing to a disease state with dual toxicities. α1-antitrypsin is a protein that inhibits the activity of elastase, which cleaves elastin in the lungs. When α1-antitrypsin is mutated, it is unable to inhibit elastase, which leads to extensive damage of the lungs. The patients develop panacinar emphysema due to the excessive destruction of elastin within the lungs by the uninhibited action of elastase. In patients who smoke, the smoke damages the epithelium, and triggers an immune response. This results in a large influx of neutrophils that respond to the cell injury by releasing increased amounts of proteases. In the case of patients with α1-antitrypsin deficiency, the effects of smoking are amplified because they have lost the ability to regulate protease function. Therefore, in patients who smoke and have α1-antitrypsin deficiency, the proteases cause excessive digestion of elastin, which compounds the risk of developing emphysema. Patients with α1-antitrypsin mutations not only present with emphysema, but also with liver damage. The mutant α1-antitrypsin is misfolded and unable to be transported to the lungs from the liver but accumulates in the endoplasmic reticulum of hepatocytes, causing cirrhosis. Such dual toxicities of the mutant α1-antitrypsin explain why these patients not only present with emphysema, but also have liver damage. The treatment of α1-antitrypsin deficiency involves enzyme replacement therapy. This therapy works well to treat the symptoms in the lungs, but it is less effective in treating the hepatic toxicities. The α1-antitrypsin protein aggregates require treatment with rapamycin or carbamazepine, drugs that increase protein degradation by enhancing autophagy.

Dominant Negative Mutations: Misfolded proteins can also compromise the function of normal, wild type protein and cause the loss of its activity, even in the context of an autosomal recessive allele in the heterozygous state. This type of mutation is known as dominant-negative mutation. Mutations in the p53 can be dominant negative since they can affect the function of the wild type version of the protein. p53 mutations are reviewed in chapters 14 and 53.

Amyloid Accumulation: Amyloid, which consists of insoluble fibril deposits, can be derived from a variety of misfolded proteins. The aggregates of fibrils are insoluble and deposit in tissues, causing organ damage. Amyloid deposition, or amyloidosis, is the cause of diverse clinical conditions including neurodegenerative disorders such as Alzheimer's, Parkinson, and spongiform encephalopathies. A familial form of amyloidosis is transthyretin-related amyloidosis. Normally, transthyretin (TTR) functions as a thyroxine and retinol transporter in the blood. However, TTR gene mutations cause misfolding of the TTR protein and its accumulation as amyloid in the nervous system and cardiac tissue.

Extracellular accumulations of amyloid β peptide in the cerebral cortex and basal ganglia is observed in **Alzheimer's disease**. The amyloid β peptide is derived from the beta-amyloid precursor protein (APP). Down syndrome patients and patients who have mutations in the APP gene, located on chromosome 21, produce excess APP protein. This excess

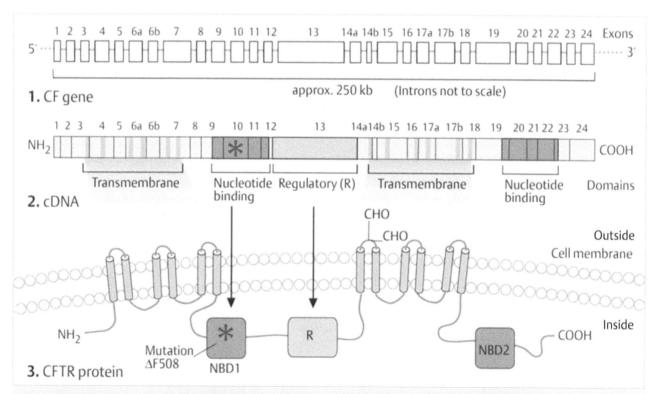

Fig. 18.8 The CFTR gene and its protein The most common mutation of the CFTR gene is the deletion of phenylalanine in the nucleotide binding domain of the CFTR protein. (Source: Passarge E, ed. Color Atlas of Genetics. 4th Edition. New York, NY. Thieme; 2012.)

Cystic fibrosis (Mucoviscidosis)

Severe progressive disease of the bronchial system and gastrointestinal tract

Disturbed function of a chloride ion channel by mutations in the *CFTR* gene

Autosomal recessive

Gene locus 7q31.2

Disease incidence approx. 1:2500

Heterozygote frequency approx. 1:25

Mutation ΔF508 in approx. 70%

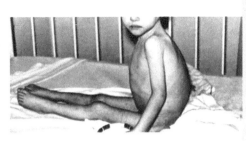

Fig. 18.9 Cystic fibrosis patients suffer from thickened mucosal secretions in the lungs and gastrointestinal system. In contrast to this, the mutant chloride channel causes increased chloride secretion in the sweat. (Source: Passarge E, ed. Color Atlas of Genetics. 4th Edition. New York, NY. Thieme; 2012.)

production of APP results in excess amyloid β peptide. Therefore, Down syndrome patients and individuals who have APP mutations develop early-onset Alzheimer's disease.

Parkinson disease is a neurological disorder caused by degeneration of the dopamine producing neurons in the substantia nigra. A few proteins have been identified as being involved in the heritable forms of Parkinson disease. These proteins are α-synuclein, parkin, and ubiquitin. Intracytoplasmic inclusion bodies of these proteins in the neurons are called Lewy bodies. Intracellular accumulation of α-synuclein causes increased oxidative stress due to mitochondrial dysfunctions.

Creutzfeldt-Jakob disease (CJD) is a type of **spongiform encephalopathy** that is found in two three major forms in humans. Familial CJD is caused by mutation of the prion gene, which is inherited in an autosomal dominant manner. Sporadic CJD results from spontaneous misfolding of endogenous prion. The variant CJD is caused by exposure to prions from bovine spongiform encephalopathy.

All prion diseases are caused by accumulation of abnormal or pathological prion protein (PrPSc). Although the wild type prion (PrPC) and pathological prion protein have the same amino acid sequence, the secondary structure of wild type prion protein has mostly alpha helices, whereas the secondary structure of pathological prion protein is dominated by beta-sheets. Since the pathological prion protein is protease resistant and insoluble, it forms amyloid fibrils in the nerve cells. This aggregated protein causes neurodegeneration and sponge-like holes.

Amyloid shows apple-green birefringence under polarized light when stained with Congo red due to the cross-β-pleated conformation. This feature has been used as the gold standard for diagnosis of amyloid related diseases.

References

[1] Mechanisms of protein-folding diseases at a glance Julie S. Valastyan and Susan Lindquist. The Company of Biologists Ltd |. Dis Model Mech. 2014; 7:9–14

[2] Chaudhuri TK, Paul S. Protein-misfolding diseases and chaperone-based therapeutic approaches. FEBS J. 2006; 273(7):1331–1349

[3] https://emedicine.medscape.com/article/335301-overview#a6

Review Questions

1. The side chain of which of the following amino acids has optimal buffering capacity at physiological pH?
   A) Arginine
   B) Aspartic acid
   C) Histidine
   D) Glutamic acid
   E) Lysine

2. Increased concentrations of which of the following compounds shifts the equilibrium of the hemoglobin conformations from T state to R state?
   A) 2,3 BPG
   B) CO_2
   C) H^+ ions
   D) Fe^{2+}
   E) $O2$

3. Which of the following statements is correct regarding the pathological prion protein?
   A) Familial CJD is acquired from mad cow disease
   B) The structure of pathological prion protein is dominated by alpha-helices
   C) The structure of pathological prion protein is dominated by beta-sheets
   D) Pathological prion is susceptible to proteases in the cells

4. Which of the following intermediates/products of anaerobic glycolysis would most likely increase the release of O2 from hemoglobin?
   A) 1,3 BPG
   B) ATP
   C) Lactate
   D) NADH
   E) Pyruvate

Answers

1. **The correct answer is C.** Buffers work best within one pH unit of their pKa values. Since the side chain of histidine has the pKa value of ~6.0, histidine is a good buffer at the physiological pH of 7.4. The side chains of aspartic acid and glutamic acid have pKa values ~4.0, and the side chains of arginine and lysine has pKa value of ~9.0. Therefore, at physiological pH, the amino acids in answer choices A, B, C, and D do not have the best buffering capacity.

2. **The correct answer is E.** Increased concentrations of O2 favor the oxygenation of hemoglobin and, therefore, shift the equilibrium of hemoglobin conformations towards the oxygenated R state. Increased concentrations of 2,3BPG, CO_2, and H^+ ions increase the release of oxygen from hemoglobin. Therefore, these compounds shift the equilibrium of hemoglobin conformations from oxygenated R state to deoxygenated T state. Hence, answer choices A, B, and C are incorrect. Fe^{2+} is important for binding of oxygen to the heme group, but Fe^{2+} is a prosthetic group of heme and does not dissociate from heme itself, therefore changing the Fe^{2+} concentrations in the environment do not affect the release of oxygen.

3. **The correct answer is C.** The structure of pathological prion protein is dominated by beta-sheets. The structure of wild type prion protein is dominated by alpha-helices. Pathological prion protein is protease resistance and insoluble. The familial CJD is caused by mutant prion gene, which is inherited in an autosomal dominant manner. The variant CJD is acquired from bovine spongiform encephalopathy.

4. **The correct answer is C.** Increased CO_2 and H^+ concentrations or decreased pH facilitates the release of oxygen from hemoglobin. The acidic environment (H^+ ions) reduces the affinity of hemoglobin for oxygen. Therefore, any acidic metabolite, such as lactate, increases the production of H^+ ions and consequently facilitates release of oxygen from hemoglobin. Answer choice A is incorrect because 2,3 BPG facilitates oxygen release, not 1,3 BPG. Answer choices B, D, and E have no effect on oxygen binding or the release.

19 Molecular and Biological Techniques

At the conclusion of this chapter, students should be able to:

- Identify the components required for a PCR reaction, describe the process of PCR, and explain how PCR can be used to identify the presence of specific infectious agents
- Describe the mechanism of action of restriction enzymes and how they can be used to create recombinant DNA, and explain how restriction enzymes can be used in the techniques of restriction length polymorphism analysis and DNA fingerprinting
- Explain the processes of Southern blotting and northern blotting and how the Southern blotting technique can be used to identify specific genetic diseases
- Describe the process of chain-termination sequencing (Sanger method) and demonstrate the ability to read sequences using this method
- Explain the process of western blotting, ELISA, and FISH, and identify their diagnostic uses

There is a multitude of molecular and biochemical techniques utilized to manipulate DNA, RNA, and proteins for clinical and research use. We will discuss some of the most common ones you will encounter. Refer to Chapters 2, 3, and 4 for an introduction to DNA, RNA, and protein synthesis.

19.1 Gel Electrophoresis

Gel electrophoresis is a technique used to separate macromolecules, such as DNA, RNA, or proteins, of differing size by administration of an electrical field. To separate the DNA by size, a sample is placed in a well at one end of an agarose gel while submerged in a buffer solution. Agarose is a polysaccharide molecule extracted from seaweed and forms a porous sponge when hardened into gel. An electrical current is applied with the negative electrode near the well and the positive electrode on the distal end (▶ Fig. 19.1). Since the backbone of DNA is negatively charged due to the phosphate molecules, the fragments will migrate toward the positive end of the gel. The smaller the DNA fragments, the easier they are to navigate through the porous agarose gel, and the faster they travel towards the other end. One usually mixes the original DNA sample prior to separation with a fluorescent tag such as ethi-

dium bromide. This allows the separated DNA bands to be visible under ultraviolet light and photographed.

19.2 Polymerase Chain Reaction

Polymerase chain reaction, or PCR, allows for amplification of one DNA fragment into vast quantities (▶ Fig. 19.2). A test based on this principal can be used for clinical diagnosis such as detecting a virus in cerebrospinal fluid or in the evaluation of urethral discharge. To amplify a specific fragment one must mix together a sample containing certain ingredients. First, a sample containing the DNA of interest must be collected. Since the PCR reaction involves the addition of heat to "melt" the two strands of DNA apart, a heat-stable DNA polymerase must be used. Taq polymerase, isolated from bacteria found in hot springs and near hydrothermal vents, is used for the reaction. Primers, which are made up of short nucleotide sequences, are needed for the DNA polymerase to build upon (as you recall DNA polymerase requires a nucleotide with a free 3'OH group to start the elongation process). A pool of deoxynucleotide triphosphates (dATP, dTTP, dCTP, and dGTP) must be added as well. All of these components are mixed, in a special thin-walled tube, with a buffer solution providing optimal pH and containing magnesium which acts as a cofactor for the DNA polymerase. The mixture is then placed inside a machine called a thermocycler, which has the ability to make wide temperature changes rapidly.

The process of PCR amplification is shown in ▶ Fig. 19.2 and involves 3 different steps.

1. Denaturation – High temperatures are used to denature the DNA, resulting in two separate strands.
2. Annealing – Cooling allows primers to anneal to complementary sequences on the desired segment of DNA.
3. Elongation – The heat-stable DNA polymerase adds nucleotides to replicate the DNA sequence that lies between the two primers.

These steps are repeated multiple times, resulting in the exponential amplification of the specific DNA segment. This has proven to be a powerful technique for generating detectable amounts of a specific DNA sample from a limited source. For example, a single DNA molecule can be amplified to 68 billion copies in only 36 cycles of the reaction. Typically, the resulting

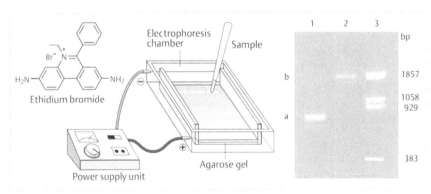

Fig. 19.1 DNA electrophoresis. (Source: Koolman J, Röhm K, ed. Color Atlas of Biochemistry. 3rd Edition. New York, NY. Thieme; 2012.)

III

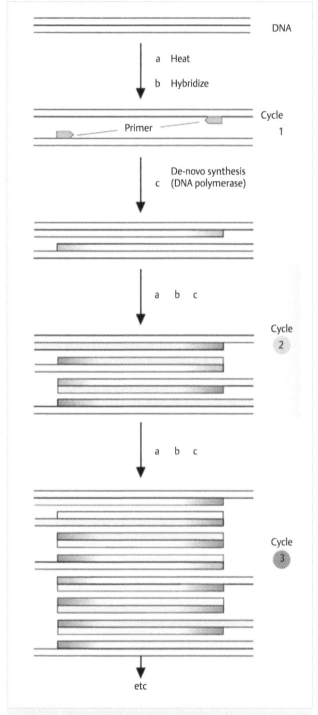

Fig. 19.2 Polymerase chain reaction (PCR). (Source: Koolman J, Röhm K, ed. Color Atlas of Biochemistry. 3rd Edition. New York, NY. Thieme; 2012.)

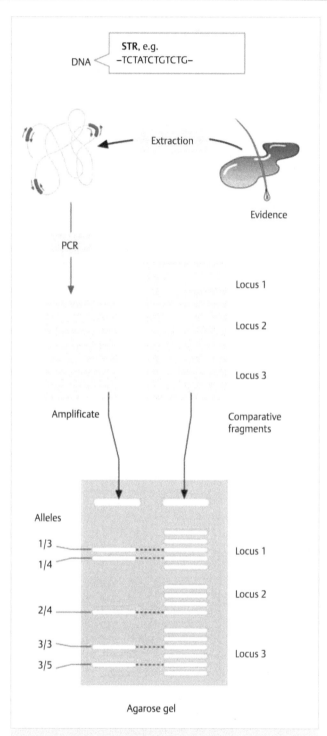

Fig. 19.3 DNA typing. (Source: Koolman J, Röhm K, ed. Color Atlas of Biochemistry. 3rd Edition. New York, NY. Thieme; 2012.)

product of PCR amplification is separated using gel electrophoresis and stained with a dye to allow for its visualization.

PCR diagnostic tests are used to identify the presence of an infectious agent by determining if certain DNA is present in bodily fluids (▶ Fig. 19.3). Primers specific for the DNA sequences of various pathogens such as *Neisseria gonorrhoeae*, herpes simplex virus, or human immunodeficiency virus (HIV) can be used to identify an active infection in the

human body. Because of the amplification capabilities of PCR, a very low level of infection can be detected. Another advantage of using PCR to identify an infectious agent is that the microorganism can be detected even during the "window period", defined as the time between the initial infection and the presence of detectable antibodies against the infectious agent. As opposed to ELISA and western blot, discussed later in the chapter, PCR does not rely on the presence

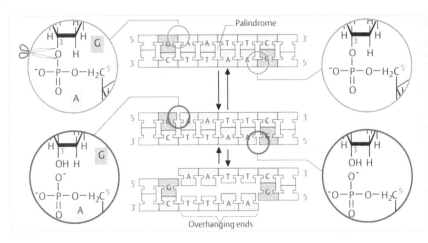

Fig. 19.4 Restriction endonucleases. (Source: Koolman J, Röhm K, ed. Color Atlas of Biochemistry. 3rd Edition. New York, NY. Thieme; 2012.)

III

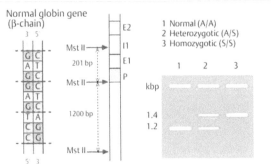

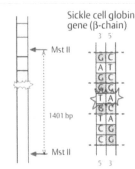

Fig. 19.5 Diagnosis of sickle-cell anemia using RFLP. (Source: Koolman J, Röhm K, ed. Color Atlas of Biochemistry. 3rd Edition. New York, NY. Thieme; 2012.)

of these antibodies and, thus, can yield a positive result much earlier on in the infection.

19.3 Restriction Enzymes

Restriction enzymes, also referred to as restriction endonucleases, are proteins commonly derived from bacteria that cleave double-stranded DNA at specific nucleotide sequences (▶ Fig. 19.4). These DNA sites are commonly 6 nucleotides long and often contain palindromes where the DNA sequence is the same when read in the 5' to 3' direction. The different enzymes are given specific names, based on the bacteria they are isolated from, with common ones including *Bam*HI, *Hind*III, and *Eco*RI. The restriction enzyme can cut the DNA in two different ways. The enzyme can act as a sort of scissor by either cutting the DNA strand to generate blunt ends or, more commonly, the restriction enzyme can cut in a staggered fashion leaving a few unpaired nucleotides on either end. This is termed "sticky ends" and is seen in ▶ Fig. 19.4. These sticky ends are complementary and can be rejoined using a DNA ligase.

A technique called restriction fragment length polymorphism (RFLP) utilizes the principal behind restriction enzymes to identify variations in homologous DNA samples (▶ Fig. 19.5). A DNA sample is "digested" with a restriction enzyme and the resulting restriction fragments are separated by gel electrophoreses based on length. This method can be used to identify the presence of various genetic diseases, including sickle cell. In sickle cell disease, there is a single nucleotide mutation in the beta globin gene from base A to T resulting in an amino acid substitution in which glutamate is replaced with valine. This specific mutation in the

beta globin gene lies in a restriction enzyme recognition site known as *Mst*II. The mutation prevents the enzyme MstII from cutting this site. Therefore, the DNA fragments generated from cutting with MstII will be different lengths compared to the normal gene. A technique called southern blot, discussed later in the chapter, is used to identify the presence or absence of the mutation based on the size of the bands that result (▶ Fig. 19.5). Although this technique was largely important in determining genetic disorders and even paternity testing, it has fallen out of favor due to new technologies such as DNA sequencing.

DNA fingerprinting was first developed for use in 1985 and has been used since in criminal cases to determine if a sample obtained from a crime scene matches a suspect. This concept is based on using restriction endonucleases to cut genomic DNA into fragments and using gel electrophoresis to separate them according to length. Genomic DNA contains multiple regions of repeated sequences, known as variable number of tandem repeats or VNTRs, which vary in length among different individuals. These VNTRs are flanked by restriction enzyme recognition sites that can be cut with their corresponding restriction enzymes to produce a distinct banding pattern when separated using gel electrophoresis. Only monozygotic twins, also known as identical twins, will have identical VNTR and banding patterns. ▶ Fig. 19.6 shows a graphic representation of DNA sequencing.

19.4 Recombinant DNA and DNA Cloning

A recombinant DNA molecule is one which contains fragments of DNA from two or more independent sources that have been

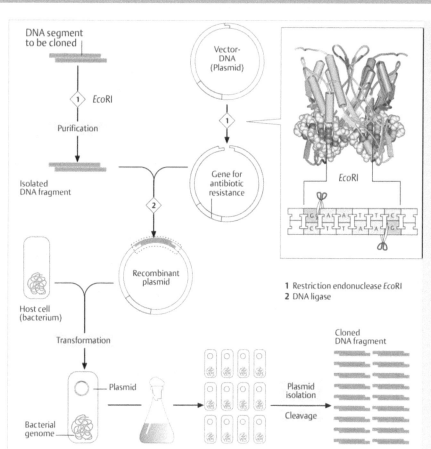

Fig. 19.6 DNA cloning. (Source: Koolman J, Röhm K, ed. Color Atlas of Biochemistry. 3rd Edition. New York, NY. Thieme; 2012.)

molecularly combined. Typically, it is formed through a process in which a DNA fragment is incorporated into a circular DNA molecule known as a plasmid (▶ Fig. 19.6). The procedure involves the cutting of a plasmid with a restriction enzyme at a specific site yielding a linear DNA sequence with "sticky" ends. A gene of interest with flanking restriction sites, identical to the one in the plasmid, is also cut with the specific restriction enzyme. When mixed in a solution these two separate DNA strands combine, joining their "sticky" ends and yielding a recombinant DNA plasmid.

DNA cloning is a process in which bacteria is used to produce a large amount of a specific DNA molecule, most commonly by amplification of a recombinant plasmid. A solution containing a recombinant DNA plasmid with the gene of interest is mixed with a solution of bacteria, most often *E. coli*. The bacteria will uptake the plasmid and begin making identical copies during cell division through exponential amplification. This chimeric DNA plasmid may then be harvested and purified for use in other experiments. The bacteria may also be induced to make protein from the DNA of interest located on the plasmid (▶ Fig. 19.7). This protein can be harvested and purified and is one of the processes for synthesizing large quantities of therapeutic proteins. Some notable therapeutic proteins that have been synthesized this way are insulin for treating diabetes mellitus, tissue plasminogen activator for treating ischemic strokes, and erythropoietin for treating certain forms of anemia.

19.5 Fluorescent Proteins

Some animal species have been found to have the ability to emit a bioluminescence signal in presence of certain wavelengths of light. This ability is due to the production of specialized proteins, called fluorescent proteins that absorb specific wavelengths of light and emit a different wavelength of light. The jellyfish *Aequorea victoria* produces a green florescent protein (GFP) which emits a green light and is popular for use in biomedical research. This protein can be fused to a cellular protein of interest, expressed in the cell, and visualized under a fluorescent microscope to follow the movement of the protein in the cell or to identify specific structures within living cells. ▶ Fig. 19.8 shows different components of the cell that can be visualized with the help of fluorescent reagents. A common fusion fluorescent protein used in biomedical research is α-tubulin-GFP. First, α-tubulin cDNA is mixed with a plasmid containing the GFP gene and a promoter sequence. These are cut with the same restriction endonuclease. The sticky ends of the α-tubulin gene are then combined with the sticky ends of the plasmid in between the promoter and GFP resulting in an α-tubulin-GFP fusion gene which can be transfected into the cell. Inside the cells, this fusion gene is made into a fusion protein, which is incorporated into the microtubule cytoskeleton, allowing for visualization of microtubule dynamics in live cells.

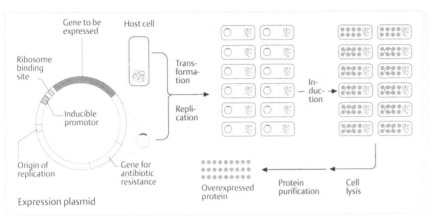

Fig. 19.7 Overexpression of proteins. (Source: Koolman J, Röhm K, ed. Color Atlas of Biochemistry. 3rd Edition. New York, NY. Thieme; 2012.)

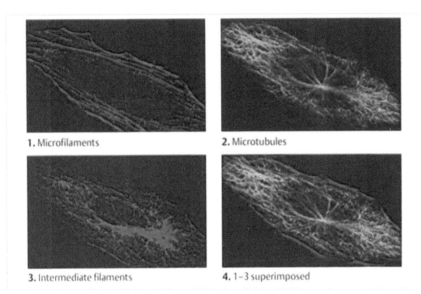

Fig. 19.8 Components of the cytoskeleton. (Source: Koolman J, Röhm K, ed. Color Atlas of Biochemistry. 3rd Edition. New York, NY. Thieme; 2012.)

19.6 Nucleic Acid Probes

Nucleic acid probes are short fragments of single stranded DNA or RNA connected to a fluorescent molecule or radioactive nucleotide. These are used to identify the presence, or lack thereof, of a specific gene within a chromosome. In addition, they can be used to determine if a detrimental chromosomal break or rearrangement has occurred. In this procedure, a double stranded DNA sample, such as a human chromosome, is heated to allow for separation of the two strands. A nucleic acid probe is then added to the solution and the temperature is lowered so that the probe can anneal to the DNA sample. The probe only hybridizes with its complimentary sequence, and the gene of interest can be visualized by its tag. This is the basis of the molecular technique known as fluorescence in situ hybridization (FISH) (▸ Fig. 19.9). In this case, fluorescence-labeled single stranded DNA probes are synthesized to recognize the DNA sequence of a chromosomal gene of interest. The gene of interest may have a mutation or contain some other alteration. These fluorescent probes are allowed to hybridize to chromosomal DNA. The location and number of genes can be directly viewed under a fluorescent microscope. This technique can be used to determine the gain or loss of genes within the chromosome such as proto-oncogenes and tumor suppressor genes

respectively. Other chromosomal alterations such as translocations, amplifications, and deletions may also be detected using FISH (▸ Fig. 19.10). Some notable examples of diseases detected using FISH are *ERBB2* gene amplifications in certain kinds of breast cancer and *BCR-ABL* translocations in chronic myelogenous leukemia.

19.7 Southern Blot

Southern blot is a technique used to visualize specific DNA sequences in a DNA sample. An unlabeled DNA sample is separated according to size on agarose gel using electrophoresis. After separation, the agarose gel is sandwiched in between a sponge and nitrocellulose membrane underneath a stack of absorptive paper towels while in a buffer solution. The separation of DNA is "blotted" onto the nitrocellulose membrane by suction, or wicking, of buffer through the gel and paper. The nitrocellulose membrane, which is much stronger than the agarose gel, can be incubated with a solution containing a fluorescent or radiolabeled probe complimentary to the DNA sequence of interest. The labeled probe is hybridized to the separated DNA sequences on the nitrocellulose membrane and, in the case of a radiolabel, can be visualized by exposing the membrane to an undeveloped x-ray film in a dark room

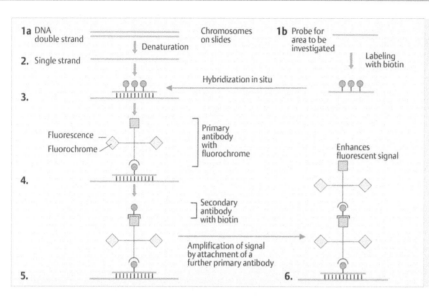

Fig. 19.9 Principle of fluorescence in situ hybridization. (Source: Passarge E, ed. Color Atlas of Genetics. 4th Edition. New York, NY. Thieme; 2012.)

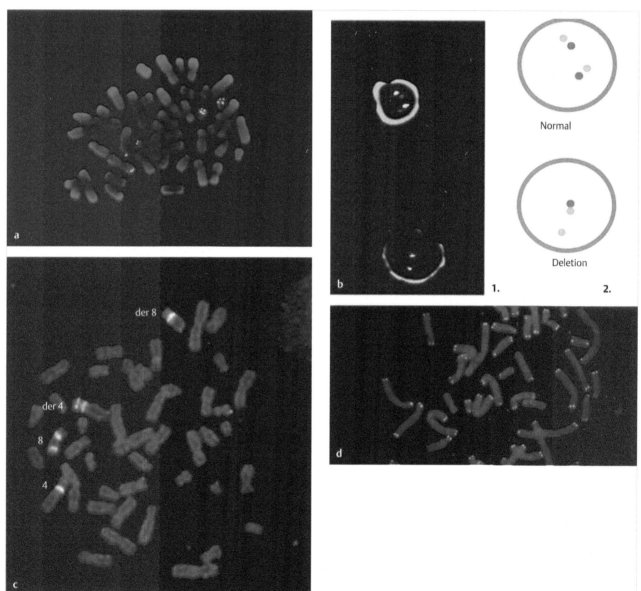

Fig. 19.10 **(a)** Example of FISH analysis in metaphase. **(b)** Interphase FISH analysis. **(c)**Translocation 4;8. **(d)** Telomere sequences in metaphase chromosomes. (Source: Passarge E, ed. Color Atlas of Genetics. 4th Edition. New York, NY. Thieme; 2012.)

(▶ Fig. 19.11). Southern blotting can be used to detect genetic diseases with known mutations. One example of this is sickle cell disease. Sickle cell disease is a result of a point mutation in the β globin gene which results in a sequence that encodes the amino acid valine at position 6 instead of the normal glutamic acid. This point mutation was found to eliminate a recognition site for the restriction enzyme MstII. A normal β globin gene has 3 MstII sites whereas an abnormal β globin gene in sickle cell disease has only 2. This allows a southern blot to detect for the presence or absence of sickle cell disease based on the difference in both the number of bands and the sizes of the individual bands generated when a genomic sample is digested with the MstII enzyme. Since the normal β globin gene has 3 MstII sites, DNA cut with this restriction enzyme will yield 2 small fragments on a southern blot compared to 1 larger fragment in someone with sickle cell disease (▶ Fig. 19.12). A northern blot procedure is performed in a similar way as the southern blot except it is used to detect RNA as opposed to DNA.

19.8 DNA Sequencing

DNA sequencing is the process of determining the precise order of nucleotides within a DNA molecule. One of the most widely used methods was the Sanger (chain-termination) method (▶ Fig. 19.13). Developed in the 1970s, this technique utilizes dideoxynucleotides to inhibit DNA propagation. Recall from chapter 2 that a deoxyribonucleoside triphosphate (dNTP) contains a hydroxyl (-OH) group in the 3' position, which is required by the DNA polymerase in order to add nucleotides to the newly-synthesized DNA chain. Dideoxyribonucleoside

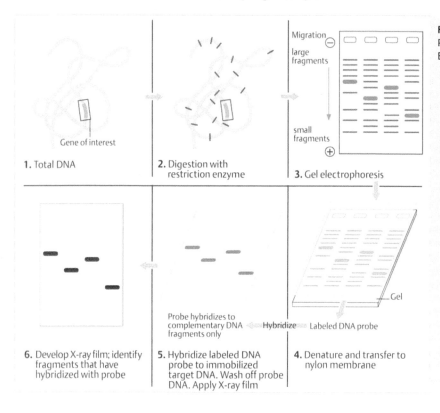

Fig. 19.11 Southern blot hybridization. (Source: Passarge E, ed. Color Atlas of Genetics. 4th Edition. New York, NY. Thieme; 2012.)

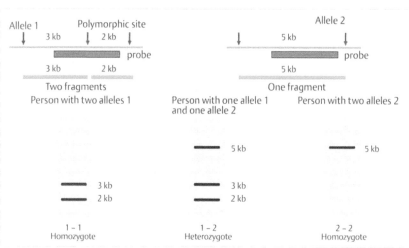

Fig. 19.12 Restriction fragment length polymorphism (RFLP). (Source: Passarge E, ed. Color Atlas of Genetics. 4th Edition. New York, NY. Thieme; 2012.)

III

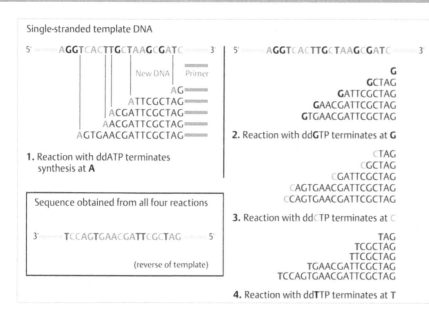

Single-stranded template DNA

5'——— AGGTCACTTGCTAAGCGATC ——— 3'

New DNA | Primer

AG
ATTCGCTAG
ACGATTCGCTAG
AACGATTCGCTAG
AGTGAACGATTCGCTAG

1. Reaction with ddATP terminates synthesis at **A**

Sequence obtained from all four reactions

3'——— TCCAGTGAACGATTCGCTAG ——— 5'

(reverse of template)

5'——— AGGTCACTTGCTAAGCGATC ——— 3'

G
GCTAG
GATTCGCTAG
GAACGATTCGCTAG
GTGAACGATTCGCTAG

2. Reaction with dd**G**TP terminates at **G**

CTAG
CGCTAG
CGATTCGCTAG
CAGTGAACGATTCGCTAG
CCAGTGAACGATTCGCTAG

3. Reaction with dd**C**TP terminates at **C**

TAG
TCGCTAG
TTCGCTAG
TGAACGATTCGCTAG
TCCAGTGAACGATTCGCTAG

4. Reaction with dd**T**TP terminates at **T**

Fig. 19.13 Dideoxy DNA sequencing. (Source: Passarge E, ed. Color Atlas of Genetics. 4th Edition. New York, NY. Thieme; 2012.)

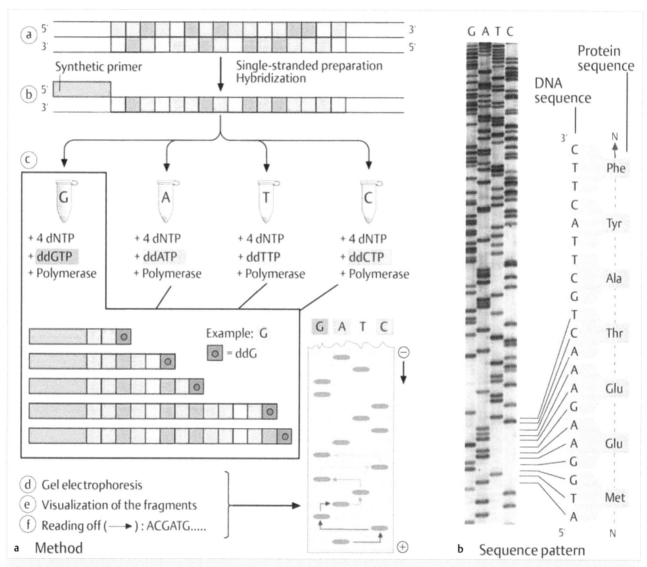

Fig. 19.14 (a) Sequencing of DNA – Method. **(b)** Sequencing of DNA – Sequence pattern. (Source: Koolman J, Röhm K, ed. Color Atlas of Biochemistry. 3rd Edition. New York, NY. Thieme; 2012.)

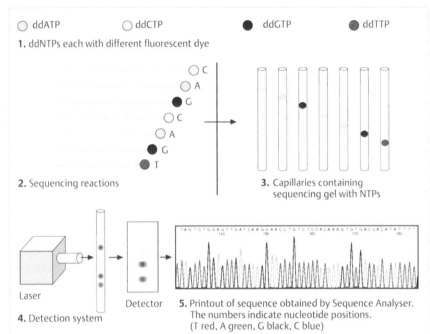

Fig. 19.15 Automated DNA sequencing. (Source: Passarge E, ed. Color Atlas of Genetics. 4th Edition. New York, NY. Thieme; 2012.)

1. ddNTPs each with different fluorescent dye
   - ○ ddATP ○ ddCTP ● ddGTP ● ddTTP

2. Sequencing reactions

3. Capillaries containing sequencing gel with NTPs

4. Detection system — Laser, Detector

5. Printout of sequence obtained by Sequence Analyser. The numbers indicate nucleotide positions. (T red, A green, G black, C blue)

triphosphates (ddNTP) do not contain this hydroxyl group and, therefore, do not allow chain extension on the 3' end. When a solution containing dNTPs (dATP, dTTP, dCTP, dGTP) is mixed with a small amount of ddNTPs (ddATP, dTTP, ddCTP, ddGTP) the rare incorporation of a ddNTP by DNA polymerase blocks further elongation of the DNA strand. Because these ddNTPs are incorporated randomly as the DNA is synthesized, numerous fragments of differing lengths are generated (▶ Fig. 19.14**a**). Using agarose gel electrophoreses, a set of 4 mixtures is separated side by side. Each mixture includes the DNA in question, a labeled primer, excess dNTPs, and a DNA polymerase. A small amount of specific ddNTPs are added to only one mixture each. One mixture contains ddATP, another ddGTP, and so on. After using gel electrophoresis to separate the various sizes of reaction products, the sequence is read from the bottom of the agarose gel and moving upward, which corresponds to the 5' to 3' DNA sequence (▶ Fig. 19.14**b**). Automated DNA sequencing method uses the same principle with the Sanger method. In the automated DNA sequencing, different colored fluorophores are used for the fluorescence labeling of DNA (▶ Fig. 19.15).

19.9 Protein Techniques

19.9.1 Gel Electrophoresis

The process of polyacrylamide gel electrophoresis (PAGE) is similar to that of agarose gel electrophoresis but is used to separate proteins based on size instead of nucleic acids. A protein sample is heated with sodium dodecyl sulfate (SDS) which acts as an anionic surfactant coating the protein in a negative charge and reducing agent such as β-mercaptoethanol which separates any of the protein subunits held together by disulfide bonds. These samples are loaded onto a polyacrylamide gel and migration through the porous matrix is mediated by an electrical current. Over time, the smaller proteins move faster towards the bottom of the gel (▶ Fig. 19.16).

19.10 Western Blot

A western blot is used to visualize specific proteins in a sample and is analogous to visualizing DNA with gel electrophoresis followed by southern blot. This technique comprises a multistep process much similar to southern or northern blotting. First, proteins are separated using PAGE. The polyacrylamide gel is then sandwiched in between a nitrocellulose membrane and filter paper. An electrical current is applied allowing the negatively charged proteins to transfer from the gel to the membrane in the same orientation. The membrane containing the protein is then incubated with a primary antibody directed against a protein of interest. A second incubation is performed with secondary antibodies directed against the primary antibodies. These secondary antibodies are conjugated to an enzyme that produces light when a substrate is added. Upon addition of the enzyme substrate, a light will be detected if the sample contained the protein of interest (▶ Fig. 19.16).

Western blotting is used in many diagnostic tests. One example of this is in human immunodeficiency virus (HIV) testing. When a patient becomes infected with HIV, their body creates antibodies against viral proteins that become detectable in a process known as seroconversion. Known HIV proteins are separated using PAGE and the patient's serum is used as a source of primary antibodies for the western blot. If HIV$^+$, the patient's antibodies will bind to the viral proteins allowing secondary antibody binding and diagnosis of HIV infection. This technique can be used to identify and diagnose other infectious diseases as well.

19.11 ELISA

Enzyme-linked immunosorbent assay (ELISA) is a sensitive method for detecting the presence of an antibody or other protein in a relatively short period of time. An antigen of interest is

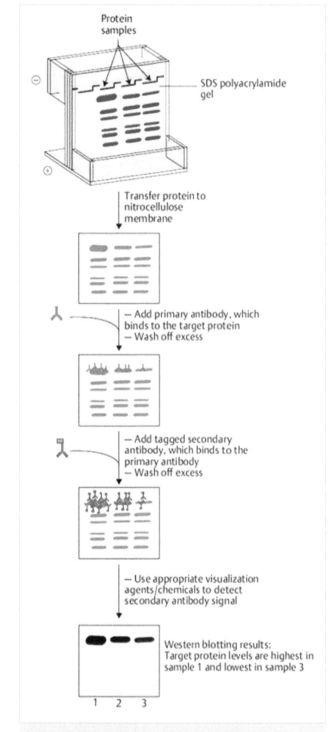

Fig. 19.16 Western blot. (Source: Panini S, ed. Medical Biochemistry-An Illustrated Review. New York, NY. Thieme; 2013.)

followed by a western blot for confirmation due to the prevalence of false-positive results. Many home HIV test kits operate using this technology.

19.12 Southwestern

A southwestern blot is a technique used to identify protein-DNA interactions. Made up of a combination of southern and western blot procedures, a sample is run using SDS-PAGE to separate proteins based on size. Separated proteins are then transferred to a membrane similar to a western blot. Double stranded DNA probes of a specific sequence are then added to the blot. If the probes bind to a protein on the blot, a DNA-protein interaction exists.

19.13 Microarray

A microarray, also known as DNA chip or gene chip, can be used to compare gene expression patterns between cells exposed to two different conditions or between a healthy cell and diseased cell. Single stranded DNA molecules representing all expressed genes in a cell are fixed to a glass slide or silicone film and used as a reference. The mRNA is isolated from each of the sets of cells and reverse transcriptase is used to make a DNA copy of the mRNA that is labeled with a dye. This DNA copy is known as complimentary DNA or cDNA. One sample may contain a red dye labeled cDNA and the other sample a green dye labeled cDNA. These labeled cDNAs are then hybridized to the glass slide containing the reference DNA. If a spot is red, that means that specific gene was expressed at a higher level in the red-labeled samples. If a spot is green, that means that specific gene was expressed at a higher level in the green-labeled sample

If the spot is yellow, that means that specific gene was expressed at the same level in both samples. In this way, the expression, or lack thereof, of specific genes can be matched to the tested condition or disease (▶ Fig. 19.18).

19.14 Proteomics

Proteomics is a technique used to compare the differences in expression of proteins between two samples. Proteins from two different samples are isolated and each sample is labeled with a different fluorescent dye. The proteins are then separated by two-dimensional (2D) gel electrophoresis. The first dimension separates based on charge and the second separates based on size. A computer program aligns the spots and determines which proteins are upregulated or downregulated based on the intensity of the spot.

19.15 Gene Therapy

Gene therapy is a method for introducing normal copies of defective genes into a genome. This technique does not replace the defective gene, it just adds a normal copy. A copy of the normal gene is incorporated into a viral particle and the virus is used to infect the patient's cells (▶ Fig. 19.19). Once inside the cell the gene is processed into a functional protein utilizing the cells normal transcriptional and translational system. There are

bound to the bottom of a welled plate. The patient's serum is added to the wells and washed out. If antibodies are present, they will bind to the antigens in the wells. An enzyme-linked secondary antibody is added, followed by the enzyme substrate. Detection of a light or color change confirms presence of reactive antibody and, thus, infection (▶ Fig. 19.17**a**). This is commonly used as a screening test for HIV (▶ Fig. 19.17**b**),

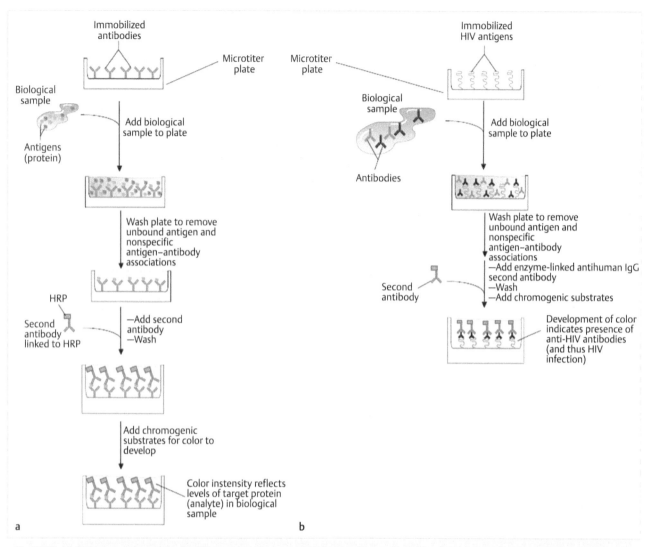

Fig. 19.17 (a) ELISA: general principle. **(b)** ELISA used to diagnose HIV infection. (Source: Panini S, ed. Medical Biochemistry- An Illustrated Review. New York, NY. Thieme; 2013)

two common viral vectors used to infect cells. One is a retrovirus and the other is an adenovirus. Retroviruses are RNA viruses that use reverse transcriptase to make a double stranded DNA copy of their RNA genome. They can stably integrate their DNA into a cells' nuclear DNA, thereby making treatment essentially permanent. Some problems with this technique include random incorporation into the genome, possibly disrupting normal functioning genes, and needing actively dividing cells to infect. Adenoviruses are DNA viruses that do not integrate into the cellular genome. They can carry larger genes than retroviruses and can infect both dividing and nondividing cells. However, since they do not integrate into the genome, therapy has to be repeated periodically.

19.16 Transgenics and Knockout Technologies

Transgenic technology makes it possible to produce organisms in which a specific gene has been replaced with a mutated version or in which the function of a specific gene is completely removed. This technology can be used to create an animal model of a human disease that is caused by mutation of a specific gene. A common animal used is known as knock-out mice (▶ Fig. 19.20). For example, mice that have been created to produce a defective version of the gene *Xpd* exhibit symptoms of premature aging, similar to those of humans with the condition trichothiodystrophy.

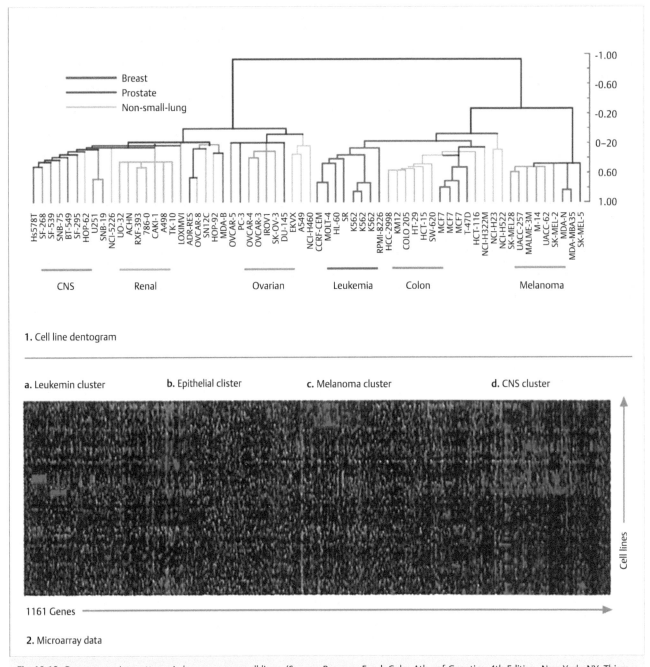

1. Cell line dentogram

a. Leukemin cluster   **b.** Epithelial clister   **c.** Melanoma cluster   **d.** CNS cluster

1161 Genes

2. Microarray data

Fig. 19.18 Gene expression patterns in human cancer cell lines. (Source: Passarge E, ed. Color Atlas of Genetics. 4th Edition. New York, NY. Thieme; 2012.)

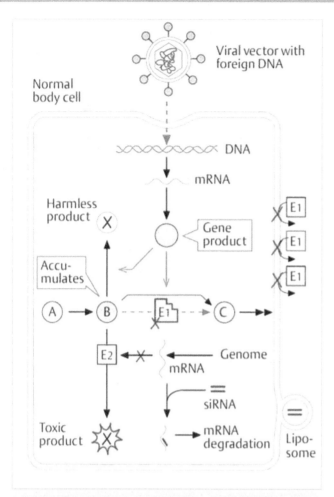

Fig. 19.19 Gene therapy. (Source: Koolman J, Röhm K, ed. Color Atlas of Biochemistry. 3rd Edition. New York, NY. Thieme; 2012.)

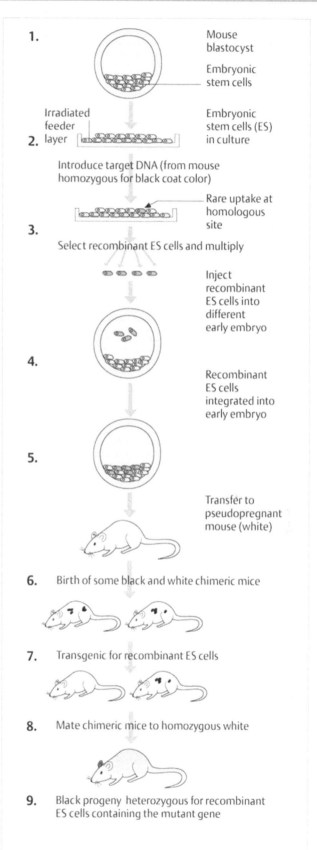

1. Mouse blastocyst

Embryonic stem cells

2. Irradiated feeder layer — Embryonic stem cells (ES) in culture

Introduce target DNA (from mouse homozygous for black coat color)

Rare uptake at homologous site

3. Select recombinant ES cells and multiply

Inject recombinant ES cells into different early embryo

4. Recombinant ES cells integrated into early embryo

5. Transfer to pseudopregnant mouse (white)

6. Birth of some black and white chimeric mice

7. Transgenic for recombinant ES cells

8. Mate chimeric mice to homozygous white

9. Black progeny heterozygous for recombinant ES cells containing the mutant gene

Fig. 19.20 Transgenic mice with targeted gene disruption. (Source: Passarge E, ed. Color Atlas of Genetics. 4th Edition. New York, NY. Thieme; 2012.)

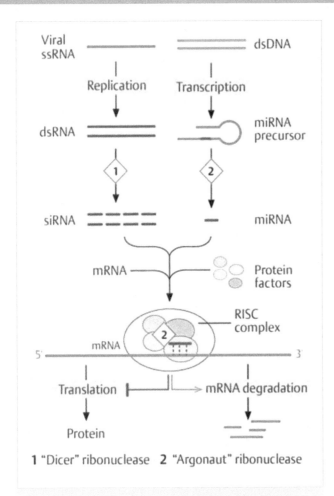

1 "Dicer" ribonuclease 2 "Argonaut" ribonuclease

Fig. 19.21 RNA interference. (Source: Koolman J, Röhm K, ed. Color Atlas of Biochemistry. 3rd Edition. New York, NY. Thieme; 2012.)

19.17 RNA Interference

Gene expression can be regulated at both the transcriptional and the translational level. Small RNA molecules known as small interfering RNAs (siRNAs) regulate translation of protein by binding to specific mRNAs to inhibit their translation which ultimately leads to their degradation, thereby preventing the protein from being made.

Based on this principle, the RNA interference method can be used to generate specific gene knockouts in tissue culture or in model organisms (▶ Fig. 19.21).

Review Questions

1. Which of the following blotting techniques can be used to identify specific sequences of RNA?
   A) Northern blot
   B) Southwestern blot
   C) Southern blot
   D) Western blot

2. A patient presents to the clinic requesting to be tested for HIV. What method would be best to rule out infection?
   A) DNA sequencing
   B) Southern blot
   C) RFLP
   D) ELISA

3. What would be the best follow-up test to rule out a false positive ELISA result?
   A) Northern blot
   B) Southwestern blot
   C) Southern blot
   D) Western blot

4. Which of the following techniques would be most appropriate for monitoring the movement of intracellular vesicles within a live cell?
   A) Green fluorescent protein (GFP) fusion
   B) Microarray
   C) Proteomics
   D) RNA interference

Answers

1. **The correct answer is A.** A northern blot is used to identify specific sequences of RNA in a sample. A Southern blot (answer choice C) identifies DNA, a western blot (answer choice D) detects protein, and a southwestern blot (answer choice B) can determine DNA-protein interactions.

2. **The correct answer is D.** ELISA is a sensitive test used frequently for the screening of HIV and is a rapid test for the detection of antibodies to viral proteins. Since HIV is a retrovirus and, thus, has an RNA genome, the viral RNA would have to be converted to complimentary DNA (cDNA) with the enzyme reverse transcriptase before it can be detected by DNA sequencing (answer choice A). This process would be more time consuming than ELISA and would therefore not be the best answer choice. Southern blot (answer choice B) and restriction fragment length polymorphisms (RFPL) (answer choice C) are used to detect differences in genomic DNA and are used to identify the presence of a genetic disease.

3. **The correct answer is D.** Western blot is a process used for the identification of protein in a sample. This technique is frequently used as a confirmatory test for HIV when the ELISA screening test is positive. Northern blot (answer choice A) is a technique used to identify RNA molecules in a sample and is typically used to assay for the presence of specific mRNA molecules. Southwestern blot (answer choice B) is used to study interactions between DNA and protein and is commonly used to test the ability of transcription factors to bind to specific sequences of DNA. Southern blot (answer choice C) is used to diagnose specific genetic diseases.

4. **The correct answer is A.** Green fluorescent fusion proteins allow the visualization under a fluorescent microscope of real-time activities within living cells. Microarrays (answer choice B) are used to compare gene expression between experimental and control groups, whereas proteomics (answer choice C) can be used to compare the abundance of specific proteins between two samples. RNA interference (answer choice D) is a method in which a short RNA molecule contains complimentary sequence with a specific mRNA and, upon binding, prevents the translation of the mRNA into protein.

III

20 Plasma Proteins and the Diagnostic Use of Enzymes

At the conclusion of this chapter, students should be able to:
- Compare the basic components of plasma and serum
- Identify the major plasma proteins and their function in blood
- Define electrophoretic separation of plasma proteins
- Describe the clinical importance of changes in the amount of plasma proteins
- Evaluate the diagnostic use of serum enzymes and tumor markers in pathological conditions

20.1 Composition of Blood

Blood is a mixture of cells, proteins, and molecules. The cells of the blood include erythrocytes, white blood cells (also known as leukocytes), and platelets. Among the white blood cells are the T and B lymphocytes, monocytes, neutrophils, eosinophils, basophils. Plasma and serum are two terms that refer to the liquid component of blood, which makes up approximately 55% of the total blood volume. The difference between serum and plasma is the absence or presence of the clotting factor fibrinogen. Fibrinogen is found in plasma, but is absent in serum. This difference determines how each is collected. Plasma is obtained in the presence of anticoagulant, whereas serum is obtained without anticoagulant. Plasma is also considered as the intravascular component of the extracellular fluid.

Plasma is mostly comprised of water but also contains other necessary components such as electrolytes (e.g. Na^+, K^+, Ca^{2+}, Mg^{2+}, HCO_3^-, and Cl^-), nutrients, metabolites (e.g. glucose, urea, uric acid, lipids, bilirubin, creatinine, acetoacetic acid, acetone, ammonia), hormones (e.g. insulin, epinephrine, glucagon, T3, cortisol), and other dissolved proteins (▶ Fig. 20.1).

20.2 Plasma Proteins and Their Functions

Albumin, immunoglobulins, and fibrinogen constitute the biggest percentage of all plasma proteins. Most of the plasma proteins are synthesized and degraded in the liver, except for immunoglobulins which are synthesized by B lymphocytes, and hormones which are synthesized by endocrine glands.

Plasma proteins have important functions including transport of molecules, maintaining osmotic balance, hemostasis, coagulation and immune defense.

1. Maintaining Blood Volume

Although all plasma proteins are involved in maintaining colloidal osmotic pressure, albumin is the main force holding water in the blood. The fluid homeostasis or balanced distribution of water is mainly maintained by albumin because albumin is the most abundant protein in the plasma. Since plasma proteins are necessary for maintaining proper intravascular volume, decreased concentrations of plasma proteins, mainly decreased levels of albumin, cause leakage of fluid from the intravascular space into interstitial space. This clinical condition is known as edema. Edema can develop due to a variety of reasons ranging from liver or kidney failure to protein malnutrition. Liver failure may result in hypoalbunemia due to insufficient production of albumin. Hypoalbunemia can also develop due to diseases that cause protein loss, such as kidney failure or gastrointestinal diseases. Insufficient dietary protein intake or malnutrition can also result in hypoalbunemia.

2. Transport

Plasma proteins transport nutrients, metabolites, and hormones throughout the body. One of the most important transport proteins in the body is albumin. It is responsible for the binding and transport of free fatty acids, calcium, zinc, copper, magnesium, bilirubin, heme, steroid hormones, and many drugs. Another plasma protein, transferrin, is responsible for transporting iron in plasma. Transferrin can bind to Fe^{3+} but not Fe^{2+}. Therefore, another plasma protein ceruloplasmin, which has ferroxidase activity, is required for oxidation of Fe^{2+} to Fe^{3+}. The transferrin-Fe^{3+} complex is taken up into cells via the transferrin receptor. Cells that have high iron demand express the transferrin receptor in order to absorb iron from the blood. Expression of the transferrin receptor is regulated at the level of translation by changing the mRNA stability with iron-response elements (IRE) found in the mRNA of the transferrin receptor (reviewed in chapter 4).

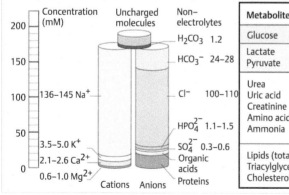

Metabolite	Concentration	
Glucose	3.6–6.1	mM
Lactate	0.4–1.8	mM
Pyruvate	0.07–0.11	mM
Urea	3.5–9.0	mM
Uric acid	0.18–0.54	mM
Creatinine	0.06–0.13	mM
Amino acids	2.3–4.0	mM
Ammonia	0.02–0.06	mM
Lipids (total)	5.5–6.0 g . L^{-1}	
Triacylglycerols	1.0–1.3 g . L^{-1}	
Cholesterol	1.7–2.1 g . L^{-1}	

Fig. 20.1 Blood plasma: composition. (Source: Koolman J, Röhm K, ed. Color Atlas of Biochemistry. 3rd Edition. New York, NY. Thieme; 2012.)

Lipoproteins function by transporting lipid molecules throughout the body in the blood. Lipoproteins and lipid transport is reviewed in the blood lipoproteins chapter 43.

3. Coagulation function of plasma proteins

There are many clotting factors, along with some plasma proteins, that are responsible for the formation and regulation of a blood clot. These components of the so-called clotting cascade are activated following an injury to blood vessels.

20.3 Clotting Cascade

After an injury to blood vessels, clot formation is initiated with the assembly of a temporary platelet plug. Trauma to endothelial cells of the blood vessels exposes collagen and the membrane protein von Willebrand factor (vWF) to the blood vessel lumen. Once exposed, these proteins cause adhesion and aggregation platelets. The platelets adhere to collagen through the platelet membrane protein glycoprotein VI on the platelet membranes. As platelets adhere to exposed collagen and to each other, granules are released from platelets. These granules contain ADP, thromboxane A2, phospholipase A2 and Ca^{2+}. The thromboxane A2 and ADP activate platelet glycoprotein IIb/IIIa,

which causes platelet aggregation. All of these molecules increase vasoconstriction and platelet aggregation.

After the initial platelet plug formation, a blood clot forms. The most important step in clot formation is conversion of fibrinogen to fibrin, a protein that forms a net-like matrix to catch erythrocytes and stop bleeding (▶ Fig. 20.2). Formation of fibrin from fibrinogen requires thrombin, which is normally synthesized and present in the blood as prothrombin, the zymogen form. The conversion of prothrombin to thrombin requires active Factor X. The activation of Factor X can be accomplished via either an intrinsic or extrinsic pathway. The intrinsic pathway is triggered by vascular injuries that result in the exposure of collagen. The vascular injury also exposes a membrane protein called tissue factor III, which triggers the extrinsic pathway of the clotting cascade. Both intrinsic and extrinsic pathways activate Factor X, which converts prothrombin to thrombin. Thrombin converts fibrinogen to fibrin.

20.3.1 Linked to Pathology

Both **hemophilia A** and **hemophilia B** are bleeding disorders. While hemophilia A is due to genetic deficiency of Factor VIII, hemophilia B is caused by deficiency of Factor IX. Both Factor VIII and Factor IX genes are located on the X-chromosome.

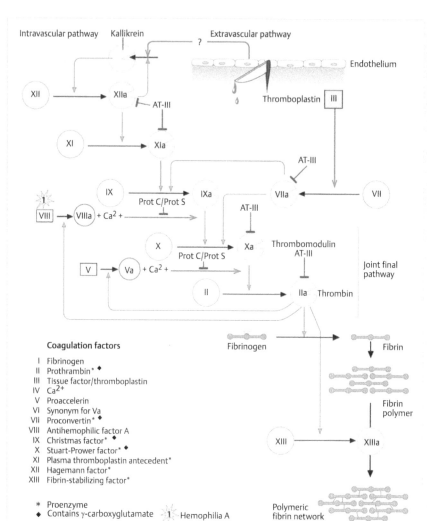

Fig. 20.2 Blood clotting. (Source: Koolman J, Röhm K, ed. Color Atlas of Biochemistry. 3rd Edition. New York, NY. Thieme; 2012.)

Coagulation factors

I   Fibrinogen
II   Prothrambin* ◆
III Tissue factor/thromboplastin
IV Ca^{2+}
V   Proaccelerin
VI   Synonym for Va
VII Proconvertin* ◆
VIII Antihemophilic factor A
IX   Christmas factor* ◆
X   Stuart-Prower factor* ◆
XI   Plasma thromboplastin antecedent*
XII Hagemann factor*
XIII Fibrin-stabilizing factor*

* Proenzyme
◆ Contains γ-carboxyglutamate   1 Hemophilia A

Therefore, these bleeding diseases are inherited in an X-linked recessive manner and commonly affect males.

20.3.2 Link to Pathology

Genetic deficiency of von Willebrand factor causes von Willebrand disease, which is a mild bleeding disorder. Patients generally experience frequent nosebleeds, easy bruising, and gum bleedings.

20.3.3 Link to Pharmacology

Aspirin, dipyridamole, clopidogrel, ticlopidine, abciximab, eptifibatide, and tirofiban are antiplatelet drugs. Aspirin inhibits COX enzymes, preventing the formation of thromboxane A2. Dipyridamole inhibits release of ADP from platelets and decreases platelet adhesion, while clopidogrel and ticlopidine inhibit ADP-induced platelet-platelet aggregation and the binding of platelets to fibrinogen. Abciximab, eptifibatide, and tirofiban are glycoprotein IIb/IIIa receptor antagonists. These drugs compete with fibrinogen and von Willebrande factor for binding to the glycoprotein IIb/IIIa receptors found on platelets.

20.4 Fibrinolysis

Plasma proteins are not only involved in the creation of a clot, but also in the dissolution of clots. Fibrin dissolution, or fibrinolysis, requires plasmin, which is synthesized as the zymogen plasminogen. Plasminogen is activated by plasminogen activators. In vascular endothelial cells, the plasminogen activator is tissue plasminogen activator (tPA). In the kidney, the plasminogen activator is called urokinase plasminogen activator (uPA). (▶ Fig. 20.3). Fibrinolysis is inhibited by α_2-antiplasmin, which is a plasma protein that inhibits plasmin.

20.4.1 Link to Pharmacology

Streptokinase and urokinase are used as thrombolytic drugs, especially in patients having acute myocardial infarctions and stoke. The enzyme streptokinase is purified from β-hemolytic streptococci, whereas urokinase is obtained from urine or cultured kidney cells. Streptokinase and urokinase converts plasminogen into plasmin.

Thrombolytic drugs such as altaplase, reteplase, and tenecteplase are recombinant tissue plasminogen activators (tPA) that also convert plasminogen into plasmin.

20.5 Anticoagulation

Clotting and fibrinolysis are regulated processes. While disorders of clotting cause bleeding, disorders of excess clotting can cause thrombosis or embolism. Embolism can be treated by anticoagulants or drugs that increase fibrinolysis. Antithrombin III is a plasma protein that binds to thrombin. Once this complex forms, thrombin becomes inactivated. This interaction between thrombin and antithrombin III is induced by heparin. Heparin is a glycosaminoglycan, which is synthesized in the mast cells and connective tissue. It is a natural anticoagulant, but can also be used as anticoagulant agent *in vitro*. Other *in vitro* anticoagulant reagents such as the calcium chelator EDTA or citrate can be used to prevent blood clotting (▶ Fig. 20.4).

20.5.1 Link to Pharmacology

Vitamin K antagonists, such as **warfarin** and **coumarins**, inhibit the γ-carboxylation of clotting factors II, VII, IX and X during translation. The γ-carboxylation of glutamate residues of these clotting factors provides a Ca^{2+} binding site, which is necessary for activation of these clotting factors and subsequently for the clot formation. Prevention of the γ-carboxylation by vitamin K antagonists is useful for the inhibition of the clotting pathway, especially in patients who have increased risk of embolism.

4. Plasma proteins defend the body against infections

Immune cells and plasma proteins such as immunoglobulins and the complement system defend the body from invading microorganisms and foreign cells. This coordinated respond against infections is called the immune response of the body. The immune response initially starts with the nonspecific, or innate, immune system, which recognizes microorganisms and distinguishes between endogenous and exogenous cells and molecules. Cellular components of the innate immune system include epithelial cells of the skin, macrophages, granulocytes, and natural killer cells. The humoral elements of the innate immune system are the components of the complement system and cytokines. The complement system is composed of ~30 different proteins, which are produced in the liver and secreted into the blood. Proteins involved in the early phase of the complement system are found in the blood as inactive zymogens. Upon activation, these complement proteins lyse bacteria. Cytokines are hormone-like peptides or proteins that are synthesized by immune system cells or other type of cells. Cytokines are involved in non-specific immune defenses, inflammatory reactions, apoptosis, and many

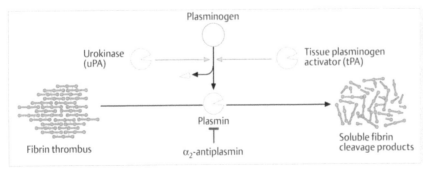

Fig. 20.3 Fibrinolysis. (Source: Koolman J, Röhm K, ed. Color Atlas of Biochemistry. 3rd Edition. New York, NY. Thieme; 2012.)

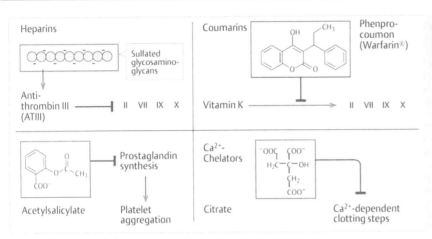

Fig. 20.4 Artificial anticoagulation. (Source: Koolman J, Röhm K, ed. Color Atlas of Biochemistry. 3rd Edition. New York, NY. Thieme; 2012.)

cellular signaling processes. There are many different types of cytokine molecules. These include interleukins, lymphokines, interferons, and colony-stimulating factors.

The acquired, or adaptive, immune response is more specific to invading microorganisms or molecules than innate immunity. This is due to the fact that the adaptive immune response is directed against specific antigens. The cellular components of the adaptive immune response are the B-lymphocytes and T-lymphocytes. B-lymphocytes, or plasma cells, produce antibodies, also known as immunoglobulins. T-lymphocytes, on the other hand, recognize cells that are infected with pathogens/microorganisms.

The immunoglobulins of the adaptive immune system and the complement proteins of the innate immune system are both plasma proteins that they are involved in the defense against infections. There are five classes of immunoglobulins; IgA, IgD, IgE, IgG, and IgM (▶ Fig. 20.5). The general structure of immunoglobulins contains two heavy chains and two light chains, each of which contain variable regions that are used to bind to specific antigens. The antibody-antigen complexes activate the complement proteins in the plasma. The complement system can also be activated by bacterial cell polysaccharides. Activation of the complement system starts a proteolytic cascade that results in the release of biologically active peptides and polypeptides. These biologically active polypeptides mediate the inflammatory response.

20.5.2 Link to Pathology: α-1 antitrypsin and emphysema

Proteolytic damage induced during the inflammatory response is limited by the synthesis of protease inhibitors. The protease inhibitor α1-antitrypsin functions to limit the damage caused by activated neutrophils, a type of white blood cell. These activated neutrophils release an enzyme called elastase which causes degradation of the protein elastin. The protease inhibitor α1-antitrypsin essentially puts the breaks on elastase's destructive power. Genetic mutations and variations in the α1-antitrypsin gene (*SERPINA1*) cause production of abnormal levels of α1-antitrypsin, ranging from a reduced level to none. Among the allelic variations of the *SERPINA1* gene, the Z and S variants are most common. While variant M is the normal form

of *SERPINA1*, the Z variant is the most severe form, resulting in only 10% α1-antitrypsin activity. Patients with the Z variant of this gene are at high risk for lung diseases such as early-onset of COPD or emphysema, as well as liver disease. The liver effects are due to misfolding and aggregation of the Z variant α1-antitrypsin protein in the endoplasmic reticulum of hepatocytes. This aggregation eventually may lead to liver failure and cirrhosis. The S variant is less severe than the Z variant. Due to the minimized or absent elastase inhibition, these patients are at increased risk for developing pulmonary conditions such as early-onset emphysema, especially in patients who are tobacco smokers. The underlying molecular reason for this is that cigarette smoking oxidizes a methionine residue in the elastase binding site of α1-antitrypsin, preventing the interaction that is necessary for elastase inhibition.

20.6 Plasma Proteins in the Diagnosis of Diseases

Plasma contains ~ 100 different proteins with concentrations ranging between ~ 6.0 – 8.0 gram/dL. The quantities of some plasma proteins can be specifically measured using different techniques. These measurements can be used to diagnose some diseases.

Electrophoresis on cellulose acetate can be used to separate plasma proteins, producing five specific bands. These bands are albumin, α1-, α2-, β- and γ-globulins (▶ Fig. 20.6). The most abundant band is albumin. Within the α1 band, the major protein is α1-antirypsin, whereas the proteins found within the α2 band are haptoglobin, α2 macroglobulin, and HDL. Transferrin, LDL, and the complement protein C3 migrate with the β-band. The γ-band contains fibrinogen, C-reactive protein, and immunoglobulins. This is one of the reasons why immunoglobulins, or antibodies, were historically called gamma-globulins.

These separated bands can also be quantitated by densitometer. Scanning the density of each band gives quantitative information about protein fractions and can be used for diagnostic purposes. For example, hypoalbuminemia, or low serum albumin, will show a comparatively lower density in the albumin band. This condition may be caused by malnutrition, liver disease, or protein loss from the urinary or gastrointestinal system.

Although we can determine the levels of major plasma proteins by electrophoresis, abnormalities in minor plasma

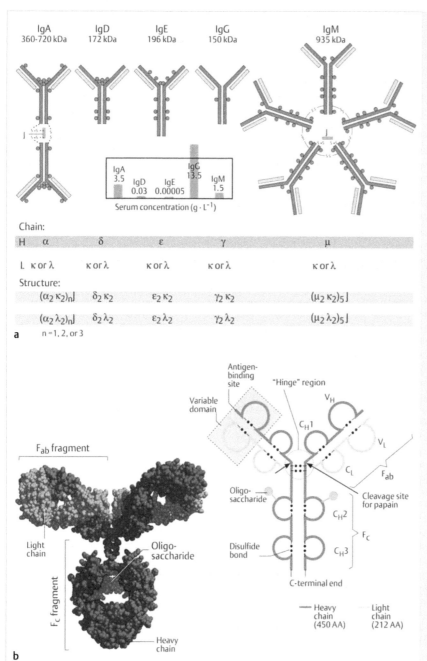

Fig. 20.5 (a) Immunoglobulin classes. **(b)** Structure of immunoglobulin G. (Source: Koolman J, Röhm K, ed. Color Atlas of Biochemistry. 3rd Edition. New York, NY. Thieme; 2012.)

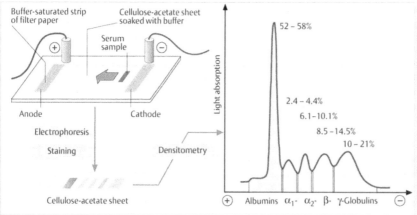

Fig. 20.6 Plasma proteins can be separated by electrophoresis. (Source: Koolman J, Röhm K, ed. Color Atlas of Biochemistry. 3rd Edition. New York, NY. Thieme; 2012.)

Table 20.1 Marker enzymes in plasma and serum

Enzyme (Acronym)	Tissue (s) of origin
Aspartate transaminase (AST, GOT)	Liver (cirrhosis, fatty liver) Heart (infarction)
Alkaline phosphatase (AP)	Liver, bile duct Bone (fractures, osteomalacia)
Amylase	Pancreas (pancreatiti)
Alanine aminotransferase (ALT, GPT)	Liver (cirrhosis, fatty liver)
Cholinesterase (CE)	Liver (cirrhosis, fatty liver)
Creatine kinase (CK, isoenzyme MB)	Heart (infarction)
Glutamate dehydrogenase (GLDH)	Liver (necrosis)
γ-Glutamyl transpeptidase (GGT)	Liver (toxins, hepatitis) Bile duct (cholestasis)
Lactate dehydrogenase (LDH)	Liver (shock), muscle, heart (infarction)
Pancreatic lipase (PLP)	Pancreas (pancreatis)

Source: Koolman J, Röhm K, ed. Color Atlas of Biochemistry. 3rd Edition. New York, NY. Thieme; 2012.

proteins cannot be revealed in this way. Instead, the quantity of minor plasma proteins must be measured by more sensitive immunological methods such as ELISA. An example of this is the detection of α-fetoprotein in amniotic fluid or in mother's plasma for the diagnosis of neural tube defects and/or Down syndrome in a fetus.

20.7 Clinical Importance of Plasma Enzymes

The presence of certain enzymes or proteins in plasma can be indicative of tissue damage or organ failure. Tissue damage causes release of intracellular enzymes into the blood. Therefore, measuring the concentrations of these enzymes in the plasma is useful for diagnostic purposes. Cell death is the most common cause of elevated plasma enzyme levels and the specific enzyme detected can often be a clue as to which cell type was damaged. Common enzymes used for clinical diagnosis can be found in ▸ Table 20.1.

Alanine transaminase (ALT) is an enzyme of amino acid metabolism. This enzyme is present in hepatocytes and leaks into the blood if the liver is damaged. Serum ALT becomes dramatically elevated in acute liver damage such as viral hepatitis and acetaminophen overdose. In viral hepatitis, ALT plasma levels increase 20–100 times above the upper limit of the normal range. **Aspartate transaminase (AST)** is similar to ALT in that it is another enzyme associated with liver cells. It is raised in acute liver damage, but is also present in red cells, cardiac, and skeletal muscle and is therefore not specific to the liver.

Pancreatic amylase and lipase are produced by the pancreas and delivered to the small intestine via the pancreatic duct. Although increased levels of amylase or lipase is a diagnostic value for pancreatitis, lipase levels along with amylase are also taken into consideration for the differential diagnosis of severe abdominal pain of sudden onset. Amylase levels can also increase in individuals who have paramyxovirus infections of the salivary glands, such as mumps. Also, serum lipase can increase due to intestinal infarction and perforation or in severe peptic ulcer disease. Therefore, the diagnosis of acute pancreatitis is best supported by an increase in both amylase and lipase levels in the serum.

Creatine kinase (CK), also called creatine phosphokinase (CPK), is found in brain, skeletal muscle, and cardiac muscle cells. CK is a dimer with two different monomers (M and B subunits). These two subunits can ultimately form 3 different isozymes: CK1 (BB) (brain), CK2 (MB), and CK3 (MM). Different isoforms of CK can be used for the diagnosis of muscle diseases including muscular dystrophies, dermatomyositis, polymyositis, rhabdomyolysis, and even myocardial infarction. After a myocardial infarction (MI), proteins and enzymes that were contained in cardiac cells leak out into the bloodstream. The appearance of myoglobin, CKMB, AST, LDH and cardiac troponin in the plasma is indicative of MI.

CKMB is specific for MI and appears 4–8 hours following onset of chest pain, reaches a peak activity within 24 hours, and returns to normal baseline in 48–72 hours. Cardiac troponin I is also specific for MI and appears in plasma within 4–6 hours after an MI, peaks 8–28 hours and remains elevated 3–10 days. This allows CKMB to be useful in diagnosing a re-infarction within a 3 to 10-day window. For example, a patient who was admitted for a myocardial infarction and received a heart stent presents to the emergency room 4 days later with chest pain. Cardiac troponin elevation would be expected but the presence of CKMB at that time would suggest another heart attack.

20.8 Tumor Biomarkers

Tumor biomarkers are secreted proteins produced by cancer cells that are detectable in the serum. They may or may not have a known function depending on the type of cancer involved. These specific proteins can be used for population screening, help with diagnosis, prognostic factors, monitoring the course of disease or treatment, and diagnosis of relapse. The minimum requirements for a protein to be considered a tumor marker include having a reliable, quick, cheap detection assay with high sensitivity (>50%) and specificity (>95%), and with a high predictive value of positive and negative results. Some tumor markers can be elevated in patients with non-cancerous diseases. For example, the tumor marker carcinoembryonic antigen (CEA) is found in patients with colon cancer, but may also be found in patients with ulcerative colitis, liver cirrhosis, and even in heavy smokers, so interpretation of a positive result should be made with caution. For this reason, the main use of tumor markers is for monitoring the effectiveness of treatment and to detect early recurrences. Some of the common tumor markers that are useful in clinical diagnostics are CA125 for ovarian cancer, CA19–9 for pancreatic cancer, prostate specific antigen (PSA) for prostate cancer, CA15–3 for breast cancer, and β2-microglobulin for myeloma. On-going research in proteomics may provide new biomarkers with increased sensitivity and specificity to allow improved cancer surveillance.

III

III

Review Questions

1. Electrophoretic analysis of plasma reveals a decreased density in the α1 band. Which of the following conditions is most likely expected in this patient?
   A) Chronic inflammation
   B) Edema
   C) Emphysema
   D) Excessive bleeding

2. Which of the following directly activates the conversion of fibrinogen to fibrin?
   A) Factor X
   B) Platelet aggregation
   C) Prothrombin
   D) Thrombin
   E) von Willebrand factor

3. Which of the following drugs inhibits the aggregation of platelets?
   A) Aspirin
   B) Heparin
   C) Urokinase
   D) Warfarin

4. Which of the following drugs is a plasminogen activator?
   A) Aspirin
   B) Heparin
   C) Streptokinase
   D) Warfarin

5. Which of the following enzymes is elevated in the blood in patients with muscular dystrophy?
   A) Alanine transaminase
   B) Amylase
   C) Creatine Kinase
   D) Lipase
   E) Troponin I

Answers

1. **The correct answer is C.** A decrease in the density of the α1 band could indicate a reduction in α1-antitrypsin. Genetic variations of *SERPINA1* gene can cause this reduction and may result in early-onset emphysema, especially in tobacco smokers. Chronic inflammation (answer choice A) is incorrect because inflammation activates the immune response and can increase the amount of antibodies, which migrates to the γ-band on the plasma electrophoresis. Edema (answer choice B) develops with a decrease in albumin levels. Excessive bleeding (answer choice D) is incorrect because a deficiency of coagulation factors cannot be detected by plasma electrophoresis.

2. **The correct answer is D.** Thrombin converts fibrinogen to fibrin. Both intrinsic and extrinsic pathways activate Factor X, which converts prothrombin to thrombin. Therefore, answer choices A and C are incorrect. Answer choices B and E are incorrect because von Willebrand factor, a membrane protein, causes adhesion and aggregation of platelets.

3. **The correct answer is A.** Aspirin is an irreversible inhibitor of cyclooxygenase (COX) enzymes. Inhibition of the COX pathway reduces the formation of thromboxane A2, which is an eicosanoid that activates platelet aggregation. Therefore, reducing the synthesis of thromboxane A2 inhibits platelet aggregation.

   Answer choice B is incorrect because heparin activates antithrombin III, which forms a complex with thrombin and inhibits its activity. Inhibition of thrombin prevents the formation of fibrin from fibrinogen. Answer choice C is incorrect because urokinase is a fibrinolytic drug. Urokinase and streptokinase can activate the conversion of plasminogen to plasmin, which dissolves the fibrin mesh. Answer choice D is incorrect because warfarin is a vitamin K antagonist. Vitamin K is a coenzyme for γ-carboxylase, which catalyzes the γ-carboxylation of glutamyl residues of the coagulation factors II, VII, IX, and X. Inhibition of γ-carboxylation prevents these factors from becoming active in the clotting cascade, therefore the conversion of fibrinogen to fibrin is inhibited by vitamin K antagonists.

4. **The correct answer is C.** Streptokinase and urokinase are fibrinolytic drugs, which activate the conversion of plasminogen to plasmin. Plasmin is a protease that degrades fibrin network and, therefore, dissolves the clot. While streptokinase is a fibrinolytic drug, aspirin, heparin, and warfarin are involved in prevention of fibrin, therefore answer choices A, B, and D are incorrect.

5. **The correct answer is C.** Elevated levels of creatine kinase can be found in patients with muscular dystrophy. CK can be used for diagnoses of muscle diseases including muscular dystrophies, dermatomyositis, polymyositis, rhabdomyolysis, and even myocardial infarction. Elevated amylase and lipase is found in acute pancreatitis, thus answer choices B and D are incorrect. Elevated alanine transaminase is found in individuals with liver injury. Therefore answer choice A is incorrect. Troponin I is specific for cardiac muscle and increases in the plasma in patients who have had a recent myocardial infarction, thus answer choice E is incorrect.

Review Questions

1. A 14-year-old male complaints of dark urine. Physical examination uncovers periorbital and bilateral leg edema. Microscopic analysis of his urine reveals hematuria, and blood tests show elevated BUN and serum creatinine. His medical history shows that he was treated for a streptococcal infection about three -weeks ago. A decreased concentration of which of the following plasma proteins is most likely causing his edema?
   A) 1-antitrypsin
   B) Albumin
   C) Complement protein C3
   D) Immunoglobulins
   E) Transferrin

2. A 58-year-old male presents to the emergency room with chest pain and shortness of breath. He tells the physician that he had a myocardial infarction and treated with balloon angioplasty three days ago. Which of the following serum enzymes is the best diagnostic test to determineif this patient is having a new infarct?
   A) Aspartate aminotransferase
   B) Cardiac troponin
   C) CKMB
   D) Lactate Dehydrogenase
   E) Lipase

3. A 55-year-old female complains of shortness of breath and chronic cough. Her history reveals that she used to smoke two packs a day, but she quit about five years ago. Complete blood count is within normal range, but plasma ALT, AST, and LDH levels are elevated. Lung spirometry results suggest an obstructive pattern. Which of the following is most likely causing her liver damage?
   A) Accumulation of elastase
   B) Aggregation of 1-antitrypsin
   C) Decreased activity of elastase
   D) Decreased affinity between elastase and 1-antitrypsin

4. A 28-year-old male presents to the emergency room with venous thromboembolism. His history reveals that his brother and his aunt both suffered from pulmonary embolism. Results of an antithrombin III assay show a deficiency in antithrombin III.Increased activity of which of the following blood clotting factors is most likely causing his embolism?
   A) Factor VIII
   B) Factor X
   C) Plasmin
   D) Thrombin
   E) von Willebrand factor

5. A two month old male is brought to the emergency room with severe dehydration due to chronic diarrhea, as well asconcerns of failure to thrive. A stool sample is found to have an abnormally high concentration of glucose, indicating a defect in glucose absorption in the intestines. The results of what procedure would be the most informative regarding the cause of this child's illness?
   A) Western blottingfor ciliary dynein
   B) Northern blottingfor occludin mRNA
   C) Sanger sequencing of the SGLT1 gene
   D) Southern blottingforthe GLUT1 transporter

6. A 17-year-old male presents with complaints of blurred vision and muscle weakness in his legs following strenuous activity. Myasthenia gravis is suspected. What technique would be most informative in confirming his diagnosis?
   A) Southern blot
   B) Northern blot
   C) FISH
   D) ELISA
   E) PAGE

7. A 61-year-old male presents with neurological symptoms consistent with Huntington's disease. Which of the following procedures would be most appropriate for confirming his diagnosis?
   A) FISH
   B) Microarray
   C) PCR
   D) DNA fingerprint
   E) Southern blot

8. A 48-year-old female with an18 pack-year smoking history presents with persistent cough, shortness of breath and a feeling of tightness in her chest. She is also found to have extensive cirrhosis of the liver. Which of the following techniques would best identify the genetic mutation that is causing her disease?
   A) Southern blot
   B) Western blot
   C) Northern blot
   D) Immunofluorescence
   E) DNA sequencing

9. A researcher wants to perform a southern blot using a probe with the sequence 5'-TCCAGCCT-3'. Before he begins, he uses the Sanger method to confirm that his target DNA sample contains the complementary sequence. Which of the following is the most likely result of this procedure?
   A) 5'-AGGCTGGA-3'
   B) 5'-TCCAGCCT-3'
   C) 5'-TCCGACCT-3'
   D) 5'-AGGTCGGA-3'

Answers

1. **The correct answer is B**. His diagnosis is post-streptococcal glomerulonephritis. After a streptococcal infection, acute glomerulonephritis may develop. This condition is an immune complex disorder. Deposition of streptococcal antigen on the glomeruli damages the glomerulus. Therefore, hematuria and proteinuria develop. Urinary loss of excess plasma proteins causes edema and hypertension. Since albumin is the main force for maintaining proper intravascular volume, urinary loss of albumin causes the edema. If the patient is not treated, glomerular damage would result in the loss of large proteins such as immunoglobulins and complement. However, loss of immunoglobulins and complement proteins leads to recurrent infections, thus answer choices C and D are incorrect. Answer choices A and E are incorrect because loss of transferrin affects iron transport, and loss of a1-

III

antitrypsin would cause lung tissue damage due to the inability to inhibit elastase.

Reference: https://emedicine.medscape.com/article/244631-overview

2. **The correct answer is C.** In a myocardial infarction, the serum CKMB isozyme reaches a peak in 24 hours and falls back to baseline in 48 to 72 hours. This allows CKMB to be useful in diagnosing re-infarction if it occurs between 3 to 10 days. Cardiac troponin (answer choice B) takes 3 to 10 days to return to baseline. Therefore, cardiac troponin would not be the best test for diagnosing the new infarct in this patient. Aspartate aminotransferase (answer choice A) and lactate dehydrogenase (answer choice D) are specific for liver damage but non-specific for myocardial tissue. Lipase (answer choice E) is useful in diagnosing acute pancreatitis.

3. **The correct answer is B.** Symptoms and history of this patient suggest that she is suffering from COPD/emphysema, which presents with shortness of breath and fatigue. Some patients with certain genetic variants of 1-antitrypsin gene (piZZ) suffer from both COPD/emphysema and liver damage. The 1-antitrypsin is a plasma protein and functions as an anti-protease, which binds and inhibits serine proteases such as elastase. Increased activity of elastase due to the deficiency of ?1-antitrypsin causes degradation of elastin. This increased proteolysis of elastin particularly effects the lungs, causing COPD/emphysema. Smoking exacerbates the effects of elastase, because cigarette smoke oxidizes the methionine residue of ?1-antitrypsin, making it ineffective as a protease inhibitor. The ?1-antitrypsin is synthesized in the hepatocytes. Patients who are homozygous for the Z variant of the ?1-antitrypsin gene may also suffer from liver damage, because misfolding of the ?1-antitrypsin causes accumulation and aggregation of the protein in the endoplasmic reticulum of the hepatocytes.

4. **The correct answer is D.** Antithrombin III is a plasma protein that binds to thrombin. Once this complex forms, thrombin becomes inactivated. Normally, thrombin converts fibrinogen to fibrin. Normally, antithrombin III prevents excessive thrombin activity to limit the formation of fibrin or clot formation. Genetic deficiency in antithrombin III causes an increase in the activity of thrombin and consequently excessive formation of fibrin. Therefore, patients with antithrombin III deficiency suffer from recurrent embolism. Factor VIII (answer choice A) is a plasma protein that circulates in the blood in an inactive form bound to von Willebrand factor (answer choice E). An injury to the blood vessel activates Factor VIII and dissociates it from von Willabrand factor. Active Factor VIII activates Factor IX. The active Factor IX activates Factor X (answer choice B), which converts pro-thrombin to thrombin. Plasmin (answer choice C) is required for dissolution of fibrin.

Reference: https://emedicine.medscape.com/article/954688-overview

5. **The correct answer is C.** The SGLT1 gene encodes the sodium-glucose transporter that is present on the apical surface of the enterocytes of the small intestines. This transporter is responsible for absorbing glucose from the lumen of the intestines using the electrochemical gradient of sodium to move glucose into the cell. Sequencing of this gene would reveal any mutations that would affect the functioning of this transporter, which is likely defective in this child. The enterocytes do not have ciliary dynein, so western blotting for this protein (answer choice A) would not be informative. Since occludin forms the tight junctions between adjacent cells within the intestinal lining, but does not function in transporting glucose, northern blotting for this mRNA (answer choice B) would not be beneficial. Southern blotting for the GLUT1 transporter would not be informative because the GLUT1 transporter allows the release of glucose from the enterocytes on the basal surface, but is not involved in the absorption of glucose from the intestinal lumen.

6. **The correct answer is D.** Myasthenia gravis is an autoimmune disease that is caused by the body's immune system making antibodies that recognize the acetylcholine receptors present on muscle fibers at neuromuscular junctions. Once these antibodies bind to the receptor, the receptor no longer responds to acetylcholine signaling and, therefore, the muscle does not receive the signal to contract. Since the diagnosis for this disease is the presence of these autoimmune antibodies, an assay that identifies the antibodies in the patient's blood is required. Of the answer choices given, ELISA is the only technique that identifies the presence of antibodies. Southern blot (answer choice A) would be used for analyzing DNA, and northern blot (answer choice B) would be used for RNA. FISH (answer choice C) uses fluorescence labeled probes to identify chromosomal aberrations associated with disease. PAGE (answer choice E) is a technique used to separate proteins based on size.

7. **The correct answer is E.** Huntington's disease is a genetic disorder that causes the progressive breakdown of nerve cells in the brain. This disease is caused by an expansion of the trinucleotide sequence CAG within the HTT gene. The number of these CAG repeats determines whether or not an individual has the disease, as well as the severity of the disease. Southern blot analysis is the standard method for determining the repeat number as the size of the resulting band can be used to calculate how many repeats are present and, therefore, the diagnosis. FISH (answer choice A) uses fluorescence labeled probes to identify chromosomal aberrations associated with disease. Microarray (answer choice B) is typically used to provide information about the level of expression of specific genes. PCR (answer choice C) is a technique that allows for the amplification of a specific fragment of DNA. DNA fingerprinting (answer choice D) is not used for identifying the presence of a disease, but rather to distinguish the DNA of one individual from another.

8. **The correct answer is E.** The symptoms presented in this patient are consistent with ? 1 antitrypsin deficiency. This disease is caused by a mutation in the SERPINA1 gene that encodes for the elastase inhibitor ? 1 antitrypsin. This protein is normally produced by the hepatocytes in the liver and released into the bloodstream. The mutation of this gene results in mis-folding of the protein, which accumulates in the liver and leads to liver damage. In addition, there is no ? 1 antitrypsin available to the lungs to prevent the elastase-mediated breakdown of lung tissue,

resulting in the lung conditions seen in this patient. DNA sequencing would show the presence of the mutation in SERPINA1 and, thus, allow for diagnosis of this disease. Southern blot (answer choice A), western blot (answer choice B), and northern blot (answer choice C) would not give information about the presence of the causative mutation. Immunofluorescence (answer choice D) can be used to visualize the location of a specific protein inside the cell, but would not be informative in this case.

9. **The correct answer is A**. According to the Watson-Crick base pairing rules, in a double stranded DNA molecule, adenosine pairs with thymine and guanine pairs with cytosine. Also recall that the two strands of the DNA molecule are oriented in an antiparallel fashion. Therefore, the sequence in answer choice A is the only one that satisfies these conditions.

III

Review Questions

1. Two new drugs both inhibit the same enzyme. Drug A binds at the same site as the substrate. Conversely, Drug B binds the enzyme at a separate location from where the substrate binds. The binding interactions of both of thesedrugs with the enzyme are found to be reversible. Which of the following is correct with respect to the Michaelis-Menten constant of the reaction between the enzyme and its substrate when drug A or drug B is present?

Choice	Km In the presence of drug A	Km In the presence of drug B
A	↓	No Change
B	↓	↓
C	No Change	No Change
D	↑	No Change
E	↑	↑

2. An investigator is studying the effects of drug Z on a viral enzyme V. The kinetics of enzyme V and its substrate are studied first and illustrated with the red line (curve 1). The investigator then adds the drug Z and its effects on the reaction between the enzyme and its substrate are illustrated by the blue line (curve 2). Which of these statements is true about drug Z?

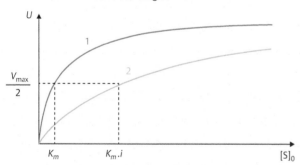

A) It disrupts the interactions between the enzyme and its coenzyme

B) It is a reversible competitive inhibitor of enzyme V

C) It is a reversible noncompetitive inhibitor of enzyme V

D) It is an irreversible inhibitor of enzyme V

E) It is a suicide inhibitor

3. A researcher is studying two different molecules that work as agonists of the serotonin receptor. Molecule X has a lower affinity for the receptor compared to molecule Y. However, both molecule X and Y have higher affinity for the receptor than serotonin. Which of the following statements is true?

A) Serotonin has the lower Km value compared to molecule X

B) Serotonin has the lower Km value compared to molecule Y

C) Serotonin has the lowest Km value compared to molecules X and Y

D) Serotonin has the highest Km value compared to molecules X and Y

E) Serotonin has the higher Km value compared to molecule X

F. Serotonin has the higher Km value compared to molecule Y

4. A 3-day-old newborn male presents with blue skin and purple lips at the local hospital. Sepsis, congenital heart disease and ingestion of oxidants are excluded. DNA testing reveals homologous mutations at the cytochrome b_5 methemoglobin reductase gene. Increased levels of which type of Hb is the cause of his conditions?

A) HbA

B) HbA2

C) HbF

D) HbM

E) HbS

5. A 12-old-male presents with dyspnea and jaundice. Physical examination reveals expiratory wheezing and clubbing. X-ray demonstrates flattened diaphragm and hyperlucency. Blood tests show abnormal hepatic enzymes and a very low level of ?1-antitrypsin. Which of the following is the most likely explanation for his liver damage?

A) Damaged endothelium triggers immune responses

B) Excessive destruction of elastin by a mutant enzyme

C) Large influx of neutrophils that release increased amounts of proteases

D) Loss of function of chloride channel of hepatocytes

E) Misfolded protein accumulates in the endoplasmic reticulum of hepatocytes

6. The affinity of hemoglobin for carbon monoxide (CO) is ~300-times higher than for oxygen. Although binding of CO to hemoglobin is reversible, exposure to CO causes hypoxia. Based on this information, which of the following is the most likely result of CO poisoning?

A) Hemoglobin-oxygen dissociation curve shifts to left

B) Hemoglobin-oxygen dissociation curve shifts to right

C) Decreased oxidation of iron in hemoglobin

D) Increased oxidation of iron in hemoglobin

E) Polymerization of ?-globin of hemoglobin

7. A 37-year-old female with Down syndrome presents with memory impairments. Her mother reports that she has become more aggressive and angryoverlittle things. During physical examination the physician notes that she did not remember her doctor's name. The expression of which of the following proteins is most likely increased in this patient?

A) ?-synuclein

B) ?-amyloid precursor protein

C) p53

D) Prion

E) Transthyretin

Answers

1. **The correct answer is D**. The Km increases in the presence of drug A, since drug A is a competitive inhibitor, which competes with substrate for the substrate binding site. Therefore, in the presence of drug A, the enzyme's affinity for its substrate decreases (Km increases). Whereas, drug B is a noncompetitive inhibitor since it does not bind to the substrate binding site. Therefore, the Km of the enzyme for its substrate remains unchanged. Since it is found that drug B inhibits the enzyme, it most likely decreases the Vmax of the reaction.

2. **The correct answer is B**. A comparison of the kinetic parameters of enzyme V in the presence and absence of drug Z indicates that there is a change in the Km value but not in the Vmax value. This indicates that drug Z binds to the substrate binding site of enzyme V, thus competing with the substrate and lowering the enzyme's affinity for its substrate. Exploration of curve 1 and 2 of the graph shows that the Vmax value did not change, therefore answer choice C is incorrect. Answer choice A is incorrect because enzymes cannot catalyze reactions if the coenzyme is not present. Answer choices D and E are incorrect because irreversible inhibitors are also called suicide inhibitors and they do not affect the Km of the enzymes. Instead, they irreversibly inactivate the enzymes.

3. **The correct answer is D**. Since both molecules X and Y have higher affinity for the receptor than serotonin, serotonin has the highest Km value for the receptor. The key to remember is the lower the Km, the higher the affinity, and the better the binding interactions.

4. **The correct answer is D**. Blue skin and purple lips indicate cyanosis or deoxygenated blood. Due to the mutations at the cytochrome b_5 methemoglobin reductase gene the ferric iron Fe^{3+} cannot be repaired or reduced to Fe^{2+}. The hemoglobin with Fe^{3+} is called methemoglobin, which cannot bind to oxygen. Deficiency of this enzyme leads to the accumulation of HbM, resulting in cyanosis. Answer choices A and B are incorrect because HbA and HbA2 are hemoglobins normally found in adults. Answer choice C is incorrect because HbF is also normal hemoglobin found in the fetus and it can bind to oxygen. HbS, also known as sickle hemoglobin, can bind to oxygen. However, when it is deoxygenated it forms a polymer with other HbS molecules and causes sickling of erythrocytes.

5. **The correct answer is E**. Physical examination and X-ray indicate emphysema which is caused by excessive destruction of elastin by elastase in this case. Elastase is overactive due to lack of inhibition by ?1-antitrypsin as indicated by very low level of ?1-antitrypsin. The mutant ?1-antitrypsin cannot be transported to the lungs but instead accumulate in the liver resulting in cirrhosis.

6. **The correct answer is A**. Although both CO and oxygen bind to heme iron and both are the substrates for hemoglobin, the affinity of hemoglobin for CO is much higher than for oxygen. This is why oxygen cannot be delivered to tissues unless the oxygen pressure is much lower than normal. This reduced release of oxygen is displayed in the hemoglobin-oxygen dissociation curve as a left-shift. Since CO poisoning decreases oxygen delivery to the tissues, the tissues that are affected the most are cardiac muscle and brain since these tissues have more demand for oxygen. Therefore, patients present with hypotension, headache, and a bright red color of the skin. A right-shift represents easier release and increased delivery of oxygen, thus answer choice B is incorrect. CO poisoning does not affect the heme iron, nor does it affect the interactions of the ?-globin subunits of the hemoglobin, therefore answer choices C, D, and E incorrect.

7. **The correct answer is B**. This patient most likely has Alzheimer's disease. There are many factors involved in the development of Alzheimer's disease, one of which is the accumulation of the amyloid ? peptide, which is derived from the beta-amyloid precursor protein (APP). Down syndrome patients and patients who have mutations in the APP gene, located on chromosome 21, produce excess APP protein. This excess production of APP results in excess amyloid ? peptide. The excess amyloid ? peptide accumulates in the basal ganglia and cerebral cortex, causing the formation of senile plaques.

   Proteins ?-synuclein and parkin are involved in the development of Parkinson disease, which is a neurological disorder caused by degeneration of the dopamine producing neurons in the substantia nigra. Parkinson disease presents with resting tremor and difficulty and rigidity in walking. Therefore answer choice A is incorrect. Since p53 is a tumor suppressor protein, genetic mutations of p53 are associated with cancer, thus answer choice C is incorrect. All prion diseases are caused by accumulation of abnormal or pathological prion protein (PrPSc). The wild type prion (mostly alpha helices) and pathological prion protein (mostly beta-sheets) both have the same amino acid sequence. Since the pathological prion protein is protease resistant and insoluble, it forms amyloid fibrils in the nerve cells. This aggregated protein causes neurodegeneration and sponge-like holes. A familial form of amyloidosis is transthyretin-related amyloidosis. Normally, transthyretin functions as a thyroxine and retinol transporter in the blood. However, TTR gene mutations cause misfolding of the TTR protein, resulting in its accumulation as amyloid in the nervous system and cardiac tissue.

Review Questions

1. A 14-year-old male complaints of dark urine. Physical examination uncovers periorbital and bilateral leg edema. Microscopic analysis of his urine reveals hematuria, and blood tests show elevated BUN and serum creatinine. His medical history shows that he was treated for a streptococcal infection about three -weeks ago. A decreased concentration of which of the following plasma proteins is most likely causing his edema?
   A) 1-antitrypsin
   B) Albumin
   C) Complement protein C3
   D) Immunoglobulins
   E) Transferrin

2. A 58-year-old male presents to the emergency room with chest pain and shortness of breath. He tells the physician that he had a myocardial infarction and treated with balloon angioplasty three days ago. Which of the following serum enzymes is the best diagnostic test to determineif this patient is having a new infarct?
   A) Aspartate aminotransferase
   B) Cardiac troponin
   C) CKMB
   D) Lactate Dehydrogenase
   E) Lipase

3. A 55-year-old female complains of shortness of breath and chronic cough. Her history reveals that she used to smoke two packs a day, but she quit about five years ago. Complete blood count is within normal range, but plasma ALT, AST, and LDH levels are elevated. Lung spirometry results suggest an obstructive pattern. Which of the following is most likely causing her liver damage?
   A) Accumulation of elastase
   B) Aggregation of 1-antitrypsin
   C) Decreased activity of elastase
   D) Decreased affinity between elastase and 1-antitrypsin

4. A 28-year-old male presents to the emergency room with venous thromboembolism. His history reveals that his brother and his aunt both suffered from pulmonary embolism. Results of an antithrombin III assay show a deficiency in antithrombin III.Increased activity of which of the following blood clotting factors is most likely causing his embolism?
   A) Factor VIII
   B) Factor X
   C) Plasmin
   D) Thrombin
   E) von Willebrand factor

5. A two month old male is brought to the emergency room with severe dehydration due to chronic diarrhea, as well asconcerns of failure to thrive. A stool sample is found to have an abnormally high concentration of glucose, indicating a defect in glucose absorption in the intestines. The results of what procedure would be the most informative regarding the cause of this child's illness?
   A) Western blottingfor ciliary dynein
   B) Northern blottingfor occludin mRNA
   C) Sanger sequencing of the SGLT1 gene

   D) Southern blottingforthe GLUT1 transporter

6. A 17-year-old male presents with complaints of blurred vision and muscle weakness in his legs following strenuous activity. Myasthenia gravis is suspected. What technique would be most informative in confirming his diagnosis?
   A) Southern blot
   B) Northern blot
   C) FISH
   D) ELISA
   E) PAGE

7. A 61-year-old male presents with neurological symptoms consistent with Huntington's disease. Which of the following procedures would be most appropriate for confirming his diagnosis?
   A) FISH
   B) Microarray
   C) PCR
   D) DNA fingerprint
   E) Southern blot

8. A 48-year-old female with an18 pack-year smoking history presents with persistent cough, shortness of breath and a feeling of tightness in her chest. She is also found to have extensive cirrhosis of the liver. Which of the following techniques would best identify the genetic mutation that is causing her disease?
   A) Southern blot
   B) Western blot
   C) Northern blot
   D) Immunofluorescence
   E) DNA sequencing

9. A researcher wants to perform a southern blot using a probe with the sequence 5'-TCCAGCCT-3'. Before he begins, he uses the Sanger method to confirm that his target DNA sample contains the complementary sequence. Which of the following is the most likely result of this procedure?
   A) 5'-AGGCTGGA-3'
   B) 5'-TCCAGCCT-3'
   C) 5'-TCCGACCT-3'
   D) 5'-AGGTCGGA-3'

Answers

1. **The correct answer is B**. His diagnosis is post-streptococcal glomerulonephritis. After a streptococcal infection, acute glomerulonephritis may develop. This condition is an immune complex disorder. Deposition of streptococcal antigen on the glomeruli damages the glomerulus. Therefore, hematuria and proteinuria develop. Urinary loss of excess plasma proteins causes edema and hypertension. Since albumin is the main force for maintaining proper intravascular volume, urinary loss of albumin causes the edema. If the patient is not treated, glomerular damage would result in the loss of large proteins such as immunoglobulins and complement. However, loss of immunoglobulins and complement proteins leads to recurrent infections, thus answer choices C and D are incorrect. Answer choices A and E are incorrect because loss of transferrin affects iron transport, and loss of a1-

III

antitrypsin would cause lung tissue damage due to the inability to inhibit elastase.
Reference: https://emedicine.medscape.com/article/244631-overview

2. **The correct answer is C**. In a myocardial infarction, the serum CKMB isozyme reaches a peak in 24 hours and falls back to baseline in 48 to 72 hours. This allows CKMB to be useful in diagnosing re-infarction if it occurs between 3 to 10 days. Cardiac troponin (answer choice B) takes 3 to 10 days to return to baseline. Therefore, cardiac troponin would not be the best test for diagnosing the new infarct in this patient. Aspartate aminotransferase (answer choice A) and lactate dehydrogenase (answer choice D) are specific for liver damage but non-specific for myocardial tissue. Lipase (answer choice E) is useful in diagnosing acute pancreatitis.

3. **The correct answer is B**. Symptoms and history of this patient suggest that she is suffering from COPD/emphysema, which presents with shortness of breath and fatigue. Some patients with certain genetic variants of 1-antitrypsin gene (piZZ) suffer from both COPD/emphysema and liver damage. The 1-antitrypsin is a plasma protein and functions as an anti-protease, which binds and inhibits serine proteases such as elastase. Increased activity of elastase due to the deficiency of ?1-antitrypsin causes degradation of elastin. This increased proteolysis of elastin particularly effects the lungs, causing COPD/emphysema. Smoking exacerbates the effects of elastase, because cigarette smoke oxidizes the methionine residue of ?1-antitrypsin, making it ineffective as a protease inhibitor. The ?1-antitrypsin is synthesized in the hepatocytes. Patients who are homozygous for the Z variant of the ?1-antitrypsin gene may also suffer from liver damage, because misfolding of the ?1-antitrypsin causes accumulation and aggregation of the protein in the endoplasmic reticulum of the hepatocytes.

4. **The correct answer is D**. Antithrombin III is a plasma protein that binds to thrombin. Once this complex forms, thrombin becomes inactivated. Normally, thrombin converts fibrinogen to fibrin. Normally, antithrombin III prevents excessive thrombin activity to limit the formation of fibrin or clot formation. Genetic deficiency in antithrombin III causes an increase in the activity of thrombin and consequently excessive formation of fibrin. Therefore, patients with antithrombin III deficiency suffer from recurrent embolism. Factor VIII (answer choice A) is a plasma protein that circulates in the blood in an inactive form bound to von Willebrand factor (answer choice E). An injury to the blood vessel activates Factor VIII and dissociates it from von Willabrand factor. Active Factor VIII activates Factor IX. The active Factor IX activates Factor X (answer choice B), which converts prothrombin to thrombin. Plasmin (answer choice C) is required for dissolution of fibrin.
Reference: https://emedicine.medscape.com/article/954688-overview

5. **The correct answer is C**. The SGLT1 gene encodes the sodium-glucose transporter that is present on the apical surface of the enterocytes of the small intestines. This transporter is responsible for absorbing glucose from the lumen of the intestines using the electrochemical gradient of sodium to move glucose into the cell. Sequencing of this gene would reveal any mutations that would affect the functioning of this transporter, which is likely defective in this child. The enterocytes do not have ciliary dynein, so western blotting for this protein (answer choice A) would not be informative. Since occludin forms the tight junctions between adjacent cells within the intestinal lining, but does not function in transporting glucose, northern blotting for this mRNA (answer choice B) would not be beneficial. Southern blotting for the GLUT1 transporter would not be informative because the GLUT1 transporter allows the release of glucose from the enterocytes on the basal surface, but is not involved in the absorption of glucose from the intestinal lumen.

6. **The correct answer is D**. Myasthenia gravis is an autoimmune disease that is caused by the body's immune system making antibodies that recognize the acetylcholine receptors present on muscle fibers at neuromuscular junctions. Once these antibodies bind to the receptor, the receptor no longer responds to acetylcholine signaling and, therefore, the muscle does not receive the signal to contract. Since the diagnosis for this disease is the presence of these autoimmune antibodies, an assay that identifies the antibodies in the patient's blood is required. Of the answer choices given, ELISA is the only technique that identifies the presence of antibodies. Southern blot (answer choice A) would be used for analyzing DNA, and northern blot (answer choice B) would be used for RNA. FISH (answer choice C) uses fluorescence labeled probes to identify chromosomal aberrations associated with disease. PAGE (answer choice E) is a technique used to separate proteins based on size.

7. **The correct answer is E**. Huntington's disease is a genetic disorder that causes the progressive breakdown of nerve cells in the brain. This disease is caused by an expansion of the trinucleotide sequence CAG within the *HTT* gene. The number of these CAG repeats determines whether or not an individual has the disease, as well as the severity of the disease. Southern blot analysis is the standard method for determining the repeat number as the size of the resulting band can be used to calculate how many repeats are present and, therefore, the diagnosis. FISH (answer choice A) uses fluorescence labeled probes to identify chromosomal aberrations associated with disease. Microarray (answer choice B) is typically used to provide information about the level of expression of specific genes. PCR (answer choice C) is a technique that allows for the amplification of a specific fragment of DNA. DNA fingerprinting (answer choice D) is not used for identifying the presence of a disease, but rather to distinguish the DNA of one individual from another.

8. **The correct answer is E**. The symptoms presented in this patient are consistent with ? 1 antitrypsin deficiency. This disease is caused by a mutation in the *SERPINA1* gene that encodes for the elastase inhibitor ? 1 antitrypsin. This protein is normally produced by the hepatocytes in the liver and released into the bloodstream. The mutation of this gene results in mis-folding of the protein, which accumulates in the liver and leads to liver damage. In addition, there is no ? 1 antitrypsin available to the lungs to prevent the elastase-mediated breakdown of lung tissue,

resulting in the lung conditions seen in this patient. DNA sequencing would show the presence of the mutation in *SERPINA1* and, thus, allow for diagnosis of this disease. Southern blot (answer choice A), western blot (answer choice B), and northern blot (answer choice C) would not give information about the presence of the causative mutation. Immunofluorescence (answer choice D) can be used to visualize the location of a specific protein inside the cell, but would not be informative in this case.

9. **The correct answer is A**. According to the Watson-Crick base pairing rules, in a double stranded DNA molecule, adenosine pairs with thymine and guanine pairs with cytosine. Also recall that the two strands of the DNA molecule are oriented in an antiparallel fashion. Therefore, the sequence in answer choice A is the only one that satisfies these conditions.

Part IV

IV

21 Digestion and Absorption of Carbohydrates

At the conclusion of this chapter, students should be able to:
- Describe the classifications of dietary carbohydrates
- Explain the role of digestive enzymes and the processes of carbohydrate digestion
- Evaluate the role of glucose transporters in the absorption from the intestinal epithelial cells into the bloodstream
- Discuss clinical tests and diseases related to carbohydrate digestion and absorption

Carbohydrates are a type of macromolecule consumed in our diet to provide our cells with a source of energy. The energy production from carbohydrates occurs through the glycolysis pathway, which is discussed in chapter 22. The basic building blocks of carbohydrates are monosaccharides, also referred to as simple sugars. Monosaccharides can be linked together through *O*-glycosidic bonds to form disaccharides, as well as larger oligo- or poly-saccharides (▶ Table 21.1).

Monosaccharides: The most important monosaccharides in human metabolism are glucose, fructose and galactose. They can be acquired through the consumption of monosaccharide-containing foods or through the ingestion of food sources containing complex carbohydrates. In this case, the complex carbohydrates are subsequently broken down into their monosaccharide constituents in the body. Monosaccharides are readily absorbed at the brush border of the small intestine through transporters are the apical surface of the enterocytes. Glucose and galactose are taken up into enterocytes from the intestinal lumen via the sodium-glucose symporter, also called the sodium-glucose linked transporter-1 (SGLT1). Fructose, on the other hand, is taken up by enterocytes via the GLUT5 transporter. Absorbed glucose, galactose ultimately exit the enterocyte via facilitative transporters GLUT2 on the basal side of the cell and enter the bloodstream. Fructose exits the enterocyte via GLUT5 on the basal side of the enterocyte (▶ Fig. 21.1).

21.1 Link to Pharmacology: Gliflozins Are SGLT2 Inhibitors

The sodium-glucose linked transporters (SGLTs) are a family of transporters that mediate the concerted transfer of glucose and Na^+ across the plasma membrane. The member of this family known as SGLT1 is found in intestinal epithelial cells, whereas SGLT2 is mainly expressed in the kidneys and functions in reabsorbing glucose in the proximal renal tubules. The SGLT2 can reabsorb all of the filtered glucose when blood glucose concentrations are up to 180 – 200 mg/dL. If the blood glucose concentration exceeds the capacity of the SGLT2 transporters (> 200 mg/dL), then glucose is excreted into the urine. Since SGLT2 mainly functions in the reabsorption of glucose from the kidneys, inhibitors of SGLT2 transporters (gliflozins) are used for treatment of type 2 diabetes. Inhibition of the reabsorption increases the glucose excretion into the urine. Thus, gliflozins

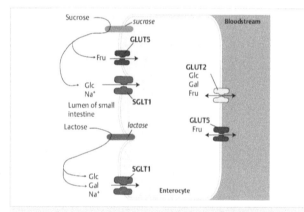

Fig. 21.1 Absorption of fructose, galactose, and glucose. (Source: Panini S, ed. Medical Biochemistry—An Illustrated Review. New York, NY. Thieme; 2013.)

Table 21.1 Classification of carbohydrates

Class	Examples	Composition and Relevant Information
Monosaccharides (1 sugar unit)	Glucose (Glc)	$C_6H_{12}O_6$; most abundant simple sugar
	Galactose (Gal)	$C_6H_{12}O_6$; obtained from mammalian milk
	Fructose (Fru)	$C_6H_{12}O_6$; obtained from fruit and honey
	Ribose/deoxyribose	$C_5H_{10}O_5$/$C_5H_{10}O_4$; components of ribonucleic/deoxyribonucleic acids
Disaccharides (2 sugar units)	Maltose	Glc + Glc; found in malt sugars and hydrolyzed by maltase
	Lactose	Glc + Gal; found in mammalian milk and hydrolyzed by lactase
	Sucrose	Glc + Fru; found in table sugar, fruits, and honey and hydrolyzed by sucrase
Oligosaccharides (3–10 sugar units)	Polymers composed of a mixture of sugar units	Found attached to lipids (glycolipids) and proteins (glycoproteins)
Polysaccharides (> 10 sugar units)	Cellulose	Polymer of Glc units: found in plants but indigestible by humans
	Starch	Polymer of Glc units: the stored form of glucose in plants
	Glycogen	Polymer of Glc units; the stored form of glucose in animals
	Glycosaminoglycans	Polymers of disaccharides; components of proteoglycans

Source: Panini S, ed. Medical Biochemistry- An Illustrated Review. New York, NY. Thieme; 2013.

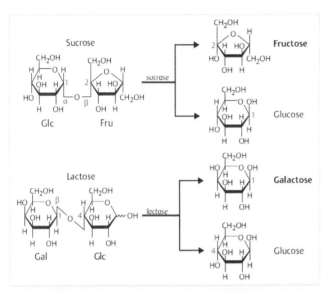

Fig. 21.2 Breakdown of sucrose and lactose. (Source: Panini S, ed. Medical Biochemistry—An Illustrated Review. New York, NY. Thieme; 2013.)

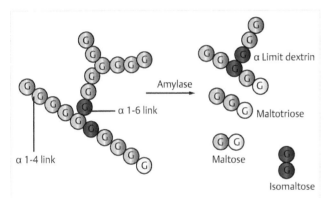

Fig. 21.3 Digestion of carbohydrates (polysaccharides) into disaccharides and monosaccharides. (Source: Digestive Functions. In: Michael J, Sircar S, ed. Fundamentals of Medical Physiology. Thieme; 2010.)

help to maintain normal blood glucose concentrations in type 2 diabetics.

Disaccharides: Disaccharides result from linkage between two monosaccharides via O-glycosidic bonds. Glucose and galactose can combine in a β-1,4 linkage to form the disaccharide *lactose*, a carbohydrate found in milk and dairy products; glucose and fructose can combine in an α-1,2 linkage to form the disaccharide *sucrose*, a carbohydrate found in table sugar; and two glucose molecules can combine in an α-1,4 linkage to form *maltose*, a disaccharide found in grain products and starch (▶ Fig. 21.2). Disaccharide digestion takes place at the brush border of the duodenum in the small intestine. It is mediated by enzymes synthesized and secreted by enterocytes. These enzymes are referred to as disaccharidases and include lactase, sucrase and maltase. As the names suggest, lactase digests lactose into glucose and galactose; sucrase digests sucrose into fructose and glucose; and maltase digests maltose into two glucose molecules. The resultant monosaccharides are then absorbed by their respective transporters.

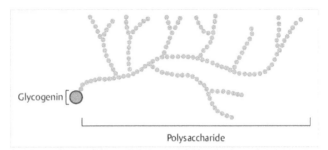

Fig. 21.4 Structure of glycogen. Glycogen is composed of repeating glucose subunits that are linked in linear chains via α-1,4 linkages, and branched points via α-1,6 linkages. (Source: Panini S, ed. Medical Biochemistry—An Illustrated Review. New York, NY. Thieme; 2013.)

Polysaccharides: There are three main polysaccharides: *starch*, *glycogen* and *cellulose*, all of which are composed of repeating glucose subunits. There are two main starches: amylose and amylopectin. Their function is to serve as an energy reserve in plant cells through the storage of glucose. Amylose and amylopectin both consist of repeating linear α-1,4 glucose subunits. Amylopectin, however, also contains glucose branch points which are joined to the linear glucose chain by an α-1,6 linkage. Following the ingestion of plant-based foods, the amylose and amylopectin are primarily digested by the enzyme amylase, which is secreted from salivary glands of the mouth and the exocrine pancreas.

Amylase is an endoglycosidase and catalyzes the cleavage of α-1,4 glucose linkages, but is unable to cleave α-1,6 glucose branch points. The breakdown of starch by amylase results in the formation of three smaller carbohydrates, and includes: maltose, maltotriose and dextrin (▶ Fig. 21.3).

Dextrin can be described as a shortened polysaccharide chain. These are subsequently broken down into individual glucose subunits by the enzymes maltase and isomaltase. These enzymes are referred to as α-glucosidases because they cleave glucose subunits joined together in an α-configuration. Maltase cleaves the remaining α-1,4 glycosidic bonds, and isomaltase cleaves the α-1,6 glycosydic branch points.

The function of glycogen is to store glucose in human hepatocytes and muscle cells. Like amylopectin, glycogen is composed of repeating glucose subunits via α-1,4 linkages. However, glycogen has more extensive α-1,6 branch points (▶ Fig. 21.4).

Cellulose is a linear polysaccharide composed of repeating glucose subunits via β-1,4 linkages and is found in plant cell walls. Human are unable to digest cellulose because we lack the β-glucosidase enzyme required to break β-1,4 glycosidic bonds (▶ Fig. 21.5). Therefore, cellulose is a non-digestible carbohydrate, otherwise known as dietary fiber. It has no direct nutritional value to humans, but can influence the health and integrity of our GI tract. The health benefits of fiber are discussed in chapter 45.

21.2 Link to Pharmacology

α-glucosidase inhibitors, **acarbose and miglitol,** inhibit the digestion of disaccharides, significantly sucrose and maltase. Inhibition of their digestion reduces the absorption of glucose.

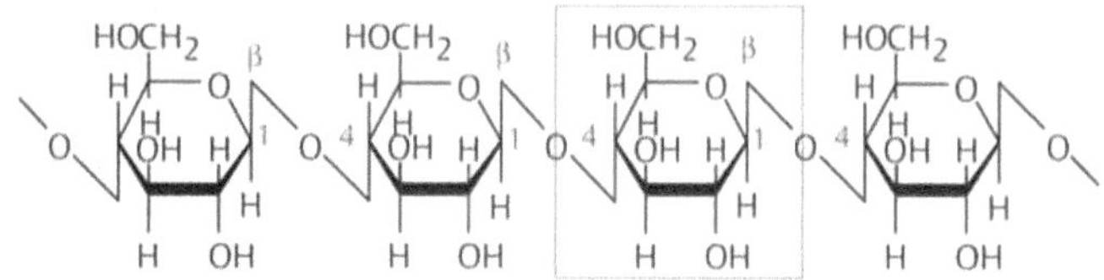

Cellulose–unbranched homopolymer (plants)

Fig. 21.5 Structure of cellulose Repeating. glucose subunits are linked via via β-1,4 linkages. (Source: Koolman J, Röhm K, ed. Color Atlas of Biochemistry. 3rd Edition. New York, NY. Thieme; 2012.)

Therefore these drugs help postprandial blood glucose concentrations in type 2 diabetics.

21.3 Link to Pathology: D-xylose Absorption Test:

The purpose of the D-xylose absorption test is to determine the etiology of malabsorption, which can be triggered by damage to the gastrointestinal mucosa (e.g. inflammatory bowel disease), an overgrowth of bacteria (e.g. Whipple disease), or secondary to malfunction of another organ involved in food breakdown or absorption (e.g. insufficiency of pancreas or gallbladder).

D-xylose is a monosaccharide that is absorbed by the gastrointestinal mucosa without the need of digestive enzymes. In patients with a normal functioning gastrointestinal mucosa, D-xylose is freely absorbed and will appear in the patient's urine in significant quantities.

High amounts of D-xylose in urine indicate that the gastrointestinal mucosa is intact and the absorption is functional. This also indicates that the malabsorption in these patients is likely secondary to dysfunction of another digestive organ (e.g. chronic pancreatic insufficiency such as cystic fibrosis).

If the amount of D-xylose in urine is significantly reduced, malabsorption may be due to bacterial overgrowth or damage to the gastrointestinal mucosa. These causes can be distinguished by giving patients a course of antibiotics and then repeating the test. If, when repeated, the urine contains a significant amount of D-xylose, it is assumed that malabsorption was due to bacterial overgrowth. If D-xylose is still significantly low in the urine following antibiotics, the cause of malabsorption is likely due to conditions that damage the gastrointestinal mucosa (e.g. Crohn's disease).

21.4 Link to Pathology: Crohn's Disease

Crohn's disease is an inflammatory bowel disease, which develops most often due to abnormal immune responses. The chronic inflammation damages the bowel mucosa of the distal ileum and this may lead to abdominal pain, cramping, diarrhea, fatigue, weight loss, and malnutrition.

21.5 Link to Pathology: Lactose Intolerance

Lactose intolerance refers to an inability to digest lactose into glucose and galactose secondary to insufficient lactase production and secretion by enterocytes. Common symptoms include diarrhea, flatulence and abdominal cramping following the consumption of lactose-containing products. The inability to digest lactose results in its fermentation by colonic bacteria into methane gas and hydrogen. This is responsible for the flatulence and abdominal cramping experienced by these patients. Lactose also attracts water, and its excess in the gastrointestinal tract results in osmotic diarrhea.

Lactose deficiency is most commonly caused by decreased expression of the lactase-persistence allele. It may also present in infants with a congenital defect in lactase expression, known as congenital lactase deficiency. Furthermore, patients may experience transient lactose intolerance following an episode of gastroenteritis secondary to injured enterocytes. Patients with lactose intolerance due to decreased expression of the lactase-persistence allele or a congenital lactase deficiency will have normal villi under microscopy. In contrast, patients with lactose intolerance secondary to gastroenteritis will have inflamed injured villi.

Lactose intolerance is a clinical diagnosis but can be supported by a positive hydrogen breath test. Following the consumption of pure lactose, the hydrogen breath content upon exhalation is measured. A positive test is a > 20ppm rise in hydrogen from baseline.

Review Questions

1. The digestion of which of the following dietary carbohydrates is significantly inhibited by acarbase?
   A) Brown rice
   B) Milk
   C) Table sugar
   D) Oat
   E) Wheat
2. Which of the following glucose transporters is involved in absorption of glucose and galactose from the apical side of enterocytes?
   A) GLUT 2
   B) GLUT 4
   C) GLUT 5
   D) SGLT-1
   E) SGLT-2
3. Which of the following laboratory tests would support the diagnosis of lactose intolerance?
   A) Blood glucose test
   B) D-xylose absorption test
   C) Hydrogen breath test
   D) Immunological test

Answers

1. **The correct answer is C**. Acarbose and miglitol are α-glucosidase inhibitors. These drugs inhibit the digestion of **disaccharides**, such as sucrose (table sugar) and maltase. It is reported that the relative affinity of acarbose for sucrase and maltase is higher than for isomaltase and lactase. Inhibition of their digestion reduces the absorption of glucose. Therefore, these drugs help to maintain postprandial blood glucose concentrations in type 2 diabetics.

The answer choices A, D, and E are **polysaccharides.** These carbohydrates are digested by endoglycosidases, such as α-amylase. The answer choice B, milk, contains lactose sugar.

The glucose and galactose that make up lactose are linked by β-glycosidic bond. Acarbose does not inhibit the enzymes responsible for breaking β-glycosidic bonds, in this case lactase.

   Reference: Heiner Laube, Department of Internal Medicine, Universität Giessen, Giessen, Germany Clin Drug Invest. 2002;22(3) (https://www.medscape.com/viewarticle/432744_2)

2. **The correct answer is D**. SGLT-1 is a sodium-glucose symporter found on the apical side of the enterocytes in the intestine and is involved in absorbing glucose and galactose from the intestinal lumen into the enterocytes. .

Answer choice A is not correct because the GLUT 2 transporter is found on the basal side of the enterocytes and transports glucose and galactose out of the enterocytes and into the bloodstream. Answer choice C is not correct because GLUT 5 is a fructose transporter. Answer choice E is not correct because SGLT-2 is expressed in proximal tubule cells of the kidney, not in the enterocytes of the intestines.

3. **The correct answer is C**. Lactose intolerance is due to insufficient lactase production and secretion by enterocytes. In the absence of lactase, the lactose sugar from milk is fermented by colonic bacteria into methane gas and hydrogen. Generally, lactose intolerance can be diagnosed clinically thorough medical, family, and diet history. However, the diagnosis can be supported by a positive hydrogen breath test. Following the consumption of pure lactose, the hydrogen breath content upon exhalation is measured. A positive test is a > 20ppm rise in hydrogen from baseline.

Answer choice A is not correct because lactase insufficiency does not affect blood glucose levels. Answer choice B is not correct because D-xylose is freely absorbed without the need for digestion. Answer choice D would not provide the information necessary to confirm the patient's diagnosis of lactose intolerance.

IV

22 Glycolysis

At the conclusion of this chapter, students should be able to:
- Explain the tissue specificity and specific functions of facilitative glucose transporters
- Describe the pathway and the functions of glycolysis in carbohydrate metabolism
- Explain in which tissues/conditions the anaerobic and aerobic glycolysis occur
- Analyze the factors involved in excess production and utilization of lactate
- Explain the clinical causes and consequences of accumulation of lactate in the body
- Explain the regulation of glycolysis and how it differs in hepatic vs. extra-hepatic tissues
- Distinguish the clinical signs and symptoms of conditions in which enzymes of glycolysis are genetically deficient or inhibited by poisons

All cells in our body need "energy" to function. The form of energy ultimately used is adenosine triphosphate (ATP) and is generated in our cells through the oxidation of glucose, fatty acids and proteins. The metabolic pathways that generate ATP are glycolysis, decarboxylation of pyruvate, the TCA cycle, and the electron transport chain. In order to function properly, all of these activities are controlled by complex, highly regulated processes.

Glycolysis → Decarboxylation of pyruvate → TCA Cycle → Oxidative Phosphorylation

In this chapter, we will review the *initial* steps of glucose oxidation known as **Glycolysis**. Although glucose is the main monosaccharide oxidized in glycolysis, the dietary sugars fructose and galactose are also metabolized in the glycolytic pathway. In order for glucose to be oxidized in the glycolytic pathway, it must first enter the cell. Uptake of glucose from blood into the cells is accomplished by tissue-specific facilitative glucose-uptake transporters ("GLUT"). The type of GLUT transporter used depends on the type of cell that is uptaking the glucose. (▸ Table 22.1)

Glucose uptake in most tissues (e.g. erythrocytes, brain, central and peripheral neuronal cells) is mediated by **GLUT 1** and **GLUT 3**.

GLUT 2 is found in hepatocytes, pancreatic β cells and the kidney. GLUT 2 is also known as an insulin (or glucose) sensor in the pancreatic β cells, as it is required for glucose entry and subsequent insulin release into the bloodstream (▸ Fig. 14.2).

Table 22.1 Location of various GLUT transporters

Glucose Transporter	Location
GLUT 1	RBCs, blood brain barrier
*GLUT 2	B-cells of pancreas, liver, kidney
GLUT 3	Neurons, placenta
*GLUT 4	Adipose tissue, skeletal and heart muscle
*GLUT 5	Spermatocytes, GI tract

After mixed meals, blood glucose concentrations increase, and therefore entry of glucose into the pancreatic β cells increases as well. Elevated glucose in pancreatic β cells induces the expression of the insulin gene. Additionally, available glucose is oxidized by the glycolytic pathway to produce ATP. The increased ATP leads to the closing of ATP-sensitive K^+ channels on the cell membrane, thereby preventing K^+ efflux from the β cells of the pancreas. This results in a depolarization of the plasma membrane and subsequent influx of Ca^{2+} through voltage-gated Ca^{2+} channels. The increased intracellular Ca^{2+} promotes release of insulin from storage vesicles into the blood to control blood glucose levels.

22.1 Link to Pathology

Neonatal Diabetes Syndrome is caused by activating mutations of the *KCNJ11* gene, which encodes a subunit of the ATP-sensitive K^+ channel. This activating mutation eliminates the ATP-dependent regulation of the channel and results in the production of a perpetually opened channel. The inability to close the channel ultimately prevents the release of insulin from pancreatic β cells in response to high blood glucose levels. This disease is also known as **maturity onset diabetes of young** (**MODY**).

Secretion of insulin from the β cells of the pancreas leads to an increased insulin/glucagon ratio in the blood. The insulin then binds to its receptor, a member of the tyrosine kinase receptor (TKR) family, on the surface of target cells and initiates the insulin signaling pathway inside the cell. One of the downstream effects of insulin signaling in skeletal muscle, cardiac muscle, and adipocytes is to increase the number of **GLUT 4** transporters on the plasma membrane (▸ Fig. 22.2). As insulin levels rise, GLUT4 containing cytoplasmic vesicles fuse with plasma membrane, thereby increasing the rate of glucose transport into the cell. Once glucose is taken up into the cells with the help of insulin signaling, it enters into the glycolytic pathway, which occurs in the cytosol of all cells.

22.2 Pathways of Glycolysis: Aerobic vs Anaerobic Glycolysis

Although the oxidation of glucose to pyruvate does not require molecular oxygen, the fate of pyruvate in the glycolytic pathway does depend on the presence of oxygen, as well as the mitochondria. Hence, the oxidation of glucose can occur under either anaerobic or aerobic conditions.

When oxygen and mitochondria are present, pyruvate is further oxidized by the pyruvate dehydrogenase complex to acetyl-CoA, which enters the tricarboxylic acid (TCA) cycle (described in chapter 26). When pyruvate enters the TCA cycle for oxidation it is referred to as aerobic glycolysis (▸ Fig. 22.3a).

Cells without mitochondria (i.e. mature erythrocytes) or oxygen (i.e. rigorously exercising muscle, tissues with poor blood supply) convert pyruvate to lactate via lactate dehydrogenase. This is referred to as anaerobic glycolysis (▸ Fig. 22.3b). The formation of lactate by lactate dehydrogenase under anaerobic

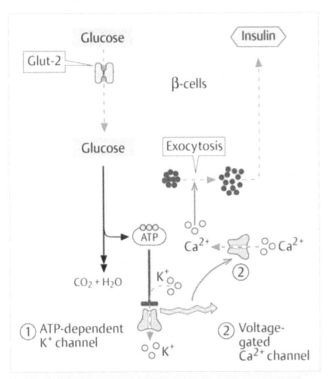

conditions provides the regeneration of NAD^+, which is required for the activity of glyceraldehyde-3-phosphate dehydrogenase. Therefore, the oxidation of NADH to NAD^+ by lactate dehydrogenase in the tissues that lack O_2 or mitochondria allows glycolysis to continue. Lactate, the final product of anaerobic glycolysis, is transported through the blood into the liver, where the lactate can enter into gluconeogenesis to resynthesize glucose. This resynthesized glucose can be released into the blood to be used in the extrahepatic tissues. This cycle of lactate and glucose is known as the Cori cycle (▶ Fig. 22.4).

22.3 Link to Pathology: Lactic acidosis

Overproduction or underutilization of lactate causes accumulation of lactate in the blood and consequently lactic acidosis. The underlying molecular reason for the overproduction of lactate is an increased $NADH/NAD^+$ ratio. An increase in NADH prevents entry of pyruvate into the mitochondria, leading to production of lactate. The reasons for an elevated $NADH/NAD^+$ ratio can be due to excess alcohol, hypoxic conditions, and genetic mutations or poisons that inhibit either the TCA cycle or electron transport chain.

Excess alcohol: The metabolism of ethanol by alcohol dehydrogenase and acetaldehyde dehydrogenase generates NADH. Therefore, excess alcohol intake increases the $NADH/NAD^+$ ratio, which shifts the equilibrium toward lactate formation. The increased $NADH/NAD^+$ ratio also inhibits

Fig. 22.1 Regulation of insulin secretion from the beta cells of the pancreas. Source: Koolman J, Röhm K, ed. Color Atlas of Biochemistry. 3rd Edition. New York, NY. Thieme; 2012.)

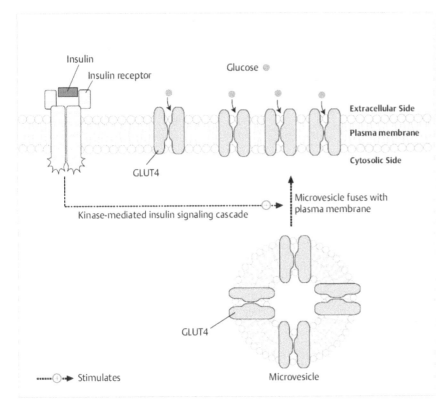

Fig. 22.2 Regulation of cell surface expression of glucose transporter 4 (GLUT4) by insulin. Insulin hormone increases the number of GLUT4 transporters in the skeletal and cardiac muscle cells, and adipocytes. This increases the glucose uptake into these cells. (Source: Panini S, ed. Medical Biochemistry—An Illustrated Review. New York, NY. Thieme; 2013.)

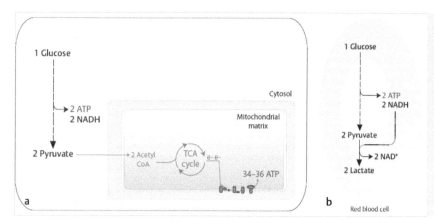

Fig. 22.3 (a) Aerobic glycolysis Cells with mitochondria and oxygen convert pyruvate to acetyl-CoA by pyruvate dehydrogenase. Acetyl CoA enters the TCA cycle for oxidation. (b) Anaerobic glycolysis Cells without mitochondria or oxygen convert pyruvate to lactate by lactate dehydrogenase. (Source: Panini S, ed. Medical Biochemistry—An Illustrated Review. New York, NY. Thieme; 2013.)

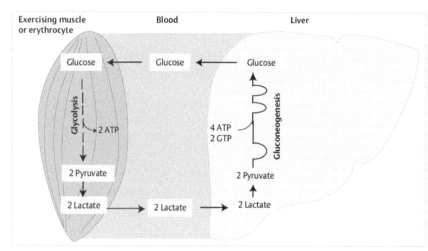

Fig. 22.4 Cori cycle. Lactate from anaerobic glycolysis is transported via blood into the liver, where lactate can enter into gluconeogenesis for glucose synthesis. This resynthesized glucose is released into the blood to be used in the extrahepatic tissues. (Source: Panini S, ed. Medical Biochemistry—An Illustrated Review. New York, NY. Thieme; 2013.)

gluconeogenesis. Thus, with excess alcohol intake, the liver becomes a lactate producer.

Hypoxic conditions can increase lactate production in the tissues. With a decreased oxygen supply, cells switch to anaerobic glycolysis to produce ATP, resulting in generation of excess amounts of lactate. Lactic acidosis can be fatal as it decreases blood vessel responsiveness to the catecholamine norepinephrine, which normally maintains blood pressure. This results in a rapid decrease in blood pressure and tissue perfusion, which further perpetuates anaerobic glycolysis and lactic acidosis. This is a vicious cycle and a highly feared complication, which may cause death in patients with sepsis.

Genetic mutations or poisons that inhibit either the TCA cycle or electron transport chain may increase lactate production. An impaired TCA cycle causes accumulation of pyruvate and, consequently, lactate formation. Since the regeneration of NAD$^+$ takes place in the ETC, any deficiency in the ETC results in an increase in the NADH/NAD$^+$ ratio, thus increasing pyruvate to lactate conversion. Deficiencies of ETC can be genetic mutations such MELAS or poisons such as CO or cyanide that inhibits the oxidative phosphorylation.

22.4 Reactions and Regulation of Glycolysis

The first step of glycolysis is catalyzed by either hexokinase or glucokinase. Both of these enzymes have the same function.

However, glucokinase is found in hepatocytes and pancreatic β-cells, while hexokinase is found in all other cells of the body. Both enzymes utilize ATP to phosphorylate glucose, thereby generating the molecule glucose-6-phosphate (G6P). The phosphate is negatively charged, which "traps" this form of glucose inside the cytoplasm. The main differences between glucokinase and hexokinase lie in their enzyme kinetics and regulation (▶ Table 22.2).

Glucokinase has a higher (Km ~ 8–10 mM) and, thus, lower affinity for glucose. Intuitively, it makes sense that when blood glucose is elevated, such as after meals, the cells in the body will oxidize glucose to make useful energy (ATP). However, during fasting states or exercise, hepatocytes do not metabolize glucose for energy. Under these conditions, hepatocytes instead spare glucose for other tissues whose primary energy source is glucose, especially erythrocytes and neurons. This is accomplished, in part, by the different kinetics of hexokinase and glucokinase. The low affinity of glucokinase for glucose ensures that hepatocytes will oxidize glucose only when glucose is plentiful in the blood.

Insulin is only secreted when β-cells sense high levels of glucose in the blood. This results in the generation of ATP from glucose oxidation. The timely secretion of insulin is accomplished by the low affinity of glucokinase for glucose in the β-cells of the pancreas. Therefore, the kinetic properties of glucokinase ensure that the pancreas will secrete insulin only when there is an adequate blood glucose level, i.e. after mixed meals.

Table 22.2 Differences between glucokinase and hexokinase

	Glucokinase	Hexokinase
Location	Liver, B-cells	All cells except liver/B-cells
Km	Low	High
Vmax	High	Low
Induced by Insulin?	Yes	No
Inhibited by G6P?	No	Yes

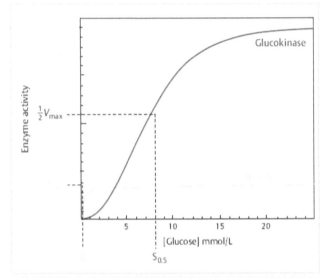

Fig. 22.5 Kinetic properties of hexokinase and glucokinase. While glucokinase has low affinity for glucose, hexokinase has high affinity for glucose. (Source: Panini S, ed. Medical Biochemistry—An Illustrated Review. New York, NY. Thieme; 2013.)

In contrast, many cell types that utilize hexokinase, such as erythrocytes and neurons, have a very high demand for ATP and thus require glucose regardless of the blood glucose level. To achieve this, the affinity of hexokinase for glucose in these cells is increased (Km ~0.1 mM) (▶ Fig. 22.5).

22.4.1 Glucokinase has a Higher Vmax

When blood glucose levels are high, such as after meals, the liver uptakes and stores excess glucose as **glycogen** via the process called **glycogenesis**. This is best achieved by glucokinase having a high Vmax, or ability to "act quickly on the available glucose".

In the pancreas, glucokinase also has a high Vmax. After meals, when the blood glucose is elevated, glucose is metabolized at a proportionally high rate. This is to ensure that the proper amount of insulin is secreted by β-cells of the pancreas into the blood.

Glucokinase is not inhibited by glucose-6-phosphate (G6P) in the hepatocytes. Glucose-6-phosphate is an intermediate of both glycolysis and glycogenesis. In order to proceed to glycogenesis to synthesize glycogen, glucokinase is *not* inhibited by its product, G6P, in the hepatocytes.

The rate-limiting-step of glycolysis is determined by phosphofructokinase-1 (PFK1), which is regulated by allosteric and hormonal (phosphorylation) mechanisms (▶ Fig. 22.6).

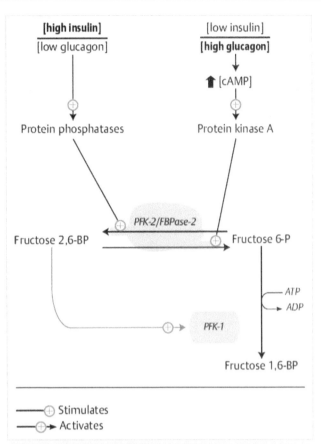

Fig. 22.6 Hormonal regulation of phosphofructokinase-1 (PFK-1) and phosphofructokinase-2 (PFK-2). After meals increased insulin hormone activates the protein phosphatase-1 (PP1). The active PP1 dephosphorylates the PFK-2, making it active. Active PFK-2 increases the concentrations of fructose 2,6-BP. The fructose 2,6-BP as an positive allosteric modulator of PFK-1 increases the activity of PFK-1, the rate-limiting step of glycolysis, thus the rate of glycolysis increases. During fasting, increased levels of glucagon hormone activates the protein kinase A (PKA), which phosphorylates the PFK-2. Phosphorylated PFK-2 becomes inactive. This causes a decrease in fructose 2,6-BP concentrations. Due to the absence of positive allosteric modulator of PFK-1, the glycolytic rate decreases. (Source: Panini S, ed. Medical Biochemistry—An Illustrated Review. New York, NY. Thieme; 2013.)

The major role of glycolysis is to generate ATP. Thus the main factor for regulation of glycolysis is the energy charge (i.e. ATP/ADP ratio) of the cell: adenosine monophosphate (AMP) is a positive allosteric modifier of PFK-1 and ATP is an allosteric inhibitor of PFK-1. Besides ATP, PFK-1 is also inhibited by citrate, a TCA cycle intermediate. When the ATP/ADP ratio is high, isocitrate dehydrogenase, the key regulatory enzyme of the TCA cycle, is inhibited. This results in accumulation of citrate in the mitochondria. This elevated citrate passes through the citrate shuttle into the cytoplasm, where it functions as allosteric inhibitor of PFK-1.

Other than allosteric regulation by the ATP/ADP ratio, the liver PFK-1 is also regulated by insulin and glucagon signaling (▶ Fig. 22.6). When blood glucose is elevated, insulin stimulates many enzymes involved in glucose metabolism, two of which

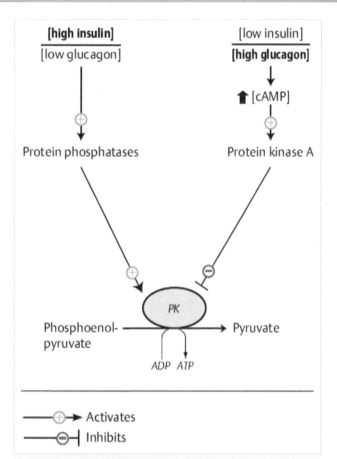

Fig. 22.7 Hormonal modulation of pyruvate kinase activity. After meals with the increased levels of insulin hormone, protein phosphotase-1 (PP1) becomes active. The active PP1 dephosphorylates the pyruvate kinase. The dephosphorylated pyruvate kinase becomes active. During fasting with the increased glucagon hormone activates the protein kinase A (PKA). Active PKA phosphorylates and consequently activates the pyruvate kinase. (Source: Panini S, ed. Medical Biochemistry—An Illustrated Review. New York, NY. Thieme; 2013.)

Table 22.3 Signs of hemolytic anemia in patients with pyruvate kinase deficiency

Sign	Reason
Fatigue, tachycardia	Low O2 carrying capacity of the blood
Jaundice, scleral icterus	Increased unconjugated bilirubin in the blood
Gallstones	Increased bilirubin in biliary system
Splenomegaly	RBC removal by splenic macrophages due to their abnormal shape; may visualize "burr cells" on peripheral blood smear
Decreased: RBC count, hemoglobin, haptoglobin **Increased:** LDH, unconjugated bilirubin	RBC lysis in the blood and release of contents

and therefore inhibition, of pyruvate kinase. This represents the switch to gluconeogenesis, where PEP enters the gluconeogenesis pathway to generate glucose.

22.4.2 Link to Pathology: Genetic Deficiencies in Pyruvate Kinase Can Cause Hemolytic Anemia

Pyruvate kinase is an important enzyme in the ATP-producing process of glycolysis. Inherited deficiencies of this enzyme, therefore, result in low ATP production. Due to the absence of mitochondria, erythrocytes are completely reliant on anaerobic glycolysis to produce ATP. The most important function of ATP in the erythrocytes is to maintain the function of the Na^+/K^+ ATPase pump in the plasma membrane. This transporter requires ATP to pump three Na^+ out of the cell in exchange for bringing two K^+ into the cell. In patients with pyruvate kinase deficiency, dysfunctional pyruvate kinase results in an insufficient amount of ATP production and causes a net accumulation of Na^+ inside the cells, which subsequently attracts water. This results in swelling of the cells and hemolysis. These patients typically present between birth and early childhood with signs of hemolytic anemia, as outlined in ▶ Table 22.3.

22.4.3 Link to Toxicology: Arsenic Poisoning

Arsenic is an element that can be found in contaminated water and pesticides. It is an inhibitor of multiple enzymes in carbohydrate metabolism - notably glyceraldehyde-3-phosphate dehydrogenase and pyruvate dehydrogenase. Inhibition of these enzymes results in decreased ATP production. In addition, inhibition of pyruvate dehydrogenase also prevents pyruvate from entering the TCA cycle and results in its conversion to lactic acid via lactate dehydrogenase, resulting in metabolic lactic acidosis. Signs of arsenic poisoning may include abdominal pain, nausea, vomiting, diarrhea and garlic breath. Treatment involves the administration of a chelating agent, such as dimercaprol.

are PFK-1 and PFK-2. Stimulation of PFK-1 results in formation of fructose-1,6-biphosphate (F1,6BP), which in turn stimulates pyruvate kinase (the last step of glycolysis). Stimulation of PFK-2 results in formation of fructose-2,6-biphosphate (F2,6BP), which in turn stimulates PFK-1. The net result is an increase in the activity of PFK-1 to promote glycolysis.

In contrast, glucagon (a hormone secreted from the pancreas which has the opposite effect of insulin), stimulates fructose-2,6-biphosphatase (F26BPase), which results in the conversion of F2,6BP to Fructose 6-Phosphate. This is to prevent the acceleration of glycolysis. It is important to note that PFK-2 and F26BPase are two separate subunits of a single polypeptide.

The final step in glycolysis is completed by **pyruvate kinase**, an enzyme that converts phosphoenolpyruvate (PEP) to pyruvate. Pyruvate kinase activity is upregulated by increased concentrations of fructose 1,6-bisphosphate. This is the feedforward regulation of pyruvate kinase. Pyruvate kinase is also regulated by insulin and glucagon signaling in the liver (▶ Fig. 22.7). While dephosphorylation by insulin signaling increases the activity of pyruvate kinase, glucagon signaling results in phosphorylation,

IV

22.4.4 Link to Toxicology: Mercury Poisoning

Mercury is a heavy metal that can be ingested by eating certain types of fish. It inhibits the glycolytic enzyme glyceraldehyde-3-phosphate dehydrogenase, thereby decreasing cellular ATP production. Signs of arsenic poisoning may include numbness, tingling, weakness and mental status changes. Treatment is with chelating agents such as dimercaptosuccinic acid (DMSA).

Review Questions

1. Which of the following is a downstream effect of insulin signaling in skeletal muscle?
   A) Closing of ATP-sensitive K^+ channels on the cell membrane
   B) Formation of lactate from pyruvate
   C) Increased ATP generation by oxidative phosphorylation
   D) Increasing GLUT4 concentrations on the cell membrane
   E) Phosphorylation and activation of pyruvate kinase

2. Which of the following is the main function of glycolysis in the erythrocytes?
   A) Generate ATP for osmotic balance
   B) Oxidize glucose to generate acetyl CoA
   C) Produce lactate for Cori cycle
   D) Produce substrate for LDH
   E) Reduce NAD^+ to NADH

3. Which one of following best explains the underlying molecular mechanism for formation of lactate in hypoxic tissues?
   A) Absence of nutrients
   B) Increased intracellular pH
   C) Increased CO_2 pressure
   D) High NADH/NAD^+ ratio
   E) Presence of mitochondria

4. Genetic deficiency of which of the following glycolytic enzymes causes hemolytic anemia?
   A) Hexokinase
   B) Glyceraldehyde-3-phosphate dehydrogenase
   C) Lactate dehydrogenase
   D) Phosphofructokinase-1
   E) Pyruvate kinase

Answers

1. **The correct answer is D**. Upon binding of insulin to its receptor, the insulin signaling pathway is initiated. One of the downstream effects of insulin signaling in skeletal muscle, cardiac muscle, and adipocytes is to increase the number of GLUT 4 transporters on the plasma membrane. This increase in GLUT4 transporters enhances the rate of glucose uptake.

Insulin signaling does not affect ATP-sensitive K + channels, therefore A is incorrect. Answer choices B and C are also incorrect. Although insulin signaling increases the rate of glycolysis in **hepatocytes** by increasing the activity of PFK-1, the rate of both oxidative phosphorylation and lactate formation are <u>not</u> affected by the insulin signaling in **skeletal muscle cells**. Answer choice E is also incorrect because the insulin signaling pathway activates protein phosphatases, which dephosphorylate pyruvate kinase, making it active only in the hepatocytes.

2. **The correct answer is A**. The main function of glycolysis in erythrocytes is to generate energy (ATP) that is used by the Na^+/K^+ ATPase pump to maintain the osmotic balance of the cell. This pump uses ATP to pump three Na^+ out of the cell in exchange for bringing two K + into the cell. Since mature erythrocytes do not have mitochondria, these cells can generate ATP only from anaerobic glycolysis.

Answer choice B is not correct because conversion of pyruvate to acetyl CoA by pyruvate dehydrogenase (PDH) takes place in the mitochondria, which are absent in erythrocytes.

Answer choices C, D, and E are not correct. Although lactate is the final product of anaerobic glycolysis, the main function of glycolysis is not to produce lactate. Production of lactate by LDH regenerates the NAD^+ for the glyceraldehyde dehydrogenase reaction, so glycolysis is sustained within the erythrocytes.

3. **The correct answer is D**. The oxygen supply is decreased in hypoxic tissues. Therefore cells switch to anaerobic glycolysis to produce ATP, resulting in the generation of excessive amounts of lactate. The underlying molecular reason for the overproduction of lactate in hypoxic tissues is that the lack of oxygen prevents the ETC from functioning since oxygen is necessary to pass electrons between the various complexes of the ETC. Impaired ETC causes an increased NADH/NAD^+ ratio, which prevents entry of pyruvate into the mitochondria, leading to production of lactate.

The absence of nutrients (answer choice A) or increased intracellular pH (answer choice B) does not increase lactate production. In hypoxic conditions, CO_2 pressure may increase, but this does not cause a depletion in the oxygen, hence answer choice C is not correct. In the absence of mitochondria, cells oxidize glucose to lactate, but if mitochondria and oxygen are present, cells oxidize glucose to pyruvate. The pyruvate then enters mitochondria to be oxidized further.

Genetic deficiency of pyruvate kinase results in an insufficient amount of ATP production and can cause hemolysis of the erythrocytes. Since erythrocytes without ATP cannot maintain the function of the Na^+/K^+ ATPase pump in the plasma membrane. This results in a net accumulation of Na^+ inside the cells, which subsequently attracts water causes hemolysis. Signs and symptoms of hemolytic anemia due to pyruvate kinase deficiency are increased concentrations of 2,3-bisphosphoglycerate and enlarged spleen. Other than pyruvate kinase, genetic deficiency of glucose 6-phosphate dehydrogenase (G6PD) also a common cause of hemolytic anemia. G6PD deficiency is reviewed in chapter 23.

The enzymes in the answer choices A, B, C, and the D are not known to have genetic deficiencies nor that of a cause of hemolysis of the erythrocytes.

IV

23 Metabolism of Fructose, Galactose and the Pentose Phosphate Pathway (HMP Shunt)

At the conclusion of this chapter, students should be able to:
- Explain how fructose and galactose are metabolized in the glycolytic pathway
- Identify diseases associated with fructose and galactose metabolism
- Explain the physiological role of the sorbitol pathway and the pathologies due to sorbitol accumulation
- Describe the functions, products and regulation of the pentose phosphate pathway
- Explain the functions of NADPH in biochemical reactions
- Identify the symptoms of glucose-6-phosphate dehydrogenase deficiency and be able to explain the underlying disease pathology
- Describe the function of the transketolase enzyme and its importance as a diagnostic test

23.1 Metabolism of Other Sugars

In addition to glucose, fructose and galactose are dietary sugars oxidized in the glycolytic pathway. In contrast to glucose, the metabolism of fructose and galactose is insulin-independent. More importantly, unlike glucose, both fructose and galactose are metabolized mainly in the hepatocytes and neither fructose nor galactose stimulates the secretion of insulin.

23.2 Fructose Metabolism

Fructose is common in the diet and found in table sugar (sucrose), many fruits, and honey. Following absorption, dietary fructose is metabolized mainly in the liver by glycolysis (▶ Fig. 23.1a). Upon entering hepatocytes, fructose is phosphorylated by *fructokinase* to form fructose-1-phosphate (F1P), which is subsequently cleaved into di-hydroxy-acetone-phosphate (DHAP) and glyceraldehyde by the enzyme *aldolase-B*. There are no other regulatory or rate-limiting steps in the metabolism of fructose. Because of this, the rate of fructose metabolism is more rapid than glucose metabolism. The products of fructose metabolism, DHAP and glyceraldehyde, can be converted to glucose or glycogen. The glyceraldehyde can be reduced to glycerol and used for the backbone of triacylglycerol in the esterification of the fatty acids (▶ Fig. 23.1b). Therefore, excessive fructose intake may result in its conversion into triglycerides in the liver. This may lead to increase in VLDL formation and subsequently fatty-liver disease. This is a concern in individuals whose intake of high-fructose corn syrup is in excessive amount.

23.3 Disorders of Fructose Metabolism

Essential Fructosuria is a benign asymptomatic autosomal recessive disorder due to mutation in the fructokinase *KHK* gene. Because of the insufficient fructokinase activity, fructose cannot be converted to fructose 1-phosphate (▶ Fig. 23.1a). Essential fructosuria is a benign disease because unlike glucose there is no threshold for the excretion of fructose from the kidneys. As soon as fructose is elevated in the blood, it is excreted from the kidneys. Thus fructose accumulates in the urine, fructosuria. Since it is excreted, individuals with this condition do not accumulate any toxic metabolites in their body.

Hereditary Fructose Intolerance is caused by a deficiency of aldolase B due to the genetic mutation of *ALDOB* gene. The disease is inherited in an autosomal recessive manner, therefore only homozygotes are affected. The estimated incidence is about 1:20,000 live births. Aldolase B converts the initially formed fructose 1-phosphate to glyceraldehyde and glycerol 3-phosphate (▶ Fig. 23.1a). Lack of aldolase B enzyme activity causes accumulation of high levels of fructose 1-phophate in the hepatocytes; this depletes the inorganic phosphate (Pi) in the hepatocytes and also in the blood (hypophosphatemia). Due to an insufficient amount of Pi, glycolysis is inhibited, and so is ATP production.

The clinical signs and symptoms experienced by these patients are right upper quadrant pain, hepatomegaly, jaundice, nausea and vomiting following the consumption of fructose. Due to insufficient ATP production gluconeogenesis and glycogenolysis are inhibited, therefore patients present with hypoglycemia following fructose consumption.

Patients are frequently presented around 6-months of age when caregivers begin to introduce juice and fruits into a child's diet. Chronic ingestion leads to hepatomegaly and irreversible liver damage. To prevent liver failure and possible death, fructose is removed from the diet.

23.4 Sorbitol (polyol) Pathway: Fructose Production

Sorbitol, a type of polyol, is a sugar generated from glucose by the enzyme *aldose reductase*. The purpose of this reaction is to "trap" glucose in the cell (similar to glucokinase/hexokinase trapping glucose as glucose-6-phosphate). Sorbitol can then be metabolized into fructose by the enzyme *sorbitol dehydrogenase* (▶ Fig. 23.2). The polyol or sorbitol pathway naturally occurs to generate fructose from glucose in seminal vesicles, ovaries, and hepatocytes. Fructose production in the seminal vesicles is important because sperm cells use fructose as their only source of energy.

23.4.1 Link to Pathology

Sorbitol accumulation causes retinopathy, nephropathy, and neuropathy in uncontrolled diabetes. The enzyme *aldose reductase* is present in the seminal vesicles, liver, ovaries, lens and retina, kidney, and Schwann cells of peripheral nerves. In contrast, the *sorbitol dehydrogenase* is found only in seminal vesicles, liver, and ovaries. The lack of sorbitol dehydrogenase in the lens, retina, kidney and Schwann cells of peripheral nerves causes problems in diabetic patients whose blood glucose levels are uncontrolled and chronically elevated.

IV

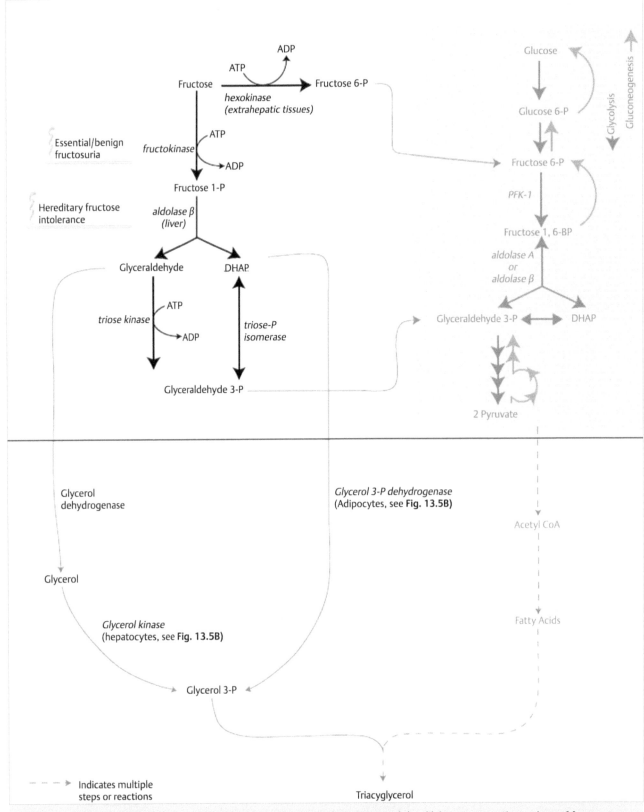

Fig. 23.1 Fructose metabolism. Fructose is metabolized in the liver by the fructokinase and the aldolase enzymes. The products of fructose metabolism, glyceraldehyde 3-P and DHAP, are intermediates of glycolysis. These intermediates are converted in the liver to pyruvate, glucose, glycogen, or fatty acids. The glycerol 3-P is used for the synthesis of triacylglycerol. (Source: Panini S, ed. Medical Biochemistry—An Illustrated Review. New York, NY. Thieme; 2013.)

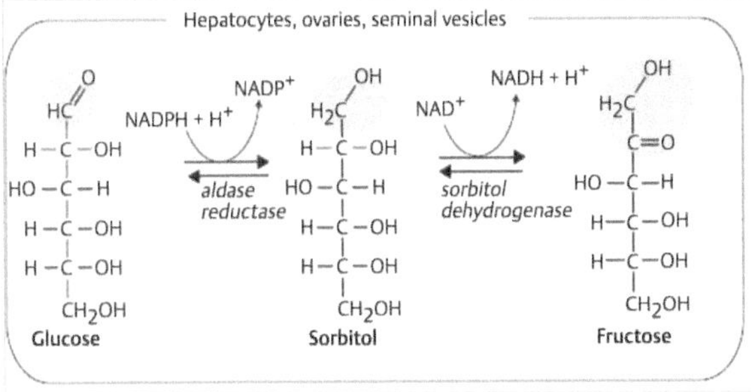

Fig. 23.2 Polyol Pathway Fructose can synthesized from glucose via the polyol (sorbitol) pathway. The aldose reductase enzyme converts glucose to sorbitol and sorbitol dehydrogenase converts sorbitol to fructose in the seminal vesicles, ovaries, and hepatocytes.

Due to increased intracellular glucose levels and the lack of sorbitol dehydrogenase, sorbitol accumulates in the lens, retina, kidney and Schwann cells of peripheral nerves. Since sorbitol is an osmolyte, it causes water retention and swelling of these cells. Therefore, uncontrolled diabetes is commonly associated with symptoms of retinopathy, cataract formation, peripheral neuropathy, and nephropathy.

23.5 Galactose Metabolism

Galactose from milk lactose (galactosyl-β-1,4-glucose) is hydrolyzed in the intestine by lactase. Once taken up into hepatocytes, β-galactose is first converted to α-galactose by galactose mutarotase, then α-galactose is phosphorylated by galactokinase to yield galactose 1-phosphate. Subsequently, galactose 1-phosphate reacts with UDP-glucose to produce UDP-galactose and glucose 1-phosphate. This reaction is catalyzed by the galactose 1-phosphate uridyltransferase (GALT) enzyme. While UDP-galactose is isomerized to UDP-glucose by epimerase, the glucose 1-phosphate can be stored as glycogen by the enzyme glycogen synthase. Glucose 1-phosphate can also undergo metabolism into glucose 6-phosphate to enter the glycolytic pathway to generate ATP (▶ Fig. 23.3).

23.6 Disorders of Galactose Metabolism

Although **GALT deficiency** is more severe than the **galactokinase deficiency**, either enzyme deficiency can cause cataracts in early life. GALT deficiency, also known as the **classic form of galactosemia,** presents with vomiting, diarrhea, and refusal to feed following the ingestion of milk. If not treated, hepatic and renal dysfunctions occur, which can lead to hyperbilirubinemia, jaundice, lethargy, hypotonia, failure to thrive, and severe intellectual disability. Galactosemia is an autosomal recessive disease, and it is included in the newborn screening test. The disease can be prevented with a lactose-free diet.

The biochemical explanation for cataract formation is due to elevated galactitol. Galactose is converted to galictitol by the enzyme aldose reductase. Galictitol is a sugar alcohol similar to sorbitol and, like sorbitol, attracts water into the lens of the eye. This water accumulation causes cell swelling, which leads to formation of bilateral cataract formation in infants.

23.7 Pentose Phosphate Pathway (PPP) or Hexose Mono Phosphate (HMP) Shunt

Depending on the needs of the cell, glucose 6-phosphate can be diverted into different metabolic pathways. If the cell requires ATP, glucose 6-phosphate is oxidized in glycolytic pathway. If the cell needs pentose sugar for nucleotide biosynthesis and/or NADPH for reducing powers, glucose 6-phosphate enters into the pentose phosphate pathway (▶ Fig. 23.4).

23.7.1 The HMP Shunt can be Divided into Two Phases.

In the oxidative phase, which is irreversible, NADPH, 5-C ribulose 5-phosphate, and CO_2 are generated. The enzymes involved in the oxidative phase are glucose 6P-dehydrogenase (G6PD) and 6-phosphogluconate dehydrogenase. The G6PD is the rate limiting step of the HMP Shunt, and deficiency of this enzyme is the most common of the inborn errors of metabolism. There are many (~300–500) known mutations, all of which cause a decrease in the production of NADPH.

The second phase of the HMP shunt, known as the non-oxidative phase, is the regenerative portion that converts excess pentose phosphates back to hexose phosphates, which can then be used for glycolysis. In this phase, the enzyme transketolase is particularly important because it requires thiamine pyrophosphate (TPP) as a coenzyme. As reviewed in chapter 26, thiamine absorption can be impaired in chronic alcoholics. A transketolase assay can be used as a diagnostic test to detect blood thiamin levels.

23.8 Roles of NADPH in Cellular Processes

The NADPH generated in the oxidative phase of the pathway provides reducing power for biosynthesis of fatty acids, cholesterol, and nitric oxide. Therefore, the HMP Shunt is particularly active in the liver, the primary organ for cholesterol and fatty acid biosynthesis (▶ Fig. 23.5a). The pathway is also active in the endocrine glands such as ovaries, testes, and adrenal cortex. These glands need to synthesize

IV

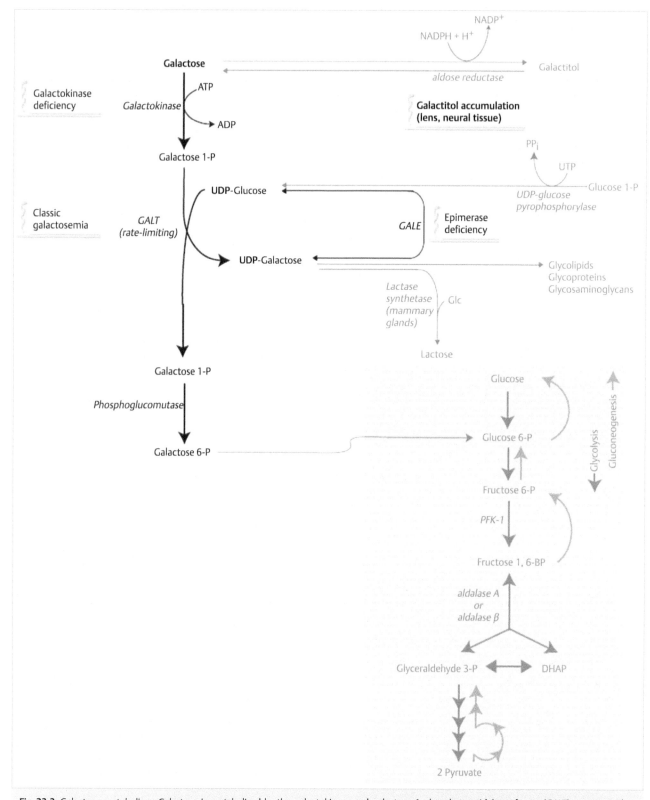

Fig. 23.3 Galactose metabolism. Galactose is metabolized by the galactokinase and galactose 1-phosphate uridyltransferase (GALT) enzymes. The product glucose 1-P can be used in glycogen synthesis or it can be isomerized to glucose 6-P and enters into glycolytic pathway. (Source: Panini S, ed. Medical Biochemistry—An Illustrated Review. New York, NY. Thieme; 2013.)

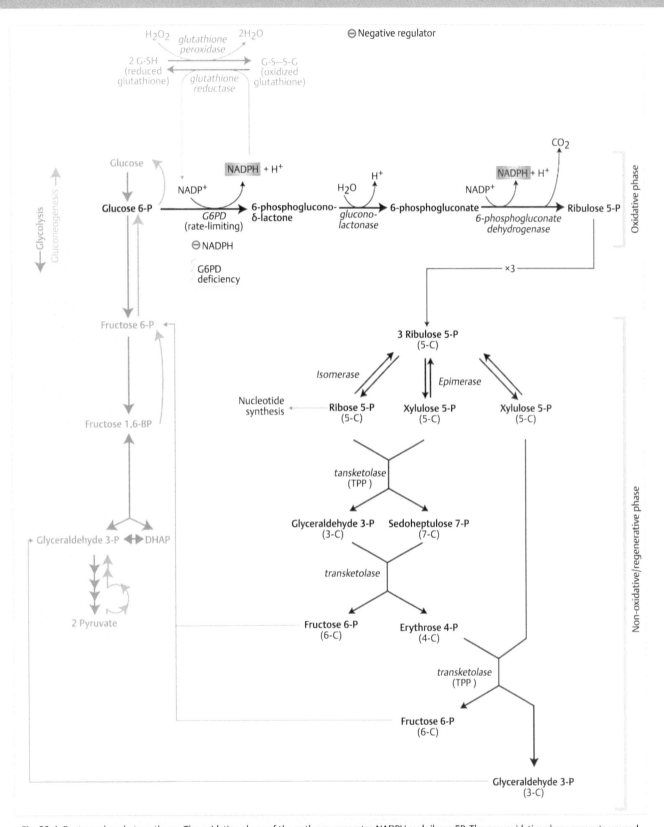

Fig. 23.4 Pentose phosphate pathway. The oxidative phase of the pathway generates NADPH and ribose-5P. The non-oxidative phase converts unused pentose-5P to fructose 6-P and glyceraldehyde 3-P. Both of which enters into glycolytic pathway. The NADPH from the pentose phosphate pathway is used by glutathione reductase to reduce the oxidized glutathione to reduced glutathione. (Source: Panini S, ed. Medical Biochemistry—An Illustrated Review. New York, NY. Thieme; 2013.)

cholesterol in order to synthesize steroid hormones, of which cholesterol is the precursor.

NADPH in phagocytes is used for the production of superoxide radicals and reactive oxygen species (ROS). These help the phagocytes to kill microorganisms and help protect against infection. (▶ Fig. 23.5b).

NADPH is also used in detoxification reactions to eliminate ROS in the erythrocytes (▶ Fig. 23.5c). Because ROS damage cell membranes and proteins, too much exposure to ROS causes hemolysis of the erythrocytes. The function of NADPH in the erythrocytes is to keep the major intracellular antioxidant, glutathione, in the reduced state (▶ Fig. 23.4). The reduced form of glutathione helps maintain the membrane integrity of red blood cells, thereby preventing hemolysis caused by exposure to ROS.

As stated earlier, the major antioxidant in cells is glutathione. In the reduced state it contains a sulfhydryl group (GSH). Reduced glutathione (GSH) helps to prevent oxidative damage to cells by reducing H_2O_2 to water. This destruction of H_2O_2 requires reduced glutathione (GSH) and the reaction is catalyzed by glutathione peroxidase. The product of the reaction is water and oxidized glutathione, which forms a dimer of two glutathione molecules (GSSG). In order to restore the oxidized glutathione back to its original reduced form, NADPH is used by glutathione reductase (▶ Fig. 23.4).

23.8.1 Link to Pathology

G6PD deficiency causes accumulation of ROS and increased risk of hemolysis. Individuals with G6PD deficiency have increased risk of hemolysis due to a decreased ability to produce NADPH and, thus, decreased ability to maintain glutathione in its reduced state in erythrocytes (▶ Fig. 23.4). In individuals with G6PD deficiency, oxidative stress causes ROS-mediated episodic hemolysis within hours. The most common triggers of oxidative stress are viral and bacterial infections, specific drugs such as antimalarial drugs and sulfonamide antibiotics, and certain foods such as fava beans. These common triggers cause reactions with O_2, either enzymatically or nonenzymatically, to produce ROS. Extensive hemolysis can ultimately result in hemolytic anemia.

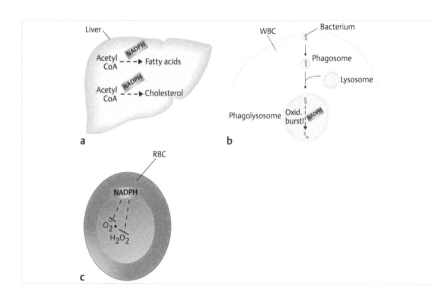

Fig. 23.5 Roles for NADPH in cellular processes. **(a)** Fatty acid and cholesterol synthesis. **(b)** Protection against infection. **(c)** Protection against oxidizing agents. (Source: Panini S, ed. Medical Biochemistry—An Illustrated Review. New York, NY. Thieme; 2013.)

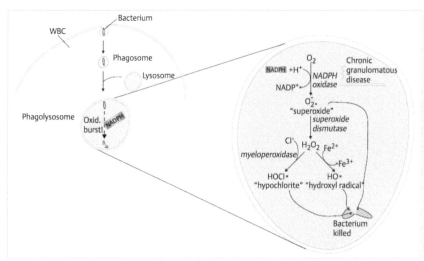

Fig. 23.6 Respiratory burst. After engulfment of bacterium by phagocytic cells, the phagosome fuses with lysosome forming phagolysosome. The NADPH oxidase generates superoxide radical within the phagolysosome. This superoxide radical can be converted to other radicals, which help to destroy the engulfed bacteria. (Source: Panini S, ed. Medical Biochemistry—An Illustrated Review. New York, NY. Thieme; 2013.)

G6PD deficiency is an X-linked recessive disorder. Therefore hemizygous males or homozygous females are affected. It is most common in Mediterranean populations.

The hemolytic anemia can be acute or chronic depending on the level of the G6PD enzyme deficiency. Many individuals with G6PD deficiency are asymptomatic unless an additional oxidative stress (infection, fava beans, and anti-malarial drugs) generates additional ROS. Patients with very low levels of G6PD activity most likely present with chronic hemolytic anemia.

23.9 Classification of G6PD Deficiency

Since there are more than 300 allelic variants, the level of G6PD activity is classified based on the severity of the disease. Class I variants have the most severe deficiency of the enzyme. These patients present with chronic anemia even in the absence of oxidative stress.

G6PD activity in Class II variants is less than 10% of normal levels. This class is also known as Mediterranean variant.

Class III variants have 10–60% of normal enzyme activity. Signs and symptoms in this class are less noticeable under ordinary conditions because only the oldest erythrocytes have decreased G6PD activity. This class may cause a shorter lifespan of erythrocytes and mild hemolysis.

Class IV variants have normal G6PD activity and Class V variants have increased activity of the G6PD enzyme.[1]

23.9.1 Heinz Bodies

Glutathione also helps to maintain a reduced state of sulfhydryl (SH) groups within intracellular proteins. Oxidation of SH groups leads to formation of denatured proteins. In the absence of glutathione, as seen in G6PD deficiency, hemoglobin proteins are oxidized and clump together to form insoluble masses called "Heinz bodies" within red blood cells.

Reference

[1] WHO Working Group. Glucose-6-phosphate dehydrogenase deficiency. Bull World Health Organ. 1989; 67(6):601–611

IV

Review Questions

1. Consumption of which of the following foods will most likely cause liver damage in patients with an aldolase B deficiency?
   A) Bread
   B) Meat
   C) Milk
   D) Sorbitol
   E) Table sugar

2. Which of the following biomolecules is the most likely cause of blurred vision problems in patients with uncontrolled type 2 diabetes?
   A) Fructose
   B) Galactose
   C) Galactitol
   D) Sorbitol

3. Which of the following enzymes can be used as a diagnostic test for thiamine deficiency?
   A) Aldolase
   B) Glucose 6-phosphate dehydrogenase
   C) Transaldolase
   D) Transketolase
   E) UDP-galactose epimerase

4. Which of the following is a trigger of hemolysis in individual who has genetic deficiency of G6PD?
   A) Fructose
   B) Galactose
   C) Hyperglycemia
   D) Hypoglycemia
   E) Primaquine

Answers

1. **The correct answer is E.** Patients with aldolase B deficiency cannot metabolize fructose. Insufficient aldolase B activity causes accumulation of high levels of fructose 1-phophate in the hepatocytes. This depletes the inorganic phosphate (Pi). Because of an insufficient amount of Pi, glycolysis is inhibited, and so is ATP production. Patients with this condition present with failure to thrive, hypoglycemia, and hepatomegaly. Chronic ingestion of fructose-containing foods leads to hepatomegaly and irreversible liver damage. Because table sugar (sucrose) is a disaccharide composed of glucose and fructose, individuals with an aldolase B deficiency should avoid the consumption of table sugar. None of the other answer choices contain fructose. Bread (answer choice A) is a starch and contains only glucose. Milk (answer choice C) has lactose, which is composed of glucose and galactose. Sorbitol (answer choice D) is not a dietary sugar. It is synthesized naturally from glucose in the cells of certain tissues. Meat (answer choice B) also does not contain fructose.

2. **The correct answer is D.** Sorbitol, a type of polyol, is a sugar generated from glucose by the enzyme *aldose reductase* in the seminal vesicles, liver, ovaries, lens and retina, kidney, and Schwann cells of peripheral nerves. In seminal vesicles, ovaries, and hepatocytes, sorbitol is converted to fructose by *sorbitol dehydrogenase*. Due to increased intracellular glucose levels and the lack of sorbitol dehydrogenase, sorbitol accumulates in the lens, retina, kidney and Schwann cells of peripheral nerves. Since sorbitol is an osmolyte, it causes water retention and swelling of these cells, leading to cellular damage. Therefore, uncontrolled diabetes is commonly associated with symptoms of retinopathy, cataract formation, peripheral neuropathy, and nephropathy.

   Answer choices A and B are incorrect because fructose or galactose is mainly metabolized in the liver, thus these sugars do not cause blurred visions in the uncontrolled diabetic patients.
   Galactitol (answer choice C) can cause cataract formation and, thus, blurred visions. However, galactitol is produced in patients who have either galactokinase or GALT deficiency, which are diagnosed during infancy.

3. **The correct answer is D.** The enzyme transketolase functions in the non-oxidative phase of the HMP shunt. To be functional, this enzyme requires thiamine pyrophosphate (TPP) as a coenzyme. Therefore, the transketolase assay can be used as a diagnostic test to detect blood thiamin levels. The enzymes in the answer choices A, B, C, and E are not used as diagnostic tests. Other enzymes that require thiamine pyrophosphate (TPP) as a coenzyme are Pyruvate dehydrogenase, α-ketoglutarate dehydrogenase, and branched-chain α-keto acid dehydrogenase.

4. **The correct answer is E.** In individuals with G6PD deficiency, oxidative stress such as viral and bacterial infections, antimalarial drugs (primaquine), sulfonamide antibiotics, and fava beans causes free radical-mediated hemolysis within hours. The oxidative stress produces either enzymatically or non-enzymatically ROS that causes damage to erythrocytes membranes ultimately leading to hemolytic anemia. The reason of this ROS-mediated hemolysis is insufficient production of NADPH from the pentose phosphate pathway. This causes cells unable to reduce the oxidized glutathione. Due to insufficient amount of reduced glutathione, the intracellular concentrations of H_2O_2 and other free radicals increase.

The answer choices A and B are not a cause of hemolysis. Hyperglycemia (answer choice C) can damage the endothelial cells by oxidation and can oxidize the intracellular proteins; however, it does not cause hemolysis of the erythrocytes. The answer choice D is not a cause of hemolysis.

24 Glycogen Metabolism and Regulation

At the conclusion of this chapter, students should be able to:
- Describe the structure and function of glycogen in the body.
- Outline the metabolic pathways for synthesis and degradation of glycogen.
- Describe the regulatory mechanisms of glycogen metabolism in the liver and in the muscle in response to insulin, glucagon, and epinephrine.
- Identify abnormalities of glycogen metabolism and the underlying molecular defects that lead to glycogen storage diseases.
- Explain the biochemical disruptions that lead to newborn hypoglycemia.

Glycogen is the storage form of glucose in animal cells. It consists of a highly branched polymer of glucose molecules linked by α-1,4 glycosidic bonds in the main chain and with α-1,6 glycosidic bonds at the branch points (▶ Fig. 24.1). The branch structure provides numerous non-reducing ends, which increases the solubility of the glycogen and serves as substrates for multiple glycogen phosphorylase enzymes. The protein glycogenin is found at the core of the glycogen and functions as a primer and catalyst in the synthesis of the first 8–10 glucose residues of glycogen. The cytosol of liver and skeletal muscle cells contains the greatest amount of stored glycogen.

The function of the liver glycogen is to maintain blood glucose concentrations for up to ~18 hours of fasting. If fasting continues, the glycogen stores become depleted and the gluconeogenesis pathway kicks in and becomes the main pathway for maintenance of blood glucose concentrations (reviewed in chapter 25). The function of glycogen in muscle cells is to release glucose for ATP generation to provide energy necessary for the muscle's function. Therefore, glucose arising from glycogen degradation in the skeletal muscle remains in the muscle cells and is rapidly metabolized by glycolysis to generate ATP for muscle contraction.

24.1 The Pathways of Glycogenesis and Glycogenolysis

Glycogen synthesis, or glycogenesis, requires ATP and UTP. The synthesis begins with glucose being phosphorylated to form glucose-6-phosphate (▶ Fig. 24.2a). This reaction is catalyzed by glucokinase in the liver and by hexokinase in other tissues. Glucose-6-phosphate is then converted to glucose-1-phosphate by phosphoglucomutase. In order for glucose to be added to the glycogen polymer, glucose must be "activated" by reacting with UTP to form UDP-glucose. The high-energy bond between UDP and glucose makes glucose a donor to a glycogen primer for the biosynthesis of glycogen. Glycogen synthase

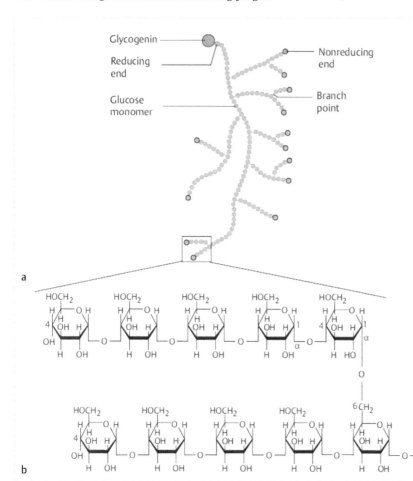

Fig. 24.1 Structure of glycogen. (a) Glycogen is a polymer of glucose that is linked via α-1,4 glycosidic bonds in the main chain and with α-1,6 glycosidic bonds at the branch points. (b) the α-1,4 glycosidic bonds and α-1,6 glycosidic bonds at the branch points are shown. (Source: Panini S, ed. Medical Biochemistry—An Illustrated Review. New York, NY. Thieme; 2013.)

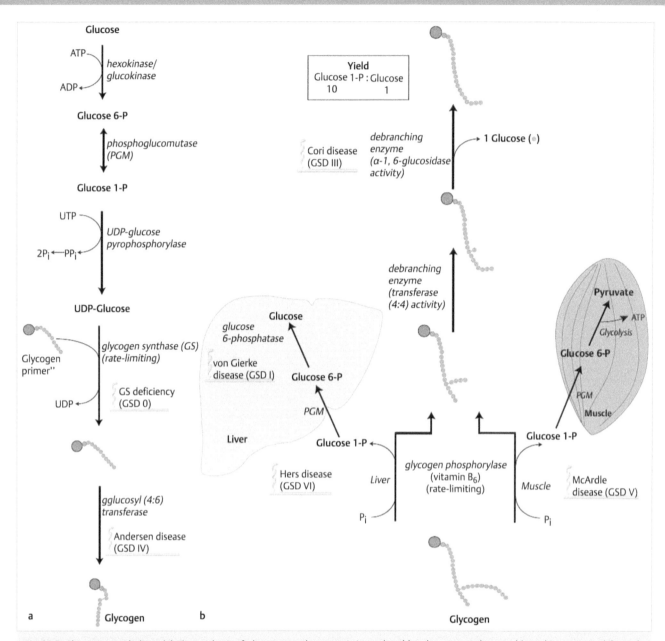

Fig. 24.2 Glycogen metabolism. **(a)** The synthesis of glycogen or glycogenesis is catalyzed by glycogen synthase and branching enzyme (glucosyl (4;6) transferase. **(b)** The degradation of glycogen or glycogenolysis is catalyzed by glycogen phosphorylase and debranching enzyme. Glucose 6-phosphatase is found only in the liver that catalyzes removal of phosphate moiety from the glucose 6-phosphate to generate free glucose in order to release into blood. (Source: Panini S, ed. Medical Biochemistry—An Illustrated Review. New York, NY. Thieme; 2013.)

catalyzes the α-1,4-glycosidic bond formation by transfer of a glucose from UDP-glucose to another glucose molecule at the end of the chain. Glycogen synthase is the key regulatory enzyme for glycogen synthesis and is therefore the site of regulation.

After approximately 8–12 glucose residues have been added, the branching enzyme (α-[1,4] α-[1,6] transferase) creates a branch point by reattaching 6–8 residues of glucose. To achieve this, the branching enzyme first removes 6–8 glucose residues by breaking the α-1,4 glycosidic bond and then links these residues with an α-1,6 glycosidic bond. These shorter chains are now ready to be elongated by glycogen synthase until the non-reducing ends reach ~11 residues of glucose, after which the process of branching repeats.

24.2 Glycogenolysis

Glycogen phosphorylase catalyzes the sequential removal of glucose-1-P units from the non-reducing ends of the glycogen branches (▶ Fig. 24.2b). The key regulatory enzyme in this release of glucose from glycogen stores, also known as glycogenolysis, is glycogen phosphorylase. Glycogen phosphorylase has the ability to cleave α-1,4 glycosidic bonds and can remove glucose residues until the partially degraded glycogen gets to ~4 glucose units of a branch point. Due to its large size and spatial hindrances, glycogen phosphorylase cannot bind to regions of the glycogen molecule that are within ~4 glucose units of a branch point. Therefore total degradation of glycogen requires an additional "debranching" enzyme. Debranching enzymes

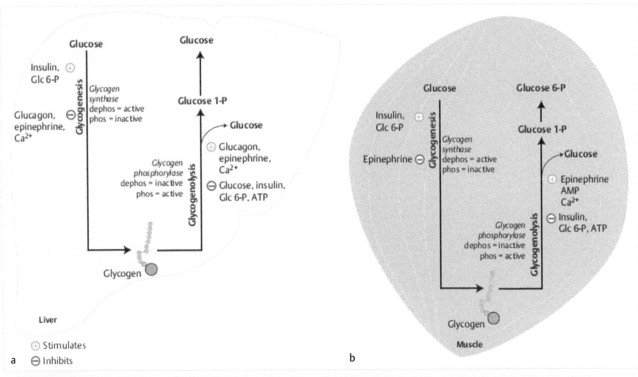

Fig. 24.3 Reciprocal regulation of glycogenesis and glycogenolysis **(a)** in the liver and **(b)** in the muscle cells. (Source: Panini S, ed. Medical Biochemistry—An Illustrated Review. New York, NY. Thieme; 2013.)

have two activities. First, α-[1,4] glucan transferase removes three of four glucose units from a branch point and transfers them to the end of another chain. This elongated chain now serves as a substrate for glycogen phosphorylase. Next, α-1,6 glucosidase removes the single glucose unit from the branch point and releases it as free glucose.

After removal of glucose 1-phosphate from the glycogen molecule, the epimerase enzyme converts glucose 1-phosphate to glucose 6-phosphate. Muscle cells use this glucose 6-phosphate for energy as it directly enters into the glycolytic pathway. However, the liver releases the glucose after removal of the phosphate moiety. Removal of the phosphate is catalyzed by glucose 6-phosphatase in the membrane of smooth endoplasmic reticulum of the hepatocytes.

24.3 Coordinated Regulation of Glycogenesis and Glycogenolysis

Glycogen synthesis and degradation are opposite pathways and are therefore not active at the same time. This *reciprocal* regulation occurs by the actions of the hormones insulin, glucagon (in the liver) and epinephrine (in the muscle). (▶ Fig. 24.3). The function of this *reciprocal* regulation is to prevent a futile, energy-wasting cycle. In addition to hormonal regulation, both glycogen synthase and phosphorylase activities are affected allosterically.

24.4 Regulation of Glycogenolysis

Glycogen phosphorylase becomes an active enzyme when it is phosphorylated, resulting in glycogen degradation. This

activation cascade is initiated by glucagon in the hepatocytes, or by epinephrine in the muscle cells and the hepatocytes (▶ Fig. 24.4).

24.5 Regulation of Glycogenolysis in the Liver

Glycogenolysis in hepatocytes can be activated by both glucagon and epinephrine/norepinephrine signaling pathways. The cell surface receptors that recognize the respective ligands are members of the G-protein coupled receptor family. Upon ligand binding, the receptors initiate signaling events inside the cell that are mediated by the $G_{\alpha s}$ subunit of the G protein complex. The downstream target of $G_{\alpha s}$ is the enzyme adenylyl cyclase, which is responsible for converting ATP to cAMP. Thus, activation of these receptors ultimately leads to increase intracellular cAMP levels. The increased concentration of cAMP activates the cAMP-dependent protein kinase -A (PKA). The PKA phosphorylates (and activates) phosphorylase kinase, which phosphorylates glycogen phosphorylase and leads to glycogen degradation and the release of glucose 1-phosphate.

Epinephrine/norepinephrine also increases the rate of glycogenolysis in the liver through α1-adrenergic receptors embedded in the plasma membrane of hepatocytes (▶ Fig. 24.5). These receptors are also members of the G-protein coupled receptor (GPCR) family. However, the α1-adrenergic receptors communicate through a different G protein subunit, $G_{\alpha q}$, to transmit the signal inside the cell. In this case, the activated $G_{\alpha q}$ activates membrane bound protein phospholipase C (PLC), which then cleaves the membrane lipid phosphatidylinositol 4,5-bisphosphate (PIP2) into a membrane-bound

IV

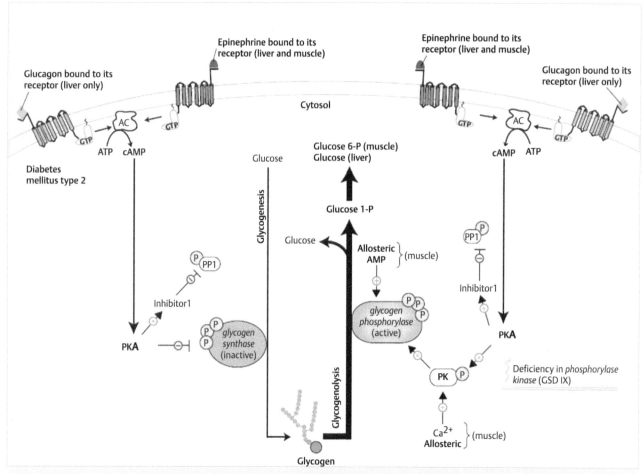

Fig. 24.4 Regulation of glycogenolysis in the liver and in the muscle. (Source: Panini S, ed. Medical Biochemistry—An Illustrated Review. New York, NY. Thieme; 2013.)

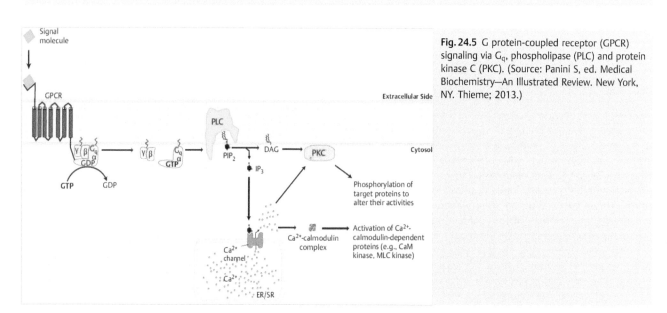

Fig. 24.5 G protein-coupled receptor (GPCR) signaling via G$_q$, phospholipase (PLC) and protein kinase C (PKC). (Source: Panini S, ed. Medical Biochemistry—An Illustrated Review. New York, NY. Thieme; 2013.)

diacylglycerol (DAG) molecule and a cytoplasmic inositol tri-sphosphate (IP3) molecule. The DAG binds to and activates protein kinase C (PKC). The active PKC phosphorylates glycogen synthase rendering it *inactive*. IP3 then travels to the membrane of the endoplasmic reticulum, where it binds a ligand-gated Ca^{2+} channel receptor, causing it to open and allowing for the release of Ca^{2+} ions into the cytoplasm. The released Ca^{2+} ions can bind to the messenger protein calmodulin, which can continue the signaling cascade by activating downstream enzymes. One of these enzymes is phosphorylase kinase. The

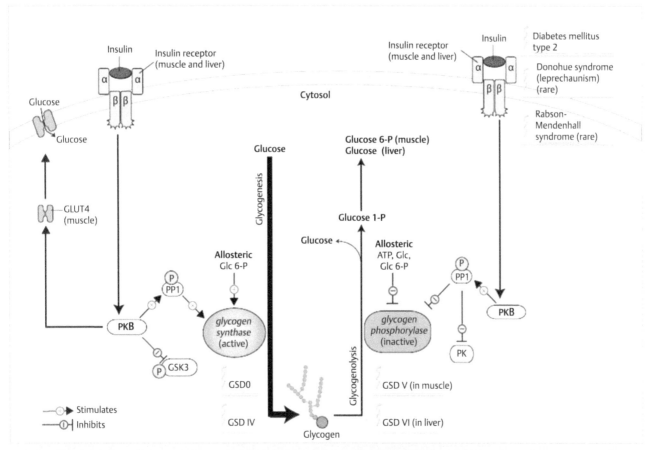

Fig. 24.6 Regulation of glycogen metabolism by insulin. (Source: Panini S, ed. Medical Biochemistry—An Illustrated Review. New York, NY. Thieme; 2013.)

activated phosphorylase kinase in turn phosphorylates glycogen phosphorylase rendering it *active.*

24.6 Regulation of Glycogenolysis in the Muscle

Degradation of glycogen in the muscle is regulated by three different mechanisms. Although the primary mode of regulation is the epinephrine signaling pathway, the accumulation of intracellular calcium and AMP can induce positive allosteric effects (▶ Fig. 24.3). One of these allosteric effectors, the calcium-calmodulin complex, binds to phosphorylase kinase and activates it directly without phosphorylation.

In skeletal muscle, the calcium that is released from the sarcoplasmic reticulum following contraction results in the activation of glycogenolysis, which provides the energy for additional contractions. This muscle contraction consumes ATP, which results in an elevation of the AMP/ATP ratio. Binding of AMP to glycogen phosphorylase results in its allosteric activation. Thus, AMP is another positive allosteric modulator of glycogen degradation through the activation of glycogen phosphorylase.

The ultimate result of these different signal transduction mechanisms is the same. They lead to the phosphorylation of glycogen synthase, which *turns it off* and puts a halt on glycogen synthesis. In addition, phosphorylation of glycogen phosphorylase occurs, which *turns it on* and allows the liver cell to break down glycogen bonds and release glucose into the blood stream.

24.7 Regulation of Glycogenesis

Glycogen synthesis occurs during the fed state via the insulin signaling pathway, which halts the degradation of glycogen and starts the synthesis of glycogen in the muscle and in the liver. One of the many downstream effects of the insulin pathway is to activate protein phosphatase-1 (▶ Fig. 24.6), which dephosphorylates, and therefore reverses, the effects of glucagon signaling on key regulatory enzymes such as PKA, phosphorylase kinase, glycogen phosphorylase and glycogen synthase. While glycogen synthase becomes an active enzyme with insulin signaling, the de-phosphorylation makes phosphorylase kinase and glycogen phosphorylase enzymes inactive.

In addition to hormonal regulation, glycogen synthase is also regulated allosterically by glucose 6-P. Increasing concentrations of glucose 6-P activates the glycogen synthase in both muscle and liver cells.

24.8 Glycogen Storage Diseases

Genetic deficiencies of enzymes in glycogen metabolism cause a group of disorders known as glycogen storage diseases. Depending on the enzyme deficiency, these disorders can be hepatic,

Table 24.1 Glycogen storage diseases

Type	Name	Affected enzyme	Symptoms
0	-	Glycogen synthase	Fatty liver, hypoglycemia when fasting
I	Von Gierke	Glucose-6-phosphatase	Hepatomegaly, hypoglycemia, lactic acidosis, ketonemia, hyperuricemia
II	Pompe	Acid $_x$- glycosidase	Hepatomegaly, muscle weakness
III	Cori-Forbes	4-$_x$-glucanotransferase	Hepatomegaly, cirrhosis, hypoglycemia, cardiac weakness
IV	Anderson	Glucan branching enzyme	Cirrhosis, muscle weakness
V	McArdle	Glycogen phosphorylase (muscle)	Muscle weakness, tendency to spasm
VI	-	Glycogen phosphorylase (liver)	Hepatomegaly, hypoglycemia when fasting, ketonemia

Source: Koolman J, Röhm K, ed. Color Atlas of Biochemistry. 3rd Edition. New York, NY. Thieme; 2012.)

myopathic or generalized types. ▶ Table 24.1 summarizes seven glycogen storage diseases with the common signs and symptoms, as well as the specific enzyme deficiency for each.

24.8.1 Link to Pathology: Neonatal Hypoglycemia

Neonatal hypoglycemia is a condition in newborns that can be due to either inadequate glycogen stores in the liver or inappropriate changes in hormone secretion, such as hyperinsulinemia. The most common cause is hyperinsulinemia, which occurs in infants whose mother had uncontrolled diabetes during pregnancy. In these cases, fetal exposure of high levels of glucose results in high glycogen stores and increased insulin levels in the fetus. The increased insulin and low glucagon suppress the mobilization of glycogen from the neonatal liver. Therefore, hyperinsulinemia may result in life-threatening hypoglycemia in the first few hours of life.

Insufficient glycogen stores can also lead to a life-threatening situation for newborns. As the maternal supply of glucose is no longer available after birth, if the infant has low blood glucose levels, epinephrine and glucagon signaling will be induced, which normally stimulates the liver to initiate glycogenolysis. If there are inadequate glycogen stores in the infant's liver, this can also contribute to a life-threatening hypoglycemic state shortly after birth.

Review Questions

1. What is the primary source of blood glucose after eight hours of fasting?
   A) Dietary carbohydrates
   B) Glycerol from adipocytes
   C) Liver glycogen degradation
   D) Muscle glycogen degradation
   E) Muscle protein degradation

2. Which of the following enzymes is shared by both gluconeogenesis and glycogenolysis?
   A) Fructose 1,6-bisphosphatase
   B) Glucose 6-phosphatase
   C) Glycogen phosphorylase
   D) Pyruvate carboxylase
   E) Pyruvate kinase

3. Increasing concentrations of which of the following intracellular secondary messengers is involved in the inactivation of glycogen synthase?
   A) cAMP
   B) PIP2
   C) PKA
   D) PKC

4. Which of the following best explains the underlying molecular mechanism of neonatal hypoglycemia?
   A) Excess liver glycogen stores
   B) Excess muscle glycogen stores
   C) High glucagon/insulin ratio
   D) Hyperinsulinemia

5. Which of the following glycogen storage disease is also classified as lysosomal storage disease?
   A) Anderson disease
   B) Cori disease
   C) Hers disease
   D) McArdle disease
   E) Pompe disease
   F) von Gierke disease

Answers

1. **The correct answer is C**. Dietary carbohydrates maintain blood glucose levels for up to ~4 hours after meals. Beyond ~4 hours following a meal, liver glycogen is used to maintain blood glucose concentrations. This is also true for fasting states up to ~18 hours. As liver glycogen stores are gradually depleted, the gluconeogenesis pathway kicks in and becomes the main pathway for maintenance of blood glucose concentrations (reviewed in chapter 25). The glucose released from muscle glycogen remains within the muscle cells to be used in glycolysis to generate ATP for muscle contraction. Therefore, the answer choices A and D are incorrect.

The glycerol from adipose tissue (answer choice B) and the amino acids from muscle protein (answer choice E) are precursors for gluconeogenesis. These carbon sources for de novo glucose production in the maintenance of blood glucose levels become significant after ~ 18 hours of fasting.

2. **The correct answer is B**. Glucose 6-phosphatase is located in the endoplasmic reticulum of hepatocytes. The enzyme catalyzes the removal of the phosphate moiety from glucose 6-phosphate, creating free glucose, which is released into the blood. This enzyme is shared by both gluconeogenesis and glycogenolysis pathways because the last intermediate of both pathways is glucose 6-phosphate.

The enzymes in answer choices of A, D, and E function in the gluconeogenesis pathway, but not the glycogenolysis pathway. Glycogen phosphorylase (answer choice C) is the key enzyme for the glycogen degradation pathway only.

3. **The correct answer is A**. The regulatory enzyme, glycogen synthase, becomes inactivated in response to glucagon and epinephrine/norepinephrine signaling. Glucagon in the liver and epinephrine in muscle cells activate adenylyl cyclase, which is responsible for converting ATP to cAMP. This increased concentration of cAMP activates the cAMP-dependent protein kinase -A (PKA). Glycogen degradation starts through the PKA pathway. PKA activates the enzyme phosphorylase kinase through phosphorylation. Phosphorylase kinase then phosphorylates and activates glycogen phosphorylase, which is responsible for the breakdown of glycogen and the release of glucose. In order to prevent the futile cycle of glycogen breakdown and subsequent glycogen synthesis, the cAMP-activated PKA directly phosphorylates and inactivates glycogen synthase, the enzyme that synthesizes glycogen.

Answer choice B is not correct because PIP2 is a membrane lipid, which is cleaved by phospholipase C into IP3 and DAG. Although this pathway is involved in the activation of glycogen phosphorylase, these second messenger do not inactivate glycogen synthase.

PKA (answer choice C) and PKC (answer choice D) are kinases that are activated through the glucagon and/or epinephrine signaling pathway, but are not secondary messengers.

4. **The correct answer is D**. Hyperinsulinemia is one of the causes of life-threatening hypoglycemia in neonates following birth. If the mother had uncontrolled diabetes during pregnancy, the fetus was exposed to high levels of glucose. This results in high concentrations of insulin secretion from the fetus' pancreas. The high insulin levels in the fetus' blood suppress the mobilization of glycogen from the neonatal liver. Since the neonate no longer has the maternal blood supply of glucose, and glycogen mobilization is suppressed by the hyperinsulinemia state, hypoglycemia occurs.

Excess liver glycogen stores (answer choice A) would not be a cause of hypoglycemia. Answer choice B is incorrect because glucose from muscle glycogen cannot be released into the blood. Answer choice C is incorrect because high insulin and low glucagon suppresses the mobilization of glycogen from the neonatal liver.

5. **The correct answer is E**. Pompe disease is due to deficiency of lysosomal α-glucosidase (acid maltase) enzyme. The deficiency of this lysosomal enzyme causes accumulation of glycogen remnants in the lysosomes. Therefore, Pompe

IV

disease is classified as both glycogen and lysosomal storage disease. The most important differentiating symptoms are cardiomegaly and muscle weakness. These symptoms are due to accumulation of glycogen remnants within the lysosomes.

The branching enzyme deficiency causes Anderson disease (answer choice A). Glycogen molecules have long outer chains with few branches. These abnormal shaped glycogen molecules accumulate within the cytoplasm and damage the liver, skeletal and cardiac muscle cells.

The Cori disease (answer choice B) is due to lack of debranching enzyme, which normally catalyzes the removal of 4 glucose residues from a branch point and then transfers these 4 glucose molecules to the end of another chain. This elongated chain now becomes substrate for glycogen phosphorylase. Without this debranching enzyme the glycogen with shorter outer branches accumulates in the liver and muscle. Patients presents with mild hypoglycemia, hepatomegaly, and muscle weakness.

While Hers disease (answer choice C) is due to the deficiency of liver isozyme of glycogen phosphorylase, McArdle disease (answer choice D) is due to deficiency of muscle glycogen phosphorylase. Since this enzyme is the key enzyme for the glycogenolysis, patients with Hers disease present with fasting hypoglycemia and hepatomegaly. The McArdle disease affects the skeletal muscles during strenuous exercise. The glycogen molecules accumulate in the cytoplasm of the skeletal muscle cells. Patients present with exercise intolerance and frequent muscle cramps.

Von Gierke disease (answer choice F) is due to either deficiency of glucose 6-phosphotase or the deficiency of translocase enzyme, which transfers the glucose 6-phosphate into the ER membrane. Since the free glucose cannot be released into the blood neither from glycogenolysis nor from gluconeogenesis, patients present with severe fasting hypoglycemia. The abnormal accumulation of glycogen within the cytoplasm of hepatocytes causes hepatomegaly. Since gluconeogenesis "backed up" the precursor lactate and pyruvate accumulate in the liver and the blood. Due to hypoglycemia the glucagon/insulin ratio increases. The elevated glucagon increases the hormone-sensitive lipase in the adipocytes, and releases the free fatty acids. Thus patients presents with hyperlipidemia.

IV

25 Gluconeogenesis

At the conclusion of this chapter, students should be able to:
• Describe the function of gluconeogenesis and the pathways involved
• Define in which physiological conditions and in which tissues gluconeogenesis is active
• Distinguish the enzymes that are unique to gluconeogenesis and the importance of these enzymes in the regulation by insulin and glucagon
• Identify the precursors for gluconeogenesis
• Explain the disorders and conditions in which the regulation of the gluconeogenesis is disrupted
• Analyze the effects of alcohol metabolism on gluconeogenesis

Gluconeogenesis is the synthesis of glucose *de novo* using carbons of lactate, glycerol and the carbon skeletons of amino acids, particularly alanine and glutamine. This *de novo* synthesis of glucose from precursors requires ATP, which is obtained in the hepatocytes from the oxidation of fatty acids. The gluconeogenesis pathway is mainly active in the liver and to a small degree in the renal cortex. The pathway is stimulated by glucagon, epinephrine, norepinephrine, and cortisol during fasting and prolonged exercise. As the liver glycogen stores become depleted under these conditions, the gluconeogenesis pathway maintains proper blood glucose levels. This ensures that the brain, central nervous system, and the red blood cells will have sufficient levels of glucose to generate ATP for necessary cellular functions.

25.1 Reactions of Gluconeogenesis

Since seven of the glycolytic pathway enzymes catalyze reversible reactions, these seven enzymes are also used in gluconeogenesis. In addition, four enzymes unique to gluconeogenesis are used to bypass the three irreversible reactions of glycolysis (▶ Fig. 25.1). These four enzymes are pyruvate carboxylase, phosphoenolpyruvate carboxykinase, fructose 1,6-bisphosphatase, and glucose 6-phosphatase (▶ Fig. 25.2).

The first step of gluconeogenesis is catalyzed by pyruvate carboxylase, which is located in the mitochondria. This enzyme carboxylates pyruvate to produce oxaloacetate (OAA). This reaction requires one ATP, biotin, and a carboxyl group from bicarbonate (HCO_3^-). Biotin serves as the coenzyme and carries the CO_2 from HCO_3^- to pyruvate to form oxaloacetate. The product OAA is reduced to malate so that it can exit the mitochondria and enter the cytoplasm. In the cytoplasm, malate is oxidized back to OAA. The cytosolic enzyme phosphoenolpyruvate carboxykinase (PEPCK) catalyzes the conversion of OAA to phos-

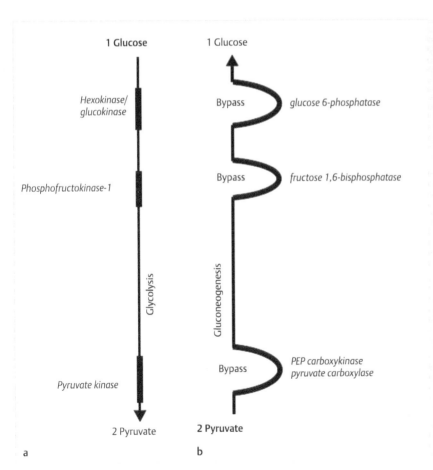

Fig. 25.1 Comparison of key features of glycolysis and gluconeogenesis: **(a)** Glycolysis (read from top to bottom); **(b)** Gluconeogenesis (read from bottom to top). (Source: Panini S, ed. Medical Biochemistry—An Illustrated Review. New York, NY. Thieme; 2013.)

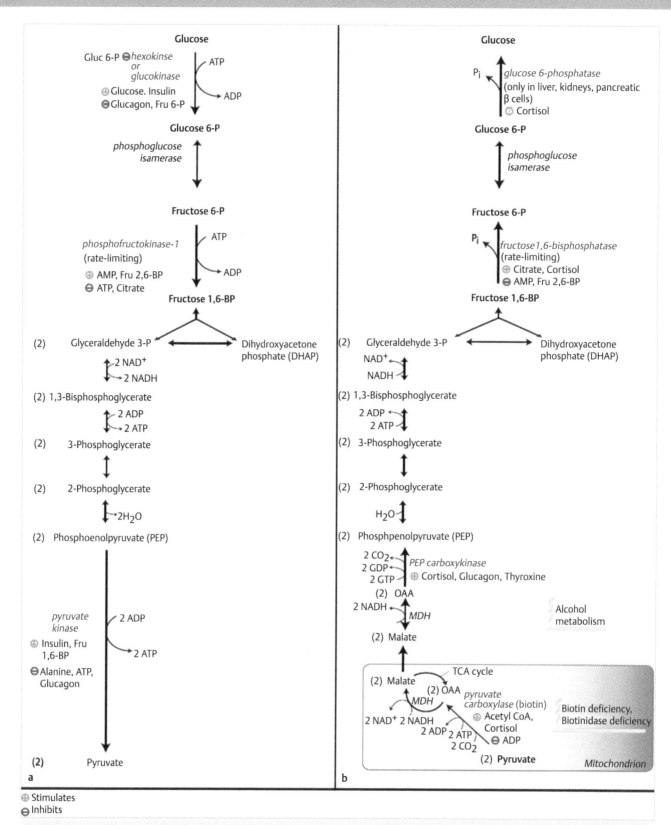

Fig. 25.2 Glycolysis versus gluconeogenesis: **(a)** Glycolysis (read top to bottom). **(b)** Gluconeogenesis (read bottom to top). (Source: Panini S, ed. Medical Biochemistry—An Illustrated Review. New York, NY. Thieme; 2013.)

phoenol pyruvate (PEP) by decarboxylating OAA and using GTP as the phosphate donor.

25.1.1 Link to Pathology

Biotin deficiency can cause hair loss, dry and scaly skin. Deficiency can also cause sleeping problems and fatigue. Deficiency of biotin develops excessive consumption of raw egg-white or can develop because of unbalanced prolonged parenteral nutrition therapy. Genetic mutations in the biotinidase enzyme prevent absorption of biotin from the intestine, thus can be a cause for biotin deficiency.

The conversion of PEP all the way to fructose 1,6 bisphosphate is catalyzed by the same enzymes that are used in glycolysis. The irreversible step of glycolysis, mediated by the activity of PFK-1, is bypassed by fructose 1,6 bisphosphates (FBPase-1), which catalyzes the conversion of fructose 1,6 bisphosphate to fructose 6-phosphate. The last step is catalyzed by glucose 6-phosphatase, which is located in the endoplasmic reticulum of the hepatocytes. The enzyme removes the phosphate group from glucose 6-P so that it can be exported from the cell as free glucose and enter the blood.

25.2 Substrates for Gluconeogenesis

Although any of the intermediates of glycolysis or the TCA cycle can serve as precursors for gluconeogenesis, the most important are lactate, glucogenic amino acids, and glycerol (▶ Fig. 25.3). The lactate formed in erythrocytes and in muscle during strenuous exercise is carried to the liver via blood and is reconverted to glucose within the hepatocytes. This glucose is then returned back to the erythrocytes and muscle to be used as an energy source via glycolysis. This cycle between lactate and glucose is known as the Cori cycle (▶ Fig. 25.4) Although lactate is an important precursor for gluconeogenesis, the main precursor for gluconeogenesis are glucogenic amino acids, especially alanine, which is derived from skeletal muscle. The conversion of alanine to glucose through the alanine cycle requires transaminases, as reviewed in chapter 33. During fasting, mobilization of triacylglycerols provides glycerol and free fatty acids. Although glycerol, the backbone of triacylglycerols, is a precursor for gluconeogenesis, free fatty acids are oxidized to acetyl CoA and cannot be a substrate for gluconeogenesis. Glycerol is produced in the adipose tissue by the hydrolysis of triacylglycerols. It is then released into the blood and taken up by the liver to be converted to glucose in the hepatocytes.

25.3 Regulation of Gluconeogenesis

The gluconeogenesis is regulated by the availability of substrates and by the hormones insulin, glucagon, epinephrine, and cortisol. During fasting and exercise, glucagon and epinephrine stimulate the availability of the substrates, such as amino acids and glycerol. Protein degradation in the skeletal muscle provides the amino acids, while triacylglycerol mobilization in the adipose tissue provides the glycerol.

The regulated steps of gluconeogenesis are the reactions that bypass glycolysis and are mediated by the enzymes pyruvate carboxylase, phosphoenolpyruvate carboxykinase, fructose 1,6 bisphosphatase, and glucose 6-phosphatase (▶ Fig. 25.2).

As we reviewed in the glycolysis chapter, pyruvate kinase is inhibited when ATP levels are elevated. Conversely, increased concentrations of ATP and acetyl CoA activate pyruvate carboxylase. This reciprocal regulation of pyruvate kinase and pyruvate carboxylase ensures that either glycolysis or gluconeogenesis is active in the hepatocytes. The phosphorylation of pyruvate kinase in the liver by glucagon signaling also turns off the enzyme, allowing the switch to gluconeogenesis. In contrast, after meals the insulin signaling pathway activates protein phosphatase-1 (PP1), which dephosphorylates the pyruvate kinase, activating it.

Phosphoenolpyruvate carboxykinase (PEPCK) is the major regulated enzyme of gluconeogenesis. It is regulated at the transcriptional level in response to the insulin/glucagon ratio. While insulin signaling inhibits PEPCK expression, the glucagon, cortisol, and thyroid hormone signaling all increase PEPCK expression.

25.3.1 Link to Pharmacology

The drug Metformin (biguanine hypoglycemic) is used to regulate blood glucose in type-2 diabetic patients. Metformin reduces hepatic glucose output by inhibiting the rate of gluconeogenesis, and also increases GLUT4 transporter concentrations on the muscle cells' surface to stimulate glucose uptake. It is also reported that insulin sensitivity in the peripheral tissues, such as muscle and adipose, is increased by Metformin. This increased insulin sensitivity is due to the drug's effects on lipid homeostasis.

An increase in the activity of AMP-dependent protein kinase (AMPK) is responsible for these effects of metformin. Under normal physiological conditions, activation of AMPK occurs when the AMP/ATP ratio is elevated within the cells. Conditions that can cause an increase in this ratio include exercise and intermittent fasting. Some of the downstream effects of AMPK are inactivation of acetyl CoA carboxylase and reduced expression of the steroid regulatory element binding protein (SREBP) in hepatocytes. AMPK also increases the number of GLUT4 transporters in the cell membrane of skeletal muscle and adipose tissue.

The fructose 1,6 bisphosphatase (FBPase-1) of the gluconeogenesis pathway and the glycolytic enzyme PFK-1 are reciprocally regulated by AMP and citrate. The FBPase-1 is activated by citrate and inhibited by AMP. Conversely, as reviewed in chapter 22, PFK-1 is inhibited by citrate and activated by AMP. During fasting, the ATP production is achieved by the oxidation of fatty acids, and this pathway generates NADH, FADH2, and acetyl CoA. The acetyl CoA molecules produced enter into the TCA cycle and ATP is generated via the electron transport chain.

The last step in generating glucose involves glucose 6-phosphatase, the enzyme that catalyzes the removal of phosphate from glucose 6-phosphate. This enzyme is also a critical step in the glycogenolysis in the generation of free glucose. The genetic deficiency of glucose 6-phosphatase is known as von Gierke disease. Since the glycogenolysis and gluconeogenesis pathways are impaired, patients with von Gierke disease suffer from severe hypoglycemia in between meals.

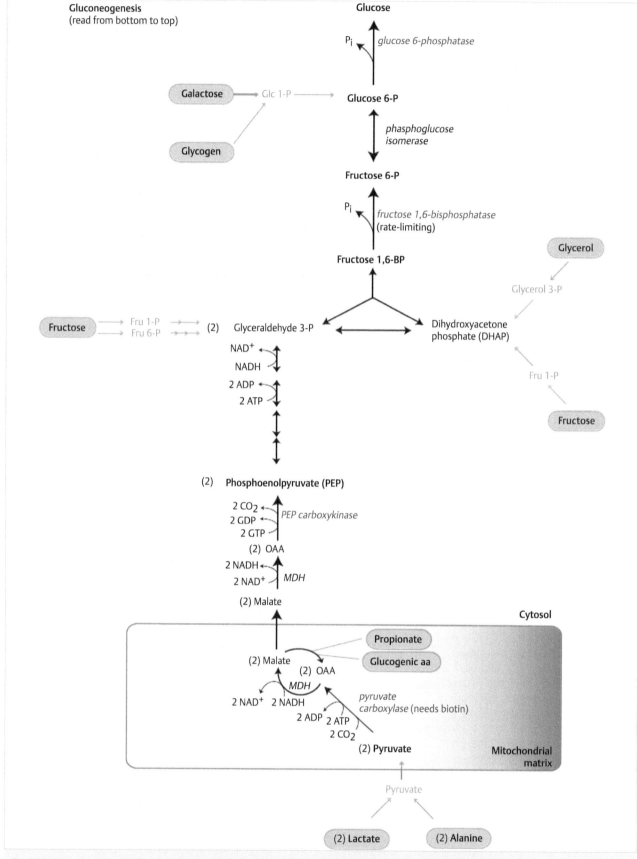

Fig. 25.3 Precursor molecules for gluconeogenesis. (Source: Panini S, ed. Medical Biochemistry—An Illustrated Review. New York, NY. Thieme; 2013.)

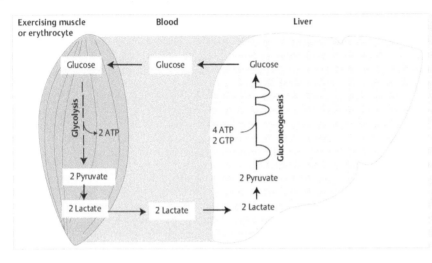

Fig. 25.4 Cori cycle. (Source: Panini S, ed. Medical Biochemistry—An Illustrated Review. New York, NY. Thieme; 2013.)

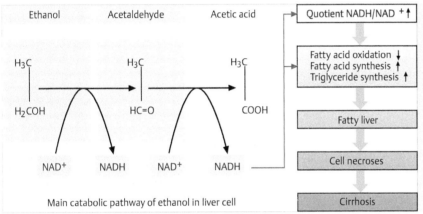

Fig. 25.5 (a, b) Ethanol metabolism. (Source: Koolman J, Röhm K, ed. Color Atlas of Biochemistry. 3rd Edition. New York, NY. Thieme; 2012.)

25.3.2 Link to Pathology

von Gierke disease is an autosomal recessive disease caused by deficiency of either glucose 6-phosphatase or the transporter of glucose 6-phosphotase. Patients present with failure to thrive, severe hypoglycemia, lactic acidosis, hyperlipidemia, hyperuricemia, and hepatomegaly. The liver enlargement is due accumulation of glycogen stores. Patients experience with severe hypoglycemia because glucose 6-phosphotase deficiency prevents the free glucose formation from glycogenolysis and gluconeogenesis, therefore the blood glucose concentrations cannot be maintained between meals. Due to impaired gluconeogenesis, lactate cannot be converted to glucose, causing lactic acidosis. Hyperuricemia is due to a buildup glucose 6-phosphate, which is the precursor and a positive effector of pentose phosphate pathway enzyme, G6PD. Increased activity of HMP shunt creates extra ribose sugar and increases the synthesis the degradation of purine nucleotides. The end product of purine catabolism is uric acid and accumulation causes hyperuricemia. The hyperlipidemia is due to high glucagon hormone. Patients' hypoglycemic state increases the glucagon secretion; this high glucagon/insulin ratio activates the hormone sensitive lipase in the adipocytes, causing release of free fatty acids.

25.3.3 Link to Pathology: Acute Alcohol Intoxication and Hypoglycemia

The metabolism of ethanol mainly takes place in the liver by the enzymes alcohol dehydrogenase (ADH) and acetaldehyde dehydrogenase (ALDH) and results in the generation of NADH (▶ Fig. 25.5a). The accumulation of NADH by ethanol metabolism interferes with many reactions in the liver (▶ Fig. 25.5b). The ensuing increases in the ratio of NADH/NAD$^+$ favors the formation of lactate, glycerol 3-phosphate, and malate, which depletes the substrates of gluconeogenesis. Thus, acute ethanol intoxication may result in severe hypoglycemia if the carbohydrate supply is insufficient.

Besides hypoglycemia, acute alcohol intoxication can result in lactic acidosis, hyperuricemia, ketoacidosis, and hyperlipidemia. All of these effects of alcohol metabolism are due to the increased NADH/NAD$^+$ ratio. Increased NADH stimulates the conversion of pyruvate to lactate, causing lactic acidosis. The increased lactate interferes with the excretion of uric acid in the kidney, causing hyperuricemia. And increased acetyl CoA (the final product of ethanol metabolism) induces ketone body and fatty acid biosynthesis. At the same time fatty acid oxidation is inhibited, due to increased NADH, contributing to more fatty acids in the blood and causing hyperlipidemia.

Review Questions

1. Which of the following molecular mechanisms prevents the futile cycle between glucose and glucose 6-phosphate in the hepatocytes?
   A) Compartmentalization of glucokinase and glucose 6-phosphatase
   B) High affinity of glucokinase for glucose
   C) Increased reaction rate of glucokinase by insulin
   D) Inhibition of glucokinase by glucose 6-phosphate

2. Which of the following is a consequence of an inhibition of the cAMP phosphodiesterase enzyme?
   A) Decreased rate of gluconeogenesis
   B) Decreased rate of glycogenolysis
   C) Increased activity of adenylate cyclase
   D) Increased activity of glycogen synthase
   E) Increased activity of PKA

3. Which of the following enzymes is the major regulated step of gluconeogenesis?
   A) Fructose 1,6 bisphosphatase
   B) Glucose 6-phosphotase
   C) Phosphoenolpyruvate carboxykinase
   D) Pyruvate carboxylase

4. Which of the following is an underlying molecular mechanism of hypoglycemia due to acute alcohol intoxication?
   A) Inhibition of ATP production
   B) Inhibition of gluconeogenesis
   C) Inhibition of glycogenolysis
   D) Inhibition of glycolysis
   E) Inhibition of NADH production

Answers

1. **The correct answer is A**. While glucokinase is located in the cytoplasm, glucose 6-phosphatase is found in the endoplasmic reticulum. After glucose 6-phosphate is taken up into the lumen of the endoplasmic reticulum, glucose 6-phosphatase removes the phosphate moiety to release free glucose into the blood.

The affinity of glucokinase for glucose is low; hence answer choice B is incorrect. Answer choice C is incorrect because insulin is not involved in the regulation of glucokinase. Glucokinase is not inhibited by its product glucose 6-phosphate. Therefore, answer choice D is not correct.

2. **The correct answer is E**. Phosphodiesterase enzymes degrade the secondary messengers cGMP and cAMP. Therefore phosphodiesterase inhibitors prevent the breakdown of cAMP and/or cGMP in the cells. There are several pharmacological compounds (i.e. drugs) that are phosphodiesterase inhibitors and are used for varying purposes. For example, sildenafil is a phosphodiesterase inhibitor that inhibits the degradation of cGMP. Theophylline is a cAMP phosphodiesterase inhibitor and is used to induce bronchodilation. Caffeine is also a cAMP phosphodiesterase inhibitor. Inhibition of cAMP degradation preserves the activity of PKA for longer periods of time than normal.

Answer choices A and B are incorrect because increased PKA activity, due to prolong concentrations of cAMP, would increase the rates of glycogenolysis and gluconeogenesis.

Answer choice D is incorrect because cAMP activates PKA, and the PKA pathway inhibits glycogen synthesis.

Phosphodiesterase inhibitors have no effect on adenylate cyclase activity. Thus, answer choice C is not correct.

3. **The correct answer is C**. All of the answer choices are the regulated steps of gluconeogenesis. The phosphoenolpyruvate carboxykinase (PEPCK) is the major regulatory enzyme and its quantity is regulated at the transcriptional level in response to the insulin/glucagon ratio. While insulin signaling inhibits PEPCK expression, the glucagon, cortisol, and thyroid hormone signaling all increase PEPCK expression.

The fructose 1,6 bisphosphatase (FBPase-1) is regulated by allosteric effectors citrate and AMP. While citrate is a positive allosteric effector, AMP is a negative allosteric effector of FBPase-1.

The activity of glucose 6-phosphatase is regulated by compartmentalization. The enzyme is located in the endoplasmic reticulum.

The pyruvate carboxylase activity is also controlled by allosteric effectors ATP and acetyl CoA. Increased concentrations of both ATP and acetyl CoA activate pyruvate carboxylase.

4. **The correct answer is B**. Metabolism of ethanol causes accumulation of NADH in the liver. The increased NADH/NAD$^+$ ratio prevents formation of pyruvate from lactate, malate from OAA, and DHAP from glycerol 3-phosphate. Thus, this inhibition principally depletes the substrates of gluconeogenesis. Therefore ethanol intoxication can cause hypoglycemia.

Answer choices A and E are incorrect because ethanol metabolism does not inhibit the production of ATP, nor the production of NADH.

The glycogen metabolism or glycolysis is not affected by ethanol metabolism, thus answer choices C and D are incorrect.

26 Pyruvate Dehydrogenase and Tricarboxylic Acid (TCA) Cycle

At the conclusion of this chapter, students should be able to:
- Explain the features and functions of PDH and the TCA cycle
- Describe the enzymatic activities of the pyruvate dehydrogenase complex, including its coenzymes and its regulation
- Identify the two intermediates (acetyl CoA and OAA) required in the first step of the TCA cycle and their metabolic sources
- Integrate the processes of glycolysis and pyruvate metabolism with the TCA cycle
- Discuss coordinated regulation of the PDH and the TCA cycle by substrate supply, allosteric effectors, and covalent modifications
- Analyze the biochemical role of thiamin in PDH and the TCA cycle and the consequences of thiamin deficiency

The TCA cycle, also known as the citric acid cycle or Krebs cycle, functions at the heart of aerobic metabolism for the oxidation of carbohydrates, lipids, and amino acids, all of which are metabolized to either acetyl-CoA or intermediates of the cycle (▶ Fig. 26.1). The function of the TCA cycle is to conserve the energy from the oxidation of acetyl-CoA in the form of reduced electron-transferring coenzymes, NADH and FAD(2H). One complete cycle yields three NADH, one FAD(2H) and one GTP molecule.

The TCA cycle reactions take place in the matrix of the mitochondrion (▶ Fig. 26.2), where six of the eight enzymes of the TCA cycle are located. The pyruvate dehydrogenase (PDH), α-ketoglutarate dehydrogenase, and succinate dehydrogenase (Complex II of Electron Transport Chain) are imbedded in the inner mitochondrial membrane, as are ADP/ATP translocase, FoFI-ATP Synthase, and the enzymes of the Electron Transport Chain (ETC).

Since the reactions of the TCA cycle degrade the 2-carbon unit of acetyl-CoA into CO_2, the cycle is primarily thought of as catabolic in nature (▶ Fig. 26.3). However, the cycle actually is the major free energy conservation system in most cells, and the many intermediates of the cycle are used as precursors for several biosynthetic reactions. Therefore the TCA cycle is **amphibolic,** being both anabolic (biosynthesis) and catabolic (degradation). When the intermediates of the cycle are used for biosynthetic reactions, the cycle serves to replenish these intermediates, therefore these reactions of the TCA cycle are known as **anaplerotic** reactions.

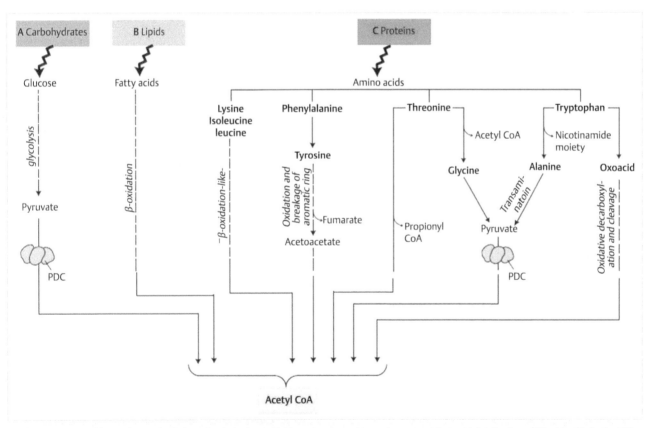

Fig. 26.1 Sources of Acetyl CoA. Oxidation of carbohydrates, lipids and amino acids produce acetyl CoA, which is further oxidized to CO_2 in the TCA cycle. (Source: Panini S, ed. Medical Biochemistry—An Illustrated Review. New York, NY. Thieme; 2013.)

IV

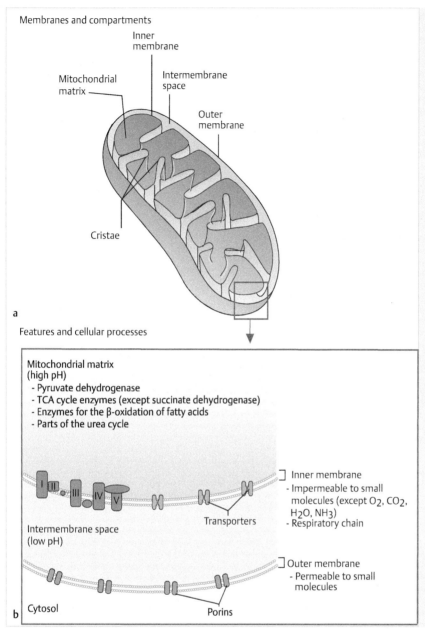

Membranes and compartments

Mitochondrial matrix

Inner membrane

Intermembrane space

Outer membrane

Cristae

a

Features and cellular processes

Mitochondrial matrix
(high pH)
 - Pyruvate dehydrogenase
 - TCA cycle enzymes (except succinate dehydrogenase)
 - Enzymes for the β-oxidation of fatty acids
 - Parts of the urea cycle

Inner membrane
- Impermeable to small
  molecules (except O_2, CO_2,
  H_2O, NH_3)
- Respiratory chain

Transporters

Intermembrane space
(low pH)

Outer membrane
 - Permeable to small
   molecules

Cytosol

Porins

b

Fig. 26.2 (a, b) Mitochondrial structure. (Source: Koolman J, Röhm K, ed. Color Atlas of Biochemistry. 3rd Edition. New York, NY. Thieme; 2012.)

26.1 TCA cycle intermediates as precursors

Citrate is a precursor for fatty acid and cholesterol biosynthesis in the cytoplasm.

Succinyl-CoA is a precursor for heme biosynthesis.

Malate is a precursor for gluconeogenesis.

26.1.1 Pyruvate Dehydrogenase Enzyme Complex

The Pyruvate Dehydrogenase Enzyme Complex (PDH) converts pyruvate to acetyl CoA and serves as a bridge between carbohydrates and the TCA cycle. The pyruvate dehydrogenase reaction is the flux generating step for the TCA cycle. With a $\Delta G = -11.5$ kcal/mol ($\Delta Go = -9.4$ kcal/mol), the reaction is essentially irreversible. This enzyme produces NADH, and releases CO_2, thereby converting pyruvate to acetyl-CoA (▶ Fig. 26.3). Acetyl Co-A enters the pathway by condensing with oxaloacetate (OAA) and it is oxidized to CO_2 in the first two steps. To be active, the PDH enzyme complex requires 5 different coenzymes: Thiamine, Lipoate, CoASH from pantothenic acid, FAD(2H) from riboflavin, and NADH from niacin. Because the PDH is a bridge to oxidize carbohydrates in the TCA cycle and to generate ATP in the ETC/oxidative phosphorylation, deficiencies in any of these vitamins affect energy metabolism, especially in aerobic tissues. Symptoms of these deficiencies include cardiac and skeletal muscle weakness, and neurologic conditions.

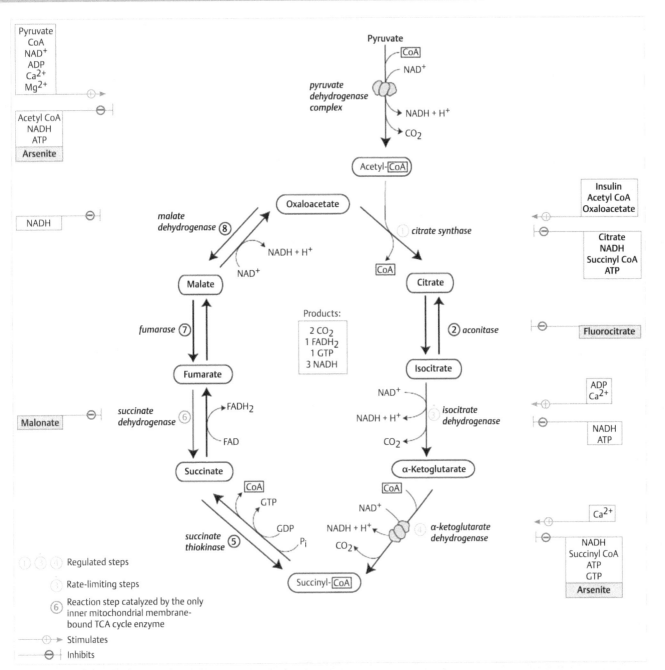

Fig. 26.3 Tricarboxylic acid cycle. (Source: Panini S, ed. Medical Biochemistry—An Illustrated Review. New York, NY. Thieme; 2013.)

26.2 Link to Pathology: Deficiency of Thiamine causes Beriberi and Wernicke-Korsakoff Syndrome

Genetic mutations of the PDH subunits or dietary thiamine deficiency impair oxidation of pyruvate and can have severe consequences, particularly in the aerobic tissues such as brain, cardiac and skeletal muscle.

Beriberi, a disease that results from a dietary deficiency of thiamine, is characterized by loss of neural function, and cardiac failure. The disease primarily is observed in populations in which

the diet is deficient in thiamin. For example, in regions where white (polish) rice, which lacks thiamine, is a major food source, there is a higher rate of the disease. Beriberi can be mild (dry beriberi) or severe (wet beriberi). Dry beriberi presents with peripheral neuropathy and ataxia. Severe deficiency of thiamine can cause wet beriberi, which presents with dry skin, peripheral neuropathy, and dilated cardiomyopathy. High-output cardiac failure in these individuals is due to decreased ATP production.

Thiamine deficiency is most common in chronic alcoholics, who habitually consume large amounts of alcohol. The combination of thiamine deficiency and alcohol toxicity causes **Wernicke-Korsakoff syndrome**. Symptoms of Wernicke encephalopathy and Korsakoff mental impairment syndrome are irritability, disordered thinking, forgetfulness, paralysis, and cardiovascular impairments. Deficiency of thiamine in chronic alcoholics develops due to two reasons. First, because of their "poor" diet, this mostly consists of empty (vitamin-free) calories of distilled spirits. And second, increased alcohol intake prevents absorption of thiamine, which occurs in the jejunum of the intestine. An elevated level of pyruvate and lactate in the blood is often an indicator of defects in pyruvate oxidation due to one of these causes.

The coenzyme form of thiamine is thiamine pyrophosphate (TPP), which is not only required for PDH activity, but is also a required coenzyme for α-ketoglutarate dehydrogenase, branched-chain keto acid dehydrogenase, and transketolase, an enzyme in the pentose phosphate pathway. Deficiency of thiamine can be diagnosed by the transketolase assay, in which the thiamine levels in the blood are measured by detecting the transketolase activity. Individuals who are suspected to be deficient in thiamine are treated by thiamine injections.

26.3 Link to Pharmacology and Toxicology: Mercury and Arsenic Poisoning

Mercury and arsenic poisoning inhibit PDH, α-ketoglutarate dehydrogenase, branched-chain keto acid dehydrogenase, and glyceraldehyde 3-P dehydrogenase. Mercury inactivates these enzymes by binding to the thiol groups in the active site. Arsenite covalently binds to sulfhydryl group of dihydrolipoic acid in the PDH enzyme complex. The resultant inactivation of lipoamide-containing enzymes, especially PDH and α-ketoglutarate dehydrogenase, brings respiration to a halt.

26.4 Link to Pathology: Genetic Mutations of the PDH Complex

Mutations in the genes encoding any one of the three subunits of PDH impair the activity of the enzyme. Deficiency of PDH is associated with accumulation of pyruvate and lactate, and, consequently, lactic acidosis. Individuals also suffer from fatigue, hypotonia, and central nervous system dysfunctions due to insufficient ATP production. Since the brain is dependent on glucose for ATP production, PDH deficiency causes intellectual disability, microcephaly, optical atrophy, and severe motor dysfunction. Starting in infancy, most children with this deficiency present with delayed development, reduced muscle tone, ataxia and seizures. To be able to support energy metabolism in most cells of the body as well as the brain, patients may benefit from a ketogenic diet consisting of high fat, low carbohydrate and low protein. Under the ketogenic diet, cells use acetyl CoA from fat metabolism, and the brain increases utilization of ketone bodies as an alternative fuel source.

26.5 Regulation of the PDH Activity

PDH enzyme activity is regulated by feedback inhibition, allosteric control, and covalent modifications (▶ Table 26.1).

Table 26.1 Modulators of pyruvate dehydrogenase complex (PDC) activity

Activators	Inhibitors
Ca^{2+}, Mg^{2+} (via allosteric activation of PDP)	Acetyl CoA, NADH (via allosteric activation of PDK)
ADP, CoA, NAD^+, pyruvate (via inhibition of PDK)	Acetyl CoA, ATP (via PDK activation), NADH
Insulin in adipose tissue, catecholamines in cardiac muscle (via increased Ca^{2+} and activation of PDP)	Arsenite (via inhibition of PDC by binding to lipoic acid in E2)

Abbreviations: ADP, adenosine diphosphate; ATP, adenosine triphosphate; CoA, coenzyme A; NAD, nicotinamide adenine dinucleotide; NADH, reduced form of nicotinamide adenine dinucleotide; PDK, pyruvate dehydrogenase kinase; PDP, pyruvate dehydrogenase phosphatase.
(Source: Panini S, ed. Medical Biochemistry—An Illustrated Review. New York, NY. Thieme; 2013.)

A high energy charge inhibits the PDH. The enzyme produces NADH directly, and ATP indirectly by the oxidative phosphorylation. Elevated concentrations of both NADH and ATP (high NADH/NAD and high ATP/ADP ratios) inhibit the activity of PDH. Acetyl-CoA, a high energy compound and the product of PDH, also inhibits PDH activity through a negative feedback loop. This negative feedback inhibition prevents the accumulation of acetyl-CoA as a metabolic intermediate.

When the levels of ATP decrease, AMP and ADP levels increases. PDH activity is stimulated by increased AMP concentrations in the cells. The activity of the PDH is also regulated by covalent modifications. PDH becomes more active when it is dephosphorylated. Insulin signaling-activated protein phosphatases (PP1) and Ca^{2+}-activated phosphatase dephosphorylate PDH, thereby activating the enzyme. When PDH is phosphorylated by PDH-kinase, it becomes inactive. The activity of the PDH-kinase is stimulated by elevated concentrations of NADH, ATP, and acetyl-CoA. In contrast, elevated pyruvate binds and allosterically inactivates PDH-kinase, preventing phosphorylation of PDH and keeping it active. Note that these factors (NADH, ATP, acetyl-CoA, and pyruvate) that regulate the PDH-kinase also allosterically regulate PDH activity.

26.6 Reactions of the TCA Cycle

The TCA cycle begins with condensation of acetyl-CoA with OAA to form citrate. The enzyme that catalyzes this condensation reaction is citrate synthase (▶ Fig. 26.3). The first three enzymes of the cycle are citrate synthase, isocitrate dehydrogenase, and α-ketoglutarate dehydrogenase, which are also the regulated and irreversible steps of the cycle. As we discussed above, α-ketoglutarate dehydrogenase shares the same coenzymes as PDH. The two-carbons from acetyl-CoA are liberated as CO_2, one in the isocitrate dehydrogenase step and the second one in the α-ketoglutarate dehydrogenase step. Succinyl-CoA, which is generated by α-ketoglutarate dehydrogenase, contains a high-energy thioester bond, and this energy is used to generate GTP by succinate thiokinase. This is the only step in the TCA cycle that synthesizes the equivalent of ATP in the substrate-level phosphorylation. Succinate dehydrogenase catalyzes oxidation of succinate to fumarate and reduces the FAD to form FAD(2H). Succinate dehydrogenase is localized in the inner-membrane of mitochondria, and it also functions in the electron transfer chain as complex-II. Malate is oxidized to OAA by malate dehydrogenase, which is the last step of the cycle.

Regulation of the TCA cycle is very similar in many aspects to the regulation of the PDH activity. Increased levels of AMP and ADP are the positive modulators of citrate synthase and isocitrate dehydrogenase as well. In parallel to PDH inactivation, citrate synthase, isocitrate dehydrogenase, and α-ketoglutarate dehydrogenase are also inhibited with increased concentrations of NADH and ATP. So the energy charge of the cell regulates both PDH activity and the enzymes of the TCA cycle, i.e. citrate synthase, isocitrate dehydrogenase, α-ketoglutarate dehydrogenase (▶ Fig. 26.3).

In addition, malate dehydrogenase activity is greatly impaired with the elevated NADH/NAD ratio. Therefore, as the increased NADH impairs formation of oxaloacetate (OAA). Because the substrates for citrate synthase are acetyl-CoA and OAA, in the absence of OAA, the TCA cycle cannot operate (▶ Fig. 26.3). This concept is usually taught in the topic of "regulation of acetyl-CoA entry into the TCA cycle".

One difference between the regulation of PDH activity and the TCA cycle is that unlike PDH, the TCA cycle is not regulated by hormonal signaling.

IV

Review Questions

1. Which of the following vitamin deficiencies inhibits pyruvate dehydrogenase enzyme activity?
   A) Ascorbic acid
   B) Biotin
   C) Folic acid
   D) Pyridoxine
   E) Thiamin

2. Which of the following TCA cycle intermediates is an allosteric regulator of glycolysis?
   A) Acetyl CoA
   B) α-ketoglutarate
   C) Citrate
   D) Fumarate
   E) Malate

3. When oxaloacetate (OAA) is used for biosynthetic reactions, the pyruvate carboxylase reaction is upregulated to replenish the OAA. Which of the following terms best describes the increased formation of OAA by pyruvate carboxylase?
   A) Amphibolic
   B) Anaplerotic
   C) Catabolic
   D) Metabolic

4. 4. Which of the following decreases the flux through the TCA cycle?
   A) Increased AMP/ATP ratio
   B) Increased glucagon levels
   C) Increased insulin levels
   D) Increased NADH/NAD ratio
   E) Increased O2 concentration

1. **The correct answer is E.** Pyruvate dehydrogenase (PDH) catalyzes the conversion of pyruvate to acetyl CoA. This enzyme complex requires five coenzymes: thiamine, lipoate, CoASH from pantothenic acid, FAD(2H) from riboflavin, and NADH from niacin. Any of these vitamin deficiencies results in the inhibition of PDH activity. Among these vitamins, the most commonly observed vitamin deficiency is thiamin. Deficiency can be due to diet that lacks thiamin, or it can be due to chronic consumption of alcohol, which prevents absorption of thiamin from the jejunum. Signs and symptoms of thiamin deficiency include a buildup of lactate and pyruvate in tissues of the body, especially in aerobic tissues such as cardiac muscle and the brain. The insufficient

generation of ATP in these tissues causes neuropathy, ataxia, and dilated cardiomyopathy. Symptoms in Wernicke-Karsakoff syndrome may be more severe, such as disorientation to time, memory loss, confusion and nystagmus.

The vitamins in answer choices A, B, C, and D are not coenzymes for the PDH complex.

2. **The correct answer is C.** The flux of the TCA cycle is coordinated with the flux of glycolysis. To ensure that intermediates of glycolysis are not going to exceed the flux of the TCA cycle, the TCA cycle intermediate citrate inhibits the rate limiting step of glycolysis, PFK-1. When citrate accumulates within the matrix, it can exit the mitochondria through the citrate shuttle and allosterically inhibit the rate limiting step of glycolysis, PFK-1, in the cytoplasm. Therefore, citrate is an allosteric regulator that modulates the energy charge of the cell.

Acetyl CoA (answer choice A) is an allosteric regulator of pyruvate carboxylase. Answer choices B, D, and the E do not function as allosteric regulators.

3. **The correct answer is B.** When the OAA is used for biosynthetic reactions, which can be for amino acid biosynthesis or can be converted to malate for gluconeogenesis, pyruvate carboxylase is upregulated to replenish the OAA. These reactions of the TCA cycle are known as **anaplerotic** reactions. The rate of the anaplerotic reactions increases as the intermediate that needs to be replenished increases. Amphibolic means being both anabolic (biosynthesis) and catabolic (degradation) at the same time. The TCA cycle is amphibolic because acetyl-CoA is catabolized to CO_2 and at the same time some of the intermediates of the cycle are used for biosynthetic reactions. Therefore, the definitions of the answer choices B, C, and D are incorrect. NADH is a negative modulator of citrate synthase, isocitrate dehydrogenase, and α-ketoglutarate dehydrogenase, the increased NADH/NAD ratio down-regulates the TCA cycle. In contrast, increased AMP/ATP ratio up-regulates the TCA cycle, since AMP is a positive regulator of these regulatory enzymes. Thus, A is incorrect. Answer choices B and C are incorrect, since the TCA cycle is not regulated by hormonal signaling. Answer choice E is incorrect as well, because O_2 is required for ETC and consequently for increasing the flux through the TCA cycle, therefore decreased concentrations of O_2 would decrease the flux through the cycle.

27 Electron Transport Chain (ETC) and Oxidative Phosphorylation

At the conclusion of this chapter, students should be able to:
- Describe the following components of the electron transport chain (ETC), the functions of each, and their locations within the mitochondria
    - Complex I (NADH dehydrogenase) "proton pump"
    - Coenzyme Q (Ubiquinone) "mobile carrier"
    - Complex II (succinate dehydrogenase) pass electrons directly to CoQ
    - Complex III (cytochrome b/c1) "proton pump"
    - Cytochrome c "mobile carrier"
    - Complex IV (cytochrome a/a3 or cytochrome oxidase) "proton pump"
- Define how the ETC and oxidative phosphorylation are coupled and how their regulation occurs
- Compare and contrast inhibitors and uncouplers of ETC and oxidative phosphorylation
- Analyze the clinical signs and symptoms associated with inhibitors and uncouplers of the ETC and oxidative phosphorylation

27.1 Purpose of the Electron Transport Chain (ETC), Localization and its Components

The ETC is the final step in the aerobic oxidation of carbohydrates, lipids and amino acids. The electrons from NADH and FADH2 are passed through a series of carriers to molecular oxygen to form water. At specific steps of the ETC, NADH and FADH2 are oxidized to NAD and FAD, respectively. These oxidation reactions release a large amount of free energy, which is captured by oxidative phosphorylation in the form of ATP.

The ETC and oxidative phosphorylation take place in the matrix of the mitochondria. Therefore, cells without mitochondria (mature erythrocytes, lens and cornea) cannot utilize the ETC. The components of the ETC consist of four complexes (complex-I, II, III, and IV), ubiquinone (coenzyme Q) and cytochrome c (▶ Fig. 27.1). While complex II is imbedded in the matrix side of the inner mitochondrial membrane, complex I, III, and IV span the entire inner mitochondrial membrane. Complex II also functions in the TCA cycle as succinate dehydrogenase. Coenzyme Q and cytochrome-c are mobile components of the ETC and these proteins freely move within the lipid bilayer of the inner membrane. As electrons flow through the components of the ETC, energy is released, and this energy is used to pump the protons to the inner-membrane space. This results in the generation of a proton gradient between the inner-membrane space and the matrix (▶ Fig. 27.2).

The complexes I (NADH dehydrogenase), III (cytochrome b/c1) and IV (cytochrome a/a3), pump protons into the inner-membrane space. The F_0F_1-ATP synthase macromolecular complex is utilized for return of these protons from the inner-membrane space back into the matrix. The return of these protons through the channel of the F_0F_1-ATP synthase complex drives the synthesis of ATP. This mechanism of ATP production is known as the chemiosmotic hypothesis.

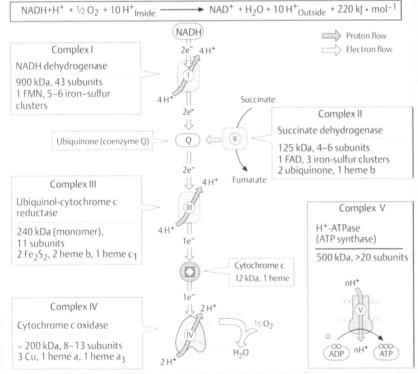

$$NADH+H^+ + \tfrac{1}{2} O_2 + 10 H^+_{Inside} \longrightarrow NAD^+ + H_2O + 10 H^+_{Outside} + 220\ kJ \cdot mol^{-1}$$

NADH

Complex I
NADH dehydrogenase
900 kDa, 43 subunits
1 FMN, 5–6 iron–sulfur clusters

$2e^-$ $4H^+$

$4H^+$

$2e^-$

Ubiquinone (coenzyme Q)

Q

Succinate

Complex II
Succinate dehydrogenase
125 kDa, 4–6 subunits
1 FAD, 3 iron-sulfur clusters
2 ubiquinone, 1 heme b

$2e^-$ $4H^+$ Fumarate

Complex III
Ubiquinol-cytochrome c reductase
240 kDa (monomer), 11 subunits
2 Fe_2S_2, 2 heme b, 1 heme c_1

$4H^+$

$1e^-$

Cytochrome c
12 kDa, 1 heme

$1e^-$

Complex IV
Cytochrome c oxidase
≈ 200 kDa, 8–13 subunits
3 Cu, 1 heme a, 1 heme a_3

$2H^+$ $\tfrac{1}{2} O_2$

H_2O

$2H^+$

Complex V
H^+-ATPase (ATP synthase)
500 kDa, >20 subunits

nH^+

V

ADP nH^+ ATP

Proton flow
Electron flow

Fig. 27.1 Components of the respiratory chain. (Source: Koolman J, Röhm K, ed. Color Atlas of Biochemistry. 3rd Edition. New York, NY. Thieme; 2012.)

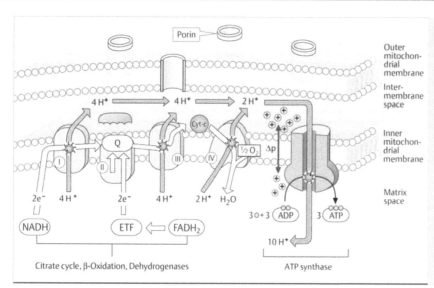

Fig. 27.2 Mechanism of oxidative phosphorylation. Transfer of electrons and hydrogens from NADH and FADH2 to O2 is facilitated by pairs of the ETC components. During this transfer of electrons, protons are pumped into the inner-membrane space that establishes a proton gradient. Return of protons to the matrix through ATP synthase accomplishes phosphorylation of ADP that produces ATP. (Source: Koolman J, Röhm K, ed. Color Atlas of Biochemistry. 3rd Edition. New York, NY. Thieme; 2012.)

27.2 Regulation of the Rates of ETC and Oxidative Phosphorylation

The need for ATP and the availability of molecular O_2 are the most important regulatory mechanisms controlling the rate of the TCA cycle, the ETC, and oxidative phosphorylation. Besides the ATP/ADP ratio and molecular O_2, the rate of the TCA cycle, the ETC, and oxidative phosphorylation is directly proportional to the number of mitochondria within the cell. The rate of respiration is higher in the tissues that have large numbers of mitochondria (see below for a link to genetic mutations in mitochondrial DNA and red-ragged fibers, LHON, MELAS, and Leigh syndrome). Tissues that are rich in mitochondria include the brain, heart and skeletal muscle. If these tissues are well-oxygenated, then the availability of ADP is the rate-limiting factor. When the ATP/ADP ratio is low, ADP allosterically activates isocitrate dehydrogenase, therefore increasing the rate of the TCA cycle and generating NADH and FADH2. The availability of NADH and FADH2 increases the rate of the ETC and oxidative phosphorylation, and ultimately production of ATP. When tissues are depleted of O_2, however, the ETC and oxidative phosphorylation cannot operate. This results in an increased in the NADH/NAD ratio, which causes the inhibition of regulatory enzymes of the TCA cycle and PDH (See the regulation of PDH and the TCA cycle in chapter 26).

27.2.1 Link to Pathology: Ischemic Tissue Damage due to Hypoxia and Lactic Acidosis

Hypoxia is one of several conditions that can lead to lactic acidosis. The decreased oxygenation of the tissues that occurs in hypoxia causes a decrease in the rate of ETC and oxidative phosphorylation, which consequently impedes the production of ATP. Due to the decreased ATP/ADP ratio, the activity of PFK-1, the rate-limiting step of glycolysis, is enhanced. However, without enough O_2, lactate becomes the final product of glycolysis (reviewed as anaerobic glycolysis in chapter 22). The resulting increased production and accumulation of lactate decreases the pH of the cells.

The insufficient production of ATP also causes impaired functioning of ATP-dependent ion channels on the plasma membranes. The normal function of these channels is to maintain osmotic balances by sustaining the correct amount of intra- and extra-cellular ion concentrations. The decreased ATP production causes osmotic imbalances due to increased intracellular Na^+ ions. This results in water retention, swelling of the cells and damaging to both the plasma membrane and the organelle membranes inside the cell. The membrane damage allows cellular proteins to leak out, intracellular Ca^{2+} to increase, and lysosomal enzymes to leak into the cytoplasm. The leaked lysosomal enzymes autolysis cellular proteins, while increased Ca^{2+} activates proteases and the acidic environment provided by increased lactate concentrations causes damage to proteins and membranes. All of these factors contribute to cell damage and consequently to cell death in ischemic tissues.

27.2.2 Ischemic Myocardial Infarction

Ischemic myocardial infarction (MI), also known as a "heart attack", is characterized by necrosis of myocardial tissue secondary to decreased blood flow to the heart. The classic clinical presentation is an elderly male complaining of severe, sub-sternal chest pain and pressure radiating to the neck or arm. It may be accompanied by diaphoresis, nausea or vomiting. It is important to note that diabetics and elderly women may experience a heart attack with vague complaints e.g. stomach pain and/or shortness of breath.

The majority of ischemic MI occurs secondary to rupture of a thrombus in the left anterior descending artery (LAD). As a result, blood flow is decreased and cardiac myocytes are unable to generate ATP via the ETC. When this occurs, the heart attempts to generate ATP via anaerobic glycolysis. The amount of ATP generated, however, is insufficient, and myocardial contractility is compromised. Additionally, anaerobic glycolysis results in excessive lactic acid production, which decreases the responsiveness of the myocardium to epinephrine. Decreased responsiveness to epinephrine further compromises myocardial contractility. Severe, prolonged ischemia eventually results in necrosis of myocytes. The biochemical process of necrosis is discussed in chapter 13.

Troponin, a protein involved in muscle contraction, is released into the blood following an MI. Elevated blood troponin is the most sensitive marker for diagnosis of a recent MI. Its levels rise in the blood after approximately 4 hours and return to baseline after 10–14 days. This is clinically important, as a patient who presents to the hospital with an MI may have normal troponin levels if it occurred within the preceding 4 hours. Creatine-kinase-MB, or CK-MB, is a myocardial enzyme released into the blood following an MI. Its levels also rise in the blood approximately 4 hours post-MI and return to baseline after 3–4 days. A feared complication in patients hospitalized for an MI is re-infarction. If re-infarction has occurred, CK-MB will be acutely elevated. For example, on day 1 of post-MI a patient will have elevated troponin and CK-MB. If this same patient is recovering normally, on day 5 the level of CK-MB should be normal and troponin will remain elevated. If, however, troponin and CK-MB are elevated, re-infarction should be suspected.

27.3 Genetic mutations in the mitochondrial DNA

27.3.1 Leber Hereditary Optic Neuropathy (LHON)

Leber Hereditary Optic Neuropathy (LHON) is a mitochondrial disorder characterized by rapidly progressive vision loss. It frequently occurs in persons 20–30 years of age. The classic clinical presentation is a younger man complaining of blurry vision or decreased visual acuity secondary to bilateral central vision loss (central scotoma). It is caused by gene mutations in cytochrome reductase of the mitochondrial electron transport chain. The gene is located within mitochondrial DNA (mtDNA), and therefore LHON follows a mitochondrial inheritance pattern. Mitochondrial inheritance patterns are discussed in chapter 49.

27.3.2 Mitochondrial Encephalomyopathy, Lactic Acidosis and Stoke-Like Episodes (MELAS Syndrome)

MELAS syndrome is a mitochondrial disorder characterized by encephalopathy, myopathy, lactic acidosis and stroke symptoms, particularly in children and adolescents. The classic clinical presentation is an adolescent complaining of one or more of the following: muscle pain, weakness, twitching, seizures or stroke-like symptoms (ie, vision loss, hemiparesis). It is caused by gene mutations in NADH-dehydrogenase, or complex I, of the ETC. Therefore, these individuals have a defect in aerobic ATP generation, and often depend on anaerobic glycolysis for

energy production during times of increased oxygen demand. This, in turn, results in elevated blood lactate levels, a byproduct of anaerobic metabolism. Elevated blood lactate results in metabolic acidosis, which may manifest clinically as muscle pain, nausea and vomiting. The gene for NADH-dehydrogenase is located within mitochondrial DNA (mtDNA), and therefore MELAS syndrome follows a mitochondrial inheritance pattern. Mitochondrial inheritance patterns are discussed in chapter 49.

27.3.3 Leigh Syndrome

Leigh Syndrome is a mitochondrial disorder characterized by severe neurological degeneration, particularly in the first year of life. The classic clinical presentation is a young child with a history of difficulty swallowing, vomiting, failure to thrive, and weak muscle tone with gross movement or balance problems. Weakness may also occur in muscles of eye movement and respiration. Consequently, affected individuals usually die by age 3 due to respiratory failure. Leigh syndrome is caused by mutations in ATP6 gene that is located in mtDNA, and therefore follows a mitochondrial inheritance pattern.

27.4 Sources of NADH and the Entry of NADH Electrons into the ETC

Most of the NADH is produced by the enzymes in the PDH complex, the TCA cycle, and the beta oxidation of lipids. All of which takes place in the matrix of the mitochondria. A small portion of the NADH is produced in the cytoplasm by the glycerol 3-phosphate dehydrogenase reaction of glycolysis, reviewed in chapter 22.

The transfer of the electrons from NADH into the ETC starts with complex-I (NADH dehydrogenase). Following this reaction, NADH is oxidized to NAD^+ and the electrons are transferred to Coenzyme-Q (ubiquinone). By accepting electrons, the Coenzyme-Q becomes reduced and transfers these electrons to complex-III (cytochrome b/c1). Next, complex-III passes electrons to cytochrome c, and the reduced cytochrome c transfers electrons to complex-IV (cytochrome c oxidase). Finally, complex-IV transfers electrons to molecular O_2 to produce water. The free energy that was created by transferring electrons from complexes I, III, and IV is used to pump protons into the inner-membrane space, generating a proton gradient.

When NADH is produced in the cytoplasm, transferring its electrons to the ETC requires tissue specific shuttle systems, since the mitochondrial membrane is impermeable to NADH. The **malate-aspartate shuttle** and the **glycerophosphate shuttle** are used to transfer electrons from NADH into the mitochondria. The malate-aspartate shuttle, active in the liver, heart, and kidneys, transfers electrons to mitochondrial NADH (▶ Fig. 27.3). Since electrons from NADH are moved from cytoplasm into mitochondria as NADH, there is no ATP loss in this shuttle system. The glycerophosphate shuttle is active in the skeletal muscle and the brain. In this shuttle system, the electrons from cytosolic NADH are passed onto the ETC through coenzyme-Q and incorporated into mitochondrial FADH2. Since

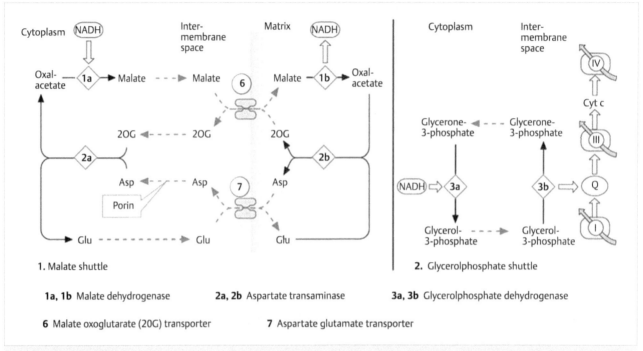

Fig. 27.3 Malate and glycerophosphate shuttle. (Source: Koolman J, Röhm K, ed. Color Atlas of Biochemistry. 3rd Edition. New York, NY. Thieme; 2012.)

the electrons entering the ETC by bypassing the complex-I, less protons will be pumped into the inner-membrane space, ultimately 1 less ATP will be generated.

27.5 Inhibitors and Uncouplers of the ETC and Oxidative Phosphorylation

Inhibitors of the respiratory chain directly bind to components of the ETC and block their function, resulting in an abrupt shut-down of oxidation/reduction reactions and the transfer of electrons. Uncouplers, on the other hand, break the link between the rate of ETC and the synthesis of ATP. In the presence of uncouplers, the ETC creates the proton gradient between the inner-membrane space and the matrix of the mitochondria as normal. However, the uncouplers allow protons to reenter the matrix without involvement of ATP synthase (▶ Fig. 27.4). The respiration rate remains fast, but the ATP production is reduced due to the reduced proton gradient. Due to the flow of the protons, the uncouplers generate heat instead of ATP. Uncoupling agents can either cause mechanical damage to the inner-membrane (e.g. cytosine arabinoside and azidothymidine) or they can be lipid-soluble substances (e.g. 2,4-dinitrophenol and high dose aspirin). Some uncouplers, such as the brown adipose tissue uncoupling protein-1 (UCP-1), or thermogenin, are naturally occurring. UCP-1 forms an ion-channel in the inner-membrane of mitochondria and functions to generate heat, especially in hibernating animals and newborns. In addition to UCP-1, several other isoforms exist. The activity of each of these is controlled by hormone signaling, especially through the thyroid hormone thyroxine. Thyroxine stimulates the synthesis of many different proteins and enzymes in the ETC and oxidative phosphorylation, including UCP. As a result, hyperthyroidism causes increased production of body heat and weight loss. By contrast, patients with hypothyroidism feel colder than usual and they gain weight easily.

27.6 Inhibitors and Genetic Mutations that Disrupt the Respiratory Chain

Disruptions in the respiratory chain can be caused by inhibitors (mainly toxic compounds that bind to specific proteins in the ETC) or by mutations in the genes that encode components of the ETC (▶ Fig. 27.5). The result of either of these disruptions is the same: a reduction in the generation of ATP. ETC inhibitors and their mechanisms are summarized in ▶ Fig. 27.5, and gene mutations that cause defects in oxidative phosphorylation are summarized in ▶ Table 27.1.

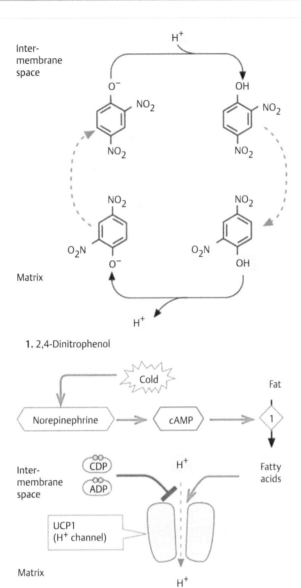

1. 2,4-Dinitrophenol

1 Hormone-sensitive lipase

2. UCP1 (thermogenin)

Fig. 27.4 Uncouplers allow protons to reenter the matrix from the inner-membrane space. Some lipid soluble agents e.g. 2,4-dinitrophenol (DNP) or agents that disrupt integrity of the membranes e.g. AZT, or some proton channels e.g. UCP-1 or thermogenin transport protons across the inner-membrane without generation of ATP. (Source: Koolman J, Röhm K, ed. Color Atlas of Biochemistry. 3rd Edition. New York, NY. Thieme; 2012.)

IV

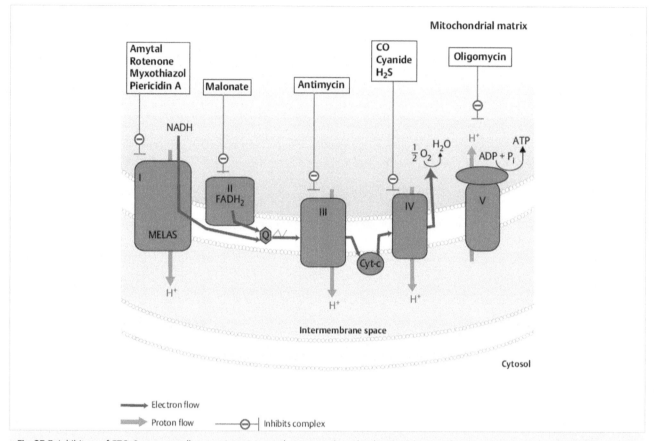

Fig. 27.5 Inhibitors of ETC. Some naturally occurring compounds or some drugs bind and inhibit specific components of the ETC. Inhibition of ETC results in decreased production of ATP. (Source: Panini S, ed. Medical Biochemistry—An Illustrated Review. New York, NY. Thieme; 2013.)

Table 27.1 Disorders associated with mutations in mitochondrial genes

Disorder	Affected Mitochondrial Gene	Affected ETC Complexes	Signs, Symptoms, Laboratories
Mitochondrial encephalopathy, lactic acidosis, andstroke (MELAS)	tRNA^{Leu}	Complex I, IV	Muscle weakness and pain, accumulation of lactic acid, vomiting, loss of appetite, seizures, strokelike episodes. Onset: Infancy to early childhood
Kearns-Sayre syndrome	Single large deletion of 1.1–10 kb of mtDNA affecting multiple genes	All	Eye pain, ophthalmoplegia, ptosis, pigmentary retinopathy, defects in cardiac conduction, ragged red fibers. Onset: before age 20
Leber hereditary optic neuropathy (LHON)	ND1, ND4, Cyt b	Complex I, III	Degenerated optic nerve, continued loss of central vision. Onset: early adulthood
Leigh syndrome	ND5, cytochrome-c oxidase	Complex I, IV	Difficulty swallowing, weak motor skills, vomiting, lesions in basal ganglia and brainstem. Onset: first year of life
Myoclonic epilepsy and ragged-red fiber disease (MERRF)	tRNA^{Lys}	Complex I, IV	Muscle spasms, hearing loss, dementia
Hypertrophic cardiomyopathy (exercise intolerance)	Cyt b, cytochrome-c oxidase	Complex III, IV	Enlargement and degeneration of cardiac muscle. Onset: birth to young adulthood
Pearson syndrome	Single large deletion of 4977 bp of leukocyte mtDNA	All	Sideroblastic anemia, exocrine pancreas dysfunction
Hypertension andhypercholesterolemia	tRNA^{Ileu}, tRNA^{Met}, tRNA^{Lys}	Complex I, IV	BP > 140/90 mm Hg, total serum cholesterol > 200 mg/dL, hypomagnesemia (< 1.7 mg/dL)
Diabetes with deafness	tRNA$^{Leu(UUR)}$	Complex I, IV	Maternal inheritance, defective ATP production andinsulin release in pancreatic β cells (different from glucokinase mutation seen in MODY2), lack of insulin resistance, impaired hearing; may respond to ubiquinone supplements

Abbreviations: ATP, adenosine triphosphate; ETC, electron transport chain; MODY, maturity onset diabetes of the young; ND1 and ND4, NADH dehydrogenase genes.
(Source: Panini S, ed. Medical Biochemistry—An Illustrated Review. New York, NY. Thieme; 2013.)

IV

Review Questions

1. What best explains the underlying molecular mechanisms of lactic acidosis due to hypoxia?
   A) High ATP/ADP ratio inhibits the glycolysis
   B) High ATP/ADP ratio inhibits the hexokinase
   C) Low ATP/ADP ratio activates phosphofructokinase-1
   D) Low ATP/ADP ratio activates pyruvate kinase
   E) Low ATP/ADP ratio inhibits lactate dehydrogenase

2. Which of the following outcomes is caused by carbon monoxide poisoning?
   A) Decreased oxygen delivery to the tissues
   B) Decreased rate of substrate level phosphorylation of GTP
   C) Increased ATP synthase activity
   D) Increased NADH dehydrogenase activity
   E) Increased reduction of oxygen

3. Which of the following is the most likely finding in individuals who have a genetic mutation in the mitochondrial NADH dehydrogenase?
   A) Hemolysis
   B) Hypocalcemia
   C) Free radical damage
   D) Increased ATP production
   E) Ragged-red fibers

4. Under normal physiological conditions, which of the following primarily affects the rate of oxidative phosphorylation?
   A) ATP/ADP ratio
   B) CO_2 pressure
   C) Complex II
   D) Mitochondrial pH

5. Which of the following processes is inhibited by uncouplers?
   A) Entry of FAD(2H) into the ETC
   B) FAD(2H) production
   C) Flux through the TCA cycle
   D) NADH production
   E) Mitochondrial ATP production

Answers

1. **The correct answer is C**. Hypoxic conditions reduce the rate of ETC and oxidative phosphorylation, especially in cardiac muscle, brain and exercising skeletal muscle. As the rate of ETC and oxidative phosphorylation decreases, so does the production of ATP. The decreased ATP/ADP ratio allosterically activates the rate limiting step of glycolysis, PFK-1. Since oxygen is not available to the tissues, the ETC and oxidative phosphorylation cannot operate. This results in an increased NADH/NAD ratio, which inhibits the pyruvate dehydrogenase (PDH) activity and increases the anaerobic glycolysis. The end product of anaerobic glycolysis is lactate, accumulation of which causes lactic acidosis. Answer choice A is not correct because under hypoxic conditions ATP levels do not increase. The answer choices B, D, and E are incorrect because these enzymes are not regulated by changes in the ATP/ADP ratio.

2. **The correct answer is A**. Although carbon monoxide (CO) is an inhibitor of complex IV (also known as cytochrome C oxidase or cytochrome a/a3) of the ETC, it has much higher affinity for hemoglobin. In fact, the affinity of CO for hemoglobin is ~300-times higher than for oxygen. Therefore, CO poisoning decreases oxygen delivery to the tissues and causes hypoxia. The tissues that are affected the most are cardiac muscle and brain since these tissues have more demand for oxygen. Patients present with hypotension, headache, and a bright red color of the skin. When CO binds to complex IV of the ETC, it inhibits the passing of electrons to oxygen. Therefore, answer choice E is incorrect. Answer choice B is not correct because substrate level phosphorylation increases in order to generate ATP from anaerobic glycolysis. The ATP synthase (also known as complex V) activity (answer choice D) does not increase with CO poisoning since the respiration rate, as well as the pumping of protons into the inner membrane space of the mitochondria, decreases.

3. **The correct answer is E**. The genes that encode the NADH dehydrogenase subunits or complex-I of ETC are located in the mitochondrial DNA. Mutations of the NADH dehydrogenase causes dysfunctions in many tissues, but especially in the tissues that are rich in mitochondria and have demand for ATP generation such as neuronal cells, heart and skeletal muscle. The function of NADH dehydrogenase, as the first enzyme of the ETC, oxidizes NADH and passes electrons to ubiquinone. The ubiquinol (reduced form of ubiquinone) transfers these electrons to complex-III (cytochrome b/c1). The reduced cytochrome c transfers electrons to complex-IV (cytochrome c oxidase). Finally, complex-IV transfers electrons to molecular O_2 to produce water. The free energy that was created by transferring electrons from complexes I, III, and IV is used to pump protons into the inner-membrane space, generating a proton gradient. Complex V, or ATP synthase, uses this proton-motive force to produce ATP from ADP. Since NADH dehydrogenase is the first step in the ETC, it functions as the rate-limiting step in ATP generation. Inhibition of this complex inhibits ATP production in aerobic cells. To compensate for the reduced ATP production, mitochondrial proliferation increases within these cells. Accumulation of this abnormal mitochondria causes aggregates and the "ragged" appearance when stained with Gomori Trichrome.

4. **The correct answer is A.** The most important modulator of the rate of the TCA cycle, the ETC, and oxidative phosphorylation is the ATP/ADP ratio in the cells. When the ATP/ADP ratio is low, ADP allosterically activates isocitrate dehydrogenase, therefore increasing the rate of the TCA cycle and generating NADH and FADH2. The availability of NADH and FADH2 increases the rate of the ETC and oxidative phosphorylation, and ultimately production of ATP. Although the availability of molecular O_2 is necessary for the ETC and oxidative phosphorylation, CO2 pressure does not affect the rate of oxidative phosphorylation, hence answer choice B is incorrect. Although the availability of NADH and FADH2 increase the rate of the ETC and oxidative phosphorylation, the complex II does affect the rate of oxidative phosphorylation. Again, with similar logic, pH of the mitochondria is important, however, does affect the rate of oxidative phosphorylation. Therefore answer choices C and D are incorrect.

5. **The correct answer is E.** Uncouplers disjoins the ETC from oxidative phosphorylation therefore inhibits the mitochondrial ATP production. Uncouplers reduce proton gradient by allowing protons to reenter the matrix without involvement of ATP synthase. The rate of the TCA cycle, therefore the production of NADH and FADH2 remains the same, or even becomes faster. In addition to the rate of the TCA cycle, the rate of ETC either remains the same or increases. Therefore, uncouplers reduce the ATP production by reducing proton gradient.

IV

28 Oxygen Toxicity and Antioxidants

At the conclusion of this chapter, students should be able to:
- Explain how O_2 can be both essential to life and toxic to the cell
- Describe the toxic effects of reactive oxygen species (ROS)
- Describe the underlying molecular mechanisms of diseases associated with ROS damage
- Describe how antioxidants and cellular scavenging enzymes protect cells against radical damage

Molecular O_2 is necessary for cells to produce a sufficient amount of ATP for normal cellular processes (reviewed in chapter 27). Molecular O_2 is also the source for free radicals, such as reactive oxygen species (ROS), in our cells. Free radicals are extremely powerful oxidizing agents that damage DNA, proteins, membrane lipids and lipoproteins. Thus, extensive exposure to ROS can result in substantial injury to our cells. Free radical damage is associated with many diseases, including cancer, atherosclerosis, coronary artery diseases, cerebrovascular disorders, ischemia/reperfusion injury, alcohol-induced liver disease, diabetes, aging, neurodegenerative disorders, Alzheimer disease, Parkinson disease, and Amyotrophic lateral sclerosis.

28.1 Generation of Superoxide Radical (O_2^-)

The generation of free radicals from O_2 is a natural process that occurs on a daily basis. The first step for generation of a free radical starts with the formation of a superoxide radical (O_2^-). The superoxide radical is the precursor for other ROS including the very reactive and most dangerous radical, hydroxyl radical ($OH\bullet$). The formation of superoxide radical in our body most often occurs in the ETC through coenzyme-Q (ubiquinone). The Coenzyme-Q occasionally loses an electron in the ETC, forming $CoQH\bullet$, which freely moves in the inner-membrane of the mitochondria and transfers an e- to dissolved O_2, thereby forming a superoxide radical (O_2^-) (▶ Fig. 28.1).

The cytochrome p450 (CYP) enzymes are involved in catalyzing transfer of electrons to organic substrates in the presence of O_2. As $CoQH^{\bullet}$ in ETC loses an electron (e-) to molecular O_2, the CYP enzymes can donate an e- to molecular O_2, as a result, generating the superoxide radical.

28.2 Link to Pharmacology and Toxicology

The function of the CYP isozymes in the liver is to metabolize alcohol and many drugs. Thus, chronic alcohol and drug abuse leads to an increase in the number of CYP enzymes, as well as an increase in the activity of each CYP. This increased CYP enzyme activity results in more superoxide formation. Therefore, chronic alcohol or drug intake is associated with increased free radical injury of the cells.

28.3 Link to Pathology: Myocardial Reperfusion Injury

Myocardial reperfusion injury is characterized by myocardial injury and inflammation after restoration of blood flow to recently ischemic myocardium. This may occur following cardiac catheterization or administration of clot busters, such as alteplase, for treatment of a myocardial infarction (MI).

When previously ischemic tissue undergoes reperfusion, abundant oxygen, white blood cells and nutrients are delivered. Although oxidative phosphorylation and ATP generation are recovered, the transient and rapid increase in the ETC results in the production of hydroxyl and superoxide free-radicals. Free-radical production results in damage to cellular DNA, proteins

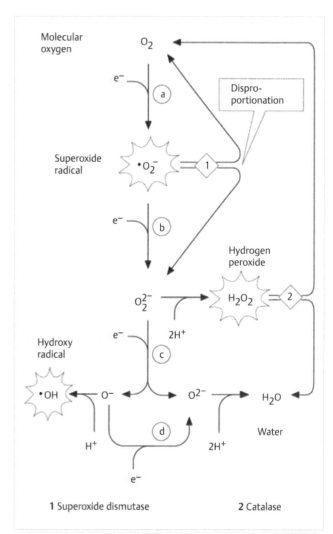

Fig. 28.1 Formation of reactive oxygen species (ROS) from oxygen. Generation of ROS starts with superoxide radical, which is a precursor for H2O2 and hydroxyl radical. The H2O2 can be reduced to water by the intracellular enzyme superoxide dismutase (SOD) or by the peroxysomal enzyme catalase. (Source: Koolman J, Röhm K, ed. Color Atlas of Biochemistry. 3rd Edition. New York, NY. Thieme; 2012.)

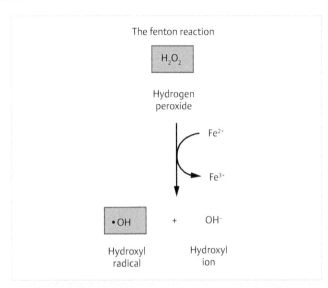

The fenton reaction

Fig. 28.2 **(a)** The Fenton reaction. **(b)** The Haber-Weiss reaction.

and the cell membrane. The end result is inflammation and damage to previously viable myocardium. This is manifested clinically by elevation in cardiac troponin and CK-MB levels following MI treatment.

28.4 Generation of Hydroxyl Radical (OH•)

The superoxide radical forms when molecular oxygen accepts a single electron (e-) from any electron carrier. The production of superoxide radical from molecular O_2 is due to its diradical nature. Since O_2 contains two unpaired electrons, when it takes up an extra electron the highly reactive superoxide radical is produced (▶ Fig. 28.1). After one more electron (e-) acceptance and binding to protons, superoxide becomes H_2O_2. Although H_2O_2 is not a free radical itself, it can diffuse through the membranes and into the cell, where it can serve as a precursor for the hydroxyl (**OH•**) radical. The conversion of H_2O_2 to the hydroxyl radical (**OH•**) occurs through several mechanisms. Two of these involves accepting electrons through non-enzymatic reactions such as Fenton and Haber-Weiss reactions. In the **Fenton reaction**, free Fe^{2+} or Cu^+ serve as single electron donors for H_2O_2, which results in formation of a hydroxyl (**OH•**) radical (▶ Fig. 28.2). In the **Haber-Weiss reaction**, H_2O_2 reacts with a superoxide radical to generate the hydroxyl (**OH•**) radical (▶ Fig. 28.2b).

28.4.1 Link to Pharmacology: Bleomycin Forms Complex with Iron and Oxygen

The molecular oxygen and iron generates free radicals, which causes single strand DNA breaks in the G2 phase of the cell cycle. Since this DNA fragmentation inhibits the cell cycle, bleomycin is used as antineoplastic drug especially for Hodgkin's lymphoma, head and neck cancer, and squamous cell carcinomas.

28.5 Generation of Superoxide Radical by Enzymatic Reactions

In addition to non-enzymatic reactions, superoxide radicals (O_2^-) can also be generated by metabolic enzymes. Examples of these include xanthine oxidase, NADPH oxidase, monoamine oxidase, and peroxisomal fatty acid oxidase.

The formation of superoxide radicals from molecular O_2 is the first step in generating ROS by both non-enzymatic and enzymatic reactions. Unlike non-enzymatic reactions, in the enzymatic reactions the molecular O_2 acts as a substrate and the product is superoxide (O_2^-).

Neutrophils and other phagocytic cells use respiratory bursts of free radicals to destroy engulfed bacteria (reviewed in chapter 23). The NADPH oxidase in phagocytes transfers electrons from NADPH to O_2 to form superoxide, which is the precursor of other reactive oxygen species (ROS). As reviewed below, free radicals destroy the membranes and proteins, therefore, the respiratory burst in the phagocytes is a key component their antimicrobial defense.

28.5.1 Link to Microbiology and Immunology

NADPH oxidase is deficient in chronic granulomatous disease (CGD), which presents with recurrent infections, chronic inflammatory lesions in the skin, lungs, and lymph nodes. Due to the genetic mutations in the NADPH oxidase gene, phagocytic cells cannot kill catalase-positive bacteria, causing recurrent, persistent infections and formation of tissue granulomas. Some patients may response to treatment with gamma interferon.

28.6 How Do Free Radicals Cause Cell Injury?

Free radicals extract an e- from other molecules, such as membrane lipids, cellular proteins and DNA, to complete their orbitals. When a free radical extracts an e- from biomolecules, they initiate chain reactions that form organic radicals and lipid peroxides. The most reactive free radical in attacking biological molecules is the hydroxyl radical (**OH•**).

Damage to lipids: Attacking membrane lipids (especially polyunsaturated membrane lipids) causes peroxidation and the generation of lipid radicals within the membranes. These unstable lipid radicals attack other lipids within the membrane, causing a self-propagating production of lipid radicals. These damaged lipid peroxyl radicals and lipid peroxides are unstable and, therefore, degraded easily within the cell membranes. Increased damage and degradation of membrane lipids results in the loss of integrity of the membrane, which causes increased cell permeability, increased influx of Ca^{2+} (an important cofactor for proteases) and ultimately cell swelling and necrosis. These reactions can be terminated by vitamin E and other lipid-soluble antioxidants.

Peptide and protein damage: Free radicals can react with iron and sulfur moieties of peptides and proteins. Oxidation of sulfhydryl group in cysteine and methionine amino acids may

IV

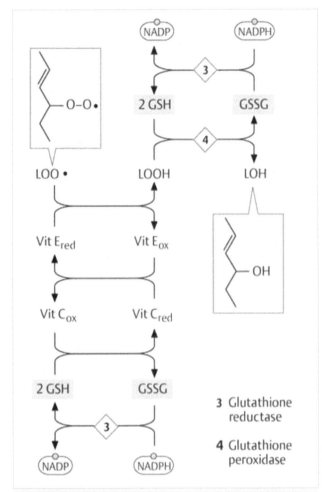

Fig. 28.3 Membrane protection. (Source: Koolman J, Röhm K, ed. Color Atlas of Biochemistry. 3rd Edition. New York, NY. Thieme; 2012.)

cause cross-linking of the protein fragments. Aggregated and cross-linked proteins are subject to immature protein degradation in the cells. The oxidation of proteins can be prevented by Glutathione, which is the major intracellular anti-oxidant molecule.

Damage to DNA: Free radicals can react with nucleotide bases of DNA. The most common target is the nucleotide guanine, which forms 8-oxo-guanine. This oxidized base may mispair during DNA replication, which ultimately causes a G-C → T-A substitution. The oxidized base may also cause breaks in the deoxyribose backbone of the DNA.

28.7 Cellular Defenses against Radical Damage

Although generation of ROS and damage to cellular compartments is unavoidable, the cell has several mechanisms to defend itself against this damage. These cellular defenses against radical damage can be split into two categories: defense

enzymes such as antioxidant scavenging enzymes, and non-enzymatic antioxidants (▶ Fig. 28.4).

1. **The defense enzymes** superoxide dismutase (SOD,) catalase, glutathione peroxidase and glutathione reductase directly or indirectly interact with ROS to convert them to non-toxic products.

The reduction of superoxide to water by defense enzymes occurs in a stepwise process. First, superoxide dismutase (SOD) reduces superoxide to hydrogen peroxide (H_2O_2). Then, glutathione peroxidase reduces the H_2O_2 to water. In this reaction, two molecules of reduced glutathione (GSH) donate two hydrogens to H_2O_2 to form water. In the process, the two glutathione molecules become oxidized glutathione (GSSG). The oxidized glutathione molecules must then be converted back to the reduced form of glutathione (GSH). This is accomplished by glutathione reductase, which uses NADPH for reducing power. Thus, NADPH is essential for protection against free radical injury. As reviewed in chapter 23, the pentose phosphate pathway also plays an essential role in the protection of erythrocytes from free radical-induced hemolysis (also see G6PD deficiency and hemolytic anemia in chapter 23).

While glutathione peroxidase is the major defense enzyme against H_2O_2 in the cytosol and in the mitochondria, the enzyme responsible for this activity in peroxisomes is catalase. After superoxide dismutase (SOD) reduces superoxide to hydrogen peroxide (H_2O_2), the peroxisomal enzyme catalase reduces H_2O_2 to water in the peroxisomes, which lack glutathione.

28.7.1 Link to Pathology

Amyotrophic lateral sclerosis (ALS), also known as Lou Gehrig disease, is caused by degeneration of neurons that mainly affects motor neurons. There are both environment and genetic factors that contributes to the development of the disease. A genetic form of the disease is caused by mutations in superoxide dismutase (*SOD1*) gene. It is reasonable to think that individuals with this gene mutation have difficulty in disposing of superoxide radicals, which can lead to accumulation of excessive ROS. This excessive amount of ROS can cause irreversible cellular degeneration especially in the motor neurons.

2. **Non-enzymatic antioxidants,** or free radical scavengers, are compounds that are either naturally occurring in the body or come from the diet. These compounds dispose, suppress, or oppose the actions of free radicals and are referred to as antioxidants. Most antioxidants terminate free radical chain reactions in the cells. Some of these antioxidants are dietary compounds. For example, vitamin E (α-tocopherol), vitamin C (ascorbic acid), carotenoids, and vitamin A are all classified as antioxidants. Examples of endogenous antioxidants that are produced in our body include melatonin, uric acid, ferritin, bilirubin, and L-carnitine (▶ Fig. 28.4).

Cellular compartmentalization, which is the physical separation of organelles involved in ROS generation, can be considered as having an antioxidant effect in the cell. Metal sequestration is also antioxidant because the protein-bound metals cannot react in Fenton or Haber-Weiss reaction.

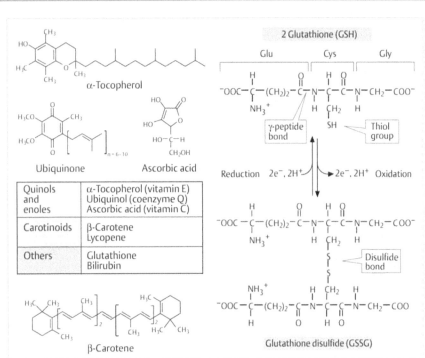

α-Tocopherol

Ubiquinone   Ascorbic acid

Quinols and enoles	α-Tocopherol (vitamin E) Ubiquinol (coenzyme Q) Ascorbic acid (vitamin C)
Carotinoids	β-Carotene Lycopene
Others	Glutathione Bilirubin

β-Carotene

2 Glutathione (GSH)

Glu   Cys   Gly

γ-peptide bond

Thiol group

Reduction   2e⁻, 2H⁺   2e⁻, 2H⁺   Oxidation

Disulfide bond

Glutathione disulfide (GSSG)

Fig. 28.4 Biological antioxidants and Glutathione. (Source: Koolman J, Röhm K, ed. Color Atlas of Biochemistry. 3rd Edition. New York, NY. Thieme; 2012.)

IV

Review Questions

1. Which of the following best explains the underlying molecular reason that patients with chronic granulomatous disease suffer from recurrent bacterial infections?
   A) Due to decreased NADPH production
   B) Due to genetic mutations in SOD
   C) Due to increased damage from ROS
   D) Due to insufficient generation of superoxide
   E) Due to lack of reduced glutathione

2. Which of the following enzymes that defends against radical damage requires NADPH for its activity?
   A) Catalase
   B) Glutathione peroxidase
   C) Glutathione reductase
   D) Glucose 6-phosphate dehydrogenase
   E) Superoxide dismutase

3. Which of the following best explains the generation of DNA strand breaks by bleomycin?
   A) Bleomycin activates DNA replication
   B) Bleomycin acts as an antioxidant
   C) Bleomycin causes cellular toxicity
   D) Bleomycin facilitates free radical formation
   E) Bleomycin inhibits NADPH production

4. Formation which of the following molecule(s) is/are the cause for myocardial reperfusion injury?
   A) ATP
   B) Antioxidants
   C) Free radicals
   D) Reducing powers
   E) NADPH

Answers

1. **The correct answer is D**. Chronic granulomatous disease is caused by a genetic mutation in the NADPH oxidase gene. Due to inadequate amounts of functional NADPH oxidase, phagocytes cannot generate sufficient amounts of superoxide radical and consequently these cells cannot generate enough ROS during the respiratory burst. Therefore, macrophages or neutrophils are unable to destroy certain bacteria, especially catalase-positive pathogens, through oxygen-dependent killing.

Answer choice A is incorrect because NADPH is produced from G6PD in the HMP shunt. G6PD deficiency causes hemolytic anemia. Mutations in the SOD gene (answer choice B) cause accumulation of ROS, which leads to cellular damage. A genetic form of the ALS disease is caused by mutations in the SOD gene. Answer choices C and E are incorrect because increased damage from ROS does not cause chronic granulomatous disease, nor does the lack of reduced glutathione.

2. **The correct answer is C**. Oxidized glutathione molecules (GSSG) are converted to the reduced form of glutathione (GSH) by glutathione reductase, which uses NADPH for reducing power. Therefore, glutathione reductase is a defense enzyme and the product of this enzyme is used in reducing the H_2O_2 to water. Answer choice A is not correct because catalase converts H_2O_2 to water and molecular oxygen in the peroxisomes without a need of reducing power (NADPH). The glutathione peroxidase, answer choice B, uses two molecules of reduced glutathione (GSH) to donate two hydrogens to O_2 to form water. This enzyme also does not need NADPH as a reducing coenzyme. The enzyme glucose 6-phosphate dehydrogenase, answer choice D, uses NADP to generate NADPH in the HMP shunt and is not involved in the defense against radical damage. Answer choice E is not correct either because superoxide dismutase (SOD) reduces superoxide to H_2O_2 without a need for NADPH.

3. **The correct answer is D**. Bleomycin and ferrous ions form complexes and in the presence of oxygen, these complexes generate free radicals. These free radicals cause DNA strand breaks during G2 phase of the cell cycle, thereby inhibiting cell growth. Answer choice C may be considered correct as bleomycin causes cellular toxicity. However, it is not the best explanation for how bleomycin causes DNA strand breaks.

Answer choice A is incorrect because bleomycin does not affect DNA replication, which happens during S phase of the cell cycle. Answer choices B and E are incorrect since bleomycin is not an antioxidant, nor does it have any effect on pentose phosphate pathway for the NADPH production.

4. **The correct answer is C**. The occlusion of a coronary artery results in ischemia. This ischemic tissue cannot generate enough ATP. The resulting increased AMP/ATP ratio activates anaerobic glycolysis and consequently lactate production. Due to decreased ATP, cells cannot maintain the activity of Na-K-ATPase on the cell membrane. This causes accumulation of intracellular Na and water retention. Due to damaged organelle membranes, intracellular Ca^{2+} increases, which activates proteases and phospholipases. During reperfusion, the ischemic myocardium receives O_2 and this allows the cells to recover the ETC and oxidative phosphorylation. This increased activity of the ETC also increases the chance of transferring electrons from CoQ· to O_2 and the generation of superoxide and free radical formation. The increased free-radicals from the ETC results in damage to cellular DNA, proteins and the cell membranes. This damage is called myocardial reperfusion injury. Answer choices A, B, D, and E are not a cause of either cellular or the reperfusion injury.

29 Oxidation of Fatty Acids and Ketogenesis

At the conclusion of this chapter, students should be able to:
- Explain the basic steps involved in lipolysis and mobilization of fatty acids in adipose tissue
- Describe the steps involved in transport of fatty acids into the mitochondria and metabolic disorders associated with each step
- Summarize the four steps in the β-oxidation spiral and explain the regulation of the rate-limiting step
- Describe the basis of MCAD deficiency and the underlying molecular mechanisms of the signs and symptoms of the MCAD disease
- Understand the alternate routes of fatty acid oxidation, and the deficiencies of these routes including α- and β-oxidation in peroxisomes and ω-oxidation in microsomes
- Describe how medium chain, unsaturated, and odd-chain length fatty acids are oxidized
- Explain the body's response to fasting/starvation through production of ketone bodies
- Explain how and why ketone bodies form and how they are used for energy
- Analyze the symptoms and the causes of ketosis and diabetic ketoacidosis

The main focus of this section is fatty acid degradation through β-oxidation and the production of ketone bodies. In addition to β-oxidation, we will also discuss alternate routes of fatty acid oxidation, as some of which are clinically important.

Fatty acids are stored in the form of triacylglycerol in the adipose tissue. Triacylglycerols are used for energy storage for two main reasons. The first being that the carbon, typically a CH_2, is almost completely reduced, therefore, its oxidation will yield the most energy possible. Secondly, the fatty acids are not hydrated as mono- and polysaccharides are, so they can pack more closely in storage tissues. While we already know that glycolysis provides quick energy for the cells, 80% of the heart and liver's energy needs are met by the breakdown of fatty acids from their major storage form, triacylglycerols. Typically, during states of fasting and exercise, the release of fatty acids from adipose tissue is triggered through the action of three major hormones: glucagon, epinephrine, and cortisol.

During a low insulin and high glucagon state, the breakdown of triglycerides into free fatty acids and glycerol is facilitated by **hormone-sensitive lipase (HSL)**. Glucagon secretion increases when blood glucose is low, this high glucagon/insulin ratio favors fatty acid release and β-oxidation through increased concentrations of intracellular cAMP. In the adipocytes, elevated intracellular cAMP activates protein kinase A (PKA), which phosphorylates and activates HSL (▶ Fig. 29.1). The active HSL hydrolyses triacylglycerols to release the free fatty acids (FFAs), which are then carried in the blood by serum albumin. When the FFAs reach the mitochondrial matrix of the skeletal muscle, cardiac muscle, and the liver, they undergo β-oxidation to release energy for the cells' functions. During prolonged starvation or exercise or in uncontrolled diabetic patients (reviewed in chapter 31) excessive fatty acid oxidation in the liver results

in acetyl CoA accumulation. This excess acetyl CoA is used for biosynthesis of ketone bodies, which are transported to extrahepatic tissues for oxidation. In this chapter, we will first review the steps of fatty acid oxidation and its regulation, and then we will discuss ketogenesis and the physiological importance of ketone bodies.

29.1 Fatty Acid Uptake

When fatty acids enter the cytosol of hepatocytes or myocytes, fatty acyl CoA synthetase or thiolase catalyzes the formation of fatty acyl CoA (▶ Fig. 29.2). Formation of a high-energy thioester bond between the fatty acid and coenzyme A (CoA) activates the free fatty acid. This formation of fatty acyl CoA requires hydrolysis of one ATP. Formation of fatty acyl CoA also traps the fatty acid in the cell. This process is energetically expensive and just barely breaks even with ATP hydrolysis. However, the subsequent hydrolysis of pyrophosphate (PP_i) drives the reaction strongly forward.

The fatty acyl CoA must then be transported into the mitochondrial matrix, the location where β-oxidation will occur. Short-chain (C_4-C_6) and medium-chain (C_8-C_{10}) fatty acids are taken up by a monocarboxylate transporter and immediately activated to acyl CoA in the mitochondrial matrix. Long-chain fatty acids (C_{12}-C_{18}) require the carrier carnitine, in order to cross the inner mitochondrial membrane. The short-, medium- and long-chain fatty acids are then broken down by β-oxidation. In contrast, very long-chain fatty acids ($>C_{18}$) and branched-chain fatty acids are transported into the peroxisomes where they are metabolized (reviewed later in this chapter).

The long-chain fatty acids are transported into the mitochondrial matrix through a transport system called the carnitine shuttle, the rate-determining step in mitochondrial fatty acid degradation (▶ Fig. 29.2). The carnitine shuttle is facilitated by two important enzymes: carnitine palmitoyl transferase I (CPT-I) and carnitine palmitoyl transferase II (CPT-II). The carnitine shuttle process begins as CPT-I catalyze the removal of the CoA and the addition of carnitine to the fatty acyl on the outer mitochondrial membrane. It is important to note that CPT-I is the rate-limiting enzyme and is inhibited by malonyl-CoA, a product of fatty acid synthesis. This ensures that fatty acid synthesis and oxidation are not occurring simultaneously. Next, acylcarnitine is transferred across the inner mitochondrial membrane into the mitochondrial matrix through a carnitine-acylcarnitine antiporter (also known as carnitine translocase). Inside the matrix, carnitine is removed and recycled back. The CPT-II transfers the CoA back onto the fatty acid **to** form fatty acyl CoA, allowing for the initiation of the β-oxidation reactions.

29.1.1 Link to Pathology

Carnitine deficiencies may occur if carnitine is not readily available and would prevent β-oxidation from occurring. While ~25% of carnitine is synthesized by the liver and kidneys from lysine, the majority of carnitine is derived from dietary intake through meat and dairy. The synthesis in the liver and kidneys

IV

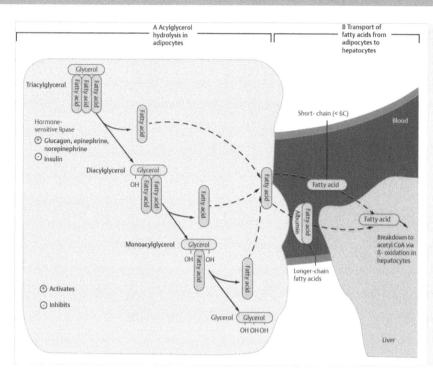

Fig. 29.1 Lipases that catalyze the hydrolysis of acyglycerols. (Source: Panini S, ed. Medical Biochemistry—An Illustrated Review. New York, NY. Thieme; 2013.)

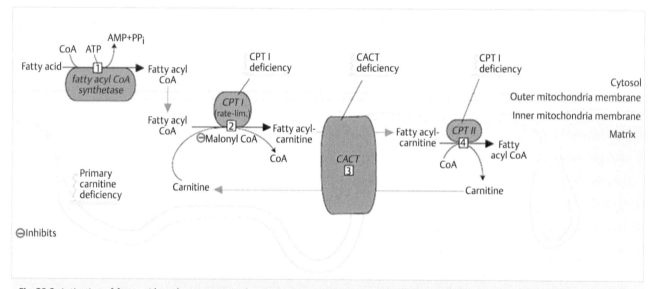

Fig. 29.2 Activation of fatty acids and transport into the matrix via carnitine shuttle. Free fatty acids (FFAs) are activated by the fatty acyl CoA synthetase enzyme in the cytoplasm. The carnitin palmitoyltransferase-I (CPT-I) in the inner-membrane of the mitochondria removes the CoA from the fatty acyl CoA and transfers fatty acyl onto carnitin, forming fatty acyl-carnitin. The fatty acyl-carnitin is transported into the matrix through carnitin-acylcarnitin translocase (CACT). The CPT II enzyme in the matrix exchanges carnitin with the CoA to re-form the fatty acyl CoA. Carnitin is transported back into the inner-membrane space. (Source: Panini S, ed. Medical Biochemistry—An Illustrated Review. New York, NY. Thieme; 2013.)

requires the cofactors s-adenosylmethionine (SAM) and vitamin C. A faulty plasma membrane carnitine transporter, either systemic or specifically in the muscle, would result in a primary deficiency of carnitine. A secondary deficiency may occur due to other metabolic disorders, such as organic acidemias, or in preterm newborns. The age of onset of carnitine deficiency varies between 1-month to 7-years of age. A cardinal sign of carnitine deficiency is **hypoketotic hypoglycemia,** as gluconeogenesis cannot be supported by fatty acid oxidation. Other signs and symptoms include: skeletal myopathies, cardiomyopathies, encephalopathy, and hepatomegaly.

CPT-I deficiency is an autosomal recessive defect in the CPT-I gene. It primarily affects fatty acid oxidation in the liver. Symptoms will usually present after a period of fasting or after a gastrointestinal illness. A patient will present with hypoglycemia and hypoketosis, lethargy, seizures and/or coma.

CPT-II deficiency (the myopathic form) will typically present in patients soon after puberty. The patient may have red or cola-colored urine, myoglobinuria, and will complain of recurrent episodic myalgia and muscle stiffness. Lab values will present a marked elevation of serum creatine kinase (CK) causing rhabdomyolysis. The patient will also complain that the

symptoms are exacerbated by prolonged exercise, stress, infections, fasting and/or by a high-fat, low-carb diet. Common misdiagnoses with these presentations include muscular dystrophy, fibromyalgia, and chronic fatigue syndrome.

29.2 β-oxidation of Fatty Acids

β-oxidation is the repeated removal of 2-carbon fragments from the fatty acid chain through a series of oxidative reactions. It is the exact opposite of the elongation process of fatty acid synthesis, as the 4 steps are reversed. This method of degradation of the fatty acids occurs through a step-by-step process called the beta-oxidation spiral. It is called a spiral because the four steps of β-oxidation are repeated on a loop until oxidation is complete, cleaving 2-carbon acetyl groups per round. For palmitic acid, a 16-carbon fatty acid, the beta-oxidation spiral occurs seven times and yields seven acetyl Co-A molecules. The 4 main steps of beta-oxidation can be described as: oxidation, hydration, oxidation, cleavage (▶ Fig. 29.3).

Step1 – oxidation: Hydrogens are removed and accepted by the coenzyme FAD between the alpha and beta carbons, creat-

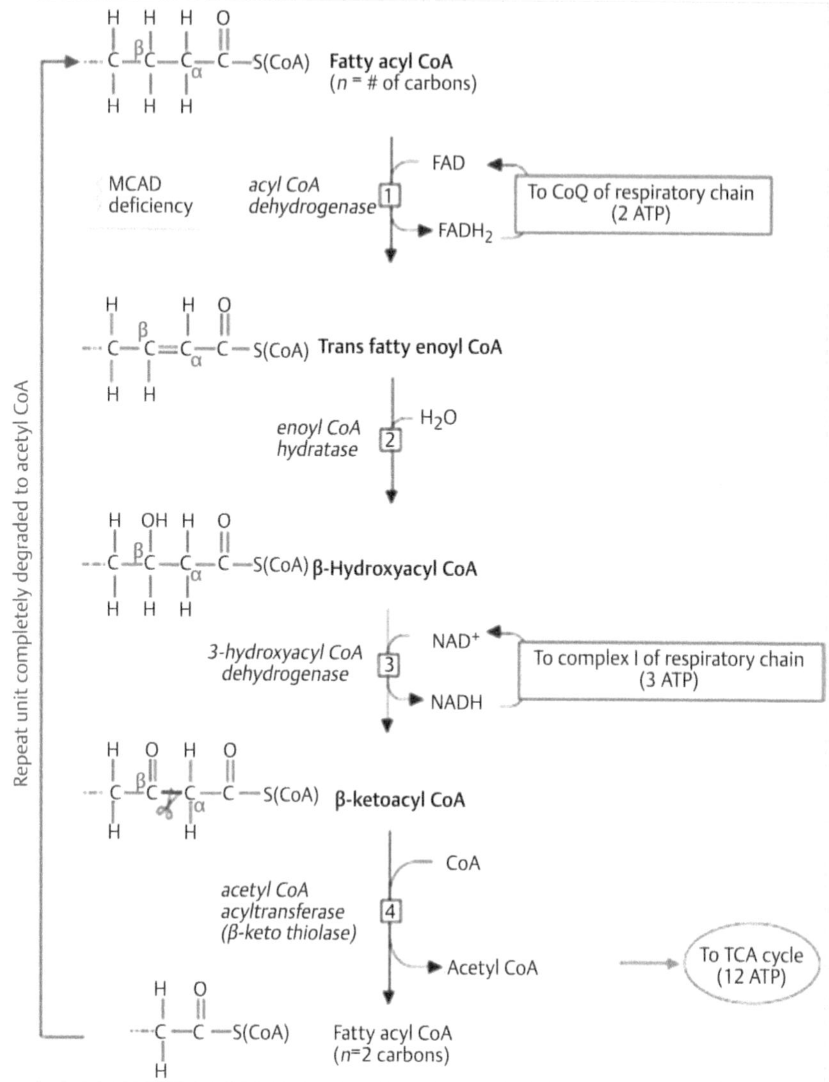

Fig. 29.3 Oxidation of saturated even-chain fatty acids. The even-chain saturated fatty acids are oxidized in the matrix by beta-oxidation, generating FADH2, NADH, and acetyl CoA. (Source: Panini S, ed. Medical Biochemistry—An Illustrated Review. New York, NY. Thieme; 2013.)

ing a double bond. This step is catalyzed by **acyl-CoA dehydrogenase** and yields a trans-delta 2-enoyl CoA. There are 4 acyl-CoA dehydrogenases that correspond to the different fatty acid chain lengths. The very long chain acyl CoA dehydrogenase (VLCAD) works in the peroxisomes, the long chain acyl CoA dehydrogenase (LCAD) in the mitochondria, and the medium chain acyl CoA dehydrogenase (MCAD) also in the mitochondria. A genetic deficiency of MCAD is well known.

Step2 – hydration: A water/OH group is added to the alpha carbon via the enzyme enoyl-CoA hydratase, creating an β alcohol. This step yields 3-L-hydroxyacyl CoA.

Step3 – oxidation: This second oxidation step involves dehydrogenating 3-L-hydroxyacyl CoA into β-keto acyl-CoA via the enzyme 3-L-hydroxyacyl-CoA dehydrogenase. The coenzyme NAD is reduced to NADH and is released.

Step4 – cleavage: The final step in beta oxidation uses the enzyme thiolase and cleaves the β-keto ester. The products of this reaction are an acetyl-CoA and a fatty acid that is two carbons shorter. Another CoA-SH comes in and reacts with the β-carbon creating a new CoA capped end, and now the process can repeat.

It is important to note that each repetition yields one acetyl CoA, one FADH$_2$, and one NADH. Thus one palmitic acid molecule yields a total of eight acetyl-CoAs. Each acetyl CoA generates ten ATP in the TCA and ETC, so a total of 80 ATP are produced just from the acetyl CoA. One FADH$_2$ and one NADH yield four ATP from ETC, so a total of 28 ATP are produced from those products. So at the completion of the β-oxidation of one palmitic acid, a total of 106 ATP molecules are produced.

29.2.1 Link to Pathology

MCAD deficiency is the most frequently diagnosed fatty acid oxidation disorder. It is an autosomal recessive disease that causes an impairment of oxidation of medium-chain fatty acids. The incidence of this disorder is 1 in 20,000, and approximately 1 in 80 healthy people are carriers of this trait. The disorder is most common in Caucasians of northern European descent. Clinical symptoms typically present between the second month and second year of life, following a period of prolonged fasting (even overnight) or following an illness. Due to this reason, MCAD deficiency accounts for about 1 in 100 sudden infant death syndromes (SIDS). Symptoms include prolonged fasting hypoglycemia, low to no ketones, hepatomegaly, dicarboxylic acidemia, vomiting, lethargy, muscle weakness, coma, and cardiopulmonary arrest. Some of these symptoms may ultimately contribute to the patient's death.

29.2.2 Link to Pharmacology/ Toxicology

Jamaican Vomiting Sickness is caused by consumption of the unripe fruit from the Jamaican ackee tree. This fruit contains the toxin hypoglycin, which inhibits both the short and medium chain acyl-CoA dehydrogenase, thus inhibiting β-oxidation. A major symptom associated with ingestion of this toxin is nonketotic hypoglycemia.

29.3 Regulation of β-oxidation

The rate limiting step of β-oxidation is the transport of fatty acids into the mitochondria, facilitated by the carnitine shuttle system. If fatty acids cannot get into the mitochondria, they cannot be degraded, thus the β-oxidation process is prevented. The rate limiting enzyme in this process is CPT-1. The enzyme CPT-1 is inhibited by malonyl-CoA, a product of acetyl-CoA carboxylase in the fatty acid biosynthesis pathway. The hormones that facilitate β-oxidation are glucagon and glucocorticoids. During the fed-state, insulin inhibits β-oxidation, as the intracellular malonyl CoA concentration is elevated when the cells are actively synthesizing fatty acids. This increased concentration of malonyl CoA inhibits the CPT-1 thereby preventing substrate cycling between fatty acid synthesis and oxidation.

Additionally, the ATP/ADP ratio and the availability of substrates and the cofactors NAD and FAD also have a regulatory effect in β-oxidation. Elevated AMP activates AMP-dependent protein kinase (AMPK), which phosphorylates and inactivates acetyl CoA carboxylase (the rate-limiting step of fatty acid synthesis), and subsequently decreases the malonyl CoA concentration. As a result, decreased malonyl CoA increases the rate of CPT-1 and fatty acid oxidation. Lastly, the rate at which acetyl CoA is processed in the TCA cycle will also show regulatory effects.

29.3.1 Link to Pathology

Fatty liver disease occurs due to alterations in mitochondrial β-oxidation of fatty acids. This leads to accumulation of triglycerides in the liver, also known as steatosis. There are multiple causes of fatty liver disease including alcoholism, obesity, diabetes, Reye's syndrome, and acute fatty liver of pregnancy.

29.4 Oxidation of Odd-Carbon Fatty Acids

The β-oxidation spiral successfully degrades even-chain fatty acids into two carbon fragments. In contrast, odd-chain carbon fatty acids are metabolized normally until it reaches a 2-C acetyl CoA and a 3-C propionyl-CoA. The propionyl-CoA is then converted to succinyl-CoA and is fed into the TCA cycle as an anaplerotic intermediate (▶ Fig. 29.4). The conversion of propionyl CoA to succinyl CoA starts with propionyl-CoA carboxylase which converts propionyl CoA to methylmalonyl CoA. The enzyme propionyl CoA carboxylase requires an ATP and uses biotin as cofactor, with the carboxyl group coming from CO_2. Methylmalonyl CoA mutase rearranges this product into succinyl CoA. The methylmalonyl CoA mutase requires vitamin B12 as coenzyme. This enzyme is one of the two enzymes in metabolism that requires B12 as coenzyme. The other enzyme that requires B12 as coenzyme is methionine synthase, which is in the one-carbon metabolism pathway (reviewed in chapter 38).

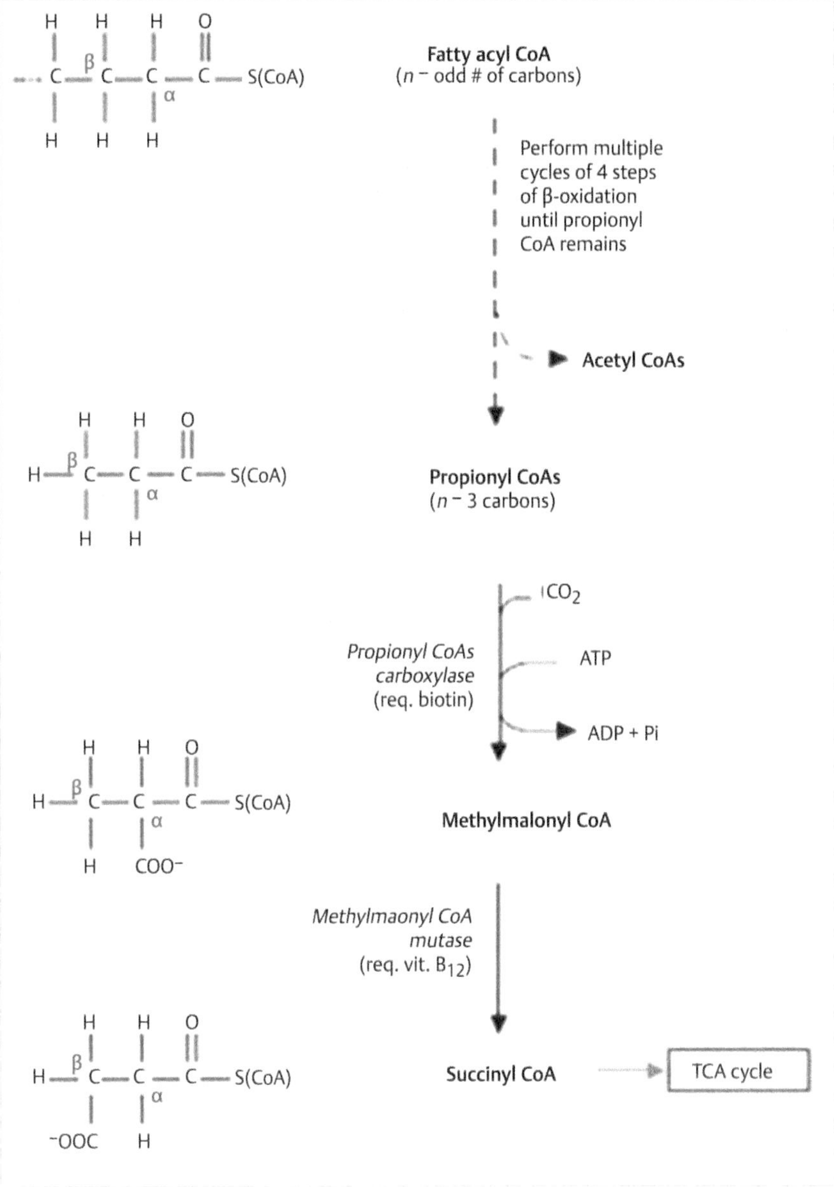

H H H O
⋯ C — C — C — C — S(CoA)
    |β | |α
H H H

Fatty acyl CoA
(n – odd # of carbons)

Perform multiple
cycles of 4 steps
of β-oxidation
until propionyl
CoA remains

→ **Acetyl CoAs**

H H O
H — C — C — C — S(CoA)
    |β | |α
H H

Propionyl CoAs
(n – 3 carbons)

CO_2

**Propionyl CoAs
carboxylase**
(req. biotin)

ATP

ADP + Pi

H H O
H — C — C — C — S(CoA)
    |β | |α
H COO⁻

Methylmalonyl CoA

**Methylmaonyl CoA
mutase**
(req. vit. B_{12})

H H O
H — C — C — C — S(CoA)
    |β | |α
⁻OOC H

Succinyl CoA ——▶ TCA cycle

Fig. 29.4 Oxidation of odd-chain fatty acids. The last step of the odd-chain fatty acids produces a three carbon propionyl CoA. Propionyl CoA is converted to succinyl CoA by the enzymes propionyl CoA carboxylase and methylmalonyl CoA mutase. (Source: Panini S, ed. Medical Biochemistry—An Illustrated Review. New York, NY. Thieme; 2013.)

IV

29.5 Oxidation of Mono-Unsaturated Fatty Acids

Diet-derived fatty acids such as oleic acids and palmitoleic fatty acids are monounsaturated fatty acids. Normal β-oxidation occurs until a cis-delta3 acyl-CoA compound is formed, which cannot be utilized by acyl-CoA dehydrogenase. Enoyl-CoA isomerase converts this to trans-delta2 acyl-CoA. This compound is then hydrated by enoyl-CoA hydratase and then fed into β-oxidation.

29.6 Oxidation of Poly-Unsaturated Fatty Acids

The dietary fatty acid linoleic acid contains two double bonds and must go through special steps in between the β-oxidation spirals (▶ Fig. 29.5). After the first three spirals, once a double bond is reached, the enzyme enoyl-CoA isomerase converts the cis double bond into a trans double bond. This bond then undergoes another round of β-oxidation. Then the enzyme 2, 4-dienoyl-CoA reductase uses NADPH and reduces the trans-delta2 cis-delta4 structure, generating a trans-delta3. Enoyl-CoA isomerase then converts this to trans- delta2 acyl CoA, allowing for the last β-oxidation spirals to occur.

29.7 Peroxisomal β-oxidation of Very Long-Chain Fatty Acids (VLCFA)

This oxidation shares some similarities to mitochondrial β-oxidation. Once short and medium-chain fatty acids are produced they are transported to the mitochondria to complete the oxidation (▶ Fig. 29.6). The VLCFA are first activated into an

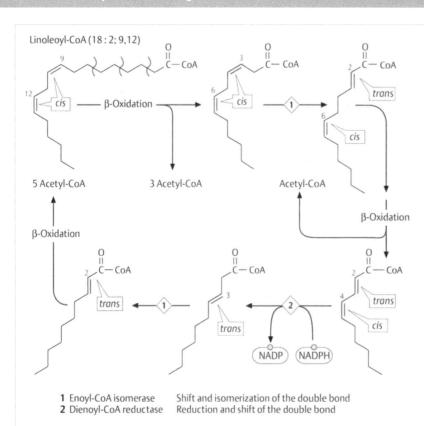

Fig. 29.5 Degradation of unsaturated fatty acids. (Source: Koolman J, Röhm K, ed. Color Atlas of Biochemistry. 3rd Edition. New York, NY. Thieme; 2012.)

acyl-CoA by a long chain fatty acyl CoA synthetase enzyme on the peroxisomal membrane. The activated compound is then transported into the peroxisome in a carnitine independent fashion. The first step in peroxisomal oxidation requires acyl-CoA oxidase, a FAD containing oxidase that forms the initial double bond. The electrons, however, go to O_2 rather than FAD, producing H_2O_2. This process continues until the fatty acid molecules reach 6–10 carbons, which are then transported to the mitochondria for completion of β-oxidation.

29.7.1 Link to Pathology: Peroxisomal disorders

Zellweger syndrome occurs in patients with mutations in the *PEX* genes, leading to a lack of peroxisomal biosynthesis. The result is defective production of ether-phospholipids and plasmalogens. These patients suffer from an accumulation of very long-chain fatty acids and branched chain fatty acids due to defective oxidation.

Adrenoleukodystrophy can originate in either an autosomal or X-linked form. The X-linked form of this disease occurs due to mutations in the ATP-binding cassette, subfamily D1 gene (*ABCD1*), which encodes a transporter present in the membrane of peroxisomes. The normal function of this transport is to allow for the entry of very long chain fatty acids into the peroxisome. Therefore, peroxisomal oxidation is impaired as a result, and the patients suffer from an accumulation of very long-chain fatty acids.

29.8 Peroxisomal α-oxidation of Branched Chain Fatty Acids

Branched chain fatty acids such as pristanic acid and phytanic acids (two of the most common) are derived from dietary sources such as leafy green vegetables (▶ Fig. 29.7a). They contain methyl (CH_3) groups at odd-number carbons, making them poor substrates for β-oxidation and necessitating their breakdown via α-oxidation. These methyl branched-chain fatty acids occur as catabolic products of chlorophyll degradation. Although humans cannot produce phytanic acid directly from chlorophyll, it is obtained from diets containing dairy products and meat. The enzyme phytanic acid α-hydroxylase hydroxylates the methyl-carbon creating a carboxyl group. Phytanic acid α-oxidase decarboxylates with oxidation at the α position. β-oxidation then proceeds past the branch.

29.8.1 Link to Pathology

Refsum's disease is an inherited disease caused by a mutation in phytanoyl-CoA hydroxylase, an important enzyme in α-oxidation. As the methyl branched group prevents beta oxidation, the oxidation of these long branched chain fatty acids is impaired, and phytanic acid accumulates in the blood and tissues. This accumulation of phytanic acid causes neurologic symptoms.

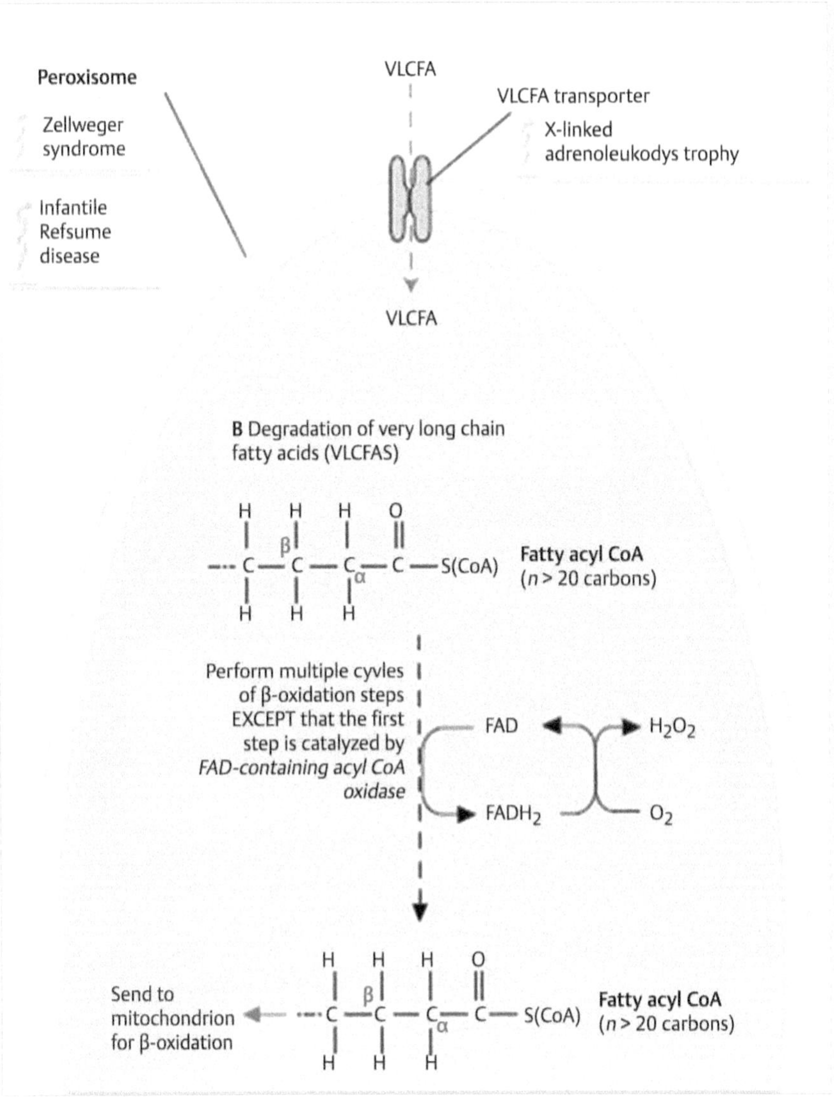

Fig. 29.6 Degradation of very long chain fatty acids (VLCFAS). (Source: Panini S, ed. Medical Biochemistry—An Illustrated Review. New York, NY. Thieme; 2013.)

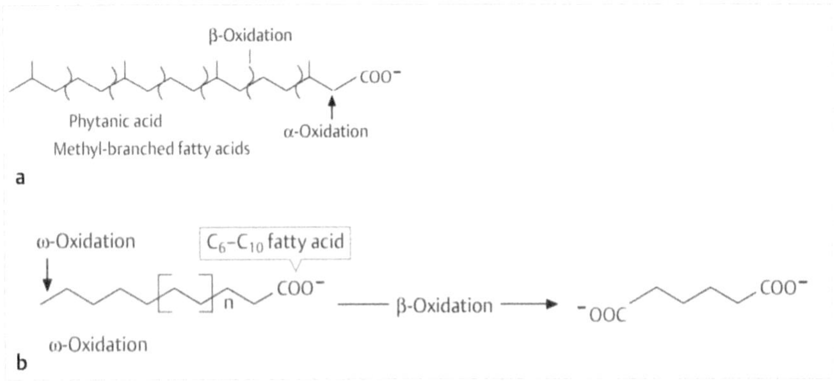

IV

Peroxisome

Zellweger syndrome

Infantile Refsume disease

VLCFA

VLCFA transporter

X-linked adrenoleukodys trophy

VLCFA

B Degradation of very long chain fatty acids (VLCFAS)

Fatty acyl CoA ($n > 20$ carbons)

Perform multiple cyvles of β-oxidation steps EXCEPT that the first step is catalyzed by *FAD-containing acyl CoA oxidase*

FAD

FADH$_2$

H$_2$O$_2$

O$_2$

Send to mitochondrion for β-oxidation

Fatty acyl CoA ($n > 20$ carbons)

Fig. 29.7 **(a)** Other fatty acid degradation pathways – Methyl-branched fatty acids. **(b)** Other fatty acid degradation pathways – ω-Oxidation. (Source: Koolman J, Röhm K, ed. Color Atlas of Biochemistry. 3rd Edition. New York, NY. Thieme; 2012.)

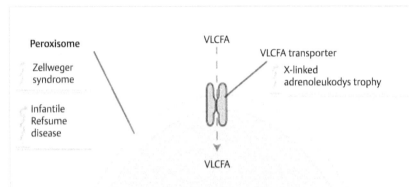

β-Oxidation

COO$^-$

Phytanic acid
Methyl-branched fatty acids

α-Oxidation

a

ω-Oxidation

C$_6$–C$_{10}$ fatty acid

COO$^-$

β-Oxidation

$^-$OOC

COO$^-$

ω-Oxidation

b

29.9 ω-oxidation of Fatty Acids

ω-oxidation involves the oxidation of the methyl-end of fatty acids, commonly referred to as the ω-carbon (▶ Fig. 29.7b). This takes place in ER microsomes and converts fatty acids into dicarboxylic acids. Dicarboxylic acids can either be oxidized by β-oxidation or can be deposited in the blood. A buildup of dicarboxylic acids in the blood can lead to a condition called **dicarboxylic aciduria**. Dicarboxylic aciduria can precipitate from any condition that leads to an increase in ω-oxidation of fats. An increase in ω-oxidation can occur as a result of impaired β-oxidation from deficiencies with CPT or MCAD enzymes, or defects in peroxisome function. It is important to note that the alternate processes of fatty acid oxidation are not

regulated with a feedback mechanism, thus the pathways are triggered by the availability of their substrate.

29.10 Ketone Bodies

The synthesis of ketone bodies occurs in the mitochondria of the liver in the presence of large amounts of acetyl CoA (▶ Fig. 29.8). Acetyl CoA is most abundant during a high rate of fatty acid oxidation in the liver. When a large amount of acetyl CoA can no longer entered into the TCA cycle, ketone bodies are formed as an alternate source of energy for tissues. It is important to note that although the liver synthesizes ketone bodies, it is incapable of utilizing it as a fuel source as it lacks the necessary enzyme for keto-oxidation, **succinyl CoA acetoacetate**

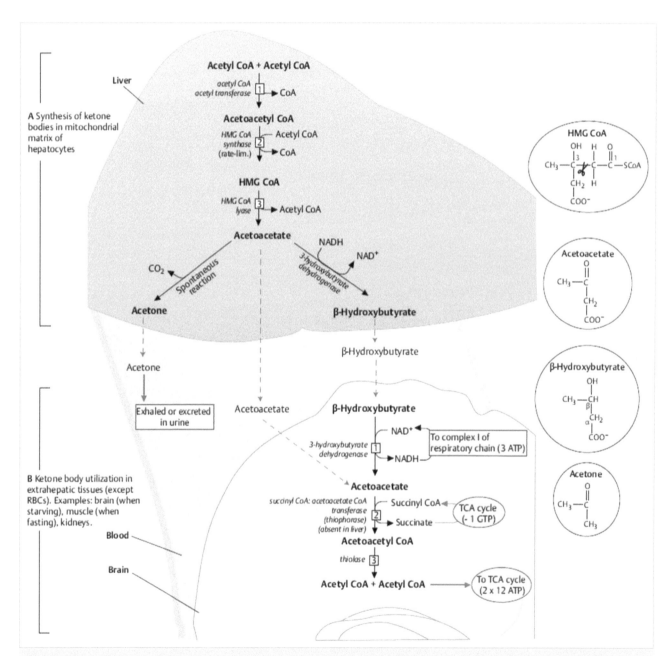

Fig. 29.8 Metabolism of ketone bodies. (Source: Panini S, ed. Medical Biochemistry—An Illustrated Review. New York, NY. Thieme; 2013.)

CoA transferase. The three major ketone bodies formed by the liver are acetoacetate, β-hydroxybutyrate and acetone. Ketone bodies are fuel for the brain, heart and muscle. They are especially important for the brain during times of starvation.

29.11 Ketogenesis

Ketogenesis only occurs in the mitochondrial matrix of the liver (▶ Fig. 29.8). Thiolase, which cleaves the 2-carbon acetyl-CoA compound in β-oxidation, reversibly condenses two acetyl CoA molecules into an acetoacetyl CoA compound. The acetoacetyl CoA then reacts with another acetyl CoA molecule via the enzyme HMG CoA synthase to produce HMG CoA 3-hydroxyl-3-methyl glutaryl CoA. In the presence of HMG CoA lyase, an acetyl CoA is cleaved off to produce acetoacetate. Acetoacetate is stable enough to have three fates. It can either enter the blood directly and be transferred to the tissue to be used as an energy source, or can be reduced to β-hydroxybutyrate by β-hydroxybutyrate dehydrogenase. β-hydroxybutyrate is a non-volatile ketone body and can also easily be transported through the blood to tissues. Acetoacetate also can be spontaneously decarboxylated into acetone and CO_2. Acetone is volatile so it cannot be transported through the blood; rather it is expelled out of the lungs and will give off a fruity smell. These reactions are mitochondrial analogues of the first two steps of cholesterol synthesis which occurs in the cytosol.

Oxidation of fatty acids is not the only source of ketone body generation. Ketone bodies can also be derived from the catabolism of amino acids. The amino acids that can produce ketone bodies are termed ketogenic amino acids. These amino acids include phenylalanine, tyrosine, isoleucine, leucine and lysine.

The process of ketone bodies synthesis occurs in a state of sustained fasting. As the fasting period increases, the body utilizes the ketone bodies for fuels. The brain prefers the use of glucose as its main energy source, however, in the absence of glucose for about 2 to 3 days, β-hydroxybutyrate levels are high enough and will cross the blood brain barrier and be used as an alternate source of energy in neuronal tissues. The muscle tissues also use β-hydroxybutyrate as an alternate source of energy to prevent muscle wasting that occurs in the early stages of starvation. Skeletal muscles, intestinal mucosa and nervous tissues are some of the tissue types that utilize this energy source.

29.11.1 Link to Pathology: Alcohol Intake during Fasting May Cause Alcoholic Ketoacidosis

Ketoacidosis is most commonly observed in uncontrolled type 1 diabetic patients as a result of insulin deficiency. The low insulin/glucagon ratio mobilizes triacylglycerols in the adipocytes, making the free fatty acids available as substrates for fatty acid oxidation and ketogenesis in the hepatocytes. The excessive fatty acid oxidation leads to overproduction of acetyl CoA. The TCA cycle cannot metabolize this excess acetyl CoA since the OAA is diverted to malate and to the gluconeogenesis pathway. Due to overproduced acetyl CoA from fatty acid oxidation and the inability of the TCA cycle to utilize it, these acetyl CoAs, consequently, are used for ketogenesis. The ketone

bodies acetoacetate and β-hydroxybutyrate accumulate in the blood and are excreted into the urine.

The most common etiology of alcoholic ketoacidosis is the intake of ethanol during starvation. The high glucagon/insulin ratio releases free fatty acids from the adipocytes, and these fatty acids are oxidized in the liver. The excess acetyl CoA from the fatty acid oxidation leads to increased ketogenesis. The high NADH production from the metabolism of the ethanol impairs the conversion of lactate to pyruvate and also impairs gluconeogenesis (reviewed in Themes 22 and 25). Therefore, patients with alcoholic ketoacidosis present with hypoglycemia. In addition to hypoglycemia, the high NADH favors the formation of β-hydroxybutyrate from acetoacetate. Therefore, the result of a urine dipstick test may show negative or slightly elevated ketones. Since this test only detects acetoacetate levels in the urine.

29.12 Oxidation of Ketone Bodies

Ketone body oxidation is simply the reverse of its synthesis in the tissues they are being utilized in (▶ Fig. 29.8). β-hydroxybutyrate and acetoacetate travel to various tissues and are transported across the cells into the mitochondrial matrix, where β-hydroxybutyrate is dehydrogenated back into acetoacetate via the enzyme β-hydroxybutyrate dehydrogenase. Acetoacetate then reacts with succinyl CoA acetoacetate CoA transferase and results in an activated acetoacetyl CoA. It is important to note that this enzyme is only available in extrahepatic tissues, therefore although the liver produces ketone bodies, it does not have the ability to degrade them and hence use them for energy. The acetoacetyl CoA compound is then cleaved by thiolase into two acetyl CoA compounds. Therefore, one acetoacetate compound has twice the energy power as one acetyl CoA compound after oxidation in the TCA cycle.

29.13 Regulation of Ketogenesis

In the fasting/starvation state, the lack of dietary glucose inhibits the release of insulin, and glucagon release is stimulated. Because of the absence of glucose and low ATP and NADH levels, the body responds by oxidizing fats. Free fatty acids are precursors of ketone bodies in the liver. Therefore the factors regulating the mobilization of FFAs from adipose tissue are also important in controlling ketogenesis. These include CPT-I, malonyl CoA, and acetyl CoA carboxylase. The partitioning of acetyl CoA between the TCA cycle and ketogenesis pathway is highly regulated. A fall in the concentration of oxaloacetate diverts acetyl CoA to ketogenesis, and an increased NADH/NAD ratio moves the equilibrium toward malate formation.

29.13.1 Link to Pathology: Diabetic Ketoacidosis (DKA)

Diabetic ketoacidosis occurs more commonly in uncontrolled type-1 diabetics, due to low insulin, high glucagon and an increase in fatty acid oxidation. In this low insulin, high glucagon state, the rates of fat mobilization, fatty acid oxidation, gluconeogenesis, and ketone biosynthesis increase. Due to excess

IV

fatty acid oxidation, the acetyl CoA accumulates and is diverted to ketone body biosynthesis. The fatty acid oxidation also generates NADH, thus increasing the NADH/NAD ratio. The high NADH diverts OAA towards malate formation in the TCA cycle. The malate is transported out from the matrix of the mitochondria into the cytoplasm for gluconeogenesis. As the OAA concentrations decrease, the TCA cycle cannot take up large quantities of acetyl CoA. Therefore, the excess acetyl CoA is used in the liver to synthesize ketone bodies. If the ketone body production exceeds the ability of the peripheral tissues to oxidize them, acetoacetic acid and β-hydroxybutyrate ($pK_a \sim 3.5$) release H^+ ions, which lower the blood pH and causes metabolic acidosis.

Low blood pH can affect protein structure and function, and can impair the ability of hemoglobin to bind O_2. Kidneys excrete these excess acids by balancing with equivalent amounts of cations, such as sodium and potassium, from cellular stores. Hence excess ketosis may deplete the body's potassium stores. Therefore, in the treatment of diabetic ketoacidosis it is advised to include potassium along with insulin.

Early symptoms of DKA include extreme thirst, constant urination, extreme weight loss, lethargy, fruity smell to breath, agitation, irritation, aggression, confusion, and muscle wasting. The ketone bodies excreted in the urine, known as ketonuria, can also be present alongside a characteristic fruity smell in the breath that is due to the blowing off excessive acetone. This characteristic smell is described as "over-ripe apples". Late symptoms of DKA include loss of appetite, extreme weakness, lethargy, apathy, vomiting, abdominal pain, flu-like symptoms, confusion, unconsciousness, and coma.

Review Questions

1. Which of the following subcellular locations is the site for the beta-oxidation of long-, medium-, and short-chain fatty acids?
   A) Cytoplasm
   B) Endoplasmic reticulum
   C) Matrix of the mitochondria
   D) Mitochondrial intermembrane space
   E) Peroxisomes

2. The oxidation of which of the following fatty acids would be impaired by a deficiency of carnitine?
   A) Long chain fatty acids
   B) Medium chain fatty acids
   C) Short chain fatty acids
   D) Very long chain fatty acids

3. In which of the following diseases is the oxidation of very long chain fatty acids impaired?
   A) Adrenoleukodystrophy
   B) Dicarboxylic aciduria
   C) Fatty liver disease
   D) Jamaican Vomiting Sickness
   E) Refsum disease

4. Which of the following diseases/conditions is the most likely cause of ketotic hypoglycemia?
   A) Acute alcohol intoxication
   B) CPT-I deficiency
   C) Jamaican Vomiting Sickness
   D) MCAD deficiency
   E) Untreated type-1 diabetes

5. Which of the following diseases is NOT a peroxisomal disorder?
   A) Adrenoleukodystrophy
   B) Carnitin deficiency
   C) Refsum disease
   D) Zellweger syndrome

Answers

1. **The correct answer is C.** The mitochondrial matrix is the location of β-oxidation for short-chain (C_4-C_6), medium-chain (C_8-C_{10}), and for long-chain fatty acids (C_{12}-C_{18}). Answer choice E is incorrect because only very long-chain fatty acids ($>C_{18}$) and branched-chain fatty acids are oxidized in the peroxisomes. Answer choice B is incorrect because ω-oxidation, not theβ-oxidation, takes place in ER microsomes.

2. **The correct answer is A.** Long-chain fatty acids (C_{12}-C_{18}) require the carrier carnitine in order to cross the inner mitochondrial membrane. Short-chain (C_4-C_6) (answer choice C) and medium-chain (C_8-C_{10}) (answer choice B) fatty acids are taken up by a monocarboxylate transporter and immediately activated to acyl CoA in the mitochondrial matrix. Very long-chain fatty acids ($>C_{18}$) (answer choice D) are metabolized in the peroxisomes. Carnitine deficiency can be either dietary or due to deficiency of enzymes and/or coenzymes in the synthesis pathway of carnitine. Patients with carnitine deficiency present with hypoketotic hypoglycemia, skeletal myopathies, cardiomyopathies,

encephalopathy, and hepatomegaly. Patients' urine may be dark red, due to myoglobinuria.

3. **The correct answer is A.** Due to genetic mutations in the ATP-binding cassette, subfamily D1 gene (*ABCD1*), the entry of very long chain fatty acids into the peroxisome is impaired in individuals with adrenoleukodystrophy. Therefore the patients with this disease cannot oxidize very long chain fatty acids.

Dicarboxylic aciduria (the answer choice B) can occur as a result of impaired β-oxidation from deficiencies in CPT or MCAD enzymes, or defects in peroxisome function. The impaired β-oxidation can lead to an excessive ω-oxidation and buildup of dicarboxylic acids in the blood, a condition called dicarboxylic aciduria.

Answer choice C is incorrect because fatty liver disease occurs due to alterations in mitochondrial β-oxidation, in which the short-, medium-, and long-chain fatty acids are oxidized. Jamaican Vomiting Sickness is caused by consumption of the toxin hypoglycin, which inhibits β-oxidation of short and medium chain acyl-CoA dehydrogenase. Thus, answer choice D is incorrect.

Refsum disease is caused by a mutation in phytanoyl-CoA hydroxylase, an important enzyme in α-oxidation. As the methyl branched group prevents β-oxidation, the oxidation of these long branched chain fatty acids is impaired, and therefore answer choice E is incorrect.

4. **The correct answer is A.** Acute alcohol intoxication causes elevated ketones in the blood and hypoglycemia. Impaired gluconeogenesis, due to high NADH production from ethanol metabolism, results in the hypoglycemia. Although an increased NADH/NAD ratio downregulates the rates of fatty acid oxidation and ketogenesis in the liver, alcoholic ketoacidosis can occur. The most common etiology of alcoholic ketoacidosis is the intake of ethanol during fasting or starvation. The high glucagon/insulin ratio releases free fatty acids from the adipocytes, and these fatty acids are oxidized in the liver. The excess acetyl CoA from the fatty acid oxidation leads to increased ketogenesis. Answer choices B, C, and D cause nonketotic hypoglycemia, because individuals with either CPT-I deficiency or MCAD deficiency and individuals who are poisoned with Jamiacan ackee tree fruit (hypoglycin), cannot oxidize the fatty acids, therefore fatty acid oxidation and ketogenesis are inhibited in these diseases/conditions. These patients may also have hypoglycemia during fasting. Answer choice E is incorrect because untreated type 1 diabetes causes ketotic hyperglycemia due to insufficient insulin.

5. **The correct answer is B.** Either primary or secondary carnitine deficiencies prevent β-oxidation from occurring, therefore it affects the mitochondria, not the peroxisomes. Adrenoleukodystrophy is due to genetic mutations in the ATP-binding cassette, which prevents the entry of very long chain fatty acids into the peroxisome. Refsum disease is caused by a mutation in phytanoyl-CoA hydroxylase, an important enzyme in α-oxidation. Genetic mutations in the *PEX* genes result in lack of peroxisomal biosynthesis leading to Zellweger syndrome.

IV

30 Acid Base Chemistry and Acid Base Disorders

At the conclusion of this chapter, students should be able to:
- Define and explain pH, pK_a, buffers and the Henderson-Hasselbalch Equation
- Apply the Henderson-Hasselbalch Equation to determine the charge of a molecule at a given pH
- Explain the bicarbonate buffer system and the respiratory and metabolic components of this buffer system
- Explain how respiratory and metabolic changes affect the bicarbonate buffer system
- Differentiate among the acid-base disorders and the fundamental relationships between pH, $[H^+]$, $[HCO_3^-]$, and P_{CO2}
- Differentiate between primary respiratory and metabolic acidosis and alkalosis
- Explain the compensatory mechanisms of acid base disorders
- Explain the plasma anion gap
- Differentiate between anion gap metabolic acidosis and non-anion gap metabolic acidosis

30.1 Acids, Bases, and Buffers

In this chapter we will review the basic ideas behind acids, bases, and buffers and how biological fluids maintain a normal pH value. We will review these basic concepts and apply them to clinical medicine. We will also learn to differentiate among acid base disorders by reviewing the fundamental relationships

between arterial pH, the partial pressure of carbon dioxide (P_{CO2}) and bicarbonate (HCO_3^-) levels.

30.2 Understanding the pH Scale

The hydrogen ion concentration in any given solution determines the pH. The pH is thus a convenient way of representing small concentrations of hydrogen ions.

Pure water dissociates into H^+ and OH^-, and although the extent of dissociation is not appreciable, the constant ion product of water is $Keq = [H^+][OH^-] = 10^{-14} M^2$. Since pure water has equal concentrations of H^+ and OH^-, then $[H^+] = [OH^-] = 10^{-7} M$. Thus, pure water contains H^+ concentrations of $10^{-7} M$. To write this H^+ concentration in a convenient way, we can write it in the $-\log_{10}$. The $-\log_{10}$ of 10^{-7} is simply 7. Therefore the pH of pure water is 7, which is also called neutral pH.

30.3 Definition of Acids and Bases

In general, acids release hydrogen ions (H^+), and bases combine with hydrogen ions (H^+). Acidic solutions produce more hydrogen ions than what is produced with the ionization of water, and the alkaline (basic) solutions produce fewer hydrogen ions than the ionization of water. There is a reciprocal relationship between pH and H^+ concentrations. The lower the pH, the higher the H^+ concentration. Therefore, solutions with higher H^+ concentration than pure water are said to be acidic. Solutions with lower H^+ concentration than pure water are called alkaline (▶ Fig. 30.1).

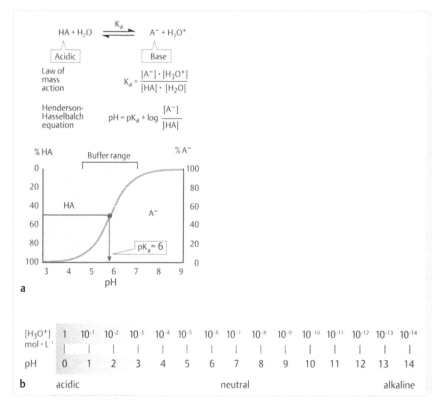

Fig. 30.1 Definition of acids and bases. **(a)** In aqueous solution, the acid HA donates a proton to the solvent. Bases accept H^+ ions from water molecules. **(b)** Pure water contains H^+ concentrations of $10^{-7} M$. Therefore, the pH of the pure water is 7, which is also called neutral pH. Aqueous solutions with a higher H^+ concentrations than water are acidic, and those with a lower H^+ concentrations are alkaline or basic. (Source: Koolman J, Röhm K, ed. Color Atlas of Biochemistry. 3rd Edition. New York, NY. Thieme; 2012.)

30.3.1 Link to Physiology: Hydrogen Ion Concentrations in the Body

The normal value of blood pH is kept constant within a range of 7.35 – 7.45. A blood pH of 7.4 corresponds to H^+ concentrations of ~ 40 nM (4×10^{-8} M). The pH value of the cytoplasm of a cell is between 7.0 – 7.3, while lysosomes have acidic pH (4.5 – 5.5), which is equates to several hundred times more H^+ than in the cytoplasm. Stomach acid has a pH value of ~2, while the environment of intestines has the pH value of ~8. The pH changes between the stomach and the intestine is due to excretion of HCO_3^- from the pancreas into the small intestine. The kidney also plays an important role in the acid-base balance of the body by controlling the absorption and excretion of H^+ and HCO_3^-, thus, the urine has wide range of pH (4.8–7.5) value (▶ Fig. 30.2).

30.3.2 Buffers Maintain the pH of the Biological Fluids

Buffers maintain the pH value of solutions by creating a perfect system of equilibrium even in the event of excess acids or bases. Buffers are made up of a weak acid and its conjugate base. In equilibrium, the buffers dissociate partially, i.e. contribute to or take away a few hydrogen ions (H^+). For example: $HA + H_2O \leftrightarrow H_3O^+ + A^-$

In this equation, HA is a weak acid and is able to dissociate into its conjugate base, A^- and H^+. The dissociation constant of this weak acid (K_a) can be written as in ▶ Fig. 30.1.

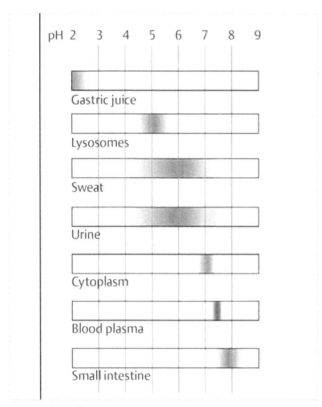

$$\text{pH} \quad 2 \quad 3 \quad 4 \quad 5 \quad 6 \quad 7 \quad 8 \quad 9$$

Gastric juice

Lysosomes

Sweat

Urine

Cytoplasm

Blood plasma

Small intestine

Fig. 30.2 pH values in the human body. (Source: Koolman J, Röhm K, ed. Color Atlas of Biochemistry. 3rd Edition. New York, NY. Thieme; 2012.)

$$K_a = \frac{[A^-]\,[H_3O^+]}{[HA]\,[H_2O]}$$

From this equation, the dissociation constant (K_a) can be described as a tendency of a weak acid to donate H^+. A higher K_a value shows a tendency to donate H^+ and indicates a stronger acid. As we did for the pH, if we convert the K_a value to – $\log_{10}$ of K_a, now it is called pK_a. As for the pH, strong acids have low pK_a values and weak acids have high pK_a values.

The **Henderson-Hasselbalch Equation** is a convenient way of viewing the relationship between the pH of a solution and the relative concentrations of conjugate acid and base of a buffer.

$$\text{pH} = \text{pK}_a + \log\frac{[A^-]}{[HA]}$$

Recall that the $[A^-]$ is the conjugate base and the [HA] is the weak acid. Therefore, the equation can be viewed as:

$$\text{pH} = \text{pKa} + \log\frac{[H + \text{acceptor or conjugate base}]}{[H + \text{donor or conjugate acid}]}$$

The Henderson-Hasselbalch equation is useful for determining the charge of a small molecule (commonly a drug) in a given pH value. In this type of problem, it is practical to remember that when the pH of the solution is less than the pKa value of an ionizable group (HA), then the group will mostly be in HA (protonated) form in the solution.

When pH < pKa, the acid (HA) is dominant or the ionizable group is protonated.

Contrary to this; when the pH is higher than the pKa value, then the group (HA) will be mostly in the deprotonated (A^-) form.

When pH > pKa, the base (A^-) is dominant or the ionizable group is deprotonated.

When the pH of the solution is equal to the pKa value, then the protonated (HA) and deprotonated (A^-) forms will be equal in a 1:1 ratio.

30.4 Plasma Buffer Systems

Buffers are important in human physiology as they allow the human body to maintain normal pH levels (▶ Fig. 30.3). Perhaps the most important human buffer system is the carbon dioxide-bicarbonate buffer system, which demonstrates the overall equilibrium between carbon dioxide and bicarbonate. $CO_2 + H2O \leftrightarrow H_2CO_3 \leftrightarrow H^+ + HCO_3^-$

Carbonic anhydrase catalyzes the reaction between CO_2 and H_2O, forming carbonic acid (H_2CO_3). The same enzyme catalyzes the conversion of carbonic acid to H^+ and HCO_3^-.

30.5 Metabolic and Respiratory Components of the Acid-Base Balance

The main players of acid-base disorders are carbon dioxide (CO_2), measured as the partial pressure of carbon dioxide (P_{CO2}), and the bicarbonate (HCO_3^-) levels. The P_{CO2} is regulated by respiration and by the levels of bicarbonate (HCO_3^-). The HCO_3^- level is regulated by the kidneys and by metabolism (▶ Fig. 30.4).

IV

IV

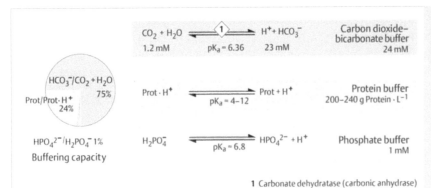

Fig. 30.3 Buffer systems in the plasma. (Source: Koolman J, Röhm K, ed. Color Atlas of Biochemistry. 3rd Edition. New York, NY. Thieme; 2012.)

1 Carbonate dehydratase (carbonic anhydrase)

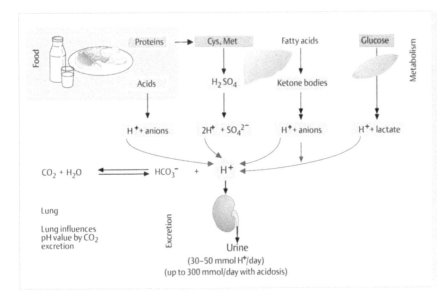

Fig. 30.4 Regulation of body pH. Human body continuously produces large quantities of acids, such as carbonic acid, sulfuric acid, acetoacetic acid, beta-hydroxybutyric acid, and lactic acid. Carbonic acid dissociates to CO_2 and breathed out by the lungs. Organic acids, lactic acid and ketones, are normally oxidized to CO_2 in the metabolism. Kidneys excrete H^+ ions into urine. Excretion of H^+ ions from the kidneys is important in the regulation of body pH. (Source: Koolman J, Röhm K, ed. Color Atlas of Biochemistry. 3rd Edition. New York, NY. Thieme; 2012.)

In order to determine the respiratory and metabolic components of pH changes, the **Henderson-Hasselbalch** equation can be reduced to the relationship between the pH, the concentration of bicarbonate [HCO_3^-], and the partial pressure of carbon dioxide (P_{CO2}). In the equation, the conjugate base (A^-) is replaced with bicarbonate (HCO_3^-) and the acid (HA) is replaced with the partial pressure of carbon dioxide (P_{CO2}). This equation now shows the relationship between blood pH and respiratory and the metabolic components of this buffer system.

$$pH = \frac{[HCO_3-]}{PCO_2}$$

30.6 Primary Acid-Base Disorders

The normal arterial pH range is 7.35 to 7.45. When the blood pH is lower than 7.35, it is determined to be acidotic. When the blood pH is higher than 7.45, it is determined to be alkalotic. Besides pH, measurements of P_{CO2} and bicarbonate levels are also necessary to determine the cause of a patient's acidosis or alkalosis.

If a change in the blood pH is due to a change in P_{CO2}, then the acid-base disorder is either **respiratory acidosis** or **respiratory alkalosis.** In healthy individuals, the normal value of P_{CO2} is 36 to 44 mmHg. **The pH is inversely proportional with the** P_{CO2}; when there is a decrease in P_{CO2} (< 36 mmHg), there is an increase in the pH; therefore, the patient is in a state of respiratory alkalosis. When there is an increase in P_{CO2} (> 44 mmHg), there is a decrease in the pH. Therefore, the patient is said to be in a state of respiratory acidosis.

If a change of the blood pH is due to a change in HCO_3^-, then the acid-base disorder is either **metabolic acidosis** or **metabolic alkalosis.** In healthy individuals, the normal value of arterial bicarbonate level (HCO_3^-) is 22 to 26 mmol/L. **The pH is directly proportional with the [HCO_3^-].** When there is an increase in HCO_3^- (> 26 mmol/L), there is an increase in the pH and the patient is in a state of metabolic alkalosis. When there is a decrease in HCO_3^- (< 22 mmol/L), there is a decrease in the pH and the patient is in a state of metabolic acidosis.

30.7 Compensatory Mechanisms of the Acid-Base Disorders

Compensatory mechanisms exist to prevent drastic changes in the acid-base status. The following equation of the carbon dioxide-bicarbonate buffer system is useful to understand the biochemistry and pathophysiology of the compensatory mechanisms of the acid-base disorders.

$$CO_2 + H2O \leftrightarrow H_2CO_3 \leftrightarrow H^+ + HCO_3^-$$

The basic foundations of the compensatory mechanisms of acid-base disorders can be explained by **Le Chatelier's Principle**, which states that "if a chemical system at equilibrium experiences a change in concentration, temperature, or pressure, the equilibrium will shift in order to minimize that change". For example, in respiratory acidosis there is an increase in P_{CO2}. In order to reduce the amount of P_{CO2}, the equilibrium of the carbon dioxide-bicarbonate buffer system will shift to the right. This results in an increase in the concentrations of H^+ and HCO_3^-. Therefore, if the P_{CO2} increases, there will be an increase in the H^+, which lowers the value of the pH, thus causing a respiratory acidosis. Bicarbonate will be increased through its reabsorption by the kidneys as a compensatory mechanism. The kidneys also decrease HCO_3^- excretion to sustain the HCO_3^-/CO_2 ratio. In this case, the kidneys increase the net acid excretion in the form of NH_4^+ (reviewed in chapter 33).

It is important to note that the body cannot fully compensate for acid-base disorders. Even in chronic acid-base disorders, the pH approaches normal values, but it never completely returns to normal. Because of this, pH changes are useful in identifying primary acid-base disorders and their associated compensatory responses. One caveat exists, however. If metabolic and respiratory abnormalities are both present, they can produce a normal-appearing pH level. It is important to note that this is not due to a compensatory mechanism, but rather a mixed acid-base disorder.

30.8 Respiratory Acidosis

Respiratory acidosis occurs when the lungs cannot effectively remove CO_2 from the body. The most common causes are diseases of the lungs and airways such as COPD and asthma, chest scoliosis, diseases of nerves and muscles, and drugs that affect breathing, such as narcotics. There are two types of respiratory acidosis: acute respiratory acidosis and chronic respiratory acidosis. **Acute respiratory acidosis**, in most cases, is a medical emergency. Conditions that lead to an acute respiratory acidosis include acute lung disease, pneumonia or pulmonary edema, central nervous system depression from opioids, narcotics, or sedatives, acute airway obstruction, weakening of respiratory muscles, flail chest, and neuromuscular disorders. In contrast, **chronic respiratory acidosis** is often a stable condition. Conditions leading to chronic respiratory acidosis include obstructive or restrictive chronic lung disease, chronic neuromuscular disorders, and chronic central hypoventilation.

30.9 Respiratory Alkalosis

Respiratory alkalosis is due to any factor that increases the patient's respiratory drive. Respiratory alkalosis is also referred to as primary hypocapnia, or hyperventilation. Patients present with tachypnea (rapid breathing), which may be due to stimulation in the central nervous system, such as fever from an infection, subarachnoid hemorrhage, or pain. Tachypnea can also be secondary to hypoxemia, which can be due to high altitude, pneumonia, pulmonary edema, or pulmonary embolism. Aspirin toxicity also presents with a respiratory alkalosis, although aspirin toxicity in the early stages presents with metabolic acidosis.

30.10 Metabolic Acidosis

Metabolic acidosis occurs either by excess production or underexcretion of acids. This increased acid results in a reduction in bicarbonate levels. The reduction of bicarbonate may be due to an increased extracellular buffering of the acid or due to loss of bicarbonate ions in the urine. Therefore, a low bicarbonate level is an indicator of metabolic acidosis.

Most often, the cause of metabolic acidosis is diabetic ketoacidosis, which is due to the buildup ketones that occurs in uncontrolled type-1 diabetics. Severe diarrhea can also cause metabolic acidosis due to excessive loss of bicarbonate from the body. Accumulation of lactic acid also results in metabolic acidosis. Conditions leading to lactic acidosis include excess alcohol consumption, prolonged intense exercise, and poisoning by salicylates, ethylene glycol, or methanol. Lactic acidosis is also caused by hypoxic conditions such as heart failure or severe anemia.

Metabolic acidosis can be further broken down into two categories: **Anion Gap Metabolic Acidosis** and **Non-Anion Gap Metabolic Acidosis**. To understand these, we need to first examine the concept of anion gap.

30.11 Anion Gap

The plasma anion gap is the calculated difference between the sodium cations and the chloride and bicarbonate anions in the plasma.

$$Plasma\ Anion\ Gap = Na^+ - (Cl^- + HCO_3^-)$$

A normal anion gap ranges between 8–12 mEq/L. The anion gap represents the concentrations of all the unmeasured anions (acetoacetate, lactate, sulfate, phosphates, and other organic anions such as proteins) in the plasma. An increased anion gap (> 12 mEq/L) occurs when the concentrations of unmeasured anions increases, and the concentration of measured anions (chloride and bicarbonate) decreases. Bicarbonate, the measured anion, decreases because the hydrogen ions react with bicarbonate to form carbon dioxide, which is then excreted via the lungs.

Patients with an anion-gap metabolic acidosis will have an increased calculated anion gap (> 12 mEq/L). The specific causes of **anion-gap metabolic acidosis** are methanol or ethylene glycol poisoning, diabetic or alcoholic ketoacidosis, lactic acidosis, excess use of salicylate, and renal failure, which presents with uremia.

If the calculated anion gap is within the normal range, this metabolic acidosis is called the **non-anion gap metabolic acidosis.** The two main causes of non-anion gap metabolic acidosis are diarrhea and renal tubular acidosis. These two conditions can be distinguished based on the patient's history alone.

30.11.1 Link to Physiology

Plasma albumin levels affect the anion gap. Albumin is the major unmeasured anion and contributes nearly the entire value of the anion gap. In every one-gram decrease in albumin, the anion gap decreases ~2.5 to 3 mmoles. Thus, if a patient has hypoalbunemia, it may appear as a non-anion gap acidosis, even though the patient actually has high anion gap acidosis. This is particularly relevant in intensive care unit (ICU) patients, where lower albumin levels are common. A lactic acidosis in a hypoalbuminemic ICU patient will commonly be associated with a normal anion gap.

30.11.2 Link to Physiology

Winters Formula is used to determine the patients expected P_{CO2}. The result can be interpreted to determine whether the body is using a respiratory compensatory mechanism or if there is another existing primary respiratory acidosis or respiratory alkalosis present. The expected P_{CO2} can be calculated by: $P_{CO2} = (1.5 \times HCO_3^-) + 8 \pm 2$

Once this expected P_{CO2} is calculated, it is then compared to the patient's actual P_{CO2} recording. If the expected P_{CO2} and actual P_{CO2} values correspond, then there is an appropriate respiratory compensation. If the expected P_{CO2} is greater than the actual P_{CO2}, then there is a primary respiratory acidosis. If the expected P_{CO2} is less than the actual P_{CO2}, then there is a primary respiratory alkalosis present.

30.12 Metabolic Alkalosis

Metabolic alkalosis occurs due to either loss of H^+ from the body or a gain in HCO_3^-. Therefore, in metabolic alkalosis, serum HCO_3^- levels increase (generally greater than 35 mEq/L).

The most common causes of metabolic alkalosis are loss of gastric acid secretions due to excessive vomiting, which results in loss of HCl and H^+, and the use of diuretics (e.g. thiazides), which increases K^+ and H^+ secretion from kidneys. These conditions also cause low urinary chloride levels, generally less than 20 mEq/L. If the patient's clinical history and physical do not indicate vomiting or past diuretic use, then the urinary chloride levels can be used to evaluate the cause of metabolic alkalosis. Conditions that cause normal or high urinary chloride levels (>20 mEq/L) include any type of excess mineralocorticoid activity such as Cushing's syndrome, Conn's syndrome, exogenous steroids, increased renin states, and Bartter's syndrome.

30.12.1 Link to Pathology

Differentiation between primary metabolic alkalosis and respiratory acidosis with metabolic compensation. As reviewed in the above example, an elevated P_{CO2} level in the setting of a primary metabolic alkalosis indicates a compensatory response of the lungs that relies on the retention of carbon dioxide. In order to achieve this increase in P_{CO2}, metabolic alkalosis causes hypoventilation, which consequently corrects the pH. However, an increased P_{CO2} greater than 55 mmHg suggests that the patient may have primary respiratory acidosis. Therefore, the P_{CO2} level, and its degree of elevation, should be considered carefully.

Review Questions

1. Arterial blood gas analysis shows:

pH: 7.32 (N= 7.35–7.45)
   P_{CO_2}: 39 mmHg (N= 36–44 mmHg)
   HCO_3^-: 19 mEq/L (N= 22–26 mEq/L)
   Which of the following is the most likely diagnosis?
a) A. Metabolic acidosis with no respiratory compensation
b) B. Metabolic acidosis with respiratory compensation
c) C. Metabolic acidosis with renal compensation
d) D. Metabolic acidosis with no renal compensation

2. Arterial blood gas analysis shows:

pH: 7.27 (N= 7.35–7.45)
   P_{CO_2}: 51 mmHg (N= 36–44 mmHg)
   HCO_3^-: 23 mEq/L (N= 22–26 mEq/L)
   Which of the following is the most likely diagnosis?
a) Metabolic acidosis
b) Metabolic alkalosis
c) Respiratory acidosis
d) Respiratory alkalosis

3. Arterial blood gas analysis shows:

pH: 7.32 (N= 7.35–7.45)
   P_{CO_2}: 53 mmHg (N= 36–44 mmHg)
   HCO_3^-: 29 mEq/L (N= 22–26 mEq/L)
   Which of the following is the most likely diagnosis?
a) Acute Metabolic acidosis
b) Acute Respiratory acidosis
c) Chronic Metabolic acidosis
d) Chronic Respiratory acidosis

4. Arterial blood gas analysis shows:

pH: 7.49 (N= 7.35–7.45)
   P_{CO_2}: 33 mmHg (N= 36–44 mmHg)
   HCO_3^-: 19 mEq/L (N= 22–26 mEq/L)
   Which of the following is the most likely diagnosis?
a) Metabolic alkalosis with respiratory compensation
b) Metabolic alkalosis with no compensation
c) Respiratory acidosis with no compensation
d) Respiratory alkalosis with renal compensation

5. Arterial blood gas analysis shows:

pH: 7.32 (N= 7.35–7.45)
   P_{CO_2}: 39 mmHg (N= 36–44 mmHg)
   HCO_3^-: 19 mEq/L (N= 22–26 mEq/L)
   Which of the following is the most likely diagnosis?
a) Metabolic acidosis with respiratory compensation
b) Metabolic acidosis with no compensation
c) Respiratory acidosis with no compensation
d) Respiratory alkalosis with renal compensation

Answers

1. **The correct answer is A.** The patient's blood is acidemic with a pH < 7.35. The bicarbonate is decreased, but the P_{CO_2} level is within normal range. Therefore, this patient has an acute primary metabolic acidosis with no respiratory

compensation. There is no respiratory compensatory mechanism occurring because the P_{CO_2} level is within normal range.

If respiratory compensation is occurring, the body responds by increasing the ventilation to help blow off CO_2 and allow the reaction to shift to the left, which helps consume hydrogen ions and essentially raises the pH level towards normal. In this example, the P_{CO_2} is within normal range. Therefore, there is no respiratory compensation occurring. This means that answer choices B, C, and D are incorrect.

2. **The correct answer is C.** The low pH indicates that the patient is in an acidotic state. Elevated P_{CO_2} and a slightly elevated bicarbonate level suggest that this patient is in respiratory acidosis with *partial* metabolic compensation. The elevation in P_{CO_2} results from a decrease in ventilation, which shifts the reaction of the bicarbonate buffer system to the right as the body is trying to consume CO_2 and form more H^+. This right shift of the reaction slightly elevates bicarbonate. The increase in hydrogen ions leads to the decrease in pH levels. In this example, the patient most likely has **acute respiratory acidosis** since the bicarbonate is only slightly elevated, meaning there is no metabolic compensation occurring. Metabolic compensation is achieved by the kidneys and it is a slower process than respiratory compensation by the lungs. The body responds to the decrease in pH by trying to increase the plasma HCO_3^- concentration, which is achieved by reabsorption of the HCO_3^- from the kidneys.

3. **The correct answer is D**. This patient most likely has chronic respiratory acidosis because the pH is just slightly low, but is still acidotic. Both P_{CO_2} and HCO_3^- levels are elevated. The elevated P_{CO_2} is most likely responsible for the acidosis because the elevated HCO_3^- does not cause an acidosis. However, the elevated HCO_3^- with low pH indicates that this is most likely metabolic compensation. Under chronic conditions, it will take 2–3 days for the body to reach maximum renal compensation. If the pH level were read as normal, 7.35 – 7.45, then we would have to consider a mixed acid-base disorder.

4. **The correct answer is D**. The high pH indicates that the patient is in an alkalotic state. The primary disorder is respiratory alkalosis as there is a decrease in P_{CO_2}. The bicarbonate is also decreased, which is most likely a result of renal compensation. Since a decrease in bicarbonate cannot result in an alkalotic state, the primary disorder is most likely due to respiration.

The decrease in P_{CO_2} shifts the reaction of the bicarbonate buffer system to the left, consuming hydrogen ions in the process and resulting in an increase in pH level.
   $$\downarrow CO_2 + H_2O \leftarrow H_2CO_3 \leftarrow \downarrow H^+ + HCO_3^-$$
The compensatory response for the respiratory alkalosis is to increase the renal excretion of bicarbonate. This increased excretion of bicarbonate shifts the reaction back to the right, which will assist in the formation of more hydrogen ions and bring the pH down towards normal.

5. **The correct answer is B.** The patient's blood is acidemic with a pH < 7.35. The bicarbonate is decreased, but the P_{CO_2}

level is within normal range. Therefore, this patient has an acute primary metabolic acidosis with no respiratory compensation.

If respiratory compensation is occurring, the body responds by increasing the ventilation to help blow off CO_2 and shift the re-action of the bicarbonate buffer system to the left. This mechanism helps consume hydrogen ions and essentially raise the pH level towards normal.

31 Glucose Homeostasis and Maintenance of Blood Glucose Concentrations

At the conclusion of this chapter, students should be able to:
- Explain the three metabolic fuels and their roles in ATP production
- Outline the action of insulin in the regulation of the metabolic pathways that are active during the fed state
- Outline the action of glucagon and epinephrine in the regulation of metabolic pathways that are active during fasting and exercise
- Define how blood glucose levels are regulated after a meal, fasting state, and during prolonged fasting
- Distinguish key differences between type1 and type2 diabetes mellitus

31.1 Metabolic Fuels

Carbohydrates, fats and proteins are the three primary sources of energy to fuel cellular processes. Although their metabolism is converged at a central intermediate, acetyl CoA, their roles in the production of ATP are entirely different.

Carbohydrate serves as the principal energy source in the fed state. Carbohydrates can be readily broken down into glucose, which can be either used immediately by any cell as a fuel source or can be stored as glycogen in the liver or muscles. Complete oxidation of glucose produces ATP through glycolysis, pyruvate oxidation, the TCA cycle, and oxidative phosphorylation. Although glycogen is a rapidly accessible source of glucose during fasting, both the liver and muscles have limited capacity for glycogen storage, with a maximum of 2000 calories worth of energy. Such limited energy storage is compensated by fat, which has a much greater capacity for energy storage. This is because fat has a higher efficiency for storing energy per unit of body weight than does glycogen, and contains almost two-times more energy compared to carbohydrates and proteins (9 calories/gram versus 4 calories/gram). Fat can also be stored without water, which is not true for the body's other sources of energy.

Although proteins are primarily used for maintaining the structures and functions of cells, muscle protein also provides about 5% of the utilized energy under normal circumstances. This utilization of protein as a fuel source increases during prolonged exercise, starvation, or with inadequate intake of carbohydrates. Unlike fatty acids, amino acids can be used for precursors of gluconeogenesis to maintain blood glucose concentrations.

31.2 Glucose Homeostasis

Glucose is the most efficient energy source in the body and, therefore, its levels in the blood must be tightly controlled so that it does not rise too high after a meal or fall too low between meals. A high concentration of glucose in the blood results in tissue dehydration due to the osmotic effect of glucose. For example, severe hyperglycemia can cause hyperosmolar coma because of dehydration of the brain. The opposite condition, hypoglycemia, can also cause coma in severe cases, but this is due to lack of ATP rather than dehydration. In healthy individuals, blood glucose levels are so well maintained that it does not drop dramatically even during prolonged fasting. Homeostasis of blood glucose is primarily achieved by concerted actions of insulin and its antagonistic hormone, glucagon. However, glucagon is not the only hormone that balances the effects of insulin. Other significant insulin counter-regulatory hormones including cortisol, epinephrine, and norepinephrine take action under the effects of neuronal signals during stress or hypoglycemia (▶ Fig. 31.1).

In the normal state, blood glucose levels fluctuate within a narrow range between the fed and the fasting states. The level of blood glucose reaches its peak (120–140 mg/dL) within thirty minutes to an hour after a high carbohydrate meal, and falls back to a fasting level (80–100 mg/dL) about two hours after the meal. Usually, blood glucose levels do not exceed 140 mg/dL as long as the tissues are taking up the glucose to oxidize or store it. The maintenance of a relatively constant level of glucose in circulation is achieved by two hormones: insulin and glucagon. These hormones regulate glucose absorption, cellular utilization, and synthesis.

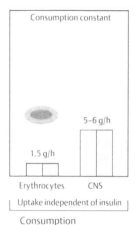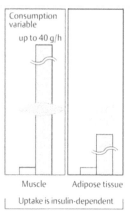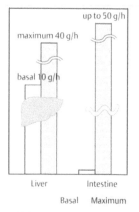

Fig. 31.1 Glucose turnover in organs. (Source: Koolman J, Röhm K, ed. Color Atlas of Biochemistry. 3rd Edition. New York, NY. Thieme; 2012.)

IV

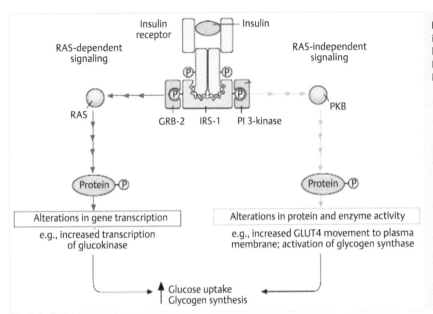

Fig. 31.2 RAS-dependent and -independent insulin signaling through receptor tyrosine kinase. (Source: Panini S, ed. Medical Biochemistry—An Illustrated Review. New York, NY. Thieme; 2013.)

The levels of these two hormones change correspondingly with the level of blood glucose.

In general, tissues absorb and utilize glucose when it is abundant in circulation and switch to other fuels when glucose supply diminishes. The circulating glucose is absorbed from a balanced diet during the fed or absorptive state, which occurs during the meal up until about four hours after a meal. During the fasting state, more than four hours after a meal, the liver glycogen stores are first mobilized to become the primary supplier of glucose in circulation, followed by gluconeogenesis in the liver.

The carbons for gluconeogenesis are provided by lactate, amino acids, and glycerol. Lactate comes to the liver via blood from exercising muscles or red blood cells, while amino acids are provided from degraded muscle proteins. Glycerol is released from triglycerides in the adipose tissue. As glycogen stores are depleted, the rate of gluconeogenesis increases. Within about 18–24 hours, all glycogen stores have been exhausted. At this point, gluconeogenesis takes over and becomes the primary source of circulating glucose.

After about three days of fasting, the rate of gluconeogenesis decreases slightly, while lipolysis and fatty acid oxidation increase. The fatty acids are either completely oxidized to CO_2 and H_2O or partially oxidized and converted to ketone bodies in the liver. With the availability of ketones, energy production in the brain switches from glucose to ketone bodies. This way, proteins are less likely to be degraded for gluconeogenesis, and thus preserved. Due to this slight decrease in protein degradation, urea production also decreases compared to the early stages of fasting.

Insulin affects the metabolism of all major fuels including carbohydrates, lipids and proteins. In general, insulin keeps blood glucose levels from rising too high by either promoting the use of glucose as an immediate energy source via glycolysis, or through its storage as an energy reserve in the form of glycogen or fat.

Insulin stimulates the transport of glucose, triglycerides and amino acids from the circulation to peripheral tissues and increases the rate of glycolysis in the liver, muscles and adipose tissues. As an anabolic hormone, insulin stimulates glycogen synthesis in the liver and muscles. When the liver reaches its limited capacity for glycogen storage, which is about 5% of liver mass, the excess glucose is shunted to the fatty acid synthesis pathway. Therefore, insulin stimulates the synthesis of fatty acids, triglycerides, cholesterol, and VLDL. Synthesis of protein in the liver and extrahepatic tissues is stimulated by insulin.

In contrast, insulin decreases the rate of glycogenolysis and fatty acid oxidation in the liver and muscles, gluconeogenesis in the liver, and lipolysis in adipose tissues.

Insulin induces these metabolic changes by binding to its receptor, a receptor tyrosine kinase, on the cell surface. This binding triggers a series of pathways within the cell, ultimately leading to the appropriate cellular response (▶ Fig. 31.2). For example, in muscle tissue and adipocytes, activation of protein kinase B (PKB) or Akt stimulates translocation of the glucose transporter GLUT4 to the cell membrane.

31.3 Action of Insulin and the Regulation of Blood Glucose Levels During the Fed State

The secretion of insulin is stimulated by increased levels of blood glucose after a high-carbohydrate meal. At the same time, secretion of glucagon is inhibited by insulin.

31.3.1 Link to Physiology

In addition to stimulating the uptake of glucose and amino acids, insulin also induces the transportation of electrolytes such as potassium, magnesium and phosphate ions into the cells. This explains why the injection of insulin can cause hypokalemia.

31.4 Action of Glucagon and Epinephrine in the Maintenance of Blood Glucose Levels During Fasting and Exercise

Glucagon is a peptide hormone made of 29 amino acids. Its secretion by α cells of the pancreas is stimulated directly by decreased insulin levels and indirectly by the fall of blood glucose levels a few hours after a meal. Therefore, a high protein, low carbohydrate diet and increased exercise can stimulate the secretion of glucagon. Glucagon then acts through a cAMP-mediated signal transduction pathway to phosphorylate downstream target enzymes (▶ Fig. 31.3).

In opposition to insulin, glucagon is primary a catabolic hormone. Glucagon prevents blood glucose levels from falling too low by promoting glycogenolysis and gluconeogenesis in the liver. About four hours after a meal, the liver becomes the primary blood glucose supplier through both glycogenolysis and gluconeogenesis. As the secretion of insulin decreases, the amount of glucagon secreted increases. The increased glucagon/insulin ratio stimulates the release of free fatty acids from adipose tissues by activating hormone-sensitive lipase. It also stimulates lipolysis and inhibits fatty acids synthesis in the liver and adipose tissue. These fatty acids serve as the primary fuel for muscles. Muscle glycogen is also used for ATP production for a limited time. While muscle can utilize both its glycogen and fatty acids for ATP generation, the brain continues to use glucose as the primary fuel. As the liver glycogen stores are depleted, gluconeogenesis takes over and becomes the primary source of circulating glucose.

In the liver, ATP from fatty acid oxidation drives gluconeogenesis. The precursors for gluconeogenesis are glycerol from fat mobilization, amino acids from muscle degradation, and lactate from red blood cells. Fatty acid oxidation generates acetyl CoA, NADH, and ATP. The acetyl CoA inhibits pyruvate dehydrogenase (PDH), but activates pyruvate carboxylase. The increased activity of pyruvate carboxylase produces more oxaloacetate for gluconeogenesis. The high level of NADH from fatty acid oxidation inhibits citrate synthesis and at the same time diverts oxaloacetate to gluconeogenesis.

During exercise, blood glucose levels are maintained by glycogenolysis and gluconeogenesis. Except for muscles, uptake of glucose decreases by the liver. The increased fatty acid oxidation provides needed ATP. During exercise, fatty acid oxidation is under the regulation of epinephrine and norepinephrine. Other hormones such as thyroxine, growth hormones, and cortisol act more indirectly by supporting the roles of epinephrine or norepinephrine, which stimulate the secretion of glucagon while inhibiting the secretion of insulin. Because of increased blood supply and increased AMP levels, glucose intake by muscles is much faster during exercise despite the decreased level of insulin. As a strong glycogenolysis stimulator, epinephrine works in concert with calcium, a secondary messenger released from the smooth endoplasmic reticulum during muscle contraction, for stimulating glycogen phosphorylase. Glycogen storage is depleted quickly in short, high-intensity exercise, whereas both glycogen and fatty acids are the primary fuel sources during long-lasting exercise.

During starvation, three to five days of fasting, or prolonged exercise, liver gluconeogenesis becomes the only glucose source to maintain blood glucose levels. At this time, the acetyl CoA from fatty acid oxidation is converted to ketone bodies in the liver. While the ketones can be used by many extrahepatic tissues, glucose is used solely by red blood cells and to a limited degree by the brain, which relies more heavily on ketone bodies.

31.4.1 Link to Pathology

Diabetes mellitus is a group of metabolic disorders characterized by hyperglycemia. Inefficient insulin or impaired insulin signaling activity prevents glucose from entering muscle and fat cells. Although glucose cannot be used by muscle and adipose tissue, the liver continues to produce glucose, contributing more to the blood and, thus, causing hyperglycemia. Once the blood glucose levels reach the threshold of glucose reabsorption in the kidney, glucose is excreted in the urine. As an osmotic diuretic, glucose causes the osmosis of large amounts

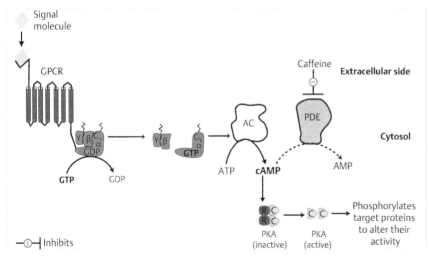

Fig. 31.3 Regulation of glycogenolysis. (Source: Panini S, ed. Medical Biochemistry—An Illustrated Review. New York, NY. Thieme; 2013.)

of water and electrolytes into the tubules, causing frequent urination in large quantities (polyuria), notably at night (nocturia) and subsequent polydipsia (excessive thirst). The diuretic effect decreases plasma volume, therefore further increasing blood glucose levels. Severe dehydration can cause cerebral dysfunction, such as the nonketotic hyperosmolar coma seen in Type II diabetic patients with extremely high levels of blood glucose. Due to the inability to take glucose into the cells, the oxidation of lipids for ATP generation is necessary.

Chronic hyperglycemia also produces pathologic effects through nonenzymatic glycation of proteins in the cell membranes, in the extracellular matrix and the serum. This process distorts protein structure, slows protein degradation, and leads to the accumulation of unwanted glycated proteins in various organs. Such glycation events contribute to long-term microvascular complications of diabetes including diabetic retinopathy, nephropathy, and neuropathy. In addition, pathological changes in the coronary artery, cerebral artery, peripheral artery, and atherosclerosis can also occur.

Fuel metabolism in uncontrolled diabetic patients is very similar to a prolonged fasting state, which explains one of the symptoms of diabetes, polyphagia (excessive hunger). As glucose cannot be used, fat is mobilized as the primary fuel source, resulting in hyperlipidemia. Fatty acid oxidation produces acetyl-CoA, which accumulates due to the lack of oxaloacetate which is diverted for gluconeogenesis. The accumulated acetyl-CoA favors ketogenesis, leading to ketoacidosis. Proteins are also degraded to allow for gluconeogenesis.

Type I diabetes is most often a result of autoimmune destruction of β-cells of the pancreas, but can be idiopathic in some cases. It is often seen in children and young adults, although diagnosis can occur at any age. About 5–10% of people with diabetes have type I, which is not related to lifestyle or diet. The destruction of β-cells starts long before the appearance of signs and symptoms. Hyperglycemia and ketoacidosis, the two classic manifestations, indicate late stages of the disease. The pathologic process is irreversible, and most patients rely on insulin to survive. Without insulin, the entry of glucose into muscles and adipose tissues is blocked, and the patient is in a constant biochemical starvation mode with high blood glucose. Glucose derived from food is not stored in the liver but remains in the blood and contributes to after-meal hyperglycemia.

Type II diabetes is much more common than type I and is often associated with older age, obesity, and a family history of diabetes. It is most common in individuals over 40 years of age who are overweight. The tendency to develop type II diabetes can be inherited, but it can also be prevented or delayed by weight loss and physical activity. Type II diabetes develops because of a loss of sensitivity to insulin, a condition known as insulin resistance. Insulin levels are higher than normal at early stages of the disease due to hyper-functional β-cells. Glucose

Table 31.1 The diagnostic criteria for pre-diabetes and diabetes

	Random Blood Glucose	Fasting Blood Glucose	OGTT
Euglycemic		< 100 mg/dL	Or, < 140 mg/dL
Pre-diabetes		> 100 but < 125 mg/dL	Or, > 140 but < 200 mg/dL
Diabetes	> 200 mg/dL, with signs and symptoms	Or, > 125 mg/dL on more than 1 occasion	Or, > 200 mg/dL after 2 hours

builds up in the blood as the β-cell compensation becomes inadequate. Weight loss and regular exercise help to restore insulin sensitivity. Together with oral medications, many people with type II diabetes can be well managed without supplemental insulin.

Gestational diabetes mellitus usually develops during pregnancy around the 24th week due to physiological insulin resistance during the second trimester. Gestational diabetes generally goes away soon after delivery. A majority of the cases of gestational diabetes can be managed by lifestyle modification alone. If this is not effective, insulin is the drug of choice to control the hyperglycemia, as insulin does not cross the placenta like other diabetic drugs. If not managed properly, gestational diabetes leads to an increased risk of fetal macrosomia (birth weight higher than 4000g), birth complications, and type II diabetes following the pregnancy.

31.4.2 Link to Pathology

Diagnosis of diabetes mellitus is based on blood glucose levels. Normally, blood glucose should be less than 100 mg/dL when fasting or less than 140 mg/dL following an oral glucose tolerance test (OGTT). The criteria for the diagnosis of diabetes and pre-diabetes are listed in ▸ Table 31.1. Pre-diabetes is known as impaired glucose tolerance and has a significant risk of progressing to diabetes.

Persistent hyperglycemia in diabetes causes the formation of glycosylated hemoglobin-Hb_{A1C} through the addition of glucose moieties to hemoglobin. Levels of Hb_{A1C} are commonly used to monitor glycemic control, with the recommendation that Hb_{A1C} be less than 7% in diabetic patients.

The measurement of the blood C-peptide levels is helpful for differentiation between type 1 and type 2 diabetes. In type1 diabetic patients, the pancreas cannot produce insulin. Thus, the C-peptide level is expected to be lower than normal. In contrast, type2 diabetics can produce insulin. However, the liver and the muscle cells develop insulin resistance. The C-peptide level is expected to be above normal or at the upper-limit range in these patients.

Review Questions

1. Which of the following is the source of blood glucose after overnight fasting?
   A) Dietary carbohydrates
   B) Fatty Acids
   C) Liver glycogen
   D) Muscle glycogen
   E) Muscle protein

2. Which of the following is the source of blood glucose after 4-days of fasting?
   A) Amino acids from muscle protein
   B) Glucose from carbohydrates
   C) Glycerol from dietary fat
   D) Lactate from erythrocytes
   E) Ketones from fat

3. The rate of which of the following pathways is most likely decreased in an untreated type 1 diabetic patient?
   A) Gluconeogenesis
   B) Glycogenolysis
   C) Glycolysis
   D) Ketogenesis
   E) Lipolysis

4. The rate of which of the following pathways is increased with an increased ratio of glucagon/insulin?
   A) Glycogenesis in the liver
   B) Ketogenesis in the liver
   C) Synthesis of triacylglycerols in the liver
   D) Transport of GLUT4 onto adipocytes membrane
   E) Uptake of glucose into the muscle cells

5. Elevated concentrations of which of the following molecules would increase the uptake of glucose in muscle cells during exercise?
   A) AMP
   B) ATP
   C) cAMP
   D) Epinephrine
   E) Glucagon

Answers

1. **The correct answer is C.** The liver glycogen stores are the first source for maintaining blood glucose levels during fasting. The liver glycogen lasts about 24–30 hours. Answer choice A is incorrect because absorption of glucose from dietary carbohydrates is completed 3–4 hours after a meal. Fatty acids are released from adipose tissue during fasting and are used to generate ATP via oxidation in the liver and muscle cells. The acetyl CoA from fatty acid oxidation cannot be converted to glucose, therefore answer choice B is incorrect. Muscle glycogen is used only in the muscle cells, as these cells lack the glucose 6-phosphatase enzyme. Thus, the answer choice D is incorrect. Muscle protein degradation provides glucogenic amino acids, which are the precursors for gluconeogenesis. Although it is a source for glucose production during fasting, it is not the primary source for maintaining blood glucose levels. Amino acids from muscle degradation become the primary source after 18–24 hours of fasting.

2. **The correct answer is A.** Blood glucose levels are maintained at a relatively narrow range. At 24–30 hours of fasting, the liver glycogen stores have been used up. As the liver glycogen stores are depleted, gluconeogenesis takes over to maintain blood glucose levels. The liver uses lactate, glycerol, and amino acids as precursors for gluconeogenesis. Although lactate and glycerol are precursors, neither are sufficient to maintain blood glucose concentrations after 2–3 days of fasting. At this point, muscle protein is degraded to provide amino acids for precursors of gluconeogenesis. Therefore, answer choice C and D are incorrect. Answer choice B is incorrect because glucose absorption is completed 3–4 hours after a meal. Answer choice E is incorrect because ketones are not the carbon source for glucose.

3. **The correct answer is C.** The rate-limiting step of glycolysis is phosphofructokinase-1 (PFK-1) activation via insulin signaling in the liver. Because type-1 diabetics lack the hormone insulin, the rate of glycolysis decreases in untreated patients. The rate of gluconeogenesis, ketogenesis, lipolysis, and glycogenolysis increase in the absence of insulin, thus answer choices A, B, D, and E are incorrect.

4. **The correct answer is B.** Increased glucagon increases the rate of glycogenolysis and gluconeogenesis in the liver. Lipolysis in the adipocytes also increases with an increased ratio of glucagon/insulin. This increased lipolysis provides free fatty acids, which are delivered to the liver and muscle for oxidation. As fatty acid oxidation increases, the precursor for ketogenesis, acetyl CoA, accumulates in the liver and leads to the production of ketones. The rate of fatty acid and triacylglycerol synthesis is increased by insulin, but not glucagon. Thus, answer choice C is incorrect. Insulin hormone also increases the rate of glucose uptake through increased numbers of GLUT4 transporters on the surface of muscle and adipose tissue. Therefore, answer choice D and E are incorrect.

5. **The correct answer is A.** Utilization of ATP in the muscle cells during exercise increases the ratio of AMP/ATP. This increased AMP activates AMP-dependent protein kinase (AMPK), which activates glycogenolysis in the muscle cells. AMPK also increases the transport of GLUT4 on the plasma membrane, which increases glucose uptake into the muscle. Both increased glucose uptake and glycogenolysis provides a continuous intracellular glucose supply to allow for muscle contraction. The uptake of glucose into the muscle is also stimulated by insulin signaling. Epinephrine binds to beta-adrenergic receptors on muscle cells and increases the cAMP levels. This increased cAMP activates PKA, which activates glycogenolysis. However, neither cAMP nor epinephrine increases GLUT4 levels. Hence, answer choices C and D are incorrect. Answer choice E is incorrect because glucagon does not affect muscle metabolism since muscle cells do not have glucagon receptors.

IV

Review Questions

1. A 14-year-old female presents with chronic abdominal pain, diarrhea and low-grade fever. Her mother reports to her doctor that she always suffered from digestive problems and diarrhea. Her medical history indicates that she has also had recurrent pulmonary infections. Physical exam reveals that her weight is in the 20th percentile and her height is the 10th percentile. Blood tests show that a metabolic panel, liver function tests, and complete blood count results are all normal. However, her serum amylase and lipase levels are slightly elevated. The results of genetic testing reveal that she is heterozygous for ?F508 and G551D mutations in theCFTR gene. Which of the followingdigestive enzymes is most likely decreased in her gastrointestinal tract?
   A) ?-glucosidase
   B) Amylase
   C) Lactase
   D) Maltase
   E) Sucrase

2. A 68-year-old female with a 25 year historyof type 2 diabetes complains of tingling, pain, and numbness inher feet. Physical examination reveals infections around her toenails. Her HbA1c level is 8.7, suggesting an average blood glucose level of 200 mg/dl. Her doctor prescribes a gliflozin. Which of the following is most likely inhibited by this drug?
   A) ATP-sensitive K$^+$ channel
   B) α-glucosidase
   C) Amylase
   D) Facilitative glucose transporter
   E) Sodium-glucose symporter

3. A 43-year-old male of Mediterranean descent complains of red-brown urine and yellowish color in his sclera. History reveals that he became a widower about 6-months ago, and since then he has been a heavy drinker. He mentions that about two days ago, he went to a party with his childhood friends and ate some food that contained fava beans. Blood tests reveal low hemoglobin levels. Which of following laboratory results most likely supports his diagnosis?
   A) Heinz bodies in his erythrocytes
   B) High blood alcohol
   C) High blood lactate
   D) High D-xylose in his urine
   E) Increased arterial pH

4. Genetic test results of a newborn reveal a mutation (A149P) in the ALDOB gene. The pediatrician counsels the mother that the newborn must not be fed with food containing table sugar, high-fructose corn syrup, fruits, honey or juice. She also advises the mother that if the baby exhibits intestinal discomforts such as nausea and vomiting, she should consult a nutritionist about food choices. Which of the following symptoms is most likely to develop if the mother feeds the baby with fructose-containing foods?
   A) Glucose excretion in the urine
   B) Hyperglycemia
   C) Hypoglycemia
   D) Lactose intolerance

5. Screening tests of a newborn suggests potential galactosemia. To avoid complications such as gastrointestinal discomfort, cataracts and mental disability, the pediatrician counsels the mother that the newborn must not be fed with milk or milk products.Which of the following biomolecules is the most likely cause of cataracts if the baby would be left untreated?
   A) Fructose
   B) Glucose
   C) Galactitol
   D) Sorbitol

6. A 69-year-old male with a history of alcoholism complainsof muscle weaknessand lethargy. He reports that he has been living by himself over two years.Physical examination shows abnormal gait and an inability to stand up straight. EEG results show evidence of encephalopathy. Blood test shows that metabolic panel, liver function tests, complete blood count results are all within normal range. Transketolase assay resultsare below normal. Which of the following nutrient deficiencies is most likely?
   A) Biotin
   B) Fructose
   C) Galactose
   D) Thiamine
   E) Vitamin C

7. A 4-month-old female presents with failure to thrive and hypotonia. Physical examination suggests hepatosplenomegaly. Liver function tests reveal hepatic injury. Although blood glucose levels are found to be normal, her creatine kinase levels are increased. Her pediatrician suspects a glycogen storage disease and orders a muscle biopsy test. The biopsy results show an altered structure of glycogen, with fewer branching points. Which of the following is her most likely diagnosis?
   A) Anderson disease
   B) Cori disease
   C) Hers disease
   D) McArdle disease
   E) von Gierke disease

8. A 47-year-old female complains of lethargy and sleeping problems. She also reports that she urinates frequently andalways feels thirsty. Physical exam reveals a BMI of 30. Urinary analysis shows slightly elevated ketones and trace amounts of glucose. A random blood glucose test is 195 mg/dl (normal is ? 100 mg/dl). Her doctor prescribes metformin to improve her glycemic control. Which of the following enzymes is most likely downregulated by this drug?
   A) Glucose 6-phosphatase
   B) Lactate Dehydrogenase
   C) Phosphoenolpyruvate carboxykinase
   D) Phosphofructokinase 1
   E) Pyruvate kinase

Answers

1. **The correct answer is B**. Since the patient has two mutations in the CFTR gene, which encodes for a chloride channel, the symptoms of this patientare most likely due to blockage of the exocrine glands, such as the pancreas and lungs. Due to the malfunctioning chloride channels, the

transport of chloride, sodium and water into the lumen of the lungs is decreased. This causes thick, viscous secretions and thick mucus in the airways, which allows bacterial growth and promotes recurrent lung infections (reviewed in Theme 18). Cystic fibrosis patients also suffer from pancreatic insufficiencies due to obstruction of the pancreatic exocrine duct. The CFTR protein secretes bicarbonate and chloride into the duodenum from the exocrine duct of the pancreas. Malfunctioning of the CFTR in the pancreatic duct cells also blocks the secretion of digestive enzymes into the intestine. Amylase is a pancreatic enzyme, therefore amylase is most likely decreased in this patient's gastrointestinal tract. Besides amylase, secretion of other pancreatic enzymes (e.g. lipase and proteases) also decreases, therefore cystic fibrosis patients suffer from digestive problems,especially diarrhea and steatorrhea. The answer choices A, C, D, and the E are incorrect because these enzymes are synthesized and secreted from the brush border of the small intestines, not the pancreas.

2. **The correct answer is E.** Gliflozins are inhibitors of the sodium-glucose co-transporter (symporter) SGLT2, which is mainly expressed in the kidneys. The SGLT2 symporter functions in reabsorbing glucose in the proximal renal tubules. The SGLT2 inhibitors (gliflozins) prevent reabsorption of glucose and increases excretion into the urine. Therefore, gliflozins help to maintain normal blood glucose concentrations in type 2 diabetics.
Gliflozins do not have any effect on the ATP-sensitive K^+ channel, facilitative glucose transporter, or the digestive en-zymes such as α-glucosidase and amylase. Therefore, answer choices A, B, C, and the D are incorrect.
The ATP-sensitive K^+ channel is found in the β-cells of pan-creas. This channel senses the increased ATP levels, which leads to the closing of ATP-sensitive K^+ channels on the cell membrane. The change in the electric activity of the plasma membrane opens up the Ca^{2+} channels. The increased intra-cellular Ca^{2+} promotes release of insulin from storage vesicles into the blood to control blood glucose levels.
Drugs that block ATP-sensitive K^+ channels are repaglinide, nateglinide, and sulfonylurea. Blocking the ATP-sensitive K^+ channels stimulates insulin secretion from the β cells of the pancreas.
Amylase and α-glucosidases are digestive enzymes for car-bohydrates. Acarbose and miglitol are drugs that specifically inhibit α-glucosidases, such as sucrase and maltase. Some studies suggest that acarbose also inhibits amylase. Facilitative glucose transporters (GLUTs) take up glucose from blood into the cells. There are five GLUTs that are tis-sue-specific transporters. GLUT1 and GLUT3 are found in most tissues, such as erythrocytes, brain, central and periph-eral neuronal cells. GLUT2 is found in hepatocytes, pancreat-ic β cells and the kidney. GLUT4 is found in adipose tissue, skeletal and cardiac muscle cells. GLUT5 is fructose trans-porter, found in enterocytes and hepatocytes.

3. **The correct answer is A.** This patient most likely has G6PD deficiency, since he is a male of Mediterranean descent. The G6PD gene is located on the X-chromosome and, therefore, males are affected more often than females. Normally the G6PD enzyme in the pentose phosphate pathway produces

enough NADPH to reduce the oxidized glutathione to the reduced form. The reduced glutathione is used to eliminate the H_2O_2 in the cells, thus preventing cells from free radical damage. Individuals with G6PD mutations (generally males), however, cannot produce enough NADPH. Therefore they are more prone to oxidative damage. The most common triggers of oxidative stress are anti-malarial drugs (primaquine), sulfonamide antibiotics (trimethoprim-sulfamethoxazole), viral and bacterial infections, and fava beans. Because of the oxidative stress, free radicals increase and damage the cell membranes, proteins and DNA. Free radicals attack the membranes of the erythrocytes and cause hemolysis. Free radicals also cause cross-linking and aggregation of intracellular proteins. These oxidized proteins clump together to form insoluble masses called "Heinz bodies" within red blood cells. Demonstration of Heinz bodies in the erythrocytes supports his diagnosis.
The answer choices B, C, D, and E are more likely incorrect because of his history. Although he likely has high alcohol levels in his blood, this doesn't support the diagnosis of his hemolytic anemia. High lactate levels are also possible in this patient. Again, the high lactate is not associated with hemo-lytic anemia either. The D-xylose test is used to identify mal-absorption, which would not be helpful for his diagnosis. If he has been drinking more than six months, his arterial pH is most likely acidic due to lactic acidosis. The arterial pH does not support hemolytic anemia either.
Reference:https://emedicine.medscape.com/article/200390-overview

4. **The correct answer is C.** The ALDOB gene encodes the aldolase B enzyme, which cleaves fructose 1-phosphate to glyceraldehyde and glycerol 3-phosphate during fructose metabolism. Mutations of this gene cause hereditary fructose intolerance. Ingestion of fructose causes digestive problems, nausea, vomiting and hypoglycemia. The biochemical cause of **hypoglycemia** is due to depletion of inorganic phosphate and consequently ATP in the hepatocytes. In the fructose metabolism pathway, the first enzyme fructokinase phosphorylates fructose, forming fructose 1-phosphate. Since the aldolase B enzyme is deficient, fructose 1-phosphate accumulates in the hepatocytes, causing depletion of inorganic phosphate. Due to unavailable inorganic phosphate, glycolysis is inhibited, resulting in a decrease in ATP production. Due to insufficient ATP production, glycogenolysis and gluconeogenesis are inhibited as well. Chronic ingestion of fructose can lead to liver damage and failure to thrive. A fructose-free diet would prevent these symptoms and is the key in treating this disease.
The answer choices A and B are incorrect because glucose metabolism is not affected in the extrahepatic tissues. An-swer choice D is incorrect since lactose is found in milk and contains galactose and glucose. The metabolism of galac-tose and glucose are most likely normal in this patient.
Reference: https://reference.medscape.com/article/944548-overview#a4

5. **The correct answer is C.** Galactokinase or GALT enzyme deficiency results in the accumulation of galactose and galactose 1-phosphate, respectively. Galactose is reduced to

IV

galactitol by aldose reductase. This sugar alcohol damages the lens of the eye and causes cataract formation.

The answer choices A and B are incorrect because fructose or glucose is not metabolized in the galactose metabolism pathway.

The sorbitol causes cataracts in uncontrolled diabetic patients due to high blood glucose concentrations. This newborn baby has the enzyme deficiency in the galactose metabolism.

6. **The correct answer is D.** The enzyme transketolase functions in the non-oxidative phase of the pentose phosphate pathway and requires thiamine pyrophosphate (TPP) as a coenzyme. As reviewed in Theme 26, thiamine absorption can be impaired in chronic alcoholics. A transketolase assay can be used as a diagnostic test to detect blood thiamin levels. Thiamine deficiency can cause beriberi disease with signs and symptoms of peripheral neuropathy, muscle weakness, dementia, short-term memory loss, and gait problems. In severe forms of beriberi, patients experience cardiac problems. The transketolase assay is a diagnostic test for thiamine deficiency, thus the answer choices A, B, C, and E are incorrect. Answer choice A is incorrect because signs and symptoms of biotin deficiency are hair loss, lethargy, and skin problems, such as dermatitis. Answer choices B and C are incorrect because neither glucose nor galactose is an essential nutrient. Vitamin C deficiency causes scurvy and presents with bleeding gums, bruising, and lethargy.

Reference: A Easter, N Katta. Thiamine Deficiency: A Case Presentation and Literature Review. Journal of Academic Hospital Medicine 2014, Volume 6, Issue 4.
http://medicine2.missouri.edu/jahm/thiamine-deficiency-case-presentation-literature-review

7. **The correct answer is A.** The patient most likely has Anderson disease due to branching enzyme deficiency. This enzyme catalyzes transfer of 8-12 glucose residues to a branch point by reattaching 6-8 residues of glucose by ?-1,6 glycosidic bonds. Deficiency of the enzyme causes abnormally shaped glycogen molecules with long outer chains and few branches. These abnormally shaped glycogen molecules accumulate within the cytoplasm and damage the liver, skeletal and cardiac muscle cells. Anderson disease patients do not present with hypoglycemia, therefore the answer choices B, C and the E are incorrect. Answer choice D is most likely incorrect because McArdle disease is due to deficiency of muscle glycogen phosphorylase, hence patients do not present with liver problems. This disease causes muscle cramps after exercise. Glycogen stores increase in the muscle. However, the structure of glycogen does not change.

Reference: https://emedicine.medscape.com/article/119690-workup

8. **The correct answer is C.** The drug Metformin helps to improve glycemic control in type-2 diabetic patients. Metformin inhibits the electron transport chain, thus increasing the AMP/ATP ratio in the hepatocytes. Increased AMP levels activate AMP-dependent protein kinase (AMPK), which downregulates the transcription of phosphoenolpyruvate carboxykinase (PEPCK), the regulatory step of gluconeogenesis. The decreased amount of PEPCK reduces hepatic glucose output by inhibiting the rate of gluconeogenesis. Through the activation of metformin, AMPK also increases the number of GLUT4 transporters on the cell membrane of skeletal muscle and adipose tissue. Answer choices A and B are incorrect because glucose 6-phosphatase and lactate dehydrogenase are regulated by the AMP/ATP ratio. Answer choices D and E are incorrect because an increased AMP/ATP ratio upregulates these enzymes.

Review Questions

1. A researcher is studying the effects of unknown compounds on the rate of ETC and oxidative phosphorylation. In his experiments, he uses isolated mitochondria and measures the ATP content, electrical and pH gradients across the inner membrane, and O_2 consumption. When he addsa lipid soluble compound into this isolated mitochondria, his data shows that ATP production, as well as the electrical and pH gradients, are decreased. However, O_2 consumption is increased. What is the most likely role of this compound?
   A) Activator
   B) Inhibitor of ATP synthase
   C) Inhibitor of complex I
   D) Modulator of isocitrate dehydrogenase
   E) Uncoupler

2. An 8-month-old male presents with poor feeding and lethargy. His mother states that he had several seizures over the last four months. His records show that his height and weight have been below the 5th percentile since birth. Physical examination reveals poor muscle tone, dystonia, abnormal eye movements, and rapid breathing. Blood tests show increased lactate and pyruvate. His pediatrician prescribes high doses of thiamine. Which of the following diets can be useful to alleviate his symptoms and the buildup of lactate?
   A) High caloric diet
   B) High carbohydrate diet
   C) Low caloric diet
   D) Low carbohydrate diet

3. A 25-year-old female who had gastric bypass surgery about 4-months ago complains of fatigue andsleep disturbances. She tells her doctor she cannot exercise because she has burning pain in her arms and legs and her muscles cramp easily. The physician notices tachycardia, and physical examination reveals edema in her lower legs. The activity of which of the following enzymes is most likely decreased in this patient?
   A) -ketoglutarate dehydrogenase
   B) Citrate synthase
   C) Isocitratedehydrogenase
   D) Succinate dehydrogenase

4. A 19-year-old female seeks genetic counseling regarding her father's recent diagnosis. She states that her father was healthy but he suddenly started to have trouble walking which included frequent tripping and an awkward gait. His symptoms quickly progressed to muscle atrophy and persistent cramping. She also reported that in less than a year, her father lost his speech and developed swallowing difficulties. Their family doctor told her that he has neurodegenerative disease which may claim his life within 1-2 years. Family history reveals that her uncle and her grandfather also had very similar symptoms and they both died in their mid-sixties. The patient should be tested for mutations in which of the following genes?
   A) *G6PDH*
   B) *MCAD*
   C) *PDH*
   D) *SOD1*

5. A 6-month-old male is brought to the emergency department in an unresponsive state. His mother informs the staff that he had a cold a few days ago but otherwise he was okay. She also reported that she wanted to wean him off of breastfeeding and didn't feed him the previous night when he was crying. Laboratory tests show his blood glucose is 35 mg/dl and negative for ketones.Which of the following urine test results is most likely to be found positive?
   A) Acetoacetate
   B) Acetone
   C) β-hydroxybutirate
   D) Dicarboxylic acid
   E) Lactic acid

6. Inhibition of which of the following pathwaysis the underlying molecular cause of hypoglycemia in individuals suffering from Jamaican Vomiting Sickness, also known as ackee poisoning?
   A) α-oxidation
   B) β-oxidation
   C) ω-oxidation
   D) Ketogenesis
   E) Oxidative phosphorylation

7. A comatose 28-year-old female is brought to the emergency department. Her friend states that she has type 1 diabetes. Physical examination shows dry mucous membranes, rapid breathing and an acetone smell in her breath. Blood glucose level is 480 mg/dl. Which of the following is the best treatment for this patient?
   A) Glucagon administration
   B) Glucagon and potassium administration
   C) Insulin administration
   D) Insulin and potassium administration

Answers

1. **The correct answer is E.** This compound is most likely an uncoupler because uncouplers allow protons to reenter the matrix without the synthesis of ATP. Since uncouplers do not affect transfer of electrons through the ETC to oxygen, the respiration rate and O_2 consumption increases but ATP production decreases. Because uncouplers destroy the proton gradient, the electrical and pH gradients also decrease across the inner membrane of mitochondria. Some uncouplers are lipid soluble compounds that create holes on the inner membrane, allowing protons to reenter the matrix. Answer choice A is not specific so it is incorrect. An inhibitor of ATP synthase (answer choice B) would inhibit ATP synthesis and would also decrease O_2 consumption. Inhibitors of complex I (answer choice C) would inhibit electron transport, ATP synthesis and O_2 consumption. Thus answer choice B and C are incorrect. Modulators of isocitrate dehydrogenase would change the rate of the TCA cycle. This change would affect ATP synthesis and O_2 consumption in the same direction, therefore answer choice D is incorrect.

2. **The correct answer is D.** This patient most likely has a genetic deficiency of the pyruvate dehydrogenase (PDH) enzyme, since he presents with neurodegenerative

IV

symptoms. Signs and symptoms of PDH deficiency are severe developmental delays, muscle weakness, and seizures. Lactate and pyruvate buildup in the blood is an important differentiating finding. The PDH enzyme complex has three different subunits: E1, E2, and E3. The most common deficiency is in the gene that encodes the E1 subunit, which is located on the X-chromosome. Thus, males are affected more often than females. Since PDH activity requires five different coenzymes, supplementation with thiamine, carnitine, and lipoic acid can help to alleviate the patient's symptoms. While thiamine and lipoic acid supplementation helps to increase PDH activity, carnitine helps the transport of long chain fatty acids into mitochondria for β-oxidation to increase ATP production. Since most pyruvate comes from glucose oxidation, , patients can be given a ketogenic diet with reduced carbohydrates to reduce the buildup of pyruvate and lactate. Answer choice B is incorrect because a high carbohydrate diet would cause increased pyruvate and, consequently, increased lactate. A high or low caloric diet (answer choices A and C) would not benefit patients who have PDH deficiency. Reference: https://emedicine.medscape.com/ article/ 948360-overview

3. **The correct answer is A**. The symptoms and history of this patient suggest thiamine deficiency, which presents with peripheral neuropathy, fatigue, dry skin, muscle cramps and cardiac problems such as high cardiac output, cardiomyopathy, and lower leg edema. The most common etiologies of thiamine deficiency are "poor" diet and malabsorption, especially in the jejunum of the intestine. The enzymes that require thiamine as a coenzyme are pyruvate dehydrogenase, -ketoglutarate dehydrogenase, branched-chain keto acid dehydrogenase, and transketolase. The diagnostic test for thiamine deficiency is transketolase assay, in which the thiamine levels in the blood are measured by detecting the transketolase activity. Answer choices B, C, and D function in the TCA cycle. However, these enzymes do not require thiamine as a coenzyme. In the first step of the TCA cycle citrate is synthesized by the citrate synthase enzyme by condensing acetyl CoA with OAA. After isomerization of the citrate to isocitrate, isocitrate dehydrogenase catalyzes the first oxidative step and forms -ketoglutarate. The next step is catalyzed by -ketoglutarate dehydrogenase to generate succinyl CoA. The -ketoglutarate dehydrogenase shares the same coenzymes as PDH. Succinyl-CoA contains a high-energy thioester bond, and this energy is used to generate GTP by succinate thiokinase. Succinate dehydrogenase catalyzes oxidation of succinate to fumarate and reduces the FAD to form FAD(2H). Succinate dehydrogenase is localized in the inner-membrane of mitochondria, and it also functions in the electron transfer chain as complex-II. After conversion of fumarate to malate by the fumarase enzyme, the last step of the TCA cycle is catalyzed by malate dehydrogenase, in which malate is oxidized to OAA. Reference: https://emedicine.medscape.com/article/ 116930-overview#a5

4. **The correct answer is D**. Based on her father's symptoms, he most likely is suffering from the disease amyotrophic lateral sclerosis (ALS). The family history suggests that this family has the genetic form of the disease, which is inherited in an autosomal dominant manner. Mutations in the *SOD1* gene, encoding for superoxide dismutase, are responsible for the familial (or genetic) form of ALS. Due to a non-functional superoxide dismutase enzyme, individuals with this gene mutation cannot dispose of superoxide radicals, leading to accumulation of superoxide radicals and increased free radical damage to tissues, especially to motor neurons. To support the diagnosis, genetic testing is recommended, however, the result of the genetic test may affect not only the patient but also family members. Answer choice A is incorrect because G6PD deficiency presents with hemolytic anemia. The onset of MCAD and PDH deficiency occurs during infancy or in the toddler years, therefore answer choice B and C are incorrect. MCAD deficiency presents with hypoketotic hypoglycemia. PDH deficiency presents with fatigue, severe developmental delays, muscle weakness and seizures early in life.
References:
https://emedicine.medscape.com/article/1170097-clinical
Saccon RA, Bunton-Stasyshyn RKA, Fisher EMC, Fratta P. Is SOD1 loss of function involved in amyotrophic lateral sclerosis? *Brain*. 2013;136(8):2342-2358. doi:10.1093/brain/ awt097.

5. **The correct answer is D**. This baby most likely has genetic deficiency of the medium chain acyl CoA dehydrogenase (MCAD) enzyme. This enzyme functions in the first step of ?-oxidation of medium chain fatty acids in the mitochondrial matrix. Most of the mutations in the MCAD gene impair the three-dimensional structure of the enzyme. Clinical situations arise when fatty acid oxidation is required, such as longer intervals between meals, as seen during fasting or illness. Gluconeogenesis is also impaired because of the decreased production of acetyl CoA, which is the positive modulator of pyruvate carboxylase. Due to decreased acetyl CoA, pyruvate carboxylase activity decreases, as well as oxaloacetate (OAA) production. Consequently, gluconeogenesis cannot operate. Since the precursor of ketones is acetyl CoA, ketogenesis is impaired, thus hypoketosis is observed. An increase in -oxidation can occur as a result of impaired -oxidation. -oxidation takes place in ER microsomes and converts fatty acids into dicarboxylic acids. Since -oxidation is impaired these dicarboxylic acids are deposited in the blood and spills into urine.
The answer choices A, B, and C are incorrect because these are ketones. Answer choice E is incorrect since patients present with hypoglycemia, and MCAD deficiency does not cause lactate production.
Reference: https://emedicine.medscape.com/ article/ 946755-overview

6. **The correct answer is B**. Ackee fruit, the national fruit of Jamaica, contains the toxin hypoglycin, which inhibits both the short and medium chain acyl-CoA dehydrogenase, thus inhibiting -oxidation. One of the major symptoms associated with ingestion of this toxin is nonketotic hypoglycemia. There are two underlying molecular reasons for the development of hypoglycemia.

First, there is an insufficient production of energy for gluco-neogenesis. The -oxidation of fatty acids in the liver provides energy for the gluconeogenesis pathway. If -oxidation is impaired, then the gluconeogenesis pathway cannot operate, thus hypoglycemia develops. Secondly, there is an insufficient production of acetyl CoA for pyruvate carboxylase activation. Due to impaired β-oxidation, acetyl CoA cannot be produced. Diminished concentrations of acetyl CoA, an activator of pyruvate carboxylase, contributes to the inhibition of gluconeogenesis.

Answer choice A is incorrect because methyl branched-chain fatty acids are oxidized by β-oxidation in the peroxisomes. Answer choice C is incorrect because -oxidationinvolves the oxidation of the methyl-end of fatty acidsin the ER microsomes. Answer D is incorrect because, although the ketogenesis is inhibited in these patients, it is not an underlying reason for hypoglycemia. Oxidative phosphorylation (answer choice E) may be slowed down in the liver due to decreased β-oxidation, which consequently decreases NADH and FADH2 production. However, inhibition of oxidative phosphorylation is not a direct underlying molecular reason for hypoglycemia.

Reference: https://emedicine.medscape.com/ article/ 1008792-overview#a5

7. **The correct answer is D**. Since type 1 diabetics cannot produce insulin, they have to receive insulin either through subcutaneous pump or injections. A failure of insulin administration results in hyperglycemia within several hours. Hyperglycemia results from an increased rate of gluconeogenesis (high glucose output from the liver) and a decreased rate of glucose uptake into the muscle and adipose tissues. High glucagon/insulin stimulates fatty acid release from adipose tissue and also stimulates fatty acid oxidation in the liver, leading to acetyl CoA accumulation. Excessive amounts of acetyl CoA is diverted to ketone body (acetoacetate) biosynthesis. Hyperglycemia induces osmotic diuresis, which results in dehydration and dried mucous membranes. Blood volume decreases, causing a decrease in blood pressure and a subsequent acceleration of pulse. The kidneys excrete extra Na^+ to neutralize organic acids, which causes withdrawal and excretion of K^+ from the cells and loss in the urine. Although K^+ deficiency develops intracellularly, serum K^+ levels are normal or even high. To bring the blood glucose concentration down insulin is required. However, insulin alone would not replace either K^+ or the electrolytes. In fact insulin causes K^+ to be pumped back into the cells. This may result in a sudden drop in blood K^+ levels and, therefore, cardiac arrhythmias. Thus, the best treatment for this patient is administration of both insulin and potassium.

Review Questions

1. A 40-year-old female presents with a history of chronic kidney disease. Her blood test results show albumin levels within the reference range, BUN and serum creatinine levels are elevated, but bicarbonate level is low. Which of the following is the most likely result of her arterial blood gas analysis?
   A) Increased anion gap and decreased pH
   B) Increased anion gap and normal pH
   C) Normal anion gap and decreased pH
   D) Normal anion gap and increased pH

2. A 19-year-old female is brought to the emergency department by her friend who found her in a semi-consciousstate in their apartment. Her friend states that the patient self-induces vomiting whenever she eats a large meal. Physical examination reveals she is pale and her BMI is 21. Blood arterial gas analysis shows pH: 7.50 (ref. range 7.35-7.45), PCO_2: 51 mmHg (ref. range 36-44 mmHg), and HCO3- : 40 mmol/L (ref. range 22-26 mEq/L). Urine chloride is 20 mEq/L (ref. range 25-40 mEq/L). Which of the following is the most likely diagnosis?
   A) Metabolic alkalosis with no compensation
   B) Metabolic alkalosis with respiratory compensation
   C) Respiratory alkalosis with no compensation
   D) Respiratory alkalosis with compensation

3. A 19-year-old male is brought to the emergency department by his friend, who found him unconscious in his home. His friend states that the patient has a history of drug abuse. Physical examination reveals pinpoint pupils and a respiratory rate below normal. Blood gas analysis shows pH: 7.26 (ref. range 7.35-7.45), PCO_2: 53 mmHg (ref. range 36-44 mmHg), and HCO3- : 22 mmol/L (ref. range 22-26 mEq/L). Which of the following is the most likely diagnosis?
   A) Metabolic acidosis with respiratory compensation
   B) Metabolic acidosis with no compensation
   C) Respiratory acidosis with renal compensation
   D) Respiratory acidosis with no compensation

4. A 25-year-old female presents to the emergency department complaining of drowsiness. She states that she has experience some unintentional weight loss over the past 2-3 months and feels tired all the time. Physical examination reveals dehydrated mucus membranes and a fruity odor on her breath. Arterial blood gas analysis showspH: 7.29 (ref. range 7.35-7.45), PCO_2: 30 mmHg (ref. range 36-44 mmHg), and HCO3- : 14 mmol/L (ref. range 22-26 mEq/L). Which of the following is the most likely diagnosis?
   A) Respiratory acidosis with no compensation
   B) Respiratory acidosis with compensation
   C) Metabolic acidosis with no compensation
   D) Metabolic acidosis with respiratory compensation

5. A hypoechoic mass measuring 1.8cm diameter was found by endoscopic pancreatic ultrasonography in a 62-year-old man. Examination of a fine-needle aspirate of the mass reveals acinar and columnar pancreatic cells with positive staining for insulin. Which of the following is the most likely laboratory finding?
   A) Hyperkalemia
   B) Hypernatremia
   C) Hypermagnesemia
   D) Hyperglycemmia

6. A 10-year-old boy presents with nausea, increased thirst, frequent urination and an unexplained weight loss of twenty pounds over the past two months. Physical examination reveals a thin, Caucasian boy with no acute distress. Laboratory studies reveal the following:Sodium 127 mmol/L (reference range: 135-145 mmol/L), Potassium 5.7 mmol/L (reference range: 3.7-5.1 mmol/L), Chloride 90 mmol/L (reference range: 95-105 mmol/L), BUN 37 mg/dL (reference range: 7-20 mg/L), glucose 370 mg/dL (reference range: 108-125 mg/dL), and serum ketones are positive. Transportation of glucose into which of the following tissues is most likely affected by his condition?
   A) Brain and muscle
   B) Muscle and adipose tissue
   C) Liver and brain
   D) Muscle and liver

7. A 15-year-old female presents to the emergency department complaining of drowsiness and nausea. Hermother states that although she hasa good appetite, she has lost about twenty five pounds over the last four months. Physical examination reveals her weight is below the 10th percentile. Her physician notes Kussmaul breathing. Laboratory studies reveal pH: 7.2 (ref. range 7.35-7.45), PCO_2: 25 mmHg (ref. range 36-44 mmHg), and HCO_3- : 14 mmol/L (ref. range 22-26 mEq/L). Accelerated synthesis of which the following is the most likely cause of her low pH?
   A) Fatty acids
   B) Glucose
   C) Ketones
   D) Lactate
   E) Pyruvate

8. A 9-year-old male is brought to the emergency department with lethargy and severe nausea. Hismother states that he seems to have lost weight recently, although he has been eating normally. Laboratory studies reveal pH: 7.1 (ref. range 7.35-7.45), PCO_2: 24 mmHg (ref. range 36-44 mmHg), and HCO3- : 15 mmol/L (ref. range 22-26 mEq/L). In addition, Kussmaul breathing is noted. Which of the following is most likely decreased in his blood?
   A) C-peptide
   B) Epinephrine
   C) Fatty acids
   D) Glucagon
   E) Glucose

Answers

1. **The correct answer is A.** The kidney maintains an acid-base balance by reclaiming most of the filtered bicarbonate. The tubular cells of the kidney also synthesize ammonia (NH_3) from glutamine. This ammonia combines with hydrogen ions to form ammonium (NH_4^+). This ammonium also serves an important role for maintaining the body's acid-base balance. Because this patient has chronic kidney disease with uremia, her kidneys most likely are not able to synthesize ammonia, regenerate bicarbonate to be returned back to the blood, and excrete hydrogen ions. Therefore,

she most likely has increased anion gap metabolic acidosis due to low concentrations of bicarbonate and decreased pH due to increased hydrogen ion concentrations. Plasma Anion Gap = $Na^+ - (Cl^- + HCO_3^-)$

Reference: Current status of bicarbonate in CDK.Dobre M, Rahman M, Hostetter TH.

J Am Soc Nephrol. 2015 Mar;26(3):515-23. doi: 10.1681/ASN.2014020205. Epub 2014 Aug 22. Review.

2. **The correct answer is B.** Increased pH and elevated bicarbonate levels indicate that the patient is in metabolic alkalosis. However, elevated PCO_2 suggests that the patient has respiratory compensation, in which the lungs retain carbon dioxide by hypoventilation. Elevated P_{CO2} cannot cause alkalosis, but shifts the bicarbonate buffer system to the right, causing the formation of carbonic acid and hydrogen ions, thus assisting in decreasing the pH level. Therefore, answer choices A, C, and D are incorrect. The most common causes of metabolic alkalosis are loss of gastric acid secretions due to excessive vomiting and the use of diuretics such as thiazides. These conditions are associated with a low (<20 mEq/L) urinary chloride level. Reference: https://emedicine.medscape.com/article/243160-overview

3. **The correct answer is D.** An elevated PCO_2 level suggests that this patient is in respiratory acidosis. Respiratory acidosis occurs when the lungs cannot effectively remove CO_2 from the body. The most common causes are diseases of the lungs and airways such as COPD and asthma, chest scoliosis, diseases of nerves and muscles, and drugs that affect breathing, such as narcotics. This patient's acute respiratory acidosis is most likely due to central nervous system depression from opioids, narcotics, or sedatives. Because the bicarbonate is not elevated, he does not have metabolic/renal compensation. Metabolic compensation is achieved by reabsorption of the HCO_3^- from the kidneys. Reference: https://emedicine.medscape.com/article/301574-clinical#b2

4. **The correct answer is D.** The patient's blood is acidemic with a pH < 7.35. The bicarbonate and P_{CO2} levels are decreased. Based on the patient's history and physical, she is most likely an undiagnosed type 1 diabetic patient. Thus, her acidosis is metabolic acidosis. In order to determine if her body is using a respiratory compensation mechanism, we can use Winter's Formula, which determines the patient's expected PCO_2. $PCO_2 = (1.5 \times HCO_3^-) + 8 \pm 2$

If calculated PCO_2 is within 2 of the actual P_{CO2} reading, then this patient has metabolic acidosis with respiratory compensation. $P_{CO2} = (1.5 \times HCO_3^-) + 8 \pm 2 = 29$.

Her calculated P_{CO2} is 29, and her actual reading P_{CO2} is 30 mmHg. Since calculated and actual P_{CO2} is within 2 units, she has metabolic acidosis with respiratory compensation. These patients generally exhibit Kussmaul breathing that decreases PCO_2 to compensate for the pH changes from the underlying metabolic acidosis.

5. **The correct answer is A.** The mass found by endoscopic pancreatic ultrasonography stains positive for insulin, indicating an insulin-secretingtumor. Abnormally high insulin levels cause hypokalemia, hypophosphatemia, andhypomagnesemia by stimulating the transportation of those ions into the cell. The normal function of insulin is to decrease blood glucose levels.Abnormally high insulin levels cause hypoglycemia instead of hyperglycemia; therefore answer choices B, C, and D are incorrect.

6. **The correct answer is B.** Based on his medical history, this patient most likely has type-1 diabetes. Insulin affects the transportation of glucose from the circulation into muscles and adipose tissues, which take up the glucose via GLUT4 transporters. The liver takes up glucose via GLUT2 and the brain takes up glucose via GLUT1 and GLUT3 transporters. The GLUT1, GLUT2, and GLUT3 transporters are not insulin sensitive.Therefore answer choices A, C, and D are incorrect.

7. **The correct answer is C.** This patient most likely has type 1 diabetes, and, therefore, cannot produce insulin. In the absence of the insulin hormone, the rate of lipolysis, gluconeogenesis, and fatty acid oxidation increases. Increased fatty acid oxidation in the liver causes accumulation of acetyl CoA, which is diverted to ketone body biosynthesis. Since ketones are acids (beta-hydroxy butyric acid, acetoacetic acid, and acetone), accumulation of ketones exceeds the body's buffering capacity, thus causing acidosis. Fatty acid (answer choice A) and glucose (answer choice B) levels are also likely increased. However, their accumulation does not cause a change in pH. Since the rate of gluconeogenesis is increased in this patient, the gluconeogenesis precursors lactate and pyruvate do not accumulate.

8. **The correct answer is A.** The presentation of this patient suggests that he most likely has type 1 diabetes and, thus, cannot produce insulin. Normally, insulin is synthesized as preproinsulin in the endoplasmic reticulum of the pancreatic beta cells. The preproinsulin contains A-chain, C-peptide, B-chain and a signal sequence. The signal sequence is removed during translation in the rough endoplasmic reticulum. After removal of the signal sequence, the molecule is called proinsulin. This proinsulin is stored in the vesicles of the Golgi apparatus, where the C-peptide is separated from the A- and the B-chain. The A- and B-chains of insulin are bound together by disulfide bonds. Insulin is secreted in equimolar amounts of C-peptide. Thus measurement of the C-peptide level is a good indicator of insulin secretion and distinguishes type 1 diabetes from type 2 diabetes.

Answer choices B, C, D, and E are incorrect because these molecules are most likely increased in this patient's blood.

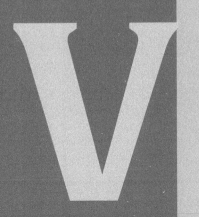

Part V

V

32 Digestion and Absorption of Proteins

At the conclusion of this chapter, students should be able to:

- Explain the concept of amino acid pool and discuss the sources and fates of the amino acids in the pool
- Describe protein digestion in the stomach and small intestine
- Name the proteolytic enzymes that digest proteins and discuss their activation from zymogens and state their specific functions in protein digestion
- Describe the absorption of amino acids by enterocytes
- Compare and contrast transportation of amino acids at the apical and basolateral membranes of enterocytes
- Differentiate diseases due to deficiency of protein-energy malnutrition
- Discuss the lysosome-autophagy and ubiquitin-proteasome pathways involved in intracellular protein degradation

Protein metabolism can be divided into two major topics: synthesis and degradation of proteins and synthesis and degradation of the building blocks of proteins.

The synthesis of proteins consists of transcription and translation, which are reviewed in chapter 2 and chapter 3. The degradation of proteins consists of both the digestion and absorption of dietary proteins and the degradation of existing body or cellular proteins, such as structural proteins, enzymes, hormones, antibodies, and hemoglobin.

Degradation of body or cellular proteins and digestion and absorption of dietary proteins results in the production of the amino acids that make up the "amino acid pool" within the cell and throughout the body (▶ Fig. 32.1). The amino acids in the pool can be "recycled" to make new proteins, new amino acids, amino acid derivatives such as heme, nucleotides, neurotransmitters, or other non-protein compounds including glucose, glycogen and lipids. They can also be catabolized to carbon dioxide and urea while generating energy. The cycle of protein degradation and synthesis, together with dietary protein intake and nitrogen-containing waste excretion, forms a dynamic balance of the amino acid pool.

In this chapter, we will review the digestion and absorption of dietary proteins and the intracellular degradation pathways of proteins. We will also review the dietary needs of proteins, since humans cannot synthesize basic (arginine, lysine, methionine, and threonine), branched chain (valine, isoleucine, leucine), and aromatic ring (tryptophan, histidine, and phenylalanine) amino acids. These amino acids that need to be supplied by the diet are called essential amino acids. Chapter 33 will review nitrogen metabolism, specifically the urea cycle. If the carbons of amino acids are used for catabolism or for storage as glycogen or lipid, then nitrogen needs to be expelled from the body as urea. The last theme of chapter 9 explains the *de novo* synthesis of non-essential amino acids and catabolism of the carbon skeletons of amino acids.

32.1 Protein Digestion in the Stomach

Minimum protein digestion is initiated in the stomach primarily by pepsin, an active form of pepsinogen, which is secreted by the chief cells of the gastric mucosa. Pepsinogen cleaves itself to become the active protease pepsin. The autocleavage of pepsinogen is aided by HCl, which is secreted by the parietal cells of the gastric mucosa. HCl also facilitates the digestive process by denaturing the ingested proteins. Pepsin, as an endopeptidase, hydrolyzes peptide bonds between amino acids, generating smaller polypeptides.

32.2 Protein Digestion in the Small Intestine

In the small intestine, protein digestion and absorption includes the following processes: intraluminal digestion by proteases, terminal digestion by peptidases at the brush border (of microvilli) or inside the enterocytes, and transepithelial transport of small peptides and amino acids.

A majority of protein digestion occurs in the small intestine by the proteases secreted from the pancreas, which also secretes bicarbonate to neutralize the acidic chyme from the stomach. Just like pepsin, the pancreatic proteases are first secreted by the exocrine acinar cells as inactive zymogens, which include trypsinogen, chymotrypsinogen, proelastase and procarboxypeptidase (▶ Fig. 32.2).

The activation of trypsinogen to trypsin is catalyzed by enteropeptidase (also known as enterokinase), which is synthesized by the duodenal mucosa. This active trypsin triggers the activation of other zymogens by cleaving short peptides from chymotrypsinogen, proelastase and procarboxypeptidase to produce active chymotrypsin, elastase and carboxypeptidase, respectively. Trypsin, chymotrypsin and elastase act as endopeptidases to further hydrolyze the polypeptides to smaller oligopeptides. Carboxypeptidase, on the other hand, acts as an exopeptidase to remove one amino acid at a time from the carboxyl end of polypeptides. Another exopeptidase secreted by the enterocytes of the small intestine is aminopeptidase, which removes one amino acid from the amino end of the oligopeptides. The exopeptidases digest proteins at the microvilli or inside the enterocytes.

32.2.1 Link to Pathology: Pancreatic Insufficiency

Exocrine cells of the pancreas not only secrete proteases for protein digestion, but also secrete amylase and lipases to digest carbohydrates and lipids. Any conditions resulting in a decrease or blockage of pancreatic secretions by stones, tumors or concretions (i.e. cystic fibrosis) can cause **pancreatic insufficiency**, which is most often seen in **chronic pancreatitis** and **cystic fibrosis** (reviewed in chapter 18). These patients develop indigestion as well as malnutrition. Exocrine pancreatic insufficiency can be detected by increased triglyceride levels in the stool (fecal fat test), which can be the first sign of pancreatic insufficiency yet is nonspecific, as it is really a test for malabsorption. Stool chymotrypsin and fecal elastase-1 tests can be done to evaluate exocrine pancreatic function more specifically, though with less

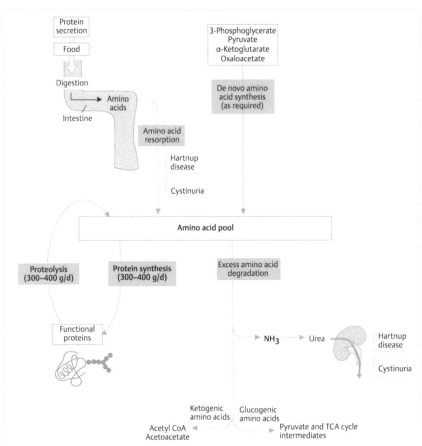

Fig. 32.1 Amino acid pool. Available amino acids in the cells or in the organism derived from de novo synthesis and from dietary sources is known as amino acid pool. The amino acids in this pool is used for the synthesis of proteins and other nitrogen containing compounds or for the energy. (Source: Panini S, ed. Medical Biochemistry—An Illustrated Review. New York, NY. Thieme; 2013.)

V

sensitivity. A negative stool chymotrypsin test or decreased levels of fecal elastase indicate pancreatic insufficiency.

32.3 Absorption of Amino Acids and Small Peptides by Enterocytes

Small peptides and amino acids are absorbed into the enterocytes (also known as absorptive cells) of the small intestine by specific amino acid transporters, which are located at the apical membrane of the enterocytes. These transporters, or Na^+/amino acid symporters, are powered by the gradient of Na^+, which is maintained by the Na^+-K^+ ATPase on the basolateral membrane of the enterocytes. Once absorbed, the amino acids are then transported into the blood via facilitated transfusion at the basolateral membrane of the enterocytes.

32.3.1 Link to Pathology: Cystinuria

Cystinuria is an autosomal recessive disorder caused by defective transportation of cysteine and basic amino acids including lysine, arginine and ornithine at both the small intestine and the kidney. Absorption and resorption of those amino acids at the small intestine and the renal tubules are both affected. However, these patients typically show no clinical signs and symptoms of amino acid deficiency, as the affected amino acids are mostly nonessential amino acids, although they can develop urolithiasis (renal calculi) due to the insolubility of cysteine.

32.3.2 Link to Pathology: Hartnup Disease

Hartnup disease is another autosomal recessive disorder caused by defective amino acids transporters at both the small intestine and the kidney. In this case, the affected transporter proteins have specificity for transporting nonpolar amino acids such as tryptophan. High levels of neutral amino acids are found in the urine of these individuals. Since tryptophan is a precursor of niacin, tryptophan and niacin deficiencies present with similar signs and symptoms. Therefore, Hartnup patients develop pellagra-like symptoms due to deficiency of tryptophan.

32.3.3 Link to Immunology: Celiac Disease

Disruption of any of the digestive or absorptive process in the small intestine leads to malabsorption. In celiac disease, gliadin, a polypeptide of 33 amino acids and one of the digestive products of gluten, enters the circulation via intestinal epithelium and triggers a series of immune reactions that damage the mucosal lining of the small intestine. In Crohn disease, an inflammatory bowel disease affecting the small intestine or colon, the mucosal structure is distorted by transmural inflammation caused by a combination of a variety factors including genetic susceptibility, defective epithelium, intestinal microbiota, and mucosal immunity.

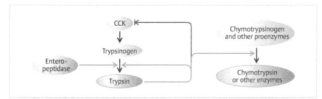

Fig. 32.2 Activation of trypsin and other zymogen proteases. The pancreatic enzyme trypsinogen is activated by the enteropeptidase, which is secreted by the duodenal mucosa. Ones it is active, trypsin activates other zymogens, e.g. chymotrypsinogen, procarboxypeptidases, proelastase, and trypsinogen. (Source: Koolman J, Röhm K, ed. Color Atlas of Biochemistry. 3rd Edition. New York, NY. Thieme; 2012.)

32.4 Essential Amino Acids

Essential amino acids cannot be synthesized in human body, and therefore must be supplied by the diet. These amino acids are Phenylalanine, Valine, Threonine, Tryptophan, Isoleucine, Methionine, Histidine, Arginine, Leucine, and Lysine.

32.4.1 Link to Pathology: Protein-Energy Malnutrition; Kwashiorkor and Marasmus

Kwashiorkor is life-threatening protein-energy malnutrition (PEM) which develops due to an insufficient intake of protein. It is commonly seen in children living in regions suffering from famine. Affected children are on a diet made up exclusively of carbohydrates. These children develop generalized edema caused by a decreased plasma protein concentration, also known as hypoalbuminemia. Besides the characteristic "flaky paint" appearance of the skin, they often have a distended abdomen due to ascites and an enlarged fatty liver. The triacyl-glycerols synthesized in the liver cannot be packed as VLDL to be secreted into the blood due to a decreased synthesis of apo-proteins caused by a marked protein deprivation. Kwashiorkor must be differentiated from another PEM syndrome, **marasmus,** in which calorie deprivation is more severe than protein deprivation. Muscle wasting and loss of subcutaneous fat, which are usually absent in Kwashiorkor, are typical in marasmus. Muscle proteins are broken down to supply the amino acid pool, which can be used either as an energy source or to synthesize the proteins that are more crucial for survival. Marasmic children present with emaciated extremities and a big head instead of big abdomen. Their serum albumin levels are relatively normal, thus these patients do not develop edema.

32.5 Intracellular Protein Degradation

Unneeded or damaged proteins inside the cell, known as "junk proteins" are disposed of by either the lysosome-autophagy (autophagocytosis) or the ubiquitin-proteasome pathways. The amino acids from such intracellular protein degradation supply the amino acid pool.

In autophagy, the unwanted cellular proteins are packed within vesicles and delivered to the lysosomes, where the ingested proteins are degraded by proteases of the cathepsin family. The key regulator of autophagy is a protein kinase called mammalian target of rapamycin or mTOR. mTOR and autophagy is controlled by the energy level of the cell. Anabolic processes, cell growth and proliferation are promoted by mTOR. In contrast, mTOR inhibits catabolic processes such as autophagy by sensing the energy levels. One of the most important stimuli during starvation is the intracellular AMP/ATP ratio. During starvation with high AMP/ATP ratio, the AMP activated protein kinase (AMPK) becomes active. This active AMPK inhibits the mTOR and consequently autophagy becomes active. In contrast, nutrients such as branched-chain amino acids, insulin, and growth factors activate mTOR signaling, and the autophagy pathway is inhibited.

In the ubiquitin-proteasome pathway, short-lived intracellular proteins are first tagged by poly-ubiquitin chains through a cascade of enzymatic reactions catalyzed by E1 activating, E2 conjugating, and E3 ligating enzymes. The poly-ubiquitinated proteins are then delivered to proteasomes, where proteases degrade the proteins. Ubiquitin is an evolutionarily conserved and ubiquitous small protein that exists in almost all tissues of eukaryotes. Besides its function in the proteolytic degradation of proteins, ubiquitin has a plethora of roles in cell development and differentiation. Dysfunctional ubiquitination is related to tumorigenesis, immune pathologies, neurodegenerative diseases, and metabolic syndromes, as well as muscle-wasting disorders.

32.5.1 Link to Pharmacology

Rapamycin and its analogs induce autophagy by inhibiting the mTOR complex. Originally, rapamycin was used as an immunosuppressant in renal transplantation due to its ability to inhibit the proliferation of T cells and the reactions to certain cytokines. As more and more roles for lysosomes in the cell cycle and cellular growth have been discovered, extended therapeutic uses of rapamycin have been explored. In recent years, rapamycin analogs (temsirolimus and everolimus) have been approved in the treatment of advanced renal cell carcinoma and advanced breast cancer.

Bortezomib (BTZ) is the first proteasome inhibitor anticancer drug. It inhibits degradation of ubiquitinated proteins by competitively binding to the 26S proteasome. The accumulation of those proteins activates certain pro-apoptotic factors and induces apoptosis of tumor cells. Bortezomib has been used in the treatment of multiple myeloma and mantle cell lymphoma. **Carfilzomib** is a more potent proteasome inhibitor that binds to proteosomal proteases with more specificity. Carfilzomib is used in the treatment of multiple myeloma that has developed resistance to bortezomib.

Suggested Reading

mTOR: a pharmacologic target for autophagy regulation. Young Chul Kim, Kun-Liang Guan Published January 2, 2015 Citation Information: J Clin Invest. 2015;125 (1):25–32. doi:10.1172/JCI73939

Review Questions

1. Which of the following protease zymogens is activated by HCl?
   - A) Chymotrypsinogen
   - B) Pepsinogen
   - C) Procarboxypeptidase
   - D) Proelastase
   - E) Trypsinogen

2. Which of the following is a role of HCl in the digestion of proteins?
   - A) Activation of the pancreatic proteases
   - B) Aiding the absorption of amino acids
   - C) Denaturation of the dietary proteins
   - D) Removal of propeptide from trypsinogen

3. Which of the following is an upstream inducer of the autophagy pathway?
   - A) AMPK
   - B) Bortezomib
   - C) Carfilzomib
   - D) Insulin
   - E) mTOR

4. Which of the following proteins is a suspected trigger of Celiac disease?
   - A) Elastin
   - B) Gliadin
   - C) Pepsin
   - D) Trypsin
   - E) Ubiquitin

5. Which of the following amino acid deficiencies causes pellagra-like symptoms in patients suffering from Hartnup disease?
   - A) Alanine
   - B) Aspartate
   - C) Glutamate
   - D) Tyrosine
   - E) Tryptophan

Answers

1. **The correct answer is B.** Pepsinogen is secreted by gastric chief cells as an inactive protease zymogen. HCl in the stomach is involved in the removal of the propeptide from pepsinogen, making it the active protease pepsin. Chymotrypsinogen, procarboxypeptidase, proelastase, and trypsinogen are pancreatic zymogens and they are secreted into the duodenum. Enterokinase, which is secreted by duodenal cells, converts trypsinogen to the active protease trypsin. The active trypsin activates the rest of the pancreatic zymogens by removing specific peptides from each zymogen. The zymogens in answer choices A, C, D, and E are activated by trypsin.

2. **The correct answer is C.** HCl lowers the pH of the stomach content. The low pH denatures proteins. Because denatured proteins are more accessible to proteases, HCl functions as aiding in the digestion of proteins. Answer choice A is incorrect because pancreatic proteases are activated by

other enzymes. For example, trypsinogen is activated by enterokinase. The active trypsin activates other pancreatic zymogens, thus D is incorrect. Answer choice B is incorrect because HCl does not have a function in the absorption. Instead, amino acids are absorbed into the enterocytes by specific Na^+ amino acid symporters. These absorbed amino acids are then transported into the blood via facilitated transporters.

Some studies suggest that the HCl or low pH content of the stomach facilitates secretion of enterokinase from the duodenal cells. This increased secretion of enterokinase by the low pH content expedites the digestion of proteins.

3. **The correct answer is A.** Active AMPK induces autophagy by inhibiting the mTOR pathway. Nutrients, especially branched-chain amino acids, insulin, and growth factors activate mTOR signaling, and consequently the activated mTOR inhibits catabolic pathways and autophagy. Therefore, answer choices D and E are incorrect.

Answer choices B and C are incorrect because they are proteasome inhibitors and these drugs do not have any effect on the autophagy pathway. Rapamycin is an inducer of the autophagy pathway. Rapamycin inhibits autophagy by inhibiting the mTOR signaling pathway.

4. **The correct answer is B.** Gliadin, a fraction of gluten, is found in wheat and wheat products and can cause an inflammatory response in the digestive tract. This inflammatory response of the immune system damages the mucosa and can result in malabsorption. This condition is known as celiac disease or gluten-sensitive enteropathy. Elastin is an extracellular matrix protein found in many connective tissues, hence A is incorrect. Answer choices C and D are digestive proteases, which aid in the digestion of dietary proteins in the stomach (pepsin) and in the small intestine (trypsin). Answer choice E is incorrect because ubiquitin is an intracellular small protein that exists in almost all tissues of eukaryotes. Proteins which are targeted for the proteasome-dependent degradation pathway are tagged with polyubiquitin chains.

5. **The correct answer is E. Hartnup disease** is inherited in an autosomal recessive manner. This disorder is caused by impairment of specific amino acid transporters in the apical brush border membrane of the small intestine and proximal tubule of the kidney. These amino acid transporters specifically transport neutral amino acids including tryptophan. Since tryptophan is a precursor of niacin and NAD and NADP coenzymes are synthesized from niacin, Hartnup patients develop pellagra-like skin problems due to deficiency of tryptophan.

Answer choices A, B, C, and D are incorrect because these amino acids are non-essential, meaning humans can synthesize these amino acids even if they are not present in the diet. The essential amino acids are Phenylalanine, Valine, Threonine, Tryptophan, Isoleucine, Methionine, Histidine, Arginine, Leucine, and Lysine.

33 Nitrogen Metabolism

At the conclusion of this chapter, students should be able to:
- Define nitrogen balance
- Explain positive and negative nitrogen balance using specific examples
- Discuss the fates of nitrogen of amino acids
- Describe the nitrogen sources of the urea cycle
- Describe the urea cycle and discuss its significance in maintaining nitrogen hemostasis and amino acid biosynthesis
- Discuss the regulation of the urea cycle by substrate availability as well as enzyme regulation and induction
- Describe the fate of urea synthesized in the liver
- Explain the role of glutamate in nitrogen metabolism
- Discuss the role of alanine and glutamine as carriers of nitrogen in circulation
- Name the three enzymes that are capable of "fixing" ammonia

33.1 Nitrogen Balance

Nitrogen balance is the net difference between nitrogen intake and nitrogen excretion. Most nitrogen intake is acquired from dietary proteins via effective absorption. Nitrogen is not stored in the body. It is either used to make nitrogenous compounds or excreted as nitrogenous waste primarily as urea, ammonia and creatinine in the urine.

Under normal conditions nitrogen hemostasis, or zero nitrogen balance, is achieved by having equivalent nitrogen intake and nitrogen excretion. Positive nitrogen balance occurs when nitrogen intake is greater than nitrogen excretion and is associated with tissue repair or with the active growth seen in children or pregnancy. Negative nitrogen balance occurs when nitrogen intake is less than nitrogen excretion and is associated with malnutrition, starvation or the wasting diseases such as marasmus and kwashiorkor that are reviewed in chapter 32.

33.2 Nitrogen Entering the Urea Cycle

Most of the nitrogen in amino acids is converted to urea via the urea cycle, primarily in the liver or to a lesser extent in the kidneys, preventing accumulation of toxic ammonia in the circulation. Basically, the liver "detoxifies" ammonia to nontoxic urea that is uncharged, water soluble and can be easily excreted in urine without taking away any electrolytes, only water.

In order to enter the urea cycle, the nitrogen needs to be removed from an amino acid via transamination, in which an amino group of an amino acid is delivered to an α-ketoglutarate forming glutamate. In this process, the original amino acid becomes an α-keto acid (▶ Fig. 33.1).

Transamination reactions are catalyzed by transaminases. These enzymes use pyridoxal phosphate (PLP), the active form of vitamin B_6, as a coenzyme. PLP is also used in the decarboxylation and dehydration reactions of amino acid metabolism. The most important transaminases are alanine transaminase (ALT) and aspartate transaminase (AST).

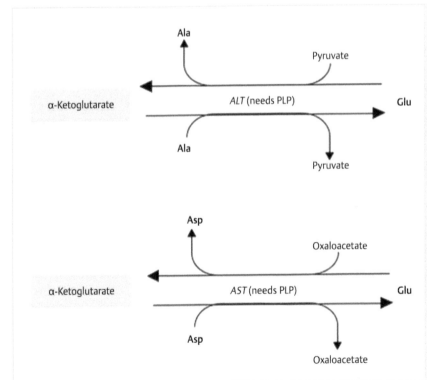

Fig. 33.1 Transamination reactions are catalyzed by transaminases. During transamination, the nitrogen of an amino acid is delivered to an In this process, the original amino acid becomes an acid and the becomes glutamate. The two most important transaminases are alanine amino transferase (ALT) and aspartate amino transferase (AST). All transaminases require pyridoxal phosphate (PLP) as coenzyme.

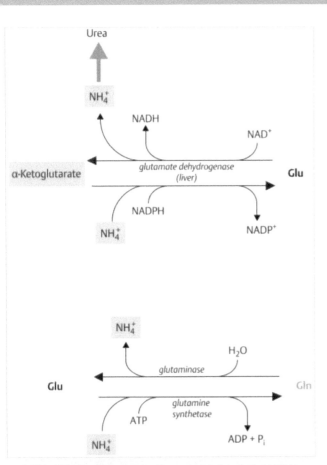

Fig. 33.2 Deamination reactions. The reactions that produce NH4+ entering the urea cycle include deamination of glutamate by glutamate dehydrogenase and deamination of glutamine by the glutaminase enzyme.

The amino group of glutamate can then be removed via deamination and is catalyzed by dehydrogenase to produce NH_4^+ entering the urea cycle (▶ Fig. 33.2). Other reactions that produce NH_4^+ entering the urea cycle include dehydratase reactions of serine and threonine, deamination reactions of glutamine, asparagine and histidine, the purine nucleotide cycle in the brain and the muscle, as well as amino acids and urea degradation mediated by bacteria in the digestive tract.

33.3 Urea Cycle

The urea cycle is first started by the nitrogen from $NH4^+$, the first source of nitrogen, which reacts with bicarbonate and ATP to produce carbamoyl phosphate. This reaction requires ATP and is catalyzed by carbamoyl phosphate synthetase I (CPSI) that exists mostly in mitochondria of the liver and the intestine (▶ Fig. 33.3). CPSI should be differentiated from carbamoyl phosphate synthetase II (CPSII), which is located in the cytosol and catalyzes the formation of carbamoyl phosphate for the pyrimidine synthesis pathway, reviewed in chapter 37. Next, carbamoyl phosphate reacts with ornithine to form citrulline. The formation of citrulline occurs in the mitochondrial matrix

and is catalyzed by ornithine transcarbamylase (OTC). Citrulline is transported out of mitochondria in exchange for ornithine. Once in the cytosol, citrulline combines with aspartate that carries the second source of nitrogen to make argininosuccinate. This reaction is catalyzed by argininosuccinate synthetase and requires ATP. Argininosuccinate is then cleaved by argininosuccinate lyase to produce fumarate and arginine. Notice fumarate is also an intermediate of the TCA cycle. Finally, arginine is cleaved by arginase to produce urea and ornithine. Ornithine is transported back to mitochondria in exchange for citrulline. Urea is transported through the blood to the kidneys and excreted in the urine.

33.4 Regulation of the Urea Cycle

The urea cycle is regulated by substrate availability and enzyme activities. The regulation of the urea cycle is "feed-forward", meaning more nitrogen intake stimulates the urea cycle. The CPSI, the rate limiting enzyme of the urea cycle, is allosterically activated by N-acetylglutamate (NAG), which is synthesized from acetyl CoA and glutamate. Production of NAG is stimulated by arginine, which also increases the level of ornithine, resulting in a higher rate of the urea cycle. The increased protein metabolism seen in a high-protein diet or in a prolonged fasting state induces enzymes of the urea cycle and moves the cycle more rapidly.

Most urea synthesized in the liver is transported via blood to the kidney and filtered through the glomerulus. About half of the filtered urea is reabsorbed by the renal tubules via passive transport, and the rest is excreted in the urine. Some urea is excreted into the intestine where, it is converted to ammonia by bacteria. That ammonia, in combination with the ammonia produced by other bacterial reactions, is absorbed by the hepatic portal vein and enters the urea cycle in the liver.

Besides urea, arginine is also produced from the urea cycle, which explains why arginine is a conditionally essential or semi essential amino acid required by preterm infants but not by most healthy people.

33.5 Nitrogen in Circulation

Some of the nitrogen of amino acids is removed as ammonia (NH_3), which exists mostly as ammonium ion (NH_4^+) under physiological pH. NH_3 can cross cell membranes while NH_4^+ cannot (▶ Fig. 33.4). In the mitochondria, NH_3 can be readily incorporated or removed from glutamate, which acts as a "central station" of nitrogen that goes either to an anabolic or catabolic pathway. The nitrogen collected by glutamate via transaminase or glutamate dehydrogenase (GDH) can be used to synthesize new amino acids, or it can enter the urea cycle as NH_4^+ or aspartate.

NH_3 is especially toxic to the central nervous system and is carried primarily by alanine and glutamine in circulation. Alanine transports the nitrogen from the muscle to the liver. Once reaching the liver, alanine releases nitrogen which enters the urea cycle via transamination and becomes the pyruvate that is used to make glucose via gluconeogenesis. The glucose is then exported out of the liver and taken back by the muscle to be

V

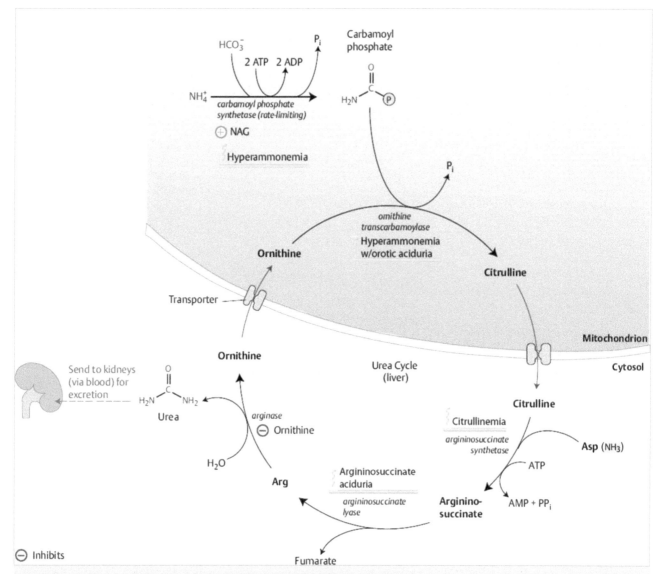

Fig. 33.3 Reactions and regulation of urea cycle. The nitrogen is eliminated from the body as urea, which is synthesized in the liver and excreted in the urine. (Source: Panini S, ed. Medical Biochemistry—An Illustrated Review. New York, NY. Thieme; 2013.)

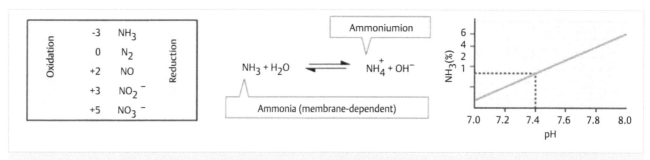

Fig. 33.4 Nitrogen containing molecules are mainly found as NH_3, with exception of NO. Ammonia (NH_3) is gaseous and is in equilibrium with its conjugated base ammonium ion (NH_4^+). At the physiological pH of 7.4, the NH_4^+ predominates. The NH_4^+ cannot pass through the membranes, only neutral NH_3 passes. Excretion of NH_3 becomes important in metabolic acidosis, because most NH_3 diffuse through the tubule membrane into the urine and it neutralizes H^+ ions to form NH_4^+. This charged NH_4^+ ions cannot return into the cells and excreted in the urine. (Source: Koolman J, Röhm K, ed. Color Atlas of Biochemistry. 3rd Edition. New York, NY. Thieme; 2012.)

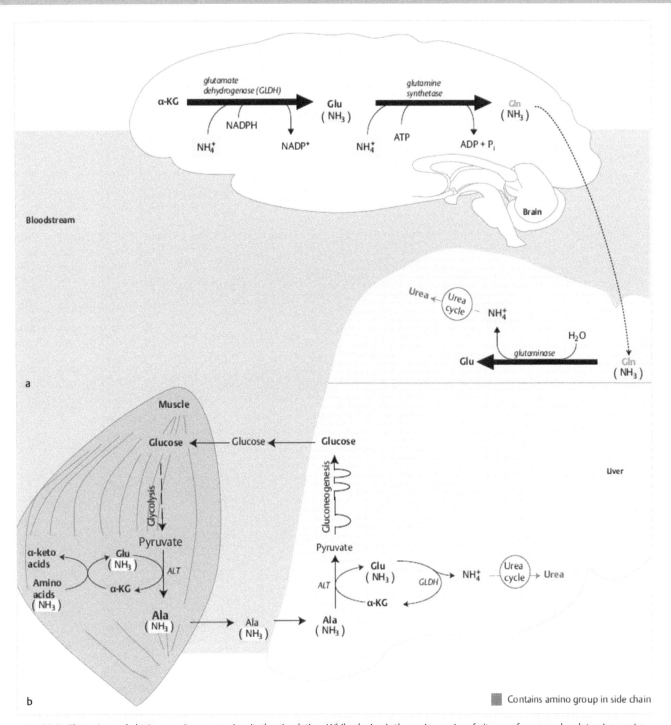

Fig. 33.5 Glutamine and alanine are nitrogen carriers in the circulation. While alanine is the major carrier of nitrogen from muscle, glutamine carries nitrogen from brain and other peripheral tissues to the liver, kidney, and intestine. In the liver, the nitrogen of glutamine is released as NH_4+ to make urea. **(b)** Nitrogen from alanine is transferred to alpha-ketoglutarate to generate pyruvate and glutamate. Gutamate dehydrogenase removes NH_4+, which enters to urea cycle. The glucose-alanine cycle and urea. (Source: Panini S, ed. Medical Biochemistry—An Illustrated Review. New York, NY. Thieme; 2013.)

used as an energy source. Such exchange of nitrogen and glucose between the liver and the muscle is called glucose-alanine cycle or alanine cycle (▶ Fig. 33.5). As a matter of fact, alanine is a major source of carbon for gluconeogenesis in the liver during fasting. The transportation of alanine from peripheral tissues to the liver is stimulated by glucagon during a fasting state.

Glutamine, a primary carrier of nitrogen in the circulation, carries nitrogen from both the muscle and other peripheral tissues to the liver and other organs including the kidney and the intestine. In the liver, the nitrogen of glutamine is released as NH_4+ to make urea. In the kidney, ammonia is released from glutamine in the tubular cells and diffuses into the tubular lumen, where it combines with hydrogen ions to form ammonium and serves an important role in maintaining acid-base equilibrium. In the intestine, glutamine is oxidized to be used as a major fuel, and the released nitrogen is ultimately taken up by the liver.

Due to neurotoxicity of NH_3, three enzymes have been designated to harness and fix NH_3 into organic molecules: carbamoyl phosphate synthetase I (CPSI) to synthesize carbamoyl phosphate from NH_4^+, glutamate dehydrogenase (GDH) to synthesize glutamate from α-ketoglutarate, and glutamine synthetase to synthesize glutamine from glutamate.

33.5.1 Link to Pathology

Blood urea nitrogen (BUN) is a measurement of the amount of blood nitrogen coming from urea. As nitrogen is removed as urea made in the liver and excreted by the kidney, impaired renal function causes an increase of BUN. Therefore, BUN is used to assess the function of the kidney. BUN is often examined together with creatinine, a breakdown product of creatine phosphate that is excreted in urine by the kidney (reviewed in chapter 35). Elevation of BUN and creatinine is defined as azotemia and is related to a variety of disorders. The ratio between BUN and creatinine is a useful indication of the causes of azotemia. The normal reference range of the BUN-to-creatinine ratio is 10:1 to 20:1. An increased ratio (i.e. high BUN but normal creatinine) indicates prerenal causes, such as high protein diet or gastrointestinal bleeding. An increased ratio with an increased creatinine (i.e. both BUN and creatinine are high) level indicates post renal causes, mostly due to obstruction. A decreased ratio (i.e. BUN is normal but creatinine is increased) indicates damage of the kidney itself (e.g., acute tubular necrosis), severe liver disease, starvation, or low protein intake.

33.5.2 Link to Pathology

The level of ammonia in the blood must be kept at very low levels due to the high toxicity of ammonia in the nervous system. **Hyperammonemia**, defined by an increased level of ammonia in the blood, occurs when the urea cycle is compromised, which can be acquired or inherited. The accumulated ammonia passes through the brain-blood barrier easily, attacks both neurons and astrocytes of the nervous system, and adversely affects neurotransmitter activities. The signs and symptoms vary depending on the onset and the age of the patients. Hyperammonemia is usually seen in hepatic encephalopathy, Reye syndrome or genetic deficiency of urea cycle enzymes.

Hepatic encephalopathy is a common complication of end-stage liver disease, namely cirrhosis, in which the liver cannot function in many metabolic processes including converting ammonia to urea. It manifests as deterioration of brain function, with a wide spectrum of neurological symptoms ranging from subtle behavior changes to deep coma.

Rye syndrome is an acute encephalopathy mostly seen in children receiving aspirin for viral infection. The exact cause is unknown. The patient develops fatty liver and swelling of the brain. Hyperammonemia, hypoglycemia and prolonged prothrombin time (PT) are common findings on lab tests in these individuals. Because of an increased risk of Rye syndrome, aspirin is not recommended for any child or teenager who is having fever, unless the child has Kawasaki disease or thrombophilia.

Genetic deficiency of any of the enzymes of the urea cycle causes increased blood levels of ammonia (hyperammonemia), which can cause permanent brain damage. The most common urea-cycle enzyme deficiency is **ornithine transcarbamylase (OTC) deficiency**, an X-linked genetic disorder. Such enzyme deficiencies cause accumulation of carbamoyl phosphate, which spills out of mitochondria to become a substrate of CPSII in the pathway of pyrimidine synthesis. As more carbamoyl phosphate is shunted to make pyrimidine, an intermediate of pyrimidine synthesis, orotic acid accumulates and becomes detectable. The affected individual often presents with lethargy, vomiting or irritability. Lab studies demonstrate increased blood ammonia and urinary orotic acid, which differentiates this condition with deficiency of CPSII in the pyrimidine synthesis pathway (reviewed in chapter 37).

Review Questions

1. Decreased activity of N-acetylglutamate synthase would cause which of the following conditions?
   A) Azotemia
   B) Hyperammonemia
   C) Increased BUN
   D) Increased creatinine
   E) Orotic aciduria

2. Which of the following is a product of transamination reactions?
   A) Acetyl CoA
   B) α-keto acid
   C) Ammonia
   D) PLP
   E) Urea

3. Which of the following amino acids is the carrier of nitrogen in the blood from peripheral tissues to the liver?
   A) Aspartate
   B) Citrulline
   C) Glutamate
   D) Glutamine
   E) Ornithine

4. Which of the following statements best explains the underlying molecular mechanisms of neurotoxicity of ammonia (NH3)?
   A) Because ammonia is a gas and can pass through the membranes
   B) Because ammonia can be converted to NH_4^+ ions
   C) Because ammonia depletes a-ketoglutarate, so the TCA cycle is inhibited
   D) Because ammonia and NH_4^+ ions causes hyperammonemia in the neurons

Answers

1. **The correct answer is B**. The rate limiting enzyme of the urea cycle is upregulated by N-acetylglutamate (NAG), which is synthesized from acetyl CoA and glutamate by NAG synthase. A high protein diet and arginine up-regulates NAG synthase and the rate of the urea cycle increases. Decreased activity of NAG synthase results in decreased NAG concentrations and consequently the rate of the urea cycle decreases. A decreased rate of the urea cycle can result in the accumulation of ammonia, which is called hyperammonemia. Answer choice C is incorrect, because decreased NAG levels decrease the rate of the urea cycle, thus the BUN levels decrease. Elevation of BUN and creatinine is defined as azotemia, therefore for the same reason answer choices A and D are incorrect. Orotic aciduria is caused by a genetic deficiency of ornithine transcarbamylase (OTC) in the urea cycle. Hence answer choice E is incorrect.

2. **The correct answer is B**. Transaminases, such as ALT and AST, catalyze the transamination reactions that transfer the amino group of an amino acid to α-ketoglutarate to form glutamate, while the original amino acid becomes an α-keto acid. Transaminases use pyridoxal phosphate (PLP), the active form of vitamin B_6, as a coenzyme. Therefore answer choice D is incorrect. Deamination reactions produce ammonia, not transaminases, thus C is incorrect. Urea is formed by arginase, the last enzyme in the urea cycle. Hence answer choice E is incorrect. Acetyl CoA is produced by pyruvate dehydrogenase through the decarboxylation of pyruvate, therefore answer choice A is incorrect.

3. **The correct answer is D**. Glutamine is a primary carrier of nitrogen in the circulation from peripheral tissues to the liver. In order to deliver the NH_3 in a "safe mode" to the liver, glutamine has to be formed in the peripheral tissues. First, aminotransferase transfers the amino group from an amino acid to α-ketoglutarate. In the reaction, α-ketoglutarate becomes glutamate and the amino acid becomes α-keto acid. The carbons of this α-keto acid now can be oxidized for energy production in the peripheral tissues. Glutamine synthetase can incorporate NH_4^+ to glutamate to synthesize glutamine. In the liver, the glutaminase enzyme releases the nitrogen of glutamine as NH_4^+ to make urea and glutamate. The glutamate's nitrogen can be transferred to OAA by aspartate transaminase (AST) to form aspartate, which enters the urea cycle as a second nitrogen carrier in the liver. Note that aspartate is not a carrier of nitrogen in the blood. It is formed in the liver by AST.

Citrulline and ornithine are the intermediates of the urea cycle, they are not found in the circulation, and thus answer choices B and E are incorrect. Another nitrogen carrier in the blood is alanine, which mainly transports the nitrogen from the muscle to the liver. Therefore answer choices A and C are incorrect.

4. **The correct answer is C**. Although all of the statements are correct, only answer choice C explains the neurotoxic effects of ammonia. Ammonia (NH_3) is a strong base and can be hydrogenated under physiological pH to form ammonium (NH_4^+) ions. Ammonia can diffuse through the membranes but not ammonium ions. Diffusing of ammonia from the tubule membranes into the urine is an important function of ammonia as a buffer, since it neutralizes H^+ ions to form NH_4^+, which is excreted in the urine. Ammonia is toxic because increased NH_3 in the blood (hyperammonemia) can pass through blood brain barrier, and once in the neurons the ammonia can deplete α-ketoglutarate, an important intermediate of the TCA cycle. This depletion of the TCA cycle intermediate consequently inhibits the cycle and energy production in the neurons.

V

34 Amino Acid Metabolism

At the conclusion of this theme, students should be able to:
- Explain the difference between essential and nonessential amino acids
- Identify and discuss the carbon sources of nonessential amino acids for synthesis
- Differentiate the causes of hyperhomocysteinemia and homocysteinuria
- Explain the difference between gluconeogenic and ketogenic amino acids
- Discuss the mechanisms of degradation of amino acids that are used as intermediates in glycolysis
- Discuss the mechanisms of degradation of amino acids that are used as intermediates of TCA cycle
- Discuss the mechanisms of degradation of amino acids that are used to produce acetyl CoA and acetoacetate
- Discuss the mechanisms of degradation of branched-chain amino acids and explain their significance
- Analyze the metabolism of tyrosine and phenylalanine
- Discuss the importance of the coenzymes (vitamins) of amino acid metabolism and state their functions

34.1 Synthesis of Nonessential Amino Acids

Of the twenty amino acids, ten of them are considered essential amino acids, as they cannot be synthesized by the body and have to be acquired from the diet. The ten essential amino acids are phenylalanine, valine, threonine, tryptophan, isoleucine, methionine, histidine, arginine, leucine, and lysine. The other ten amino acids are called nonessential amino acids since they can be synthesized by the body.

The carbons of amino acids are essentially derived from glucose. The exception is tyrosine, which is synthesized from the essential amino acid phenylalanine. The carbons of the other nine nonessential amino acids are derived from the intermediates of either glycolysis or the TCA cycle (▶ Fig. 34.1).

34.2 Serine, Glycine, Alanine and Cysteine are Derived from the Intermediates of Glycolysis

Serine is synthesized from 3-phosphoglycerate via oxidation, transamination and dephosphorylation. The majority of glycine is synthesized from serine via the one carbon transfer reaction catalyzed by serine hydroxymethyl transferase (▶ Fig. 34.1). This enzyme requires pyridoxal phosphate (PLP) and tetrahydrofolate (FH$_4$) as coenzymes (reviewed in Chapter 35). A small amount of glycine is derived from threonine degradation. Alanine is synthesized from pyruvate via transamination catalyzed by alanine aminotransferase (ALT).

Synthesis of cysteine requires serine and methionine (▶ Fig. 34.1). While serine provides the carbon and nitrogen, methionine provides the sulfur. Methionine first has to be converted to homocysteine. Cystathionine β-synthase combines homocysteine and serine to form cystathionine. Cystathionine is then cleaved to cysteine and α-ketoglutarate by cystathionase. The enzymes catalyzing both reactions (cystathionine β-synthase and cystathionase) require PLP as a coenzyme. Deficiency of either of these two enzymes causes accumulation of homocysteine both in the blood and in the urine.

34.2.1 Link to Pathology: Hyperhomocysteinemia and Homocysteinuria

Genetic deficiency of cystathionine β-synthase or methionine synthase (homocysteine methyl transferase) in the methionine metabolic pathway causes accumulation of homocysteine in the blood and in the urine. This condition is called homocysteinuria (▶ Fig. 34.2).

This condition is also called hyperhomocysteinemia, which is also caused by the genetic deficiency of the methylene tetrahydrofolate reductase (MTHFR) enzyme, reviewed in Chapter 35. Hyperhomocysteinemia, the accumulation of homocysteine in the blood can be due to genetic deficiency of either methionine synthase, or cystathionine synthase, or methylene tetrahydrofolate reductase. The symptoms of hyperhomocysteinemia due to either enzyme deficiencies are more severe and develops in early ages compared to the hyperhomocysteinemia due to nutritional deficiency of either B12, or folate, or vitamin B6."

Patients with either **hyperhomocysteinemia** or **homocysteinuria** are at increased risk of cardiovascular diseases, thrombosis, neurodegenerative diseases, and fractures. The increased risk of cardiovascular and neurodegenerative diseases are likely due to the nature of homocysteine, which is a potent oxidizing agent.

34.3 Glutamate, Glutamine, Proline, Arginine, Aspartate and Asparagine are Derived from the Intermediates of the TCA Cycle

Glutamate, glutamine, proline, and arginine are related to α-ketoglutarate (▶ Fig. 34.1). Glutamate is synthesized from α-ketoglutarate either by transaminase or glutamate dehydrogenase. Two of the ways to fix free ammonia in the body are through the formation of glutamate from α-ketoglutarate by glutamate dehydrogenase or the formation of glutamine from glutamate by glutamine synthase (see Theme 33). Limited quantities of arginine synthesized from the urea cycle are enough for adults but not for growing children, which makes arginine an essential amino acid in children but not in adults.

The other two amino acids derived from the TCA cycle, aspartate and asparagine are related to oxaloacetate. Aspartate is synthesized from oxaloacetate via transamination. Asparagine is synthesized from aspartate by acquiring the amide nitrogen from glutamine (▶ Fig. 34.1).

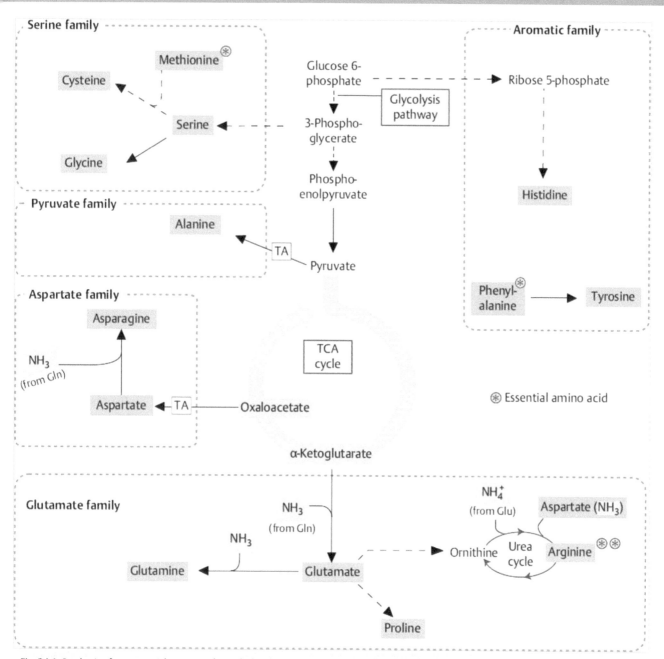

Fig. 34.1 Synthesis of nonessential amino acids. With the exception of tyrosine, all nonessential amino acids are derived from glucose. Nonessential amino acids; serine, glycine, alanine, and cysteine are synthesized from glycolytic intermediates. Nonessential amino acids; glutamate, glutamine, proline, arginine, aspartate, and asparagine are synthesized from TCA cycle intermediates. (Source: Panini S, ed. Medical Biochemistry—An Illustrated Review. New York, NY. Thieme; 2013.)

34.4 Amino Acids Degradation

When amino acids are degraded, the nitrogen can either be used to synthesize other nitrogen containing compounds, including nucleic acids, or excreted as nitrogen containing waste, such as urea. Energy can be generated through the oxidation of the carbons of amino acids to CO_2. These carbons can also be used to make glucose or ketone bodies.

The amino acids degraded to intermediates of glycolysis or the TCA cycle can ultimately make glucose via gluconeogenesis. Therefore, these amino acids are categorized as glucogenic.

Ketogenic amino acids are those that can be degraded to acetyl CoA, the precursor of ketone bodies. Among the twenty amino acids, leucine and lysine are exclusively ketogenic. Tryptophan, threonine, phenylalanine, tyrosine, and isoleucine are both ketogenic and glucogenic. The other thirteen amino acids are exclusively glucogenic (▶ Fig. 34.3).

The amino acids that are degraded to intermediates of glycolysis are alanine, serine, and glycine. Glycine can be cleaved by the glycine cleavage enzyme. This reaction generates CO_2, ammonia and methyl- tetrahydrofolate, a molecule generated by the transfer of one carbon unit from glycine to

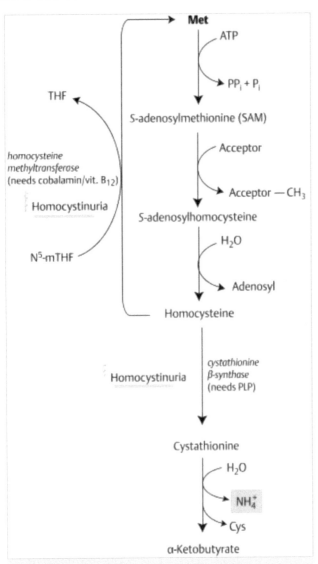

Fig. 34.2 Synthesis of methionine and cysteine. Methionine is synthesized from homocysteine, which can be converted to cysteine or back to methionine. The conversion of homocysteine to methionine requires vitamin B12 and folate, while the conversion of homocysteine to cysteine requires vitamin B6. Deficiency of any of these vitamins (either vitamin B12, or folate, or vitamin B6) causes accumulation of homocysteine, which is called homocystinuria or hyperhomocysteinemia.

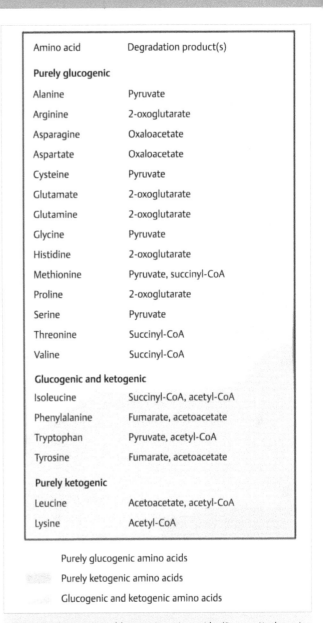

Amino acid	Degradation product(s)
Purely glucogenic	
Alanine	Pyruvate
Arginine	2-oxoglutarate
Asparagine	Oxaloacetate
Aspartate	Oxaloacetate
Cysteine	Pyruvate
Glutamate	2-oxoglutarate
Glutamine	2-oxoglutarate
Glycine	Pyruvate
Histidine	2-oxoglutarate
Methionine	Pyruvate, succinyl-CoA
Proline	2-oxoglutarate
Serine	Pyruvate
Threonine	Succinyl-CoA
Valine	Succinyl-CoA
Glucogenic and ketogenic	
Isoleucine	Succinyl-CoA, acetyl-CoA
Phenylalanine	Fumarate, acetoacetate
Tryptophan	Pyruvate, acetyl-CoA
Tyrosine	Fumarate, acetoacetate
Purely ketogenic	
Leucine	Acetoacetate, acetyl-CoA
Lysine	Acetyl-CoA

  Purely glucogenic amino acids

  Purely ketogenic amino acids

  Glucogenic and ketogenic amino acids

Fig. 34.3 Glucogenic and ketogenic amino acids. (Source: Koolman J, Röhm K, ed. Color Atlas of Biochemistry. 3rd Edition. New York, NY. Thieme; 2012.)

tetrahydrofolate. Glycine can also be converted to glyoxylate. Glyoxylate is then oxidized to oxalate, which is eliminated by the kidney. Oxalate has a very low solubility in the body. Normally, oxalate is converted back to glycine by alanine-glyoxylate aminotransferase. If the alanine-glyoxylate aminotransferase enzyme is deficient, then oxalate levels increases, causing hyperoxaluria. When oxalate combines with calcium, it forms crystals and precipitates in the kidneys.

34.4.1 Link to Pathology: Hyperoxaluria and Kidney Stones

Calcium oxalate is the primary constituent of the most common renal calculi. The oxalate part of the stone is the end degradative

product of glycine and can only be eliminated by the kidney. Deficiency of **alanine-glyoxylate aminotransferase** (*AGXT* gene) results in accumulation of oxalate and formation of calcium-oxalate crystals in the kidneys. This enzyme deficiency is inherited in an autosomal recessive manner and the disease is called primary **hyperoxaluria type 1.** Patients with primary hyperoxaluria type-1 often develop urolithiasis and further renal failure.

The carbon part of cysteine is converted to pyruvate, while the sulfur is converted to sulfate. Methionine is used to synthesize homocysteine, which can be converted to cysteine or back to methionine (▶ Fig. 34.2). The conversion of homocysteine to methionine requires vitamin B_{12} and folate, while the conversion of homocysteine to cysteine requires vitamin B_6. Deficiency of any of the three vitamins (vitamin B_{12}, folate, vitamin

B_6) or the enzyme methylene tetrahydrofolate reductase, which catalyzes the production of 5-methyltetrahydrofolate, a co-substrate for re-methylation of homocysteine to methionine. This results in hyperhomocysteinemia.

34.5 Amino Acids Degraded to Intermediates of the TCA Cycle

Glutamate is converted to α-ketoglutarate by transaminase or glutamate dehydrogenase. The α-ketoglutarate can be converted to malate in the liver and then to glucose via gluconeogenesis. Glutamate can be converted to glutamine by glutamine synthetase. Glutaminase converts glutamine back to glutamate.

Glutamate, glycine, and cysteine are used in glutathione biosynthesis via the γ-glutamyle cycle. Reduced glutathione (GSH) has a strong reducing power and acts as an important antioxidant to protect cells from oxidative damage caused by reactive oxygen species (reviewed in Chapter 28).

Proline and arginine are also derived from glutamate. Proline is first converted to glutamate semialdehyde, which is oxidized to glutamate. Arginine is only converted to glutamate semialdehyde when ornithine is in excess, otherwise it is cleaved in the urea cycle to form ornithine and urea. The essential amino acid histidine is also converted to glutamate when degraded.

Aspartate and asparagines, which are derived from oxaloacetate, can be converted back to oxaloacetate when degraded. Asparagine is first converted back to aspartate by asparaginase. Aspartate can also be degraded to form fumarate when acting as a nitrogen source in the urea cycle (reviewed in Chapter 33) or in the purine nucleotide cycle.

34.6 Methionine, Threonine, Valine and Isoleucine are Degraded to Propionyl CoA.

These amino acids enter the TCA cycle through propionyl CoA, which is converted to succinyl CoA, an intermediate of the TCA cycle (▶ Fig. 34.4 (a, b)). Succinyl CoA can be converted to pyruvate and enter the pathway of gluconeogenesis. When degraded via homocysteine to propionyl CoA, methionine is first converted to S-adenosylmethionine (SAM), a methyl group donor. Regeneration of methionine from homocysteine requires both folate and vitamin B_{12} (reviewed in Chapter 38). Deficiency of either vitamin causes accumulation of homocysteine. Valine and isoleucine serve as important energy sources, especially in the muscles, when degraded.

Amino acids degraded to acetyl CoA or acetoacetate are tryptophan, threonine, phenylalanine, tyrosine, isoleucine, lysine and leucine. These amino acids are known as ketogenic amino acids but can also be glucogenic, with the exception of lysine and leucine, which are strictly ketogenic.

Tryptophan can be oxidized to alanine. Tryptophan is also used for the synthesis of niacin in the liver. This process requires iron and other vitamins including riboflavin and vitamin B_6. The majority of threonine is converted to succinyl CoA as discussed previously. Only a small amount of threonine is converted to acetyl CoA.

The branched-chain amino acids isoleucine, leucine and valine are all first converted to α-keto analogs by transaminase, and then go through oxidative decarboxylation catalyzed by branched-chain α-keto acid dehydrogenase (▶ Fig. 34.5). This reaction generates NADH and FADH2, which are the primary sources of energy in muscles, as the muscle has the highest levels of activity of the first two enzymes in the pathway of branched-chain amino acid degradation. Among the three branched-chain amino acids, valine is glucogenic, as it forms the intermediate propionyl CoA which is converted to succinyl CoA. Isoleucine can be converted to either succinyl CoA or acetyl CoA, so it is both glucogenic and ketogenic. Leucine is strictly ketogenic as it can only be converted to acetyl CoA or acetoacetate, but not succinyl CoA.

The conversion of propionyl CoA to succinyl CoA starts with propionyl-CoA carboxylase, which converts propionyl CoA to methylmalonyl CoA. The enzyme propionyl CoA carboxylase requires an ATP and uses biotin as a coenzyme, with the carboxyl group coming from CO_2. Methylmalonyl CoA mutase rearranges this product into succinyl CoA. The methylmalonyl CoA mutase requires vitamin B12 as coenzyme.

34.6.1 Link to Pathology

Maple syrup urine disease (MSUD) is a recessive genetic disorder caused by gene mutations that produce a defective branched-chain α-keto acid dehydrogenase complex. Such enzyme deficiencies cause an accumulation of α-keto acids of leucine, isoleucine and valine, resulting in metabolic acidosis. Ketoacidosis is an important laboratory finding in affected infants. Their urine or ear wax has a characteristic sweet odor that smells similar to maple syrup or fenugreek due to the presence of sotolon, a lactone produced when branched-chain amino acids are not degraded properly.

34.7 Phenylalanine and Tyrosine Metabolism

Phenylalanine is hydroxylated to tyrosine by phenylalanine hydroxylase, which requires the cofactor tetrahydrobiopterin (BH_4). Tyrosine can be converted to homogentisic acid, which can be further degraded to produce both fumarate and acetoacetate (▶ Fig. 34.6).

34.7.1 Link to Pathology: Phenylketonuria (PKU)

Phenylketonuria (PKU) is a recessive genetic disorder characterized by an accumulation of phenylalanine, which cannot be degraded properly. The mutations of the *PAH* gene causes an absence or defect of phenylalanine dehydrogenase, a rate limiting enzyme that catalyzes conversion of phenylalanine to tyrosine. Therefore, the phenylalanine dehydrogenase deficiency causes accumulation of phenylalanine in the blood. Phenylalanine is a large neutral amino acid. Other amino acids in this group are valine, leucine, isoleucine, tryptophan, and tyrosine. All of these large neutral amino acids cross the blood brain barrier via a single transporter. Thus the accumulated

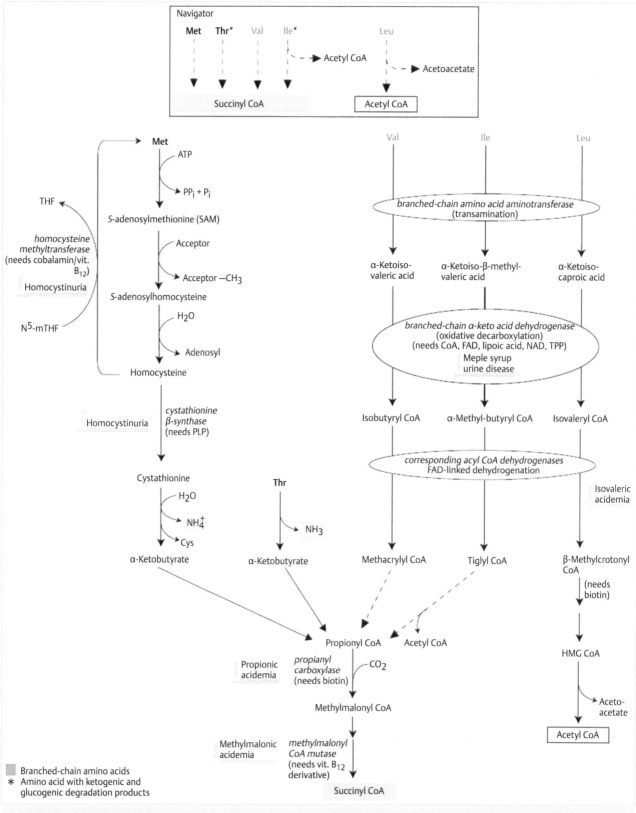

Fig. 34.4 (a, b) Degradation of methionine, valine, isoleucine, and leucine. Methionine is converted to S-adenosylmethionine (SAM), a one-carbon donor, which is converted to homocysteine after donating its methyl group. Homocysteine can be converted back to methionine or can be used for the synthesis of cysteine. The branched chain amino acids, valine, isoleucine, and leucine are degraded to propionyl CoA via the pathway that requires the branched-chain alpha-keto acid dehydrogenase enzyme. Propionyl CoA is then converted to succinyl CoA by the enzymes propionyl CoA carboxylase and methylmalonyl CoA mutase. Dashed arrows indicate multiple reactions.

Name (frequency)	Affected enzyme(s)	Symptoms
Phenylketonuria (1:10 000)	Phenylalanine hydroxylase	Delayed intellectual development, neurological defects
Alkaptonuria (1:25 000)	Homogentisate dioxygenase	Dark urine, arthritis
Albinism (1:20 000)	Tyrosinase	Absent melanization
Homocystinuria (1:150 000)	Cystathionine β-synthase methionine synthase	Thrombosis tendency, lens ectopia, intellectual retardation
Maple syrup urine disease (1:100 000)	Branched-chain dehydrogenase	Failure to thrive, lethargy, neurological defects
Methylmalonate acidemia (1:30 000)	Methylmalonyl-CoA mutase	Acidosis, failure to thrive

Fig. 34.5 Degradation of amino acids to succinyl coenzyme A (succinyl CoA), acetoacetate, and acetyl CoA. (Source: Panini S, ed. Medical Biochemistry—An Illustrated Review. New York, NY. Thieme; 2013.)

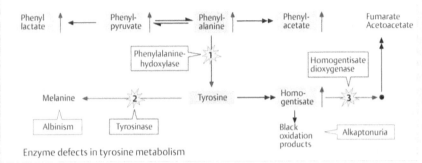

Enzyme defects in tyrosine metabolism

phenylalanine saturates this transporter and consequently causes intellectual disability and other neuropsychological disturbances. In addition, the urine, skin, breath, and ear wax have a mousy or musty odor.

Increased phenylalanine also inhibits tyrosinase, an enzyme catalyzing the production of melanin from tyrosine, which explains why PKU patients often have fair skin and hair. The neurotoxicity of phenylalanine can be prevented by a diet low in this amino acid. A rare form of PKU, known as non-classical PKU, can be caused by a deficiency of tetrahydrobiopterin (BH4) associated with mutations of a different set of genes. BH$_4$ also acts as a coenzyme for the tyrosine hydroxylase enzyme that converts tyrosine to L-dopa, which is further converted dopamine. Therefore, the level of dopamine is low in non-classical PKU but not in classical PKU. Low levels of dopamine stimulate the production of prolactin, which is not seen in classical PKU.

34.7.2 Link to Pathology: Alkaptouria

Alkaptouria is an autosomal recessive genetic disorder caused by a defect in the *HGD* gene coding for the homogentisic acid oxidase enzyme in the pathway of phenylalanine catabolism. Deficiency of homogentisic acid oxidase causes accumulation of the enzyme substrate homogentisic acid, which is oxidized to alkapton. This molecule is responsible for turning the urine a brown or black color when exposed to oxygen in the air. Homogentisic acid also accumulates in other tissues including cartilage, causing arthritis. In this type of arthritis, joints become darkened. This darkening of the joints is called ochronosis.

34.7.3 Link to Pathology

Tyrosinemia type I, II, and III are caused by deficiencies of enzymes in the phenylalanine catabolism pathway.

Tyrosinemia type I is caused by deficiency of the fumarylacetoacetate hydrolase enzyme. Deficiency of tyrosine aminotransferase in the same pathway causes **tyrosinemia type II**. **Tyrosinemia type III** is caused by a deficiency of p-hydroxyphenylpyruvate dioxygenase. All three types of tyrosinemias have increased excretion of tyrosine in the urine. This excess tyrosine in the urine gives urine a cabbage-like smell. Each of the three types of tyrosinemia exhibits a different set of clinical signs and symptoms due to the accumulation of different intermediates of tyrosine metabolism. Of the three types, type I is the most severe. Patients develop progressive liver and kidney dysfunction which may lead to death at an early age. Type III is very rare and often seen with neurologic problems. It can be easily managed by dietary restriction and with supplemental ascorbate, a coenzyme of the deficient enzyme p-hydroxyphenylpyruvate dioxygenase. The severity of type II falls in between type I and type III. Patients develop skin and eye lesions, it can be managed with a diet free of phenylalanine and tyrosine.

34.7.4 Link to Pharmacology

Asparaginase is used in treating childhood acute lymphocytic leukemia (ALL). As asparaginase hydrolyzes circulating L-asparagine to aspartate and ammonia, the cancer cells are starved to death as the cancer cells of ALL, unlike normal cells, are incapable of making asparagine themselves (▶ Fig. 34.6).

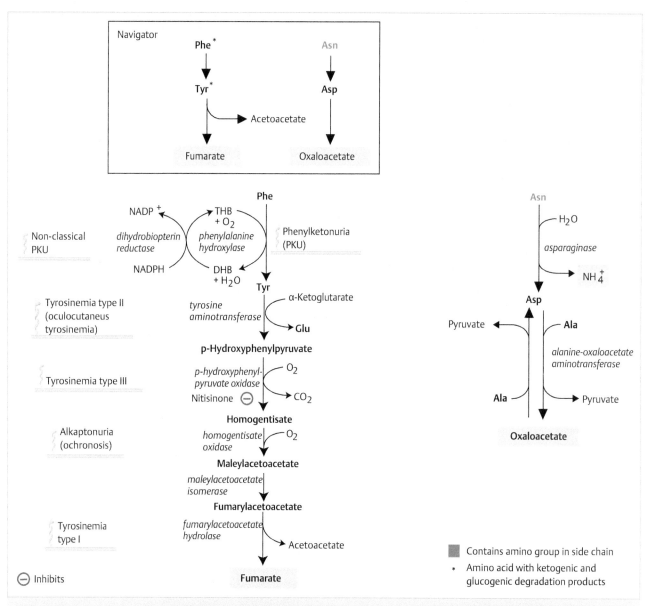

Fig. 34.6 Degradation of phenylalanine, tyrosine and asparagine. Phenylalanine, an essential amino acid, is converted to tyrosine by the phenylalanine hydroxylase enzyme, which requires tetrahydrobiopterine (THB) as coenzyme. Deficiency of this enzyme causes PKU. The ring of tyrosine is converted to fumarate by serial of enzymes. Genetic deficiencies of these enzymes cause diseases include; tysosinemia type I, II, and III, and alkaptonuria or also known as orchonosis. Asparaginase converts asparagine to aspartate, which is converted to oxaloacetate by aspartate amino transferase (AST), a TCA cycle intermediate. (Source: Koolman J, Röhm K, ed. Color Atlas of Biochemistry. 3rd Edition. New York, NY. Thieme; 2012.)

Review Questions

1. Which of the following enzyme deficiencies causes hyperoxaluria?
   A) Alanine aminotransferase
   B) Alanine-glyoxylate aminotransferase
   C) Branched-chain α-keto acid dehydrogenase
   D) Cystathionine β-synthase
   E) Cystathionase

2. In the synthesis of cysteine, which of the following intermediates is formed by combining homocysteine and serine?
   A) Cystathionine
   B) Methionine
   C) Tetrahydrofolate
   D) Pyridoxal phosphate

3. Which of the following amino acids is essential in children but not in adults?
   A) Alanine
   B) Arginine
   C) Cysteine
   D) Phenylalanine
   E) Tyrosine

4. Improper degradation of which of the following amino acids causes maple syrup urine disease?
   A) Cysteine
   B) Phenylalanine
   C) Tryptophan
   D) Tyrosine
   E) Valine

5. Which of the following coenzymes are required for the conversion of propionyl CoA to succinyl CoA?
   A) Biotin and B6
   B) Biotin and B12
   C) Biotin and PLP
   D) B6 and B12
   E) B6 and PLP

Answers

1. **The correct answer is B**. Alanine-glyoxylate aminotransferase converts oxalate to glycine. Therefore, deficiency of this enzyme causes hyperoxaluria, the accumulation of oxalate. Oxalate can be eliminated from the body by the kidneys. However, it has a very low solubility and combines with calcium to form crystals in the kidneys. Thus, hyperoxaluria is associated with kidney stones. Hyperoxaluria patients often develop urolithiasis and further renal failure.

   Answer choice A is incorrect because alanine amino transferase catalyzes transfer of an amino group between alanine and pyruvate. Deficiency of branched-chain α-keto acid dehydrogenase causes accumulation of α-keto acids of leucine, isoleucine and valine. This enzyme deficiency causes maple syrup urine dis-

ease; therefore, C is incorrect. Answer choices D and E are incorrect because deficiency of cystathionine β-synthase or cystathionase causes accumulation of homocysteine both in the blood and in the urine.

2. **The correct answer is A**. In the cysteine biosynthesis pathway, cystathionine β-synthase combines homocysteine and serine to form cystathionine. The enzyme cystathionase then cleaves cystathionine into cysteine and α-ketoglutarate. Methionine provides sulfur for cysteine biosynthesis. However, it needs to be converted to homocysteine. Therefore, answer choice B is incorrect. Answer choices C and D are coenzymes, not metabolites. In the synthesis of methionine, methyl-tetrahydrofolate is used as a methyl donor. Pyridoxal phosphate functions as coenzyme in many reactions of the amino acid metabolism.

3. **The correct answer is B**. Arginine is a conditionally essential amino acid, which is required by children but not most in healthy adults. Arginine can be synthesized in the urea cycle by argininosuccinate lyase. This enzyme cleaves arginine succinate to produce fumarate and arginine. Limited quantities of arginine synthesized from the urea cycle are enough for adults but not for growing children. Answer choices A, C, and E are incorrect because these amino acids are nonessential in both children and adults. Phenylalanine (answer choice D) is essential in both children and adults.

4. **The correct answer is E**. One of the enzymes in the degradation pathway of branched-chain amino acids (leucine, isoleucine, and valine) is the branched-chain α-keto acid dehydrogenase complex. Genetic deficiency of this enzyme causes maple syrup urine disease (MSUD), which is inherited in an autosomal recessive manner. This enzyme deficiency causes accumulation of α-keto acids of leucine, isoleucine and valine, resulting in metabolic acidosis. These α-keto acids give patients the characteristic body odor of burnt sugar or maple syrup.

Due to genetic deficiencies of the basic amino acid transporter, cysteine accumulates in the urine, which causes cystinuria. High concentrations of cysteine forms cistine, which has a very low solubility and forms crystals in the urinary tract. Thus answer choice A is incorrect.

Tryptophan (answer choice C) can be oxidized to alanine. Tryptophan is also used for synthesis of niacin in the liver. Therefore, deficiency of tryptophan can cause pellagra-like symptoms. This process requires iron and other vitamins including riboflavin and vitamin B6.

Phenylalanine is converted to tyrosine by phenylalanine hydroxylase. Deficiency of this enzyme causes PKU. Therefore, answer choice B is incorrect. Answer choice D is incorrect because diseases in the tyrosine degradation pathway are tyrosinemia type I, type II, type III, and alkaptonuria.

5. **The correct answer is B**. The conversion of propionyl CoA to succinyl CoA requires two enzymes and two coenzymes. Propionyl-CoA carboxylase converts propionyl CoA to methylmalonyl CoA. This enzyme requires biotin as a coenzyme and uses ATP to transfer a carboxyl group from

V

CO_2. The product of propionyl CoA carboxylase is metylmalonyl CoA, which is converted to succinyl CoA by methylmalonyl CoA mutase. This enzyme requires vitamin B12 as a coenzyme. Genetic deficiency of propionyl CoA carboxylase causes propionic acidemia. Methylmalonic aciduria can be due to either dietary deficiency of B12 or genetic deficiency of methylmalonyl CoA mutase. Levels of methyl malonic acid can be measured, and this is often used to diagnose B12 deficiency. Both enzyme deficiencies present with similar signs and symptoms.

35 Amino Acid Derivatives

At the conclusion of this chapter, students should be able to:
- Discuss the synthesis and inactivation pathways of the amino acid derivatives catecholamines (dopamine, norepinephrine, epinephrine), melanin, thyroid hormones, serotonin, melatonin, GABA, histamine, creatine, and NO
- Describe the role of pyridoxal phosphate, tetrahydrobiopterin, SAM, and vitamin C in the synthesis of catecholamines, serotonin, melatonin, histamine, and GABA
- Explain the physiological effects of catecholamines (dopamine, norepinephrine, epinephrine), melanin, thyroid hormones, serotonin, melatonin, GABA, histamine, creatine, and NO in metabolism
- Explain the clinical importance of creatinine measurements in the blood and in the urine
- Discuss the synthesis of thyroid hormones and the regulation processes for thyroid hormone synthesis and secretion
- Analyze the clinical presentation of Parkinson disease and schizophrenia, hypothyroidism and hyperthyroidism

35.1 Introduction

Amino acids not only serve as the building blocks of protein synthesis, but are also important for the synthesis of various biomolecules that have essential and diverse biological roles (▶ Fig. 35.1). Clinically important biomolecules derived from amino acids include porphyrins/heme, purine and pyrimidine nucleotides, creatine, nitric oxide, glutathione, thyroid hormones, catecholamines, melanin, serotonin, melatonin, histamine, and GABA.

Tyrosine is the precursor of the catecholamines dopamine, norepinephrine, and epinephrine. Tyrosine is also the precursor of thyroid hormones (thyroxine and triiodothyronine) and melanin, the pigment that gives human skin, hair, and eyes their color.

Tryptophan is important for the synthesis of 5-hydroxytryptamine (serotonin), N-acetyl methoxytryptamine (melatonin) and niacin. Many conditions such as Parkinson's disease, depression, seizures, and retinal damage result from altered synthesis of these biomolecules. Therefore the pathways leading to amino acid derived molecules have been the prime targets for pharmaceutical agents. The main focus of this chapter is the metabolism of amino acid derived biomolecules. We will review the synthesis and degradation of these clinically important amino acid derivatives. Because most of these amino acid derivatives are either neurotransmitters or hormones, we will also review their physiological actions.

35.2 Cofactors Important for the Synthesis of Amino Acid Derivatives

Monooxygenase (or mixed-function oxygenase) enzymes require the cofactor tetrahydrobiopterin for their catalysis. Phenylalanine hydroxylase, tyrosine hydroxylase, and tryptophan hydroxylase are examples of such enzymes. In the reactions catalyzed by these enzymes, one atom of O_2 is transferred to the substrate, and the other oxygen is used for formation of water. The hydrogen donor in these reactions is tetrahydrobiopterin (BH4), which is oxidized to dihydrobiopterin (BH2). Recycling of dihydrobiopterin back to tetrahydrobiopterin is essential and is catalyzed by dihydropteridine reductase (DHPR) (▶ Fig. 35.2).

The synthesis of several of amino acid derivatives involves amino acid decarboxylation by pyridoxal phosphate (PLP) dependent enzymes (▶ Fig. 35.2).

S-adenosylmethionine (SAM) is required as the methyl donor for the synthesis of epinephrine from norepinephrine and melatonin from N-acetylserotonin. As a result, S-adenoylhomocysteine (SAH) is formed by the demethylation of SAM. It is important to remember that synthesis of SAM requires both vitamin B12 and folate (discussed in Chapter 38). Therefore, deficiency in either of these vitamins may decrease the synthesis of these two important neurotransmitters (**Fig. 35.2**).

Vitamin C is required for the synthesis of norepinephrine from dopamine (▶ Fig. 35.2). This reaction is catalyzed by dopamine-β-hydroxylase, which is a Cu^{2+} containing enzyme that is localized in the presynaptic vesicles. As a coenzyme of the reaction, vitamin C prevents the oxidation of Cu^{2+} to Cu^{3+}.

35.3 Action and Fate of Neurotransmitters

Neurotransmitters are synthesized and stored in the presynaptic neurons. They are released into the synaptic cleft by Ca^{2+}-regulated exocytosis in response to stimuli. The postsynaptic neurons contain receptors for these neurotransmitters on the cell membrane, most of which are G-protein coupled receptors. Once secreted into the synaptic cleft, neurotransmitters bind to

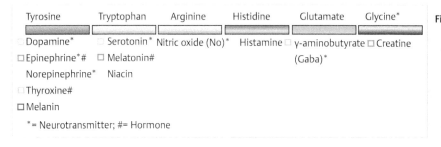

Tyrosine	Tryptophan	Arginine	Histidine	Glutamate	Glycine*
Dopamine*	Serotonin*	Nitric oxide (No)*	Histamine	γ-aminobutyrate (Gaba)*	Creatine
Epinephrine*#	Melatonin#				
Norepinephrine*	Niacin				
Thyroxine#					
Melanin					

*= Neurotransmitter; #= Hormone

Fig. 35.1 Amino Acid Derivatives.

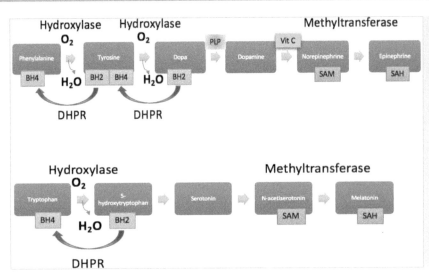

Fig. 35.2 Cofactors Important for the Synthesis of Amino Acid Derivatives. Tetrahydrobiopterin (BH4), pyridoxal phosphate (PLP from vitamin B6) and vitamin C are required for the synthesis of catecholamines (dopamine, norepinephrine, and epinephrine).

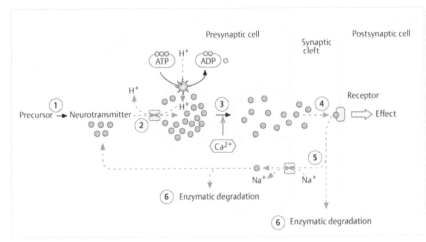

Fig. 35.3 Fate of neurotransmitters. (1) The biosynthesis of neurotransmitters occur in the cytoplasm of presynaptic neurons. (2) Neurotransmitters are stored in the storage vesicles. (3) Neurotransmitters are released into the synaptic cleft by Ca^{+2}-regulated exocytosis. (4) Neurotransmitters bind to their receptors on the post-synaptic membrane. (5) Termination of neurotransmitters action can be either by reuptake or by enzymatic degradation (6). (Source: Koolman J, Röhm K, ed. Color Atlas of Biochemistry. 3rd Edition. New York, NY. Thieme; 2012.)

their receptors on the postsynaptic neurons and trigger their actions by initiating signaling cascades. Termination of the neurotransmitters' action can occur either by reuptake into presynaptic nerve, or by enzymatic degradation (▶ Fig. 35.3).

35.3.1 Link to Pharmacology

There are many pharmaceutical agents that change the actions or fate of neurotransmitters. Drugs may affect the concentration of neurotransmitters by increasing their synthesis, reuptake, or enzymatic degradation. Some drugs may also have an effect on the receptors, acting as either agonists or antagonists.

35.4 Derivatives of Tyrosine

35.4.1 Synthesis of Catecholamines

Dopamine, norepinephrine and epinephrine are derived from tyrosine and are collectively called catecholamines (▶ Fig. 35.4). Dopamine is synthesized in the dopaminergic neurons of the substantia nigra and ventral tegmental area (VTA), which lack the enzymes for the synthesis of norepinephrine and epinephrine. While dopamine from substantia nigra regulates movement and motor functions, dopamine from the VTA controls the brain's reward and pleasure centers. It provides feelings of

enjoyment and reinforcement to motivate us to do, or continue doing, certain activities.

35.4.2 Link to Pathology

Schizophrenia and Parkinson's disease develop due to an imbalance in the dopaminergic system (▶ Fig. 35.5). Parkinson's disease develops from the loss of dopaminergic neurons in the substantia nigra. Loss of these cells can be due to genetic factors, environmental factors, and free-radical injury to substantia nigra. Regardless of etiology, deficiency in dopamine synthesis leads to Parkinson's disease. Parkinson's patients present with unintentional tremor, especially in the extremities, gait problems, rigidity, and bradykinesia. Under microscopic examination of the brain, Lewy bodies are commonly observed. In contrast to Parkinson's disease, schizophrenia is believed to be due to hyperactive dopamine transmission. Binding of dopamine to its receptors D1 and D2 initiates movement at the cortex.

35.4.3 Link to Pharmacology

Although there are several strategies for the treatment of Parkinson's disease, the most common is levodopa (L-DOPA). It is the precursor to dopamine, which is decarboxylated to

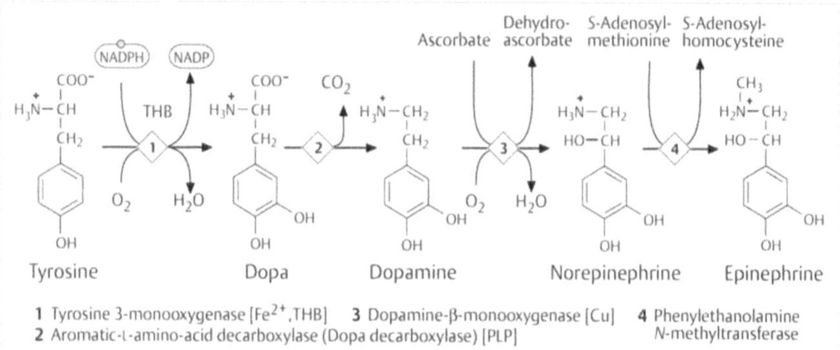

Fig. 35.4 Biosynthesis of the catecholamines. Source: Koolman J, Röhm K, ed. Color Atlas of Biochemistry. 3rd Edition. New York, NY. Thieme; 2012.)

1 Tyrosine 3-monooxygenase [Fe²⁺,THB] 3 Dopamine-β-monooxygenase [Cu] 4 Phenylethanolamine
2 Aromatic-L-amino-acid decarboxylase (Dopa decarboxylase) [PLP]   *N*-methyltransferase

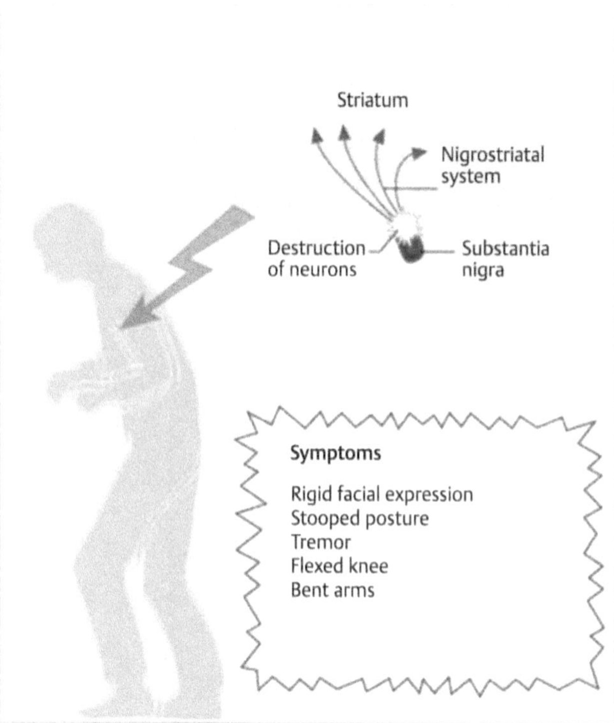

Fig. 35.5 Parkinson disease. (Source: Koolman J, Röhm K, ed. Color Atlas of Biochemistry. 3rd Edition. New York, NY. Thieme; 2012.)

dopamine in the brain. Common side effects of L-DOPA treatment are nausea and vomiting due to the peripheral conversion of levodopa to dopamine, thereby activating of peripheral dopamine receptors. Carbidopa is a decarboxylase inhibitor and is administered with levodopa to decrease its peripheral conversion to dopamine and also to prevent levodopa from being broken down before it reaches the brain.

Norepinephrine is synthesized through the oxidation of dopamine in a reaction that requires ascorbate (vitamin C) and copper (▶ Fig. 35.4). It acts as both a hormone and a neurotransmitter, depending on the site of synthesis. When released from the adrenal medulla into the blood, it acts as a hormone. However, it is a neurotransmitter when released by noradrenergic neurons in the central nervous system and sympathetic nervous system. In the peripheral nervous system, norepinephrine is the primary neurotransmitter released by the sympathetic nerves.

As a hormone, norepinephrine binds to α1, α2, and β1 receptors and mediates the fight-or-flight response, increasing heart

rate, triggering the release of glucose from energy stores, increasing blood flow to skeletal muscle, and decreasing gastrointestinal motility.

In the central nervous system, norepinephrine acts as a neurotransmitter and leads to increased alertness and arousal by regulating sleep, memory, learning, and emotions.

35.4.4 Link to Pharmacology

Norepinephrine can be used as a drug to prevent hypotension. It increases blood pressure through its vasopressor activity.

Even though the major site for **epinephrine** synthesis is the adrenal medulla, it is also synthesized in a small group of neurons in the brainstem that utilize epinephrine as their neurotransmitter. Just as norepinephrine, it mediates the fight-or-flight response which leads to increased heart rate, increased blood flow to skeletal muscle, and increased glucose mobilization from glycogen.

35.4.5 Inactivation of Catecholamines

Catecholamines in the circulation are taken up by non-neuronal tissues, especially liver, and rapidly inactivated by monoamine oxidase (MAO) and by catechol-O-methyltransferase (COMT) enzymes (▶ Fig. 35.6). The reactions catalyzed by these two enzymes can occur in either order. The resulting degradation products are vanillylmandelic acid (VMA) from norepinephrine and homovanillic acid (HVA) from dopamine. These inactive products are excreted in the urine.

35.4.6 Link to Pathology

The measurements of inactive end-products (VMA and HVA) are useful for screening for catecholamine-secreting tumors such as neuroblastoma, pheochromocytoma, or other neural crest tumors. The measurement of urine levels of these molecules may also be helpful in monitoring patient response to certain treatments. Urine HVA levels may also be lower due to MOA deficiency. On the other hand, it may be elevated due to a deficiency of dopamine beta-hydrolase, the enzyme that converts dopamine to norepinephrine.

35.4.7 Link to Pharmacology

The anti-depression drugs known as MAO inhibitors may cause hypertensive crisis if tyramine-containing foods such as

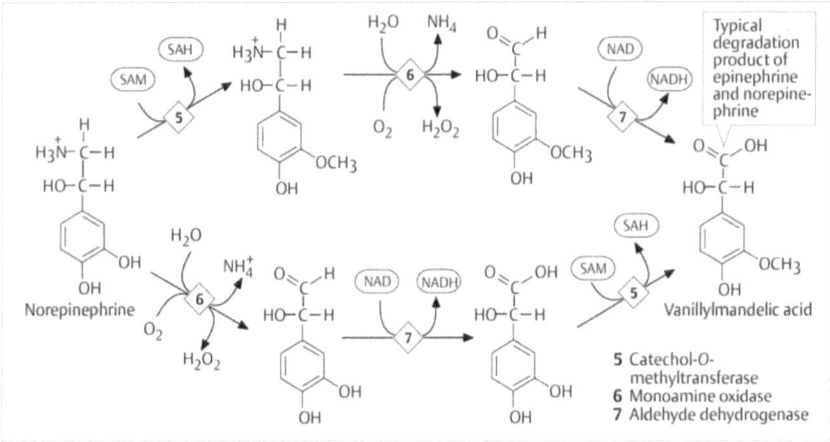

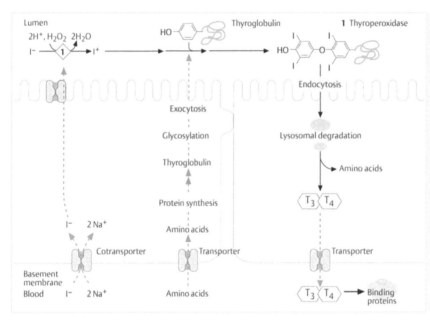

Fig. 35.6 Inactivation of norepinephrine by catechol-o-methyltransferase (COMT) (5) and monoamine oxidase (MAO) (6). (Source: Koolman J, Röhm K, ed. Color Atlas of Biochemistry. 3rd Edition. New York, NY. Thieme; 2012.)

5 Catechol-O-methyltransferase
6 Monoamine oxidase
7 Aldehyde dehydrogenase

Fig. 35.7 Biosynthesis of thyroid hormones. (Source: Koolman J, Röhm K, ed. Color Atlas of Biochemistry. 3rd Edition. New York, NY. Thieme; 2012.)

aged-cheese or wine are consumed. This is known as the "cheese effect.". Since liver MAO metabolizes the tyramine from food, MAO inhibitors inactivate the liver enzyme, which can lead to elevation of tyramine in the blood. Tyramine increases blood pressure through increasing norepinephrine and its vasopressor activity.

35.4.8 Synthesis of Melanin

Melanin is a brown-black pigment responsible for human skin, eye and hair color. Melanin protects skin cells from sun damage by absorbing ultraviolet rays and scavenging free radicals. It is synthesized by melanocytes, located in the skin, hair roots, iris and retina of the eye. These specialized cells convert tyrosine to dopaquinone by a copper-dependent tyrosinase enzyme.

35.4.9 Link to Pathology

Oculocutaneous albinism is caused by mutations in the tyrosinase gene, preventing melanocytes from synthesizing melanin.

35.4.10 Synthesis of Thyroid Hormones

Tyrosine is the precursor of the thyroid hormones T4 (thyroxine or tetraiodothyronine) and T3 (triiodothyronine). Both T3 and T4 are synthesized in the follicular cells of the thyroid gland (▶ Fig. 35.7). Follicular cells of the thyroid gland synthesize thyroglobulin protein, which is secreted into the lumen of the follicle (colloid). In the colloid, the thyroperoxidase enzyme catalyzes iodination of tyrosine residues of the thyroglobulin. The follicular cells then take up the iodinated thyroglobulin from the colloid by endocytosis. These endosomes fuse with lysosomes and the lysosomal enzymes digest the thyroglobulin. This lysosomal digestion results in iodinated tyrosine residues, producing T3 and T4. These mature thyroid hormones are released from the follicular cells into the bloodstream. In the peripheral tissues, the enzyme deiodinase converts T4 to T3. In the bloodstream, the concentration of T4 is higher than T3, although T3 has higher biological activity than T4. In the blood, both T3 and T4 are bound to thyroxine-binding globulin, transthyretin, and albumin.

35.4.11 Link to Physiology: Physiological and Caloric Effects of Thyroid Hormones

Together, T3 and T4 regulate protein synthesis, the basal metabolic rate of metabolism, and body temperature. The receptors for thyroid hormones are found inside the cell as nuclear receptors. These hormones regulate metabolism by changing the sensitivity of the hepatocytes and adipocytes to the action of epinephrine. In the liver, T3 increases the rate of glycolysis, cholesterol and bile acid synthesis. T3 also increases the rate of gluconeogenesis and glycogenolysis. In the adipocytes, T3 has a bipolar affect, in which T3 increases both lipolysis and the availability of glucose to fat cells. This is accomplished by sensitizing the adipocytes to epinephrine and increasing the secretion of insulin from the pancreas. In the muscle, T3 increases glucose uptake and protein synthesis.

T3 regulates body temperature (thermogenesis) by stimulating ATP utilization, which results in heat production. This is due to T3's ability to increase norepinephrine release from the sympathetic nervous system. The norepinephrine increases the synthesis of thermogenine (UCP) in brown adipose tissue, resulting in increased heat production through the uncoupling of the ETC (reviewed in Chapter 27).

35.4.12 Regulation of Thyroid Hormone Levels

The activity of the thyroid gland is regulated by the hypothalamus and the anterior pituitary hormones (▶ Fig. 35.8). External stimuli such as cold and stress are registered by the hypothalamus, which secretes thyroid releasing hormone (TRH). TRH stimulates the secretion of thyroid stimulating hormone (TSH) from the anterior pituitary. This TSH regulates the activity of the thyroid gland by stimulating both the synthesis and the secretion of the thyroid hormones T3 and T4. When there is adequate levels of T3 present in the blood, the release of TSH and TRH is inhibited by negative feedback regulation.

35.4.13 Link to Pathology

Dysregulation of thyroid hormones may lead to either hyperthyroidism or hypothyroidism, meaning either excess or insufficient systemic levels of thyroid hormones, respectively. These two conditions can be caused by a variety of etiologies. One of the causes of thyroid disorder is iodine deficiency, which can lead to goiter with hypothyroidism. Iodine deficiency causes hypothyroidism because T3 and T4 cannot be synthesized in the thyroid gland.

Another thyroid disorder is Graves' disease, which leads to goiter with hyperthyroidism. Gravis disease is an autoimmune disease which is caused by the binding of IgG autoantibodies to TSH receptors on the thyroid gland. This binding stimulates the thyroid gland to secrete excessive T3 and T4. The increased T3 and T4 levels decrease the TSH levels by negative feedback regulation. Therefore, in patients with Gravis disease, hyperthyroidism develops, which presents with weight loss and tachycardia due to an increased basal metabolic rate.

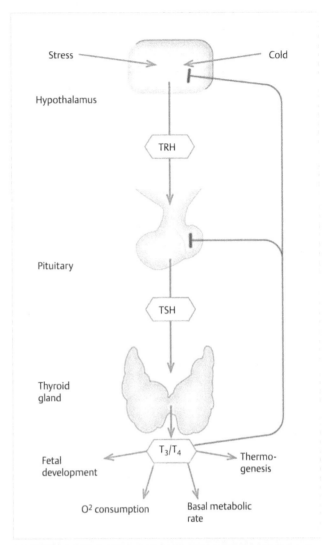

Fig. 35.8 Regulation of thyroid activity. (Source: Koolman J, Röhm K, ed. Color Atlas of Biochemistry. 3rd Edition. New York, NY. Thieme; 2012.)

35.5 Derivatives of Tryptophan

Tryptophan is the precursor for the neurotransmitter **serotonin**. Serotonin synthesis is similar to that of dopamine. Just like in the dopamine synthesis pathway, the first step is catalyzed by a tetrahydrobiopterin-dependent hydroxylase that synthesizes 5-hydroxytryptophan (▶ Fig. 35.2). In the case of serotonin, however, the enzyme is tryptophan hydroxylase. However, the second step is catalyzed by the same enzyme, DOPA decarboxylase, a pyridoxal phosphate (PLP) dependent enzyme. Here, the end product is serotonin rather than dopamine.

Serotonin is found in various types of cells such as intestinal mucosal cells, platelets, and in the serotonergic neurons of the central nervous system. Serotonin has a wide variety of physiological functions and regulates behaviors such as sleep, appetite, mood, cognitive functions, body temperature, circadian rhythmicity, and neuroendocrine functions.

Apart from its own important physiological effects, serotonin is also the precursor of the hormone **melatonin**. It is converted to melatonin in the pineal gland via acetylation and methylation using SAM as the methyl donor. Melatonin synthesis is influenced by the circadian cycle, with melatonin synthesis and secretion being reduced during daylight hours.

35.5.1 Link to Pharmacology

Selective serotonin reuptake inhibitors (SSRI) are the most widely used antidepressants. They can attenuate the symptoms of depression by blocking the reabsorption of serotonin back into the presynaptic neurons, thus increasing the levels of available serotonin in the synaptic cleft.

35.5.2 Link to Pharmacology and Toxicology

Cocaine blocks the reuptake of dopamine, norepinephrine, and serotonin. Cocaine and amphetamines cause CNS stimulation by increasing the dopamine, norepinephrine, and serotonin levels in the synaptic cleft. While cocaine blocks the reuptake mechanism, amphetamines increases the secretion of dopamine, norepinephrine, and serotonin from presynaptic vesicles into the synaptic cleft. The increased neurotransmitters bind to their receptors on the post-synaptic neurons and stimulate these neurons, therefore, cocaine and amphetamines are called CNS stimulants. Toxicity of these drugs causes tachycardia, hypertension, excessive sweating, and agitation.

35.5.3 Niacin

Even though the liver can synthesize niacin from tryptophan, the amount is thought not to be sufficient to meet the body's requirements. Therefore niacin, as vitamin B3, must be acquired from the diet to meet these needs.

35.5.4 Synthesis of Histamine from Histidine

Histamine is synthesized from histidine by a pyridoxal phosphate dependent histidine carboxylase. This enzyme is mainly expressed in the mast cells, gastric enterochromaffin-like cells (ECLCs), and histaminergic neurons of the CNS (▶ Fig. 35.9). In response to allergic reactions, histamine acts on vascular smooth muscle cells and endothelial cells, leading to vasodilation and increased vascular permeability. In the stomach, it increases gastric secretion. As a neurotransmitter, it is important for the regulation of the circadian cycle, appetite, learning, memory and the nervous system.

The wide effects of histamine are transmitted through four different types of membrane receptors. **H1 receptors** of the skin and respiratory tract mediate allergic reactions that cause bronchoconstriction, vasoconstriction or vasodilation. **H2 receptors** are involved in HCl secretion from the parietal cells of the stomach. **H3 receptors** in the CNS are involved in the regulation of brain's vigilant responses such as learning and memory. Inhibition of these receptors causes sleepiness and a decrease in alertness. **H4 receptors** found on mast cells and basophilic leukocytes are involved in regulating chemotaxis.

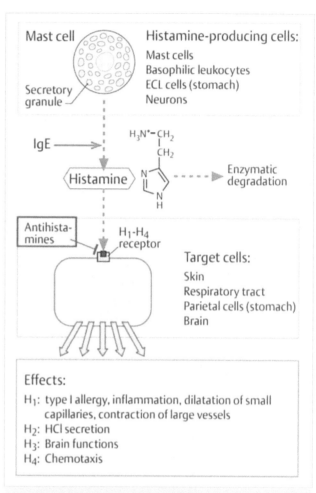

Fig. 35.9 Synthesis and the effects of histamine. (Source: Koolman J, Röhm K, ed. Color Atlas of Biochemistry. 3rd Edition. New York, NY. Thieme; 2012.)

35.5.5 Link to Pharmacology

Antihistamines are used to relieve the symptoms of allergies such as hay fever, hives, conjunctivitis and reactions to insect bites or stings. They are also used to prevent motion sickness and insomnia. Some antihistamines (e.g. diphenhydramine) used in the treatment of allergic reactions can cross the blood-brain barrier and cause drowsiness. The new generation antihistamines (e.g. fexofenadine) are designed in such a way that they cannot cross the blood-brain barrier. Therefore, this family of antihistamines does not cause drowsiness.

Excessive HCl production from the parietal cells of gastric mucosa can be inhibited by H2 receptor antagonists such as cimetidine and ranitidine.

35.5.6 Synthesis of γ-aminobutyrate (GABA) from Glutamate

Glutamate and GABA are both synthesized in the brain and function as important neurotransmitters. While glutamate is an excitatory neurotransmitter, GABA is the most widely distributed inhibitory neurotransmitter in the central nervous system. Its primary role is to attenuate the excitability of

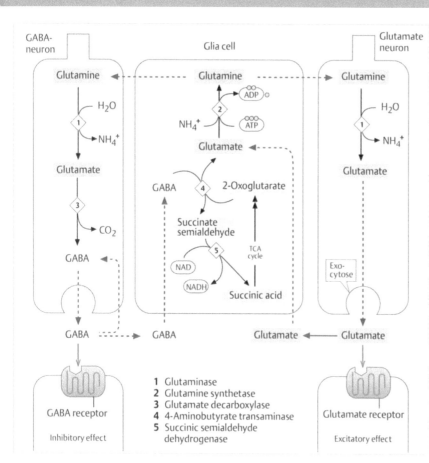

Fig. 35.10 An inhibitory neurotransmitter GABA is synthesized from glutamate. Glutamate is an excitatory neurotransmitter. (Source: Koolman J, Röhm K, ed. Color Atlas of Biochemistry. 3rd Edition. New York, NY. Thieme; 2012.)

V

1 Glutaminase
2 Glutamine synthetase
3 Glutamate decarboxylase
4 4-Aminobutyrate transaminase
5 Succinic semialdehyde dehydrogenase

neuronal activity in all areas of the brain. Excessive GABAergic stimulation leads to sedation, amnesia, and ataxia. On the other hand, attenuation of GABA signaling results in arousal, anxiety, restlessness, insomnia, and exaggerated reactivity. Although alcohol affects the brain's neurons in various ways, one of its activities occurs through binding to the GABA receptors, which increases the movement of chloride ions into the cell through its GABA-gated chloride channel. Due to these effects, alcohol causes muscle relaxation and sedation.

GABA is synthesized through decarboxylation of glutamate by the PLP-dependent enzyme glutamate decarboxylase (▶ Fig. 35.10). Glutamate decarboxylase is expressed only in the GABAergic neurons. The GABA shunt is a mechanism by which the supply of GABA in the CNS is conserved. In the GABA shunt, α-ketoglutarate is formed from glucose via glycolysis and the TCA cycle and is transaminated to form glutamate, the precursor of GABA. Once it is released into the synaptic cleft, GABA is recycled by reuptake into the presynaptic terminals and/or surrounding glia and transaminated to succinic semialdehyde, which can then be converted to succinate, a TCA intermediate, thereby completing the loop.

35.5.7 Link to Pathology and Pharmacology

GABA is the main inhibitory neurotransmitter counterbalancing neuronal excitation in the cerebral cortex. When this balance is perturbed, seizures may develop. Therefore, the GABA system has been targeted for the treatment of epilepsy and seizures. For example, benzodiazepines and barbiturates are GABA$_A$ receptor agonists and work by enhancing GABA-mediated inhibition. Drugs that increase synaptic GABA by either inhibiting its uptake or its transamination to succinic semialdehyde are potent anticonvulsants.

35.5.8 Synthesis of Nitric oxide from Arginine

Nitric oxide (NO) is the smallest known signaling molecule. Since it is a gas, it is easily diffusible within its surroundings to function either as a neurotransmitter or a vasodilatory local hormone. NO is synthesized from the amino acid arginine by nitric oxide synthase (NOS) (▶ Fig. 35.11). There are three isozymes of NOS: iNOS (inducible), eNOS (endothelial) and nNOS (neuronal). NO produced from eNOS promotes relaxation in the endothelial cells of smooth muscles through activating an intracellular guanylate cyclase enzyme, which increases the concentration of the secondary messenger cGMP. Increased cGMP activates a protein kinase called PKG, which triggers relaxation of smooth muscles and dilation of blood vessels. NO produced from iNOS is involved in inhibiting platelet aggregation, and NO produced in the brain from nNOS promotes learning processes.

35.5.9 Link to Physiology

Just like NO, atrial natriuretic peptide (ANP) reduces blood pressure through cGMP-triggered relaxation of the smooth muscles. Binding of ANP to its receptor activates the guanylyl

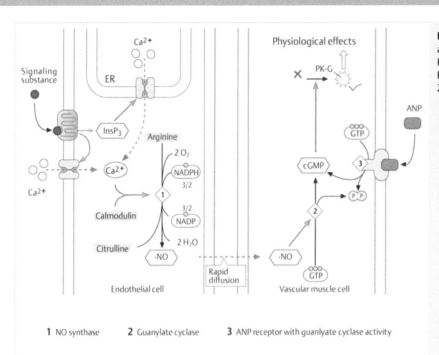

Fig. 35.11 Synthesis of NO from arginine (left) and signaling pathway of NO (right). (Source: Koolman J, Röhm K, ed. Color Atlas of Biochemistry. 3rd Edition. New York, NY. Thieme; 2012.)

1 NO synthase   **2** Guanylate cyclase   **3** ANP receptor with guanlyate cyclase activity

cyclase activity of the ANP receptor and therefore increases intracellular cGMP levels.

35.5.10 Link to Pharmacology

Isosorbide dinitrite (nitroglycerin) tablets are useful in the treatment of angina pectoris because nitroglycerin generates NO, which activates guanylyl cyclase in the vascular smooth muscle cells. Active guanylyl cyclase increases cGMP, which relaxes the blood vessels, thereby increasing cardiac muscle perfusion.

Another drug, **sildenafil** (Viagra), also increases NO levels by inhibiting the phosphodiesterase in vascular smooth muscle cells, which normally degrades cGMP. Inhibition of the phosphodiesterase enzyme keeps cGMP levels elevated, causing relaxation of the smooth muscle and allowing blood flow into the penis.

35.5.11 Link to Pathology

NO is also a radical that is used in macrophages and neutrophils to kill invading bacteria. Because NO is a radical, accumulation of NO leads to its oxidation to peroxynitrite (ONOO-) which is thought to be an important mediator of chronic inflammation and neurodegenerative diseases by virtue of its ability to nitrate and oxidize biomolecules.

35.5.12 Synthesis of Creatine from Glycine, Arginine and S-adenosylmethionine

The amino acids glycine, arginine and S-adenosylmethionine (SAM) are required for the *de novo* synthesis of creatine in the kidney and the liver (▶ Fig. 35.12). When there is excess ATP, the enzyme creatine kinase produces phosphocreatine from creatine. Phosphocreatine in the muscle serves as a high-energy molecule that can be used as a substrate to rapidly synthesize ATP from ADP at the beginning of intense muscular contraction. Creatinine is formed from phosphocreatine via non-enzymatic hydrolysis in muscle. Creatinine is the waste product of this reaction and is excreted in the urine.

35.5.13 Link to Pathology

Creatinine clearance is an indicator of the renal glomerular filtration rate. Accumulation of creatinine in the blood is an indication of renal glomerular damage. Thus, renal function can be monitored by determining the levels of creatinine and protein in the urine.

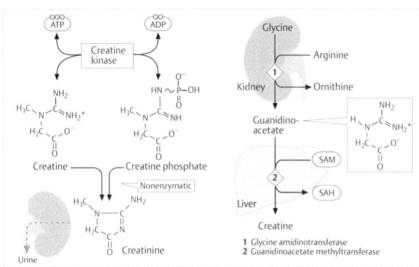

Fig. 35.12 Synthesis of creatine from glycine, arginine and S-adenosylmethionine (SAM) (right). Creatine is phosphorylated by creatine kinase to form creatine phosphate (left). Creatinine is the waste product of creatine. (Source: Koolman J, Röhm K, ed. Color Atlas of Biochemistry. 3rd Edition. New York, NY. Thieme; 2012.)

V

Review Questions

1. Which of the following enzymes requires pyridoxal phosphate as a coenzyme?
   A) Dopamine-β-hydroxylase
   B) DOPA carboxylase
   C) Monoamine oxidase
   D) Tyrosinase
   E) Tyrosine hydroxylase

2. Which of the following amino acids is the precursor for melanin?
   A) Arginine
   B) Glycine
   C) Histidine
   D) Tryptophan
   E) Tyrosine

3. Which of the following secondary messengers is increased by NO signaling?
   A) cAMP
   B) Ca^{2+}
   C) cGMP
   D) DAG
   E) PIP2

4. Synthesis of which of the following hormones is most likely to increase in patients who have nutritional deficiency of iodine?
   A) T3 and TSH
   B) T4 and TSH
   C) T3 and T4
   D) TSH
   E) TSH and TRH

5. Which of the following histamine receptors is inhibited by anti-heartburn medications such as cimetidine?
   A) H1
   B) H2
   C) H3
   D) H4

Answers

1. **The correct answer is B**. Dopa decarboxylase catalyzes the synthesis of dopamine from dopa, as well as the synthesis of serotonin from 5-hydroxytryptophane. This enzyme requires pyridoxal phosphate (PLP), which is the coenzyme form of vitamin B6.

Dopamine-β-hydroxylase (answer choice A) is a Cu^{2+} containing enzyme and that requires vitamin C as a coenzyme for its function. Monoamine oxidase (answer choice C) requires FAD as coenzyme. Tyrosinase (answer choice D) is a copper containing enzyme that catalyzes the synthesis of melanin, the colored pigment of the skin and hair. Tyrosine hydroxylase (answer choice E) requires tetrahydrobiopterin (BH4) as coenzyme in the synthesis of dopa.

2. **The correct answer is E**. Melanin is the pigment of hair, skin, and iris of the eyes and is synthesized in melanocytes from the amino acid tyrosine. Arginine and glycine (answer choices A and B) are precursors for creatine synthesis. Histamine is synthesized from histidine (answer choice C). Tryptophan (answer choice D) is a precursor for serotonin, melatonin and niacin.

3. **The correct answer is C**. NO diffuses into the cells through membranes and activates intracellular guanylate cyclase, which produces cGMP from GTP. Increased cGMP activates protein kinase G. Activation of the PK G relaxes endothelial smooth muscle cells, dilating the blood vessels and allowing increased blood flow and reducing blood pressure.

The answer choices A, B, D, and E are incorrect because although these are secondary messengers as well, NO signaling does not increase the levels of these secondary messengers.

4. **The correct answer is E**. Iodine is needed for the synthesis of the thyroid hormones T3 and T4. Therefore, deficiency of iodine decreases the synthesis of these hormones. Since the T3 levels are low, absence of negative feedback inhibition of TSH and TRH secretion will cause a stimulation of the pituitary and the hypothalamus, causing an increase in TSH and TRH levels.

5. **The correct answer is B**. Histamine binds to H2 receptors in the gastric parietal cells and stimulates the secretion of HCl. Anti-heart burn medications such as cimetidine are H2 receptor antagonists or H2 blockers that inhibit the secretion of HCl. Medications that block H1 receptors (answer choice A) are used for alleviating allergic reactions in the skin and upper respiratory tract. H3 receptors (answer choice C) in the CNS are involved in regulation of the brain's vigilant responses such as learning and memory. Inhibition of these receptors causes sleepiness and decreases alertness. H4 receptors (answer choice D) found on the mast cells and basophilic leukocytes and are involved in regulating chemotaxis.

V

36 Porphyrin Heme Metabolism and Iron Homeostasis

At the conclusion of this theme, students should be able to:
- Discuss the major classes of heme containing proteins and the basic structural features of porphyrins
- Identify the major tissues and subcellular compartments in which heme biosynthesis occurs
- Describe the pathway of heme synthesis and its regulation
- Explain the biochemical basis of lead poisoning and discuss its diagnosis and treatment options
- Discuss the biochemical and genetic basis, as well as treatment options for porphyria cutanea tarda, acute intermittent porphyria and erythropoietic protoporphyria
- Describe how bilirubin is derived from heme and how it is handled in the body
- Distinguish among the biochemical and genetic basis, diagnosis, and the treatment options of the neonatal, hemolytic, hepatocellular, and obstructive forms of jaundice
- Explain normal iron metabolism
- Discuss the pathology related to iron metabolism and its diagnosis and treatment options

Protoporphyrin IX chelated with iron is called heme. Heme is a prosthetic group found in many proteins. The heme iron facilitates systemic oxygen transfer via hemoglobin. Heme is also a component of various oxidoreductases, peroxidases, monooxygenases, and numerous cytochrome P450 enzymes. The oxidoreductases, which include cytochrome oxidase and cytochrome c, participate in the mitochondrial electron transport chain. Cytochrome P450 enzymes are involved in detoxification, degradation, and biosynthetic reactions. Some of the compounds detoxified by cytochrome p450s include xenobiotics such as alcohol, drugs, and carcinogens. The synthesis of cholesterol, vitamin D, and bile acids also requires the activity of some members of the cytochrome p450 family.

In this chapter, first we will review the synthesis of heme and disorders of the heme synthesis pathway. Next, heme degradation pathway and causes of hyperbilirubinemia or jaundice will be reviewed. And finally, iron hemostasis will be reviewed.

36.1 Heme Synthesis

Although heme can be synthesized in all cells, the predominant sites of heme synthesis are the erythroid cells of the bone marrow and the liver. In the liver, heme is mainly used by the cytochrome P450 enzymes. Therefore, based on the demand for heme proteins, the rate of heme biosynthesis changes accordingly. In erythroid cells, heme is used for hemoglobin synthesis. Since there is a continuous turnover of the erythroid cells in the body, the synthesis of heme is constant.

Heme synthesis starts with two simple molecules: succinyl CoA, which is produced as part of the TCA cycle in the mitochondria, and glycine. Heme synthesis starts in the mitochondria, continues in the cytoplasm, and ends in the mitochondria (▶ Fig. 36.1). The reason that this process begins in the mito-

chondria is because this is the only place within the cell that succinyl CoA is found.

The first reaction involves the condensation of one glycine with one succinyl CoA by the pyridoxal phosphate-containing enzyme, δ-amino levulinic acid synthase (ALAS), which is the rate-limiting step of the heme biosynthesis pathway. There are two isoforms of this enzyme: ALAS1 and ALAS2. Each is produced by a different gene and controlled by different mechanisms (see below for tissue specificity and regulation).

The next step is catalyzed by ALA dehydratase, also known as porphobilinogen synthase, which is cytoplasmic and requires zinc for its activity, although it is inhibited by heavy metals such as lead. Subsequently, a condensation reaction involving four porphobilinogens and resulting in the elimination of ammonia to generate hydroxymethylbilane occurs. This condensation reaction is catalyzed by porphobilinogen deaminase, also known as either uroporphyrinogen I synthase or hydroxymethylbilane synthase. Deficiency of porphobilinogen deaminase causes **acute intermittent porphyria**.

The next step is the ring closure of the linear hydroxymethylbilane, which is catalyzed by uroporphyrinogen III synthase and creates uroporphyrinogen III. Uroporphyrinogen III decarboxylase then converts uroporphyrinogen III to coproporphyrinogen III. Genetic deficiency of uroporphyrinogen III decarboxylase causes **porphyria cutanea tarda**.

Coproporphyrinogen III translocates from the cytoplasm into the mitochondria and is further decarboxylated to form protoporphyrinogen IX. Oxidation of this product results in the synthesis of protoporphyrin IX. The final step is catalyzed by ferrochelatase, a mitochondrial enzyme, which inserts ferrous ion (iron in oxidation state + 2) into the protoporphyrin IX to form heme. Just like ALA dehydratase, ferrochelatase is also inhibited by heavy metal poisoning, such as lead. Genetic deficiency of ferrochelatase causes **erythropoietic protoporphyria**.

36.1.1 Regulation of Heme Synthesis

As mentioned previously, ALAS is the rate-limiting step of the heme synthesis pathway. The two isoforms of ALAS, ALAS1 and ALAS2, are produced by different genes and controlled by different mechanisms. While the *ALAS1* gene is located on chromosome 3, the gene encoding *ALAS2* is located on the X-chromosome. The ALAS1 enzyme is expressed in non-erythroid tissues, whereas the ALAS2 enzyme is expressed in erythroid specific cells. Consequently, the activity of these two isoforms is regulated by different mechanisms.

Hepatic Regulation of Heme Synthesis: ALAS1

The liver isozyme of ALAS is regulated both allosterically and transcriptionally. The final product of the pathway, heme, is the negative allosteric regulator of the ALAS1 enzyme, and can provide a negative feedback mechanism for the regulation of its synthesis. In addition, heme also represses the transcription of ALAS1 mRNA, which is the most effective means of regulating the activity of ALAS1. However, physiological concentrations of

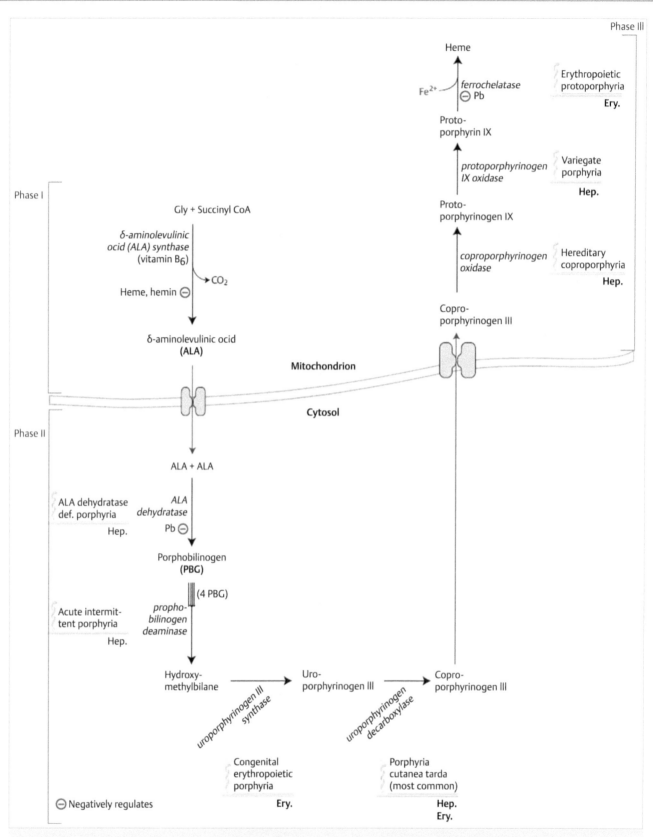

Fig. 36.1 Biosynthesis of heme. (Source: Panini S, ed. Medical Biochemistry- An Illustrated Review. New York, NY. Thieme; 2013.)

heme is not sufficient to provide negative feedback inhibition to regulate ALAS1. Therefore, pharmacological agents, such as hematin and hemin, have been developed to take advantage of this control mechanism.

Transcriptional regulation of ALAS1 is subject to regulation by diverse factors, which are frequently associated with xenobiotic, hormone, and drug metabolism.

Medication such as barbiturates, some antibiotics, alcohol, and xenobiotics induce the transcription of ALAS1 mRNA, because the metabolism of these compounds induces the synthesis of cytochrome P450 enzymes. Since the P450 enzymes contain heme, the synthesis of heme along with the synthesis of P450 enzymes must be upregulated. As a result, de-repression of transcription increases the ALAS1 mRNA levels and consequently the synthesis of ALAS1 increases.

Several other factors also increase the ALAS1 mRNA levels. These include blood loss due to menstruation, an accident, or surgery, malnutrition, and maladies such as sickness, infection, or stress. Since the transcription of ALAS1 mRNA is indirectly regulated by blood glucose concentrations, low blood glucose levels cause an upregulation of *ALAS1* gene expression.

Erythroid Regulation of Heme Synthesis: ALAS2

ALAS2 expression is regulated by the amount of iron present. The 5'UTR of ALAS2 mRNA contains an iron responsive element (IRE), which binds to iron responsive binding protein (IRP). If iron levels are low, IRP binds to IRE. This interaction between IRE and IRP represses the translation of ALAS2 mRNA. In contrast, high iron levels prevent the binding of IRP to IRE, thereby increasing ALAS2. The availability of iron parallels erythroid heme biosynthesis. Therefore, ALAS2 translation is up-regulated when iron is abundant. In addition to ALAS2, expressions of other proteins that are involved in iron homeostasis, such as transferrin receptor and ferritin, are also regulated by IRE and IRP interactions. Transferrin receptor mRNA has a IRE in its 3' UTR, which binds to IRP when iron levels are low. This IRE-IRP interaction protects transferrin receptor mRNA from degradation, making iron available and taken up for heme biosynthesis. In contrast to transferrin receptor expression, ferritin mRNA expression is downregulated by IRE-IRP interactions in its 5' UTR when iron levels are low. [1]

36.1.2 Disorders of Heme Biosynthesis

Disorders of heme biosynthesis can be congenital or acquired and can lead to either sideroblastic anemias or porphyrias, or in some cases both.

Genetic deficiencies of ALAS2 give rise to **X-linked sideroblastic anemia**, while genetic deficiency of porphobilinogen deaminase causes **acute intermittent porphyria**, and genetic deficiency of uroporphyrinogen III decarboxylase causes **porphyria cutanea tarda**. Deficiency of ferrochelatase causes **erythropoietic protoporphyria,** which is characterized by the presence of both porphyria and sideroblastic anemia. The defective in this enzyme results in overproduction, accumulation and excretion of toxic precursor compounds, such as ALA and porphobilinogen or protoporphyrin IX. With the exception of X-linked sideroblastic anemia, most of the diseases of heme

metabolism have an autosomal dominant inheritance pattern, resulting in a 50% reduction in activity of the affected enzyme.

Besides the genetic deficiencies of ALAS2, nutritional deficiency of vitamin B6 also affects the activity of both ALAS1 and ALAS2 enzymes. Heme biosynthesis also is affected by lead poisoning, which causes an acquired deficiency of ferrochelatase or ALA dehydratase enzymes.

36.1.3 Sideroblastic Anemias

Sideroblastic anemias are a heterogeneous group of disorders that share two main characteristics: impaired heme biosynthesis causing microcytic-hypochromic erythrocytes, and the emergence of ringed sideroblasts, seen as nucleated erythrocytes containing iron-laden mitochondria, in the bone marrow. A sideroblast is show in ▶ Fig. 36.2**a.**

There are various causes of sideroblastic anemias with the most common being congenital X-linked sideroblastic anemia, sideroblastic anemia due to isoniazid treatment, or lead poisoning.

Congenital X-linked sideroblastic anemia is due to mutations in the *ALAS2* gene that impair the enzyme's activity and cause the disruption of the normal production of heme in the developing red blood cells. A reduction in the amount of heme prevents these cells from making enough hemoglobin. Because almost all of the iron transported into erythroblasts is normally incorporated into heme, the reduced production of heme leads to a buildup of excess iron in these cells.

36.1.4 Link to Pharmacology

Nutritional deficiency of vitamin B6, also known as pyridoxine phosphate in the active form, causes inactivation of ALAS enzymes since these enzymes require pyridoxine phosphate as a coenzyme for their activity. As a consequence, this deficiency may lead to sideroblastic anemia. Isoniazid used in tuberculosis treatment and chronic consumption of ethanol impairs the absorption of vitamin B6 and, therefore, may also lead to development of sideroblastic anemia by decreasing the availability of pyridoxal phosphate.

Lead poisoning inhibits heme synthesis. Cosmetics, jewelry, toys, pottery, paint chips from lead-based paint, water contaminated by lead leaching from pipes, soil contaminated with lead, and occupational exposure are just a few sources of lead. Lead inhibits two enzymes of heme biosynthesis: ALA dehydratase and ferrochelatase. ALA dehydratase inhibition by lead is stronger than ferrochelatase. Another enzyme inhibited by lead is pyrimidine 5'nucleotidase, which inhibits RNA

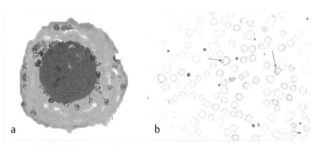

Fig. 36.2 (a) sideroblast photograph. **(b)** Basophilic stippling of erythrocytes. Arrows shows basophilic stippling of erythrocytes.

translation, leading to the accumulation of RNA in erythroblasts and giving the appearance of numerous small dots, hence the name basophilic stippling (▶ Fig. 36.2**b**).

The symptoms of lead poisoning differ depending on the degree of poisoning. In mild lead poisoning, lethargy, anorexia, abdominal discomfort and arthralgia may develop. In moderate poisoning, anemia, headache, abdominal cramps, peripheral neuropathies and gingival and long bone lead lines may be expected. High levels of lead poisoning may lead to convulsion, coma, encephalopathy and renal failure. The diagnosis is based on the blood lead concentration in association with compatible clinical symptoms. The diagnostic tests for lead poisoning include the accumulation of Zn-protoporphyrin, sideroblastic immature erythrocytes in bone marrow smears, hypochromic and microcytic mature erythrocytes with basophilic stippling in peripheral smear.

Treatment options for lead poisoning include chelators such as desferrioxamine, sodium calcium edetate (sodium calcium EDTA), and pencillamine. These molecules bind to lead, resulting in its excretion in urine.

36.1.5 Porphyrias

Hereditary or acquired deficiencies of the enzymes in the porphyrin synthesis pathway give rise to a group of disorders called porphyrias (▶ Table 36.1). Porphyrias are classified as hepatic or erythropoietic porphyrias. While an overproduction or accumulation of intermediates of the pathway in the liver results in hepatic porphyrias, accumulation of the intermediates in the erythroid cells or in the bone marrow results in erythropoietic porphyrias.

Porphyrias are also categorized according to the presenting symptoms: neuropathy and psychiatric symptoms or photosensi-

Table 36.1 Porphyrias

Defective Enzyme	Associated Porphyria	Inheritance and Features	Type of Porphyria
δ-ALA dehydratase	δ-ALA dehydratase deficiency porphyria	Autosomal recessive Acute attacks of abdominal pain and neuropathy	Hepatic
PBG deaminase (in liver)	Acute intermittent porphyria	Autosomal dominant Deficiency leads to excessive production of ALA and PBG Periodic attacks of abdominal pain and neurologic dysfunction	Hepatic
Uroporphyrinogen III synthase (in erythrocytes)	Congenital erythropoietic porphyria	Autosomal recessive Deficiency leads to accumulation of uroporphyrinogen I and its red-colored, air oxidation product uroporphyrin I Photosensitivity; red color in urine and teeth; hemolytic anemia	Erythropoietic
Uroporphyrinogen decarboxylase	Porphyria cutanea tarda (PCT)	Autosomal dominant Deficiency leads to accumulation of uroporphyrinogen III, which converts to uroporphyrinogen I and its uroporphyrin oxidation products *Most common porphyria in the United States*; photosensitivity resulting in vesicles and bullae on skin of exposed area; wine red–colored urine	Hepatoerythropoietic
Coproporphyrinogen oxidase	Hereditary coproporphyria	Autosomal dominant Photosensitivity and neurovisceral symptoms (e.g., colic)	Hepatic
Protoporphyrinogen IX oxidase	Variegate porphyria	Autosomal dominant Photosensitivity and neurologic symptoms and developmental delay in children	Hepatic
Ferrochelatase	Erythropoietic protoporphyria	Autosomal dominant Photosensitivity with skin lesions after brief sun exposure. Patients may have gallstones and mild liver dysfunction	Erythropoietic

Abbreviations: ALA, δ-aminolevulinic acid; PBG, porphobilinogen.

Note: All porphyrias cause neurologic symptoms, whereas those that lead to the accumulation of cyclic tetrapyrroles (coproporphyrinogen III and beyond) result in photosensitivity. Urine that turns dark or reddish on standing is seen with all porphyrias except ALA dehydratase deficiency porphyria.

Source: Panini S, ed. Medical Biochemistry—An Illustrated Review. New York, NY. Thieme; 2013.)

tivity. Neuropathy and psychiatric symptoms are mostly seen in acute porphyrias, in which porphobilinogen and ALA accumulate. Photosensitivity is caused by sun-light, which oxidizes the porphyrinogens to porphyrins in the skin. These photo-excited porphyrins mediate oxidative damage causing cutaneous lesions.

36.1.6 Acute intermittent porphyria (AIP)

Porphobilinogen deaminase deficiency causes acute intermittent porphyria. In addition to enzyme deficiency, other factors must also be present for the disease to exhibit symptoms. The precipitating factors for the acute attacks in AIP patients are medications such as barbiturates, some antibiotics, alcohol, and xenobiotics, blood loss due to menstruation, accident, or surgery, malnutrition, and maladies. As hepatic ALAS activity increases, the synthesis of ALA and porphobilinogen also increase. Since the porphobilinogen deaminase activity is reduced to ~50% of normal, ALA and porphobilinogen accumulate. During acute attacks, urinary porphobilinogen and ALA levels markedly increase. The color of urine in a patient with AIP during crisis turns a dark-brown color resembling the color of port wine due to porphobilinogen oxidation.

Treatment and/or prevention options include the avoidance of precipitating factors and the adherence to a high-carbohydrate diet. Heme and hematin treatments also reduce the ALAS1 activity in the liver. Clinical presentation includes abdominal pain and neurologic dysfunction.

Porphyria cutanea tarda (PCT) is the most common porphyria and it occurs due to a decreased activity of uroporphyrinogen decarboxylase, which may be low due to genetic or sporadic changes in the enzyme activity. The clinical symptoms of PCT usually develop after the age of 30 and its onset in childhood is rare. Environmental factors contributing to PCT include alcohol abuse, iron overload in the liver, exposure to sunlight, hepatitis B, hepatitis C, HIV infections, and estrogen treatment used for either oral contraceptives or for prostate cancer treatments. Clinical presentation is usually confined to the skin and consists of blistering skin lesions in exposed areas, especially on the backs of the hands. Hypertrichosis and hyperpigmentation on the face are more prominent in females, bringing them to the dermatologist. There are no neurologic symptoms in PCT.

Plasma, urinary, and fecal porphyrin levels are useful in differentiating PCT from other porphyrias. Diagnosis of PCT can be made by detecting uroporphyrin and heptacarboxylate porphyrin in urine, or isocoproporphyrin in feces.

Treatment options include avoidance of environmental exposures such as alcohol, tobacco, and estrogen. Sunscreen may help for the photosensitivity. Iron chelation therapy or phlebotomy to remove the excess iron load which indirectly inhibits uroporphyrinogen decarboxylase activity may also be helpful. Removal of iron activates heme synthesis, thereby alleviating some of the symptoms of PCT.

Erythropoietic Protoporphyria (EPP) is the most common erythropoietic porphyria and is inherited in an autosomal recessive manner. It is due to decreased ferrochelatase activity, the last enzyme of heme biosynthesis. Less commonly, EPP is due to increased activity of ALAS2 and in this case it is termed the X-linked form of EPP. Painful photosensitivity starts early in childhood and includes the following symptoms: burning pain, redness, and itching occurring within minutes of sunlight exposure. EPP can greatly impair quality of life, with chronic liver disease developing in later stages. Treatment options are limited in EPP. Avoiding sunlight exposure and wearing clothing designed to provide protection is the mainstay of management of EPP, and is necessary throughout life.

36.2 Heme Degradation

When erythrocytes reach their lifespan of ~120 days, they are phagocytosed by macrophages of the spleen, liver, and bone marrow (known collectively as the reticuloendothelial system). The hemoglobin component of the cells is broken down to its constituent amino acids and heme. Heme is further degraded to produce bilirubin by the macrophages of the reticuloendothelial system. In the first reaction of this process, heme is degraded by heme oxygenase, releasing green-pigmented biliverdin, iron and carbon monoxide (CO) (▶ Fig. 36.3). Biliverdin is then converted to red-orange-colored bilirubin by biliverdin reductase.

Because bilirubin is poorly soluble in the aqueous medium of the plasma, it must be transported to the liver by albumin. After entering the hepatocytes, bilirubin is solubilized by conjugation to glucuronic acid, which yields bilirubin monoglucuronide and diglucuronide. The conjugation of glucuronic acid to bilirubin is catalyzed by UDP-glucuronyl transferase.

The conjugated bilirubin is excreted into bile, which drains into the duodenum. Conjugated bilirubin passes through the proximal small bowel and when it reaches the distal ileum and colon, it is hydrolyzed to unconjugated bilirubin by bacterial glucuronidases. The unconjugated bilirubin is reduced to urobilinogens by the normal gut bacteria. About 80–90% of urobilinogens are excreted in feces, either unchanged or oxidized to brown-orange derivatives called urobilins or stercobilins, giving feces its characteristic color. The remaining ~10% of the urobilinogens is reabsorbed into the enterohepatic circulation. From there, some of the urobilinogen is taken up by liver, but a small amount of urobilinogen escapes hepatic uptake, filters across the renal glomerulus, and is excreted in urine, giving urine its characteristic color.

36.2.1 Link to Pathology: Intravascular Hemolysis can cause Hemoglobinuria

Intravascular hemolysis of erythrocytes can occur in sickle cell disease, thalassemia, and in the deficiencies of the G6PD and the pyruvate kinase. In these conditions, the membranes of erythrocytes rupture and their hemoglobin leaks into the plasma. This hemoglobin in the plasma breaks down into hemoglobin dimers that are then bound by the plasma proteins haptoglobin or α-2 macroglobulin. If intravascular hemolysis continues, the hemoglobin dimers exceed the binding capacity of haptoglobin and α-2 macroglobulin in the plasma, which leads to hemoglobinemia. This unbound free hemoglobin in the plasma can be readily filtered through the glomerulus of kidneys, causing hemoglobinuria.

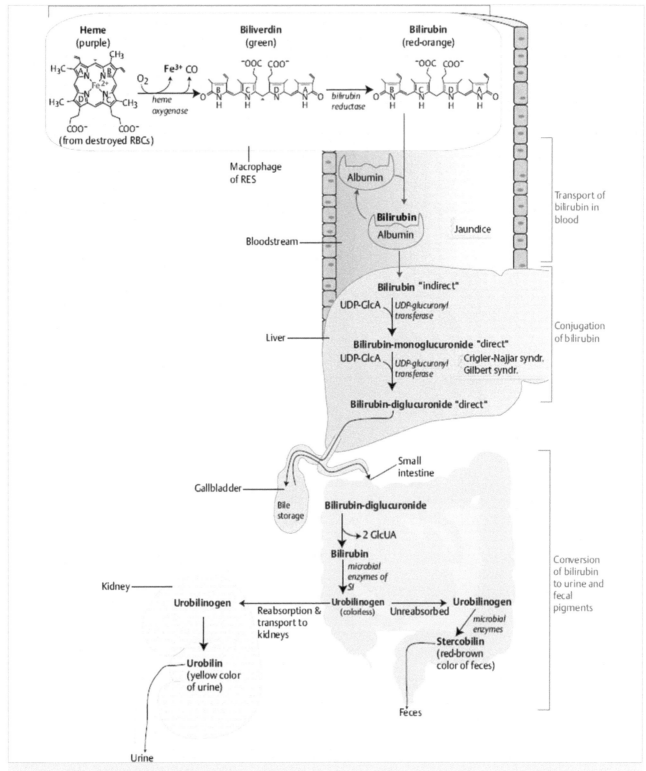

Fig. 36.3 Heme degradation to urobilin and stercobilin. (Source: Panini S, ed. Medical Biochemistry—An Illustrated Review. New York, NY. Thieme; 2013.)

36.2.2 Jaundice or Hyperbilirubinemia

Jaundice (icterus) develops when there is an imbalance between bilirubin production and clearance that leads to increased levels of bilirubin deposition in the tissues.. This condition is characterized by yellow discoloration of the skin, conjunctivae, and mucous membranes. Due to its neurotoxicity, the body must maintain the ability to excrete this compound. Both unconjugated bilirubin and conjugated bilirubin may accumulate. However, because of poor solubility of unconjugated bilirubin and tight binding to albumin, it cannot be excreted in urine, whereas conjugated bilirubin is soluble and is excreted in the urine.

36.2.3 Link to Pathology: Measurement of Serum Bilirubin

The terms direct- and indirect-reacting bilirubin are derived from the van den Bergh reaction. This assay is still used in laboratories to determine the serum bilirubin level. The direct fraction reacts with diazotized sulfanilic acid in the absence of an accelerator substance such as alcohol and provides an approximate determination of the conjugated bilirubin in serum. The total serum bilirubin is the amount that reacts after the addition of alcohol. The indirect fraction is the difference between the total and the direct bilirubin and provides an estimate of the unconjugated bilirubin in serum.

The causes of jaundice can be classified as prehepatic, hepatic and post-hepatic.

36.2.4 Pre-Hepatic Jaundice

Pre-hepatic jaundice arises due to extensive release of heme from hemolysis that exceeds the capacity of hepatic conjugation of unconjugated bilirubin. Subsequently, unconjugated bilirubin accumulates in the blood and the peripheral tissues. Common causes of hemolysis are G6PD deficiency and sickle cell crisis.

36.2.5 Hepatic Jaundice

Hepatic jaundice is caused by liver damage such as cirrhosis, viral hepatitis, or drugs used to treat hepatocarcinoma. In hepatic jaundice, both unconjugated and conjugated bilirubin accumulates in the blood.

36.2.6 Post-Hepatic Jaundice

Post-hepatic jaundice, also known as obstructive jaundice, develops from a mechanical obstruction of the bile duct that is cause by conditions such as gallstones or tumors, thus preventing the secretion of conjugated bilirubin into the intestines. The feces of afflicted individuals have a characteristic pale clay color due to the relative absence of the brown colored stercobilin.

36.2.7 Neonatal Jaundice

Newborns, particularly those that are premature, may develop neonatal jaundice due to low levels of the hepatic enzyme UDP-glucuronyl transferase at birth. It takes about two weeks until the required enzyme levels are reached. As a result, unconjugated bilirubin accumulates during this time, which can create a serious condition known as kernicterus (chronic bilirubin encephalopathy), a permanent disabling of neurons. In order to avoid the accumulation of unconjugated bilirubin, newborns receive phototherapy to convert the unconjugated bilirubin to a more soluble derivative which can be filtered in the urine.

36.2.8 Hereditary Jaundice

Several genetic defects of the UDP-glucuronyl transferase enzyme can cause Crigler-Najjar syndrome type I and type II and Gilbert syndrome, all of which present with unconjugated hyperbilirubinemia. Mutations in the promoter region of the UDP-glucuronyl transferase gene result in Gilbert syndrome, while mutations in the coding region of the UDP-glucuronyl transferase gene result in Crigler-Najjar syndrome (Type 1 and 2).

36.2.9 Gilbert Syndrome

Gilbert Syndrome is relatively common in the US population and may go unnoticed for years. It is a benign condition caused by a mutation in the 5' promoter region of the *UDPGT* gene, leading to reduced levels of the enzyme. The enzyme levels fluctuate with situations that are conducive to stress such as illness, strenuous exercise, or fasting.

36.2.10 Crigler-Najjar Syndrome

Crigler-Najjar syndrome type I is caused by a complete deficiency of *UDPGT*, leading to very high levels of unconjugated bilirubin, thereby kernicterus ensues. Due to the absence of conjugated bilirubin, the feces of the Crigler-Najjar syndrome type 1 patient is colorless or clay colored. Because of the severity of this disease, it is fatal without liver transplantation.

Unlike Crigler-Najjar syndrome type I, the type II form has some UDP- glucuronyl transferase activity. As a result, this syndrome is benign. Patients are responsive to phenobarbital induction therapy, which increases the UDP- glucuronyl transferase enzyme activity.

36.2.11 Dubin-Johnson Syndrome

Dubin-Johnson syndrome develops as a result of mutations in the *MRP2* gene, which is responsible for the secretion of conjugated bilirubin into the bile. This leads to chronically elevated levels of conjugated bilirubin. Dark pigmented granules develop in the liver in response to the accumulation of conjugated bilirubin that is deposited there. Most patients are asymptomatic except for recurrent jaundice.

36.2.12 Rotor Syndrome

Rotor syndrome is also a relatively mild condition with predominantly conjugated bilirubin in the blood. Just like in Gilbert syndrome and Dubin-Johnson syndrome, jaundice develops intermittently. Both organic anion transporting polypeptide 1B1 (*OATP1B1*) and organic anion transporting polypeptide 1B3 (*OATP1B3*) are responsible for the uptake of

Table 36.2 Disorders associated with abnormal levels of iron

	Disorder Description and Causes	Symptoms
Iron deficiency	Generally due to insufficient dietary uptake Iron *depletion*, however, can only occur through bleeding (e.g., traumatic blood loss, abnormal menstruation, colon cancer)	Fatigue, anemia
Iron excess/overload	Iron poisoning	Hemorrhagic gastritis and liver necrosis
	Hemochromatosis is an autosomal recessive disease that causes accumulation of iron in the liver, heart, pancreas, and skin. It is caused by unregulated duodenal reabsorption of iron due to low hepcidin levels	Cirrhosis, heart failure, diabetes mellitus, bronzed skin, malabsorption
	Hemosiderosis is a condition in which ferritin is degraded to insoluble hemosiderin, which may accumulate. It is observed in the tissues of alcoholics and patients who have blood transfusions (e.g., hemolytic anemia)	Alveolar hemorrhage
	Sideroblastic anemia is caused by impaired heme synthesis, which results in iron buildup in mitochondria	Pale skin, fatigue, enlarged spleen

Source: Panini S, ed. Medical Biochemistry—An Illustrated Review. New York, NY. Thieme; 2013.

bilirubin by the liver. Therefore, mutations in *OATP1B1* gene and/or *OATP1B3* gene cause the development of Rotor syndrome. Both of these proteins are found in the hepatocytes, where they transport bilirubin and other compounds from the blood into the liver so that they can be cleared from the body. In the liver, bilirubin is dissolved in digestive fluid called bile and then excreted from the body.

36.3 Iron Homeostasis

Due to the detrimental effects of both excess and deficiency of iron, the proper control over iron homeostasis is critical. Iron deficiency anemia is one of the most prevalent nutritional problems in the world (▶ Table 36.2). In the case of iron excess, which may occur through either accidental poisoning or due to hemochromatosis, toxicity develops because the free Fe^{+2} reacts with molecular oxygen in the Fenton reaction and causes accumulation of hydrogen peroxide, which further generates OH radical. In order to prevent this oxidative stress, iron is shielded from molecular oxygen and surrounding milieu by binding to proteins. This is the reason that iron availability is tightly regulated both at cellular and at systemic levels. Many proteins such as ferritin, transferrin, iron regulatory proteins (IRPs), and hepcidin play important roles in iron homeostasis.

While IRPs exert their control at the cellular level, hepcidin works at the systemic level. IRPs regulate the transcription and translation of transferrin receptor and ferritin (reviewed above in this Theme). Hepcidin regulates duodenal absorption of dietary iron and iron storage in macrophages. The absorption of

iron is dependent on the body's iron stores, hypoxia and the rate of erythropoiesis.

36.3.1 Absorption, Transport, and Storage of Iron

Iron is obtained through the diet and excreted from the body via bleeding and shedding of cells through the feces, sweat, and urine. Approximately, 70% of the total body iron is associated with hemoglobin (▶ Fig. 36.4). To maintain the required rate of hemoglobin synthesis, a daily intake of 20–25 mg of iron is needed.

Dietary iron is found in either the heme or non-heme forms (▶ Fig. 36.5). In the acidic pH of the stomach, heme is dissociated from hemoproteins. At the same time, the reducing environment of the stomach converts free iron to the ferrous form, which is the only form that can be absorbed in the duodenum. For this reason, vitamin C and other acidic components derived from the diet can increase iron absorption. The remaining iron that is in the ferric form is reduced to ferrous by intestinal brush border ferri-reductases and transported into the intestinal enterocytes by the divalent metal transporter 1 (DMT1). Heme iron is transported across the brush border of the intestines by the heme carrier protein (HCP1). Once in the cells, it is degraded to release ferrous iron. From here on, the metabolism of iron from the diet, either as heme or non-heme, is identical.

Once iron is absorbed into the enterocyte, its fate depends on the iron pool within the cell. Ferrous iron can either be stored in the enterocyte bound to ferritin or released to the circulation by ferroportin. In order for the iron to bind to transferrin, it needs to be in the ferric form. Hephaestin, bound to the basolateral side of enterocytes, is a ferroxidase that oxidizes the ferrous form (Fe + 2) of iron to the ferric from (Fe + 3) before it is incorporated into serum transferrin.

Transferrin is the main protein involved in iron transport in plasma. Once in the circulation, transferrin-bound iron is distributed mainly to the liver but also to muscle, bone marrow and to some extend other tissues. Transferrin-iron complexes are taken up by hepatocytes from the blood by binding to the transferrin receptors, which are then internalized via receptor mediated

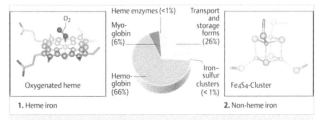

Fig. 36.4 Distribution of iron. (Source: Koolman J, Röhm K, ed. Color Atlas of Biochemistry. 3rd Edition. New York, NY. Thieme; 2012.)

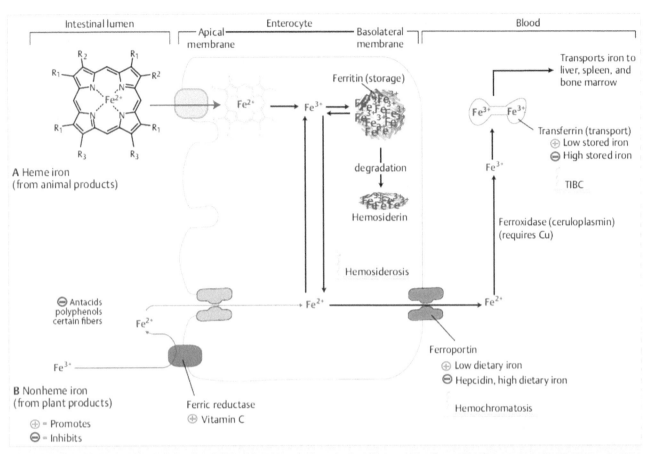

| Intestinal lumen | Enterocyte | Blood |

Fig. 36.5 Iron absorption, storage, and transport. (Source: Panini S, ed. Medical Biochemistry- An Illustrated Review. New York, NY. Thieme; 2013.)

endocytosis. In the endosome inside the cell, iron is released from transferrin due to the acidic environment of the endosome and is reduced to the ferrous form by endosomal ferri-reductase. The transferrin-receptor complex is then recycled back to the cell surface, where the unbound transferrin dissociated from its receptor and returns to the blood to bind to another iron molecule. In addition to the transferrin-bound iron, free iron in the circulation is also taken up by hepatocytes, in this case using DMT1, the transferrin-independent iron transporter.

Once in the cell, iron is stored bound to ferritin, which is the major intracellular iron storage protein. Each ferritin molecule can bind to approximately 4,500 iron atoms. Ferritin also has enzymatic activity that allows it to reduce ferric iron to ferrous iron. When iron concentrations become abnormally high, the liver converts ferritin into another storage protein called hemosiderin. Storing excess iron in hemosiderin protects the body against the damage that free iron can cause. Hemosiderin is most commonly found in macrophages and is especially abundant in situations following hemorrhage, suggesting that its formation may be related to phagocytosis of red blood cells and hemoglobin.

Ceruloplasmin, a plasma ferroxidase, plays an essential role in mobilizing iron from stores by oxidizing the ferrous form to the ferric form. This allows iron to bind to transferrin so that it can be delivered to other tissues of the body.

Hepcidin is a hormone secreted by the liver that controls iron levels in the circulation by inhibiting the dietary absorption in the intestine, as well as the release of iron from macrophages and hepatocytes. It does so by binding to ferroportin and causing its degradation. When iron levels are low, hepatocytes decrease hepcidin synthesis, allowing more iron to enter circulation. However, when iron is abundant, hepcidin synthesis increases to limit iron absorption and the release from stores.

36.3.2 Link to Pathology: Iron Deficiency and Overload

Iron deficiency is one of the most prevalent nutritional challenges in the world. Populations at risk include menstruating or pregnant females, infants and children, and those with malabsorption diseases such as celiac disease. If the iron deficiency is not due to blood loss, it may be due to low intake or lack of bioavailable iron. This can be seen in infants deprived of iron-fortified formula or cereal after the age of 6 months. It can also be seen in vegans since the non-heme iron in their diet contains many factors that decrease the bioavailability of iron.

Symptoms of iron deficiency are fatigue, feelings of faintness, and cold or abnormal sensations of the extremities, shortness of breath, immunosuppression, low IQ, and behavioral problems. These symptoms can be explained by the roles of iron in the body. The first four symptoms are due to iron-deficiency anemia. Iron-deficiency anemia is characterized by microcytic and hypochromic erythrocytes. Iron deficiency develops in stages. Iron stores first diminish, followed by a decrease in circulating iron, and finally a subsequent fall in hemoglobin production. Blood tests used to detect iron deficiency include measurements of hemoglobin, hematocrit (the percentage of blood volume made up by red blood cells), ferritin, transferrin saturation and total iron-binding capacity (TIBC). Low hemoglobin, hematocrit, ferritin levels and transferrin saturation with a concomitant increase in total iron-binding capacity (TIBC) are indicative of iron deficiency.

Iron overload can be due to accidental poisoning or frequent blood transfusions. However, another cause of iron overload is the genetic disease **hemochromatosis (HH).**

The most prevalent form of this disease in Caucasians, HH-HFE, is an autosomal recessive disorder, caused by mutations in the *HFE* gene. The name of the hemochromatosis gene, *HFE*, stands for "**H**igh **I**ron (Fe)".

The HFE protein controls the production of hepcidin. Low levels of hepcidin leads to increased ferroportin activity. Subsequently, iron uptake from the diet via enterocytes, iron release from macrophages into the circulation, and deposition of the excess of iron in parenchymal cells of tissues all increase, leading to a condition of systemic iron overload. Chronic hemochromatosis may cause bronzed pigmentation to the skin and damage to the liver and pancreas, potentially leading to diabetes mellitus. For this reason, this disease is sometimes called "**bronze diabetes**".

Reference

[1] Chiabrando D, Mercurio S, Tolosano E. Heme and erythropoiesis: more than a structural role. Haematologica. 2014; 99(6):973–983

Review Questions

1. Deficiency in which of the following enzymes causes sideroblastic anemia?
   A) ALAS 1
   B) ALAS2
   C) Porphobilinogen deaminase
   D) Uroporphyrinogen III decarboxylase

2. Which of the following enzymes is inhibited by lead?
   A) ALAS1
   B) ALAS2
   C) ALA dehydratase
   D) Porphobilinogen deaminase
   E) Uroporphyrinogen III synthase

3. Deficiency in which of the following enzymes most likely causes hemolytic jaundice?
   A) Galactose 1-phosphate uridyl transferase
   B) Glucose 6-phosphatase
   C) Glucose 6-phosphate dehydrogenase
   D) UDP-glucuronyl transferase

4. In which of the following conditions is associated with increased direct bilirubin levels in the blood?
   A) Crigler-Najjar syndrome type 1
   B) Crigler-Najjar syndrome type 2
   C) Gilbert Syndrome
   D) Dubin-Johnson syndrome

5. Which of the following proteins is involved in the regulation of iron levels at the systemic level?
   A) Ferritin
   B) Hepcidin
   C) Iron regulatory proteins
   D) Transferrin

Answers

1. **The correct answer is B.** The ALAS2 enzyme is expressed in erythroid specific cells, therefore genetic deficiencies of ALAS2 causes sideroblastic anemia. Because the *ALAS2* gene is located on the X-chromosome, this disease commonly effects males. Sideroblastic anemia is also caused by genetic deficiency of the ferrochelatase enzyme and nutritional deficiency of vitamin B6, which is the coenzyme for ALAS enzymes.

Answer choice A is incorrect because the ALAS1 enzyme is involved in the synthesis of heme for non-erythroid cells, mostly for cytochrome P450 enzymes.

Answer choices C and D are incorrect because genetic deficiency of porphobilinogen deaminase causes acute intermittent porphyria, and genetic deficiency of uroporphyrinogen III decarboxylase causes porphyria cutanea tarda.

2. **The correct answer is C.** Lead inhibits two enzymes of heme biosynthesis: ALA dehydratase and ferrochelatase, although its inhibition of ALA dehydratase is stronger. The diagnostic tests for lead poisoning include the accumulation of Zn-protoporphyrin, sideroblastic immature erythrocytes in bone marrow smears, and hypochromic and microcytic mature erythrocytes with basophilic stippling in peripheral smear.

The answer choices A, B, D, and E are incorrect because lead does not inhibit ALAS1, ALAS2, porphobilinogen deaminase, and uroporphyrinogen III synthase. ALAS1 and ALAS2 enzymes are the rate limiting step of the heme synthesis pathway. The ALAS1 isozyme is found in non-erythroid cells, while ALAS2 is present in erythroid cells.

3. **The correct answer is C.** G6PD deficiency in the pentose phosphate pathway may cause hemolysis of erythrocytes, especially during increased oxidative stress and reactive oxygen species. Individuals with G6PD deficiency cannot generate sufficient amounts of NADPH to reduce oxidized glutathione. Due to insufficient amounts of reduced glutathione, free radicals accumulate. These increased free radicals damage the cellular membranes, especially in the erythrocytes, therefore causing hemolysis. Increased hemolysis increases the degradation of heme and the formation of bilirubin. This increased bilirubin exceeds the capacity of the liver for conjugation, thus unconjugated bilirubin increases, leading to jaundice.

Although genetic deficiency of galactose 1-phosphate uridyl transferase (GALT) causes jaundice, it does not cause hemolysis of the erythrocytes. The UDP-galactose is required to generate UDP-glucose, which is used for conjugation of bilirubin in the liver. Conjugated bilirubin is excreted in the bile and disposed of mostly through the feces. The deficiency of GALT enzyme causes galactosemia, therefore answer choice A is incorrect. Galactosemia patients present with jaundice, hepatosplenomegaly, and cataracts.

Answer choice B is incorrect because deficiency of glucose 6-phosphatase causes von Gierke's syndrome, which presents with severe hypoglycemia due to the inability to maintain blood glucose concentrations.

Answer choice D is incorrect because deficiency of the UDP-glucuronyl transferase enzyme causes hereditary jaundice, including Crigler-Najjar syndrome type I and type II and Gilbert syndrome, all of which present with unconjugated hyperbilirubinemia. Mutations in the promoter region of the UDP-glucuronyl transferase gene result in Gilbert syndrome, while mutations in the coding region of the UDP-glucuronyl transferase gene result in Crigler-Najjar syndromes.

4. **The correct answer is D.** Dubin-Johnson syndrome develops as a result of mutations in the *MRP2* gene, which is responsible for the secretion of conjugated bilirubin into the bile. This leads to chronically elevated levels of conjugated bilirubin. Conjugated bilirubin is also called direct bilirubin since it reacts with diazotized sulfanilic acid in the absence of an accelerator substance such as alcohol and provides an approximate determination of the conjugated bilirubin in serum.

Answer choices A, B, and C are incorrect because deficiency of the UDP-glucuronyl transferase enzyme causes hereditary jaundice, such as Crigler-Najjar syndrome type I and type II and Gilbert syndrome, all of which present with unconjugated hyperbilirubinemia.

5. **The correct answer is B.** Although all of these proteins play important roles in iron homeostasis, the hepcidin hormone regulates iron absorption from the intestine, and therefore

controls iron levels at the systemic level. Increased hepcidin hormone decreases the absorption of iron by stimulating degradation of ferroportin, which is responsible for transporting ferrous iron from enterocytes to hepatocytes and macrophages.

Iron regulatory proteins (answer choice C) exert their control at the cellular level by regulating the transcription and translation of transferrin (answer choice D) receptor and ferritin (answer choice A).

37 Nucleotide Metabolism

At the conclusion of this chapter, students should be able to:
- Discuss the role of the pentose phosphate pathway in the synthesis of 5' phosphoribosyl pyrophosphate (PRPP) and explain its central role in nucleotide metabolism
- Discuss the biosynthesis of purine and pyrimidine nucleotides, with emphasis on the key regulatory steps
- Describe the purine salvage pathway and explain its role in pathologies
- Discuss the pyrimidine salvage pathway and its role in pharmacotherapy for the treatment of cancer and herpes infections
- Discuss purine degradation to uric acid, and the pathophysiology and treatment options of hyperuricemia
- Differentiate between the de novo and salvage pathways in maintaining steady-state purine and pyrimidine nucleotide levels
- Discuss the role of adenylate kinase (myokinase) in generating AMP
- Describe the importance and regulation of ribonucleotide reductase in cancer chemotherapy and in adenosine deaminase deficiency
- Differentiate the causes of orotic aciduria
- Compare and contrast the effects of 5-flurouracil (5-FU) and methotrexate (MTX) on the synthesis of thymidine

Nucleotides are composed of a heterocyclic base, a five-carbon sugar (deoxyribose or ribose), and phosphate(s) (▶ Fig. 37.1). DNA and RNA each have four different bases. The purine bases are adenine and guanine and the pyrimidine bases are cytosine, uracil and thymine. Purines and pyrimidines are essential for replication, transcription, and cellular metabolism. Furthermore, nucleotides also serve an important role in coupling endergonic reactions with exergonic reactions to drive otherwise unfavorable reactions. There are several disorders involving nucleotide metabolism. For example, disorders of purine metabolism include gout, Lesch-Nyhan syndrome, and adenosine deaminase deficiency and purine nucleoside phosphorylase deficiency. A disorder of pyrimidine synthesis is orotic aciduria. Nucleotide metabolism has also been targeted for antibiotic treatment and chemotherapy.

37.1 Do Novo Synthesis versus Salvage Pathways of Purines and Pyrimidines

Nucleotide synthesis can occur through two different pathways: de novo synthesis, referring to the synthesis of nucleotides from simple molecules such as amino acids, folates, and ribose, and recycling of nucleotides after their partial degradation to free bases, such as seen in the salvage pathways. Although the de novo synthesis is important for both purines and pyrimidines, the salvage pathway is important for only purines.

37.2 Synthesis of 5-Phosphoribosyl-Pyrophosphate (PRPP)

The sugar molecule of nucleotides, ribose-5-phosphate, is synthesized by the pentose phosphate pathway. Activation of the ribose-5-phosphate is achieved by phosphoribosyl-pyrophosphate (PRPP) synthetase (▶ Fig. 37.2).

De novo synthesis of purines and pyrimidines use different strategies to link the ribose-5-phosphate to the base. In purine nucleotide synthesis, the parent base is synthesized first before the addition of PRPP. In contrast, pyrimidines use the activated sugar moiety PRPP as a backbone to build the parent base. In salvage pathways, phosphoryl transfer from PRPP converts adenine, hypoxanthine, and guanine to their mononucleotides.

37.3 Synthesis of Deoxyribonucleotides (dNDP) by Ribonucleotide Reductase (RR)

The end products of both purine and pyrimidine nucleotide synthesis are ribonucleotides that are used for RNA synthesis and for cellular metabolism. However, they cannot be used for DNA synthesis. The conversion of ribonucleotides to deoxyribonucleotides is catalyzed by **ribonucleotide reductase (RR)** by reduction of ribonucleotide diphosphates (NDPs) to dNDPs (▶ Fig. 37.3). Ribonucleotide reductase requires reduced thioredoxin. Maintenance of reduced thioredoxin is achieved through coupling of this reaction with thioredoxin reductase, using NADPH as a source of reducing equivalents. NADPH for this reaction is generated by the pentose phosphate pathway. Therefore, the pentose phosphate pathway is intimately linked to nucleotide metabolism, providing both the ribose-5-phosphate and NADPH for nucleotide synthesis.

Ribonucleotide reductase is highly expressed in actively proliferating cells and is active during S-phase of the cell cycle. The activity of RR is tightly regulated to ensure balanced production of dNTPs for synthesis of DNA. RR is a heterodimeric tetramer composed of two identical subunits: a regulatory unit and an active unit. The regulatory subunits mediate the activity and

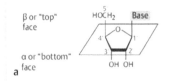

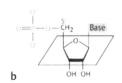

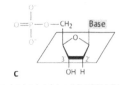

Fig. 37.1 (a-c) Nucleoside, nucleotide, and deoxynucleotide structures and their numbering system. (Source: Panini S, ed. Medical Biochemistry- An Illustrated Review. New York, NY. Thieme; 2013.)

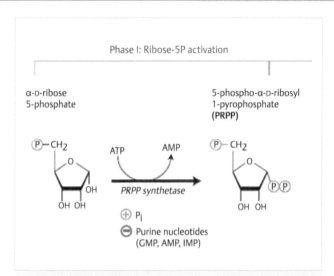

Fig. 37.2 Synthesis of phospho-ribosyl pyrophosphate (PRPP) by PRPP synthetase. The ribose sugar is converted to PRPP. (Source: Panini S, ed. Medical Biochemistry—An Illustrated Review. New York, NY. Thieme; 2013.)

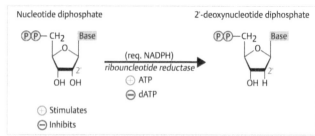

Fig. 37.3 Reduction of nucleoside diphosphates by ribonucleotide reductase. (Source: Panini S, ed. Medical Biochemistry—An Illustrated Review. New York, NY. Thieme; 2013.)

substrate specificity of the RR enzyme complex. dATP is a very potent inhibitor of RR. Therefore binding of dATP to the active site inhibits the enzyme and prevents formation of all dNTPs. In contrast to this, binding of ATP to the active site activates the enzyme.

37.3.1 Link to Pharmacology

Hydroxyurea is a potent inhibitor of RR, therefore hydroxyurea has been used in the treatment of cancer for many years, as it effectively arrests the cells in S phase. It is widely used for the treatment of chronic myelogenous leukemia, polycythemia vera and essential thrombocythemia. It is also used in the treatment of other non-neoplastic pathologies such as sickle cell anemia or refractory psoriasis.

37.4 De Novo Purine Synthesis

The liver is the major site of purine synthesis in the body, and this biosynthetic process requires energy in the form of ATP. The purine ring is formed using glycine, glutamine, aspartate, CO_2 and reduced tetrahydrofolate derivatives (▶ Fig. 37.4). Purine biosynthesis is tightly regulated in response to physiological needs. The central molecule that regulates purine synthesis is PRPP. There are four major phases for the synthesis of purine ribonucleotides, with the first being the activation of ribose-5-P to form PRPP. This is common to both purine and pyrimidine nucleotide synthesis. The second phase is the committed step in which PRPP is converted to phosphoribosylamine. Phase three involves the synthesis of the parent purine IMP. AMP and GMP are then generated from IMP in the fourth phase.

As discussed above, PRPP is the activated sugar intermediate for both the de novo and salvage pathways of purine and pyrimidine nucleotide synthesis. PRPP synthetase is activated by inorganic phosphate and is inhibited by purine-5'-ribonucleotides, especially by AMP and GMP. Increased concentrations of AMP and GMP signify that nucleotide synthesis is sufficient in the cell.

After the formation of PRPP, its pyrophosphate moiety is replaced with an amide group from glutamine to generate phosphoribosylamine by PRPP-amidotransferase. This is the rate limiting step of de novo purine nucleotide synthesis and if, therefore, tightly regulated. This enzyme is activated as the concentration of PRPP increases. There is also allosteric feedback inhibition by IMP, AMP, and GMP.

After the formation of phosphoribosylamine, a series of nine reactions generate IMP, which is the parent purine. The formation of IMP is an energy consuming process and also requires glycine, aspartate, CO_2 and two formate molecules. The donors of the formyl moieties are reduced tetrahydrofolate (FH4) derivatives in the form of 10-methenyl tetrahydrofolate (FH4. CH_2) and 10-formyl tetrahydrofolate (FH4.COH).

The purine pathway branches at IMP to generate AMP and GMP in a two-step, energy requiring process. While the conversion of IMP to AMP requires GTP, the conversion of IMP to GMP requires ATP. The synthesis of AMP and GMP from IMP is regulated in a reciprocal manner. High levels of GTP favor IMP conversion to AMP. On the other hand, high levels of ATP favor IMP conversion to GMP. This reciprocity of regulation allows the cell to balance the synthesis of ATP and GTP by decreasing the synthesis of one purine nucleotide if there is a deficiency in the other nucleotide.

37.4.1 Link to Pharmacology: Methotrexate and Pemetrexed are Antifolates

Antifolates, which are structural analogues of folates, such as methotrexate and pemetrexed, have been used to target purine synthesis as antineoplastic agents in the clinic. Antifolates inhibit the enzyme GARFT (glycinamide ribonucleotide formyltransferase). It is important to note that although methotrexate can inhibit GARFT, the primary target of methotrexate is **dihydrofolate reductase (DHFR),** which will be discussed later in this chapter (synthesis of thymine nucleotides).

37.4.2 Link to Pharmacology: Mycophenolic Acid and Ribavirin are Inhibitors of IMP Dehydrogenase

Near the end of GMP synthesis, IMP is converted to the intermediate xanthosine monophosphate (XMP) by the enzyme IMP dehydrogenase. This reaction is the rate limiting step in the de

V

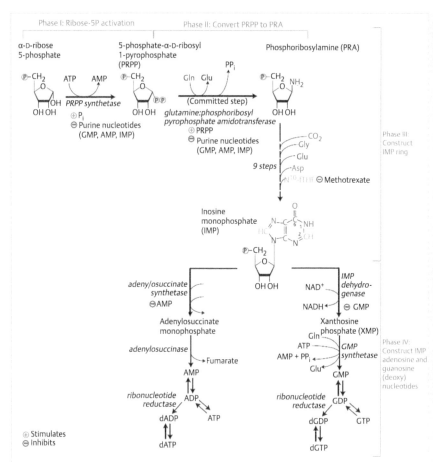

Fig. 37.4 Phases of the de novo synthesis of purine nucleotides and deoxynucleotides. (Source: Panini S, ed. Medical Biochemistry—An Illustrated Review. New York, NY. Thieme; 2013.)

novo synthesis of GTP. Its inhibition depletes intracellular pools of guanine nucleotide. Due its importance in cell division, inhibitors of the IMP dehydrogenase have been used in immunosuppressive chemotherapy (mycophenolic acid) and antiviral chemotherapy (ribavirin).

37.5 Regulation of De Novo Purine Synthesis

Purine nucleotide synthesis is regulated in the first two steps: PRPP synthetase and PRPP amidotransferase (▸ Fig. 37.4). PRPP synthetase is activated by inorganic phosphate, while PRPP amidotransferase is activated by accumulation of PRPP. Both enzymes are inhibited by AMP and GMP.

In the final phase, the generation of AMP and GMP are also regulated reciprocally. In the branch leading from IMP to AMP, the first enzyme is feedback-inhibited by AMP and the corresponding enzyme in the branch from IMP to GMP is feedback-inhibited by GMP.

37.6 Purine Salvage Pathway

De novo purine synthesis occurs mainly in the liver. In other tissues, however, de novo synthesis is not sufficient to meet normal physiological needs. Therefore, extrahepatic tissues supplement their nucleotide needs by using the salvage pathway (▸ Fig. 37.5).

Degradation of RNA generates free purine bases (adenine, guanine, and hypoxanthine). That are resynthesized by the salvage pathway to their corresponding nucleotides by enzymes called phosphoribosyltransferases. Adenine phosphoribosyl transferase (APRT) converts adenine to AMP, while hypoxanthine-guanine phosphoribosyl transferase (HGPRT) recognizes both guanine and hypoxanthine as substrates to catalyze the formation of GMP and IMP, respectively. The role of APRT is not as significant as the role of HGPRT in synthesizing the purine nucleotides from free bases.

37.6.1 Link to Pathology: Lesch-Nyhan Syndrome is due to Deficiency of HGPRT

Lesch-Nyhan Syndrome is an X-linked recessive disorder caused by a complete deficiency of the HGPRT enzyme, leading to the inability to salvage hypoxanthine and guanine. Inhibition of the salvage pathway results in the accumulation of PRPP, which activates PRPP-amidotransferase. This consequently activates de novo purine nucleotide synthesis. Due to absence of the salvage pathway, hypoxanthine and guanine bases build up. These purine bases are destined to the purine degradation pathway, which generates excessive amounts of uric acid. Therefore, the deficiency of HGPRT enzyme results in hyperuricemia, and neurological deficits including spasticity, intellectual disability, aggression, and self-mutilation.

A plausible biochemical explanation for the neurological symptoms is due to the significant reliance of the purine

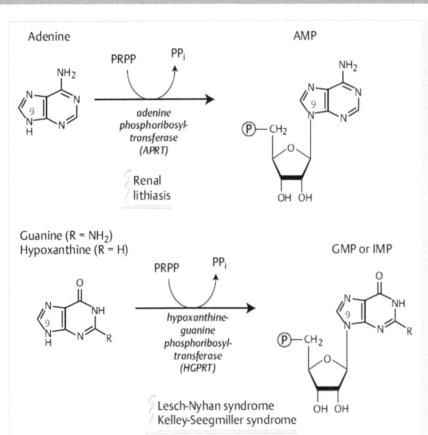

Fig. 37.5 Purine salvage pathway. Free purine bases adenine, guanine, and hypoxanthine, can be salvaged to nucleotides by phosphoribosyltransferases. APRT converts adenine to AMP and HGPRT converts guanine and hypoxanthine to GMP and IMP, respectively by transfering PRPP to guanine and hypoxantine. (Source: Panini S, ed. Medical Biochemistry—An Illustrated Review. New York, NY. Thieme; 2013.)

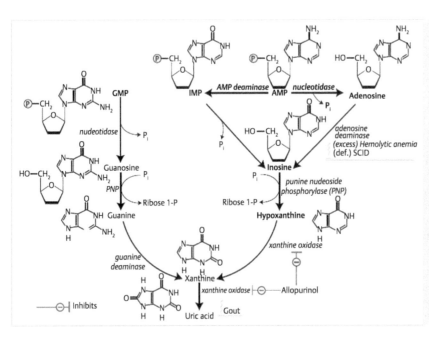

Fig. 37.6 Purine nucleotide catabolism. (Source: Panini S, ed. Medical Biochemistry—An Illustrated Review. New York, NY. Thieme; 2013.)

salvage pathway rather than the de novo pathway. As a result, GTP levels are low in brain. GTP is important for signal transduction through the dopamine receptor-agonist interaction. In addition, GTP is used in the first and rate-limiting step in the synthesis of tetrahydrobiopterin, which is the cofactor for the synthesis of catecholamine transmitters and serotonin.

37.7 Purine Degradation

As with synthesis, the degradation of purines mainly occurs in the liver. The degradation process starts with the conversion of nucleotides to nucleosides by nucleotidases (▶ Fig. 37.6). In humans, adenosine and guanosine are degraded to the intermediate xanthine, which is metabolized to uric acid. Adenine nucleotide degradation starts with AMP. While the deamination of AMP yields IMP, removal of phosphate from AMP yields adenosine. These two intermediates, IMP and adenosine, are converted to inosine. Adenosine to inosine conversion requires adenosine deaminase (ADA). Inosine, through hypoxanthine, is converted to xanthine. Guanine nucleotide degradation starts with the conversion of GMP to guanosine. Guanosine, through guanine, is converted to xanthine. The final step in purine degradation is the formation of uric acid from xanthine by xanthine oxidase, the same enzyme that is used for the conversion of hypoxanthine to xanthine.

37.7.1 Link to Immunology and Pathology: Severe Combined Immunodeficiency Disease (SCID)

The conversion of adenosine to inosine is catalyzed by adenosine deaminase (ADA). Genetic deficiency of ADA is an autosomal recessive condition that impairs the immune system and causes severe combined immunodeficiency disease (SCID). ADA deficiency results in the accumulation of adenosine, which is converted to AMP and ADP by kinases. Ribonucleotide reductase (RR) generates dADP from this accumulated ADP. The elevated dADP in turn inhibits the RR since it is a very potent inhibitor of RR. This inhibition of RR prevents the formation of other dNTPs, which ultimately impairs DNA synthesis.

Because ADA is most active in lymphocytes, ADA deficiency impairs lymphocyte proliferation, differentiation and maturation and also leads to death of lymphocytes in the thymus. Patients with SCID cannot develop effective immune protection from foreign invaders such as bacteria, viruses, and fungi. Since there are a decreased number of lymphocytes to fight off infections, these patients are prone to repeated and persistent infections that can be very serious or even life-threatening.

37.7.2 Link to Pathology: Hyperuricemia and Gout

The primary risk factor for developing gout is high levels of uric acid in serum, referred to as hyperuricemia.

Under physiological conditions, uric acid is found as sodium urate in the plasma. Both uric acid and sodium urate have poor solubility, leading to their precipitation in urine and synovial fluid. In the presence of high levels of uric acid, acute gout may develop when the resulting sodium urate crystals cause a severe inflammatory response. This results in excruciating joint pain, swelling, warmth, redness, and tenderness. In the chronic condition, deposits of sodium urate may form in the tissues, especially at the metatarsophalangeal joint, knees, ankles, and wrists.

Hyperuricemia may develop due to genetic causes or due to an acquired disorder. It may arise either from increased urate formation or from decreased excretion. In some cases, there is overproduction and under secretion of urate at the same time. While approximately 90% of gout cases are due to under secretion, only 10% is due to increased production.

Examples of causes of overproduction of uric acid include defective HGPRT (Lesch-Nyhan syndrome), gain of function in PRPP synthetase, deficiency of glucose-6-phosphatase (Von Gierke disease), hereditary fructose intolerance, and increased turnover of nucleic acids in proliferative diseases and in chemotherapy.

Examples of causes of under secretion of uric acid include chronic ethanol intake, diabetic ketoacidosis, and starvation. In the latter case, accumulation of lactate and ketone bodies competes with renal uric acid excretion. Therefore, diseases associated with metabolic syndrome such as hypertension, diabetes and cardiovascular disease are also associated with hyperuricemia.

37.7.3 Link to Pharmacology: Allopurinol is Hypoxanthine Analog

Allopurinol, an analog of hypoxanthine, is used in the treatment of gout. It inhibits xanthine oxidase in a rather complicated fashion. Xanthine oxidase catalyzes the conversion of allopurinol into alloxanthine (also known as oxypurinol), which is a potent irreversible inhibitor of xanthine oxidase. Subsequently, synthesis of uric acid is blocked. This leads to the accumulation of xanthine and hypoxanthine, which are more soluble than uric acid. Allopurinol is also converted by HGPRT to allopurinol ribonucleotide. This nucleotide serves as a feedback inhibitor of PRPP-amidotransferase, the key enzyme in de novo purine synthesis, resulting in decreased purine synthesis and degradation.

37.7.4 Link to Pharmacology: Use of Colchicine, Indomethacin, and Probenecid in Gout

Acute attacks of gout initially are treated with the drugs colchicine or indomethacin to reduce the inflammation. If the hyperuricemia is due to under-excretion of uric acid, then these patients are treated with an uricosuric drug (probenecid). However, gout that is due to overproduction of uric acid is treated with allopurinol or febuxostat, which is a nonpurine inhibitor of xanthine oxidase.

37.8 Purine Nucleotide Cycle in Muscle

The purine nucleotide cycle is a sequence of biochemical reactions that generate fumarate plus free ammonia from aspartate. Deamination of AMP to IMP releases the ammonia. AMP is resynthesized from IMP by two enzymes of the de novo purine pathway. In skeletal muscle, AMP deaminase plays an essential role in energy metabolism in several ways. 1) The purine nucleotide cycle has an anapleuretic role in replenishing the TCA cycle intermediates through fumarate, which is converted to oxaloacetate. As a result, there is an increase in ATP production for the exercising muscle. 2) Adenylate kinase, also known as myokinase in the muscle, coupled with AMP deaminase regulates the adenylate energy charge in the muscle. Adenylate kinase is a reversible enzyme and catalyzes the formation of ATP and AMP from two ADP molecules, thereby maintaining equilibrium among AMP, ADP, and ATP. A slight increase in AMP levels activates AMP-dependent kinase (AMPK), which acts as a signal transducer for metabolic adaptations by

V

responding to an altered cellular energy status. AMPK inhibits biosynthetic pathways and activates catabolic pathways to restore cellular energy stores. However, the continual accumulation of AMP also halts ATP production. AMP deaminase ensures that the adenylate kinase reaction favors ATP production. 3) Ammonia produced in the AMP deaminase reaction buffers the H^+ produced during ATP hydrolysis and also neutralizes the tubular urine from weak acids such as lactate, pyruvate, and ketone bodies that are produced during exercise. The NH_3 condenses with glutamate for the synthesis of glutamine, which is secreted into the circulation. In the kidney, glutamine is converted back to NH_3 to maintain the neutrality of the urine.

37.8.1 Link to Pathology: Myoadenylate Deaminase (Muscle AMP Deaminase) Deficiency and Exercise Intolerance

Myoadenylate deaminase deficiency is an autosomal recessive disorder that is associated with myalgia, cramps, and fatigue following vigorous or even moderate exercise due to reduced ATP replenishment for the reasons described above. Patients with this deficiency do not develop muscle wasting, and the histological appearance of muscle is unremarkable. Measurement of venous ammonia from the forearm ischemic exercise test is a useful screening test.

37.9 De Novo Pyrimidine Synthesis

In de novo pyrimidine synthesis, the pyrimidine ring is assembled first and then is linked to ribose-1-phosphate (▶ Fig. 37.7). Building the pyrimidine ring requires aspartate, glutamine, CO_2, glycine, and reduced tetrahydrofolate derivatives. The parent ring molecule in pyrimidine synthesis is orotate, from which UMP forms. Although the formation of UMP is relatively simple, the generation of the other pyrimidine nucleotides from UMP is a more complicated process.

Pyrimidine biosynthesis requires energy and is tightly regulated. The first reaction of pyrimidine synthesis starts with the synthesis of carbamoyl phosphate by carbamoyl synthetase II (CPS II) in the cytosol, where the majority of regulation occurs. CPSII is allosterically activated by PRPP and is inhibited by UTP. The CPSII is different from the mitochondrial carbamoyl phosphate synthase I of urea synthesis. The different localizations of these two isozymes generate independent pools of carbamoyl phosphate for these two pathways. Generation of carbamoyl phosphate in pyrimidine synthesis occurs from the condensation of the amide group of glutamine with carbon dioxide, with the expenditure of two ATP molecules. CPS I in the urea cycle uses free ammonia rather than glutamine. The regulation of these isozymes is also different. While CPS I is allosterically regulated by N-acetylglutamate, CPS II is allosterically regulated by PRPP, ATP and UTP. While PRPP and ATP are activators of the enzyme, UTP is an inhibitor.

Orotate is formed by a multienzyme complex. After the synthesis of orotate, and the pyrimidine nucleotide UMP is synthesized by UMP synthetase, another multienzyme complex. UDP

and UTD are synthesized from UMP by ATP-dependent uridylate kinase and nucleoside diphosphate kinase. CTP is synthesized by the amination of UTP. CTP can be converted to dCDP by ribonucleotide reductase (RR). The dCDP is then phosphorylated to dCTP for DNA synthesis.

In order to synthesize deoxythymidine nucleotides for DNA synthesis, first the dUMP is synthesized by RR. Thymidylate (dTMP) is synthesized from dUMP by thymidylate synthase using methylene-tetrahydrofolate (FH4-CH2) as a coenzyme. In this reaction, FH4-CH2 is converted to dihydrofolate (FH4) by donating the methyl (CH_3) group to the pyrimidine ring. As discussed in the one-carbon metabolism section, Chapter 38, reduction of dihydrofolate (FH2) to tetrahydrofolate (FH4) by dihydrofolate reductase (DHFR) using NADPH as the reducing power is essential to re-populate the reduced folate pools. Due to the essential role of DHFR in thymidylate synthesis, it has been a target for the development of antibiotics, antiparasitics, and antineoplastic agents (see below).

37.9.1 Link to Pathology: Hereditary Orotic Aciduria and Megaloblastic Anemia

Hereditary orotic aciduria can develop due to UMP synthase (UMPS) deficiency. The UMPS is a bifunctional enzyme with orotate phosphoribosyl transferase and orotidylate decarboxylase activities. Mutations in either domain can lead to accumulation of orotate in urine. Babies with UMP synthase deficiency present early in life with megaloblastic anemia. These patients' megaloblastic anemia cannot be treated with either folate or vitamin B12. The impairment in pyrimidine synthesis may lead to defective erythrocyte membrane synthesis due to the reduced availability of nucleotide-lipid cofactors in the bone marrow. For example, CTP is required for cytidine diphosphocholine, which is an essential molecule for membrane phospholipid biosynthesis. Pyrimidine replacement with oral uridine results in hematological remission with reduced urinary excretion of orotic acid.

37.9.2 Link to Pathology: Hereditary Orotic Aciduria due to OTC Deficiency

Orotic aciduria may also develop due to deficiency of the ornithine transcarbamylase (OTC) enzyme in the urea cycle. Decreased carbamoyl phosphate utilization in the mitochondria results in the leakage of mitochondrial carbamoyl phosphate into the cytosol and leads to the stimulation of pyrimidine synthesis. Uracil and orotic acid accumulation are useful markers of urea cycle disorders. While uracil accumulates due to increased degradation of pyrimidines, orotic acid accumulates due to limited activity of UMPS due to the depletion of PRPP.

37.9.3 Link to Pharmacology: 5-Fluorouracil and Methotrexate are Antimetabolites

Synthesis of thymidylate (TMP) from UMP has been targeted for the development of antineoplastic agents. As discussed

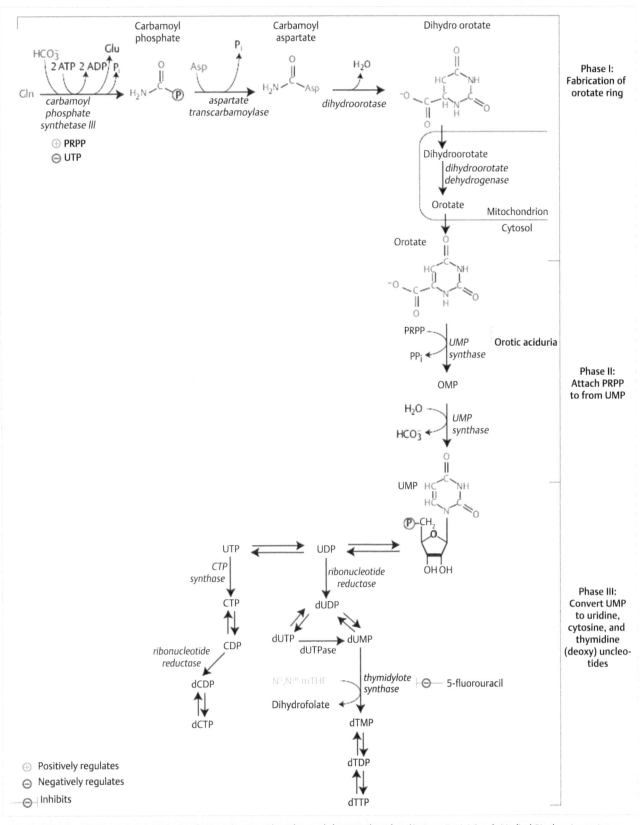

Fig. 37.7 Phases of the de novo synthesis of pyrimidines, nucleotides, and deoxynucleotides. (Source: Panini S, ed. Medical Biochemistry—An Illustrated Review. New York, NY. Thieme; 2013.)

above, hydroxyurea inhibits RR, which normally reduces the NDPs to dNDPs. An important antimetabolite known as 5-Flurouracil (5-FU) is a pro-drug, this is metabolized to three different antimetabolites. These antimetabolites are: 5-fluorouridine-5'-triphosphate (5-FUTP), 5-fluoro-2'-deoxyuridine-5'-triphosphate (5-FdUTP), and 5-fluoro-2'-deoxyuridine-5'-monophosphate (5-FdUMP). Since 5-FUTP and 5-FdUTP are incorporated into RNA and DNA, these metabolites disrupt RNA and DNA synthesis, respectively. FdUMP binds covalently to the nucleotide-binding site of thymidylate synthase, forming a stable ternary complex with the enzyme and its coenzyme methyltetrahydrofolate (FH4-CH3). As a result, dTMP levels drop leading to depletion of dTTP, hence thymineless cell death occurs.

Methotrexate, a folate analog, competitively inhibits dihydrofolate reductase (DHFR), therefore causing the depletion of active folate, FH4, in the cells. Since FH4 is a required coenzyme and a one-carbon donor for both purines and thymidine nucleotide synthesis, methotrexate inhibits DNA replication, especially in rapidly dividing cells (▶ Fig. 37.8).

Antimetabolites not only kill the cancer cells but also kill rapidly dividing normal cells such as in the gastrointestinal tract and hematopoietic cells. Therefore, antimetabolites are commonly used for treatment of intestinal cancers and leukemia.

37.10 Pyrimidine Salvage Pathways

In the pyrimidine salvage pathway, the nucleosides (cytidine, uridine, and thymidine) that are generated during DNA and RNA degradation can be converted back to nucleotide monophosphates. This allows them to return to the pyrimidine synthesis pathway (▶ Fig. 37.9). Therefore, unlike the purine salvage pathway, the pyrimidine salvage pathway occurs at the nucleoside rather than the base level. The uridine kinase and thymidine kinase enzymes of the pyrimidine salvage pathway regenerate UMP and dTMP, respectively.

It is important to recognize that uridine can give rise to all of the pyrimidine nucleotides, as uridine becomes UMP by uridine kinase and UMP can be converted to TMP and CMP. This is the basis for the treatment of orotic aciduria with uridine.

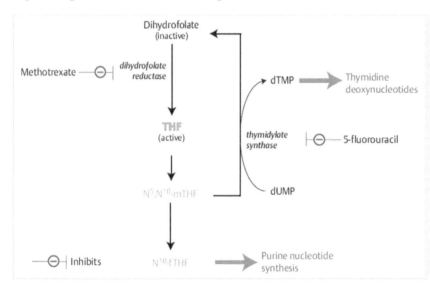

Fig. 37.8 Antineoplastic agents and folate derivatives. (Source: Panini S, ed. Medical Biochemistry—An Illustrated Review. New York, NY. Thieme; 2013.)

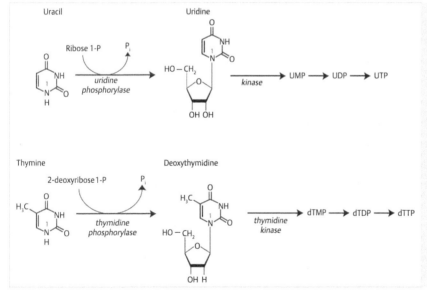

Fig. 37.9 Pyrimidine nucleotide salvage pathway. (Source: Panini S, ed. Medical Biochemistry—An Illustrated Review. New York, NY. Thieme; 2013.)

Another important difference between purines and pyrimidines is that dietary pyrimidines can be readily absorbed from enterocytes and secreted into the circulation as pyrimidine nucleosides, whereas dietary purines cannot be absorbed. Instead, they are degraded to uric acid.

37.10.1 Link to Pharmacology: Acyclovir is a Substrate for Viral Thymidine Kinase

Acyclovir is a guanosine analog and is an antiviral drug used for herpes virus infections such as genital herpes, cold sores, shingles, and chicken pox. Acyclovir is preferentially phosphorylated by viral thymidine kinase, but not by human thymidine kinase. Therefore only the infected human cells can metabolize acyclovir. Viral DNA polymerase, which is less selective than human DNA polymerase, incorporates the phosphorylated acyclovir into viral DNA, resulting in premature chain termination.

37.11 Degradation of Pyrimidines

Unlike purine degradation, which results in the single end product uric acid, pyrimidine degradation does not converge to a single molecule. The cleavage of uridine to uracil and the cleavage of thymidine to thymine by uridine phosphorylase initiate the catabolic pathway of pyrimidines in humans. Uracil and cytosine nucleotides are degraded to uridine and deoxyuridine, which are further metabolized to β-alanine. Thymidine on the other hand is degraded to thymine, which is further converted to β-aminoisobutyrate. These two end products of pyrimidine degradation are converted into intermediates of the citric acid cycle.

V

Review Questions

1. Which of the following enzymes is inhibited by dATP?
   A) Adenosine deaminase
   B) Dihydrofolate reductase
   C) PRPP synthetase
   D) Ribonucleotide reductase
   E) Thymidylate synthase
2. Which of the following metabolites is a central molecule in the de novo synthesis of purines and pyrimidines?
   A) Carbamoyl phosphate
   B) dNTP
   C) IMP
   D) PRPP
   E) UMP
   F) Xanthine
3. Deficiency in which of the following enzymes causes orotic aciduria?
   A) Adenosine deaminase
   B) Dihydrofolate reductase
   C) Hypoxanthine Guanine Phosphoribosyl Transferase
   D) Thymidylate synthase
   E) UMP synthase
4. Which of the following drugs inhibits the formation of uric acid?
   A) Acyclovir
   B) Allopurinol
   C) Colchicine
   D) 5-Fluoro uracil
   E) Methotrexate
5. Increased activity in which of following enzymes cause hyperuricemia?
   A) Glucose 6-phosphatase
   B) HGPRT
   C) PRPP synthetase
   D) UMP synthase

Answers

1. **The correct answer is D**. dATP is a potent inhibitor of ribonucleotide reductase (RR), preventing the formation of all dNTPs. In contrast to this, binding of ATP to the active site activates the enzyme. Ribonucleotide reductase is highly expressed in actively proliferating cells because the activity of RR is required for the generation of deoxynucleotides (dNTPs) needed for DNA replication. The drug hydroxyurea is also a potent inhibitor of RR. Therefore hydroxyurea been used in the treatment of cancer, especially for the treatment of chronic myelogenous leukemia, polycythemia vera and essential thrombocythemia. It is also used in the treatment of other non-neoplastic pathologies such as sickle cell anemia and refractory psoriasis.

Adenosine deaminase deficiency causes severe combined immunodeficiency syndrome (SCID). In this syndrome adenosine and deoxyadenosine accumulates. Accumulated adenosine can be phosphorylated by adenosine kinase, which produces ATP. Ribonucleotide reductase (RR) reduces the ribose of ATP to form dATP, which inhibits the RR. This inhibition prevents syn-

thesis of other deoxynucleotides (dCTP, dTTP, dGTP). Answer choice A is incorrect because dATP does not inhibit the adenosine deaminase.

Dihydrofolate reductase is inhibited by the folate analog methotrexate, thus answer choice B is incorrect. Answer choices C and E are incorrect because neither PRPP synthetase nor thymidylate synthase activity is affected by dATP. Thymidylate synthase is inhibited by 5-fluorouracil.

2. **The correct answer is D**. Phosphoribosyl-pyrophosphate (PRPP) is the activated ribose sugar which is used in both purine and pyrimidine nucleotide synthesis. The de novo synthesis of purines and pyrimidines, however, utilize different strategies to link the ribose-5-phosphate. In purine nucleotide synthesis, the parent base is synthesized first before the addition of PRPP. In contrast to this, pyrimidines use the activated sugar moiety PRPP as a backbone to build the pyrimidine base.

Answer choice A is incorrect because carbamoyl phosphate is an intermediate in the urea cycle and in the de novo synthesis of pyrimidines, but not in purines. Answer choice B is incorrect because dNTP stands for deoxynucleotide triphosphates, which includes dATP, dGTP, dCTP, dTTP. While IMP is the parent purine for ATP and GTP biosynthesis, UMP is the parent pyrimidine in UMP, CMP, and dTMP biosynthesis. Therefore, answer choices C and E are incorrect. Answer choice F is incorrect because xanthine is an intermediate in the purine degradation pathway.

3. **The correct answer is E**. Hereditary orotic aciduria can develop due to deficiency of either UMP synthase (UMPS) or the urea cycle enzyme ornithine transcarbamoylase (OTC). Genetic mutations in UMPS lead to accumulation of orotate in urine because orotic acid is the substrate of UMPS, which converts orotic acid to UMP. Due to the deficiency of uridine synthesis, CPSII, the rate limiting step of pyrimidine pathway, is upregulated. Treatment with oral uridine helps to downregulate the CPSII and, therefore, decreases orotic acid synthesis.

Answer choices A, B, C, and D are incorrect because deficiencies of these enzymes do not cause accumulation of orotic acid. Adenosine deaminase (ADA) deficiency causes SCID and Hypoxanthine Guanine Phosphoribosyl Transferase (HGPRT) deficiency causes Lesch-Nyhan syndrome.

4. **The correct answer is B**. Allopurinol is an analog of hypoxanthine that inhibits xanthine oxidase and subsequently the synthesis of uric acid formation. Forms of gout that are due to overproduction of uric acid are treated with allopurinol or febuxostat, which is a nonpurine inhibitor of xanthine oxidase. Colchicine or indomethacin, answer choice C, is also used in the treatment of gout. However, colchicine reduces the inflammation that is produced by uric acid crystals.

Acyclovir, answer choice A, is a guanosine analog that is used as an antiviral drug for herpes virus infections. Acyclovir inhibits viral thymidine kinase, but not human thymidine kinase.

5-Fluoro uracil (answer choice D) inhibits thymidylate synthase and, as a result, dTMP levels drop. This leads to depletion of dTTP, hence thymineless cell death occurs. Answer choice E is incorrect because methotrexate, a folate analog, competitively inhibits dihydrofolate reductase (DHFR), therefore causing the depletion of active folate, FH4, in the cells.

5. **The correct answer is C.** Increased activity due to gain of function mutations in the PRPP synthetase enzyme causes hyperuricemia and, consequently, gout. PRPP allosterically activates the rate-limiting step, PRPP amidotransferase, of the purine synthesis pathway. Increased synthesis of PRPP activates purine synthesis. Excess purine nucleotides are degraded, which result in overproduction of uric acid, or hyperuricemia.

Answer choices A and B are incorrect because decreased activity of glucose 6-phosphatase and decreased activity of HGPRT causes hyperuricemia. Answer choice D is incorrect because UMP synthase is involved in pyrimidine biosynthesis, not purine biosynthesis. The degradation products of the pyrimidines are β-alanine and β-aminoisobutyrate.

V

38 Vitamins in One-Carbon Metabolism

At the conclusion of this chapter, students should be able to:
- Explain the absorption of folate and vitamin B12
- Explain the factors and diseases that interfere with the absorption of folate and vitamin B12
- Distinguish between the synthesis of tetrahydrofolate (FH4) in humans and bacteria
- Compare the mechanism of action of antimetabolites for antineoplastic drugs vs the antibiotics trimethoprim and sulfonamides
- Discuss the role of FH4 in purine and thymidylate synthesis
- Discuss the synthesis of S-adenosylmethionine (SAM) and its role in one-carbon metabolism
- Describe the mechanism of the folate trap due to deficiency of vitamin B12
- Explain the basic pathophysiology of both vitamin B12 and folate deficiency
- Interpret laboratory data to distinguish between primary and secondary folate deficiency

Many biochemical reactions require the transfer of an activated one-carbon group from a donor to an acceptor compound. Donations of the one-carbon groups (methyl, methylene, and formyl) are important for the biosynthesis of serine, methionine, glycine, choline, thymidylate and the purine nucleotides. Folate (vitamin B9), an essential nutrient in the human diet, functions as a carrier of one-carbon groups. Tetrahydrofolate (FH4) is the active form of the folate and can receive one-carbon groups from a variety of different molecules and then transfer them to unique recipients.

While S-adenosylmethionine (SAM) functions as a methyl (CH_3) donor for more than 30 reactions in our body, both vitamin B12 and active folate have a role in methylation reactions and both of these coenzymes are required for the formation of SAM from homocysteine.

In this chapter, first we will review the absorption of folate, the synthesis of functional folate, and the functions of folate as a one-carbon donor in nucleotide synthesis. Next, we will review the functions of B12 in metabolism, with a focus on the function of B12 in SAM biosynthesis. We will also review the absorption of B12 and the causes of its deficiency as well as the diseases associated with B12 deficiency.

38.1 Absorption of Folate

Folate is found in dark leafy vegetables such as spinach, broccoli, and legumes, and citrus fruits. Liver is also rich in folate and so are the fortified foods such as bread and cereal. Folic acid has no biological activity unless it is converted into its coenzyme form tetrahydrofolates (FH4).

The structure of folate contains the ring structure pteridine, an amino acid glutamate and a bridge molecule para-aminobenzoic acid (PABA) between the pteridine ring and glutamate (▶ Fig. 38.1).

Folates are absorbed in the jejunum (▶ Fig. 38.2). In order for transport into the intestinal cells, folate polyglutamates (folate

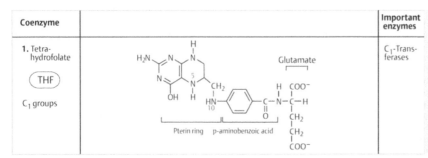

Coenzyme		Important enzymes
1. Tetra-hydrofolate THF C_1 groups	(structure of tetrahydrofolate: Pterin ring, p-aminobenzoic acid, Glutamate)	C_1-Trans-ferases

Fig. 38.1 Structure of tetra-hydrofolate Folate contains a pterin ring, a para-amino benzoic acid, and a glutamate. (Source: Koolman J, Röhm K, ed. Color Atlas of Biochemistry. 3rd Edition. New York, NY. Thieme; 2012.)

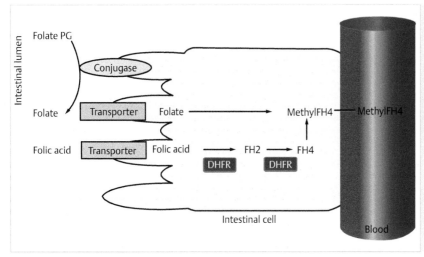

Fig. 38.2 Absorption of folate in the jejunum. Fruits and vegetables contain folate as folate polyglutamates (folate PG), which is hydrolized to folate by the conjugase enzyme. Folate and folic acid are absorbed trough transporters into the enterocytes where it is converted to tetra-hydrofolate by dihydro folate reductase (DHFR) enzyme.

PG) are first hydrolyzed to folate monoglutamates by an intestinal mucosal enzyme conjugase (also known as gamma glutamyl hydrolase). Folic acid is reduced to tetrahydrofolate by dihydrofolate reductase (DHFR). In the enterocytes, folate is converted to 5-methyltetrahydrofolate (FH4-CH3), which is the major circulating form of folate in the body. The liver stores a 3 to 4 month supply of folates.

38.1.1 Link to Pathology and Pharmacology

Patients with Celiac disease or Crohn's disease may have reduced conjugase activity due to the erosion of enterocytes. This reduced conjugase activity affects the absorption of folate and, therefore, leads to folate malabsorption. Conjugase activity is also inhibited by phenytoin, an anti-seizure medication, and by oral contraceptives.

38.2 Tetrahydrofolate Derivatives Function as One-Carbon Donors in Nucleotide Synthesis

A major role of tetrahydrofolate is to provide one-carbon groups for purine and thymidylate synthesis (reviewed in Chapter 37). Synthesis of both AMP and GMP requires the donation of formyl groups from formyl-tetrahydrofolate (FH4-CHO) in two separate reactions. Thymidylate synthesis from dUMP is catalyzed by thymidylate synthase and requires the donation of a methyl group from methylene-tetrahydrofolate (FH4-CH2). In these reactions, tetrahydrofolate (FH4) is converted to dihydrofolate (FH2), which is an inactive form of folate. In order to regenerate the active form of folate, dihydrofolate reductase (DHFR) reduces dihydrofolate (FH2) to tetrahydrofolate (FH4). The reducing power and the coenzyme of the DHFR enzyme is NADPH. Because the active form of folate is required for nucleotide biosynthesis and for DNA replication, enzymes in folate metabolism have been important targets for antineoplastic drugs and also for antibiotics.

38.2.1 Link to Pharmacology: Inhibition of DHFR in Bacteria and in Eukaryotes

There are several drugs belonging to the antifolate family that bind to the DHFR enzyme and competitively inhibit the conversion of dihydrofolate to tetrahydrofolate. Since there is significant variation in the human and bacterial DHFR, the antibiotics sulfonamides and trimethoprim are able to preferentially inhibit bacterial DHFR with little or no effect on human DHFR. Combination treatment of sulfonamides and trimethoprim is synergistic and is used widely for treatments of infections as well as prevention of opportunistic infections.

Methotrexate and aminopterin are also antifolates, but unlike trimethoprim and sulfonamides, methotrexate binds to human DHFR very tightly. As a result, there is extensive thymidylate depletion and dUMP accumulation, causing thymineless cell death. Although thymidylate depletion is the major outcome of methotrexate inhibition, it also impairs purine synthesis by

inhibiting the formyl transfer from FH4-CHO. Inability to synthesize both DNA and RNA due to methotrexate treatment results in the inhibition of cancer cell proliferation. Methotrexate has been used effectively for the treatment of various cancers such as acute lymphocytic leukemia, Burkitt lymphoma in children, breast cancer, bladder cancer, and head and neck carcinomas. However, methotrexate is toxic to rapidly proliferating normal cells, leading to myelosuppression, oral and intestinal mucositis, and neurotoxicity. A strategy called leucovorin rescue is used to avoid toxicity and still provide clinical benefit. Leucovorin (5-formyltetrahydrofolate) is a stable, reduced form of folate that is converted to tetrahydrofolate without requiring DHFR, thus bypassing the inhibition of tetrahydrofolate synthesis by methotrexate.

38.2.2 Link to Pathology: Megaloblastic Anemia and Neural Tube Defects due to Folate Deficiency

Folate deficiency may develop either due to nutritional deficiency or inadequate absorption arising from enterocyte dysfunction, such as in patients with Celiac disease and Crohn disease. Folate deficiency may also develop when there is an increased demand by the growing fetus during pregnancy. Some situations that lead to folate deficiency are the cancer therapy methotrexate, alcoholism, and extensive use of antacids. Alcohol and antacids may hinder the absorption of folates by increasing the pH of the upper intestine.

Folates are important for the synthesis of DNA, RNA and some amino acids. Therefore, they are critical for rapidly dividing cells such as embryos, red blood cells, and epithelial cells. Folate deficiency affects blood-forming cells in the bone marrow, leading to macrocytic anemia, which presents with weakness, fatigue, headache, heart palpitations, and shortness of breath.

Folate deficiency during pregnancy causes neural tube defects such as spina bifida. In 1991, folate fortification of grains was initiated in an attempt to reduce neural tube defects due to folate deficiency in expecting mothers. Now, women who are planning to be pregnant are advised to take folic acid at least a month before the pregnancy and continue during pregnancy. Folate supplementation is also advised for nursing mothers.

38.3 Functions of Vitamin B12

Vitamin B12 functions as coenzyme for two enzymes in metabolism: methylmalonyl CoA mutase and methionine synthase.

Methylmalonyl CoA mutase is the second enzyme in the propionyl CoA pathway (▶ Fig. 38.3). Propionyl CoA pathway is involved in the metabolism of odd-chain fatty acids and certain amino acids, converting them into succinyl CoA, a TCA cycle intermediate. In the case of vitamin B12 deficiency, the activity of methylmalonyl CoA mutase is compromised, leading to accumulation of plasma and urine concentrations of methylmalonic acid, a degradation product of methylmalonyl CoA. Therefore, measurement of methylmalonyl CoA is a diagnostic test for vitamin B12 deficiency.

Methionine synthase (also known as homocysteine methyltransferase) is a B12-dependent enzyme that transfers a methyl group from FH4-CH3 to homocysteine to synthesize methionine

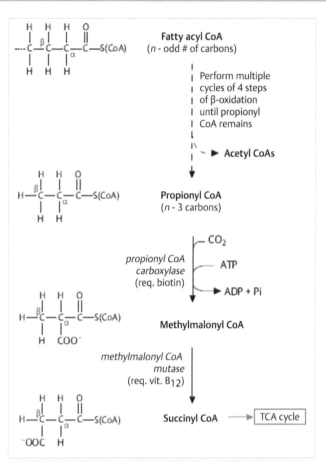

Fig. 38.3 Propionyl CoA pathway. Propionyl CoA, final product of odd-chain fatty acid oxidation, is converted to succinyl CoA by the propionyl CoA carboxylase and the methylmalonyl CoA mtase enzymes. In the pathway methylmalonyl CoA mutase requires vitamin B12 for its activity. (Source: Panini S, ed. Medical Biochemistry- An Illustrated Review. New York, NY. Thieme; 2013.)

(see Chapter 34). This reaction accomplishes two goals: regeneration of FH4, and the synthesis of S-adensylmethionine (SAM). SAM is the major methyl-group donor in one-carbon metabolism, as this methyl group can be transferred to a variety of acceptor molecules, such as DNA, histones, and norepinephrine in the synthesis of epinephrine (reviewed in Chapter 35).

38.4 Folate Trap Hypothesis

A deficiency in B12 results in accumulation of FH4-CH3 and, hence, decreased levels of FH4. This absence of FH4 due to deficiency of B12 results in impaired DNA synthesis. As reviewed above and in Chapter 37, de novo synthesis of purine and thymidine nucleotides requires FH4-CHO and FH4-CH2. Therefore, B12 deficiency produces folate deficiency by trapping the folate in its reduced methyl form (FH4-CH3). Since all the folate (FH4) is trapped in its methyl form (FH4-CH3), functional folate is not available in the form of FH4-CHO or FH4-CH2 for the synthesis of purines and thymidine, respectively.

38.4.1 Link to Pathology: Folate or B12 Deficiency Causes Megaloblastic Anemia

As reviewed above, B12 deficiency depletes the functional folate and, therefore, impairs DNA synthesis. Because of this, either folate deficiency or vitamin B12 deficiency presents with megaloblastic anemia. Megaloblast formation is due to impaired DNA synthesis, which results in a blocked cell cycle in those cells. Even though cells cannot enter into mitosis, cell growth and hemoglobin synthesis continue. This leads to an increase in the size of erythrocytes (macrocytosis) with normal hemoglobin production (normochromic). While large erythroid precursors with immature nuclei and giant band neutrophils are present in the marrow, macrocytic erythrocytes and hyper-segmented neutrophils with more than five nuclear lobes are seen in peripheral blood. Other rapidly growing tissues are also affected, particularly in the gastrointestinal track. For example, a painful, smooth, beefy red tongue due to atrophic glossitis is a common presentation in folate and vitamin B12 deficiency.

38.4.2 Link to Pathology: B12 Deficiency Causes Neurological Dysfunctions

In addition to megaloblastic anemia, vitamin B12 deficiency leads to peripheral neuropathy with numbness, tingling, and pain in the hands. This is because vitamin B12 deficiency decreases the synthesis of SAM, which is an important methyl donor for sphingomyelin synthesis. Since the maintenance of the myelin sheet of the neurons requires sphingomyelin synthesis, B12 deficiency results in abnormal myelination and nerve transmission. As reviewed in Chapter 35, SAM is also required for the synthesis of several neurotransmitters such as serotonin, norepinephrine and dopamine. Therefore, the involvement of SAM in these reactions may explain neurological manifestations due to B12 deficiency.

It is essential to distinguish megaloblastic anemia caused by folate deficiency from megaloblastic anemia caused by vitamin B12 deficiency in order to prevent the development of neurological symptoms. An important consideration in the diagnosis of vitamin B12 deficiency is that folate supplementation may mask its associated early hematological symptoms. While the hematological symptoms of folate deficiency are reversible, vitamin B12 deficiency may lead to irreversible neurologic damage if the deficiency is prolonged. Therefore, patients with megaloblastic anemia should either be given both folic acid and vitamin B12 for corrective action, or the levels of methylmalonic acid should be measured to rule out B12 deficiency.

38.4.3 Link to Pathology: HyperhomoCysteinemia is a Multifactorial Condition

Nutritional deficiency of folate, vitamin B12 or vitamin B6 causes the elevation of homocysteine levels. Homocysteine levels are also elevated in patients who have genetic deficiency in the methylenetetrahydrofolate reductase (MTHFR) enzyme. This enzyme converts methylenetetrahydrofolate (FH4-CH2) to methyltetrahydrofolate (FH4-CH3). Genetic deficiency of the methionine synthase or cystathionine synthase also results in hyperhomocysteinemia.

Hyperhomocysteinemia is more severe in patients who have genetic deficiency in either of those aforementioned enzymes, compared to nutritional deficiencies. Genetic hyperhomocysteinemia presents early in life with developmental delays, intellectual disability, severe cardiovascular problems such as thrombosis, arteriosclerosis, and stroke, and downward dislocated lenses (lens ectopia). The cardiovascular problems are most likely due to the oxidizing nature of homocysteine, which inactivates nitric oxide by oxidizing it, and also occur in the non-genetic forms of hyperhomocysteinemia. Hyperhomocysteinemia also increases the risk of thrombosis such as venous thromboembolism. Hyperhomocysteinemia is also called homocysteinuria because accumulated homocysteine in the serum is excreted into the urine.

38.5 Absorption of Vitamin B12

Vitamin B12 (cobalamin) is found in animal products and is exclusively synthesized by bacteria, such as those present in the rumen of cattle. Absorption of vitamin B12 requires many factors such as R-binders and intrinsic factor (IF). R-binders, also known as haptocorrin and transcobalamin-I, prevent the degradation of the acid sensitive B12 in the stomach. R-binders are released by the salivary glands and gastric mucosa in response to the ingestion of food. In addition, both R-binders and IF are secreted from the parietal cells of the gastric mucosa (▶ Fig. 38.4). Once vitamin B12 is released from the food proteins in the stomach through the action of gastric acid, it binds to as the R-binders.

Vitamin B12 bound to R-binders move into the duodenum, where pancreatic proteases digest the R-binders, thus enabling vitamin B12 to preferentially bind to intrinsic factor (IF). From the duodenum to the ileum, vitamin B12 remains bound to IF and thus protected from digestion. In the ileum, IF receptors allow internalization of the vitamin B12-IF complex.

Within the enterocytes of the ileum, vitamin B12 dissociates from IF. Transcobalamin-II binds to vitamin B12 when it is released into the bloodstream. Transcobalamin-II transports and releases vitamin B12 to liver, bone marrow, and other tissues. If secreted into bile, vitamin B12 can be efficiently taken up again through the enterohepatic circulation. The human body stores vitamin B12 in the liver at levels that are sufficient for several years.

38.5.1 Link to Pathology: Causes of Vitamin B12 Deficiency

Vitamin B12 deficiency can occur through several mechanisms. In strict vegans, deficiency is due to an absence of animal products in their diet and, hence, inadequate intake. Vitamin B12 deficiency can develop in elderly individuals due to poor diet or development of absorptive defects such as declined production of IF due to aging. The socioeconomic risk and lack of access to adequate nutrition also leads to deficiency of B12.

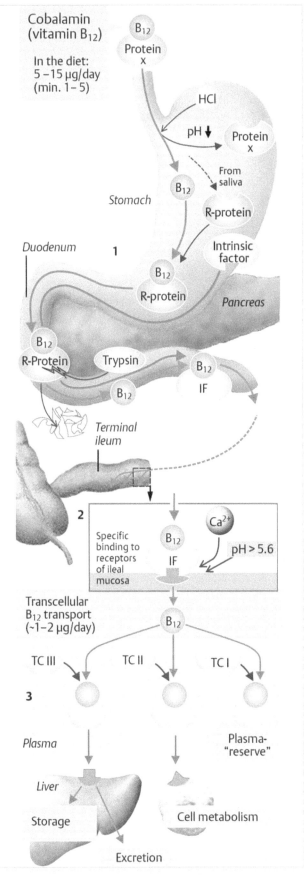

Fig. 38.4 Digestion and Absorption of B12.

Vitamin B12 digestion and absorption is a rather complex process, therefore impairment at any of the stages of its digestion and absorptive process may lead to vitamin B12 deficiency. As previously discussed, the gastric parietal cells are essential for the synthesis of R-binders and IF. Destruction of these cells therefore prevents the absorption of B12. Increased hydrogen ions in the terminal ileum also affect the B12 absorption. Pancreatic insufficiency or the absence of pancreatic proteases causes ineffective digestion of R-binders; therefore B12 cannot be released from R-binders and cannot be absorbed.

Additional causes for B12 deficiency include excess use of antacids and proton pump inhibitors such as omeprazole, which reduces stomach acid production and secretion of R-binders and IF from the parietal cells. Bariatric bypass surgery and surgical removal of the terminal ileum, such as with Crohn's disease, cause insufficient absorption of B12. In addition, a genetic deficiency of transcobalamin-II would prevent the normal transport of B12 in the blood and produce a B12 deficiency.

Reference

[1] Palmer AM, Kamynina E, Field MS, Stover PJ. Low vitamin B12 impairs nuclear dTMP synthesis. Proceedings of the National Academy of Sciences 2017, 114 (20) E4095 E4102

Review Questions

1. Insufficient synthesis of which of the following molecules is most likely due to the deficiency of B12?
   A) Functional folate
   B) Hemoglobin
   C) Intrinsic factor
   D) R-binders
   E) Thymidine

2. Deficiency of which of the following vitamins/molecules causes methylmalonic aciduria?
   A) Folate
   B) SAM
   C) Vitamin B6
   D) Vitamin B12

3. Which of the following enzymes is inhibited by methotrexate?
   A) Conjugase
   B) Dihydrofolate reductase
   C) Methionine synthase
   D) Methylmalonyl CoA mutase
   E) Thymidylate synthase

4. Folate malabsorption can result from which of the following?
   A) Deficiency of transcobalamin-II
   B) Excessive use of antacids
   C) Insufficient production of IF
   D) Omeprazole
   E) Pancreatitis
   F) Phenytoin
   G) Strict vegan diet

Answers

1. **The correct answer is E**. Deficiency of B12 traps the folate in the form of methyltetrahydrofolate (FH4-CH3). This creates deficiency in methylenetetrahydrofolate (FH4-CH2), which is required for de novo synthesis of thymidine (dTMP). Vitamin B12 is the coenzyme of methionine synthase, which catalyzes re-methylation of homocysteine to methionine and regenerates functional folate (FH4) from FH4-CH3.

   Answer choice A is incorrect because deficiency of B12 traps the functional folate in the form of FH4-CH3. Dihydrofolate reductase catalyzes the synthesis of functional folate (FH4), therefore B12 deficiency does not affect its synthesis, and however it traps it in its methyl form.
   Answer choices B, C, and D are incorrect because B12 does not have any role in the synthesis of these proteins.

2. **The correct answer is D**. Vitamin B12 is the coenzyme of methylmalonyl CoA mutase in the propionyl CoA pathway. In this pathway, propionyl CoA is converted to methylmalonyl CoA by propionyl CoA carboxylase. The methylmalonyl CoA is then converted to succinyl CoA, which enters the TCA cycle to be oxidized. Measurements of methylmalonyl CoA is a diagnostic test for B12 deficiency. Folate and B12 deficiency presents with mild

hyperhomocysteinemia and hematological symptoms of megaloblast formation. This diagnostic test therefore is an important test to differentiate between the two major causes of megaloblastic anemia. Vitamin B6 is the coenzyme for cystathionine β-synthase, which converts homocysteine to cystathionine. Therefore, B6 deficiency also causes mild hyperhomocysteinemia. SAM functions as a one-carbon donor in more than 30 reactions in metabolism. Folate, SAM and vitamin B6 do not have any role in the propionyl CoA pathway. Therefore methylmalonyl CoA does not increase and thus answer choices A, B, and C are incorrect.

3. **The correct answer is B**. Methotrexate is a structural analog of folate and competitively inhibits the dihydrofolate reductase enzyme, which catalyzes synthesis of dihydrofolate and tetrahydrofolate, a functional folate. Conjugase (answer choice A) is an intestinal mucosal enzyme, which catalyzes conversion of folate polyglutamates to folate monoglutamates. The folate monoglutamates can be absorbed into the enterocytes. Methotrexate does not inhibit the conjugase enzyme. Methionine synthase (answer choice C) catalyzes the transfer of a methyl group from B12–CH3 to homocysteine to synthesize methionine. This enzyme is not inhibited by methotrexate either. Vitamin B12 is also the coenzyme for methylmalonyl CoA mutase, which catalyzes the conversion of propionyl CoA to succinyl CoA. Methotrexate does not inhibit either methionine synthase or methylmalonyl CoA mutase. Thymidylate synthase transfers a methyl group from methylenetetrahydrofolate (FH4–CH2) to dUMP to synthesize dTMP. This enzyme can be inhibited by 5-fluorouracyl, but methotrexate does not inhibit thymidylate synthase, hence answer choice E is incorrect.

4. **The correct answer is F**. The antiepileptic drug phenytoin inhibits the activity of the conjugase enzyme in the small intestine. Since the conversion of folate polyglutamates to folate monoglutamates is necessary for the absorption of dietary folate, and conjugase is responsible for this conversion, this inhibition causes the malabsorption of folate. In addition to the drug phenytoin, erosion of enterocytes due to celiac or Crohn disease also reduces the intestinal conjugase, therefore causing folate malabsorption.

Answer choices A, B, C, D, E, and G are incorrect because all of these conditions/drugs inhibit the absorption of B12. Excess use of antacids (answer choice B) and proton pump inhibitors such as omeprazole (answer choice D) reduce stomach acid production and secretion of R-binders and IF from the parietal cells. A genetic deficiency of transcobalamin-II (answer choice A) causes impaired transport of B12 in the blood, thus resulting in a deficiency. Patients with pancreatitis (answer choice E) may have reduced proteases, which impair the digestion of R-binders. Therefore B12 cannot be freed from R-binders to bind to IF. A vegan diet (answer choice G) does not result in folate deficiency, because folate is found in fruits and vegetables. A strict vegan diet, however, may cause B12 deficiency since B12 is found only in animal products.

V

Review Questions

1. A 16-year-old male presents to the emergency department with a severe sudden onset of pain in his lower abdomen. He also reports that he feels nauseated. Physical examination of his abdomen is unremarkable. Urinary tests show hematuria, signs of a urinary tract infection, and hexagonal crystals. His physician directs himto drink a large volume of fluids. Which of the following is his most likely his diagnosis?
   A) Cystinuria
   B) Homocysteinuria
   C) Hyperhomocysteinemia
   D) Hyperoxaluria

2. A 36-year-old man complainsof constant upper abdominal pain forthe last three months. He reports that the pain radiates to the back and gets worse whenever he eats and drinks. Sitting up and leaning forward helps to reduce the pain. A complete blood count shows increased white blood cells. Although a stool chymotrypsin test is negative,afecal elastase-1 test demonstrates decreased levels of elastase. In addition, numerous fat droplets are found uponfecal fat testing. Dysfunction of which of the following organs is the most likely cause of his current condition?
   A) Large intestine
   B) Small intestine
   C) Pancreas
   D) Stomach

3. A48-year-old renal transplantation patient is being treated with an immunosuppressant to inhibit T-cell proliferation and organ rejection. Which of the following is also inhibited by this treatment?
   A) Absorption of amino acids
   B) Autophagy pathway
   C) Digestion of proteins in the small intestine
   D) mTOR pathway
   E) Proteosomal degradation of proteins

4. A researcher has discovered amolecule that induces apoptosis in cultured bone marrow cells. It is also found that under these conditions the ubiquitination of proteinsis increased.Which of the following is most likely inhibited by this molecule?
   A) Degradation of proteins through autophagy
   B) Degradation of proteins by the 26S proteasome
   C) mTOR dependent proteolysis
   D) Ubiquitination of cellular proteins

5. A 5-week-old male presents with frequent vomiting and failure to thrive. His mother states that he doesn't sleep well and cries a lot. His pediatrician notes hypotonia and tachypnea. Blood tests show decreased BUN and citrulline, but elevated ornithine, glutamine and alanine. Urine analysis reveals orotic aciduria. A nutritionist orders discontinuation of protein intakeand a high carbohydrate and lipid diet. Which of the following is most likely to develop if the patient is fed protein-containing diet?
   A) Hepatic encephalopathy
   B) Hyperammonemia
   C) Kwashiorkor
   D) Marasmus
   E) Rye syndrome

6. A 46-year-old man with a long history of cirrhosis is found in a coma. Laboratory tests reveal extremely high levels of serum ammonia. Which of the following amino acids is increased in his plasma?
   A) Asparagine
   B) Aspartate
   C) Glutamate
   D) Glutamine

7. A-three-month-old female is brought for her well-child visit. The baby's mother states that everything is fine except that she cannot breast-feed her baby anymore because of her poor suckling. Her physician notes hypopigmentationof her skin and hair. Her weight and height is below the 10th percentile. Neurological examination reveals delay in her moto-neural development. Review of her genetic screening tests reveal normal genetic screening results. Which of the following supplementationswould most likely alleviate this baby's symptoms?
   A) Pyridoxal phosphate (PLP)
   B) Tetrahydrobiopterin (BH4)
   C) Tetrahydrofolate

8. A one-week-old female presents with fatigue and vomiting. Physical examination reveals abnormal muscle tone and unresponsiveness. Her physician notes an odor of burned sugar. Urine analysis show ketones. Which of the following is most likely the diagnosis?
   A) Alkaptonuria
   B) Maple syrup urine disease
   C) Phenylketonuria
   D) Tyrosinemia

Answers

1. **The correct answer is A.** The hexagonal shape of the crystals indicates that the patient has cysteine crystals. These crystals form in the urinary tract due to a genetic deficiency of the basic amino acid transporter. In the proximal tubule, this transporter normally removes cysteine and basic amino acids (arginine, lysine, and histidine) from the filtrated urine back into the blood. However, with the genetic deficiency of this transporter, cysteine levels increase in the urine and crystals form. Patients develop kidney stones in early teenage years and they suffer throughout their life with frequent kidney stones. Hydration and avoidance of methionine rich foods (meat and animal products) is the key for prevention of stone formation. Hyperoxaluria (answer choice D) also causes formation of kidney stones. Oxalate and calcium precipitates and forms crystals in the kidney. Microscopic analysis of the calcium ox-alate crystals reveals a characteristic "picket fence" or "enve-lope" shape. Therefore, answer choice D is incorrect. Homocysteinuria and hyperhomocysteinemia do not present with kidney stones, therefore answer choices C and D are incorrect.

2. **The correct answer is C.** Based on the clinical symptoms and the lab tests, he is most likely to have chronic pancreatitis that needs to be confirmed by imaging studies. Increased white blood cells indicates inflammation. Positive

V

fecal fat test is a sign of malabsorption. Both chymotrypsin and elastase are digestive enzymes that are secreted by the pancreas and both are present in stool under normal conditions. Absence of chymotrypsin and a decreased level of elastase indicate pancreatic insufficiency.

Answer choices A, B, and D are incorrect because chymotrypsin and elastase are secreted by the pancreas.

3. **The correct answer is D.** This patient is most likely being treated with rapamycin. As an immunosuppressant, rapamycin inhibits T-cell proliferation and also inhibits the reaction of T-cells to certain cytokines. Rapamycin's antiproliferative activity is due to its ability to inhibit the mTOR pathway. Inhibition of mTOR activates the autophagy pathway. Therefore, answer choice B is incorrect. Rapamycin does not affect the digestion and absorption of proteins in the small intestine. Thus answer choices A and C are incorrect. Proteasomal degradation is ubiquitin-dependent, which is not affected by rapamycin. Therefore, answer choice E is incorrect.

4. **The correct answer is B.** Because the treated cells have more ubiquitinated proteins than the untreated cells, this molecule must be either a proteasome inhibitor or it increases the rate of protein ubiquitination by increasing the activity of E3 ligases in the ubiquitin pathway. The drug bortezomib (BTZ) inhibits degradation of ubiquitinated proteins by competitively binding to the 26S proteasome. The proteasome-dependent degradation pathway is important for the cells to maintain the balance of regulatory proteins involved in many cellular processes, such as control of growth and the cell cycle. This dysregulation due to inhibition of proteasome degradation pathway induces apoptosis and cell death.

Answer choices A and C are not ubiquitin-dependent proteasome pathways. The autophagy pathway does not involve the proteasome.

Reference: Field-Smith A, Morgan GJ, Davies FE. Bortezomib (VelcadeTM) in the Treatment of Multiple Myeloma. *Therapeutics and Clinical Risk Management.* 2006;2(3):271-279.

5. **The correct answer is B.** This patient most likely has inherited ornithine transcarbamoylase (OTC) deficiency. The OTC gene is located on the X-chromosome. Therefore, males are affected more often than females. Because of this genetic deficiency, the urea cycle does not operate. Thus hyperammonemia occurs. Ammonia can pass through membranes and inhibit ATP generation by inhibiting the TCA cycle. Therefore, ammonia must be kept at very low levels to prevent neuronal toxicity. A low protein diet prevents ammonia formation in these patients. Therefore, a low protein diet is recommended to prevent brain damage. Due to a deficiency of OTC, carbamoyl phosphate accumulates in the mitochondria and finds its way to the cytoplasm. Since carbamoyl phosphate is an intermediate of the pyrimidine biosynthesis pathway, orotic acid increases. Decreased plasma levels of citrulline and arginine are also consistent with OTC deficiency.

Hepatic encephalopathy (answer choice A) and Rye syndrome (answer choice E) can also result in hyperammonemia. However, this patient's disease is not caused by either of these conditions. Kwashiorkor and Marasmus are protein energy malnutrition diseases, which do not present in early life. These diseases develop later life time. Also, these two conditions are treated with a balanced diet, containing sufficient amounts of proteins, carbohydrates, and lipids.

Reference: https://emedicine.medscape.com/article/950672-overview

6. **The correct answer is D.** The patient most likely has hepatic encephalopathy as indicated by cirrhosis and coma. Because of high levels of serum ammonia, glutamate is consumed in fixing excess ammonia to produce glutamine by glutamine synthase. Therefore, the plasma level of glutamate is decreased while the plasma level of glutamine is increased. Considering glutamine as a primary nitrogen carrier in the circulation, increased plasma levels of glutamine is consistent with hyperammonemia of hepatic encephalopathy.

7. **The correct answer is B.** This baby's signs and symptoms indicate phenylketonuria (PKU). However, newborn screening results were negative and did not detect a genetic deficiency in the phenylalanine hydroxylase (PAH) gene. Since the patient has PKU-like symptoms, it may be non-classical (aka secondary) PKU. Deficiency in the synthesis of tetrahydrobiopterin (BH4), the coenzyme of phenylalanine hydroxylase, causes the non-classical form of PKU. Either phenylalanine hydroxylase or BH4 deficiency leads to accumulation of phenylalanine. High concentrations of phenylalanine in the blood saturates the neutral amino acid transporter in the blood-brain barrier. This transporter transports phenylalanine, branched-chain amino acids (isoleucine, leucine, and valine), tyrosine and tryptophan. When this transporter is saturated with phenylalanine, these amino acids cannot pass into the brain and this results in psychomotor developmental delays. To prevent progress of the disease, a diet low in phenylalanine is necessary for these patients. Reports also show that BH4 supplementation significantly lowers blood phenylalanine levels in secondary PKU patients. Hypopigmentation of skin and hair is also common in PKU patients because increased phenylalanine inhibits tyrosinase, an enzyme catalyzing the production of melanin from tyrosine.

Answer choices A, C, and D are incorrect because these coenzymes are not required for the phenylalanine hydroxylase reaction.

Reference: https://emedicine.medscape.com/article/949470-overview

8. **The correct answer is B.** Genetic deficiency of the branched-chain ?-keto acid dehydrogenase enzyme causes maple syrup urine disease (MSUD), which is inherited in an autosomal recessive manner. Because of this enzyme deficiency, branched-chain amino acids are not degraded properly. This results in accumulation of ?-keto acids of leucine, isoleucine and valine, and consequent metabolic acidosis. Although ketoacidosis is an important laboratory finding in affected infants, the positive ketones in the urine is not due to acetoacetate and ?-hydroxybutyrate. It is due, instead, to accumulation of -keto derivatives of branched chain amino acids. These ?-keto acids give patients the characteristic body odor of burned sugar or maple syrup.

V

Alkaptonuria (answer choice A) is due to the accumulation of homegentisic acid, which is excreted in urine and turns a brown-black color when it is exposed to air. Phenylketonuria (answer choice C) patients also develop metabolic acidosis; however, their body odor is described as mousy or musty.

Tyrosinemia type I, II, and III are caused by deficiencies of enzymes in the phenylalanine catabolism pathway. All three types of tyrosinemias have increased excretion of tyrosine in the urine. This excess tyrosine in the urine gives urine a cabbage-like smell.

V

Review Questions

1. A 25-year-old woman complains of severe headache and unusual sweating. Physical examination reveals heart palpitations and high blood pressure. Twenty-four-hour urine analysisshowselevated levels of vanillylmandelic acid (VMA). Which of the following is the most likely explanation for the elevated levels of VMA in her urine?
   A) Degradation product of ?-aminobutyrate
   B) Degradation product of norepinephrine
   C) Decreased serotonin levels
   D) Increased thyroid hormones

2. A 19-year-old male presents with tachycardia, hypertension, and excessive sweating. Physical examination reveals dilated pupils and irregular breathing. He reports that he tried an unknown substance that his friend gave him in an effort to increase his energy and improve his mood.Increased levels of which of the following neurotransmitters are responsiblefor this patient's condition?
   A) ?-aminobutyrate, glutamate, and serotonin
   B) Dopamine, norepinephrine, and serotonin
   C) Epinephrine, norepinephrine, and serotonin
   D) Nitric oxide, norepinephrine, and dopamine

3. A 33-year-old male presents to the emergency department with chest discomfort. He reports that while playing soccer he suddenly felt pressure and heaviness in his chest. His medical history reveals that he has hypercholesterolemia, and his BMI is over 30. Angina pectoris is suspected, and he is given a sublingual nitroglycerin. Activation of which of the following enzymes alleviates this patient's discomfort?
   A) Adenylyl cyclase
   B) Membrane-bound guanylyl cyclase
   C) Phosphodiesterase
   D) Soluble guanylyl cyclase

4. A 45-year-old female presents with weight loss and agitation. She also complains of heat intolerance. Physical examination reveals tachycardia and an enlarged thyroid gland. Blood tests show increased T4 but decreased TSH levels. Her physician prescribes a corticosteroid. Activity of which of the following enzymesis expected to be decreased with her medication?
   A) Deiodinase
   B) Lysosomal enzymes
   C) PEPCK
   D) Thermogenin
   E) Thyroperoxidase

5. A 30-year-old female presents with fatigue, weight loss, and fever. Physical examination shows no abnormalities except skin pallor. A peripheral blood smear shows ring sideroblasts that are microcytic and hypochromic. Her history reveals that she hasbeen on isoniazid for the last six months. Decreased activity in which of the following enzymesis the most likely cause of her symptoms?

   A) ALA synthase
   B) ALA dehydratase
   C) Ferrochelatase
   D) Porphobilinogen deaminase
   E) Uroporphyrinogen decarboxylase

6. A 1-week-old infant is brought to the physician by his parents because of yellowish-colored skin. Physical examination revealsno apparent abnormalities. Laboratory tests show hemoglobin18 g/dL (N = 14 to 24 g/dL), reticulocytes 1.0% (N = 0.1-1.3%). This patient's blood most likely has an excess amount of which of the following?
   A) - aminolevulinic acid
   B) Bilirubin
   C) Biliverdin
   D) Porphobilinogen
   E) Stercobilin

7. A two-year-old male is brought to the pediatric dentist by his parents because of severe and repeated lip chewing and aggressive tongue biting. Urine analysis reveals a significant increase of uric acid production. Physical examination reveals that his head control is poor, and the child shows delayed motor skills. Intellectual disability is also evident, and he displays self-destructive behavior. This patient most likely has a deficiency in which of the following enzymes?
   A) AMP deaminase
   B) HGPRT
   C) PRPP synthetase
   D) Ribonucleotide reductase
   E) Xanthine oxidase

8. A 4-year-old child is brought to the physician by his mother because of fatigue and weakness. The child's skin appears pale, and physical examination further shows pale conjunctiva. Laboratory studies show hemoglobin of 6 g/dL and high levels of orotate in the urine. Aperipheral blood smear reveals thepresence of abnormally large erythrocytes. Which of the following is the most appropriate pharmacotherapy for this child?
   A) Allopurinol
   B) Folate
   C) 5-Fluorouracil
   D) Uridine
   E) Vitamin B12

9. A 32-year-old female presents with pelvic pain and a persistent urge to urinate. She is diagnosed with a urinary tract infection and is prescribed a combination drug containing trimethoprim and sulfamethoxazole. Which of the following is the most likely mechanism of action of this drug combination?
   A) Inhibition of transcription of dihydrofolate reductase
   B) Inhibition of transport across bacterial cell walls
   C) Inhibition of bacterial synthesis of cobalamin
   D) Inhibition of bacterial synthesis of tetrahydrofolate
   E) Inhibition of synthesis of phospholipids in bacteria

V

10. A 62-year-old male who had a recent gastrectomy presents with memory loss and weakness. The physician performs a broad workup and discovers that the patient is anemic and has a vitamin B12 deficiency. Which of the following findings is most likely associated with this patient's symptoms?

	Mean Corpuscular Volume (80-100 µm³)	Homocysteine	Methylmalonic acid	S-adenosylmethionine
A.	90	elevated	elevated	normal
B.	105	elevated	normal	elevated
C.	110	elevated	elevated	low
D.	120	normal	elevated	normal
E.	120	elevated	normal	low
F.	110	elevated	elevated	elevated

Answers

1. **The correct answer is B**. Vanillylmandelic acid (VMA) is the degradation product of norepinephrine. This patient most likely has pheochromocytoma, which is a norepinephrine/epinephrine secreting tumor. Patients generally presents with headache, palpitations, diaphoresis, and severe hypertension. The measurements of VMA and HVA are useful for the screening of catecholamine-secreting tumors such as neuroblastoma, pheochromocytoma, or other neural crest tumors. Homovanillic acid (HVA) is the degradation product of dopamine. Answer choice A is incorrect because GABA is not degraded. Instead, it enters into the GABA shunt. The degradation product of serotonin is 5-hydroxyindoleacetic acid (5-HIAA), thus answer choice C is incorrect. The degradation of thyroid hormones T3 and thyroxine (T4) does not produce any metabolite for urinary excretion, hence answer choice D is incorrect.

2. **The correct answer is B**. The patient is displaying symptoms of drug overdose, most likely cocaine or amphetamine. Both of these drugs increase the levels of dopamine, norepinephrine, and serotonin in the synaptic cleft. While cocaine blocks the reuptake mechanisms of these neurotransmitters, amphetamines increase their release. Norepinephrine binds to adrenergic receptors of the cardiac muscle cells, thus increasing heartbeat and causing tachycardia. Hypertension and excessive sweating (diaphoresis) are side effects of increased levels of norepinephrine, while dilated pupils are due to increased serotonin levels. CNS stimulation (enhanced mood and exhilaration) is due to increased levels of dopamine and serotonin.

   Answer choice A is incorrect because GABA is an inhibitory neurotransmitter. Answer choice C is incorrect because epinephrine is not released from the CNS. It is mainly released from the adrenal medulla. Serotonin by itself would not cause exhilarations. Answer choice D is incorrect because NO is a smooth muscle relaxant, thus the patient would not have hypertension with an increase in NO levels.

3. **The correct answer is D**. Nitroglycerin tablets are useful in the treatment of angina pectoris because nitroglycerin generates NO, which is a gas and thus can pass through the membranes. NO activates intracellular guanylyl cyclase in the vascular smooth muscle cells, resulting in an increase in cGMP, which relaxes the blood vessels and increases cardiac muscle perfusion.

   Answer choice A is incorrect because adenylyl cyclase generates cAMP, which activates PKA. The membrane bound guanylyl cyclase (answer choice B) is activated by atrial natriuretic peptide (ANP), which also reduces blood pressure through cGMP-triggered relaxation of the smooth muscles. However, ANP is a peptide hormone and cannot diffuse through the cell membrane. It binds to its receptor and activates the guanylyl cyclase activity of the receptor and therefore increases intracellular cGMP levels. Answer choice D is incorrect because inhibition of phosphodiesterase in vascular smooth muscle cells keeps cGMP levels elevated, causing relaxation of the smooth muscle.

4. **The correct answer is A**. The thyroid gland produces both T3 and T4. In the bloodstream, the concentration of T4 is higher than T3, although T3 has higher biological activity than T4. The T4 (thyroxine) is deiodinated to T3 by the deiodinase enzyme, which is mainly found in the liver and kidney. This patient most likely has Graves' disease, which leads to goiter with hyperthyroidism. Graves' disease is an autoimmune disease which is caused by the binding of IgG autoantibodies to TSH receptors on the thyroid gland. This binding stimulates the thyroid gland to secrete excessive T3 and T4. The increased T3 and T4 levels decrease the TSH levels by negative feedback regulation. Therefore, patients with Graves' disease present with weight loss and tachycardia due to an increased basal metabolic rate. The administration of corticosteroids helps these patients because corticosteroids inhibit the deiodinase enzyme, therefore decreasing the conversion of T4 to T3.

   Answer choices B and E are incorrect because there isn't any known effect of corticosteroids on these enzymes. The thyroperoxidase enzyme catalyzes iodination of tyrosine residues of the thyroglobulin within the thyroid gland cells. Answer choice C is incorrect because corticosteroids increase the activity of PEPCK by increasing the transcription of the PEPCK gene and therefore PEPCK synthesis increases. Thermogenin (answer choice D) is not an enzyme, it is a natural uncoupling protein, and its expression is increased in patients with hyperthyroidism.

5. **The correct answer is A**. The patient most likely has sideroblastic anemia due to isoniazid treatments. Long term use of isoniazid may lead to sideroblastic anemia by decreasing the availability of pyridoxal phosphate, the active form of vitamin B6. The vitamin B6 is the coenzyme of ALA synthase. Therefore, if there is not enough pyridoxal phosphate to bind ALA synthase, its activity drops. Answer choices B, C, D, and E are incorrect because none of these enzymes requires pyridoxal phosphate as a coenzyme.

6. **The correct answer is B**. This patient is showing the symptoms of newborn jaundice. The immature liver of this patient is unable to perform the conjugation of each bilirubin molecule with two molecules of glucuronic acid. Hence unconjugated bilirubin accumulates and causes physiological newborn jaundice.
   Answer choice A is incorrect because ALA is found in the mitochondria of pre-erythroid cells. Answer choice C is incorrect because biliverdin is found within the macrophages of the reticuloendothelial system. Biliverdin is the intermediate of the heme degradation pathway. Heme is degraded to biliverdin by heme oxygenase enzyme. Biliverdin is then converted to bilirubin by bilirubin reductase. Bilirubin is carried to the liver bound to albumin. Porphobilinogen (answer choice D) is an intermediate of the heme synthesis pathway. Urobilinogen (answer choice E) is a degradation product of bilirubin, but it forms in the colon by colonic bacteria from urobilinogen.

7. **The correct answer is B**. This patient most likely has Lesch-Nyhan syndrome, a rare inborn error of metabolism. It is characterized by intellectual disability and self-destructive behavior resulting in self-mutilation through biting and scratching. Hypoxanthine-guanine phosphoribosyl transferase (HGPRT) is deficient in Lesch-Nyhan syndrome and leads to the inability to salvage hypoxanthine or guanine. Inhibition of the salvage pathway results in the accumulation of PRPP, which activates PRPP-amidotransferase. This causes the activation of de novo purine nucleotide synthesis. At the same time, there is a buildup of hypoxanthine and guanine due to the absence of the salvage pathway. These purine bases are destined to purine degradation to generate excessive amounts of uric acid.
   Answer choice A is incorrect because genetic deficiency of AMP deaminase (ADA) causes severe combined immunodeficiency syndrome (SCID), which presents with recurrent infections. There is no known disease with the genetic deficiencies of PRPP synthetase (answer choice C), ribonucleotide reductase (answer choice D) or the xanthine oxidase enzymes (answer choice E).

8. **The correct answer is D**. This patient most likely has orotic aciduria due to a deficiency in UMP synthase (UMPS), which is a bifunctional enzyme with orotate phosphoribosyltransferase and orotidylate decarboxylase activities. Mutations in either domain can lead to accumulation of orotate in the urine. Patients with this disease present with megaloblastic anemia, which cannot be treated with either folate or vitamin B12 (answer choice E). Pyrimidine replacement with oral uridine results in hematological remission with reduced urinary excretion of orotic acid.
   Allopurinol (answer choice A) is used for the treatment of gout and inhibits the xanthine oxidase enzyme in the purine degradation pathway. 5-fluorouracil (answer choice C) is an anti-neoplastic drug, which inhibits the thymidylate synthase enzyme and, therefore, causes decreased synthesis of dTMP.

9. **The correct answer is D**. The sulfonamides such as sulfamethoxazole are structural analogues of paraaminobenzoic acid (PABA) that inhibit the do novo synthesis of folates. Trimethoprim, an antifolate, preferentially inhibits bacterial DHFR with little or no effect on human DHFR. Combination treatment of sulfonamides and trimethoprim is synergistic and is used widely for the treatment of infections due to susceptible bacteria, as well as prevention of opportunistic infections.

10. **The correct answer is C**. This patient has vitamin B12 deficiency. Gastrectomy leads to reduced intrinsic factor (IF) secretion by the parietal cells, thereby reducing the absorption of vitamin B12 in the terminal ileum. In vitamin B12 deficiency, the size of the red blood cells is enlarged. Due to the inhibition of tetrahydrofolate synthesis, folate is trapped in the form of FH4-CH3. S-adenosylmethionine (SAM) levels are low and homoctsyteine levels are elevated. Vitamin B12 deficiency also leads to the reduction of methylmalonyl mutase activity. Thus, accumulation of methylmalonic acid is a diagnostic test for this patient.

Part VI

VI

39 Digestion and Absorption of Lipids

At the conclusion of this chapter, students should be able to:
- Classify the types of lipids and their biological roles in the body
- Explain the nomenclature and classification of fatty acids
- Understand the importance of essential and non-essential fatty acids in nutrition
- Explain the digestion processes of triacylglycerols (TG) to free fatty acids (FA) and 2-monoacylglycerol (2-MG)
- Describe the role of bile acids and salts in the digestion of lipids
- Explain the synthesis of bile acids and salts
- Describe how micelles enter epithelial cells and are reconverted to TG
- Explain how TG, cholesterol plus apoproteins, and other lipids form nascent chylomicrons that exit intestinal epithelial cells

In this chapter of digestion and absorption of lipids, we will explain how dietary lipids are hydrolyzed into their absorbable components and how the dietary lipids are transported in the body. Since lipids are structurally and functionally diverse molecules, we will start with the nomenclature and classification of lipids. There are four classes of medically important lipids: triacylglycerols (fats), phospholipids, sphingolipids, and cholesterol.

Triacylglycerols are composed of a glycerol backbone esterified to three fatty acids. Triacylglycerols, also called fats, are the major form of storage and the transport form of fatty acids. There are two major groups of phospholipids: glycerophospholipids and sphingophospholipids. The difference between glycerophospholipids and sphingophospholipids is their backbone, with glycerophophoslipids having a glycerol backbone and sphingophospholipids having a sphingosine backbone. The sphingosine backbone is also found in glycolipids, in which phosphate and the choline head group of sphingomyelin is replaced by a carbohydrate group. Both glycerophospholipids and sphingophospholipids function as major structural components of biological membranes. Cholesterol's structure and function is reviewed in Chapter 42.

These four classes of lipids can also be categorized based on their biological roles (▶ Table 39.1). Lipids in our body can be used for energy production, energy storage, structural components of cell membranes, and synthesis of some hormones. For example, cholesterol is a precursor for steroid hormones

(reviewed in Chapter 44), and arachidonic acid is a precursor for eicosanoids (reviewed in Chapter 41).

39.1 Fatty Acids Nomenclature and Classification

Fatty acids are carboxylic acid molecules with hydrocarbon chain and serve as important components of many lipids. In addition to fatty acids being esterified with glycerol to form triacylglycerols (fat), fatty acids are also esterified with sphingosine to form sphingomyelin and with cholesterol to form cholesterol ester. Although there are some odd-number fatty acids, most naturally occurring fatty acids have an even number of carbon atoms (▶ Fig. 39.1). The synthesis of 16-carbon palmitic acid (16:0), a saturated fatty acid without a double bond, occurs in the liver. While saturated fatty acids have no double bond, unsaturated fatty acids contain one or more double bonds. Fatty acids with one double bond are called monounsaturated (e.g. oleic acid 18:1), and fatty acids with more than one double bond are called polyunsaturated (e.g. linoleic acid 18:2 and linolenic acid 18:3).

For polyunsaturated fatty acids, the location of the double bond is important. In this classification the most important fatty acids are ω-3 (alpha linolenic acid ALA, eicosapentaenoic acid EPA, docosahexaenoic acid DHA) and **ω -6 fatty acids** (linoleic acid, arachidonic acid).

The numbering of the fatty acids starts with the carboxyl group being #1. Along with numbering, the fatty acids' carbons are also designated with Greek letters. In this nomenclature, the α-carbon (C-2) is next to the carboxyl carbon (C-1), the β-carbon is C-3 and so on. Regardless of the fatty acid carbon number, the last carbon is always called the ω-carbon. Based on this nomenclature, if the first double carbon is at the third position from the omega carbon, then the fatty acid is called ω-3 fatty acid. The same rule applies for ω-6 and ω-9 fatty acids.

Fatty acids are also classified as **essential** and **non-essential fatty acids**. Essential fatty acids have to be in the diet since humans cannot synthesize them. Linoleic acid (18:2), a ω-6 fatty acid, and linolenic acid (18:3), a ω-3 fatty acid, are two essential fatty acids that must be obtain from the diet. Although the human body can elongate arachidonic acid to synthesize EPA (20:5) and DHA (22:6), it is recommended that these lipids should be taken from the diet since elongation and desaturation

Table 39.1 Biological roles

	Substance	Function
Fuel	Fats, fatty acids	Store and provide energy
Cell components	Phospholipids, sphingolipids, and cholesterol	Membrane components
Insulation	Phospholipids, sphingolipids, and cholesterol	Mechanical cushion and thermal insulator Electrical insulator
Special tasks	Steroid hormones, glycerolipids, fatty acids, eicosanoids	Signal function: hormone, mediator, second messenger
	Fatty acids, isoprenoids	Membrane anchor
	Isoprenoids	Cofactor for enzymes
	Retinal	Visual pigment

Source: Koolman J, Röhm K, ed. Color Atlas of Biochemistry. 3rd Edition. New York, NY. Thieme; 2012.

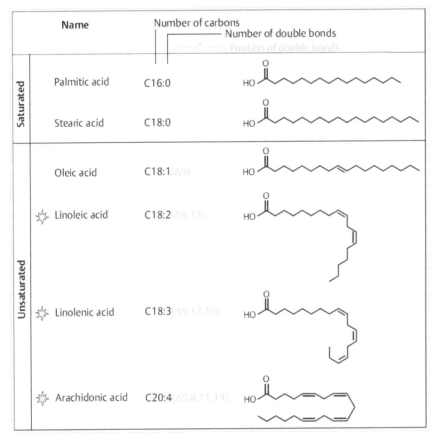

Name	Number of carbons		

Fig. 39.1 Saturated and unsaturated fatty acids. Saturated fatty acids have no double bond, e.g. C16:0. Unsaturated fatty acids may have one double bond (monounsaturated) or may have more than one double bond (polyunsaturated). Linoleic acid and linoleic acid are essential fatty acids. Arachidonic acid is precursor for eicosanoids. (Source: Panini S, ed. Medical Biochemistry—An Illustrated Review. New York, NY. Thieme; 2013.)

VI

processes are not as effective in humans. Arachidonic acid is not essential either, because linoleic and linolenic fatty acids can be elongated and desaturated to synthesize arachidonic acid.

39.1.1 Link to Physiology: Arachidonic Acid is a Precursor for Eicosanoids

The local hormones prostaglandins, thromboxanes, and leukotrienes are known as eicosanoids. These local hormones have important roles in many biological processes including inflammation, allergic responses, platelet aggregation, and vasoconstriction. These compounds are also widely used pharmacologically, including prevention and treatment of gastric ulcers, control and alleviation of high blood pressure and asthma, relief of allergic reactions, and labor induction. The synthesis and physiological roles of eicosanoids is reviewed in Chapter 41.

39.2 Cis vs Trans Unsaturated Fatty Acids

Usually the double bonds in the unsaturated fatty acids are in cis form. If the acyl chains are on the same side of the double bond, it is called cis configuration. The cis configuration creates a "kink" in the fatty acid's hydrocarbon chain. Fats rich with unsaturated fatty acids are liquid at room temperature. Most trans fatty acids are man-made through the process of partially hydrogenating vegetable oils. Trans fats, such as margarine, were introduced in the early 1950's. Although trans fatty acids are chemically unsaturated, they have adverse health effects

which have been studied extensively in many clinical research studies. The results of these studies repeatedly show that trans fatty acids increase the risk of cardiovascular disease.

39.3 Digestion and Absorption of Dietary Lipids

The digestion and absorption of lipids in various forms are an important part of human metabolism. When fats are ingested in the diet, a series of steps and reactions take place in order to get it to its appropriate locations in the body.

While the digestion of triacylglycerols can occur in the mouth through the actions of a small group of lingual lipases, the digestion of lipids takes place in the first part of the small intestine. The enzyme **lipase** is secreted into the duodenum from the exocrine acinar cells of the pancreas. Pancreatic lipase attacks the triacylglycerols and digests them by removing two free fatty acids from the 1st and the 3rd positions of glycerol. The products of lipase-catalyzed triacylglycerol digestion are two free fatty acids and a 2-monoacylglycerol (▶ Fig. 39.2). Pancreatic lipases are supported by co-lipase, which enhances their functions.

39.3.1 Link to Physiology and Pathology: Cystic Fibrosis May Impair the Function of the Pancreas

The pancreas secretes many digestive enzymes into the small intestine. These enzymes include amylase for carbohydrate digestion (reviewed in Chapter 21), trypsinogen,

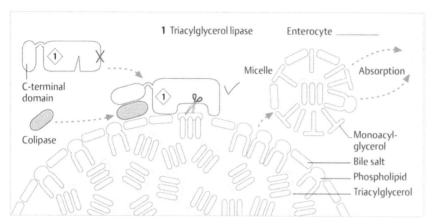

Fig. 39.2 Hydrolysis of triacylglycerol by pancreatic lipase. Pancreatic lipase with the help of colipase removes two free fatty acids from the 1st and the 3rd positions of the triacylglycerols. Bile acids and salts package fatty acids, 2-monoacylglycerols, and other lipids into a micelle. Micelles travel to the microvilli in the ileum, the lipids, but not the bile acids and salts, get absorbed into the enterocytes. (Source: Koolman J, Röhm K, ed. Color Atlas of Biochemistry. 3rd Edition. New York, NY. Thieme; 2012.)

chymotripsinogen, proelastase, procarboxypeptidase, proaminopeptidase for protein digestion (reviewed in Chapter 32), and lipase for triacylglycerol digestion. In addition to these enzymes, the pancreas also secretes bicarbonate (HCO_3^-) into the small intestine. This secreted bicarbonate, along with the digestive enzymes, helps to neutralize the acidic pH of the stomach. The neutral pH is required for the activity of the digestive enzymes, especially for lipase, which is the most pH sensitive enzyme compared to the other pancreatic enzymes. Patients with **cystic fibrosis** may experience steatorrhea (fatty stool) due to insufficient secretion of bicarbonate and digestive enzymes. An acidic pH due to insufficient bicarbonate denatures lipase and causes impaired lipid digestion and absorption, which results in fatty, foul-smelling, bulky stools. If not treated, poor digestion and absorption of lipids also impairs the absorption of lipid-soluble vitamins.

39.3.2 Link to Pathology: Chronic Pancreatitis

One of the most common impaired digestion and absorption of triacylglycerols is due to chronic pancreatitis, in which the secretion of lipase is decreased. These patients also experience steatorrhea.

39.4 Function of Bile Acids and Bile Salts in the Digestion of Lipids

Besides lipase, the digestion of triacylglycerols also requires bile acids and bile salts, which are synthesized in the liver, stored in the gallbladder, and secreted in the duodenum (the synthesis of bile acids and salts are reviewed in Chapter 42). Bile acids and salts are amphipathic molecules, which emulsifies fats like detergents. This emulsification process increases the surface areas for the pancreatic enzymes to act on the fat droplets.

Bile acids and salts package fatty acids, 2-monoacylglycerols, and other lipids into a micelle. When the micelles travel to the microvilli in the ileum, the lipids, but not the bile acids and salts, get absorbed into the enterocytes. About 95% of the bile acids and salts are returned to the liver via enterohepatic circulation.

In the endoplasmic reticulum of the enterocytes, fatty acids and 2-monoacylglycerols are re-esterified back into triacylglycerols (▶ Fig. 39.3). Fatty acyl CoA synthetase attaches a CoA molecule to the fatty acid to form fatty acyl CoA, which is called "activated" fatty acid. The "activated" fatty acids react with 2-monoacylglycerol to form diacylglycerol (DAG). DAG acyltransferase transfers another fatty acid from a fatty acyl CoA to form triacylglycerol (TG). These triacylglycerols, cholesterol esters, phospholipids, and lipoproteins are packaged into particles called **chylomicron**s. Transport of chylomicrons, carrying the dietary fats, into the blood occurs through the lymphatic system.

39.5 Synthesis and Fate of Chylomicrons

The lipid portion of chylomicron is synthesized within the endoplasmic reticulum of the intestinal epithelial cells. The protein component of the chylomicron is apolipoprotein B48 (ApoB48), which is synthesized in the rough endoplasmic reticulum (RER) of the enterocytes. First the lipid portion of chylomicron is assembled in the Golgi complex. Then the ApoB48 is attached to the surface, and finally the newly synthesized chylomicron is exocytosed from enterocytes into the lymphatic system from where they eventually enter into the blood stream. The chylomicrons are high in triacylglycerols, thus it is classified as low-density lipoproteins (other blood lipoproteins are reviewed in Chapter 43).

When chylomicrons enter into the blood stream, they are called nascent chylomicrons. In the blood stream, the chylomicrons interact with HDL particles and receive two proteins, an ApoCII and an ApoE from the HDL. At this point these complexes are called mature chylomicrons. The ApoCII activates the lipoprotein lipase (LPL) on the capillary endothelial cells of adipose and muscle cells. Activation of LPL breaks down the TG from the chylomicron molecules into FFAs and glycerol molecules that are absorbed by the adipose and muscle tissues. When chylomicrons unload most of their TG, the ApoCII is delivered back to the HDL. These chylomicrons are then called chylomicron remnants. The function of the ApoE is to guide the chylomicron remnants to the liver, where the chylomicrons are internalized and taken up by the liver through receptor-mediated endocytosis (reviewed in Chapter 43).

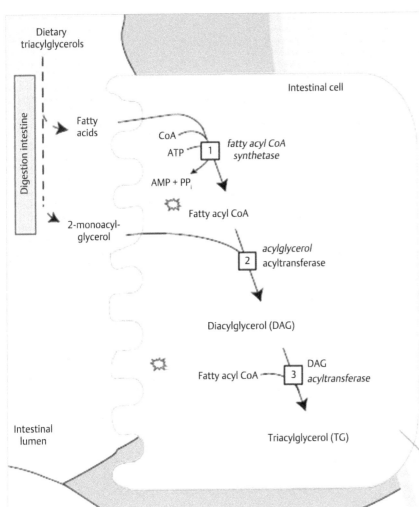

Fig. 39.3 Synthesis of triacylglycerol in the enterocytes. Fatty acids and 2-monoacylglycerols are re-esterified back into triacylglycerols In the endoplasmic reticulum of the enterocytes. (Source: Panini S, ed. Medical Biochemistry—An Illustrated Review. New York, NY. Thieme; 2013.)

VI

Review Questions

1. Which of the following is a polyunsaturated ω-6 fatty acid?
2. Which of the following fatty acids is an essential fatty acid?
   A) Alpha-linolenic acid
   B) Arachidonic acid
   C) Oleic acid
   D) Palmitic acid
3. After digestion of triacylglycerols by pancreatic lipase, which of the following products are absorbed into the enterocytes?
   A) Fatty acids and diacylglycerol
   B) Fatty acids and glycerol
   C) Fatty acids and monoacylglycerol
   D) Glycerol and diacylglycerol
   E) Monoacylglycerol and diacylglycerol
4. Which of the following tissues/cells is the site for nascent chylomicron synthesis?
   A) Blood
   B) Duodenum
   C) Enterocytes
   D) Liver
   E) Lymph nodes

Answers

1. **The correct answer is D**. Unsaturated fatty acids can be mono- or poly-unsaturated. Besides the number of double bonds, the location of the double bond is important. With the omega nomenclature system, the methyl carbon is called the omega carbon. Based on this nomenclature, if the first double bond is at the sixth position from the omega carbon, then the fatty acid is called ω-6 fatty acid.

Answer choices A, B, and C are ω-3 fatty acids. A is alpha linolenic acid (ALA), B is eicosapentaenoic acid (EPA), and C is docosahexaenoic acid (DHA).

2. **The correct answer is A**. There are two essential fatty acids that must be obtained from the diet. These essential fatty acids are ω-6 linoleic acid (18:2) and ω-3 alpha linolenic acid (18:3). Arachidonic acid (answer choice B) is not essential because linoleic and alpha linolenic fatty acids can be elongated and desaturated to synthesize arachidonic acid in the human body. Oleic and palmitic fatty acids are not essential. In fact, palmitic acid is the final product of the fatty acid synthesis pathway, reviewed in Chapter 40.

3. **The correct answer is C**. The products of lipase-catalyzed triacylglycerol digestion are two free fatty acids and a 2-monoacylglycerol. The pancreatic lipase attacks the triacylglycerols and digests them by removing two free fatty acids from the 1st and the 3rd positions of glycerol. Pancreatic lipases are supported by co-lipase, which enhances their functions.

4. **The correct answer is C**. Nascent chylomicrons are synthesized in the enterocytes. The lipid portion of chylomicron is synthesized within the endoplasmic reticulum and the protein component, ApoB48, is synthesized in the rough endoplasmic reticulum (RER) of the enterocytes. First the lipid portion of chylomicron is assembled in the Golgi complex. Then the ApoB48 is attached to the surface, and finally the newly synthesized chylomicron is exocytosed from enterocytes into the lymphatic system from which they eventually enter into the blood stream. In the blood, the nascent chylomicrons interact with HDL particles and receive two proteins, an ApoCII and an ApoE from the HDL. At this point they are called mature chylomicrons.

VI

40 Synthesis of Fatty Acids and Triacylglycerols

At the conclusion of this chapter, students should be able to:
- Identify the main substrate of fatty acid synthesis
- Outline the steps of fatty acid synthesis
- Explain the allosteric and the hormonal regulation of fatty acid synthesis
- Describe the elongation and the desaturation processes of fatty acids
- Outline the steps for the formation of triacylglycerols in the liver and the adipocytes

Fatty acid biosynthesis, an anabolic reaction, takes place in the cytosol of the hepatocytes and the adipocytes. The *de novo* synthesis of fatty acids is stimulated by insulin hormone in the liver and adipose tissue when there are excess carbons. Dietary carbohydrates, especially glucose and the carbon skeletons of proteins, are the main sources of these excess carbons.

In the pathway of fatty acid biosynthesis, acetyl CoA molecules are condensed into 16-carbon palmitic acid in the cytosol. However, since acetyl CoA is generated in the mitochondria and cannot cross the membrane, it must be shuttled into the cytosol in the form of citrate via the citrate shuttle (▶ Fig. 40.1). Citrate

only exits mitochondria during insulin signaling, when the cells have high energy levels.

The enzyme citrate lyase converts citrate back into oxaloacetate and acetyl CoA in the cytosol. Oxaloacetate (OAA) is converted to malate using cytosolic malate dehydrogenase, and this malate is then converted to pyruvate by malic enzyme. The malic enzyme reaction is an important step because the reaction produces NADPH, which provides the needed reducing power for fatty acid biosynthesis.

Acetyl-CoA, a two-carbon compound, is the main monomer for palmitate synthesis. Palmitate is a 16-carbon saturated fatty acid. Therefore, eight acetyl-CoA molecules are utilized per palmitate. The overall reaction is thus;

8 Acetyl CoA + 7ATP + 14 NADPH → Palmitic Acid

40.1 Regulation of Fatty Acid Synthesis

Fatty acid synthesis occurs during the fed state. The presence of dietary glucose is crucial and so insulin is a key regulatory hormone in this synthesis. In order for fatty acid synthesis to occur, insulin hormone activates the key regulatory enzymes, includ-

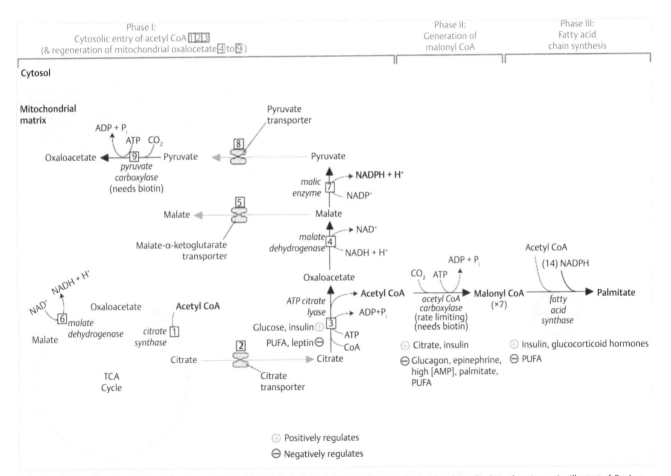

Fig. 40.1 Synthesis of fatty acids and regulatory steps in the synthesis pathway. (Source: Panini S, ed. Medical Biochemistry—An Illustrated Review. New York, NY. Thieme; 2013.)

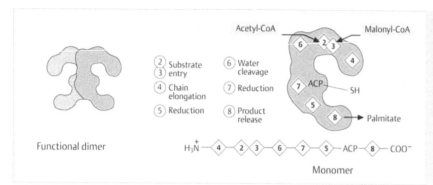

Fig. 40.2 Structure and subunits of the fatty acid synthase enzyme. (Source: Panini S, ed. Medical Biochemistry—An Illustrated Review. New York, NY. Thieme; 2013.)

ing phosphofructokinase-1 (PFK-1), pyruvate dehydrogenase (PDH), citrate lyase, and acetyl CoA carboxylase, the rate-limiting enzyme in fatty acid biosynthesis. The activity of malic enzyme and glucose 6-phosphate dehydrogenase (G6PD) are also important for the production of NADPH, the reducing power for fatty acid synthesis.

When acetyl CoA gets into the cytosol, it is converted to malonyl CoA by acetyl CoA carboxylase. The acetyl CoA carboxylase reaction requires an ATP, CO_2, and biotin as a coenzyme (▶ Fig. 40.1).

The rate limiting step of fatty acid synthesis, acetyl CoA carboxylase, is controlled by allosteric regulators and by hormones. The positive allosteric regulator is citrate and the negative allosteric regulator is palmitoyl CoA, the end product of fatty acid biosynthesis. Acetyl CoA carboxylase is activated by insulin and inhibited by glucagon. This process is somewhat intuitive as insulin is present when there is dietary glucose, and therefore the body has the ability to make and store fatty acids. During fasting, glucagon inhibits synthesis and storage of fatty acids. With glucagon signaling, the acetyl CoA carboxylase enzyme gets phosphorylated by active protein kinase A (PKA). The phosphorylated acetyl CoA carboxylase cannot polymerize and stays in an inactive dimer. The *de novo* synthesis of fatty acids is also inhibited with an increased AMP/ATP ratio in the liver. Increased AMP levels activate AMP-dependent kinase (AMPK), which phosphorylates and inhibits acetyl CoA carboxylase. Contrary to glucagon, insulin signaling stimulates protein phosphatase-1, which dephosphorylates acetyl CoA carboxylase, thereby increasing the polymerization of the dimers and activating the enzyme.

The product of acetyl CoA carboxylase is malonyl CoA, which plays an important role as a regulator of β-oxidation (reviewed in Chapter 29) by inhibiting the activity of CPT-1, the rate limiting enzyme of β-oxidation. The inhibition of fatty acid oxidation by malonyl CoA during fatty acid synthesis prevents substrate cycling and the futile use of energy.

Fatty acid synthase (FAS), a multimeric enzyme, catalyzes the assembly of palmitic acid from malonyl CoA and acetyl CoA (▶ Fig. 40.2). FAS is made up of acyl carrier protein (ACP) and three domains with a total of seven catalytic sites. The ACP requires pantothenic acid as a coenzyme, which serves as an anchor of the fatty acid chain. The four reactions within fatty acid synthesis, with the help of FAS, are condensation, reduction, dehydration and reduction.

The cycle starts with the initial charging of FAS with an acetyl moiety from acetyl CoA and a malonyl moiety from malonyl

CoA. The first step is the condensation of acetyl and malonyl moieties to form β-ketobutyryl-ACP. Reduction, dehydration and reduction reactions, in that order, produce butyryl-ACP. Both reduction reactions utilize NADPH for reducing power. In the next round, this butyryl moiety will be moved to the SH group of the condensing subunit. The freed ACP subunit is now open to accept the second malonyl group from malonyl CoA. Again, the reduction, dehydration, and reduction reactions will produce a 6-carbon fatty acid. Additional rounds of FAS complex reactions will grow the acyl chain until it reaches a 16-carbon palmitate. When the acyl chain reaches 16 carbons, the fatty acid is released from the FAS complex.

It is important to note that the condensation, reduction, dehydration, and reduction steps in fatty acid synthesis is the reverse of fatty acid oxidation. This should make sense as fatty acid oxidation is the breakdown of fatty acids while synthesis is the building of fatty acids.

40.2 Further Elongation

After palmitate has been made, it undergoes activation by addition of a CoA group. The product palmitoyl CoA can be elongated to other fatty acid molecules. The CoA group bound to fatty acid serves as an anchor during the elongation reactions. The carbon donor for the elongation is also malonyl-CoA, and the reducing power is also NADPH. The further elongation reaction takes place in the smooth endoplasmic reticulum (SER) and can be repeated multiple times for the desired fatty acid length.

40.3 Desaturation of Fatty Acids

Each fatty acid molecule can be desaturated (addition of a double bond). Desaturation requires the enzyme fatty acyl-CoA desaturase, which is a mixed function oxidase. Three substances are required for the activation of this enzyme – oxygen, NADH and cytochrome B. In humans, fatty acids cannot be desaturated past the 9th carbon, but are still required by the body. This is why we have the concepts of essential and nonessential fatty acids.

Because humans cannot synthesize double bonds past the 9th carbon, ω6 linoleic and ω3 linolenic fatty acids are required in the diet. Arachidonic acid is a ω6 non-essential fatty acid because humans can synthesize arachidonic acid from linoleic

VI

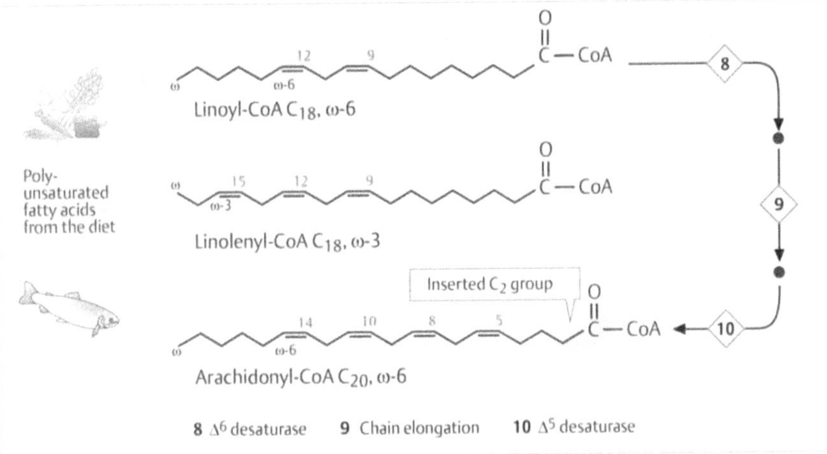

Fig. 40.3 Human can synthesize arachidonic acid from linoleic acid by desaturation (8 and 10), by elongation (9). (Source: Panini S, ed. Medical Biochemistry—An Illustrated Review. New York, NY. Thieme; 2013.)

8 Δ^6 desaturase   **9** Chain elongation   **10** Δ^5 desaturase

acid by adding a 2-carbon unit and introducing two double bonds (▶ Fig. 40.3). Hence, it is not needed in the diet.

40.4 Synthesis of Triacylglycerols

A triacylglycerol (TG) contains three fatty acids, which are esterified to a glycerol backbone. TG synthesis is active during the fed-state. The liver and adipose tissue converts excess dietary carbohydrates and the carbon skeletons of the amino acids to TG in order to store energy. The synthesis of triglycerides (TG) begins with the formation of glycerol 3-phosphate. Glycerol 3-phosphate can be synthesized in the liver and the adipose tissue, although through different pathways.

In the liver, glycerol is converted to glycerol 3-phosphate via the enzyme glycerol kinase. It is important to note that this enzyme is only present in the liver and, as such, adipose cells cannot use this metabolic pathway in the synthesis of glycerol 3-phosphate (▶ Fig. 40.4**a**).

Because of the unavailability of glycerol kinase in the adipocytes, glycerol 3-phosphate is obtained from the glycolytic intermediate dihydroxyacetone phosphate (DHAP) (▶ Fig. 40.4**b**). This further emphasizes that TG synthesis occurs in the presence of glucose in the fed state.

The next step in TG synthesis involves a reaction between two fatty acyl-CoA molecules and a glycerol 3-phosphate to form phosphatidic acid. The activation of fatty acids into fatty acyl-CoA molecules involves the enzyme fatty acyl-CoA synthetase and ATP. Phosphatidic acid is then dephosphorylated into diacylglycerol (DAG). A third and final fatty acyl-CoA is added to DAG to form TG. After synthesis, TG is then transferred from the liver to the blood in the form of VLDL (the VLDL pathway is reviewed in Chapter 43).

40.4.1 Link to Pathology

Alcoholic fatty liver disease is a common problem in chronic alcoholics. Ethanol metabolism produces NADH and acetate. While fatty acid synthesis is stimulated by a high NADH/NAD ratio, β-oxidation and gluconeogenesis are inhibited. This causes an elevation of fatty acids in the liver and favors the formation of TG. This extra fat accumulates in the liver because not only is the synthesis of fat increased, but also the transport of VLDL is inhibited. Acetaldehyde, the intermediate of ethanol metabolism, forms adducts with proteins, causing protein denaturation and preventing VLDL transport from the liver into the blood.

VI

VI

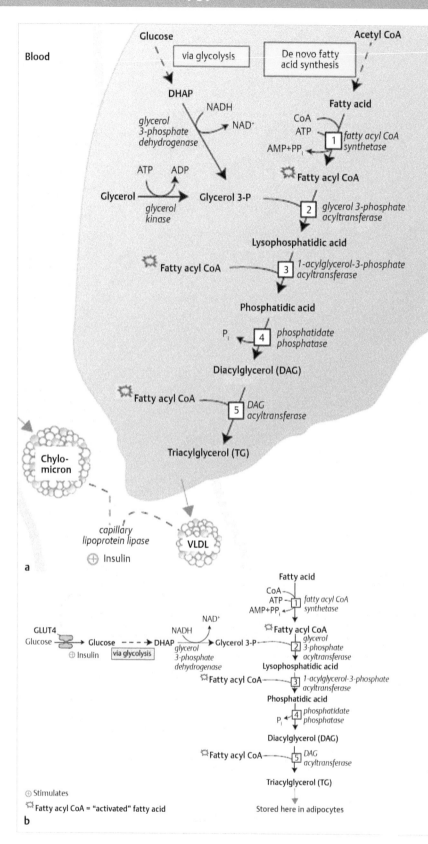

Fig. 40.4 (a) TG synthesis in hepatocytes. **(b)** TG synthesis in adipocytes. (Source: Panini S, ed. Medical Biochemistry—An Illustrated Review. New York, NY. Thieme; 2013.)

Review Questions

1. Which of the following modulators/hormones activates acetyl CoA carboxylase in the liver?
   A) AMPK
   B) Glucagon
   C) Insulin
   D) Palmitoyl CoA
2. Which of the following enzymes produce the reducing power for fatty acid biosynthesis?
   A) Glucose 6-phosphate dehydrogenase and citrate lyase
   B) Glucose 6-phosphate dehydrogenase and phosphofructokinase-1
   C) Malic enzyme and citrate lyase
   D) Malic enzyme and glucose 6-phosphate dehydrogenase
   E) Malic enzyme and phosphofructokinase-1
3. Which of the following statements best describes role of malonyl CoA in fatty acid metabolism?
   A) Malonyl CoA is an inhibitor of the rate limiting step of β-oxidation
   B) Malonyl CoA is used for the initial charging of fatty acid synthase
   C) Malonyl CoA is used for elongation of fatty acids
   D) Malonyl CoA is produced by the rate limiting step of fatty acid synthesis

Answers

1. **The correct answer is C.** Acetyl CoA carboxylase is the rate limiting step of fatty acid synthesis. The enzyme is active during the fed state. After meals, available glucose and energy induces insulin secretion from the pancreas. Insulin activates protein phosphatase-1, which dephosphorylates acetyl CoA carboxylase and activates the enzyme. Contrary to insulin signaling, glucagon (answer choice B) and AMPK (answer choice A) signaling pathways phosphorylate and inactivate acetyl CoA carboxylase and, consequently, fatty acid synthesis. The acetyl CoA carboxylase enzyme is also regulated by the allosteric modulators citrate and palmitoyl CoA. Citrate, as a positive allosteric modulator, binds to dimers of acetyl CoA carboxylase and induces the polymerization and activation of the enzyme. Palmitoyl CoA (answer choice D), as a negative allosteric modulator, causes depolymerization and, therefore, inactivates the enzyme.

2. **The correct answer is D.** The malic enzyme and glucose 6-phosphate dehydrogenase (G6PD) produce NADPH, the reducing power for fatty acid synthesis. Malic enzyme in the cytoplasm converts malate to pyruvate. G6PD is the rate limiting step of the pentose phosphate pathway and is important for the production of NADPH for biosynthetic reactions, such as fatty acid and cholesterol biosynthesis. Availability of glucose 6-phosphate induces G6PD.

Citrate lyase converts citrate back into oxaloacetate and acetyl CoA in the cytosol. Oxaloacetate (OAA) is converted to malate using cytosolic malate dehydrogenase. These reactions do not produce NADPH. Phosphofructokinase-1 (PFK-1) is the rate limiting step of glycolysis and this step does not produce NADPH either.

3. **The correct answer is A.** Although all of the statements are correct, the importance of malonyl CoA in fatty acid metabolism lies in the fact that it inactivates CPT-1, the rate limiting enzyme of β-oxidation. The inhibition of fatty acid oxidation by malonyl CoA during fatty acid synthesis prevents substrate cycling and the futile use of energy.

Malonyl CoA is produced by acetyl CoA carboxylase by carboxylation of acetyl CoA. This enzyme is the rate limiting step of fatty acid synthesis and uses ATP and requires biotin as a coenzyme. The carbon source for the elongation of fatty acids is also malonyl CoA.

VI

41 Metabolism of Membrane Lipids and Lipid Derivatives

At the conclusion of this chapter, students should be able to:
- Classify the major membrane lipids and their functions in the cell
- Outline the synthesis of glycerophospholipids
- Explain the clinical importance of surfactant in newborn babies
- Describe glycerophospholipid degradation and the important enzymes in this process
- Outline the synthesis of eicosanoids by cyclooxygenase and lipoxygenase pathways
- Describe the inhibition mechanisms of cyclooxygenases
- Describe the various functions of eicosanoids and their therapeutic uses
- Outline the synthesis of ceramide and sphingolipids
- Distinguish between examples of disease processes caused by sphingolipid synthesis or breakdown (i.e. lysosomal storage diseases)

The major lipids found in biological membranes are glycerophospholipids and sphingolipids. Glycerophospholipids are not only the major constituents of these membranes, but are also vital for blood lipoproteins and lung surfactant. Sphingolipids are important for the creation of the myelin sheath that surrounds nerve fibers, insulating them and enhancing the speed of conduction. Ether glycolipids are present in plasminogen as well as plasmin activating factor, making it crucial for the regulation of the blood clotting cascade. In this chapter, we will review glycerophospholipids first and then the eicosanoid lipid derivatives. Finally, we will review the sphingolipids.

41.1 Glycerophospholipids

Glycerophospholipids contain a polar head with a phosphate moiety and two nonpolar tails of fatty acids that are esterified to a glycerol backbone (▶ Fig. 41.1). The polar head is hydrophilic and the non-polar fatty acid tails are hydrophobic. Their amphipathic structure is crucial for their function as a regulator of water and solutes inside and outside of the cellular membrane. The polar head group can be serine, choline, ethanolamine, glycerol or inositol.

41.2 Synthesis of Glycerophospholipids

The synthesis of glycerophospholipids occurs in all cells except erythrocytes. The initial steps to phospholipid synthesis are like that of triacylglycerol. Glycerol 3-phosphate is formed from either glycerol or from the glycolytic intermediate dihydroxyacetone phosphate (DHAP) (▶ Fig. 41.2). In the liver, glycerol or DHAP is used as the starting substrate, but in adipose tissue only DHAP is utilized. Glycerol is phosphorylated by glycerol kinase, which utilizes a molecule of ATP. In the liver and adipose, glucose follows the initial steps of glycolysis to make DHAP, which is converted to glycerol 3-phosphate with the use of NADH. Glycerol 3-phosphate then reacts with two activated fatty acids to form phosphatidic acid.

The addition of the polar head group can occur in one of two distinct ways: In the first pathway, the phosphate group is cleaved off of phosphatidic acid to form diacylglycerol. A head group is then added with cytosine triphosphate (CTP) on the 3rd carbon of diacylglycerol to form glycerophospholipid. In the second pathway, a CTP is combined with diacylglycerol and then a head group is added through a condensation reaction.

41.3 Functions of Glycerophospholipids

The liver, being the major center of glycerophospholipid synthesis, utilizes phospholipids to coat the plasma lipoprotein VLDL. The phospholipids are also secreted into the lumen of the intestine from the liver through bile ducts. These phospholipids function as emulsifiers in the intestinal lumen during the digestion of lipids. Enterocytes in the intestine also produce phospholipids that coat the chylomicrons.

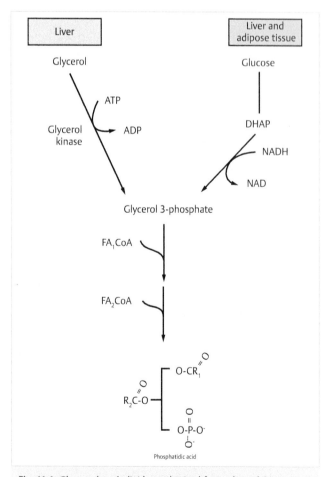

Fig. 41.1 Glycerophospholipids synthesized from glycerol 3-phosphate, serves as a backbone, two fatty acids, and a polar head with a phosphate moiety. (Source: Panini S, ed. Medical Biochemistry—An Illustrated Review. New York, NY. Thieme; 2013.)

VI

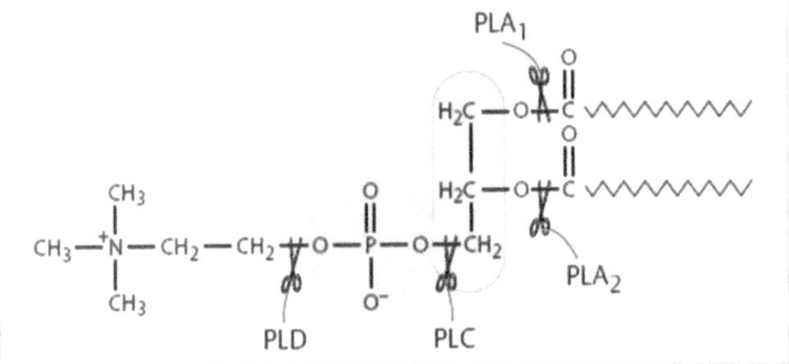

Fig. 41.2 Glycerophospholipids are degraded by phospholipases (PLA1, PLA2, PLC, and PLD). (Source: Panini S, ed. Medical Biochemistry—An Illustrated Review. New York, NY. Thieme; 2013.)

Phosphatidylinositol is used as an anchor for certain proteins in the plasma membrane. Diacylglycerol, ceramide, and lysophosphatidic acid are used as signaling molecules that activate various protein kinases.

41.3.1 Link to Physiology

Lung surfactant is composed mostly of phospholipids. The major phospholipid of surfactant is dipalmitoylphosphatidylcholine, which is complexed with surfactant proteins. The surfactant decreases surface tension within the alveoli and prevents the alveoli from collapsing during expiration. If the alveoli collapse, it takes increased inspiratory pressure to re-inflate the alveoli. Collapse of the alveoli can also cause permanent damage. Surfactant is produced late in pregnancy near the 35th week of gestation by the type II cells of the lung. Therefore, infants born prior to this time are at risk of not having formed enough surfactant in their lungs. Insufficient production of surfactant can cause neonatal respiratory distress syndrome in newborns. In this syndrome, the alveoli's surface tension is increased and, therefore, collapse more easily.

41.4 Degradation of Glycerophospholipids

Glycerophospholipids are broken down by phospholipases (▶ Fig. 41.2). Phospholipases are enzymes that can be found either in the plasma membrane or within lysosomes. Their presence in the plasma membrane is vital, as this is the site of a majority of glycerophospholipids. There are four enzymes that are involved in the degradation of glycerophospholipids. Phospholipase A1 removes a fatty acid group from the C1 carbon of the glycerol moiety, whereas the fatty acid at the C2 carbon is cleaved by phospholipase A2 (PLA2). The fatty acid at the C2 location is often arachidonic acid and its cleavage generates an eicosanoid. Phospholipase C cleaves phosphatidylinositol 4,5-bisphosphate (PIP2) into diacylglycerol (DAG) and inositol triphosphate (IP3). This step is part of a secondary messenger system of a G protein receptor. Phospholipase D cleaves the head group from phosphatidic acid.

41.5 Eicosanoid Synthesis

Eicosanoids are potent autocrine and paracrine hormones that act on adjacent cells and provide a variety of cellular responses.

These signaling molecules, which include prostaglandins, thromboxanes and leukotrienes, have a very short half-life of only a few minutes and are produced from the cleavage of arachidonic acid (▶ Fig. 41.3). Arachidonic acid can be obtained from the diet or from cellular damage that causes an increase in PLA2 activity, which results in an increase in arachidonic acid release from glycerophospholipids within the plasma membrane. Arachidonic acid is then utilized to form different eicosanoid compounds through multiple enzymes. Different cells in different regions of the body contain various enzymes that determine the type of eicosanoid produced.

41.6 Cyclooxygenase Pathway

The cyclooxygenase (COX) pathway is responsible for the production of prostaglandins and thromboxanes (▶ Fig. 41.3). The first compound formed in this process is prostaglandin H2, which serves as a precursor for other prostaglandins, prostacyclin and thromboxanes. There are two forms of the COX enzyme, COX-1 and COX-2.

COX-1 is a constitutively active enzyme that is present in cells and is responsible for the production of prostaglandins. These prostaglandins are important for many normal cellular activities, which are referred to as "housekeeping functions". These activities include the production of gastric mucous in the stomach, regulation of gastric acid, endothelial integrity, maintaining normal blood pressure, uterine contraction, platelet aggregation and bronchodilation.

COX-2 is a stimulated enzyme that is produced in higher quantities in response to injury. This enzyme plays a more active role in pain and inflammation and is stimulated by inflammatory mediators, such as growth factors and cytokines.

41.7 Functions of Prostaglandins

Prostaglandin H2, produced by COX from arachidonic acid, can be converted by the enzyme prostaglandin E synthase to produce prostaglandin E2 (PGE2). PGE2 serves many functions in the body, depending on the region in which it is acting. Systemically, it can act on the hypothalamus to increase the body's thermostatic temperature and produce fever. PGE2 is also involved in causing pain through its actions on peripheral nerves. In addition, PGE2 is a vasodilator, and acts on the arteries of the kidneys, particularly the afferent vessels, increasing blood flow to the kidneys. This action is important in the autoregulation of both blood pressure and maintaining adequate blood flow to the kidney. PGE2 also plays a role in cervical thinning and induction of labor by causing uterine contractions. PGE2 keeps the ductus arteriosus patent in utero, providing an important shunt from the pulmonary arteries to the aorta, bypassing the lungs. PGE2 also increases the production of protective gastric mucous in the stomach that helps prevent and heal ulcers.

Prostaglandin H2 can also be used to produce prostacyclin (PGI2) through the enzyme prostacyclin synthase. PGI2 is a potent vasoconstrictor that is an important piece of the clotting cascade and healing of endovascular injury.

Prostaglandin H2 is used by the enzyme thromboxane synthase to produce thromboxane (TXA2). TXA2 causes

VI

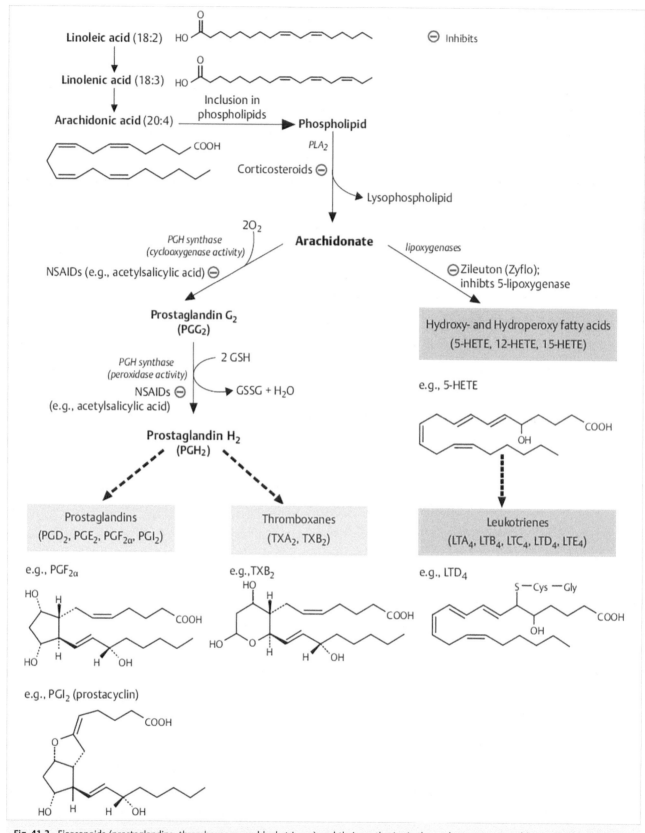

Fig. 41.3 Eicosanoids (prostaglandins, thromboxanes, and leukotrienes) and their synthesis via the cyclooxygenases and lipoxygenase pathways. (Source: Panini S, ed. Medical Biochemistry—An Illustrated Review. New York, NY. Thieme; 2013.)

vasoconstriction and platelet aggregation. TXA2 is also involved in healing of endovascular injury.

41.7.1 Link to Pharmacology

Aspirin reduces the production of prostaglandins by irreversibly inhibiting the COX enzymes. Decreasing the inflammatory mediators like prostaglandins relieves pain and reduces fever. Other non-steroidal anti-inflammatory drugs (NSAIDs) also act by decreasing the production of prostaglandins by competitively inhibiting the COX enzymes.

Low dose aspirin specifically inhibits the synthesis of thromboxanes in the platelets. This COX inhibition by low dose aspirin has been found to decrease platelet aggregation and has become a cornerstone of therapy and prevention of coronary artery disease, decreasing future cardiovascular events such as thrombus formation and undoubtedly saving lives.

NSAIDs non-selectively inhibit both COX1 and COX2 enzymes. However, inhibition of prostaglandins' housekeeping functions brings negative consequences. For example, NSAIDs have been implicated in gastric ulcer disease due to the decreased production of protective mucous in the stomach. They also lead to increased bleeding events. If taken during pregnancy, they can lead to premature closure of the ductus arteriosus in the fetus, as PGE2 is required for its patency. Also, prostaglandins are required for afferent renal artery vasodilation to increase renal blood flow. If prostaglandin synthesis is inhibited by NSAIDs, it can lead to vasoconstriction of afferent renal arteries, decreasing renal blood flow and causing renal ischemia.

The discovery of COX2 enzymes led scientists to develop selective COX-2 inhibitors that more selectively block the prostaglandins involved in the inflammatory processes. Although these selective COX-2 inhibitors do not cause stomach irritation like COX-1 inhibitors, some COX-2 inhibitors have been withdrawn from the market due to cardiovascular side effects.

41.7.2 Link to Pharmacology

Steroids are anti-inflammatory drugs. Glucocorticoids inhibit PLA2 and, therefore, decrease the release of arachidonic acid and inflammatory reactions. Anti-inflammatory corticosteroids also inhibit the transcription of COX2 enzymes.

41.8 Lipoxygenase Pathway

Arachidonic acid is utilized by the enzyme lipoxygenase to produce leukotrienes (▶ Fig. 41.3). In the first step, arachidonic acid is converted to 5-hydroxyeicosatetraenoic acid (5-HETE) by the enzyme lipoxygenase. The 5-HETE is then converted to leukotriene A4 (LTA4). Leukotriene A4 can then either be converted to leukotriene B4 or leukotriene C4. Leukotriene C4 can be converted to leukotriene D4 and further leukotriene E4. Leukotrienes are responsible for the inflammatory processes within the lungs and are strong bronchoconstrictors. They are active within leukocytes and mediate chemotaxis.

41.8.1 Link to Pathology and Pharmacology

Leukotrienes from the lipoxygenase pathway have been implicated as major players in the inflammatory process that causes asthma. Therefore, they have been targeted by drug therapy to treat this condition. The drug Zileuton is an inhibitor of lipoxygenase, whereas Montelukast is a leukotriene receptor antagonist.

41.8.2 Link to Pathology

Aspirin (or any NSAID) induced asthma is a condition which can develop in certain individuals who take aspirin. Since NSAIDs block the cyclooxygenase pathway, more arachidonic acid is produce, which drives leukotriene synthesis, causing bronchoconstriction and asthma symptoms.

41.9 Metabolism of Sphingolipids

Sphingolipids are sphingomyelin and glycolipids that are found on the cellular membrane. Sphingomyelin serve structural functions, especially in the myelin sheet. Glycolipids are utilized for cell-cell interactions and for the antigenic ABO blood groups. Cholera toxin binds to GM1 gangliosides.

The backbone of the sphingolipids is a ceramide group (▶ Fig. 41.4). Ceramide synthesis starts with the condensation of serine and palmitoyl CoA, which is called sphingosine. Addition of a fatty acid to sphingosine creates the ceramide.

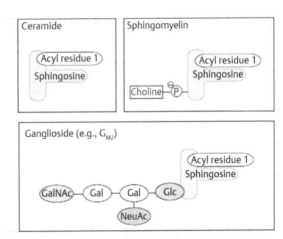

Fig. 41.4 Synthesis/degration of sphingomyelin and ganglioside. (Source: Panini S, ed. Medical Biochemistry—An Illustrated Review. New York, NY. Thieme; 2013.)

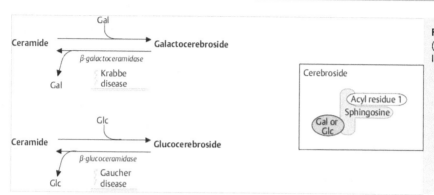

Fig. 41.5 Synthesis/degradation of cerebrosides. (Source: Panini S, ed. Medical Biochemistry—An Illustrated Review. New York, NY. Thieme; 2013.)

Table 41.1 Sphingolipidoses and their associated defective enzyme

Sphingolipidose	Defective Enzyme	Effect
Niemann-Pick disease	Sphingomyelinase	Accumulation of sphingomyelin in brain and red blood cells; leads to seizures, ataxia; fatal before 3 years
Tay-Sachs disease	Hexosaminidase A	Accumulation of gangliosides (GM$_2$) in neurons; leads to muscle atrophy, blindness (Sandhoff disease is more rapid)
Sandhoff disease	Hexosaminidase A and B	Similar to Tay-Sachs, but Sandhoff disease progresses more rapidly than the former Tay-Sachs
Krabbe disease	β-Galactosidase	Accumulation of glycolipids in oligodendrocytes; leads to complete demyelination
Gaucher disease	β-Glucosidase	Accumulation of glucocerebrosides in liver, spleen, and red blood cells; leads to skeletal erosion
Fabry disease	α-Galactosidase	Accumulation of glycolipids in brain, heart, kidney; leads to rashes, kidney failure, peripheral neuropathy
Metachromatic leukodystrophy	Arylsulfatase A	Accumulation of sulfatides in neural tissue; leads to demyelination; nervous tissue stains brownish yellow with cresyl violet dye

Source: Panini S, ed. Medical Biochemistry—An Illustrated Review. New York, NY. Thieme; 2013.

When the head group choline is transferred from phosphatidylcholine to ceramide, the lipid molecule is called sphingomyelin. Sugars can also be added to ceremide to produce cerebrosides, globosides, and gangliosides. These sphingolipids function as cellular receptors and for cellular adhesion molecules.

Degradation of these sphingolipids takes place in the lysosomes. A group of hereditary lysosomal storage diseases, also known as sphingolipidoses, are caused by a genetic deficiency of lysosomal enzymes that degrade sphingolipids (▶ Fig. 41.4, ▶ Fig. 41.5). This deficiency causes accumulation of lipid molecules in the lysosomes, leading to cell damage and subsequent failure of one or more organs. Sphingolipidoses and their associated defective enzymes are listed in the ▶ Table 41.1.

41.9.1 Link to Microbiology

GM1 gangliosides are the binding site for cholera toxin (B subunit) and the heat-labile enterotoxin of *E. coli*. The A subunit of the cholera toxin is endocytosed into the cytoplasm, where it catalyzes the ADP-ribosylation of the Gα subunit, making it constrictively active. The active Gα activates adenylate cyclase,

which increases intracellular cAMP concentrations. High intracellular cAMP activates the CFTR, causing sodium and water retention in the lumen of the intestine and leading to watery diarrhea.

41.9.2 Link to Pathology

Multiple sclerosis is a demyelinating disease of the central nervous system. Destruction of neurons due to demyelination of axons interferes with nerve conduction. Patients present with sensory loss, motor dysfunctions such as weakness of lower extremities and muscle cramping, and autonomic dysfunctions such as bladder, bowel or sexual dysfunctions.

41.9.3 Link to Pathology

Lysosomal Storage Diseases: Genetic deficiency of any lysosomal enzyme results in accumulation of the substrates in the lysosomes, thus these diseases are called lysosomal storage diseases. Although, most of the lysosomal storage diseases are lipid storage diseases (also known as sphingolipidosis), there are also few diseases that are caused by accumulation of glycogen molecule (Pompe disease reviewed in Chapter 24), or glycosaminoglycans (mucopolysaccharidosis), or I-cell disease, reviewed in Chapter 6.

The most common sphingolipidosis disease is **Gaucher disease**. Genetic deficiency of lysosomal β-glucosidase (also known as glucocerebrosidase or glucosylceramidase) causes accumulation of glucocerebrocytes within the lysosomes. Toxic levels of glucocerebrocytes kill the cells. These dead cells with accumulated lipid molecules are engulfed by macrophages. Therefore monocytes and macrophages form characteristic "wrinkled paper" appearance. The disease also affects the liver, spleen, and the bone marrow. Patients suffer from hepatosplenomegaly, which is responsible for the anemia and pancytopenia. Because the bone marrow is also affected, patients also complain of bone and joint pain. Bone deformities and fractures are commonly seen in these patients. Another common symptom is a cherry-red spot on the macula of the eyes. Gaucher disease is the most common lysosomal storage disease and it is especially prevalent in Askenazi Jews. The onset of the disease varies from early childhood to adulthood.

Tay-Sachs and Sandhoff diseases are similar. While Tay Sachs disease is due to deficiency of hexosaminidase A, Sandhoff disease is due to deficiency of both hexosaminidase A and B isoforms. In both diseases GM2 gangliosides accumulate within the lysosomes of neurons and the retina. Patients are normal at birth but present with motor developmental delays by six months of age. Besides motor weakness, patients present with an unusual startle reflex. Hepatomegaly is not common in the early stages of the disease. Both diseases are autosomal recessive and are more common in Ashkenazi Jews. Another common finding is cherry-red spots on the macula of the eyes.

Genetic deficiency of the lysosomal sphingomyelinase enzyme causes **Niemann-Pick disease**. Sphingomyelin accumulates in the lysosomes of brain and reticuloendothelial system cells. The most common symptoms are hepatosplenomegaly and cherry red spots on the macula. Differentiation between Niemann-Pick and Tay-Sachs diseases are hepatosplenomegaly in Niemann-Pick and motor developmental delays in Tay-Sachs disease.

Fabry disease affects multiple organs including the skin, gastrointestinal tract, kidneys, heart and brain. Common symptoms are skin problems such as angiokeratomas, corneal and lenticular changes. Fabry disease develops in childhood due to deficiency of the lysosomal α-galactosidase enzyme, which is an X-linked disease. Therefore, males are affected more commonly than females.

Krabbe disease develops due to deficiency of the β-galactosidase enzyme. Deficiency of this enzyme causes accumulation of glycolipids in oligodendrocytes, resulting in demyelination. Krabbe disease presents with irritability and psychomotor arrest.

Metachromatic leukodystrophy is an inherited neurodegenerative disorder. Deficiency of arylsulfatase A causes accumulation of sulfatides in neuronal tissues and consequently demyelination occurs. There are four types of this disease: late infantile, early juvenile, late juvenile, and adult. The most common symptoms are memory loss, gait disturbances, and loss of motor functions. Patients do not develop cherry-red spots on the macula.

VI

Review Questions

1. Choline is required for synthesis of which of the following lipid molecules?
   A) Glycerophospholipid
   B) Glycolipid
   C) Sphingomyelin
   D) Stearic acid
   E) Triacylglycerol

2. Thromboxane A2 is produced by which of the following enzymes?
   A) Cyclooxygenase
   B) Hexosaminidase
   C) Lipoxygenase
   D) Sphingomyelinase

3. Which of the following fatty acids is a precursor for eicosanoid biosynthesis?
   A) Arachidonic acid
   B) Oleic acid
   C) Palmitic acid
   D) Stearic acid

4. Which of the following enzymes is inhibited by cortisol?
   A) COX1
   B) COX2
   C) Lipoxygenase
   D) PLA2

5. Which of the following is the receptor for the B subunit of the cholera toxin?
   A) Cerebrosides
   B) Ceramide
   C) Ganglioside
   D) Globosides

Answers

1. **The correct answer is C.** Sphingomyelin is synthesized by transferring the head group choline from phosphatidylcholine to ceramide. Phosphatidylcholine is synthesized by transferring the head group choline from CDP-choline to a diacylglycerol (DAG). Since sphingomyelin is an important component of myelin sheets and neuronal membrane lipids, infant formulas and baby foods are supplemented with not only ω3 fatty acids, but also with choline for healthy brain development.
   Glycerophospholipids (answer choice A) are composed of a glycerol backbone with two fatty acids, a phosphate and a head group. The nomenclature of these molecules is based on the type of head group present. For example, glycerophospholipids with a serine head group are called

phosphatidylserine. Glycolipids (answer choice B) generally have ceramide backbone and sugar molecule attached to it. Stearic acid (answer choice D) is a saturated 18-carbon fatty acid. Triacylglycerol (answer choice E) is also called fat, which has a glycerol backbone esterified to three fatty acids.

2. **The correct answer is A.** Cyclooxygenase (COX) enzymes convert arachidonic acid to PGG2 and PGH2, which are the parent compounds for prostaglandins and thromboxane A2. Hexosaminidase (answer choice B) is a lysosomal enzyme that degrades ganglioside (GM2) by removing the sugar molecules. Genetic mutations of the *HEXA* gene result in the production of an ineffective enzyme, leading to an accumulation of GM2 gangliosides in the lysosomes. The disease that results from this defect is called Tay-Sachs disease and is especially destructive to motor neurons and leads to muscle atrophy. The lipoxygenase (answer choice C) pathway produces leukotrienes through an HETE intermediate. Sphingomyelinase (answer choice D) is also a lysosomal enzyme that degrades sphingomyelin in the lysosomes. Deficiency of this enzyme causes Niemann-Pick disease, in which patients present with hepatosplenomegaly.

3. **The correct answer is A.** Eicosanoids are synthesized from arachidonic acid through the cyclooxygenase and lipoxygenase pathways. The cyclooxygenase pathway synthesizes prostaglandins and thromboxanes, while the lipoxygenase pathway synthesizes leukotrienes. Monounsaturated fatty acids such as oleic acid (answer choice B), and the saturated fatty acids palmitate (answer choice C) and stearate (answer choice D) are not used for eicosanoid synthesis.

4. **The correct answer is D.** Cortisol, a glucocorticoid, is a steroid hormone which induces the synthesis of phospholipase inhibitor. Therefore, PLA2 activity decreases with increased cortisol hormone activity. Due to decreased PLA2, less arachidonic acid becomes available, which causes a loss of prostaglandin synthesis, as well as a loss of synthesis of other inflammatory eicosanoids. Because of this, cortisol is an effective anti-inflammatory drug. Answer choices A and B are incorrect because COX1 and COX2 enzymes are inhibited by NSAIDs. Answer choice C is incorrect because the drug Zileuton is an inhibitor of lipoxygenase.

5. **The correct answer is C.** The B subunit of the cholera toxin binds to GM1 ganglioside of the enterocytes and the A subunit enters the cytoplasm. Cerebrosides (answer choice A) and globosides (answer choice D) are also glycolipids, however they are not receptors for the toxin. Ceramide (answer choice B) is the backbone of all sphingolipids and it does not serve as specific receptor.

VI

42 Cholesterol Metabolism

At the conclusion of this chapter, students should be able to:
- Explain the synthesis of cholesterol and its functions
- Explain the regulatory mechanisms of cholesterol biosynthesis
- Describe the fates of cholesterol
- Describe the synthesis and enterohepatic circulation of bile acids
- Describe the role of bile acids in lipid digestion
- Apply your knowledge of cholesterol metabolism on treatment strategies for hyperlipidemias

Hypercholesterolemia is a common initiator of arterial plaque formation and is responsible for atherosclerosis and cardiovascular diseases. Due to this increased risk of cardiovascular disease, cholesterol synthesis has been an important target for medical therapy. In addition, medications that lower blood cholesterol also decrease the risk of other cardiovascular problems. Although cholesterol is often portrayed as a contributor to cardiovascular disease, it must be emphasized that it also plays very important roles in normal biological processes. For example, cholesterol is a component of the plasma membrane, and it plays a central role in the process of myelination, providing insulation for neurons and allowing for the efficient communication between distant neurons. It is also a precursor for the synthesis of bile acids, vitamin D, and steroid hormones.

Cholesterol is a 27-carbon compound (▶ Fig. 42.1) that is obtained in the diet but as can also be made from acetyl-CoA molecules. De novo cholesterol synthesis can occur in many different organs of the body, but the main site of synthesis is the liver. In addition, endocrine glands such as adrenal medulla, the testes, and the ovaries also synthesize cholesterol for steroid synthesis. In the liver, approximately one gram of cholesterol is synthesized per day. This cholesterol can be used to form hepatocyte membranes, synthesized into bile acids, or stored in the hepatocytes as cholesterol esters. Cholesterol can also be transported to extrahepatic tissues through the blood in the form of VLDL. The details of blood lipoproteins and lipid transport are discussed in Chapter 43.

42.1 Biosynthesis of Cholesterol

Cholesterol synthesis starts with the condensation of three acetyl-CoA molecules in the cytoplasm. Acetyl- CoA, formed in the mitochondria from many metabolic processes such as pyruvate dehydrogenase, oxidation of fatty acids, amino acids, and ethanol, must be transported out into the cytosol for cholesterol synthesis. Since acetyl-CoA is unable to permeate the membranes of the mitochondria, it is transported into the cytosol in the form of citrate via the citrate shuttle. Citrate is formed in the mitochondria by combining oxaloacetate with acetyl-CoA. Both of which are abundant in an energy rich state, therefore cholesterol synthesis is stimulated by insulin and occurs when the cells have high energy levels.

42.1.1 Link to Pathology

Increased cholesterol levels are associated with obesity and a high carbohydrate diet. This association can be attributed to an excess of acetyl-CoA, which serves as the substrate for cholesterol synthesis. Therefore, the carbohydrate and fat content of the diet is just as important as the cholesterol content. This condition is known as polygenic hypercholesterolemia and it is the most common cause of high blood cholesterol. The patients present with moderate elevations in the LDL, however, triglyceride levels may be within the reference range. This condition is generally caused by atherogenic diet, obesity, and sedentary life style. The lifestyle modifications, diet and exercise can be the medical therapy.

In the first step of cholesterol metabolism, one acetyl-CoA molecule is combined with another acetyl-CoA to form acetoacetyl-CoA in a reaction that is catalyzed by the enzyme thiolase (▶ Fig. 42.2). Then HMG-CoA synthase catalyzes condensation of a third acetyl-CoA with acetoacetyl-CoA to form β-hydroxyl-β-methylglutaryl-CoA (HMG-CoA). The HMG-CoA synthase of ketogenesis is made in a different way from the HMG-CoA synthase of cholesterol synthesis through the use of subcellular compartmentalization. While the HMG-CoA synthase of cholesterol synthesis is found in the smooth endoplasmic reticulum, the HMG-CoA synthase of ketogenesis is localized in the mitochondria, reviewed in Chapter 29.

The next step of the cholesterol biosynthesis is the reduction of HMG-CoA to mevalonate. This step is catalyzed by the enzyme HMG-CoA reductase and is the rate limiting step of cholesterol synthesis. This step is completed in the smooth endoplasmic reticulum.

42.1.2 Link to Pharmacology

Statin medications are competitive inhibitors of HMG-CoA reductase. HMG-CoA reductase is the target enzyme for the treatment of hypercholesterolemia and, therefore, is also cornerstone in the treatment of cardiovascular diseases. Statin

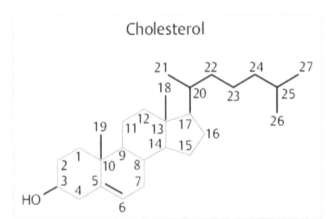

Fig. 42.1 Structural features of cholesterol. (Source: Panini S, ed. Medical Biochemistry—An Illustrated Review. New York, NY. Thieme; 2013.)

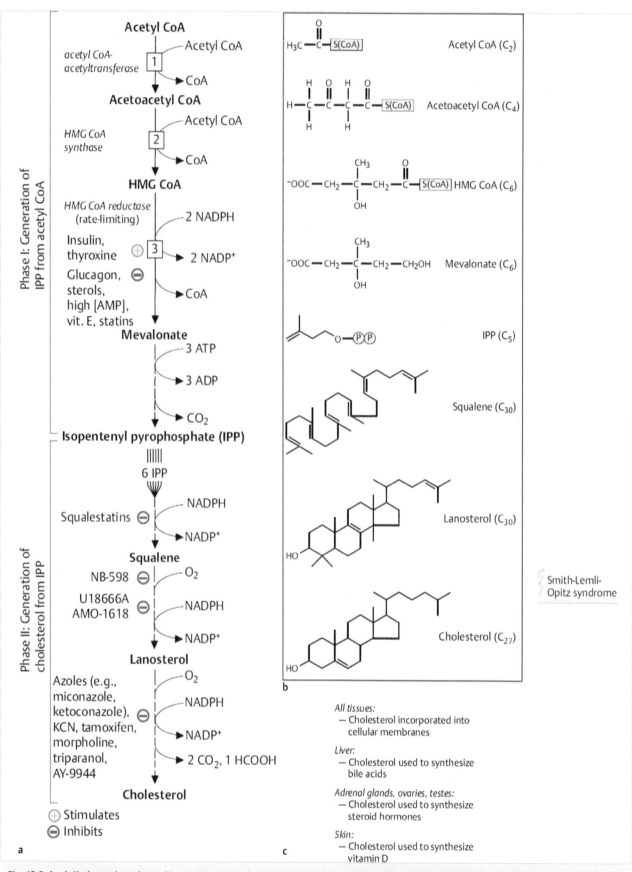

Fig. 42.2 **(a-c)** Cholesterol synthesis. (Source: Panini S, ed. Medical Biochemistry—An Illustrated Review. New York, NY. Thieme; 2013.)

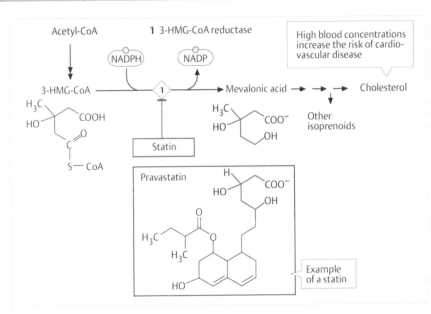

Fig. 42.3 Statins (structural analogs of HMG-CoA) are used for treatments of high blood cholesterol levels. (Source: Koolman J, Röhm K, ed. Color Atlas of Biochemistry. 3rd Edition. New York, NY. Thieme; 2012.)

VI

drugs, being structural analogs of HMG-CoA, are extensively used for the treatments of high blood cholesterol levels (▶ Fig. 42.3). The proposed mechanism for lowering blood cholesterol is that statins decrease the pool of intracellular cholesterol in the liver. In response, the hepatocytes upregulate the production of LDL receptors to increase the uptake of LDL cholesterol from the blood. This upregulation of LDL receptors further decreases the blood LDL cholesterol. Being that cholesterol is a central contributor to atherosclerotic plaque formation, this leads to decreased progression of atherosclerosis.

The next step of cholesterol synthesis, involves the conversion of mevalonate to isoprene units. The condensation of two isoprene units generates a 10-carbon geranyl pyrophosphate. Condensation of another isoprene unit with geranyl pyrophosphate produces a 15-carbon farnesyl pyrophosphate. The farnesyl pyrophosphate is an important intermediate of the cholesterol biosynthesis pathway because it is the precursor of ubiquinone, also known as coenzyme Q, of the electron transport chain. Therefore, statin medications decrease the production of coenzyme Q. This decreased Coenzyme Q production consequently decreases the capacity of oxidative phosphorylation and ATP production, which leads to muscle pain, fatigue, and lethargy.

In the last condensation step, two farnesyl pyrophosphates are joined in a reaction catalyzed by squalene synthase to produce squalene.

In the final step of cholesterol synthesis, cyclization of squalene occurs to form a lanosterol steroid nucleus. Through several complex reactions cholesterol is formed (▶ Fig. 42.2).

42.2 Regulation of Cholesterol Synthesis

The regulatory processes of cholesterol synthesis are centered on HMG-CoA reductase activity, the rate limiting step of cholesterol synthesis (▶ Fig. 42.2).

Transcriptional regulation of the HMG-CoA reductase gene, and thus synthesis of the enzyme, occurs via the activity of the sterol-regulatory element-binding protein (SREBP), a transcription factor (▶ Fig. 42.4). The SREBP is activated by SREBP cleavage-activating protein (SCAP), both of which are localized within the endoplasmic reticulum membrane. When cholesterol levels are low, SREBP interacts with SCAP. This interaction between SREBP-SCAP causes the translocation of SREBP from the membrane of endoplasmic reticulum to the golgi apparatus. Subsequently, SREBP is transported to the nucleus, where it binds to specific DNA sequences and induces transcription of the HMG-CoA reductase gene. In contrast, when cholesterol levels are elevated, cholesterol binds to SREBP and prevents the interactions between SCAP and SRBEP, therefore preventing the expression of the HMG-CoA reductase gene.

When cholesterol is at high levels, the HMG-CoA reductase enzyme is also eliminated through proteosomal degradation. This occurs because under high concentrations of cholesterol (as well as plant sterols), HMG-CoA reductase undergoes oligomerization, which makes the enzyme more susceptible to degradation. This oligomerization results in the ubiquitination of membrane bound HMG-CoA reductase by E3 ligases and its targeted proteosomal degradation.

HMG-CoA reductase can also be regulated by covalent modifications. Phosphorylation of HMG-CoA reductase inactivates the enzyme, which happens under a state of high glucagon and AMP (▶ Fig. 42.5). In contrast, insulin activates HMG-CoA reductase by dephosphorylation. This is why type-2 diabetic patients also develop cholesterolemia due to the hyperinsulinemia.

The synthesis of cholesterol is also inhibited by a negative-feedback loop. Besides cholesterol, free fatty acids, bile acids, and stanols can all inhibit the activity of HMG-CoA reductase.

42.3 Absorption of Cholesterol

Cholesterol obtained from the diet is absorbed through diffusion into enterocytes. The cholesterol within the enterocytes is esterified and incorporated into chylomicrons through the activity of the acyl-CoA cholesterol acyl transferase (ACAT)

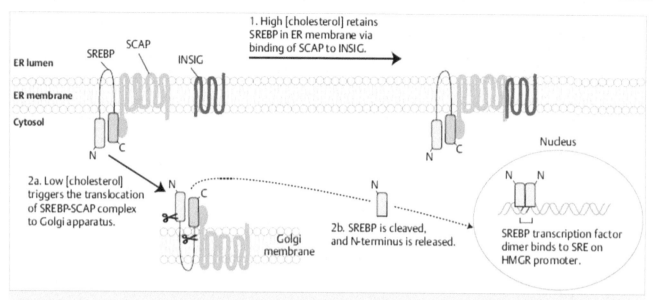

Fig. 42.4 Regulation of HMGR gene transportation. (Source: Panini S, ed. Medical Biochemistry—An Illustrated Review. New York, NY. Thieme; 2013.)

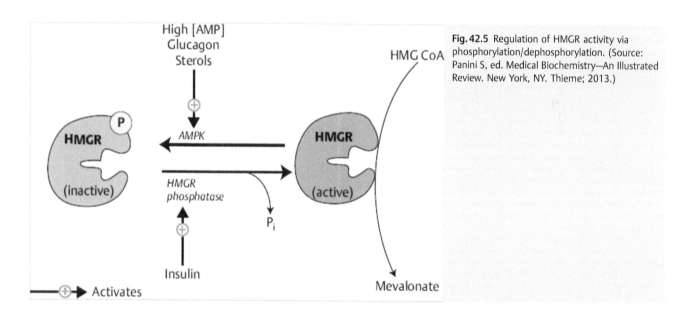

Fig. 42.5 Regulation of HMGR activity via phosphorylation/dephosphorylation. (Source: Panini S, ed. Medical Biochemistry—An Illustrated Review. New York, NY. Thieme; 2013.)

enzyme. The chylomicrons transport cholesterol and triacylglycerols to other tissues through the lymphatic system and then to the blood (transport of lipids by lipoproteins is reviewed in Chapter 43).

There is also a process to transport cholesterol back into the intestinal lumen, through an ATP binding cassette (ABC) protein. This allows for excess cholesterol to be eliminated from the cell through coupling with ATP hydrolysis. This is an important process as it prevents excess amounts of cholesterol from being absorbed through the enterocytes and into the bloodstream. The expression of the genes that code for the ABC proteins is controlled by the amount of cholesterol absorbed.

42.3.1 Link to Pharmacology

The process of absorption of cholesterol into enterocytes has been exploited for pharmacologic control of cholesterol levels.

The drug ezetimibe prevents the absorption of dietary cholesterol by inhibiting the Niemann-Pick C1L1 protein, which is involved in the transport of free cholesterol from micelles into the enterocytes. This decreased intestinal absorption of cholesterol decreases the cholesterol content of chylomicrons and consequently cholesterol levels in the liver. Therefore, the liver increases LDL-receptor gene expression, which increases LDL uptake. Ezetimibe has been utilized in patients who need better control of cholesterol, such as patients who are not optimally controlled with statins or intolerant to statins.

42.3.2 Link to Pathology

A defect in the ABC proteins, ABCG5 and ABCG8, leads to accumulation of cholesterol in enterocytes. Due to the lack of function of these proteins, excess cholesterol cannot be transported back into the intestinal lumen. Instead, this excess cholesterol

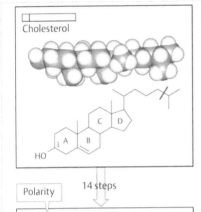

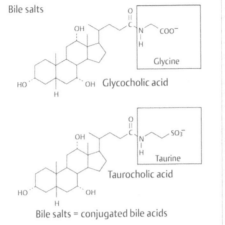

Bile acids	HO– in position		
Primary bile acids — Cholic acid — Chenodeoxycholic acid	C–3 C–3	C–7 C–7	C–12 –
Secondary bile acids — Deoxycholic acid — Lithocholic acid	C–3 C–3	– –	C–12 –

Fig. 42.6 Structure of bile acids and bile salts. (Source: Koolman J, Röhm K, ed. Color Atlas of Biochemistry. 3rd Edition. New York, NY. Thieme; 2012.)

VI

Bile salts = conjugated bile acids

is transported to the blood, which leads to an increase in blood cholesterol. This defect in ABC proteins is an autosomal recessive disorder called sitosterolemia and increases the risk of cardiovascular diseases.

42.4 Fate of Cholesterol

Cholesterol synthesized within the liver is utilized for several processes throughout the body. Within the hepatocytes, cholesterol is esterified at the hydroxyl group of the C-3 location by the enzyme acyl-CoA-cholesterol-acyl-transferase (ACAT), which places a fatty acid on the hydroxyl group. The esterification of free cholesterol with a fatty acid makes cholesterol more hydrophobic, facilitating easier transport in the core of lipoproteins and helps in the storage of cholesterol esters within the hepatocytes.

Cholesterol is also stored within the gallbladder as biliary cholesterol, which is released in response to food and aids in the absorption of lipids by forming micelles.

42.4.1 Link to Pathology

Cholesterol stored in the gallbladder may contribute to the formation of gall bladder stones, a condition known as cholelithiasis. Cholethiasis leads to a clinical emergency condition, in which the intense pain often requires surgical removal of the gallbladder. The stones can also become lodged in the biliary tract and can lead to pancreatitis. Gall bladder stones can also lodge within the intestine, causing a bowel obstruction known as gallstone ileus. Cholethiasis is often associated with obesity, biliary stasis, and infections.

Cholesterol is used for synthesis of vitamin D and steroids. Synthesis of vitamin D and steroid hormones is discussed in Chapter 44.

42.5 Bile Acid Synthesis

Most of the cholesterol in the hepatocytes is converted to bile acids and is excreted into the intestine in the form of bile products. Bile acids and salts are effective detergents due to their polar and nonpolar properties, and therefore help in the digestion and absorption of lipids (▶ Fig. 42.6). Bile acids, cholic acid, and chenodeoxycholic acid, are synthesized in the liver through hydroxylation reactions beginning with the 7α-hydroxylase enzyme, a microsomal cytochrome p450 enzyme (CYP7A1). This enzyme is the rate limiting step of the bile acid synthesis pathway and the activity of the enzyme is regulated by the concentration of bile acids and cholesterol (▶ Fig. 42.7).

When bile acids are conjugated with either the amino acid taurine or glycine, they are now called bile salts. The bile salts with lower pKa value are better emulsifiers. This allows for enhanced lipid digestion and absorption in the intestine.

The bile acids and salts recycle back to the liver for reuse. Approximately 95% of secreted bile acids are reabsorbed in the terminal ileum (▶ Fig. 42.8). Bile salts are deconjugated and dehydroxylated by the bacteria within the intestine. This leads to less solubility of the bile salts and increased reabsorption within the ileum. Without deconjugation and dehydroxylation, the bile salts cannot be readily reabsorbed, causing an increase in their excretion. Upon return to the liver, the secondary bile salts can be reconjugated but not rehydroxylated.

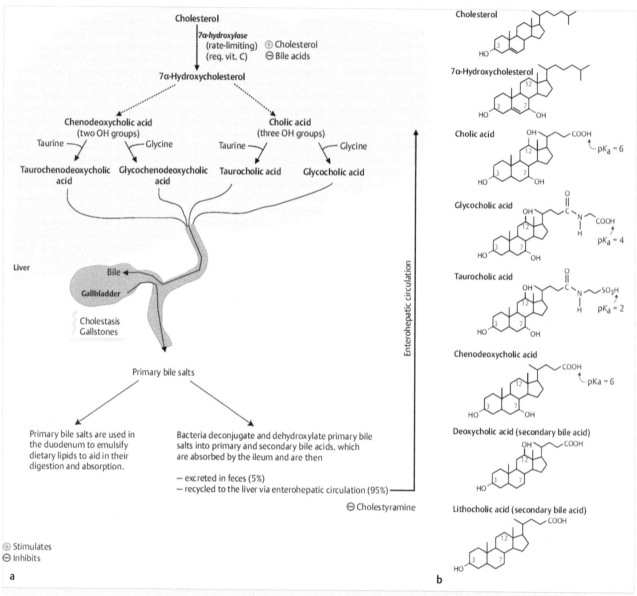

Fig. 42.7 (a, b) Synthesis and regulation of bile acids. (Source: Panini S, ed. Medical Biochemistry—An Illustrated Review. New York, NY. Thieme; 2013.)

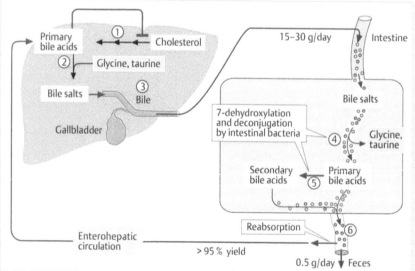

Fig. 42.8 Enterohepatic circulation of bile acids. (Source: Koolman J, Röhm K, ed. Color Atlas of Biochemistry. 3rd Edition. New York, NY. Thieme; 2012.)

Excretion of bile salts occurs only in small amounts. However, because the body cannot metabolize cholesterol, excretion of bile salts remains the only way the body can remove cholesterol.

42.5.1 Link to Pharmacology

Bile acid sequestrants like cholestyramine binds to bile salts and prevents them from being reabsorbed, which leads to elimination of large amounts of cholesterol from the body. This then causes upregulation of 7α-hydroxylase, leading to further use of the body's cholesterol for synthesis of bile acids. These drugs have been shown to greatly decrease LDL and total cholesterol levels but are not regularly utilized due to their side effect profile of GI distress, constipation and decreased lipid soluble vitamin absorption.

VI

Review Questions

1. Expression of which of the following proteins is upregulated by the use of statin drugs?
   A) Coenzyme Q
   B) HMG-CoA synthase
   C) HMG-CoA reductase
   D) LDL receptors

2. Increased concentrations of which of the following molecules/hormones decreases the synthesis of cholesterol?
   A) AMP
   B) ATP
   C) Insulin
   D) SREBP

3. Which of the following is an inhibitor of the cholesterol absorption in the enterocytes?
   A) ABCG5
   B) ABCG8
   C) Cholestyramine
   D) Ezetimibe
   E) Statin drugs

4. Activity of which of the following enzymes is expected to be increased with use of the drug cholestyromine?
   A) 7α-hydroxylase
   B) ACAT
   C) Coenzyme Q
   D) HMG-CoA synthase

Answers

1. **The correct answer is D**. Statin drugs decrease intracellular cholesterol synthesis in the liver. Therefore, the hepatocytes upregulate expression of LDL receptor gene to increase the uptake of LDL cholesterol from the blood. Statin drugs are structural analogs of HMG-CoA, therefore these drugs competitively inhibit the HMG-CoA reductase enzyme, and hence answer choice C is incorrect. Answer choice B is incorrect because statin drugs do not affect the activity or the synthesis of HMG-CoA synthase. Statin drugs reduce the synthesis of coenzyme Q by decreasing the intermediate farnesyl pyrophosphate, thus answer choice A is incorrect.

2. **The correct answer is A**. Synthesis of cholesterol increases in states of high energy. HMG-CoA reductase activity decreases during exercise and fasting. Therefore, high AMP decreases the activity of HMG-CoA reductase. In contrast to this, high ATP and insulin levels activates the enzyme. Thus, answer choices B and C are incorrect. Answer choice D is incorrect because SREBP is a transcription factor which binds to SRE on the DNA and increases the transcription of HMG-CoA reductase, the rate limiting step of cholesterol synthesis. When SREBP levels are high, synthesis of cholesterol is upregulated.

3. **The correct answer is D**. The drug ezetimibe prevents the absorption of dietary cholesterol by inhibiting the Niemann-Pick C1L1 protein, which is involved in the transport of the free cholesterol from micelles into the enterocytes.

Answer choices A and B are incorrect because ABCG5 and ABCG8 are involved in transporting excess cholesterol from enterocytes back into the intestinal lumen. Although these proteins control cholesterol levels, they do not inhibit the absorption of cholesterol into the enterocytes. Answer choice C is incorrect because cholestyramine is a bile acid sequestrants. Cholestramine inhibits the absorption of bile acids from intestinal lumen and prevents bile acid recycling. Answer choice E is incorrect because statin drugs inhibit the intracellular HMG-CoA reductase enzyme.

4. **The correct answer is A**. Cholestyramine is a bile acid sequestrant which binds to bile salts and prevents them from being reabsorbed back into the liver. This increased elimination of bile salts causes upregulation of the 7α-hydroxylase enzyme, which is the rate limiting step of bile acid synthesis. The activity of the 7α-hydroxylase enzyme is regulated by the concentration of bile acids and cholesterol in the liver.

Answer choice B is incorrect because ACAT is an intracellular enzyme that esterifies free cholesterol with a fatty acid to synthesize cholesterol ester. The esterified cholesterol can be stored intracellularly. Coenzyme Q is an ETC component, and its activity is not affected by cholestyramine, thus answer choice C is incorrect. Answer choice D is incorrect because HMG-CoA synthase is not a rate limiting enzyme.

43 Blood Lipoproteins

At the conclusion of this chapter, students should be able to:
- Describe the composition, function, and metabolism of blood lipoproteins
- Explain the importance of the VLDL to LDL pathway in health and diseases
- Describe the role of HDL in reverse cholesterol transport & apoprotein/lipid exchange
- Explain the process and molecular mechanism of receptor-mediated endocytosis
- Explain the causes of hyperlipidemias and the biochemical aspects of atherosclerosis
- Explain the role of Lp(a) in heart disease

Cholesterol, triacylglycerol, phosholipids, and lipid-soluble vitamins are transported in the blood in the form of lipoproteins. The lipoproteins are macromolecules that are hypophilic but have a hydrophobic core, which contains triacylglycerol and cholesterol ester. The hydrophilicity is provided by phospholipids and proteins, which are found at the outer shell of lipoproteins. There are several different types of lipoproteins in the plasma, each categorized according to its density. and include chylomicrons, VLDL, LDL and HDL (▶ Fig. 43.1).

43.1 Composition of Lipoproteins

The outer shell of lipoproteins is mainly composed of phospholipids and proteins, with the head-groups of the phospholipids and the amino acid side chains of the proteins giving the lipoprotein hydrophilic properties. The protein component of lipoproteins is made from apolipoprotein molecules. Apolipoproteins can function as activators of certain enzymes or as ligands for some receptors. Apolipoproteins also help stabilize the outer shell and increase the structural stability with cholesterol molecules, which are interspersed to give the lipoproteins their spherical shape. Apolipoproteins are divided into classes with the majority of them belonging to the classes of A, B, C, and E (▶ Table 43.1).

43.2 Chylomicrons

The largest and the least dense lipoprotein is the chylomicrons, which are transporters of dietary lipids (also reviewed in Chapter 39) and are produced within the endoplasmic reticulum of enterocytes.. The chylomicrons' characteristic apolipoprotein, ApoB48, is required for chylomicrons to exit the cells of the small intestine. Chylomicrons are absorbed into the lymphatic system, which transports chylomicrons to the bloodstream via thoracic duct to the left subclavian vein. Once within the bloodstream, chylomicrons receive ApoCII and ApoE through interaction with high density lipoprotein (HDL). The ApoCII activates the lipoprotein lipase, which is located on the surface of the capillary endothelial cells. This enzyme is responsible for the release of free fatty acids to the capillaries of heart, adipose, skeletal muscle and lactating mammary glands. ApoE of the chylomicrons functions as the ligand of ApoE receptors, which is vital for absorption of chylomicron remnants in the liver.

43.2.1 Link to Pathology

Hyperlipoproteinemias are genetic disorders that are characterized by deficiencies in the breakdown of chylomicrons and VLDL, or in the reabsorption of chylomicrons.

Type I hyperlipoproteinemia, also known as **hyperchylomicronemia,** is a rare condition (prevalence, 1:10,000) caused by impaired hydrolysis of chylomicron triacylglycerols. Individuals with this condition have inherited a deficiency of either lipoprotein lipase (LPL) or its activator, ApoC-II. The plasma triglyceride level in these patients can reach 1000 mg/dl, but LDL is reduced to 20% of normal or less, and most cholesterol is present in VLDL rather than LDL. Patients present with eruptive cutaneous xanthomas, abdominal pain after fatty meals, and recurrent attacks of pancreatitis, but no excessive atherosclerosis. Treatment consists of a low-fat diet.

Type III hyperlipoproteinemia, also called **dysbetalipoproteinemia** is caused by homozygosity of the ApoE2 gene, a genetic variant of ApoE. The ApoE2 has low affinity for binding to hepatic ApoE receptors and, therefore, leads to the accumulation of

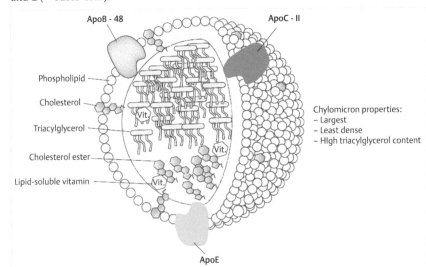

Fig. 43.1 Composition and structure of lipoproteins. (Source: Panini S, ed. Medical Biochemistry—An Illustrated Review. New York, NY. Thieme; 2013.)

Chylomicron properties:
– Largest
– Least dense
– High triacylglycerol content

VI

Table 43.1 Composition of lipoprotein complexes

Type	Apoproteins	Lipid proportion (%)			Function
		Triacyl glycerol	Cholesterol	Phospholipid	
Chylo-microns	B-48, C-II, E	80–95	2–7	3–9	Transport of dietary lipids from the intestine
Chylo-micron remnants	B-48, C-II, E				Transport of dietary lipids to the liver
VLDL	B-100, C-II, E	55–80	5–15	10–20	Transport of endo-genous lipids from the liver
IDL	B-100, C-II, E	20–50	20–40	15–25	Transport of endo-genous lipids from the liver
LDL	B-100	5–15	40–50	20–25	Provision of cholesterol to cells
HDL	C-II, E	5–10	15–25	20–30	Return of cholesterol to the liver

chylomicron remnants and IDL-like VLDL remnants in the blood. The phagocytosis of remnant particles by macrophages leads to xanthomas on palms, knees, elbows, and buttocks. Atherosclerosis shows a predilection for peripheral arteries, but the cardiovascular disease risk is increased as well. Although ~1% of the population has the offending ApoE2 genotype, only 2% to 10% of these individuals become hyperlipidemic.

Genetic variance in the alleles that code for ApoE has also been found to have a connection to Alzheimer's Disease. There are three common variants in the alleles for ApoE: ApoE2, ApoE3 and ApoE4. The ApoE4 allele has been found to be common within those who have Alzheimer's disease, making it a risk factor for its development.

43.3 VLDL to LDL Pathway

Very low density lipoproteins (VLDL) function as transporters for triacylglycerols and cholesterol and are formed in the liver. The VLDL to LDL pathway is also called the endogeneous pathway of triacylglycerol transport (▶ Fig. 43.2). When dietary carbohydrate and fat intake exceed the body's energy needs, triacylglycerol and cholesterol synthesis in the liver increases. In turn, the increased triacylglycerol and cholesterol content in the liver leads to an increase in synthesis of VLDL. Therefore, it is not only excess fat that can lead to the formation of atherosclerosis. Diets high in carbohydrates can cause high levels of fatty acid, triacylglycerol and cholesterol synthesis.

VLDLs are packed with cholesterol, cholesterol esters and triacylglycerol in the liver. The characteristic apolipoprotein of VLDL is ApoB100, which is encoded by the same gene as ApoB48. After release into the blood, VLDL interacts with HDL to receive ApoCII and ApoE. ApoCII activates lipoprotein lipase in various tissues, while triacylglycerols are released from VLDL. Approximately 50% of VLDL remnants are reabsorbed back into the liver via binding of ApoE to receptors on the hepatocytes.

43.4 IDL and LDL

The remaining VLDL remnants have more of their triacylglycerols removed and become less-dense intermediate density lipoproteins (IDL). Triacylglycerols can be further cleaved by hepatic lipase in the hepatic sinusoids, producing less-dense,

low density lipoproteins (LDL). LDLs, depleted of triacylglycerol, have a high content of cholesterol and cholesterol esters. LDLs are responsible for delivery of cholesterol to peripheral tissues. The development of LDL from the metabolism of VLDL is shown in ▶ Fig. 43.2.

Cholesterol received from LDL in the extrahepatic tissues is used for plasma membrane formation and steroid hormone synthesis. The LDL is taken up in peripheral tissues by receptors that recognize ApoB100 through a process called receptor-mediated endocytosis. When LDL levels are elevated within the bloodstream, these receptors become saturated. This leads to non-receptor mediated uptake of LDL, also known as the non-specific absorption of LDL. This non-specific uptake within arterial walls causes endothelial cell dysfunction, which increases the risk of atherosclerosis.

LDL within the vascular walls creates an inflammatory response that leads to oxidation of LDL lipoproteins. The oxidized LDL is engulfed by tissue macrophages, via the scavenger receptors SC-A1 and SC-A2, and initiating the formation of foam-cells. This uptake of oxidized LDL occurs. Eventually, the arterial lumens narrow. This is why LDL is known as "bad cholesterol", as elevations lead to adverse health effects. Atherosclerosis formation is discussed in further detail at the end of this section.

43.5 HDL

While LDL is commonly referred to as "bad cholesterol", HDL is considered "good cholesterol." The main reason for this is that HDL takes up cholesterol from the walls of blood vessels and delivers this cholesterol back to the liver, inhibiting the buildup of cholesterol within the arterial walls. Taking up cholesterol from the subintimal space of arterial walls decreases the risk of cholesterol oxidation. Thus, lower levels of oxidized LDL results in less engulfing by inflammatory cells. This is why elevated levels of HDL are found to decrease the risk of development of atherosclerosis.

HDL lipoprotein is released from the liver as an empty shell of phospholipids, with very low triacylglycerol and cholesterol within its core. This newly released HDL has a disc-like structure and is called nascent HDL. The nascent HDL also contains apoproteins AI, AII, CI, CII, D, and E. As the nascent HDL fills with triacylglycerol and cholesterol, its structure changes from disc-like to a spherical shape. This lipid-filled HDL known as mature HDL and delivers the cholesterol to the liver in a process known as reverse cholesterol transport (▶ Fig. 43.3). HDL

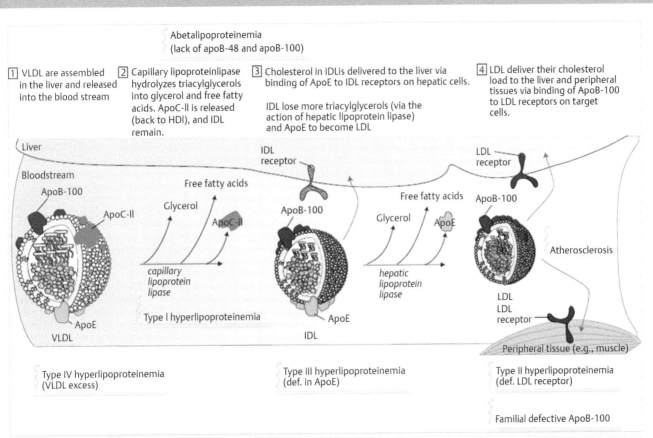

VI

Fig. 43.2 VLDL, IDL, and LDL processing. (Source: Panini S, ed. Medical Biochemistry—An Illustrated Review. New York, NY. Thieme; 2013.)

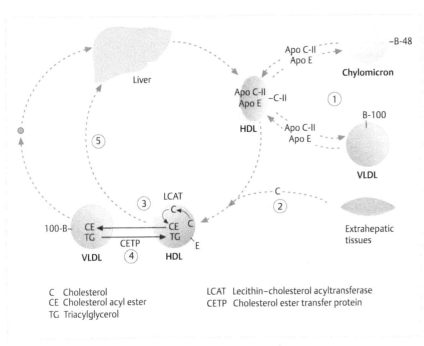

Fig. 43.3 Function and fate of HDL. 1. HDL transfers Apo C-II and Apo E to chylomicrons and VLDLs. 2. HDL collects excess cholesterol from extrahepatic tissues 3. Free cholesterol is esterified by LCAT in the blood. These esterified cholesterol is transported into the core of the HDL. 4. HDL exchanges its cholesterol esters with the triacylglycerols of VLDL. This process is facilitated by Cholesterol ester transfer protein (CETP). 5. HDL is taken up into the hepatocytes via scavenger receptors SC-B1. (Source: Koolman J, Röhm K, ed. Color Atlas of Biochemistry. 3rd Edition. New York, NY. Thieme; 2012.)

C   Cholesterol
CE Cholesterol acyl ester
TG Triacylglycerol

LCAT Lecithin–cholesterol acyltransferase
CETP Cholesterol ester transfer protein

also delivers the key apolipoproteins ApoCII and ApoE to chylomicrons and VLDL in the blood.

In peripheral tissues, cholesterol is transported to the outer surface of cell membranes by an ATP-binding cassette (ABC1) transporter. It can then be absorbed by circulating HDL. The cholesterol that is collected from the cell membranes must be esterified in order to prevent its release from HDL before delivery to the liver. The lecithin-cholesterol acyltransferase (LCAT) enzyme esterifies the cholesterol at the C-3 carbon, similar to the process of the intracellular esterification, and this esterified cholesterol can then be packed within the core of the HDL particle.

This HDL filled with cholesterol from peripheral tissues is now a mature HDL molecule and can deliver its cholesterol cargo to the liver by binding to scavenger receptor SC-B1 on hepatocytes via the ligand ApoE.

HDL also exchanges its cholesterol esters with the triacylglycerols and phospholipids of VLDL. This lipid exchange process is facilitated by a protein called cholesterol ester transfer protein (CETP). This exchange is controlled by the amount of triacylglycerol particles that are present within the blood. When triacylglycerol levels are elevated, more cholesterol and cholesterol esters are delivered to the liver via LDL, and less by HDL.

43.6 Receptor Mediated Lipoprotein Endocytosis

Cells in need of cholesterol can take up cholesterol using LDL receptors. To initiate LDL uptake, the LDL receptors bind to a binding domain on the ApoB100 component. Of note, while VLDL also contains ApoB100, it is not accessible by LDL receptors until it is transformed into LDL through the process described earlier.

The LDL receptors contain clathrin-coated pits on the cytosolic side. Upon binding to LDL, the receptor-LDL complex is internalized into the cell by endocytosis (▶ Fig. 43.4). The endocytosed vesicles fuse with lysosomes, and the high acidic content of the lysosomes causes the LDL to dissociate from the receptor. The cholesterol esters are then hydrolyzed to free cholesterol, which is either incorporated into plasma membranes or quickly reesterified by ACAT. Since excess incorporation of free cholesterol into the membrane may lead to plasma membrane damage, reesterification of free cholesterol by ACAT is an important process for minimizing membrane damage. Cholesterol esters are used in cellular processes such as steroid production or storage within the cell. The ApoB100 is

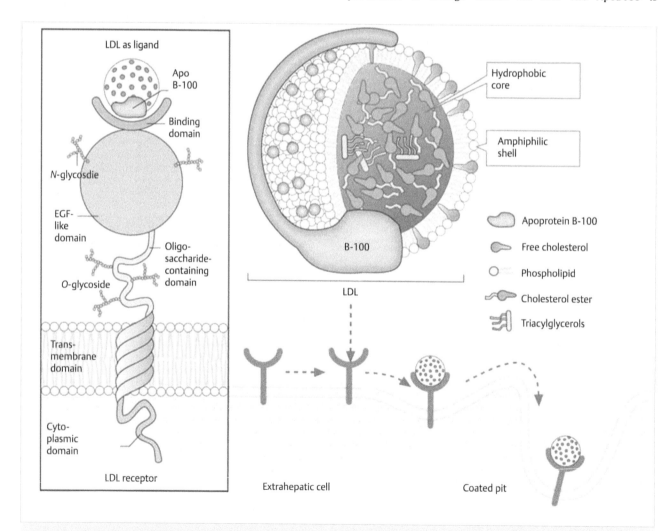

Fig. 43.4 Structure of LDL receptor and the process of receptor-mediated endocytosis of LDL. (Source: Koolman J, Röhm K, ed. Color Atlas of Biochemistry. 3rd Edition. New York, NY. Thieme; 2012.)

hydrolyzed and it's amino acids are released into the cytosol. LDL receptors are either recycled to the plasma membrane or degraded. One of the enzymes involved in degradation of LDL receptors is known as proprotein convertase subtilisin/kexin type 9 (PCSK9). Binding of PCSK9 to the LDL-LDL receptor complex initiates LDL receptor degradation.

43.6.1 Link to Pharmacology

A new class of LDL-cholesterol lowering drugs called **PCSK9 inhibitors** are monoclonal antibodies that bind to PCSK9 and inhibit the degradation of LDL receptors. Since these receptors are no longer degraded but, instead, return to the cell surface, the uptake of LDL is enhanced, which results in lower LDL levels in the blood. PCSK9 inhibitors are used to lower cholesterol and for the prevention of atherosclerosis, especially in patients who have genetic deficiencies in the LDL receptor gene (*LDLR*).

43.6.2 Link to Pathology

Familial hypercholesterolemia, also known as type II hypercholesterolemia, is charactarized by elevated blood LDL cholesterol. The disease is due to an autosomal dominant mutation within the gene encoding the LDL receptor. Heterozygous individuals (incidence ~ 1 in 500) have about a 3-fold increase in plasma LDL, which leads to an increased risk of atherosclerosis. Homozygotes often die in their early twenties or thirties from myocardial infarctions. Patients typically present with deposits of cholesterol esters within tendons, skin, and around the iris in the eye. These deposits are known as xanthomas (▶ Fig. 43.5).

43.7 Atherosclerosis

Atherosclerosis is a progressive deposition of lipids, which leads to degeneration and inflammation of arteries. Atherosclerotic plaque formation is initiated with damage of the intimal (innermost endothelial layer) of the blood vessels. Damage to the endothelial layer can be caused by increased shear stress from increased blood pressure. This damage initiates an inflammatory response that recruits neutrophils and macrophages (monocytes). These inflammatory cells are responsible for oxidizing LDL through the production of oxidizing agents such as superoxide, nitric oxide, and hydrogen peroxide. In addition, the damaged endothelial cells begin to express vascular cell adhesion molecule-1, which recruits more inflammatory cells and consequently increases oxidation of LDL.

Macrophages internalize the oxidized LDL using the scavenger receptors SR-A1 and SR-A2. These lipid-laden macrophages are called foam cells. These cells have a high affinity and low saturability for oxidized LDL, creating lipid-engorged cells. These macrophages release pro-inflammatory cytokines (interferon-gamma and tumor necrosis factor-alpha, TNF-α) that are responsible for perpetuating the inflammatory response by further recruitment of macrophages and induction of a T cell response. The foam cells also create a nidus for platelet aggregation. These platelet aggregates also release cytokines that play a role in furthering the inflammatory response, which leads to recruitment of vascular smooth muscle cells and

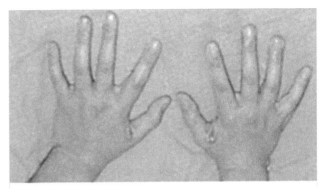
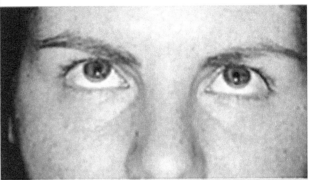

Fig. 43.5 (a) Xanthoma formation, and **(b)** Arcus lipoides. (Source: Passarge E, ed. Color Atlas of Genetics. 4th Edition. New York, NY. Thieme; 2012.)

stimulation of fibroblasts, which produce the extracellular matrix components that help form the plaque. As the atherosclerotic plaque evolves, a fibrous cap is created. The fibrous cap is composed of smooth muscle cells, cholesterol, extracellular matrix, and inflammatory cells. The thin end-regions of the fibrous cap are known as "shoulders", which are more prone to damage by the shear stress of high blood pressure. This damage can cause the shoulder of the plaque to rupture or significantly weaken. Macrophages are then stimulated by the damage to release metalloproteases that further break down the plaque, ensuring rupture. Once rupture occurs, foam cells are exposed to the bloodstream again and create a nidus for platelet aggregation. This platelet aggregation initiates thrombus formation, which is the cause of myocardial infarction.

43.8 Role of Lipoprotein (a) in Heart Disease

Lp(a) is structurally identical to LDL, with an additional Apo(a) protein that is covalently linked to ApoB100. An increased risk of coronary heart disease is associated with and increased in Lp (a), which is due to the structure of Apo(a), a structural homolog of plasminogen. Plasminogen is a precursor of plasmin, which is a protease found in plasma. Plasmin degrades fibrin to dissolve the blood clots. Due to the structural similarity between plasminogen and Apo(a), increased levels of Lp(a) competitively inhibit the conversion of plasminogen to plasmin. This decreased plasmin formation results in accumulation

VI

of blood clots and consequently leads to cardiovascular disease, particularly in patients with hypercholesterolemia. Clinical studies also suggest that diets with increased levels of trans fatty acids increase the circulating Lp(a) concentrations.[1]

Reference

[1] Anuurad E, Enkhmaa B, Berglund L. Enigmatic role of lipoprotein(a) in cardiovascular disease. Clin Transl Sci. 2010; 3(6):327–332

Review Questions

1. Which of the following apolipoproteins is the cause of dysbetalipoproteinemia?
   A) ApoA1
   B) ApoB48
   C) ApoCII
   D) ApoE2

2. Which of the following apolipoproteins is attached to VLDL during synthesis in the hepatocytes?
   A) Apo(a)
   B) ApoB48
   C) ApoB100
   D) ApoCII
   E) ApoE

3. Increased concentrations of which of the following two lipoproteins would most likely increase triacylglycerol levels in the blood?
   A) HDL-LDL
   B) HDL-Chylomicron
   C) LDL-Chylomicron
   D) LDL-VLDL
   E) VLDL-HDL
   F) VLDL-Chylomicron

4. Which of the following drugs inhibits the degradation of LDL receptors?
   A) Fibrates
   B) Niacin
   C) PCSK9
   D) Statins
   E) Thiazolidinedione

5. Which of the following receptors is involved in the uptake of HDL lipoprotein into hepatocytes?
   A) SR-A1
   B) SR-A2
   C) SR-B1
   D) LDL receptors
   E) Scavenger receptors

Answers

1. **The correct answer is D**. ApoE2 has lower affinity for hepatic ApoE receptors compared to ApoE3. Therefore, the presence of ApoE2 results in the accumulation of chylomicron remnants and IDL-like VLDL remnants in the blood. Patients with dysbetalipoproteinemia present with high blood triacylglycerol and high blood cholesterol levels. Xanthomas on palms, knees, elbows, and buttocks develop since the macrophages phagocytose the chylomicron- and VLDL-remnants.

Answer choice A is incorrect because ApoA1 is associated with HDL lipoprotein. ApoA1 activates lipoprotein lipase, therefore, it facilitates taking up cholesterol from peripheral tissues and from other lipoproteins into the core of the HDL. ApoB48, answer choice B, is required for chylomicrons to exit the cells of the small intestine. Answer choice C is incorrect because genetic deficiency of ApoCII or lipoprotein lipase (LPL) causes type I hyperlipoproteinemia or hyperchylomicronemia. These patients also develop cutaneous xanthomas.

2. **The correct answer is C**. ApoB100, the apolipoprotein characteristic of VLDL, is synthesized in the liver as VLDLs are packed with cholesterol, cholesterol esters and triacylglycerol. ApoB100 is found not only in VLDL, but also in IDL and LDL lipoproteins. The LDL receptors recognize ApoB100 for internalization of LDL into the hepatocytes.

Answer choice A is incorrect because Apo(a) is found in Lp(a), which is structurally identical to LDL. While the LDL contains ApoB100, the Lp(a) contains both ApoB100 and Apo(a), which is covalently linked to ApoB100. Answer choice B is incorrect because ApoB48 is found on chylomicrons. The ApoCII (answer choice D) and ApoE (answer choice E) are incorrect because these two apolipoproteins are transferred from HDL to VLDL and chylomicrons in the bloodstream.

3. **The correct answer is F**. Chylomicrons and VLDLs are rich in triacylglycerols, therefore increased concentrations of these lipoproteins increases triacylglycerol concentrations in the blood. Answer choices A, B, C, D, and E are incorrect because these answer choices have either HDL or LDL. Both of these lipoproteins are rich in cholesterol, but not in triacylglycerol.

4. **The correct answer is C**. PCSK9 is an enzyme that is involved in the degradation of LDL receptors in hepatocytes. Inhibitors of PCSK9 prevent the degradation of LDL receptors, therefore increasing the uptake of LDL from the blood.

Answer choice A is incorrect because fibrates stimulate the proliferation of peroxisomes by activating the PPARα (peroxisome proliferator-activated receptors) pathway. The PPAR pathway induces transcription of the lipoprotein lipase gene, as well as genes that are involved in lipid oxidation. Therefore, these drugs are used for the treatment of hypertriglyceridemia. Thiazolidinedione (answer choice E) acts similar to fibrates and activates the PPARγ pathway, which increases peroxisome proliferation and also decreases the free acid levels in the blood by increasing their oxidation in the peroxisomes. Therefore, thiazolidinedione is used in treatment of type 2 diabetes and dislipidemia (metabolic syndrome). Niacin (answer choice B) inhibits the hormone sensitive lipase in adipocytes, thus decreasing VLDL formation in the liver. Statins (answer choice D) are inhibitors of HMG-CoA reductase in hepatocytes.

5. **The correct answer is C**. Mature HDL delivers cholesterol to the liver by binding via the ApoE ligand to the scavenger receptor SC-B1 on hepatocytes.

SR-A1 and SR-A2 (answer choices A and B) are scavenger receptors that are found on macrophages, which take up oxidized LDL. Answer choice D is incorrect because LDL receptors take up LDL by receptor-mediated endocytosis. The receptors recognize ApoB100 for internalization of LDL.

VI

44 Steroid Hormones and Vitamin D

At the conclusion of this chapter, students should be able to:
- Describe the synthesis of steroid hormones in the adrenal cortex and in the gonads
- Explain the regulation of steroid hormone synthesis and secretion
- Compare and contrast signaling pathways associated with steroid and thyroid hormones
- Describe the synthesis of vitamin D
- Discuss the effects of vitamin D in calcium homeostasis

The synthesis of specific steroid hormones occurs in the mitochondria of both the adrenal cortex and the sex organs. The corticosteroids, glucocorticoids (cortisol), mineralocorticoids (aldosterone) and the androgen precursor dehydroepiandrosterone (DHEA) are synthesized in the adrenal cortex while the sex hormones testosterone and estrogen are synthesized in the testes and ovaries. Progesterone is synthesized and secreted by the corpus luteum.

Unlike protein or peptide hormones, steroid hormones are not encoded in the DNA. Through several oxidative reactions by cytochrome P450 enzymes, steroid hormones are synthesized from cholesterol. The production of the steroid hormones is controlled by other hormones that control the activity of the synthetic enzymes. When these enzymes are activated, cholesterol is taken in by cells of the adrenal cortex, ovaries, testicles or ovarian corpus luteum to produce steroids. The synthesis starts with the removal of the side chain carbons on the seventeenth carbon of cholesterol. Once formed, steroids travel to their destination bound to serum proteins, such as albumin, as their hydrophobicity would inhibit their movement in the bloodstream. Once at the destination, steroids utilize their hydrophobicity to cross the cell membrane. Once in the cell, steroid hormones bind to receptors in the cytoplasm or nucleus that create a transcription factor for DNA to induce expression of specific genes. Therefore, it is important to remember that the actions of steroid hormones are like those of transcription factors and serve to further enhance cellular processes.

One example of steroid hormone signaling that we reviewed in the gluconeogenesis section of Chapter 25 is cortisol signaling, which increases the expression of the PEPCK gene. After the hormone associates with its intracellular receptor, the receptor-hormone complex binds to regulatory elements in the genome and activates or represses the tissue-specific gene transcription.

44.1 Classes of Intracellular Receptors

There are two major classes of intracellular receptors: steroid hormone receptors and thyroid hormone receptors.

The steroid hormone receptor family includes glucocorticoids, mineralocorticoids, androgens and progesterone receptors. These receptors exist as cytoplasmic multimeric complexes that are bound to heat shock proteins (HSPs).

Binding of ligand to the receptor causes both the dissociation of receptor from the HSPs and the ligand-bound receptors to form homodimers. Ligand binding also exposes a nuclear translocation signal on the receptor, which results in translocation of the ligand-receptor homodimer to the nucleus. This ligand-receptor homodimer binds to specific DNA sequences known as Hormone Response Elements (HREs) and recruits the transcription factors necessary for the expression of specific target genes.

The thyroid hormone receptor family includes the thyroid hormone receptor, retinoic acid receptor, retinoid X receptor, estrogen receptor, peroxisome proliferator-activator receptor (PPAR), and vitamin D receptor. In contrast to the steroid hormone receptors, the thyroid hormone receptor family is mainly found in the nucleus and constitutively bound to DNA. In the absence of ligand, these receptors are bound to a histone deacetylase. When ligand binds, the receptors recruit histone acetylases, resulting in the formation of an activator complex and gene expression.

44.2 The Common Pathway of Steroid Hormone Synthesis

The synthesis of all steroid hormones begins in a similar pattern. The synthesis starts with transport of cholesterol from the cytoplasm to the inner membrane of mitochondria by the steroidogenic acute regulatory (StAR) protein. The first step for converting cholesterol to the parent compound pregnenolone is identical in the adrenal cortex, ovaries, and testes. Hydroxylations at C22 and C20 require NADPH and molecular oxygen. Cleavage of the side chain creates the 21 carbon pregnenolone. This process is catalyzed by a cytochrome P450 side chain cleavage enzyme (CYP11A/desmolase) (▶ Fig. 44.1). Pregnenolone is then converted to progesterone by the enzyme 3-β-hydroxysteroid dehydrogenase. Progesterone then can be converted to several other steroid hormones with the use of various enzymes, some of which are used in multiple pathways.

44.3 Synthesis of Cortisol and Physiological Effects of Cortisol

Cortisol is also known as a stress hormone because it mediates the body's response to physiological stresses such as pain, hypoglycemia, infections, trauma, exercise, and exposure to cold. In order to promote survival during these physiological stresses, cortisol increases the rate of gluconeogenesis in the liver and proteolysis in the extrahepatic tissues. The amino acids from proteolysis are converted to blood glucose by gluconeogenesis in the liver. This increased proteolysis and gluconeogenesis during starvation maintains the blood glucose concentrations at normal levels (Fig. 44.2). In the adipocytes, cortisol increases the lipolysis and free fatty acid levels in the blood. The effect of cortisol on carbohydrate, protein, and lipid metabolism is the opposite of the effects of insulin. Therefore, therapeutic use of cortisol can cause side effects that include increased blood glucose concentrations.

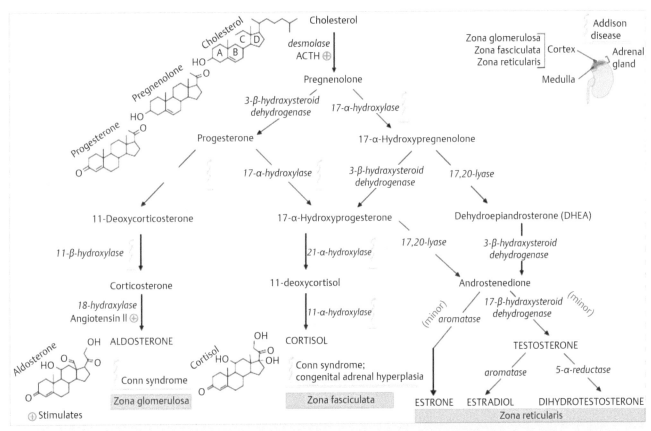

Fig. 44.1 Synthesis of steroid hormones. (Source: Panini S, ed. Medical Biochemistry—An Illustrated Review. New York, NY. Thieme; 2013.)

Cortisol also suppresses all allergy and immune responses and decreases the body's healing response. This is why cortisol in high concentrations is used as an anti-inflammatory agent (Fig. 44.2).

The synthesis and secretion of cortisol is stimulated by adrenocorticotropic hormone (ACTH), which is released from the anterior pituitary. Release of ACTH is stimulated by corticotropin releasing hormone (CRH), which is secreted from the hypothalamus in response to external stimuli such as emotional stress, cold, pain, infections, trauma, exercise, and starvation. Stimulation of the zona fasciculata of the adrenal cortex by ACTH initiates the transport of cholesterol to the mitochondria and also activates the CYP11A/desmolase enzyme to produce pregnenolone from cholesterol.

In the cortisol synthesis pathway, the 3-β-hydroxysteroid dehydrogenase enzyme converts pregnenolone to progesterone. The CYP17 (17-α-hydroxylase) enzyme, located on the smooth endoplasmic reticulum, hydroxylates progesterone to produce 17-α-hydroxyprogesterone, which is further converted to 11-deoxycortisol by the enzyme CYP21 (21-α-hydroxylase). The 11-deoxycortisol then travels back to the inner mitochondria. Here, it undergoes the third hydroxylation by CYP11 (11-β-hydroxylase) to produce cortisol (▶ Fig. 44.1). The synthesis and secretion of cortisol is inhibited by a negative feedback loop, in which increased concentrations of cortisol in the blood inhibits the release of both CRH and ACTH.

44.3.1 Link to Pathology

Insufficient production of adrenal steroids is known as adrenal deficiency syndrome. This condition can develop in patients who have been using therapeutic glucocorticoids over a long period of time. Due to the negative feedback action of cortisol, long-term exogenous glucocorticoid use inhibits the release of ACTH and CRH from the anterior pituitary and the hypothalamus, respectively. Therefore, upon abrupt withdrawal of exogenous steroids, appropriate amount of endogenous cortisol is not released from the adrenal cortex due to the absence of the ACTH and the CRH. The exogenous glucocorticoids (hydrocortisone, prednisone, etc.) are commonly prescribed for patients who suffer for medical conditions such as inflammatory diseases, chronic COPD, inflammatory bowel disease, and systemic lupus erythematous. Therefore, it is vital to taper exogenous steroids rather than abrupt withdrawal. Tapering of these exogenous steroids allows the body to begin producing CRH and ACTH again.

44.3.2 Link to Pharmacology

High concentrations of glucocorticoids are used pharmacologically for immunosuppressive and anti-inflammatory properties. A long-term administration of corticosteroids may cause exogenous Cushing syndrome. Since cortisol is an antagonist of

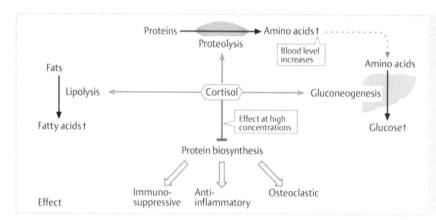

Fig. 44.2 Biochemical effects of cortisol. (Source: Koolman J, Röhm K, ed. Color Atlas of Biochemistry. 3rd Edition. New York, NY. Thieme; 2012.)

insulin, patients taking high doses of glucocorticoids may develop diabetes mellitus as a side effect. Glucocorticoids also inhibit the activity of osteoblasts in bone (reviewed later in this chapter), which may cause demineralization of bones if used for long periods of time (▶ Fig. 44.2).

44.3.3 Link to Pathology

There are several congenital adrenal hyperplasia (CAH) syndromes that result from the deficiency of enzymes in the cortisol synthesis pathway. The deficiency of 21-α-hydroxylase (the most common form of CAH) results in an absence of aldosterone and cortisol. This insufficient production of cortisol results in continual stimulation of the adrenal cortex by ACTH. This continual stimulation of the gland leads to hyperplasia of the adrenal cortex and can result in elevated levels of the precursors of androgens such as DHEA and its sulfate derivative DHEAS. Androstenedione, another weak androgen, is also produced in the zona reticulosum of the adrenal cortex by the enzyme 3-β-hydroxysteroid dehydrogenase. These adrenal androgens, DHEA and androstenedione, are converted to testosterone in the muscle and estrogen in the adipose tissue.

The total deficiency of 21-α-hydroxylase leads to excess androgen production, which causes the development of ambiguous genitalia in females at birth and early virilization in males. The lack of aldosterone and cortisol leads to adrenal crisis and salt wasting later in life. Common symptoms include an inappropriate response to illness due to cortisol insufficiency, hyponatremia, hyperkalemia and metabolic acidosis. The increased levels of 17-α-hydroxyprogesterone is considered diagnostic, as its production is increased due to the inability to produce cortisol and aldosterone from progesterone and pregnenolone.

The partial deficiency of 21-α-hydroxylase presents with milder symptoms and appears later in life generally during adolescence years. Common symptoms in females are oligomenorrhea, hirsutism, and infertility. In males, early virilization and advanced musculoskeletal growth.

44.4 Synthesis of Aldosterone and Physiological Effects of Aldosterone

Aldosterone is synthesized in a similar way as cortisol with a few exceptions. Aldosterone synthesis is also stimulated by ACTH. In the zona glomerulosa of the adrenal cortex, the progesterone is hydroxylated at C21 by CYP21 (21-α-hydroxylase) to form 11-deoxycorticosterone (▶ Fig. 44.1). CYP11 (11- β-hydroxylase) then catalyzes the conversion of 11-deoxycorticosterone to corticosterone. The corticosterone is then oxidized by the 18-hydroxylase enzyme to produce aldosterone. In this last step, 18-hydroxylase is stimulated by angiotensin II.

As the renin-angiotensin-aldosterone system is responsible for regulating blood fluid homeostasis, it is activated in response to hypovolemic or hyponatremic conditions. When low amounts of sodium are presented in the macula densa within the kidney, renin is released. Renin stimulates the production of angiotensin I, which is converted to angiotensin II in the lungs by the angiotensin converting enzyme (ACE). Angiotensin II stimulates efferent renal arterial vasoconstriction and production of aldosterone, which in turn increases sodium reabsorption, potassium secretion, and acid secretion in the distal collecting ducts of the kidney. The increased sodium increases passive water reabsorption, thereby increasing intravascular volume.

44.4.1 Link to Pathology: Conn Syndrome and Addison Disease

Excess aldosterone production from tumors of the adrenal cortex causes primary hyperaldosteronism, also known as Conn syndrome. Common symptoms include uncontrollable hypertension, hypokalemia, hypernatremia and metabolic alkalosis. In contrast to Conn syndrome, Addison disease is an autoimmune disease that results in destruction of the adrenal cortex. The destruction of adrenal cortex leads to low aldosterone production and hypotension due to excess excretion of sodium and water from the kidneys.

44.4.2 Link to Pathology

Deficiency of the enzyme 11- β-hydroxylase causes hypertension and excess production of androgens such as DHEA and androstenedione. In addition, patients with this disease cannot produce cortisol, aldosterone, and corticosterone (▶ Fig. 44.1). Insufficient cortisol production causes over-stimulation of the adrenal cortex by ACTH due to the absence of negative feedback inhibition. Over-stimulation of the adrenal cortex causes the overproduction of adrenal androgens, which results in ambiguous genitalia in females and early virilization in males. This enzyme deficiency also results in excess production of the 11-deoxycorticosterone, which has mineralocorticoid activity. Due to its aldosterone-like effects, increased 11-deoxycorticosterone suppresses the renin-angiotensin system. More importantly, this intermediate, like aldosterone, increases sodium reabsorption and fluid retention.

44.5 Synthesis of Sex Hormones

Androgens, estrogens and progesterone are produced in the sex organs of the testicles and ovaries (▶ Fig. 44.3). They are also formed to a lesser extent in the adrenal cortex. In the testicles, testosterone synthesis occurs in the Leydig cells and is stimulated by luteinizing hormone (LH), which stimulates desmolase to convert cholesterol into pregnenolone. Pregnenolone is further converted to 17-α-hydroxypregnenolone by 17-hydroxylase. The 17, 20 lyase / desmolase enzyme is then utilized to form dehydroepiandrosterone (DHEA), which is used as a substrate by 3-β-hydroxysteroid dehydrogenase to produce androstenedione. Androstenedione is then converted to testosterone by 17-β-hydroxysteroid dehydrogenase. Testosterone is then transported out of the Leydig cells. Some testosterone is converted to dihydrotestosterone by 5-α-reductase, which is a much more potent androgen. Dihydrotestosterone stimulates sperm production in the Sertoli cells and development of secondary sex characteristics. The enzyme aromatase converts testosterone to estradiol and is present in high quantities in adipose tissue. Therefore, obese males have a risk of higher estrogens in their body.

44.5.1 Link to Pharmacology

Finasteride is an inhibitor of 5-α-reductase enzyme, which catalyzes the conversion of testosterone to dihydrotestosterone. This drug is used for treatment of benign prostatic hyperplasia.

The production of estrogens occurs in the ovarian follicle and corpus luteum, with the primary estrogens being formed in the granulosa cells of the ovary, and is stimulated by follicle stimulating hormone (FSH). Estrogen, along with progesterone, is responsible for development of the endometrial lining for implantation. Estrogens are formed in the same pathway of testosterone, apart from aromatase forming estradiol or estrone from androstenedione (▶ Fig. 44.3). In the ovaries, FSH can also stimulate the conversion of testosterone to estradiol, an estrogen with a hydroxyl group at C17, by the enzyme aromatase (▶ Fig. 44.3).

44.5.2 Link to Pathology

The enzyme aromatase can be found in adipose tissue of both men and women. While in men the aromatase can cause feminine features by making estradiol from testosterone, in women it can be more pathologic. Increased exposure to estrogen is a known risk factor for endometrial and breast cancer. Increased

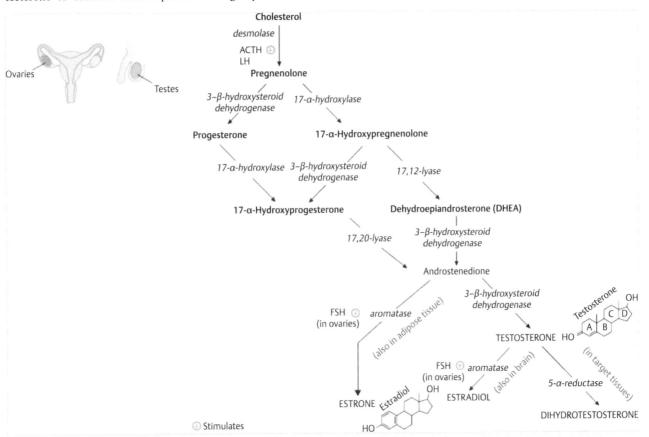

Fig. 44.3 Steroid hormone synthesis in the testes and ovaries. (Source: Panini S, ed. Medical Biochemistry—An Illustrated Review. New York, NY. Thieme; 2013.)

estrogen due to obesity is a cause for endometrial hyperplasia which is a major risk factor for endometrial cancer.

44.5.3 Link to Pharmacology

Rivizor (Vorozole) inhibits the aromatase enzyme and it is used for treatments of estrogen positive breast cancer.

Progesterone is formed in the ovarian follicle and corpus luteum and is dependent on luteinizing hormone (LH). Once pregnenolone is formed from cholesterol by the desmolase enzyme, it is converted to progesterone via the activity of 3-β-hydroxysteroid dehydrogenase. Progesterone is responsible for maintenance of the endometrial lining for implantation. If implantation occurs, human chorionic gonadotrophin (HCG) from the implanted embryo maintains the corpus luteum which continues to produce progesterone to maintain the endometrial lining. If implantation does not occur, the corpus luteum degenerates and menses occurs.

44.6 Synthesis of Vitamin D

Vitamin D is a lipid-soluble vitamin that is necessary for calcium homeostasis. Vitamin D can be taken from the diet or can be synthesized from cholesterol in the body. The synthesis is stimulated by low calcium levels and by parathyroid hormone (PTH), which is released in conditions of low calcium levels.

Vitamin D synthesis begins in the liver with the formation of 7-dehydrocholesterol, which is transported to the skin (▶ Fig. 44.4). Upon exposure to UV light in the skin, 7-dehydrocholesterol is converted to cholecalciferol (Vitamin D3). Cholecalciferol is returned to the liver, where C25 is hydroxylated by 25-α-hydroxylase to form 25-hydroxycholecalciferol. The 25-hydroxycholecalciferol is then converted to 1,25-hydroxycholecalciferol (calcitriol) in the kidney by 1-α-hydroxylase. This enzyme is the rate limiting step of calcitriol synthesis and its activity is upregulated by PTH.

The active form of vitamin D is calcitriol, also known as 1,25-hydroxycholecalciferol. Calcitriol stimulates calcium absorption in the intestine by increasing the production of the calcium-binding protein calbindin. Calbindin increases calcium absorption from the intestines and stimulates the ATP-dependent calcium pump, which transports calcium into the blood. Calcitriol also increases calcium reabsorption in the kidneys. Because calcitriol increases the deposition of calcium, the mineralization of bones is induced. Vitamin D deficiency therefore causes disruption of calcium deposition in the bones.

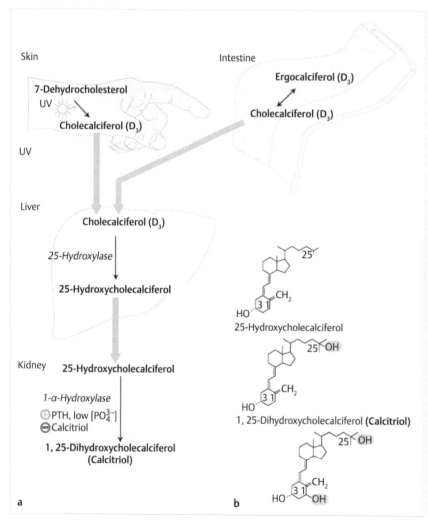

Fig. 44.4 (a, b) Synthesis and regulation of calcitriol, the active form of vitamin D. (Source: Panini S, ed. Medical Biochemistry- An Illustrated Review. New York, NY. Thieme; 2013.)

44.6.1 Link to Pathology

Vitamin D deficiency can lead to rickets in children and osteomalacia in adults. Chronic kidney or liver disease, lack of exposure to sun light and gastrointestinal problems that cause lipid malabsorption can all lead to vitamin D deficiency. Because PTH stimulates the formation of the active form of vitamin D (calcitriol), hypoparathyroidism also causes vitamin D deficiency.

44.7 Calcium Hemostasis and Bone Remodeling

Vitamin D, along with PTH and calcitonin, is involved in the regulation of blood calcium levels (▶ Fig. 44.5). While calcitriol and calcitonin increase calcium deposition and improve bone mineralization, PTH promotes mobilization of calcium from the bones. When blood calcium levels decrease, the parathyroid gland secretes PTH, which activates osteoclasts. The activated osteoclasts break down bone, releasing calcium and increasing calcium levels in the blood. In contrast, increased blood calcium decreases the secretion of PTH from the parathyroid gland.

The most important mineral of bones is apatite, a form of crystalline calcium phosphate. In adults, more than 1 kg of calcium is stored in the bones in the form of hydroxyapatite, carbonate apatite, or fluoroapatite. Through the activity of osteoclasts and osteoblasts, calcium is continuously incorporated and then removed from bones. While osteoblasts deposit collagen and incorporate calcium into bones, osteoclasts secrete $H+$ ions and collagenases to function in demineralization of the bones. These activities are regulated by calcitriol, PTH, and calcitonin. Mineralization of osteoid by osteoblasts is critical for proper development of bones.

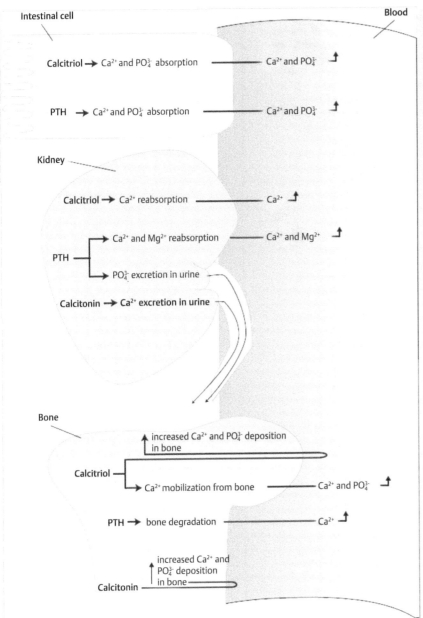

Fig. 44.5 Calcium homeostasis. Regulation of blood calcium levels is maintained by calcitriol, PTH, and calcitonin. (Source: Panini S, ed. Medical Biochemistry—An Illustrated Review. New York, NY. Thieme; 2013.)

VI

44.7.1 Link to Pathology

When osteoblasts are unable to form hydroxyapatite, due to insufficient calcium and/or phosphate, rickets occurs in growing children and in adults, osteomalacia may result. Osteomalacia (bone softness) exhibits normal amounts of collagen matrix, but there is a deficiency in mineralization. This disease is most often caused by insufficient levels of vitamin D. Treatment involves the replacement of calcium and vitamin D.

Osteoporosis, on the other hand, is a condition in which bone resorption outweighs bone formation. Principle causes of osteoporosis are decreased estrogen, vitamin D, and calcium. Since osteoporosis is due to increased activity of osteoclast cells, treatment can include bisphosphonates, which inhibit bone resorption by the osteoclasts.

VI

Review Questions

1. Which of the following is the rate limiting step in the biosynthesis of cortisol?
   A) 17-α-hydroxylase
   B) 21-α-hydroxylase
   C) Side chain cleavage enzyme
   D) Steroid Acute Regulatory Protein (StAT)

2. Which of the following enzymes is inhibited by Vorozole?
   A) 5-α-Reductase
   B) Aromatase
   C) CYP11A
   D) CYP17
   E) CYP21
   F) Desmolase

3. Synthesis of which of the following hormones is expected to be decreased in patients who have genetic deficiency in the 18-hydroxylase enzyme?
   A) Aldosterone
   B) Corticosterone
   C) Cortisol
   D) Deoxycortisol
   E) Deoxycorticosterone

4. Which of the following diseases/syndromes presents with hypotension?
   A) Addison disease
   B) Conn Syndrome
   C) 11-β-hydroxylase deficiency
   D) Cushing Syndrome

5. The synthesis of which of the following hormones is increased in patients with Cushing syndrome?
   A) Aldosterone
   B) ACTH
   C) CRH
   D) FSH
   E) LH

Answers

1. **The correct answer is D**. Cortisol synthesis is stimulated by ACTH from the anterior pituitary. Stimulation of the zona fasciculate of the adrenal cortex by ACTH activates the transport of cholesterol from the cytoplasm into the mitochondria by steroidogenic acute regulatory (StAR) protein. Although ACTH also stimulates the side chain cleavage enzyme CYP11A/desmolase, the rate limiting step of steroid hormone synthesis is the transport of cholesterol into the mitochondria by StAR. Thus, answer choice C is incorrect. The enzymes in answer choices A (17-α-hydroxylase) and B (21-α-hydroxylase) are located in the smooth endoplasmic reticulum. The CYP17 enzyme hydroxylates progesterone to produce 17-α-hydroxyprogesterone. CYP21 produces 11-deoxycortisol. Neither of these enzymes is regulated, therefore answer choices A and B are incorrect.

2. **The correct answer is B**. Rivizor (Vorozole) inhibits the aromatase enzyme. This enzyme is stimulated by FSH in the ovaries and converts testosterone to estradiol in the ovaries. Aromatase is also found in the adipose tissue of both men

and women. Therefore, obese men and women have more conversion of testosterone to estrogen. While the increased estrogen levels cause feminine features in men, obese women are at risk for development of endometrial and breast cancer. The drug Rivizor (Vorozole) is used for the treatment of estrogen receptor positive breast cancer.

Answer choice A is incorrect because vorozole is not the target of 5-α-reductase. The 5-α-reductase enzyme converts testosterone to dihydrotestosterone. This enzyme is inhibited by the drug finasteride, which is used for the treatment of benign prostatic hyperplasia. Answer choices C and F refer to the same enzyme and vorozole is not the target of CYP11A/desmolase. The CYP17 and CYP21 are not inhibited by vorozole either, thus answer choices D and E are incorrect.

3. **The correct answer is A**. In the synthesis of aldosterone, the conversion of corticosterone to aldosterone is catalyzed by 18-hydroxylase, which is upregulated by angiotensin II. Therefore, deficiency of this enzyme would decrease the synthesis of aldosterone. Answer choices B and E are incorrect because these intermediates are upstream of aldosterone synthesis and the levels would increase. Answer choices C and D are incorrect because 18-hydroxylase is the last step of the aldosterone synthesis pathway and, therefore, does not affect the cortisol pathway.

4. **The correct answer is A**. Addison disease is caused by autoimmune destruction of the adrenal cortex, which results in low aldosterone production. Since the function of aldosterone is to increase sodium reabsorption, potassium secretion, and acid secretion in the distal collecting ducts of the kidney, low levels of aldosterone results in higher sodium and water excretion, blood volume decrease, and hypotension.

Answer choice B is incorrect because Conn syndrome, also known as primary hyperaldosteronism, is due to excess aldosterone production from tumors of the adrenal cortex. These patients suffer from uncontrollable hypertension, hypokalemia, hypernatremia and metabolic alkalosis. Answer choice C is incorrect because 11-β-hydroxylase deficiency causes hypertension. Patients who have genetic deficiency of the 11-β-hydroxylase enzyme cannot produce cortisol, aldosterone, and corticosterone. This absence of cortisol production causes excess release of ACTH and overstimulation of the adrenal cortex. In the synthesis of aldosterone, the intermediate 11-deoxycorticosterone accumulates. This intermediate acts like aldosterone and, therefore, patients develop hypertension.

Answer choice D is incorrect because Cushing syndrome patients suffer from elevated levels of cortisol, which is an antagonist of insulin. Cortisol does not have effects on blood volume or sodium and water homeostasis.

5. **The correct answer is B**. Pituitary or adrenal tumors cause development of endogenous Cushing syndrome, which is associated with increased cortisol levels. The increase in cortisol results from either stimulation of the adrenal cortex or excess secretion of ACTH from the pituitary. Cushing syndrome can also develop in patients who have been treated with high levels of glucocorticoids. This condition is called exogenous Cushing syndrome.

VI

Answer choice A is incorrect because aldosterone levels do not change. CRH is secreted from the hypothalamus and Cushing syndrome patients do not have elevated levels of CRH. Thus,

Answer choice C is incorrect. Answer choices D and E are incorrect because FSH and LH are not associated with Cushing syndrome.

VI

45 Nutrition and Metabolism

At the conclusion of this chapter, students should be able to:
- Describe nutritional recommendations and the macronutrient composition of the diet
- Explain the role and health effects of dietary carbohydrates, proteins, and fats
- Describe the health benefits of dietary fiber
- Explain how the body uses energy during the resting (basal) metabolic rate, physical activity and for generating body heat
- Explain the effects of thyroid hormone in regulating the metabolic rate
- Explain mechanisms of hormones that are involved in the regulation of eating, controlling appetite, and weight homeostasis
- Explain the consequences of under-nutrition and excess energy intake (obesity)
- Analyze the mechanisms of insulin resistance in the development of metabolic syndrome and type-2 diabetes

The main focus of this chapter is to discuss the impact of nutrition on health and disease. Maintaining proper nutrition is a continuous effort for every human being and requires a balanced diet in every stage of our life. An unbalanced diet on the other hand, in the form of either under-nutrition or over-nutrition, is a concern for both individuals and their health care providers. The main question for nutrition researchers is "what is the ideal macronutrient balance for the diet?"

There are many factors that influence the ratios of carbohydrates, fat, and protein intake. Thus, nutritional recommendations can vary vastly between individuals, especially in children and pregnant woman. For example, there is a wide variability in estimating children's nutritional needs. Caloric need differs child to child and should be tailored individually for each child. During pregnancy, healthy eating habits are essential for both the mother and the baby. Key factors to consider are adequate weight gain, the consumption of unprocessed foods, appropriate vitamin and mineral supplementation, avoidance of harmful substances like alcohol and tobacco, and strict adherence to safe food handling guidelines. Sufficient intake of folate, calcium, iron, iodine and vitamin D is also very important during pregnancy.

The role of thyroid hormone in regulating the metabolic rate is also an important concept reviewed in this chapter. Another area of interest in nutrition research concerns the hormonal mechanisms necessary for the regulation of the appetite and eating. The results of these studies could potentially lead to the development of new medicines, especially for weight reduction. And finally, we review the consequences of an unbalanced diet and excess energy intake in the development of insulin resistance and type-2 diabetes.

45.1 Recommendations for Macronutrient Composition of the Diet

Dietary recommendations for adults are based on many randomized, controlled trials. In these recommendations, maintaining caloric balance is one of the most important factors. It is generally recommended that a "normal weight" individual should have a dietary intake that is equal to their energy expenditure. In addition to the amount of energy, the type of macronutrients consumed is an important factor. For example, consuming unsaturated fats from fish is more beneficial than a burger fried in trans-oil.

Good health can be maintained by many different types of diets, but in general, a diet that consists of the daily consumption of fruits and vegetables, replaces refined grains (white bread, sweetened cereals, white rice) with whole grain options (bread, cereals, brown rice), and minimizes red meat and meat products but includes the consumption of fish has proven to be the most beneficial, especially for cardiovascular health.

The USDA publishes dietary health guidelines every 5 years. Recent USDA guidelines suggest that individuals need to combine healthy choices from all food groups (fruits, vegetables, grains, dairy, oils and protein). (https://health.gov/dietaryguidelines/2015/resources/2015–2020_Dietary_Guidelines.pdf). We will review these food groups in the context of carbohydrates, proteins, and fats.

45.2 Carbohydrates and Their Health Effects

The primary role of carbohydrates is to provide energy for the cells in our body. Carbohydrates produce 4 kcal/g of energy when digested. If there is excess energy, the carbons of the carbohydrates are either stored as glycogen or converted to triacylglycerols. Both of which are stimulated by the insulin hormone.

The carbohydrates in the diet can be categorized as either simple or complex based on the number of sugar molecules that are contained in their chemical structures. While simple sugars refer to monosaccharides (glucose, galactose, and fructose) and disaccharides (sucrose, lactose, and maltose), the complex carbohydrates in the diet are polysaccharides and fiber.

Simple sugars, such as glucose, are the main source of energy in our body. Glucose is primarily stored as glycogen in the liver and the skeletal muscle. As reviewed in Chapters 24 and 31, the liver is the only organ that breaks down its stored glycogen to supply glucose to other tissues during a fasting state. The other two monosaccharides, fructose and galactose, are mainly metabolized by the liver. The metabolism of fructose and galactose is reviewed in Chapter 23.

45.2.1 Link to Pathology: Is High-Fructose Corn Syrup a Cause of Obesity?

High-fructose corn syrup (HFCS) is enzymatically processed corn syrup, which has been used as a sweetener since the early 1970's in many beverages. In this process, sucrose's glucose is converted to fructose by the enzyme isomerase. The advantages of using HFCS over traditional sweeteners are its ease of storage and its cost-effectiveness. Although fructose is also found in table sugar, clinical studies suggested that consumption of HFCS is strongly linked to obesity. The link between high fructose consumption and obesity can be explained by the metabolism of fructose. As the pathway shows (Chapter 23), fructose metabolism bypasses the rate-limiting step of glycolysis and produces intermediates that can be used in the synthesis of tri-acylglycerols in the liver.

Polysaccharides are polymers of glucose molecules. While starch and glycogen are similar in structure and digested by amylase, the fibers (cellulose, lignin, pectin, and gums) are non-digestible polysaccharides. Even though fiber cannot be digested in the gastrointestinal system, it has many health benefits. Insoluble or non-fermentable fiber absorbs water and increases the bulk of stool. This provides a softening of the stool, which helps in bowel movements, or defecation, and decreases the risk for diverticulosis. Examples of insoluble fiber include wheat bran, fruits and vegetables. Soluble or fermentable fiber also softens the stool and increases fecal bacterial mass. Examples of soluble fiber include oat bran, psyllium seeds, and fruits. Dietary fiber can bind to potential carcinogens, such as lithocholic acid, estrogens, and cholesterol, and help to excrete them in the stool.

45.3 Proteins and Their Health Effects

Most proteins in our body have either structural or functional roles, and in some cases both. The recommended daily intake of protein in adults is 0.8 – 1 gram per kilogram of body weight. While proteins such as collagen are important structural components within the body, other proteins, including enzymes, hormones, plasma proteins, transporters, and antibodies, have a more functional role. Regardless of their function, all proteins are made of amino acids, which can be categorized as either essential or non-essential. Essential amino acids are phenylalanine, valine, threonine, tryptophan, isoleusine, methionine, histidine, arginine, leucine, and lysine. Although proteins are essential to produce and maintain muscle, a portion of this muscle protein is broken down during the fasting state and resynthesized during the fed state. During the fasting state, the carbons of the amino acids are converted to either glucose or ketones, while the amino groups are excreted as urea. The intake and excretion of nitrogen that occurs during the process of body protein turnover must be equalized,, a concept known as nitrogen balance.

Nitrogen balance refers to a situation in which the amount of nitrogen incorporated into the body each day exactly equals the amount excreted. **Positive nitrogen balance** occurs when the amount of nitrogen incorporated into the body exceeds the amount excreted. Therefore, positive nitrogen balance occurs during growth, pregnancy, and the recovery phase of an injury or surgery. **Negative nitrogen balance** occurs when nitrogen loss exceeds incorporation into the body. This is associated with protein malnutrition (kwashiorkor), a dietary deficiency of just one essential amino acid, starvation, uncontrolled diabetes, and infections.

45.3.1 Link to Pathology: Kwashiorkor and Marasmus

Kwashiorkor is a form of severe protein malnutrition with normal total caloric intake. This disease predominantly affects children. These patients have normal intake of carbohydrates but deficiency in protein and essential amino acids. Visceral protein becomes decreased because the liver is unable to synthesize albumin and other proteins. The disease is primarily due to deficiency of protein intake resulting from malnutrition, and free radical damage to endogenous proteins caused by infections. These patients tend to have a higher susceptibility to fungal, parasitic and bacterial infections. Clinical features of this disorder include pitting edema, ascites (fluid in the abdomen), massive hepatomegaly, diarrhea, anemia and a 'flaky' appearance of the skin. The massive hepatomegaly is the most prominent feature and it is caused by fatty liver. Development of the fatty liver is due to decreased synthesis of ApoB100, which is required for assembly and secretion of VLDL. Decreased release of VLDL causes accumulation of fat in the liver, causing it to be fatty.

Marasmus is energy malnutrition, which also mainly affects children. In this disease, there is total calorie deprivation with a dietary deficiency of both protein and carbohydrates. In these individuals, muscle proteins are degraded for fuel, with the liberated amino acids being used for gluconeogenesis to maintain blood glucose concentrations. These patients present with severe muscle wasting, and their extremities appear like 'broomsticks'.

An increased risk of infectious diseases in both Kwashiorkor and Marasmus patients results from impaired immune response due to insufficient protein/antibody synthesis.

45.4 Fats and Their Health Effects

Fats including triacylglycerol, phospholipids, and cholesterol have unique functions that are immensely important for our health. Triacylglycerols are the storage form of fat and contains ~9 kcal/gram of energy. The types of fat we consume daily influence the level of blood lipids and blood cholesterol. The different types of fatty acids are distinguished based on their number of double bonds, location of double bonds (ω-3 and ω-6) and cis versus trans configuration of the double bonds. **Saturated fatty acids** are mainly found in dairy, meat and some vegetable oils such as coconut and palm oil. The nutritional importance of saturated fats is their ability to increase blood LDL cholesterol. They are able to do so by increasing the formation of LDL and by decreasing LDL turnover. The lower carbon length saturated fatty acids are more prone to increasing LDL cholesterol than long chain fatty acids. **Monounsaturated fatty acids** are found mainly in vegetables (olive oil) and fish. These are a great substitute to consuming saturated fats. Monounsaturated fatty acids, such as oleic acid (18:1), actually lower plasma LDL cholesterol by increasing hepatic LDL receptor number and LDL turnover. **Polyunsaturated fatty acids (PUFA)** have more than one double bond and the location of the double bond is very important (ω-3 and ω-6).

The ω-6 PUFA: The first double bond is found at the sixth carbon from the methyl end. Sources include nuts, avocado, cotton seed, and corn oil. While linoleic acid (18:2) is an essential fatty acid arachidonic acid (20:4) can be synthesized in our body by elongation and desaturation processes (reviewed in Chapter 40).

The ω-3 PUFA: The alpha-linolenic acid (ALA - 18:3) is an essential ω-3 fatty acid that is found in flax seed, walnut, and spinach. The eicosapentaenoic acid (EPA - 20:5) and the docosahexaenoic acid (DHA - 22:6) are found in fish. The body can attempt to convert a small amount of ALA to EPA or DHA, but this is not a very efficient process in human metabolism. Therefore, a diet rich in EPA and DHA containing food, such as fish, is recommended. The health benefits of EPA and DHA include the prevention of arrhythmia and a lowering of the risk of depression, dementia, and arthritis.

45.5 The Ratio of ω-3/ ω-6 PUFA in Diet

The health benefits of ω-3 fatty acids differ from those of ω-6. The ω-3 PUFAs decrease the risk of heart diseases by decreasing platelet aggregation, inflammation, and arrhythmia. The ω-3 PUFAs also increase endothelial relaxation. The ω-6 PUFAs, on the other hand, can actually increase the risk of thrombotic events. This effect of ω-6 is due to the fact that arachidonic acid is the precursor for the pro-aggregating agent thromboxaneA2 (TXA2). Clinical studies have shown that diets rich in ω-3 PUFAs significantly decrease the risk of sudden cardiac death. Therefore, a diet with a high ratio ω-3 to ω-6 is recommended.

Trans-fatty acids are chemically classified as unsaturated, but behave like saturated fats. These fatty acids are made by the hydrogenation of liquid vegetable oils, also known as margarine. Trans-fats not only elevate serum LDL cholesterol, but also lower serum HDL cholesterol. Furthermore, as reviewed in Chapter 43, clinical studies have found that increased trans-fat intake not only increased blood LDL levels but also cause the elevation of Lp(a). Thus, trans-fat intake increases the risk of coronary artery diseases and has negative impact on health. For these reasons, it is recommended that all trans-fats be eliminated from the diet.

45.6 Cholesterol and Its Health Effects

Blood cholesterol levels (LDL, HDL, and the total cholesterol) are the most important indicators of coronary health, and there are many factors that influence these levels. Elevated LDL cholesterol is the primary risk factor for heart disease and atherosclerosis. Increased cholesterol production in the liver, an unhealthy diet (such as excess energy and trans-fat intake), and genetic factors contribute to the elevation of LDL cholesterol.

The process of elevated LDL-cholesterol causing atherosclerosis is mostly due to the oxidation of LDL. As we reviewed in Chapter 43, oxidized LDL is recognized as a foreign particle and taken up by macrophages. The macrophage receptors SR-AI & SR-A2 are involved in mediating the uptake of the LDL lipoproteins. Macrophages that are engorged with lipids transform into "foam cells", which participate in the formation of an atherosclerotic plaque.

The entire process starts with endothelial injury of muscular and elastic arteries by oxidized LDL, which allows monocytes to adhere to endothelial cells and move to the subendothelium (intima) and transform into macrophages. Once these monocytes transform into macrophages, they consume excess oxidized lipoproteins, causing them to become foam cells.

Accumulated foam cells release growth factors, cytokines, and matrix metalloproteinases, all of which induce the migration of smooth muscle cells from the media to the intima. In later stages, the matrix components collagen, proteoglycans, and elastin are produced, resulting in the formation of a mature fibrous plaque with a necrotic core. This structure is indicative of the pathognomonic lesion of atherosclerosis.

45.6.1 Link to Pathology and Pharmacology: Treating Hypercholesterolemia by Diet

The type of fat and the amount of carbohydrate (especially simple sugar) consumed has a more prominent effect on controlling plasma cholesterol levels than a sole reduction in dietary cholesterol intake. In fact, it is important to note that negative feed-back inhibition of HMG-CoA reductase can be accomplished by adequate amounts of dietary cholesterol. Therefore, a cholesterol-free diet may have little or no influence on plasma cholesterol levels. However, avoiding trans-fats, reducing dietary saturated fats, and increasing unsaturated fat intake has been found to be more helpful in reducing the risks of cardiovascular diseases. In addition, a diet rich in vegetables and fruits, which are rich in fiber, plant sterols, and stanols helps to reduce absorption of dietary cholesterol.

45.7 Thyroid Hormones Regulate the Metabolic Rate

Daily energy expenditure (DEE) is the amount of energy that is used by the body in one day. The three primary factors that determine how many calories are needed in one day include: the basal metabolic rate (BMR), physical activity, and the thermic effect of food.

Basal metabolic rate (BMR) is defined as the rate at which the body uses energy to support all of the involuntary activities that are necessary to sustain life. BMR attributes to 60% of the DEE. Examples of these involuntary activities include respiration, circulation, cardiac muscle contraction, and brain and neuronal functions. BMR changes with age, body weight, level of physical activity, and gender. Lean muscle mass increases BMR, because muscle cells are more metabolically active compared to adipocytes.

The other 40% of the DEE depends on the degree of physical activity and the thermic effect of food. DEE increases with increased physical activity. Individuals with a sedentary life style use an amount of energy that is almost equal to the BMR. The thermic effect of food, also known as diet-induced thermogenesis, occurs naturally after meals during the digestive process. Although the metabolic rate increases after meals, it constitutes only a small portion (~10%) of the DEE.

Thyroid hormones (T3 and T4) are the main modulators of the BMR. T3 and T4 are secreted from the thyroid gland upon stimulation by thyroid stimulating hormone (TSH). Free T3 is the active form of thyroid hormone in our body. Free T4 (FT4) is considered a "prohormone" because it is converted to free T3 (FT3) peripherally by the outer ring deiodinase. Although T4 is not an active hormone, it provides negative feedback for the release of TSH from the anterior pituitary and thyrotropin-releasing hormone (TRH) from the hypothalamus.

Thyroid hormones regulate energy expenditure, the rate of lipolysis, and protein degradation. They also regulate the rate of triacylglycerol and protein synthesis. These bipolar effects of thyroid hormones are accomplished by increasing insulin secretion from the pancreas, and by increasing secretion of norepinephrine from the sympathetic nervous system.

Norepinephrine stimulates ATP utilization and heat production by increasing synthesis of the uncoupling protein (UCP) thermogenin in brown adipose tissue. Norepinephrine in the adipocytes also increases lipolysis. In contrast to the norepinephrine, insulin increases glucose uptake, glycolysis, and triacylglycerol synthesis in adipose tissue.

In muscle cells, norepinephrine increases protein degradation. In contrast, insulin acts as an anabolic hormone to increase glucose uptake and protein synthesis.

In the liver, while the rates of glycolysis and cholesterol biosynthesis are increased by insulin, norepinephrine causes the rate of both glycogenolysis and gluconeogenesis to increase.

Increased release of norepinephrine by T3 consequently affects heart muscle through beta-adrenergic receptors and causes increased cardiac output and "nervousness".

45.7.1 Link to Pathology: Hyperthyroidism and Hypothyroidism

Thyroid hormone is the main regulator of the basal metabolic rate in the human body. Therefore, effects on the levels of these hormones can result in significant changes in the body. Individuals with too much thyroid hormone experience anxiety, heart palpitations, sweating, and heat-intolerance and can even develop atrial fibrillation. This condition is called hyperthyroidism. On the other end of the spectrum, individuals with a deficit of thyroid hormone feel lethargic, tired, fatigued, and experience constipation. This is called hypothyroidism and can cause cold-intolerance and weight gain.

45.8 Hormones that Regulate Eating, Appetite, and Weight Homeostasis

While thyroid hormones have huge effects on the regulation of metabolism, there are other hormones and factors that regulate eating, appetite and, consequently, weight homeostasis. For example, appetite is satisfied by chewing, swallowing and a full stomach. GI-derived hormones indirectly affect fuel metabolism through their ability to influence the digestion, absorption and assimilation of ingested nutrients.

Gastrin, secreted by stomach, induces gastric acid secretion. **Ghrelin**, a GI-derived hormone, functions as neuropeptide in the central nervous system and induces a feeling of hunger. **Motilin** is secreted by the M-cells in the duodenum and jejunum, and stimulates gastric and pancreatic enzyme secretion. **Pancreatic polypeptide (PP)** reduces gastric emptying and slows upper intestinal motility. **Peptide YY (PYY)** inhibits pancreatic secretion and gastric motility. **Secretin** regulates the pH of the intestine by regulating secretion of gastric acid from the stomach and secretion of bicarbonate from the pancreas. **Cholecystokinin**, a GI-derived hormone, functions in the intestine and also as a neuropeptide in the brain. Cholecystokinin, as a neuropeptide, stimulates satiety and has an appetite suppressing effect. **Incretin hormones**, glucagon-like peptide1 (**GLP-1**), and glucose-dependent insulinotropic polypeptide (**GIP**) are important in enhancing the synthesis and release of insulin. GLP-1 also inhibits glucagon secretion.

Leptin is produced by adipocytes and acts as neuropeptide. This hormone stimulates satiety, suppresses appetite, and increases energy expenditure. The paraventricular and arcuate nucleus of the hypothalamus control food intake and energy expenditure. Leptin stimulates the anorexigenic (appetite suppressant) neurons in the arcuate nucleus, and inhibits the orexigenic (appetite stimulator) neurons. Leptin activates the anorexigenic neurons in the arcuate nucleus to express proopiomelanocortin (POMC). The POMC is the precursor of α-melanocyte stimulating hormone (α-MSH), which is released from POMC expressing anorexigenic neurons of hypothalamus. The α-MSH binds to its receptors MC3-R and MC4-R at the paraventricular nucleus (PVN). The binding of α-MSH to its receptors activates the melanocortin signaling pathway, which inhibits food intake and increases energy expenditure. The most common genetic mutations of inherited obesity have been linked to the MC4-R receptor mutations, which are inherited in a co-dominant fashion. Thus the melanocortin signaling pathway is an important contributing factor to inherited obesity.

Anorexigenic neurons not only suppress appetite, but they are also heavily involved in the activation of the sympathetic nervous system. Activation of the sympathetic nervous system causes the release of cortisol and thyroid hormones. These hormones increase the rate of BMR and energy expenditure. In contrast to leptin, ghrelin activity increases appetite by stimulating orexigenic neurons.

45.9 Unbalanced Diet: Excess Energy Intake

Consumption of more energy than exertion causes weight gain and pathological conditions such as obesity. Obesity is defined as a body mass index (BMI) greater than $30 \, kg/m^2$, while overweight BMI is $> 25 \, kg/m^2$, the morbid obesity BMI is $> 40 \, kg/m^2$. The prevalence of obesity increases with aging. However, these levels off after around 60 years of age. The risk factors that can contribute to increased morbidity and mortality in obese patients are dependent upon the distribution of fat in the body. For example, fat around the waist or visceral fat increases morbidity and mortality significantly. Therefore, waist to hip ratio is considered to be a better indicator than BMI.

45.9.1 Link to Pharmacology: Weight Reduction Therapy with Pharmacological Agents

In general, weight reduction can be accomplished by exertion of more energy than consumption. However, due to environmental, cultural, or genetic factors weight reduction via caloric restriction may not be successful. Therefore, a variety of pharmacologic options have been developed. **Orlistat** is a pancreatic lipase inhibitor that greatly increases fecal fat excretion. **Lorcaserin**, a serotonin antagonist, decreases appetite and, therefore, individuals taking this drug tend to eat less. **Phentermine, a** sympathomimetic drug, functions by increasing norepinephrine in the neuronal cleft, which leads to appetite suppression.[1]

45.10 Obesity Causes Insulin Resistance and Metabolic Syndrome

The combination of obesity, insulin resistance, high triacylglycerol, low HDL and high blood pressure is known as **metabolic syndrome**. Metabolic syndrome is commonly observed in obese individuals since obesity is the leading cause of insulin resistance, leading to the other components of metabolic syndrome, dyslipidemia and hypertension.

45.11 Mechanisms of Insulin Resistance

Although obesity is the leading cause of insulin resistance, genetic mutations of insulin receptors or antibody binding to the receptors can also create insulin resistance. In addition to deficiencies of insulin receptors, a malfunction in intracellular signaling molecules can create insulin resistance. Regardless of etiology, the consequence of insulin resistance is the same: cells are unable to respond to insulin.

Insulin resistance more often develops in obese individuals because obesity can cause excess free fatty acids (FFAs) and excess adipokines (TNF-α) in the circulation. The source of excess FFAs can be either the diet or adipocytes. The larger adipocytes in obese individuals contribute more FFA release compared to normal-weight individuals. Also, visceral fat deposits have higher rates of lipolysis that contributes to elevation of FFAs in the circulation. Blood FFA levels are also increased by the action of the **adipokine hormone TNF-α,** which is secreted from adipocytes. TNF-α elevates blood FFA levels by increasing lipoprotein lipase activity. TNF-α also activates SREBP in the liver, which increases VLDL production, contributing to dyslipidemia.

Elevated levels of FFAs in the circulation accumulate in the muscle and in the liver, leading to insulin resistance in these tissues. Insulin resistance in the muscle causes decreased uptake of glucose by GLUT4, while insulin resistance in the liver not only decreases the glucose uptake by GLUT2, but also increases the rate of fatty acid oxidation. Increased fatty acid oxidation increases the NADH/NAD ratio and acetyl CoA concentrations in the hepatocytes. While an increased NADH/NAD ratio decreases the rate of pyruvate dehydrogenase activity, increased acetyl CoA increases the rate of pyruvate carboxylase. Therefore, the rate of gluconeogenesis increases and, consequently, the output of glucose into the blood increases.

Overall, decreased uptake of glucose by GLUT4 and increased production of glucose by the liver leads to hyperglycemia.

To overcome the insulin resistance the pancreas produces more insulin, resulting hyperinsulinemia. Unfortunately, hyperinsulinemia increases the endocytosis and degradation of insulin receptors. The reduced glucose utilization and increased gluconeogenesis raises blood glucose levels further and further, causing **hyperglycemia**, which stimulates the sympathetic nervous system, leading to sodium and water retention and vasoconstriction. All of these events lead to the development of hypertension.

In summary, obesity is associated with insulin resistance, which causes hyperinsulinemia, dyslipidemia, i.e. increased triglycerides and decreased HDL, and hypertension. All of which in turn increases the risk of cardiovascular diseases, hypertension, atherosclerosis, certain types of cancer, and type-2 diabetes. Hyperinsulinemia leads to development of certain types of cancers because; excess insulin can bind to insulin-like growth factor (IGF) receptors, which stimulates the MAPK pathway and cellular growth.

45.11.1 Link to Pharmacology

Although, there is a variety of pharmacological options for weight reduction, exercise and fasting remain the most effective methods due to their up-regulation of the insulin receptor. As reviewed in Chapter 30, exercise also increases glucose uptake in the muscle cells.

45.12 Type-2 Diabetes Mellitus

Type-2 Diabetes Mellitus is a metabolic disorder characterized by hyperglycemia. It is most common in individuals over 40 years of age who are overweight, but it can also be inherited through genetic factors, which are generally multifactorial. Therefore, the three common risk factors are obesity, older age, and a family history of diabetes.

The simplified mechanism of type-2 diabetes is that the pancreas cannot secrete adequate levels of insulin to overcome insulin resistance. Because the cells do not respond to insulin appropriately, glucose uptake is impaired and hyperglycemia ensues, leading to type-2 diabetes. Insulin resistance leads to development of type-2 diabetes after ~5–15 years; insulin levels become elevated further and further as an attempted compensatory mechanism by the pancreatic beta cells. However, due to degeneration of the beta-cells, pancreas cannot produce enough insulin to compensate the receptor resistance anymore; thus, the type-2 diabetes develops.

45.12.1 Link to Pathology

Diabetic complications are caused by hyperglycemia. Chronic hyperglycemia damages vascular endothelial cells, the retina, renal glomerulus, and peripheral nerve cells. As a result, diabetes is associated with atherosclerosis, cardiovascular complications, nephropathy, retinopathy, cataracts, neuropathy, decreased sensation in the feet, and skin ulcers on the heels and toes.

Increased flux of the polyol pathway, increased non-enzymatic glycation of proteins, hyperglycemia-induced activation of PKC pathway, over-modification of proteins by N-acetylglucosamine, and oxidative stress are the major underlying molecular mechanisms in the development of diabetes complications.

VI

45.12.2 Link to Pharmacology: Diabetic Medications and Mechanisms of Actions

Biguanides (metformin) is an inhibitor of NADH dehydrogenase (Complex-I) in the ETC. The inhibition of Complex-I increases the AMP levels in the cell. Increased AMP activates AMP-dependent kinase (AMPK), which increases the concentrations of GLUT4 on the muscle cells, thereby stimulating glucose entry into the muscles. In the liver, activated AMPK reduces the rate of gluconeogenesis.

Alpha-glucosidase inhibitors (acarbose) block the digestion of complex carbohydrates, resulting in a loss of glucose absorption in the intestines.

Sulfonylureas (gliclazide, glimepiride, glyburide) stimulate release of insulin by closing the ATP-sensitive K-channels in the beta-cells of the pancreas.

SGLT-2 inhibitors (dapaglifozine, empagliflozine) inhibit the reabsorption of glucose by inhibiting SGLT-2 in the kidney, resulting in the elimination of glucose through the urine.

Thiazolidinediones (rosiglitazone, pioglitazone) increase adiponectin levels through binding to peroxisome proliferator activated receptor-Υ (PPAR-Υ) in the adipocytes. Increased adiponectin decreases FFA levels in the circulation, thus decreasing insulin resistance in the muscle cells.

Glucagon-like peptide-1 (GLP-1) agonists (liraglutide, dulaglutide, exenatide) are derivatives of GLP-1. These drugs bind to GLP-1 receptors and stimulate insulin secretion.

Dipeptidyl-peptidase-4 (DPP-4) inhibitors (linagliptine, sitagliptine, saxagliptine, alogliptine) inhibit the degradation of GLP-1, thus also stimulating insulin secretion.

Reference

[1] Kang JG, Park C-Y. Anti-Obesity Drugs: A Review about Their Effects and Safety. Diabetes Metab J. 2012; 36(1):13–25

VI

Review Questions

1. Increased levels of which of the following in the circulation contributes to the development of insulin resistance?
   A) Amino acids
   B) Cholesterol
   C) Glucose
   D) Free Fatty Acids
   E) Proteins

2. Which of the following hormones decreases appetite?
   A) Gastrin
   B) Ghrelin
   C) Leptin
   D) Motilin

3. Insulin resistance in muscle cells is most likely decreased by which of the following treatment options for type-2 diabetes?
   A) Acarbose
   B) Dapaglifozine
   C) Empagliflozine
   D) Liraglutide
   E) Rosiglitazone

4. Which of the following hormones has the greatest effect on the basal metabolic rate (BMR)?
   A) Glucagon
   B) Insulin
   C) T3
   D) T4

Answers

1. **The correct answer is D**. Elevated levels of FFAs in the circulation cause insulin resistance, especially in the muscle and in the liver. The insulin resistance in the muscle causes decreased uptake of glucose by GLUT4. Insulin resistance in the liver not only decreases the glucose uptake by GLUT2, but also increases the rate of fatty acid oxidation. Insulin resistance causes elevation of blood glucose because both the muscle and the liver cannot utilize glucose for energy but, instead, generate energy from fatty acid oxidation. Thus, answer choice C is incorrect. Answer choices A and E are incorrect because blood levels of amino acids or proteins are not known to affect insulin-receptor interactions. Increased level of blood cholesterol (hypercholesterolemia) is associated with vascular damage but not with insulin resistance, thus, answer choice B is incorrect.

2. **The correct answer is C.** Leptin decreases appetite by stimulating the anorexigenic (appetite suppressant) neurons in the arcuate nucleus. In contrast to leptin, ghrelin (answer choice B) stimulates appetite through actions on the anorexigenic neurons and orexigenic neurons.

   Gastrin (answer choice A) induces gastric acid secretion, thus inducing hunger. Motilin (answer choice D) also stimulates gastric and pancreatic enzyme secretion. Therefore, answer choices A, B, and D are incorrect.

3. **The correct answer is E.** Rosiglitazone is in the class of thiazolidinediones, which bind to peroxisome proliferator activated receptor- Ƴ (PPAR- Ƴ) in the adipocytes, causing them to proliferate. The generation of new adipocytes allows the body to store fat safely in the new healthy fat cells. The activated PPAR- Ƴ pathway also increases adiponectin release from the adipocytes. Increased adiponectin decreases FFA release from the adipocytes and, therefore, FFA levels decrease in the circulation. Decreased FFAs in the circulation decrease insulin resistance in the muscle cells and, therefore, glucose uptake into the muscle is increased.

   Answer choice A is incorrect because acarbose is an alpha-glucosidase inhibitor, which inhibits the digestion of complex carbohydrates. Therefore, absorption of glucose from the intestine is inhibited. Answer choice B and C are incorrect because both of these drugs are SGLT-2 inhibitors, which inhibit the reabsorption of glucose by inhibiting SGLT-2 in the kidney, allowing for glucose elimination through the urine. Answer choice D is incorrect because liraglutide is glucagon-like peptide-1 (GLP-1) agonist, which stimulates insulin secretion.

4. **The correct answer is C.** Although both T3 and T4 are categorized as thyroid hormones, only free T3 is active. The free T4 (FT4) is considered an inactive "prohormone", which must be converted to T3 by the outer ring deiodinase in the peripheral tissues. Thus, answer choice D is incorrect. Free T3 regulates energy expenditure, rate of lipolysis, and protein degradation. It also regulates the rate of triacylglycerol and protein synthesis. These bipolar effects of thyroid hormones are accomplished by increasing insulin secretion from the pancreas, and by increasing secretion of norepinephrine from the sympathetic nervous system. Neither insulin nor glucagon by itself can have bipolar effects like the thyroid hormone T3. Therefore, answer choices A and B are incorrect.

VI

46 Vitamins and Minerals

At the conclusion of this chapter, students should be able to:
- Describe the functions of water-soluble and lipid-soluble vitamins in metabolism
- Distinguish among the disorders associated with vitamin A, D, E, and K deficiencies
- Distinguish among the disorders associated with thiamine, niacin, vitamin B6, folate, vitamin B12, and ascorbic acid deficiencies
- Describe the disorders associated with abnormal levels of copper

Vitamins are essential organic compounds that are not biosynthesized in the human body, but are required in small amounts for many metabolic reactions. Vitamins can be the precursors of coenzymes and regulatory molecules, or they can act as antioxidants. The dietary requirements for vitamins vary in each individual and are influenced by age, gender, and physiological conditions. The variations in physiological conditions can stem from pregnancy, breast-feeding, exercise, and nutrition.

Vitamins are divided into two categories: water-soluble and fat/lipid-soluble. The lipid-soluble vitamins include vitamins A, D, E, and K, while the water-soluble vitamins are all the B vitamins and vitamin C. Deficiency diseases can result from either the lack of vitamins in the diet or from malabsorption. On the other hand, vitamin toxicity can develop with excess intake of vitamin A, D, and B6. Vitamins A, D, E, and B12 can be stored in the body for relatively longer periods of time than the rest of the vitamins.

46.1 Lipid-Soluble Vitamins

Absorption of the lipid-soluble vitamins, A, D, E, and K require bile acids, dietary lipids, and micelle formation in the intestine. Therefore, absorption of these vitamins is easier when they are ingested with food (▶ Fig. 46.1).

46.2 Vitamin A

The parent compound of vitamin A is retinol, which can be further converted in the body to retinal and retinoic acid. Retinal and retinoic acid can also be produced from β-carotene, which is found in fruits and vegetables and functions as an antioxidant. Once it reaches the intestines it is cleaved to yield two molecules of retinal.

The 11-cis form of retinal is essential for vision, as it is a major component of the visual pigment of rhodopsin. Absorption of light by rhodopsin causes isomerization of 11-cis retinal to 11-trans retinal. This isomerization of retinal activates rhodopsin, which functions as a G-protein coupled receptor. Activated rhodopsin activates the G-protein transducin through the association with GTP. The GTP-bound transducin activates the phosphodiesterase enzyme, which hydrolyzes cGMP, thus decreases the intracellular cGMP concentration. A decrease in cGMP closes the sodium channels, which hyperpolarizes the membranes. Therefore, vitamin A deficiency can result in visual impairment and night blindness.

Retinoic acid is a gene regulator which is required for the maintenance of epithelial cells and for tissue differentiation. Inside the nucleus, retinoic acid binds to retinoic acid receptors and retinoid X receptors and regulates the expression of genes that affect growth, development, and differentiation of tissues. Retinoid X receptors can bind and form dimers with either the vitamin D receptor or the thyroid receptors. Therefore, vitamin A deficiency impairs the functions of vitamin D and thyroid hormones. Deficiency leads to a decrease in the growth rate in children as bone development is slowed, and can also affect spermatogenesis in males.

Retinoic acid also promotes growth and differentiation of epithelial cells and mucus secretion from columnar epithelial cells. Deficiency may result in metaplasia of the corneal epithelium, dry eyes, bronchitis, pneumonia, or follicular hyperkeratosis. As mentioned previously, an early sign of vitamin A deficiency is night blindness, with prolonged deficiency leading to irreversible loss of visual cells. Severe vitamin A deficiency can also lead to xerophthalmia, a pathologic dryness of the conjunctiva and cornea.

A deficiency of vitamin A can be treated by dietary supplements that contain retinol or its precursor. Acne, psoriasis, and skin aging can be treated with topical application of retinoic acid. Severe recalcitrant cystic acne can be treated with oral administration of 11-cis retinoic acid. However, since retinoic acid functions as a gene regulator, either deficiency or excess of vitamin A can be teratogenic, which could result in significant congenital malformations to the developing fetus.

46.3 Vitamin D

Calcitriol is the active form of vitamin D and functions to stimulate calcium absorption from the intestine. Vitamin D along with parathyroid hormone (PTH) and calcitonin regulate calcium homeostasis. This topic is reviewed in Chapter 44.

46.4 Vitamin E

Vitamin E (α-tocopherol) is an antioxidant that prevents peroxidation of unsaturated fatty acids on the membranes. Due to its antioxidant properties, vitamin E has a reputation for reducing the risk of cancer, coronary heart disease, and age related degenerative diseases. Alpha-tocopherol is the most potent form of vitamin E, which is mainly found in plants. Deficiency can result due to fat malabsorption, such as is seen in cystic fibrosis patients. Deficiency causes accumulation of free radicals, which can lead to hemolysis and neurologic problems.

46.5 Vitamin K

The dietary form of vitamin K is found in leafy vegetables and is known as phytomenadione, or K1. The active form of the vitamin K (menaquinones also known as K2) is produced by bacteria of the intestinal flora. Vitamin K is the coenzyme of the glutamate

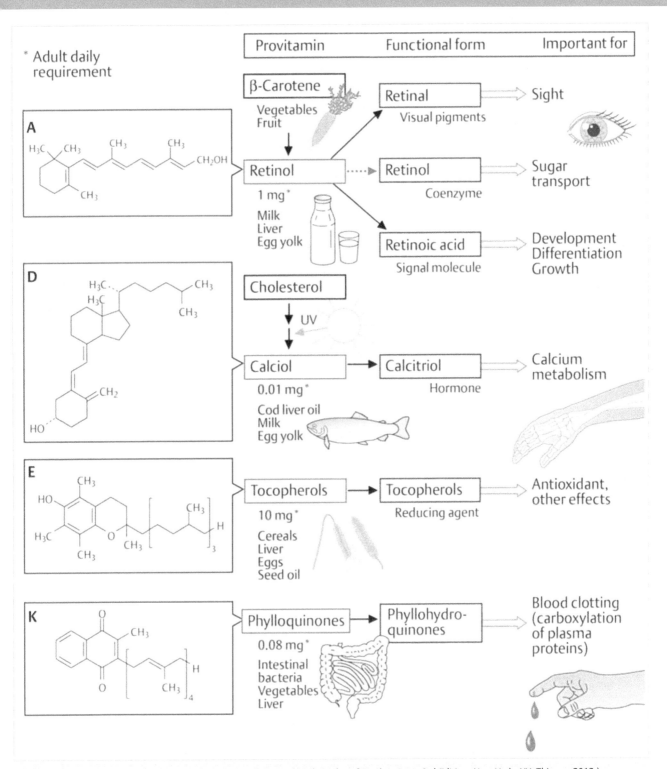

* Adult daily requirement

| Provitamin | Functional form | Important for |

A

β-Carotene
Vegetables
Fruit

Retinol
1 mg*
Milk
Liver
Egg yolk

Retinal
Visual pigments → Sight

Retinol
Coenzyme → Sugar transport

Retinoic acid
Signal molecule → Development Differentiation Growth

D

Cholesterol
↓ UV

Calciol
0.01 mg*
Cod liver oil
Milk
Egg yolk

Calcitriol
Hormone → Calcium metabolism

E

Tocopherols
10 mg*
Cereals
Liver
Eggs
Seed oil

Tocopherols
Reducing agent → Antioxidant, other effects

K

Phylloquinones
0.08 mg*
Intestinal bacteria
Vegetables
Liver

Phyllohydro-quinones → Blood clotting (carboxylation of plasma proteins)

Fig. 46.1 Lipid-soluble vitamins. (Source: Koolman J, Röhm K, ed. Color Atlas of Biochemistry. 3rd Edition. New York, NY. Thieme; 2012.)

VI

carboxylase enzyme, which catalyzes the γ-carboxylation of glutamate residues of the coagulation factors II, VII, IX, X, and proteins C and S. The γ-carboxylation of glutamate is a co-translational modification, which creates calcium binding sites in these blood clotting factors. Therefore, deficiency of vitamin K presents with hemorrhagic conditions resulting from decreased formation of γ-carboxyglutamate residues on the blood clothing factors, which leads to a non-functional coagulation pathway. This can cause increased prothrombin time and easy bruising.

Vitamin K deficiency is most common in newborns due to the inadequate intestinal flora of newborns, in conjunction with the low levels of vitamin K in breast milk. Infants with vitamin K deficiency present with abnormal bleeding in the umbilical stump. If not treated, deficiency can cause hemorrhagic disease of the newborn, which can lead to intracranial hemorrhages. Vitamin K deficiency is also associated with long-term broad spectrum antibiotic therapy or with conditions that cause fat malabsorption. The most important lab test for evaluating vitamin K status is the prothrombin time test.

46.5.1 Link to Pharmacology

The vitamin K antagonists warfarin and dicumarol inhibit blood coagulation by inhibiting γ-carboxylation of the coagulation factors during translation. Thus, they are used as anticoagulant therapy in individuals who are at high risk for thrombotic events.

46.6 Water-Soluble Vitamins

All of the water-soluble vitamins, except B12, are absorbed from the jejunum of the upper small intestine. The water-soluble vitamins include thiamine (B1), riboflavin (B2), niacin (B3), pantothenic acid (B5), pyridoxine (B6), biotin (B7), folate (B9), cyanocobalamin (B12), and vitamin C (▸ Fig. 46.2).

46.7 Thiamine (B1)

Vitamin B_1 (thiamine)'s active form is thiamine pyrophosphate (TPP). TPP is required as a coenzyme for the catalysis of pyruvate dehydrogenase (PDH) in the conversion of pyruvate to acetyl CoA, and alpha-ketoglutarate dehydrogenase in the TCA cycle for the generation of succinyl CoA. TPP is also a coenzyme for the branched-chain ketoacid dehydrogenase enzyme in the metabolism of branched-chain amino acids, and the transketolase enzyme in the pentose phosphate pathway. The activities of PDH and alpha-ketoglutarate dehydrogenase are important in energy metabolism, particularly in the cells of the nervous system.

Although deficiency can be due to inadequate intake, chronic alcoholics are more susceptible to thiamin deficiency for two reasons: alcohol inhibits the absorption of thiamine, and excretion of thiamine is increased. Moderate deficiency can result in dry beriberi, which presents with peripheral neuropathy, muscle weakness, fatigue, and shorter attention span. Severe deficiency results in wet beriberi characterized by high-output cardiac failure and edema. Patients with beriberi can present with dry skin, irritability, disordered thinking, forgetfulness, neurological disturbances, cardiac insufficiency, muscular atrophy, and paralysis. The combination of thiamine deficiency and alcohol toxicity causes Wernicke-Korsakoff syndrome. This is characterized by mental derangement, loss of memory, delirium, ataxia and paralysis of the eye muscles. These patients may also present symptoms of confusion, hallucinations, disorientation, and frenzy.

46.8 Riboflavin (B2)

Vitamin B_2 (riboflavin) is a precursor for the cofactors flavin adenine dinucleotide (FAD) and flavin mononucleotide (FMN). These molecules serve as coenzymes for several dehydrogenases and are electron carriers in oxidation/reduction reactions in the body. Deficiency in vitamin B2 can be caused by a poor diet or may be seen in infants receiving photo therapy to treat jaundice. Deficiency can cause corneal neovascularization, cheilosis, glossitis, stomatitis, and a magenta colored tongue.

46.9 Niacin (B3)

Nicotinamide is required for biosynthesis of the coenzymes nicotinamide adenine dinucleotide (NAD) and nicotinamide adenine dinucleotide phosphate (NADP). These are hydrolyzed in the gastrointestinal tract and absorbed into the small intestine. They are found in whole grains, milk, and meat. Humans are also able to generate nicotinate from tryptophan, however at a very low yield. Niacin deficiency is known as pellagra and initially presents with lethargy, weakness, and indigestion. Patients also develop rough skin and pigmented dermatitis. Other symptoms include digestive disturbances, diarrhea, dementia and depression. Severe niacin deficiency is also known as 4D disease because of the signs and symptoms: dermatitis, diarrhea, dementia, and, if not treated, death.

46.9.1 Link to Pharmacology

Very high doses of niacin can be used for the treatment of hypercholesterolemia. It has been shown to reduce blood LDL by about 5 to 25% and increase blood HDL levels. This is accomplished through inhibition of lipolysis by reducing the activity of hormone-sensitive lipase in the adipose tissue and esterification of triglycerides in the liver. This leads to reduced VLDL synthesis in the liver, which in turn decreases blood LDL levels. Niacin also increases lipoprotein lipase (LPL) activity. Side effects of high doses of niacin include flushing, itching, gastrointestinal distress, hyperglycemia, hyperuricemia, and hepatotoxicity.

46.10 Pantothenic Acid (B5)

Vitamin B_5 (pantothenic acid) is the precursor of coenzyme A. Coenzyme A contains a thiol group that carries acyl compounds as activated thiol esters. Examples include succinyl CoA, fatty acyl CoA, and acetyl CoA. Coenzyme A is also a component of the acyl carrier protein in fatty acid synthase. Deficiency of vitamin B_5 is rarely observed in humans.

46.11 Pyridoxine (B6)

Vitamin B_6 (pyridoxine and pyridoxal phosphate) is used as coenzyme in many of the reactions of amino acid metabolism. It is the coenzyme for aminotransferases (e.g. ALT and AST), decarboxylases (e.g. dopa decarboxylase), and delta-amino levulinic acid (ALA) synthase. Deficiency is mostly seen in alcoholics who present with cheilosis, stomatitis, or sideroblastic anemia. Other symptoms include peripheral neuropathy, glossitis, irritability, psychiatric symptoms, and epileptic seizures.

Unlike other water-soluble vitamins, vitamin B6 is toxic in high doses. Excess can lead to peripheral neuropathy, nerve damage that can cause weakness, and loss of sensation.

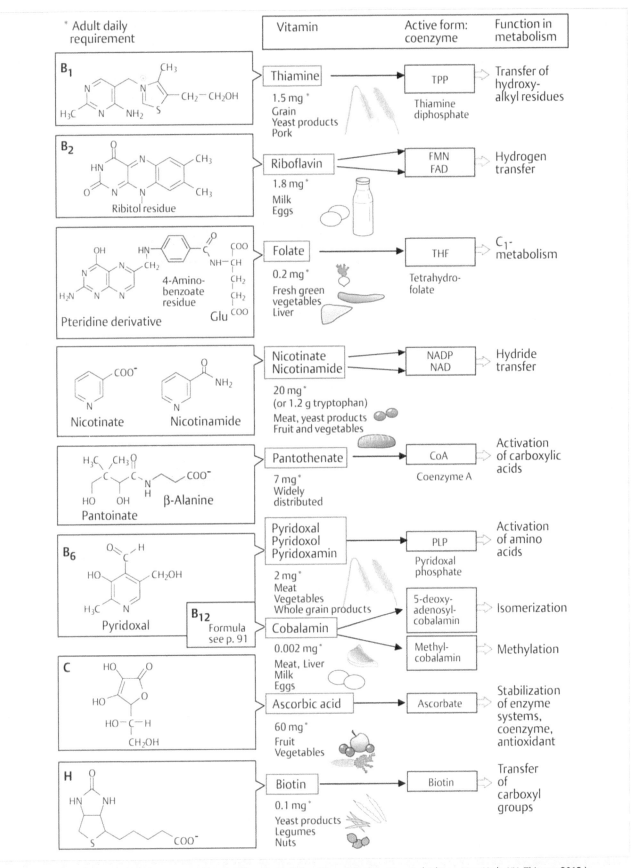

Fig. 46.2 Water-soluble vitamins. (Source: Koolman J, Röhm K, ed. Color Atlas of Biochemistry. 3rd Edition. New York, NY. Thieme; 2012.)

46.12 Biotin (B7)

Vitamin B_7 (biotin) is an important coenzyme in carboxylation reactions. Pyruvate carboxylase, acetyl CoA carboxylase, and propionyl CoA carboxylase enzymes require biotin as a coenzyme. Non-dietary biotin deficiency can be caused by the deficiency of the enzyme biotinidase, which cleaves the biotin-lysine conjugate biocytin. Therefore, biotinidase deficiency is often included in newborn screening. Another protein called avidin is found in egg whites and has high affinity for biotin and prevents its absorption. Many athletes consume uncooked egg white and, therefore, they are at risk for developing biotin deficiency. Symptoms of biotin deficiency include: scaly dermatitis, alopecia, and glossitis, loss of appetite, abdominal discomfort and nausea.

46.13 Folate (B9)

Vitamin B_9 (folic acid) is a coenzyme for thymidylate synthase and the purine synthesis enzymes. Folate is rich in leafy green vegetables and fruits such as berries. The active form of folate is tetra hydro folate (FH4), which is synthesized by dihydrofolate reductase (DHFR). Deficiency of folic acid causes megaloblastic anemia and neural tube defects. Due to insufficient nucleotide synthesis that results in impaired DNA replication, rapidly dividing cells are most affected. The functions of folate as a one-carbon carrier is reviewed in Chapter 37, and vitamin B12 in the synthesis of S-adenosyl methionine (SAM) is reviewed in Chapter 38.

46.13.1 Link to Pharmacology: Importance of Folic Acid During Pregnancy

The incidence of neural tube defects such as spina bifida, anencephaly, and myelomeningocele can be prevented if a woman takes supplemental folic acid before and after conception.

46.14 Cobalamin (B12)

Vitamin B_{12} (cobalamin) is a required coenzyme for homocysteine methyltransferase and methyl-malonyl-CoA mutase. It is exclusively found in animal products and can only be absorbed in the small intestine when the gastric mucosa secretes R-binders and intrinsic factor. Dietary deficiency of B12 is rare because the liver can store vitamin B12 in large amounts. When deficiency is present, it is usually a consequence of impaired secretion of intrinsic factor, malabsorption such as celiac disease, pancreatic insufficiency, or genetic deficiency in cobalamin receptors.

Deficiency causes pernicious anemia, which presents with two types of abnormalities: megaloblastic anemia, and neurological dysfunction with demyelination. Thus, patients with pernicious anemia present with peripheral neuropathy.

46.15 Vitamin C

Vitamin C (ascorbic acid) is the coenzyme for the prolyl and lysyl hydroxylase enzymes, which are required for the conversion of procollagen to collagen. Hydroxylated prolyl and lysyl residues are required for fibril cross-linking of collagen fibers. Therefore, vitamin C functions in the maintenance of normal connective tissue and wound healing. It is also used as a coenzyme in tyrosine degradation, bile acid synthesis, and catecholamine synthesis pathways. Vitamin C, as a reducing agent, is required for absorption of iron from the intestines. It is also required in many iron and copper containing enzymes to keep their metal in a reduced state. For example, dopamine β-hydroxylase is a copper containing enzyme, which requires vitamin C for the reduction of copper. Vitamin C, as water-soluble antioxidant, is the scavenger of free radicals in the blood.

Vitamin C is abundant in fruits and vegetables. Deficiency can develop in 2 to 3 months if the diet lacks vitamin C-containing foods. Deficiency is also known as scurvy and is characterized by dry mouth and eyes, decaying, peeling gums, loose teeth, delayed wound healing, ecchymosis and petechiae, weakness, and lethargy. The development of most of these symptoms is due to disrupted collagen structures in the blood vessel walls.

46.16 Minerals

The minerals essential for life can be divided into macroelements (daily requirement > 100 mg) and microelements, which are required in trace amounts of less than 100 mg). The macroelements are sodium, potassium, calcium, magnesium, chlorine, phosphorus, and sulfur (▶ Table 46.1). Almost all macroelements in the body function either as nutrients or electrolytes such as sodium, potassium, calcium, and magnesium. Microelements such as iodine and macroelements such as calcium function in signaling cascades either as ligands or secondary messengers. Most of the microelements such as zinc, copper, cobalt, chromium, selenium, and molybdenum serve as important cofactors for the activity of many enzymes. The functions and homeostasis of iron is reviewed in Chapter 36., while calcium homeostasis is reviewed in Chapter 44.

Zinc is found in more than 200 cellular proteins that have either structural or catalytic functions. Deficiency of zinc can arise from inadequate intake or from malabsorption and leads to hair loss, dermatitis, poor wound healing, and depressed immune function.

Copper is a cofactor of oxidases that transfer electrons to oxygen such as cytochrome c oxidase, superoxide dismutase, tyrosinase, and monoamine oxidase. Menkes syndrome results in a deficiency of copper due to mutations of an ATP-dependent membrane transporter for copper, which impairs the absorption of copper from the intestine. Menkes syndrome is inherited in an X-linked recessive manner and is characterized by microcytic hypochromic anemia, hemorrhagic vascular changes, bone demineralization, and neurological problems.

Wilson disease has the opposite effect, causing deposition of copper in the body. Autosomal recessive mutations in a kidney copper transporter impair the excretion of copper from the kidneys, which leads to excessive accumulation of copper in the brain, liver, and kidneys. Patients present with liver damage and neurological and hematological complications. It can be treated with penicillamine, which forms a soluble complex with copper and aids in excretion.

Table 46.1 Minerals

Mineral	Content[a] (g)	Major source	Daily requirement (g)	Functions/Occurrence
Water	35000–40000	Drinks Water in solid foods From metabolism (300g)	1200 900	Solvent, cellular building block, dielectric, coolant, medium for transport, reaction partner
Macroelements (daily requirement > 100 mg)				
Na	100	Table salt	1.1–3.3	Osmoregulation, membrane potential, mineral metabolism
K	150	Vegetables, fruit, cereals	1.9–5.6	Membrane potential, mineral metabolism
Ca	1300	Milk, milk products	0.8	Bone formation, blood clotting, signal molecule
Mg	20	Green vegetables	0.35	Bone formation, cofactor for enzyme
Cl	100	Table salt	1.7–5.1	Mineral metabolism
P	650	Meat, milk, cereals, vegetables	0.8	Bone formation, energy metabolism, nucleic acid metabolism
S	200	S-containing amino acids (Cys and Met)	0.2	Lipid and carbohydrate metabolism
Microelements (trace elements)			Mg	
Fe	4–5	Meat, liver, eggs, vegetables, potatoes, cereals	10	Hemoglobin, myoglobin, cytochromes, Fe/S clusters
Zn	2–3	Meat, liver, cereals	15	Zinc finger proteins, insulin storage, zinc enzymes
Mn	0.02	Found in many foodstuffs	2–5	Enzymes
Cu	0.1-.02	Meat, vegetables, fruit, fish	2–3	Oxidases
Co	<0.01	Meat	Traces	Vitamin B$_{12}$
Cr	<0.01		0.05–0.2	Not clear
Mo	0.02	Cereals, nuts, legumes	0.15–0.5	Redox enzymes
Se		Vegetables, meat	0.05–0.2	Selenium enzymes
I	0.03	Seafood, iodized salt, drinking water	0.15	Thyroxin
Requirement not exactly known				
F		Drinking water (fluoridated), tea, milk	0.0015–0.0004	Bones, dental enamel

[a]Content in the body of a 65 kg adult

Source: Koolman J, Röhm K, ed. Color Atlas of Biochemistry. 3rd Edition. New York, NY. Thieme; 2012.

VI

Review Questions

1. Which of the following vitamins is considered as antioxidant?
   A) Vitamin A
   B) Biotin
   C) Niacin
   D) Vitamin D
   E) Vitamin B12

2. Which of the following vitamins have effects on blood LDL cholesterol levels?
   A) Vitamin A
   B) Biotin
   C) Vitamin C
   D) Folate
   E) Niacin

3. Deficiency in which of the following compounds can lead to sideroblastic anemia?
   A) Cobalamin
   B) Pyridoxal phosphate
   C) Retinol
   D) Tetrahydrofolate
   E) Thiamine pyrophosphate

4. Which of the following diseases is caused by deficiency of thiamine?
   A) Beriberi
   B) Menkes syndrome
   C) Pellagra
   D) Pernicious anemia
   E) Wilson disease

5. Which of the following cellular mechanisms is most likely decreased by deficiency of B12?
   A) Degradation of proteins
   B) Formation of collagen
   C) Lipolysis
   D) Methylation of DNA
   E) Synthesis of heme

Answers

1. **The correct answer is A.** Free radicals and lipid peroxyl radicals can be quenched by carotenoids such as beta-carotene, lycopene, and lutein. Although vitamin E is considered the most effective antioxidant for scavenging peroxyl radicals of the cellular membranes, vitamin A is a lipid soluble antioxidant that can also reduce lipid peroxyl radicals. Since vitamin C is water soluble, it cannot quench lipid peroxyl radicals. However, vitamin C regenerates vitamin E, thus supporting the antioxidant properties of vitamin E. Answer choices B, C, D, and E are incorrect because these vitamins do not have any antioxidant activity.

Reference: Antioxidant functions of vitamins. Vitamins E and C, beta-carotene, and other carotenoids. Sies H, Stahl W, Sundquist AR. Ann N Y Acad Sci. 1992 Sep 30;669:7–20. Review.

2. **The correct answer is E.** A very high dose of niacin is used for treating hypercholesterolemia. It has been shown to reduce blood LDL about 5 to 25% and increase blood HDL levels. Niacin decreases the activity of hormone-sensitive lipase in the adipose tissue and esterification of triglycerides in the liver. This leads to reduced VLDL synthesis in the liver, which in turn decreases blood LDL levels. Answer choices A, B, C, and D are incorrect because these vitamins do not have any effect on lipid or cholesterol metabolism.

3. **The correct answer is B.** Vitamin B6 deficiency can lead to sideroblastic anemia because delta-amino levulinic acid (ALA) synthase, which is a component of the heme synthesis pathway, requires vitamin B6 (pyridoxal phosphate) as a coenzyme. Deficiency of cobalamin (B12) or tetrahydrofolate causes megaloblastic anemia. Therefore, answer choices A and D are incorrect. Answer choice C is incorrect because retinol (vitamin A) does not function in the synthesis of heme nor does it affect DNA replication in the pre-erythroid cells. Answer choice E is incorrect because thiamine pyrophosphate (TPP) is a coenzyme for PDH and alpha-ketoglutarate dehydrogenase, which are important enzymes in energy metabolism.

4. **The correct answer is A.** Deficiency of thiamine results in beriberi. Patients with beriberi can present with dry skin, irritability, disordered thinking, forgetfulness, neurological disturbances, cardiac insufficiency, muscular atrophy, and paralysis. The combination of thiamine deficiency and alcohol toxicity causes Wernicke-Korsakoff syndrome. This is characterized by mental derangement, loss of memory, delirium, ataxia and paralysis of the eye muscles. These patients may also present with symptoms of confusion, hallucinations, disorientation, and frenzy.

Answer choice B is incorrect because Menkes syndrome is caused by deficiency of copper due to insufficient absorption of copper from the intestines. Pellagra is caused by deficiency of niacin, thus answer choice C is incorrect. Pernicious anemia is caused by deficiency of vitamin B12. Therefore answer choice D is incorrect. Wilson disease is the accumulation of copper in the liver and brain, which is due to mutations in the kidney copper transporter. Thus, answer choice E is incorrect.

5. **The correct answer is D.** Vitamin B12 is a required coenzyme in the synthesis of S-adenosyl methionine (SAM), which is the one-carbon methyl (CH_3) donor in more than 35 reactions in the metabolism and is involved in the methylation of DNA. Therefore, deficiency of B12 decreases the synthesis of SAM and the methylation of DNA. Answer choice B is incorrect because vitamin C is a coenzyme in the collagen synthesis pathway. Lipolysis is not affected by B12, thus answer choice C is incorrect. Synthesis of heme requires vitamin B6 as a coenzyme, therefore answer choice E is incorrect.

Review Questions

1. A 14-year-old female presents with chronic abdominal pain, diarrhea and low-grade fever. Her mother reports to her doctor that she always suffered from digestive problems and diarrhea. Her medical history indicates that she has also had recurrent pulmonary infections. Physical exam reveals that her weight is in the 20th percentile and her height is the 10th percentile. Blood tests show that a metabolic panel, liver function tests, and complete blood count results are all normal. However, her serum amylase and lipase levels are slightly elevated. The results of genetic testing reveal that she is heterozygous for ?F508 and G551D mutations in the*CFTR* gene. Which of the following symptoms is expected to be present in this patient?
   A) ?-glucosidase deficiency
   B) Fatty liver
   C) Lactose intolerance
   D) Steatorrhea

2. A 29-year-old male with a history of chronic alcoholism presents with jaundice. He reports that he lives by himself and he never cooks at home. Physical examination shows abnormal gait and an inability to stand up straight. EEG results show evidence of encephalopathy. The liver function test results indicate that he has liver damage. A liver biopsy shows steatosis and fibrosis. Which of the following is most likely the explanation of the liver steatosis in this patient?
   A) Decreased lipase activity due to high ethanol intake
   B) Increased NADH/NAD ratio due to high ethanol intake
   C) High intake of saturated and trans fats in his diet
   D) Lack of essential fatty acids in his diet

3. A 57-year-old female with chronic Crohn's disease presents with dry, scaly skin. She also complains of hair loss. Her medical history reveals that she had her gallbladder removed about a year ago. She reports to her doctor that she has been following a low fat diet since her gallbladder surgery. Which of the following is most likely causing her skin problems?
   A) Deficiency of cholesterol
   B) Deficiency of essential fatty acids
   C) Deficiency of bile acid synthesis
   D) Deficiency of triacylglycerols

4. A 67-year-old female recovering from acute myocardial infarction is advised to take low dose (32 mg/day) aspirin to prevent future cardiovascular events. Which of the following statements best explains how low dose aspirin helps this patient?
   A) Inhibits the COX pathway
   B) Inhibits eicosanoid synthesis
   C) Inhibits the lipoxygenase pathway
   D) Inhibits prostaglandin synthesis
   E) Inhibits TXA2 synthesis in the platelets

5. A 45-year-old male with a history of Cushing syndrome complains of shortness of breath. He reports that he recently moved to the area and he thinks that pine trees around his house are causing his allergic reactions. Physical examination reveals wheezing sounds. Which of the following treatment options would be best for this patient?
   A) Aspirin
   B) Glucocorticoids

C) NSAIDs
   D) COX2 inhibitors
   E) Zileuton

6. A 3-month-old male presents with failure to thrive. His mother states that he cries a lot and it is difficult to feed him. Physical exam identifies hepatosplenomegaly. His motor development, however, is normal for his age. An ophthalmologic examination reveals cherry-red macula. Which of the following enzymes is most likely deficient in this patient?
   A) ? -galactosidase
   B) ? -glucosidase
   C) Hexosaminidase A
   D) Sphingomyelinase

7. A 34-year-old female presents with fatigue and severe joint pain. Physical examination reveals yellow fatty deposits on the sclera and yellowish-brown skin pigmentation. Her physician also notes hepatosplenomegaly. Complete blood count reveals low levels of erythrocytes, white blood cells and platelets. Microscopic analysis of the peripheral blood cells reports a 'wrinkled paper" appearance of the macrophages. Which of the following is the most likely diagnosis?
   A) Fabry disease
   B) Gaucher disease
   C) I-cell disease
   D) Krabbe disease

8. A 6-month-old male is brought to his pediatrician for a well-child visit. His mother expresses concern for his lethargy, as he is having difficulty lifting up his head and he is not able to sit up on his own. During physical examination it is noted that he is easily startled by sudden noises. His neuro motor development is slow for his age. An ophthalmologic examination reveals cherry-red macula. Which of the following is the most likely diagnosis?
   A) Metachromatic leukodystrophy
   B) Mucopolysaccharidosis
   C) Niemann-Pick disease
   D) Pompe disease
   E) Tay-Sachs disease

Answers

1. **The correct answer is D**. Since the patient has two mutations in the CFTR gene, which encodes for a chloride channel, the symptoms of this patient are most likely due to blockage of the exocrine glands, such as the pancreas and lungs. Cystic fibrosis patients suffer from pancreatic insufficiencies due to obstruction of the pancreatic exocrine duct. The CFTR secretes bicarbonate and chloride into the duodenum from the exocrine duct of the pancreas. The bicarbonate is important for neutralization of the acidic pH of the stomach contents. Due to decreased bicarbonate secretion, digestive enzymes cannot function in the low pH environment. Lipase is the most pH sensitive enzyme among the pancreatic enzymes. Malfunctioning of the CFTR in the pancreatic duct cells also blocks or decreases the secretion of digestive enzymes into the intestine. Therefore, cystic

fibrosis patients suffer from digestive problems, especially diarrhea and steatorrhea.

The answer choices A and C are incorrect because ?-glucosidase and lactase enzymes are synthesized and secreted from the brush border of the small intestines, not the pancreas. Fatty liver (answer choice B) is a very unlikely symptom of cystic fibrosis.

2. **The correct answer is B.** Metabolism of ethanol increases the NADH/NAD ratio in the liver. This high NADH increases the glycerol 3-phosphate concentrations, which results in enhanced esterification of fatty acids and formation of triacylglycerols (TG) in the liver. Fatty acids are increased because high NADH also inhibits fatty acid oxidation. Increased TG stimulates VLDL formation in the liver. However, release of VLDL into the blood is inhibited due to damage of the microtubules from acetaldehyde, an intermediate of ethanol metabolism. The increased fatty acids and TGs synthesis, decreased fatty acid oxidation, and VLDL release contributes to steatosis. Therefore, this condition is called alcohol-induced fatty liver disease or steatosis.

Answer choice A is incorrect because ethanol does not have a direct effect on pancreatic lipase activity, and also decreased lipase activity in the intestine would result in steatorrhea (fatty stool). Answer choice C and D are incorrect because a high saturated and trans fat diet increases cardiovascular disease risk. A dietary lack of essential fatty acids causes skin problems and alopecia.

3. **The correct answer is B.** Since she is suffering from Crohn's disease, which causes malabsorption, she may have deficiency of many vitamins and minerals. Also, since she doesn't have a gallbladder, she must avoid high fat meals. While pancreatic lipase and colipase digest fats into free fatty acids and monoacylglycerol, bile acids and salts pack them into micelles. Micelles are absorbed into the enterocytes. Bile acids and salts stored in the gallbladder are secreted into the duodenum after meals. Due to her condition, she is most likely deficient in essential fatty acids, which are important for membrane lipid biosynthesis. Essential fatty acid deficiency generally presents with scaly, dry skin and alopecia in adults. The deficiency of essential fatty acids causes growth problems and intellectual disability in children.

Answer choices A and D are incorrect because the body can synthesize cholesterol and triacylglycerols. Answer choice C is incorrect because bile acids are synthesized in the liver, and stored in the gallbladder.

4. **The correct answer is E.** Although the statements in answer choices A and D are correct, aspirin is an irreversible inhibitor of COX1 and COX2 enzymes, therefore prostaglandin synthesis is inhibited. However, low dose aspirin specifically inhibits the synthesis of thromboxanes (TXA2) in the platelets. This COX inhibition by low dose aspirin has been found to decrease platelet aggregation by decreasing TXA2 synthesis and preventing future coronary artery diseases, such as heart attacks and stroke. Therefore, the best statement for how low dose aspirin helps this patient is in answer choice E.

Aspirin does not inhibit the lipoxygenase pathway and nor does it affect every eicosanoid synthesis. Thus, the answer choices B and C are incorrect.

5. **The correct answer is E.** Leukotrienes are synthesized in the lipoxygenase pathway. The drug Zileuton inhibits the lipoxygenase enzyme. Zileuton would be useful for this patient because leukotrienes are responsible for the inflammatory processes within the lungs and are strong bronchoconstrictors. They are active within leukocytes and mediate chemotaxis as well. Therefore, inhibition of leukotrienes alleviates allergic reactions and asthma. Since leukotrienes are synthesized by the lipoxygenase pathway, inhibition of the COX pathway would not be beneficial. Therefore, answer choices A, C, and D are incorrect. Glucocorticoids can also alleviate inflammatory and allergic reactions because cortisol, a glucocorticoid, inhibits PLA2, which releases arachidonic acid, the precursor for all eicosanoids. Glucocorticoids would not be a good option for this patient, however, because he has a history of Cushing syndrome, which is caused by an excess of cortisol secretion from their adrenal glands.

Leukotriene receptor antagonists such as Montelukast also would be helpful for this patient.

Reference: Paggiaro P, Bacci E. Montelukast in Asthma: A Review of its Efficacy and Place in Therapy. *Therapeutic Advances in Chronic Disease*. 2011;2(1):47-58. doi:10.1177/2040622310383343.

6. **The correct answer is D.** This patient most likely has Niemann-Pick disease. Although patients are normal at birth, by age of 3-months hepatosplenomegaly develops. Motor development and hypotonia may develop after one year of age. Deficiency of lysosomal sphingomyelinase causes accumulation of sphingomyelin in the monocytes and macrophages, which leads to development of hepatosplenomegaly. Niemann Pick disease is an autosomal recessive disease and its prevalence is more common in Ashkenazi Jews. The sphingomyelinase enzyme activity can be measured in white blood cells or in fibroblast cell cultures. There are two types of Niemann-Pick disease: type A and type B. Type A has a more severe enzyme deficiency, age of onset is very early, and death occurs by 2-3 years of age. Type B is a milder form and patients live up to 20-30 years.

Niemann-Pick and Gaucher disease (answer choice B) have similar presentations. Both have hepatosplenomegaly and cherry red spots in the macula of the eyes. The differentiation is the skeletal deformities and erosion, which is present in Gaucher disease. Therefore, answer choice B is incorrect. Hexosaminidase A (answer choice C) deficiency causes Tay-Sachs disease, which presents with abnormal motor developments and cherry red spots in the macula of the eyes. Tay-Sachs patients do not present with hepatosplenomegaly but they do develop motor dysfunctions.

The ?-galactosidase enzyme deficiency (Krabbe disease) causes accumulation of glycolipids and results in demyelination. Krabbe disease presents with irritability and psychomotor arrest. Therefore, the answer choice A is incorrect.

Reference: https://emedicine.medscape.com/article/951564-overview

7. **The correct answer is B.** Common symptoms of Gaucher disease are joint pain and hepatosplenomegaly, as well as the 'wrinkled paper" appearance of the macrophages. Genetic deficiency of lysosomal ?-glucosidase causes accumulation of glucocerebrocytes within the lysosomes. The disease affects the liver, spleen, and the bone marrow. Patients suffer from hepatosplenomegaly, which is responsible for the anemia and pancytopenia. Because the bone marrow is also affected, patients also complain of bone and joint pain. Bone deformaties and fractures are commonly seen in these patients. Gaucher disease is the most common lysosomal storage disease and it is especially prevalent in Askenazi Jews. The onset of the disease varies from early childhood to adulthood.

Fabry disease (answer choice A) affects multiple organs including the skin, gastrointestinal tract, kidneys, heart and brain. Common symptoms are skin problems such as angiokeratomas, corneal and lenticular changes. Fabry disease develops in childhood due to deficiency of the lysosomal α-galctosidase enzyme, which is an X-linked disease. Therefore, males are affected more commonly than females.

I-cell disease (answer choice C) is caused by a mutation in GlcNac-1-phosphtransferase. This enzyme adds mannose 6-phsophate to lysosomal enzymes. Tagging the lysosomal enzymes with mannose 6-phosphate targets them from Golgi to lysosomes (reviewed in Theme 6). Deficiency of GlcNac-1-phosphotransferase causes accumulation of lysosomal enzymes in the extracellualr space and the blood. Because lysosomes lack the enzymes, undigested substrates accumulate within the lysosomes. I-cell disease affects multiple organs and presents with developmental delay, skeletal abnormalities, coarse facial features, hepatosplenomegaly and hypotonia.

Krabbe disease (answer choice D) develops due to deficiency of the β-galactosidase enzyme. Deficiency of this enzyme causes accumulation of glycolipids in oligodendrocytes, resulting in demyelination. Krabbe disease presents with irritability and psychomotor arrest.

Reference: https://emedicine.medscape.com/article/944157-overview

8. **The correct answer is E.** This child's neuro motor developmental delays suggest either Tay-Sachs or Sandhoff disease. While Tay Sachs disease is due to deficiency of hexosaminidase A, Sandhoff disease is due to deficiency of both hexosaminidase A and B isoforms. In both diseases GM2 gangliosides accumulate within the lysosomes of neurons and the retina. Patients are normal at birth but present with motor developmental delays by six months of age. Besides motor weakness, patients present with an unusual startle reflex. Hepatomegaly is not common in the early stages of the disease. Both diseases are autosomal recessive and are more common in Ashkenazi Jews. Another common finding is cherry-red spots on the macula of the eyes.

Metachromatic leukodystrophy (answer choice A) is an inherited neurodegenerative disorder. Deficiency of arylsulfatase A causes accumulation of sulfatides in neuronal tissues and consequently demyelination occurs. There are four types of this disease: late infantile, early juvenile, late juvenile, and adult. The most common symptoms are memory loss, gait disturbances, and loss of motor functions. Patients do not develop cherry-red spots on the macula.

Genetic deficiency of mucopolysaccharide-cleaving enzymes that degrade mucopolysaccharides or glycosaminoglycans (GAGs) in the lysosomes leads to the accumulation of these GAGs. These diseases are called mucopolysaccharidosis. Two of the mucopolysaccharidosis are Hurler and Hunter syndromes. Deficiency of iduronate sulfatase causes Hunter syndrome, which is inherited in an X-linked recessive manner. Hunter syndrome presents with heart problems and hepatosplenomegaly. Hurler syndrome is inherited in an autosomal recessive manner, and it is caused by deficiency of α-L-iduronidase enzyme. Hurler syndrome presents with spinal deformities, stiff joints, and cloudy cornea. Neither of the mucopolysaccharidosis presents with cherry-red spots on the macula of the eyes, therefore answer choice B is incorrect.

Genetic deficiency of the lysosomal sphingomyelinase enzyme causes Niemann-Pick disease. Sphingomyelin accumulates in the lysosomes of brain and reticuloendothelial system cells. The most common symptoms are hepatosplenomegaly and cherry red spots on the macula. Differentiation between Niemann-Pick and Tay-Sachs diseases are hepatosplenomegaly in Niemann-Pick and motor developmental delays in Tay-Sachs disease. Therefore, answer choice C is incorrect.

Pompe disease is both a lysosomal storage and a glycogen storage disease (reviewed in Theme 24). Deficiency of lysosomal α-glucosidase causes accumulation of glycogen remnants in the lysosomes. Common symptoms are cardiovascular problems and hypotonia. Therefore, answer choice D is incorrect.

VI

Review Questions

1. A 16-year-old female presents with pancreatitis. Her past medical history reveals that she had episodes of abdominal pain since childhood. Physical examination reveals eruptive cutaneous xanthomas and hepatosplenomegaly. Blood tests show a normal lipid panelexcept for a triglyceride level of 950 mg/dl (normal is < 150 mg/dl). Which of the following would be the best treatment option of her symptoms?
A) Fibrates
B) PCSK9 inhibitors
C) Restriction of dietary fat
D) Statin drugs
E) Thiazolidinones

2. A 38-year-old male presents to his family physician for his annual exam. His medical history is unremarkable and he has a BMI of29 and blood pressure of145/95 mmHg. A fasting lipid panel reveals atotal cholesterol of 290 mg/dL (N < 200 mg/dL), LDL of 210 mg/dL (N < 100 mg/dL), HDL of 35 mg/dL (N > 35 mg/dL), and triacylglycerids of 165 mg/dL (N < 150 mg/dL). Random blood glucose is 105 mg/dL (N = 110-140 mg/dL). Which of the following is the best treatment option for this patient?
A) Diet and exercise
B) Fibrates
C) PCSK9 inhibitors
D) Statin drugs
E) Thiazolidinones

3. A 52-year-old female presents to the emergency department with fatigue, shortness of breath and pain in her arms. Her medical history reveals that she was hospitilized about a year ago for stroke-like symptoms. She has been taking high doses of a statin drug. Physical examination reveals xanthomas around the iris of her eyes and her tendons. Elevated levels of which of the following lipoproteins is most likely causing her cardiovascular problems?
A) Chylomicrones
B) HDL
C) LDL
D) VLDL

4. A 44-year-old male presents to the office to discuss his recent muscle pain, fatigue and lack of concentration. History reveals that he has been taking statin drugs for the last 2months. His doctor changes his prescription to cholestramine. Activity in which of the following enzymes is expected to be increased in his liver with his new medication?
A) 7-α-hydroxylase
B) 17-α-hydroxylase
C) 18-hydroxylase
D) 21-α-hydroxylase

5. A 37-year-old male presents to his physician with fatigue, shortness of breath and pain in hisarms. His blood cholesterol is 570 mg/dL. Physical examination reveals multiple xanthomas. A mutation in which of the following proteins is most likely responsible for his symptoms?
A) HMG-CoA reductase
B) LDL receptor
C) PCSK9

D) SR-A1
E) SR-A2

6. A researcher is studying the effects of different compounds on the growth of breast cancer cells. His findings indicate that growth of these cells is enhanced in the presence of estrogen. Inhibition in which of the following enzymes would prevent the growth of these cells?
A) 5-α-reductase
B) Aromatase
C) 11-β-hydroxylase
D) 18 hydroxylase
E) 21-α-hydroxylase

7. A researcher is studying the effects of different compounds on the cellular proliferation of epithelial and stromal cells in the periurethral area of the prostate. Findings indicate that androgens increase the mass of these cells. Decreased concentrations of which of the following androgens would most effectively inhibit prostate growth?
A) Androstenedione
B) Dehydroepiandreosterone
C) Dihydrotestosterone
D) Testosterone

8. A-16-year-old female presents with oligomenorrhea. Although genotyping shows 46-XX, her physician notices that she is very masculine and has male-like facial hair. Physical examination indicates signs of dehydration and low blood pressure. Blood tests reveal low cortisol and aldosterone but high DHEAS and 17-?-hydroxyprogesterone levels. Which of the following enzymes is most likely deficient in this patient?
A) CYP11
B) CYP 17
C) CYP 21
D) CYP11B1

9. A 4-year-old African American girl presents for her annual exam. Her mother states that she has been complaining of pain in her legs. History reveals that she had a recent broken bone due to a fall when she was playing in the backyard. Physical examination shows that her height is below the 5th percentile but her weight is at the 50th percentile. Her dental history is remarkable for four cavities. History also shows that she is lactose intolerant. Blood tests show that her blood calcium and phosphorus levels are slightly low, but alkaline phosphatase is higher than the reference range. Which of the following additional blood test results is most likely to be found?
A) High levels of PTH
B) Low levels of PTH
C) Normal levels of PTH
D) Low levels of calcitriol
E) High levels of calcitriol

Answers

1. **The correct answer is C.** Based on the symptoms, this patient has genetic deficiency of either lipoprotein lipase (LPL) or apolipoprotein CII (ApoCII). Her most likely diagnosis is hyperchylomicronemia. In order to reduce plasma

triglyceride levels and also prevent the episodes of abdominal pain, dietary fat restriction is recommended. Answer choices B and D are incorrect because PCSK9 inhibitors and statin drugs are used for treatment of hypercholesterolemia, not hyperchylomicronemia. Answer choices A and E are incorrect because both fibrates and thiazolidinones reduce VLDL formation in the liver, which would not correct the hyperchylomicronemia in this patient. Fibrates induce the PPAR? (peroxisome proliferator-activated receptors) pathway, which results in the transcription of the lipoprotein lipase gene. Since this patient most likely has a mutation in the LPL gene, an inactive enzyme would be made. Thiazolidinones act similar to fibrates and activate the PPAR? pathway, which increases peroxisome proliferation.

2. **The correct answer is A.** Although low dose statin drugs would be useful for lowering his LDL cholesterol, the best medical therapy for this patient would be a low fat diet and aerobic exercise. Answer choices B and E are incorrect because fibrates and thiazolidinones are useful for lowering serum triacylglycerel levels, which is not necessary for this patient. Answer choice C is incorrect because PCSK9 inhibitors are useful for patients who has elevated LDL in spite of of high doses of statin treatments, which is not the case in this patient.

3. **The correct answer is C.** The most likely diagnosis of this patient is familial hypercholesterolemia, which is characterized by very high levels of blood LDL cholesterol. Patients typically present with xanthomas, which are the deposits of cholesterol esters within tendons, skin, and around the iris in the eye. Answer choice A is incorrect because hyperchylomicronemia presents in early childhood with abdominal pains. Answer choice B is incorrect because HDL is known as "good" cholesterol and elevation of this lipoprotein is inversely associated with cardiovascular events. Answer choice D is incorrect because elevated levels of VLDL is associated with high triacylglycerols, which is not reported in this patient. This condition is treated with fibrates and/or thiazolidinones, which are useful for lowering serum triacylglycerel levels.

4. **The correct answer is A.** Cholestramine is a bile acid sequestrant, which increases the excretion of bile acids through the feces. Less bile acid reabsorption from the intestine to the liver causes upregulation of bile acid synthesis. Since the rate limiting step of bile acid synthesis is 7-α-hydroxylase, this enzyme is upregulated by the decreased concentrations of bile acids in the liver. Because bile acids are synthesized from cholesterol, increased synthesis of bile acids decreases the intracellular pool of cholesterol. This decreased concentration of intracellualr cholesterol upreglates LDL receptor expression, thereby increasing LDL uptake from the blood and decreasing the amount of LDL in the blood. Answer choices B, C, and D are incorrect because these enzymes are involved in the synthesis of steroid hormones.

5. **The correct answer is B.** She most likely has very high blood LDL cholesterol due to a genetic disorder in the LDL-receptor gene (*LDLR*). This condition is known as familial hypercholesterolemia. Heterozygote patients develop coronary artery disease and xanthomas by the early 40s and

50s. Homozygous patients present with xanthomas at birth or in early childhood. Due to deficiency of hepatic LDL receptors, LDL cholesterol cannot be taken up by the liver, thus LDL accumulates in the blood. This accumulated LDL is susceptable to oxidation by reactive oxygen species. The oxidized LDL is taken up by extrahepatic tissues and macrophages via scanger receptors SR-A1 and SR-A2. This leads to foam cell formation and plaque deposition, which are the initial steps of artheroschlerosis. Therefore, answer choices D and E are incorrect. Answer choice A is incorrect because there are no reported cases of mutations in HMG-CoA reductase gene. Answer choice C is incorrect because binding of PCSK9 to the LDL receptor inhibits the recycling of the receptor and leads to LDL receptor proteolysis. Therefore, mutation in the PCSK9 gene would have an increased number of LDL receptors and thus patients would not suffer from high blood LDL cholesterol levels.

6. **The correct answer is B.** Aromatase is responsible for producing eestrogens (estrone and estradiol) from androstenedione and from testosterone. The enzyme is stimulated by follicle stimulating hormone (FSH) in the ovaries. Inhibition of this enzyme would decrease the synthesis of estrogens. Answer choice A is incorrect because 5-α-reductase converts testosterone to dihydrotestosterone. Answer choice C is incorrect because 11-β-hydroxylase converts 11-deoxycortisol to cortisol. Answer choice D is incorrect because 18 hydroxylase converts corticosterone to aldosterone. Answer choice E is incorrect because 21-α-hydroxylase is an adrenal cortex enzyme and it is involved in multiple steps in the synthesis of aldosterone and cortisol.

7. **The correct answer is C.** Dihydrotestosterone is the most potent androgen and studies show that reducing concentrations of dihydrotestosterone decrease the growth of the prostate. Therefore, the inhibitors of 5-α-reducatse, such asfinasteride, are used for treatment of benign prostate hyperplasia. Anwer choices A and B are incorrect because these hormones are weak androgens. Although testosterone is an androgen, it is not as potent as dihydrotestosterone, therefore answer choice D is incorrect.
Reference:Smith AB, Carson CC. Finasteride in the treatment of patients with benign prostatic hyperplasia: a review. *Therapeutics and Clinical Risk Management*. 2009;5:535-545.

8. **The correct answer is C.** Mild or partial deficiency of 21-α-hydroxylase enzyme (CYP21) in females presents with oligomenorrhea, hirsutism, and male-like masculinization during adolescence years. Males, however, present with early virilization, acne, deepening of the voice, and advanced skeletal maturation. If patients are not treated, low levels or the complete lack of cortisol and aldosterone lead to adrenal and salt wasting crisis. The increased levels of 17-α-hydroxyprogesterone is considered diagnostic of 21-α-hydroxylase (CYP 21) deficiency.
Answer choice D is incorrect because the 11-β-hydroxylase (CYP11B1) deficiency causes hypertension. Although, both CYP21 and CYP11B1 deficiencies cause excess production of androgens such as DHEA and androstenedione, CYP21 deficiency causes hypotension but CYP11B1 deficiency causes hypertension. This is caused by the over-stimulation of the adrenal cortex by ACTH due to absence of negative feedback

VI

inhibition by cortisol. CYP11B1 enzyme deficiency also results in excess production of the 11-deoxycorticosterone, which has mineralocorticoid activity and, therefore, increases sodium reabsorption and fluid retention.

Answer choice A is incorrect because desmolase (CYP11), also known as side chain cleavage enzyme, catalyzes the conversion of cholesterol to pregnenolone, which is the first step in steroid hormone biosynthesis. Thus with the deficiency of this enzyme, none of the corticosteroids, mineralocorticoids or sex steroids would be produced. Answer choice B is incorrect because CYP17 deficiency also presents with hyperaldosteronism and thus hypertension.

Reference: https://emedicine.medscape.com/article/919218-overview

9. **The correct answer is A**. This patient most likely has Rickets due to vitamin D deficiency. This condition is more common in children with dark skin since they need more sunlight to produce sufficient amounts of vitamin D. Since this child is lactose intolerant, she is most likely not eating dairy products, resulting in a nutritional deficiency in vitamin D and calcium. The most common symptoms of vitamin D deficiency are sore and painful bones, skeletal deformities, bowed legs, thickening of the ankles and knees (knock-knees), dental problems with delayed teeth emergence and increased cavities, poor growth, and fragile bones. Answer choices B, C, D, and E are incorrect because blood test results of Rickets patients with vitamin D deficiency consistently show increased parathyroid hormone. The increased PTH levels affect calcitriol levels, which may be normal or slightly elevated. Due to increased PTH activity, blood calcium levels can be very close to reference range or slightly lower. Vitamin D inhibits the release of PTH. If vitamin D is deficient, then PTH increases, which causes release of calcium from the bones to maintain normal blood calcium levels.

Reference: https://emedicine.medscape.com/article/985510-overview

Review Questions

1. A 45-year-old male presents with weight loss, agitation, sweating, and heat intolerance. Physical examination reveals tachycardia and an enlarged thyroid gland. Blood tests show increased T3 and T4, but decreased TSH levels. Increased secretion of which of the following hormones/neurotransmitters is most likely the cause of his weight loss?
   A) Glucagon
   B) Ghrelin
   C) Insulin
   D) Motilin
   E) Norepinephrine

2. A 39-year-old female complains of constant hunger. She states that despite taking weight loss medications she has not been successful in losing weight. Her family history reveals that most of her relatives have high BMI levels. Decreased levels of which of the following hormones/neurotransmitters is most likely responsible of her increased appetite?
   A) Ghrelin
   B) Epinephrine
   C) Leptin
   D) Motilin
   E) Norepinephrine

3. A 24-year-old morbidly obese male presents with complains of constant hunger. He states that despite of weight loss medications and psychological therapy, he cannot lose weight. He also states that he has a twin brother whohas similar weight problems. Laboratory tests reveal that his leptin levels were within reference range. Genetic mutations in which of the following receptors is most likely responsible of his weight problems?
   A) Glucagon
   B) Insulin
   C) MC4-R
   D) SR-A2
   E) Thyroid

4. A 62-year-old female presents for her annual examination with no complaints. Physical examination reveals that her BMI is 38 (N = 18-25) and her blood pressure is 180/95 (N= 140/80). Her fasting metabolic panel shows: glucose 125 mg/dL (N < 100 mg/dL), total cholesterol 300 mg/dL (N < 200 mg/dL), LDL 185 mg/dL (N < 100 mg/dL), HDL 35 (N > 35), and triglycerides 190 mg/dL (N < 130 mg/dL). Which of the following is most likely increased in her blood?
   A) Adiponectin
   B) Cholecystokinin
   C) Leptin
   D) TNF-?

5. A 49-year-old male with a BMI of 36 and a history of high blood pressure presents for his annual examination. Random blood glucose is 195 mg/dL andfasting blood glucose is 155 mg/dL. His doctor prescribes metformin to normalize his blood glucose levels. The activity of which of the following proteins/molecules will most likely be increased by his medication?
   A) Adiponectin
   B) AMPK
   C) GLP-1

D) GLUT2
E) PPAR-?

6. A 62-year-old male who had a recent gastrectomy presents with memory loss and weakness. The physician performs a broad workup and discovers that the patient has macrocytic anemia. His blood homocysteine levels are elevated and urine analysis shows methylmalonic aciduria. Supplementation of which of the following vitamins would help to alleviate his anemia?
   A) Biotin
   B) Vitamin B6
   C) Vitamin C
   D) Vitamin B12
   E) Folate

7. A 58-year-old male presents to his family physician with complaints of knee pain. Following a physical examine, his doctor suggests knee replacement surgery. His history reveals that he has a blood clotting disorder, for which he currently does not take any medications. However, he notes that he avoids green, leafy vegetables in his diet. Which of the following medications would decrease the risk of stroke during and after his surgical procedure?
   A) Niacin
   B) Statin
   C) Vitamin K
   D) Warfarin
   E) ?-3 PUFA

8. An eight-month-old female presents with diarrhea. Physical examination reveals a very skinny child with apparent wasting. Her height and weight is below the 10th percentile. Her grandmother states that the child's mother has been out of work for the last 6-months and they are struggling to pay their bills. Her grandmother also states that the child was weaned from breast milk and formula almost 4 months ago. Which of the following is the most likely clinical finding of this patient?
   A) Edema
   B) Fatty liver
   C) Hepatomegaly
   D) Muscle wasting

Answers

1. **The correct answer is E.** This patient most likely has Graves' disease, which leads to goiter with hyperthyroidism. Graves' disease is an autoimmune disease which is caused by the binding of IgG autoantibodies to TSH receptors on the thyroid gland. This binding stimulates the thyroid gland to secrete excessive T3 and T4. Thyroid hormones regulate energy expenditure, the rate of lipolysis, and protein degradation. They also regulate the rate of triacylglycerol and protein synthesis. These bipolar effects of thyroid hormones are accomplished by increasing insulin secretion from the pancreas, and by increasing secretion of norepinephrine from the sympathetic nervous system. Norepinephrine stimulates ATP utilization and heat production by increasing synthesis of the uncoupling protein (UCP) thermogenin in brown adipose tissue. Norepinephrine

VI

in the adipocytes also increases lipolysis. In contrast to norepinephrine, insulin increases glucose uptake, glycolysis, and triacylglycerol synthesis in adipose tissue. Therefore, answer choice C is incorrect. Answer choice A is incorrect because glucagon secretion is not affected by thyroid hormones. Answer choices B and D are incorrect because ghrelin induces a feeling of hunger and motilin stimulates gastric and pancreatic enzyme secretion, both of which would be associated with weight gain, not weight loss.

2. **The correct answer is C**. Leptin stimulates satiety, suppresses appetite, and increases energy expenditure. Decreased secretion of leptin from the adipocytes is responsible for an increase in appetite. In contrast to leptin, ghrelin hormone activity increases appetite by stimulating orexigenic neurons. Although increased ghrelin or motilin increases appetite, decreased ghrelin or motilin levels do not have opposite effects, therefore answer choice A and D are incorrect. Although increased levels of epinephrine and norepinephrine lead to appetite suppression, decreased levels do not have opposite effects. Therefore answer choice B and E are incorrect.

3. **The correct answer is C**. Leptin activates the anorexigenic neurons in the arcuate nucleus to express proopiomelanocortin (POMC). The POMC is the precursor of α-melanocyte stimulating hormone (α-MSH), which is released from POMC-expressing anorexigenic neurons of hypothalamus. The ?-MSH binds to its receptors MC3-R and MC4-R at the paraventricular nucleus (PVN). The binding of α-MSH to its receptors activates the melanocortin signaling pathway, which inhibits food intake and increases energy expenditure. The most common genetic mutations of inherited obesity have been linked to MC4-R receptor mutations, which are inherited in a co-dominant fashion. Thus the melanocortin signaling pathway is an important contributing factor to inherited obesity.

Answer choices A and B are incorrect because genetic muta-tions in glucagon or insulin receptors do not induce hunger. Although insulin is a satiety hormone, its effects in appetite suppression are significantly lower than the leptin and mela-nocortin signaling pathways. Answer choice D is incorrect because SR-AI & SR-A2 receptors are found in macrophages, and these receptors are involved in mediating the uptake of the LDL lipoproteins, which leads to the formation of athero-sclerotic plaques. Thyroid hormones regulate energy expen-diture, the rate of lipolysis, and protein degradation. They also regulate the rate of triacylglycerol and protein synthesis. Mutations in thyroid hormone receptors would modulate BMR, but not appetite. Therefore answer choice E is incor-rect.
Reference:Lanfray D, Richard D. Emerging Signaling Pathway in Arcuate Feeding-Related Neurons: Role of the Acbd7. *Frontiers in Neuroscience.* 2017;11:328. doi:10.3389/fnins.2017.00328.

4. **The correct answer is D**. This patient is most likely pre-diabetic due to insulin resistance. The insulin resistance is likely due to her obesity, which can cause excess release of free fatty acids (FFAs) and adipokines (TNF-α) into the circulation from adipocytes. Excess FFAs in the circulation causes insulin resistance in the muscle and in the liver. While

insulin resistance decreases glucose uptake, it increases glucose output from the liver, leading to elevation of blood glucose levels. To compensate for insulin resistance, her pancreas most likely secretes excess insulin, causing hyperinsulinemia. Hyperinsulinemia stimulates the sympathetic nervous system, leading to sodium and water retention, vasoconstriction and hypertension. Excess amounts of TNF-α activate SREBP in the liver, which increases VLDL production, contributing to her dyslipidemia, characterized by high LDL and triacylglycerols, and low HDL. In contrast to TNF-α, adiponectin levels decrease in obese individuals. Adiponectin normally decreases FFA levels in the circulation and, thus, decreases insulin resistance as well. Therefore, answer choice A is incorrect. Answer choice B is incorrect because cholecystokinin, as a neuropeptide, stimulates satiety and has an appetite suppressing effect. It is most likely that this patient does not have increased levels of cholecystokinin. Answer choice C is incorrect because leptin increases the basal metabolic rate and suppresses the appetite. Therefore, it is not likely that leptin is elevated in this patient.

5. **The correct answer is B**. Normally the insulin signaling pathway leads to an increase in the GLUT4 transporters on the cell membranes of muscle and adipose cells, thereby increasing glucose uptake. Metformin also increases GLUT4 transporters on the cell membranes of the muscle cells. However, metformin does this by inhibiting NADH dehydrogenase (Complex-I) in the ETC. The inhibition of Complex-I increases the AMP levels in the cell. Increased AMP activates AMP-dependent kinase (AMPK), which increases the concentrations of GLUT4 on the muscle cells, thereby stimulating glucose entry into the muscles. Answer choice A and E are incorrect because release of adiponectin is stimulated by thiazolidinediones (rosiglitazone, pioglitazone). These drugs binds to peroxisome proliferator activated receptor-Y (PPAR-Y) in the adipocytes. Increased adiponectin decreases FFA levels in the circulation, thus decreasing insulin resistance in the muscle cells. Answer choice C is incorrect because GLP-1 levels can be increased by dipeptidyl-peptidase-4 (DPP-4) inhibitors (linagliptine, sitagliptine, saxagliptine, alogliptine). GLP-1 is a peptide hormone and is normally secreted after meals. Binding of GLP-1 to its receptors stimulates insulin secretion. Answer choice D is incorrect because GLUT2 is not an insulin sensitive transporter and its levels do not change with the changes in insulin or glucagon levels.

6. **The correct answer is D**. Macrocytic anemia is caused by deficiency of either folate or vitamin B12. This patient most likely has vitamin B12 deficiency because he has homocysteinemia and methylmalonic aciduria, which is the diagnostic test for vitamin B12 deficiency. Therefore, answer choice E is incorrect. Answer choice B is incorrect because vitamin B6 deficiency causes microcytic hypochromic anemia. Vitamin C deficiency causes scurvy, which affects collagen synthesis. Therefore, answer choice C is incorrect. Biotin deficiency causes alopecia and abdominal discomfort, thus answer choice A is incorrect.

7. **The correct answer is D**. Warfarin and dicumarol are vitamin K antagonists, which decrease the regeneration of

vitamin K. Vitamin K as a coenzyme is required for the γ-carboxylase enzyme. This enzyme catalyzes the γ-carboxylation of the glutamate residues of blood coagulation factors (II, VII, IX, and X) during translation. Decreased ?-carboxylation decreases the calcium binding sites on these coagulation factors, thus the coagulation cascade slows down. Therefore, vitamin K antagonists are used as anticoagulant therapy in individuals who are at high risk for thrombotic events. Green leafy vegetables are rich in vitamin K, thus, individuals at high risk for thrombotic events are advised to avoid eating these vegetables. Answer choice C is incorrect because vitamin K supplementation would increase thrombosis. Answer choice A and B are incorrect because niacin and statin drugs are used for the treatment of hypercholesterolemia. Although ω-3 PUFA has antithrombotic effects and decreases platelet aggregation, the efficiency of these fatty acids would require a much longer time before his surgery; therefore answer choice E is incorrect.

8. **The correct answer is D**. This child most likely has marasmus, which is caused by deficiency in calories and energy. In these patients, muscle protein is degraded to liberate muscle amino acids for glucose production to maintain blood glucose concentrations, leading to muscle wasting. Clinical findings such as edema, fatty liver, and hepatomegaly, are not associated with marasmus. Therefore answer choices A, B, and C are incorrect. Kwashiorkor presents with edema. Since these patients have normal intake of carbohydrates but deficiency in protein and essential amino acids, the liver is unable to synthesize albumin and other proteins. Kwashiorkor also presents with massive hepatomegaly, which is caused by fatty liver. Development of the fatty liver is due to decreased synthesis of ApoB100, which is required for assembly and secretion of VLDL. Decreased release of VLDL causes accumulation of fat in the liver, causing it to be fatty.
Reference: https://emedicine.medscape.com/article/984496-clinical

Part VII

VII

47 DNA Packaging and Meiosis

At the conclusion of this chapter, students should be able to:
- Describe the organization of human genome and packaging of DNA into chromatin
- Describe the stages of meiosis
- Analyze the differences between meiosis and mitosis
- Describe the events that generate genetic diversity (independent assortment and crossing over) during meiosis

47.1 DNA Packaging

Compared to the other components of the cell, DNA molecules are very large. In fact, it has been estimated that if the entire amount of nuclear DNA of a single cell is stretched out, end to end, it would span a distance of approximately two meters (six feet). In order for this amount of material to fit into a nucleus of roughly six microns in diameter, DNA must be tightly packaged. This is accomplished through a series of increasingly higher order structures that include specific DNA packaging proteins.

The first level of packaging involves a family of proteins known as histones. Histones are basic proteins whose positive charges allow them to associate with the negatively-charged DNA. There are four major families of core histone proteins: H2A, H2B, H3, and H4. Two copies of each of these proteins assemble to produce an octamer of histones known as a nucleosome, around which the DNA is wound. Strings of nucleosomes are further wound into helical, tubular structures called solenoids. These solenoids are then packaged to form the condensed chromosomes.

This combination of histone proteins and DNA is collectively known as chromatin. Chromatin exists as two major forms in the nucleus: heterochromatin and euchromatin. Heterochromatin is a highly condensed form of chromatin and makes up approximately 10% of the chromosome. In a typical linear eukaryotic chromosome, the centromere and telomeres consist of heterochromatin, as well as some interspersed areas within other regions of the chromosome. Due to the compactness of these regions, they are devoid of transcriptional activity and, hence, gene expression does not occur in these areas. Euchromatin, by contrast, is more loosely packaged and comprises the majority of the chromosome. These are the regions in which gene expression occurs.

As described in Chapter 11, chromosomes are further condensed during mitosis, and meiosis, through the formation of condensin complexes. Recall that these structures allow for the tightly packaged sister chromatids that will be separated during both mitotic and meiotic cell divisions.

47.2 Meiosis

Meiosis is the process that creates gamete cells that will ultimately transfer genetic information from one generation to the next. The process begins with a diploid cell and results in four haploid gamete cells, each of which contains half of the genetic information of the original cell. In order to create four gametes, one cell goes through a two-part cell division process, termed a reduction division, consisting of eight stages (▶ Fig. 47.1). Many of the stages are similar to the stages of mitosis (previously discussed in Chapter 11). The differences and similarities between these two processes are discussed throughout this chapter.

As described in chapter 11, the eukaryotic cells must complete the initial cell cycle stages of interphase in preparation for meiosis or mitosis. Interphase is the stage of the cell cycle in which the cell grows and replicates its DNA to form sister chromatids. In this process, a diploid cell with 2n DNA content becomes a diploid cell 4n DNA content. The centrosome, which we will discuss shortly, is also duplicated at this stage.

Unlike mitosis where the objective is to make two identical cells, the purpose of meiosis is to produce four haploid cells which are genetically different from the parent cell. Hence, generating genetic diversity is vital. One way a cell can create genetic variation is through recombination during meiosis. **Recombination**, also called **crossing over**, can occur in **prophase I** when homologous chromosomes condense and pair together to form a tetrad (synapsis). The tetrad is held together through a structure known as the synaptonemal complex, which is composed of three parallel dense elements that holds the homologous chromosome pairs together and mediates crossing over. The crossing over is an exchange of genetic information between the homologous chromosomes, resulting in recombinant chromosomes (▶ Fig. 47.2). Therefore, recombination allows genetic information from both the individual's mother and father to be passed on via one chromosome. This ensures genetic diversity in the offspring.

A cell also generates genetic diversity during the next stage in meiosis, **metaphase I**, when the nuclear membrane breaks down. During metaphase I, the homologous pairs formed during prophase I line up, with one chromosome from each pair on either side of the equator. The manner by which homologous chromosomes line up on either side, i.e. maternal on the right and paternal on the left versus maternal on the left and paternal on the right, is an independent event termed **independent assortment**. As shown in ▶ Fig. 47.3, when there is variation between which sides the paternally-inherited chromosomes (and equally the maternally-inherited chromosomes) settle on, the daughter cells contain a mixture of paternally- and maternally-inherited genetic material. A simple calculation reveals the number of different ways the 23 pairs of chromosomes can line up: $2^{23} = 8,388,608$, which illustrates the vast amount of genetic variation that independent assortment can contribute to the gametes.

In contrast, because synapsis of homologous chromosomes does not occur during mitosis, there is no crossing over or independent assortment, meaning the cells produced are genetically identical to the parent cell. This fidelity of mitosis is required to produce the necessary cells needed for growth and repair, whereas the genetically diverse diploid daughter cells created by meiosis are important for reproduction.

The meiotic spindle forms during metaphase I, with spindle fibers developing from the centrosomes at either pole and attaching to the centromeres of the chromosomes, as seen in ▶ Fig. 47.3. Homologous chromosomes are pulled apart via the spindle fibers in **anaphase I**. Sister chromatids remain attached,

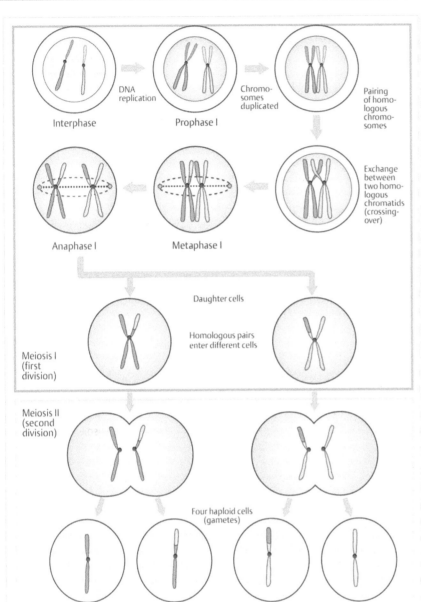

Interphase

DNA replication

Prophase I

Chromosomes duplicated

Pairing of homologous chromosomes

Exchange between two homologous chromatids (crossing-over)

Anaphase I

Metaphase I

Daughter cells

Homologous pairs enter different cells

Meiosis I (first division)

Meiosis II (second division)

Four haploid cells (gametes)

Recombinant

Nonrecombinant

Fig. 47.1 Meiosis. (Source: Passarge E, ed. Color Atlas of Genetics. 4th Edition. New York, NY. Thieme; 2012.)

however, and the cell is still considered diploid. It is not until **telophase I** when the nuclei divide into each cell and **cytokinesis,** the physical process of cell division when one cell separates into two cells, occurs that one cell becomes two genetically diverse daughter cells with equal amounts of genetic material.

Each of these newly formed cells then enters the next phase, meiosis II, which is distinct from meiosis I and closely resembles mitosis (▶ Fig. 47.1), with the major difference being that the cells are haploid upon entering meiosis II. **Prophase II** marks the beginning of meiosis II with spindle formation. However, unlike prophase I, this step does not include crossing over. The chromosomes align at the equator again during **metaphase II**, this time without their homologous chromosomes, and the sister chromatids are pulled apart during **anaphase II**. The

haploid (1n) cells are created in the final stage of **telophase II** with cytokinesis resulting in each of the two cells splitting to form four haploid gametes.

To review, meiosis is divided into two stages, with the separation of homologous chromosomes occurring during meiosis I, and separation of the sister chromatids occurring in meiosis II. When all of these stages transpire without error, four haploid granddaughter cells are produced to be used in reproduction. However, if homologous chromosomes are not pulled to opposite sides of the cell in anaphase I, or if sister chromatids do not separate in meiosis II, the resultant cells can have an abnormal number of chromosomes, known as chromosomal aneuploidy, due to **nondisjunction** (discussed in more detail in Chapter 48).

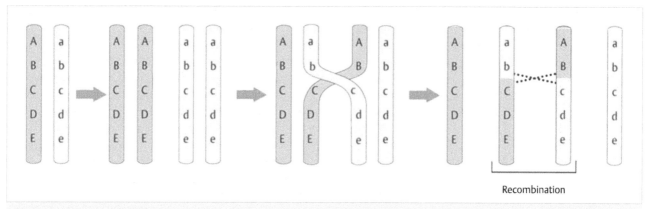

Fig. 47.2 Genetic recombination by crossing-over. (Source: Passarge E, ed. Color Atlas of Genetics. 4th Edition. New York, NY. Thieme; 2012.)

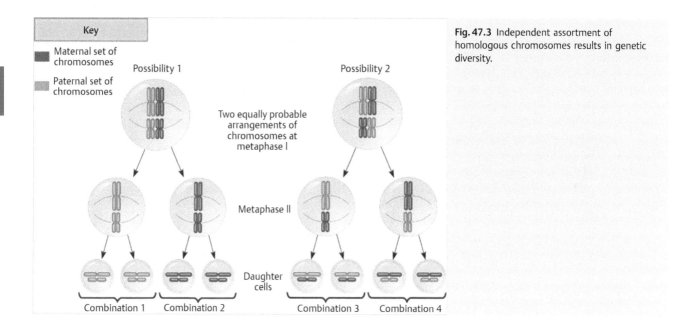

Fig. 47.3 Independent assortment of homologous chromosomes results in genetic diversity.

When sperm cells are created in males via spermatogenesis, the meiotic process results in four functional haploid sperm cells (▶ Fig. 47.4). However, during the creation of egg cells in females, a process called oogenesis, only one of the resultant granddaughter cells is functional, while the other three are nonfunctional (▶ Fig. 47.4). These three nonfunctional haploid cells are called polar bodies, the first of which is created at telophase I. Although these polar bodies are nonfunctional, their cytoplasms are squeezed into the functional egg cell to increase its volume since the sperm cells that may ultimately fertilize this egg have virtually no cytoplasm of their own. As a result, almost everything in a developing embryo's cytoplasm, including the mitochondrial DNA, is passed from the mother, with little to no contribution from the father.

When these gametes, the sperm and egg cell, come together, unique diploid cell is formed which then undergoes mitosis an inestimable number of times to eventually form a fetus. The specifics of fertilization and early embryogenesis are covered in chapter 55.

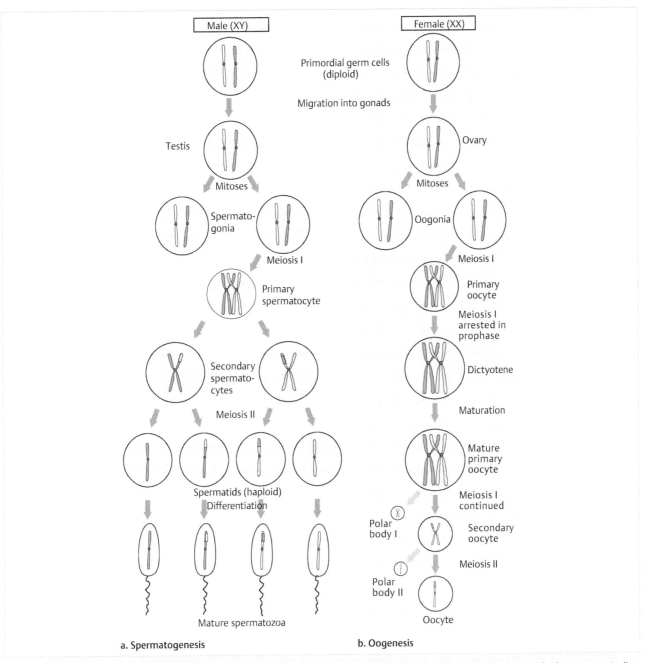

Fig. 47.4 Spermatogenesis **(a)** one primary spermatocyte forms four spermatids. Oogenesis **(b)** cytoplasm of primary oocyte divides asymmetrically in both meiosis I and meiosis II. The larger cell becomes the egg and the smaller cells becomes polar body, these polar bodies degenerate and do not develop. (Source: Passarge E, ed. Color Atlas of Genetics. 4th Edition. New York, NY. Thieme; 2012.)

Review Questions

1. Which of the following structures is produced by the assembly of eight subunits of histones?
   A) Centromere
   B) Condensin
   C) Nucleosome
   D) Solenoid

2. In which of the following phases of meiosis does recombination occur?
   A) Prophase I
   B) Metaphase I
   C) Prophase II
   D) Metaphase II

3. Which of the following is a normal process that results in three nonfunctional cells and one functional cell?
   A) Independent assortment
   B) Nondisjunction
   C) Oogenesis
   D) Spermatogenesis

4. Which of the following events is shared by both mitosis and meiosis?
   A) Formation of synaptonemal complex
   B) Recombination
   C) Reductive division
   D) Separation of sister chromatids

Answers

1. **The correct answer is C.** The DNA of an eukaryotic cell undergoes several levels of packaging to allow for approximately two meters of genetic material to fit into a nucleus with a diameter of approximately six microns. The first level of packaging involves the assembly of histone proteins to form a structure called a nucleosome. There are four major forms of histone proteins: H2A, H2B, H3, and H4. Two copies of each of these histones assembles to form an octameric structure called a nucleosome, around which the DNA is wound. Multiple nucleosomes are then organized to form helical, tubular structures called solenoids (answer choice D). The centromere (answer choice A) is a heterochromatic region on the chromosome that serves as the assembly site of the kinetochore during mitosis. Condensin (answer choice B) is a structure that aides in the condensation of chromosomes during mitosis and meiosis, but it is not composed of histones.

2. **The correct answer is A.** Recombination, or crossing over, is the exchange of genetic material between homologous chromosomes. This process is important for increasing the genetic diversity of offspring, and only occurs during prophase I of meiosis, when chromosomal tetrads are formed.

3. **The correct answer is C.** Meiosis is the process that creates haploid gamete cells from diploid cells through reductive divisions. There are two forms of gametogenesis: oogenesis in females, and spermatogenesis in males. The end result of spermatogenesis (answer choice D)is the formation of four functional sperm cells, whereas oogenesis results in one functional egg cell and three nonfunctional cells called polar bodies. Independent assortment (answer choice A) is a term that describes the random assembly of paternal and maternal chromosomes along the metaphase plate during metaphase I of meiosis. The manner by which these homologous chromosomes line up on either side, i.e. maternal on the right and paternal on the left versus maternal on the left and paternal on the right, contributes to genetic variation in the offspring. Nondisjunction (answer choice B) refers to a defect in the segregation of chromosomes during cell division, which results in aneuploidy, or an abnormal number of chromosomes in the resulting daughter cells.

4. **The correct answer is D.** In somatic cells, separation of sister chromatids occur during anaphase of mitosis. Ultimately, mitosis results in the creation of two diploid daughter cells. This process of sister chromatid separation also occurs during meiosis in germ cells, but is associated with anaphase of the second cell division (anaphase II) in this case, leading to the production of four haploid cells. The events of meiosis in which a diploid cell ultimately produces four haploid daughter cells is called reductive division (answer choice C), which doesn't occur in mitosis. Recombination (answer choice B), or the exchange of genetic material between homologous chromosomes, occurs during prophase I of meiosis and contributes to genetic variation in the resulting haploid germ cells. The synaptonemal complex (answer choice A), a structure consisting of three linear dense elements, is formed during prophase I and mediates the process of recombination. Recombination does not normally occur during mitosis, therefore answer choices A and B are incorrect.

VII

48 Cytogenetics: Chromosomal Basis of Human Diseases

At the conclusion of this chapter, students should be able to:
- Describe the structure and function of chromosomes and define Barr bodies
- Explain the types of chromosomes (telocentric, acrocentric, submetacentric, metacentric)
- Explain normal and abnormal karyotypes including chromosomal copy number changes and structural variations
- Explain mosaicism and how mosaicism affects the phenotypic expression of a chromosomal disorder
- Identify the following types of chromosomal abnormalities: triploidy, trisomy, reciprocal translocations, Robertsonian translocations, paracentric and pericentric inversions, deletions, and copy number variants
- Describe the disruptions that lead to numerical and structural chromosome abnormalities
- Determine the most appropriate diagnostic laboratory test for a given chromosomal disorder

The study of the structure and inheritance of our chromosomes is called **cytogenetics**. Each human has 23 pairs of chromosomes, or 46 total. And for each pair, one chromosome comes from the mother while the other chromosome comes from the father. Chromosome abnormalities include a change in the number of chromosomes present or alterations in the chromosome structure. While some of these chromosome changes are considered benign, chromosome abnormalities account for up to 1 in every 150 genetic conditions seen in live births. This chapter will focus on the main characteristics of a chromosome, as well as chromosome abnormalities and how they relate to human genetic disease.

48.1 Parts of a Chromosome

Chromosomes are thread-like structures composed of DNA that is tightly coiled many times around proteins called histones. Each chromosome has a **centromere**, which separates the short arm, or the **p arm**, from the long arm, or the **q arm**. Additionally, the ends of the p and q arms are referred to as the **telomeres** (▶ Fig. 48.1).

It is often the location of the centromere that gives each chromosome its own unique shape (▶ Fig. 48.2). For example, **metacentric** chromosomes are those with their centromere near the middle of the chromosome, **submetacentric** chromosomes have a centromere that is off center but still between the middle and the end of the chromosome, **acrocentric** chromosomes have a centromere that is closer to the end of the p arm, and finally **telocentric** chromosomes have centromeres located at the terminal end of the chromosome, giving this type of chromosome only one chromosome arm. While humans have the first three types, we do not possess telocentric chromosomes.

48.2 Chromosome Identification

Chromosomes can be identified by their centromere location, but also by their banding pattern and their size. But in order for chromosomes to be visible under a microscope for analysis, metaphase chromosomes must first be treated with trypsin and then treated with Giemsa stain. This staining, referred to as G-banding, will darken the heterochromatic regions of the chromosomes, which tend to be adenine and thymine rich DNA. Meanwhile, the chromatin that is less condensed, which tends to be more transcriptionally active and rich with guanine and cytosine, will absorb less stain and appear as light bands. Various combinations of these bands give each chromosome pair their own unique banding pattern. Additionally, each stained band is numbered to allow any band on the chromosomes to be identified, which is important not only for sorting the chromosomes, but also for the identification of abnormalities. ▶ Fig. 48.3 demonstrates the numbering system of the G-bands of the X chromosome, which occurs for each arm of the chromosome separately, beginning at the centromere and counting to the telomere.

Once the chromosomes are stained, they can be identified by their banding pattern and the location of their centromere, and then lined up according to size in a visual referred to as a **karyotype**. The first 22 pairs of homologous chromosomes, also known as the **autosomes,** are the same between females and males and have been assigned a number from 1 to 22 based on their size. These numbers are useful not only for ordering the chromosomes in a karyotype, but they also serve as the name of each chromo-

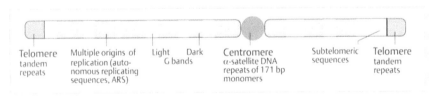

Fig. 48.1 Basic features of a eukaryotic chromosome. (Source: Passarge E, ed. Color Atlas of Genetics. 4th Edition. New York, NY. Thieme; 2012.)

Telomere tandem repeats | Multiple origins of replication (autonomous replicating sequences, ARS) | Light — Dark G bands | Centromere α-satellite DNA repeats of 171 bp monomers | Subtelomeric sequences | Telomere tandem repeats

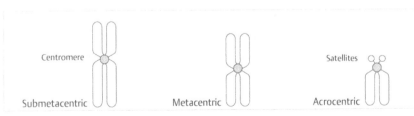

Fig. 48.2 Types of metaphase chromosome. (Source: Passarge E, ed. Color Atlas of Genetics. 4th Edition. New York, NY. Thieme; 2012.)

Centromere

Submetacentric   Metacentric   Satellites   Acrocentric

Fig. 48.3 Banding patterns and sizes of human chromosomes 1–12. (Source: Passarge E, ed. Color Atlas of Genetics. 4th Edition. New York, NY. Thieme; 2012.)

some. The 23rd pair of chromosomes is known as the **sex chromosomes** and determines whether someone is a female or a male. Females have two X chromosomes and their karyotypes are read as 46,XX, which indicates that this individual has 46 total chromosomes *including* two X chromosomes. Males, on the other hand, have one X chromosome and one Y chromosome. A male karyotype is read as 46,XY, indicating that this individual has 46 total chromosomes *including* one X chromosome and one Y chromosome. ▶ Fig. 48.4 demonstrates a complete human karyotype with the chromosomes properly sorted by banding pattern, centromere location, size, and gender.

While females have two X chromosomes, it has been found that they do not have double the quantity of X chromosome products because one X in every somatic cell of a female is inactivated, or transcriptionally silenced, during an early embryonic phase. This process, known as **lyonization**, is discussed further in Theme 50. Since this process is permanent, all cells that descend from that particular cell line will feature the same inactive X. The inactivated X-chromosome forms a structure called the **Barr body,** which is visible at interphase. There will always be one fewer Barr body than the total number of X chromosomes. Therefore, a Barr body will not be present in the 46,XY male karyotype.

48.3 Chromosome Abnormalities

Most chromosome abnormalities occur due to missegregation of chromosomes during mitosis or meiosis, or by misrepair of damaged chromosomes. We will review several types of abnormalities that can be detected with cytogenetic techniques.

As previously discussed, a typical compliment of chromosomes for a human being contains two sets of 23 chromosomes, or 46 total chromosomes, which is referred to as a diploid set. However, **polyploidy** (Greek, *poly*= "many", *ploid* = "set") refers to multiple sets of the 23 pairs of chromosomes, meaning there are typically not errors in individual chromosomes, but an error that has resulted in too many sets of chromosomes. The extra chromosomes in polyploid cases encode a surplus of genetic information and will often lead to multiple congenital anomalies and a pregnancy that will spontaneously terminate. Polyploid conceptions that do survive to term typically die shortly after birth. Polyploid chromosome sets that have been observed in humans include **triploidy**, which includes three sets of 23 chromosomes, or 69 total chromosomes (▶ Fig. 48.5), and **tetraploidy**, which includes four sets of 23 chromosomes, or 92 total chromosomes.

The most common cause of triploidy is the fertilization of one egg by two separate sperm, which results in a zygote that has received 23 chromosomes from the egg, 23 chromosomes from the first sperm, and 23 chromosomes from the second sperm. However, triploidy can also occur if an ovum and a polar body, each of which has 23 chromosomes, fuse together and are then fertilized by a sperm cell carrying an additional 23 chromosomes. Furthermore, an error in mitosis that resulted in a sperm or egg with two sets of 23 chromosomes, or a diploid sperm or egg, can also produce a triploid zygote if fertilization occurs.

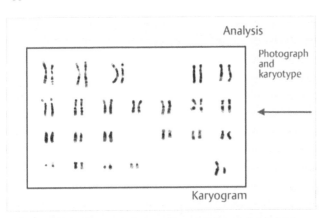

Fig. 48.4 Normal human karyotype. (Source: Passarge E, ed. Color Atlas of Genetics. 4th Edition. New York, NY. Thieme; 2012.)

Tetraploidy occurs much less frequently than triploidy and is thought to be caused by the fusion of two diploid zygotes or by a mitotic error in an early embryo which results in all of the duplicated chromosomes migrating to one of the two daughter cells.

48.4 Aneuploidies

Aside from additional entire sets of chromosomes, another numerical chromosome abnormality is **aneuploidy**. Aneuploidy refers to extra or missing chromosomes in a homologous pair. Common aneuploid karyotypes can involve either **monosomy** of a chromosome pair, meaning one copy of the homologous chromosomes is missing, or **trisomy** of a homologous chromosome pair, meaning there is an extra chromosome present.

One known cause of aneuploidy is an error called **nondisjunction,** which by definition is the failure to separate or disjoin. Aneuploidy will result whether nondisjunction occurs in meiosis I or meiosis II (▶ Fig. 48.6). When nondisjunction occurs in meiosis I, and the homologous chromosomes fail to separate, the result is that one daughter cell will receive both chromosomes in that pair. The second meiotic division will separate the sister chromatids as expected, however, due to the meiosis I nondisjunction error, two of these cells will now receive two chromosomes, instead of one, while the other two daughter cells will continue without DNA material. Fertilization with a haploid gamete will then result in either trisomy for that particular chromosome, or monosomy for that particular chromosome. In this case, none of the possible fertilization events will have the expected number of chromosomes.

On the other hand, when nondisjunction occurs in meiosis II, half of the fertilization events will lead to a disomic cell, with the expected number of chromosomes, since the first meiotic division was successful. However, nondisjunction between the sister chromatids will result in one gamete receiving two chromosomes for that particular chromosome pair, and fertilization with a haploid gamete will result in either a trisomy for that particular chromosome, or monosomy when the haploid gamete is the only chromosome present.

Trisomy 21, the chromosome composition that leads to an individual with Down syndrome, includes three copies of chromosome 21 and is one of the most common types of trisomy. ▶ Fig. 48.7 illustrates the karyotypes of three different trisomies in humans, including trisomy 21. A female with a trisomy 21 karyotype is read as 47,XX, +21, or 47 total chromosomes, with two X chromosomes indicating a female, and an additional 21st chromosome. Down syndrome occurs in 1 in every 800 live births and, along with trisomy 13 and trisomy 18, is

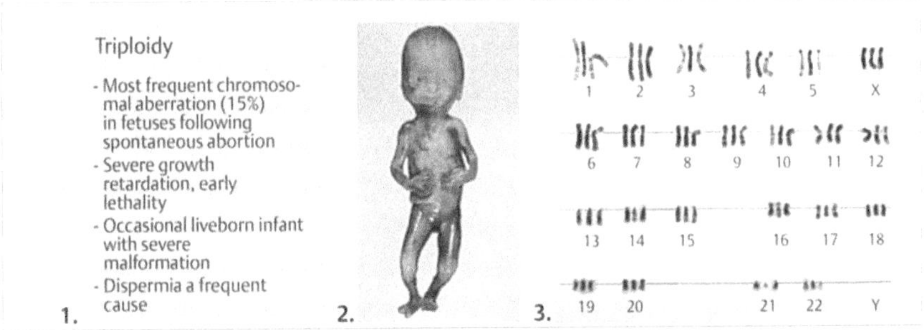

VII

Triploidy

- Most frequent chromosomal aberration (15%) in fetuses following spontaneous abortion
- Severe growth retardation, early lethality
- Occasional liveborn infant with severe malformation
- Dispermia a frequent cause

1.

2.

3.

1 2 3 4 5 X

6 7 8 9 10 11 12

13 14 15 16 17 18

19 20 21 22 Y

Fig. 48.5 Triploidy. (Source: Passarge E, ed. Color Atlas of Genetics. 4th Edition. New York, NY. Thieme; 2012.)

VII

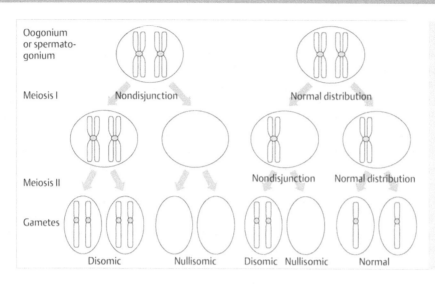

Fig. 48.6 Nondisjunction in meiosis I or meiosis II. (Source: Passarge E, ed. Color Atlas of Genetics. 4th Edition. New York, NY. Thieme; 2012.)

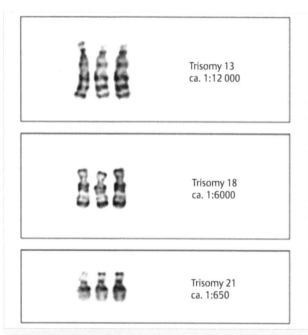

Trisomy 13
ca. 1:12 000

Trisomy 18
ca. 1:6000

Trisomy 21
ca. 1:650

Fig. 48.7 Autosomal trisomies in humans. (Source: Passarge E, ed. Color Atlas of Genetics. 4th Edition. New York, NY. Thieme; 2012.)

one of the few chromosomal trisomies that will possibly lead to a live birth.

On the other hand, monosomy of an entire chromosome is almost always lethal, with the exception of monosomy X, which is the presence of a single X chromosome. This chromosome composition leads to a female with Turner syndrome and a karyotype represented by 45,X, indicating 45 total chromosomes and only one X chromosome (▶ Fig. 48.8).

48.5 Structural Variations

Another type of chromosome abnormality, collectively known as copy number variants (CNVs), are **deletions**, or loss of chromosomal segments, and **duplications**, or gain of chromosomal

segments. Deletions and duplications can occur by chromosomal breakage, unequal crossing over between misaligned chromosomes, and abnormal segregation between translocations and inversion, examples of which will be provided in this section. When an individual has a deletion of a chromosomal segment, they are then monosomic for the homologous region (▶ Fig. 48.9), whereas an individual who has a duplication of a chromosomal segment is trisomic for that region.

Abnormalities of individual chromosomes tend to occur when chromosomes break and are then reconstructed in an abnormal manner. Such abnormalities will typically include a **derivative chromosome**, which is one that has been structurally rearranged, but retains its centromere. A chromosome rearrangement is referred to as **balanced** when the final set of chromosomes has the expected complement of chromosomal material. Balanced rearrangements typically do not have a phenotypic effect and can sometimes be passed through generations without the family members' awareness of the existence of a chromosomal rearrangement in their family.

One such balanced rearrangement is an **inversion**. Inversions occur when a single chromosome breaks in two places and is reconstructed with the segment between those two breaks inverted. There are two known types of inversions. When both breaks occur on the same chromosome arm, and hence do not include the centromere in the inversion, it is termed a **paracentric** inversion, whereas **pericentric** inversions include a break on each arm, and the inverted material includes the centromere (▶ Fig. 48.10).

One very common and well described inversion seen in humans is a small pericentric inversion of chromosome 9, known as inv(9)(p11q12). However, the inv(9)(p11q12) has not been described with any phenotypic abnormality in those individuals that carry it, and has not been associated with any significant risk of miscarriage or offspring with unbalanced karyotypes.

Although inversions do not typically have a phenotypic effect on the carrier, there is the possibility they could produce unbalanced gametes. Due to the inverted material, the chromosome pairs are forced to create a loop during meiosis I in order for homologous regions of DNA to pair up for recombination purposes. The recombination that occurs between the inverted material can result in either balanced gametes or unbalanced

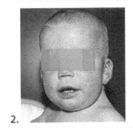
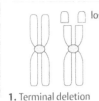
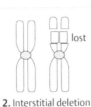
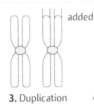

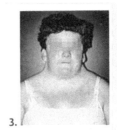

1.

2.

3.

Fig. 48.8 Monosomy X (Turner syndrome; 45, XO). (Source: Passarge E, ed. Color Atlas of Genetics. 4th Edition. New York, NY. Thieme; 2012.)

lost

lost

added

normal

cen

q

p

q

q

1. Terminal deletion   2. Interstitial deletion   3. Duplication   4. Isochromosome for the long arm (q) of the X chromosome

Fig. 48.9 Deletion, duplication, isochromosome. (Source: Passarge E, ed. Color Atlas of Genetics. 4th Edition. New York, NY. Thieme; 2012.)

VII

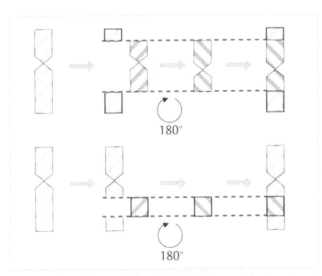

180°

180°

Fig. 48.10 Inversion. (Source: Passarge E, ed. Color Atlas of Genetics. 4th Edition. New York, NY. Thieme; 2012.)

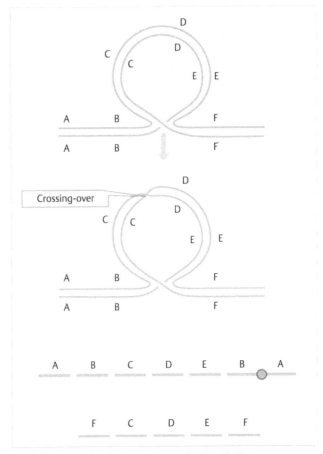

Fig. 48.11 Aneursomy by recombination. (Source: Passarge E, ed. Color Atlas of Genetics. 4th Edition. New York, NY. Thieme; 2012.)

gametes (▶ Fig. 48.11). When recombination occurs with a paracentric inversion, the resulting recombinant chromosomes are typically **acentric** or **dicentric** and, as such, typically do not lead to a live birth. What this means clinically for an individual with a paracentric inversion is that the chance of having a liveborn child with an unbalanced chromosome abnormality is actually quite low. Recombination between chromosomes with pericentric inversions, however, typically results in gametes with both duplications and deletions of chromosome material. The duplicated and deleted segments of chromosomes are those that are distal to the inversion. It has been estimated that an individual with a pericentric inversion has a 5–10% chance of having a live-birth with an unbalanced karyotype.

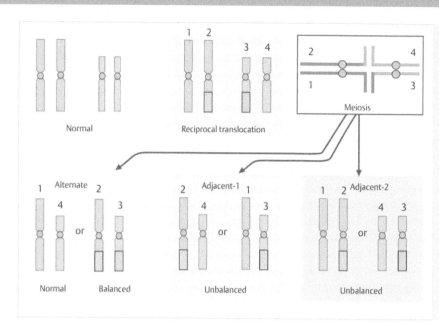

Fig. 48.12 Reciprocal translocation. (Source: Passarge E, ed. Color Atlas of Genetics. 4th Edition. New York, NY. Thieme; 2012.)

48.6 Translocations

An individual can also carry a balanced **translocation** within their own chromosomes without phenotypic effect. There are two main types of translocations. The first is a **reciprocal transloca-tion,** which involves breakage of two non-homologous chromosomes and then exchange of the broken sections (▶ Fig. 48.12). The total number of chromosomes in this case remains the same and the translocation is balanced if none of the broken segments were deleted or duplicated during the exchange.

However, individuals with balanced reciprocal translocations are at risk for offspring with unbalanced translocations. ▶ Fig. 48.13 demonstrates the possible zygotes that can occur with each pregnancy for an individual who carries a balanced reciprocal translocation between the long arm of chromosome 1 and the short arm of chromosome 20. When a gamete is fertilized with a single chromosome 1 and a single chromosome 20 that are not involved in this individual's translocation, the zygote is expected to inherit the typical 46 total chromosomes without being a carrier for the reciprocal translocation. And when a gamete is fertilized that contains only the two chromosomes involved in the translocation, that zygote will be a carrier of the balanced translocation without phenotypic effect. The clinical abnormalities for the zygote occur due to two separate possibilities. The first is when the single chromosome 1 not involved in the translocation is included in a gamete with the derivative chromosome 20 that includes missing DNA from its short arm as well as additional DNA from chromosome 1. When this gamete is fertilized, the zygote will be a carrier of an unbalanced reciprocal translocation, which will include partial trisomy of chromosome 1 and partial monosomy of chromosome 20. The second possibility for a zygote that will be expected to have clinical abnormalities is when the derivative chromosome 1 that includes missing DNA from its long arm and additional DNA from chromosome 20 is included in a gamete with a single chromosome 20 that is not involved in the translocation. The fertilization of this gamete will also result in a zygote with an unbalanced reciprocal translocation, this time with partial trisomy 20 and partial monosomy 1. Due to the unbalanced chromosome material in each of these two zygote possibilities, these pregnancies may not come to full term, and if they do, they are expected to have some physical and/or intellectual abnormalities. As most translocations are unique to a family, it is often challenging to find published literature regarding possible clinical outcomes. Therefore, it is not always feasible that a family can be given a list of clinical features or prognosis for an individual with an unbalanced translocation.

48.6.1 Link to Pathology: Translocation Carriers May Experience Multiple Miscarriages

From a clinical standpoint, translocations should be suspected in families with a history of multiple miscarriages and affected family members, as seen in ▶ Fig. 48.14. The healthy adult siblings in this family could be balanced translocation carriers who are passing on unbalanced chromosomal material, resulting in miscarriages and a child with clinical features including, in this case, developmental delay, short stature, and dysmorphic features.

The second common type of translocation is a **Robertsonian translocation**, which only occurs between two acrocentric chromosomes (chromosomes 13, 14, 15, 21, and 22). The short arms of these chromosomes are lost and fusion occurs near the centromeres, which results in a karyotype with only 45 chromosomes instead of 46. While all combinations of Robertsonian translocations have been seen, the 13q14q and 14q21q are the most commonly detected in the human population. Although the short arms are lost, the Robertsonian translocation is still considered a balanced chromosome abnormality since the short arms of all the acrocentric chromosomes carry multiple copies of genes for ribosomal RNA.

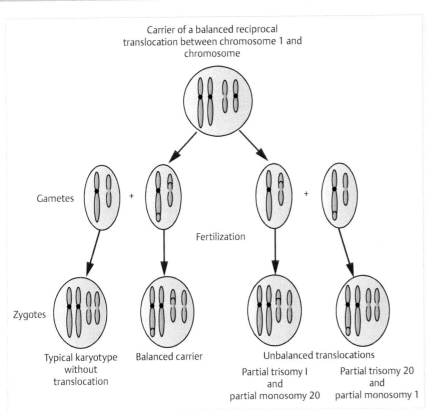

Carrier of a balanced reciprocal translocation between chromosome 1 and chromosome

Gametes

+

Fertilization

+

Zygotes

Typical karyotype without translocation

Balanced carrier

Unbalanced translocations

Partial trisomy I and partial monosomy 20

Partial trisomy 20 and partial monosomy 1

Fig. 48.13 Possible zygotes that can occur with each pregnancy for an individual who carries a balanced reciprocal translocation.

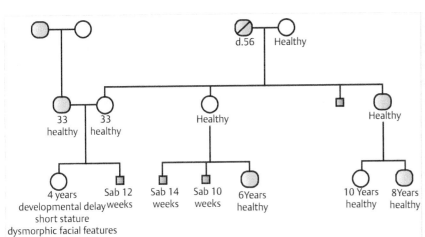

d.56   Healthy

33 healthy   33 healthy   Healthy   Healthy

4 years developmental delay short stature dysmorphic facial features

Sab 12 weeks   Sab 14 weeks   Sab 10 weeks   6 Years healthy

10 Years healthy   8 Years healthy

Fig. 48.14 A pedigree example of a family who would be appropriate to offer cytogenetic testing. Abbreviation: Sab, Spontaneous abortion.

VII

48.6.2 Link to Pathology: Robertsonian Translocation Can Cause Down Syndrome

While a balanced Robertsonian translocation does not typically have any clinical implications for the carrier, there is the possibility for unbalanced material to be passed to the next generation. ▶ Fig. 48.15 demonstrates the possible zygotes that can occur with each pregnancy for an individual who carries a 14;21 Robertsonian translocation. When a gamete with a single chromosome 14 and single chromosome 21 are fertilized, the zygote is expected to inherit the typical 46 total chromosomes without even being a carrier for the balanced translocation.

And when a gamete that contains only the 14;21 Robertsonian translocation is fertilized, that zygote will be a carrier of the balanced translocation without phenotypic effect. The clinical abnormalities for the zygote occur due to two separate possibilities. The first is when the Robertsonian translocation is included in a gamete with either a typical chromosome 14 or a typical chromosome 21, which will result in a zygote with either trisomy 14, which is a lethal condition, or trisomy 21, which is also known as Down syndrome. The second type of possibility is when a gamete includes only a single chromosome 14 and is missing the DNA material from chromosome 21, or vice versa which includes a single chromosome 21 without the DNA material from chromosome 14. Fertilization of

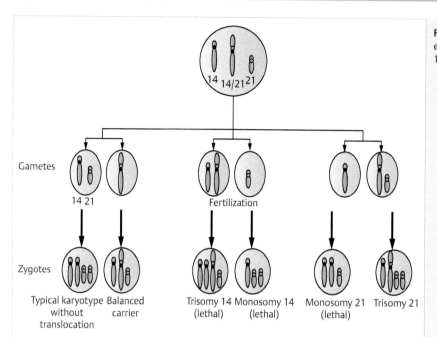

Fig. 48.15 Possible zygotes that can occur with each pregnancy for an individual who carries a 14;21 Robertsonian translocation.

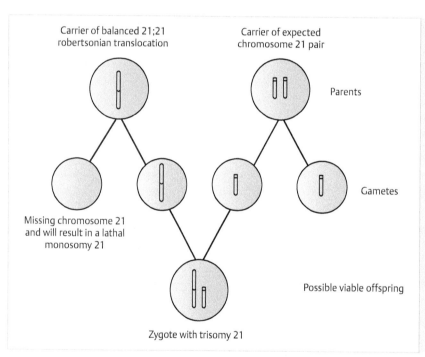

Fig. 48.16 Possible zygotes that can occur with each pregnancy for an individual who carries a 21;21 Robertsonian translocation

these gametes will result in additional lethal conditions, monosomy 14 and monosomy 21, respectively.

While considering the possible outcomes for pregnancies for an individual with a Robertsonian translocation, it is also important to consider the patient who carries a 21;21 Robertsonian translocation. This particular translocation is still considered a balanced translocation, however, since both of this individual's chromosome 21 are fused together, their options for gametes include either both number 21 chromosomes, which will always result in trisomy 21, or a gamete without chromosome 21 DNA, which will always result in the lethal monosomy 21 (▶ Fig. 48.16).

48.7 Mosaicism

No matter the chromosome abnormality, including polyploidy, aneuploidy, translocations, copy number variants, etc., if an individual has one cell line with one complement of chromosomes and a second, or even third or fourth, cell lines with a different complement of chromosomes, this is referred to as **mosaicism**. As seen in ▶ Fig. 48.17, nondisjunction during mitosis can lead to an individual with mosaicism. In this example, some of the cells will contain the expected disomy complement, while other cells will now contain a trisomy of a particular chromosome, and the monosomy cells are typically lethal. When an individual is mosaic for more than one cell line, this may lead to a slightly altered phenotype with possibly milder features than may be expected for the disease-associated cell line, due to the fact that not every cell has the chromosomal complement of the disease-associated chromosome abnormality. Unfortunately, it is currently impossible to determine which percentage of an individual's cells contains each chromosome compliment. Therefore, it can also be difficult when discussing with a family what possible clinical features to expect, as it will always be unknown what percentage of cells contains the disease-causing chromosome complement.

Germline mosaicism refers to a phenomenon in which an individual's somatic cells do not have a chromosomal abnormality, but an abnormality is present in their germline cells, such as ovum or sperm. Germline mosaicism is often discussed when a couple has a child who has been found to have a genetic abnormality, which can include a chromosomal abnormality or any other type of DNA change, not found in either parent. The child's abnormality may be *de novo*, meaning it is occurring in the family for the first time in that child, or it may be that the parent's genetic test was negative because the abnormality is located only in their germ cells. This is important to review with parents in terms of recurrence risk, since they would have a chance to pass on this abnormality again if it is present in their gametes. When a possible *de novo* change is discovered, these families are often offered prenatal diagnosis for this specific abnormality for future pregnancies since germline mosaicism cannot be ruled out (▶ Table 48.1).

48.8 Cytogenetic Clinical Laboratory Tests

There are multiple cytogenetic tests that can be ordered for the previously discussed chromosome abnormalities. A chromosomal karyotype can diagnosis large chromosome abnormalities, such as polyploidy, trisomy, or monosomy of individual chromosomes, and some inversions, translocations, and CNVs if the material involved is at least 5 megabases (Mb), typically at 550 band level and at 100x on the microscope.

Another common cytogenetic test is **fluorescence *in situ* hybridization** (FISH), which uses a small DNA sequence, called a probe, which has sequence complimentary to a specific DNA sequence on a chromosome and has attached to it a fluorescent. ▶ Fig. 48.18 demonstrates that the fluorescent signal of each specific probe can be seen under a microscope that is fitted with a filter that allows the fluorescence to be visible. Each probe will have its own fluorescent color assigned, and while it is expected to see two copies of each autosomal chromosome probe, missing or additional fluorescent probes will indicate when a CNV is present. FISH is not only useful for determining the presence or absence of a particular region of DNA, but can also determine the number of chromosomes present, as well as the organization of the chromosomes.

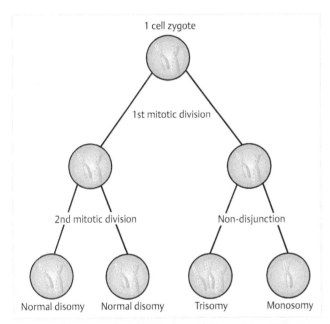

Fig. 48.17 An example of nondisjunction during a mitotic division that lead to mosaicism.

1 cell zygote

1st mitotic division

2nd mitotic division

Non-disjunction

Normal disomy   Normal disomy   Trisomy   Monosomy

Table 48.1 Nomenclature of chromosome abnormalities per the International System for Human Cytogenetic Nomenclature (ISCN)

Type of chromosome abnormality	Correct nomenclature examples per ISCN	Notes:
Triploidy	69,XXX	
Trisomy	47,XY + 18 47,XXX	Gain of an autosomal chromosome is indicated by +
Monosomy	45,X	
Mosaicism	46,XX/46,XY 46,XY/47,XXY + 21	Each cell line is separated by /
Inversion	46,XY,**inv**(9)(p11q12)	**inv** represents inversion
Reciprocal translocation	46,XX,**der**(9)t(9;14)(p24.2;q32.11)	**der** represents the derivative chromosome. This is an unbalanced translocation between the short arm of chromosome 9 and the long arm of chromosome 14
Robertsonian translocation	45,XX,**der**(14;21)(q10q10)	A balanced Robertsonian translocation between chromosome 14 and chromosome 21. q10 represents the centromere

VII

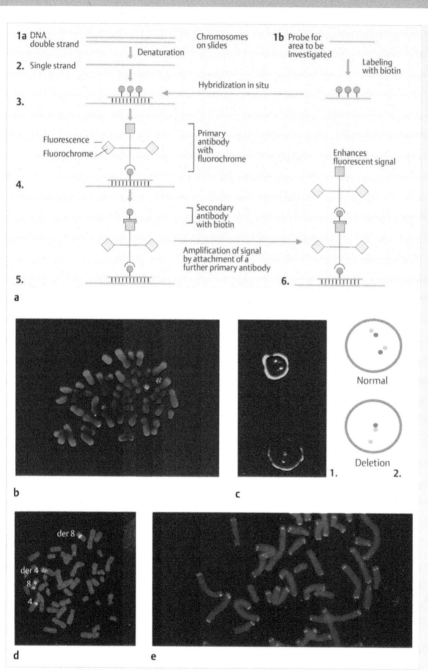

1a DNA double strand — Chromosomes on slides — **1b** Probe for area to be investigated

Denaturation

2. Single strand

Labeling with biotin

Hybridization in situ

3.

Fluorescence — Fluorochrome

Primary antibody with fluorochrome

Enhances fluorescent signal

4.

Secondary antibody with biotin

Amplification of signal by attachment of a further primary antibody

5.

6.

a

Normal

Deletion

1.   2.

b

c

der 8

der 4

8

4

d   e

Fig. 48.18 **(a)** Principle of fluorescence in situ hybridization. **(b)** Example of FISH analysis in metaphase. **(c)** Interphase FISH analysis. **(d)** Translocation 4;8 **(e)** Telomere sequences in metaphase chromosomes. (Source: Passarge E, ed. Color Atlas of Genetics. 4th Edition. New York, NY. Thieme; 2012.)

There are certain well-known genetic diseases that can be diagnosed using FISH analysis alone, such as 22q11.2 deletion syndrome (▶ Fig. 48.19) and cri du chat syndrome. FISH can also be used for parental testing when a child has been found to have a CNV. In some cases, the parents' samples can be tested using just a FISH probe for the region that has been duplicated or deleted in their child, instead of evaluating an entire karyotype.

However, since FISH analysis relies on probes that locate only a specific region of a chromosome, if the abnormality of that chromosome is located in a different region, the FISH analysis will return a normal result. In this case, additional genetic testing will need to be pursued for a diagnosis. Additional genetic testing is also required for chromosomal abnormalities that are smaller than 5 Mb, since these will not be seen on a standard karyotype. In cases such as these, **chromosomal microarray** technology can be utilized.

Chromosomal microarrays examine the genome for deletions and duplications, but at a much higher resolution than a G-banded karyotype. This array-based genomic copy number analysis, which can include array-based comparative genomic hybridization (CGH) and single nucleotide polymorphism (SNP) arrays, was designated as a first-tier clinical diagnostic test for individuals with neurodevelopmental disorders and congenital anomalies in 2010 (3). Microarray can not only detect areas of known recurrent CNVs, but it is also capable of identifying novel CNVs that may then become associated with a recurrent clinical picture and provide diagnoses for patients and their families. The use of SNP array platforms also allow for the

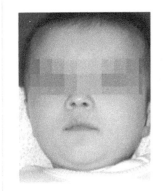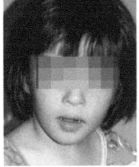

1. 5 months old   **2.** 3¼ years old

a

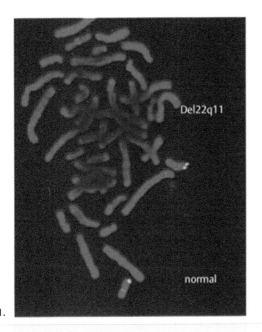

b   **1.**

Fig. 48.19 **(a)** Deletion 5p- (Cri-du-chat-syndrome). **(b)** Deletion 22q11. (Source: Passarge E, ed. Color Atlas of Genetics. 4th Edition. New York, NY. Thieme; 2012.)

detection of regions of homozygosity, which may result from uniparental disomy, which is covered in Chapter 52.

The microarray itself is a glass slide coated with probes, which in this case are short pieces of DNA called oligonucleotides,

placed across unique regions of the human genome, with the exception of centromeres, telomeres, and satellites. These probes span the human genome and have the capability to identify duplications and deletions as small as 0.3 Mb - 0.5 Mb, including intragenic abnormalities. The patient's DNA and control DNA are fluorescently labeled and hybridized to the glass slide before plotted against a map of the human genome. While a karyotype can provide G-band coordinates for regions of deletions or duplications, microarray is able to provide the expected size of the CNV, the CNV's exact coordinates, and the genes included in the region.

One such condition that has only been discovered due to microarray technology is a recurrent 593 kilobase (kb) deletion (located at breakpoints 29,606,852–30,199,855) that has now been associated with 16p11.2 deletion syndrome. Some of the common features reported in those with 16p11.2 deletion syndrome include cognitive delay, developmental delay, autistic features or autism spectrum disorder, a predisposition to being overweight or obese, and seizures (2). Although this deletion cannot be detected by routine analysis of G-banded chromosomes, once detected in a proband via microarray, FISH probes are available to test additional family members for the presence or absence of this region.

Microarray technology can only detect unbalanced chromosome abnormalities, since balanced translocations and inversions do not change the DNA copy number. However, the advantage to this is that when a translocation is detected on karyotype, microarray can confirm if it is balanced or unbalanced. Low level mosaicism and sequence variants are also not detected with this type of technology.

References

[1] Shaffer LG, McGowan-Jordan J, Schmid M, eds. An International System for Human Cytogenetic Nomenclature. S Karger: Basel; 2013

[2] Miller DT, Chung W, Nasir R, et al. 16p11.2 Recurrent Microdeletion. 2009 Sep 22 [Updated 2015 Dec 10]. In: Pagon RA, Adam MP, Ardinger HH, et al., editors. GeneReviews® [Internet]. Seattle (WA): University of Washington, Seattle; 1993–2016. Available from: https://www.ncbi.nlm.nih.gov/books/NBK11167/

[3] Miller DT, Adam MP, Aradhya S, et al. Consensus statement: chromosomal microarray is a first-tier clinical diagnostic test for individuals with developmental disabilities or congenital anomalies. Am J Hum Genet. 2010; 86(5): 749–764

VII

Review Questions

1. Which of the following chromosome types is not found in human cells?
   A) Acrocentric
   B) Metacentric
   C) Submetacentric
   D) Telocentric

2. Which of the following karyotypes represents a viable monosomy?
   A) 45, X
   B) 45, Y
   C) 47, XX + 21
   D) 47, XY + 18

3. Which of the following types of chromosomes is most likely to be involved in a Robertsonian translocation?
   A) Metacentric
   B) Submetacentric
   C) Acrocentric
   D) Telocentric

4. Which of the following laboratory tests would be most appropriate for detecting a trisomy of chromosome 21?
   A) Chromosomal microarray
   B) Comparative genomic hybridization
   C) G-banding
   D) SNP microarray

Answers

1. **The correct answer is D**. Metacentric chromosomes (answer choice B) are those with their centromere near the middle of the chromosome, while submetacentric chromosomes (answer choice C) have a centromere that is off center but still between the middle and the end of the chromosome. Acrocentric chromosomes (answer choice A) have a centromere that is closer to the end of the p arm, and telocentric chromosomes have centromeres located at the terminal end of the chromosome, meaning that this type of chromosome only has one chromosome arm. Human cells contain metacentric, submetacentric, and acrocentric chromosomes, but do not contain telocentric chromosomes.

2. **The correct answer is A**. Monosomy is a type of aneuploidy in which a cell has inherited only one copy of a chromosome instead of the normal two. This condition is most commonly a result of nondisjunction, or the missegregation of chromosomes during cell division. Although both answer choices A and B are karyotypes representative of monosomies, only answer choice A is known to produce a viable offspring. The condition that results from this type of monosomy is known as Turner syndrome. Another type of aneuploidy is trisomy, or the presence of three copies of a chromosome, which is represented by the karyotypes in answer choices C and D. Answer choice C is a karyotype of an individual with Down syndrome, whereas answer choice D is a karyotype of an individual with Edwards syndrome. Although those affected by Edwards syndrome die before birth, it is not caused by a monosomy, therefore answer choice D is not correct.

3. **The correct answer is C**. Robertsonian translocations are a common type of translocation that only occurs between two acrocentric chromosomes. The acrocentric chromosomes in humans are chromosomes 13, 14, 15, 21, and 22. In these translocations, the short arms of these chromosomes are lost and fusion occurs near the centromeres, which results in a karyotype with only 45 chromosomes instead of 46. Robertsonian translocations are usually balanced and, since the short arms of all the acrocentric chromosomes carry multiple copies of genes for ribosomal RNA, there are typically no clinical implications for the carrier of this type of translocation.

4. **The correct answer is C**. G-banding is a technique in which metaphase chromosomes are treated with Giemsa stain, allowing for the visualization and identification of all 46 chromosomes in the human cell. Any gross chromosomal abnormality, such as a trisomy of chromosome 21, would be readily identified by this method. The other listed technologies are needed for examining the genome at higher resolution to find smaller genetic abnormalities. Comparative genomic hybridization (CGH) (answer choice B), SNP microarrays (answer choice D) and chromosomal microarray technology (answer choice A) can be used to examine the genome for deletions and duplications and for the detection of regions of homozygosity.

49 Single Gene Disorders: Autosomal Dominant and Recessive Inheritance

At the conclusion of this chapter, students should be able to:

- Define the terms phenotype, genotype, heterozygote, homozygote, dominant, and recessive
- Distinguish between the characteristic features of autosomal dominant and recessive inheritance patterns
- Explain the Punnett square and illustrate its use in predicting potential genotypes and phenotypes
- Produce and interpret a pedigree of various single-gene inheritance patterns
- Use information in a pedigree to calculate probabilities of autosomal dominant and recessive diseases
- Explain family genome analysis and how genome analysis can reveal Mendelian inheritance patterns

In order to discuss genetics in relation to human disease, it is important to know the fundamental vocabulary used in clinical genetics. This chapter will focus on common inheritance patterns and how they relate to genetic disease recurrence risk for the next generation.

It is an individual's **genotype**, or genetic make-up, that determines their **phenotype**, or the observable characteristics of an individual. The human genotype consists of the genes lined along paired, or homologous, chromosomes. Due to variations in genes, called **alleles**, and considering that one chromosome in a homologous pair is inherited from the mother and the other is inherited from the father, the matching set of genes on a homologous pair may not be an identical set.

The most common, and therefore expected, allele in a population is referred to as the **wild type allele**, while a variation or permanent change from the wild type allele is referred to as a **variant**. A variant can be disease-causing, or it can be completely benign. Some variants can also be advantageous to an individual, such as that seen with the variant associated with sickle cell trait. Sickle cell disease is inherited in an **autosomal recessive** fashion, which means an individual must inherit two disease-causing, or **pathogenic**, sickle cell variants, one from each parent, in order to exhibit disease. These pathogenic variants can be **homozygous**, which means that both alleles are the same abnormal variant, or they can inherit two different pathogenic variants, termed **compound heterozygous.** However, an individual who is **heterozygous** for the sickle cell variant, meaning they have one copy of the sickle cell variant on one chromosome and a wild type allele on the homologous chromosome, will not exhibit features of sickle cell disease. In this case, being heterozygous for this condition is known to be advantageous as it provides the individual some resistance against malaria. In contrast, in the case of a disease inherited in an **autosomal dominant** fashion, an individual heterozygous for a pathogenic variant is expected to exhibit the disease phenotype associated with that gene.

49.1 Punnett Square

The **Punnett square** is a tool used in genetics to determine the probability of a specific genotype in offspring. The simplest Punnett square, which focuses on only one trait or genetic disease, uses a grid with four boxes in two columns and two rows. The genotype of one parent is indicated along the top of the grid and the genotype of the other parent is marked along the left side. The genotype of a simple Punnett square will contain only two letters as representation of the trait or disease.

For instance, if one parent has an autosomal dominant genetic disease, the allele for the disease can be represented by a capital letter, such as "D", while the wild type allele can be represented by a lower case letter, such as "d". The genotype of this individual for the purpose of the Punnett square would therefore be "Dd" and these two letters will go along the top two columns. On the other hand, the parent who is unaffected by the disease will be represented by two wild type alleles, or "dd", and these two letters will go along the left of the two rows. The empty squares are then filled in by combining the letters of the parental genotypes. For this example, the letters combine to form "Dd" and "dd" in the squares of the top row and "Dd" and "dd" in the squares of the bottom row (▶ Fig. 49.1). These four squares now represent the genotype

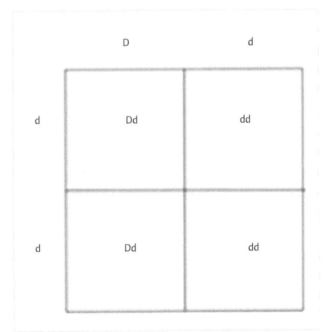

Fig. 49.1 A Punnett square representing the genotype outcomes between a genotype consisting of an autosomal dominant "D" allele and a wild type "d" allele, crossed to a genotype with two wild type alleles.

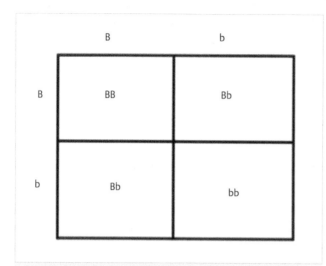

Fig. 49.2 A Punnett square representing the genotype outcomes between two genotypes consisting of a wild type "B" allele and an autosomal recessive "b" allele.

VII

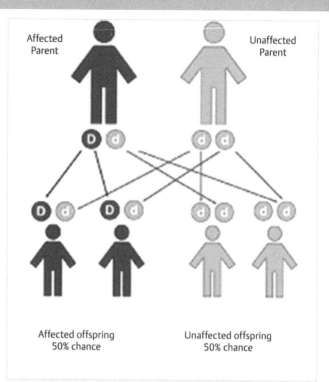

Fig. 49.3 A visual based on Punnett square logic for a couple who would like to know the recurrence risk for an autosomal dominant genetic condition. This represents the recurrence risks for a single pregnancy.

probabilities that can occur between these two parents. This also reveals that there is a 2 in 4, or 50%, chance that each offspring will inherit the dominant "D" allele and exhibit features of the autosomal dominant disease, and a 2 in 4, or 50%, chance that each offspring will inherit two wild type "dd" alleles and not exhibit features of the autosomal dominant disease.

The possible genotype outcomes between two individuals who are carriers for an autosomal recessive condition can also be represented in a Punnett square. In this case, each parent would have one wild type allele represented by an uppercase letter, such as "B", while the allele representing the autosomal recessive carrier allele would be represented by a lower case letter, such as "b". In this example, the letters combine to form "BB" and "Bb" in the squares of the top row and "Bb" and "bb" in the squares of the bottom row (▶ Fig. 49.2). This particular Punnett square reveals that there is a 1 in 4, or 25%, chance that an offspring will inherit two wild type "BB" alleles and will not exhibit features of the autosomal recessive condition, nor are they a carrier of the condition, and a 2 in 4, or 50%, chance that the offspring will inherit an autosomal recessive allele as well as a wild type allele, meaning that their genotype would be "Bb" and they would be a carrier of the condition. According to this Punnett square, there is also a 1 in 4, or 25%, chance that an offspring will inherit both autosomal recessive "bb" alleles and exhibit features of the disease.

As discussed, the Punnett square provides us with the percentile chance for each offspring genotype in the next generation. Therefore, the Punnett square can be a visual tool provided to couples who are asking about recurrence risk for a known genetic disease in their family. For instance, ▶ Fig. 49.3 provides us with the recurrence risk for an autosomal dominant condition for a couple, where the capital letter "D" represents the allele associated with the autosomal dominant disease, and the lower case letter "d" represents the wild type allele. The individual who is affected by the genetic condition in this example will be represented by "Dd" and the individual

who is not affected will be represented by "dd". Like the Punnett square, this visual also provides the couple with an opportunity to see each of their alleles paired up, although in this discussion it is important to stress that this is the recurrence risk for each individual pregnancy. In this case, the couple can see that the parent with the autosomal dominant genetic disease has a 50% chance to pass on the allele associated with disease and a 50% chance to pass on the wild type allele. When combined with their partner's two wild type alleles, this couple can visualize the possible outcomes for each pregnancy, including that they will have a 50% chance to have a child with the same autosomal dominant condition as the affected parent (Dd), and a 50% chance to have a child who is unaffected (dd).

We can also use this approach to predict the outcome of a mating between individuals who are known to be carriers for an autosomal recessive condition (▶ Fig. 49.4). Here, we allow the upper case letter "R" to represent the wild type allele and the lower case letter "r" to represent the allele associated with the autosomal recessive disease. This provides us with a genotype of "Rr" for both known carriers in this couple. Therefore, both individuals in this example will each have a 50% chance to pass on the wild type allele and a 50% chance to pass on the allele associated with the autosomal recessive condition. And if we look at the possible outcomes for a single pregnancy, we can see that there will be a 25% chance to have a child who inherits two wild type alleles (RR) and is therefore not only unaffected, but who is also not a carrier for this condition, a 50% chance to have a child who inherits both a wild type allele and a pathogenic allele (Rr), who like the parents will not be

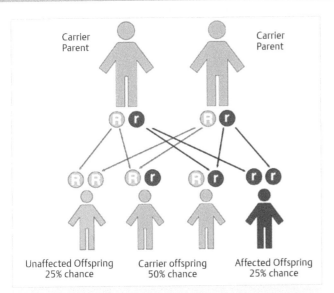

Fig. 49.4 A visual based on Punnett square logic for a couple who would like to know the recurrence risk for an autosomal recessive genetic condition when both individuals in the couple are known carriers for that condition. This represents the recurrence risks for a single pregnancy.

their children, and then dad's parents, before moving on to the other side of the pedigree to ask the same questions about the maternal family history.

Once this multiple-generation pedigree is drawn, health histories are added. While every pedigree will have a set of main questions to ask, which are determined by the practice specialty, it is also important to note that each pedigree should be tailored based on the proband's reason for referral. For instance, if a proband is referred for a discussion on the probability that their multiple café au lait spots are related to a possible diagnosis of neurofibromatosis, the medical health related questions asked about each family member will relate to signs and features of neurofibromatosis, such as (although not limited to) skin findings, eye findings, concern for developmental delay or intellectual disability, and any cancer diagnoses. Tailoring these questions for relatives will keep the focus on the proband's reason for visit and will avoid lengthy discussions about the proband's uncle's diagnosis of diabetes and hypertension, which are unrelated to the reason for referral.

affected but will be a carrier, and finally a 25% chance to have a child who inherits a pathogenic allele from each parent (rr) and will therefore be affected with this condition.

This same technique can be used in order to create visuals for couples who wish to discuss the recurrence risk for autosomal dominant conditions when both parents represent any of the possible genetic variations: both parents are affected, autosomal recessive conditions when only one parent is affected, autosomal recessive conditions when only one parent is a carrier, X-linked recurrence risks, and so on.

49.2 Pedigree Analysis

Another important visual in clinical practice is the **pedigree**, or family history. The pedigree indicates not only how additional family members are related to the **proband**, or index case in a family, but also indicates each family member's status with respect to different medical conditions. The National Society of Genetic Counselors Pedigree Standardization Task Force worked to create standardized pedigree nomenclature in the 1990s which are included in ▶ Fig. 49.5. These standardized symbols not only include basics such as circles to represent females and squares to represent males, but additional rules to indicate if someone is deceased or has an ongoing pregnancy, and shading rules in regards to labeling an individual with more than one medical condition.

To draw a pedigree, the proband, indicated by an arrow, is drawn first and then shaded appropriately based on their medical condition(s). The pedigree key can be created as the pedigree is drawn to indicate what each shaded mark is representing. The proband's siblings, with their gender and age, are added next, followed by the proband's parents. From here, it is generally a good idea to focus on one side of the family, such as the paternal family history including dad's siblings,

49.3 Pedigree Analysis of Autosomal Dominant and Recessive Diseases

As an example, ▶ Fig. 49.6 demonstrates a pedigree with a proband who was found to have congenital hearing loss. Using standardized symbols, it can be seen that the proband, a circle marked by an arrow, is a two month old female. And using the pedigree key to determine the meaning of the shading in the proband and additional relatives, it can also be seen that there is hearing loss in three direct generations in this family. The apparent transmission of a pathogenic hearing loss variant from one family member to the next in each generation is a classic example of likely autosomal dominant inheritance.

49.3.1 Link to Pathology

Diseases that are inherited in an autosomal dominant manner include familial hypercholesterolemia, Huntington disease, neurofibromatosis type1, hereditary spherocytosis, Marfan syndrome, acute intermittent porphyria, achondroplasia, and familial adenomatous polyposis.

On the other hand, the pedigree shown in ▶ Fig. 49.7 demonstrates hearing loss in two siblings, while neither parent is affected. The mother's brother also has a history of hearing loss, which suggests a possible autosomal recessive inheritance for this family's hearing loss.

49.3.2 Link to Pathology

Diseases that are inherited in an autosomal recessive manner include phenylketonuria, maple syrup urine disease, thalassemia, sickle cell anemia, cystic fibrosis, Tay-Sachs, albinism, Gaucher, Niemann-Pick, Fabry, Krabbe, metachromatic leukodystrophy, I-cell disease, and the glycogen storage diseases von Gierke, Cori's, Andersen's, Pompe's, McArdle's, and Hers' disease.

The next step for each of these families is to determine if the type of hearing loss of family members is similar in regards to

instructions:

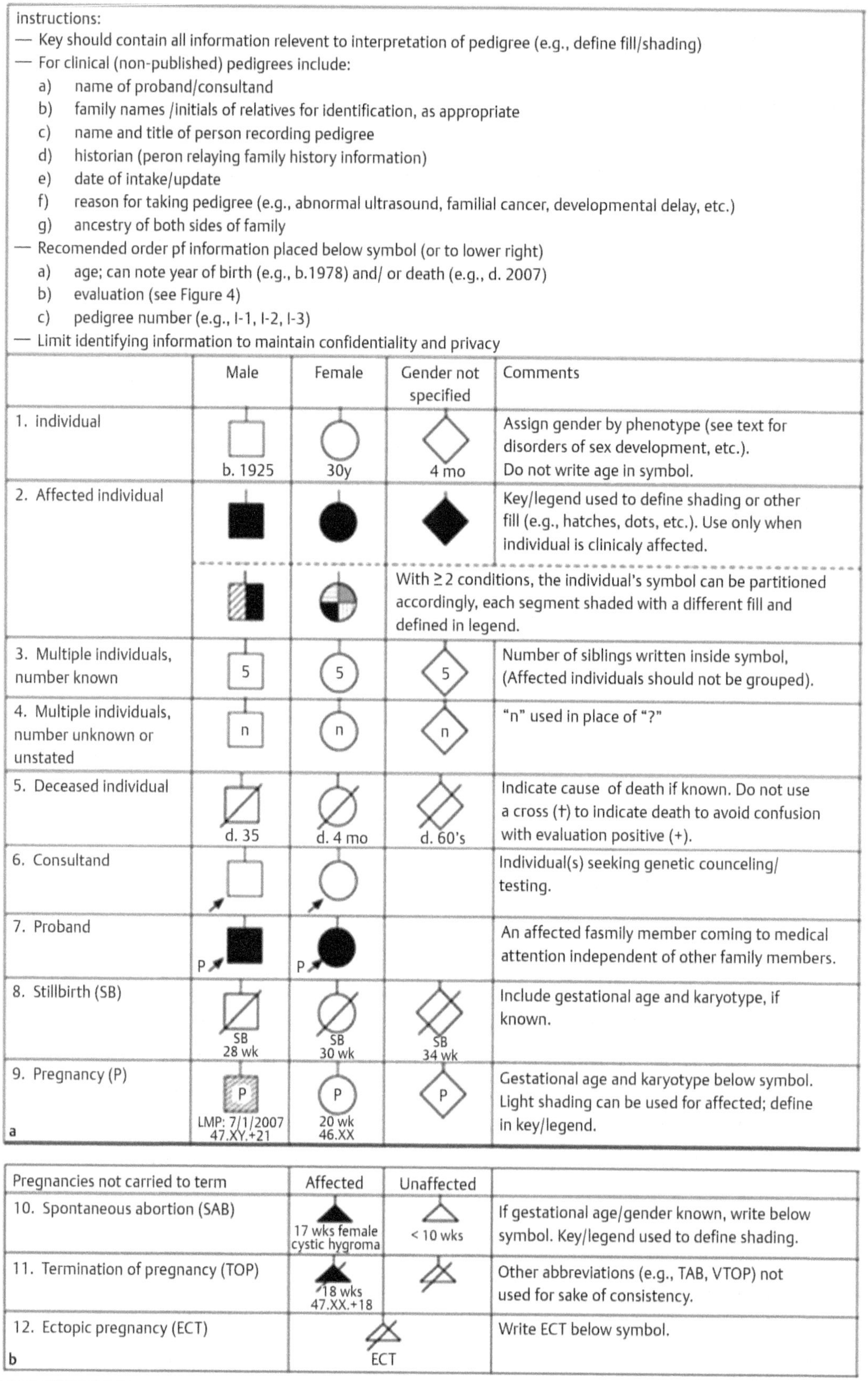

— Key should contain all information relevent to interpretation of pedigree (e.g., define fill/shading)
— For clinical (non-published) pedigrees include:
  a) name of proband/consultand
  b) family names /initials of relatives for identification, as appropriate
  c) name and title of person recording pedigree
  d) historian (peron relaying family history information)
  e) date of intake/update
  f) reason for taking pedigree (e.g., abnormal ultrasound, familial cancer, developmental delay, etc.)
  g) ancestry of both sides of family
— Recomended order pf information placed below symbol (or to lower right)
  a) age; can note year of birth (e.g., b.1978) and/ or death (e.g., d. 2007)
  b) evaluation (see Figure 4)
  c) pedigree number (e.g., I-1, I-2, I-3)
— Limit identifying information to maintain confidentiality and privacy

	Male	Female	Gender not specified	Comments
1. individual	b. 1925	30y	4 mo	Assign gender by phenotype (see text for disorders of sex development, etc.). Do not write age in symbol.
2. Affected individual				Key/legend used to define shading or other fill (e.g., hatches, dots, etc.). Use only when individual is clinicaly affected.
				With ≥ 2 conditions, the individual's symbol can be partitioned accordingly, each segment shaded with a different fill and defined in legend.
3. Multiple individuals, number known	5	5	5	Number of siblings written inside symbol, (Affected individuals should not be grouped).
4. Multiple individuals, number unknown or unstated	n	n	n	"n" used in place of "?"
5. Deceased individual	d. 35	d. 4 mo	d. 60's	Indicate cause of death if known. Do not use a cross (†) to indicate death to avoid confusion with evaluation positive (+).
6. Consultand				Individual(s) seeking genetic counceling/ testing.
7. Proband	P	P		An affected fasmily member coming to medical attention independent of other family members.
8. Stillbirth (SB)	SB 28 wk	SB 30 wk	SB 34 wk	Include gestational age and karyotype, if known.
9. Pregnancy (P)	P LMP: 7/1/2007 47.XY.+21	P 20 wk 46.XX	P	Gestational age and karyotype below symbol. Light shading can be used for affected; define in key/legend.

a

Pregnancies not carried to term	Affected	Unaffected	
10. Spontaneous abortion (SAB)	17 wks female cystic hygroma	< 10 wks	If gestational age/gender known, write below symbol. Key/legend used to define shading.
11. Termination of pregnancy (TOP)	18 wks 47.XX.+18		Other abbreviations (e.g., TAB, VTOP) not used for sake of consistency.
12. Ectopic pregnancy (ECT)		ECT	Write ECT below symbol.

b

Fig. 49.5 (a, b) Standardized pedigree symbols and definitions. (Source: Bennett RL, French KS, Resta RG, Doyle DL. Standardized human pedigree nomenclature: update and assessment of the recommendations of the national society of genetic counselors. J Genet Counsel, 2008;17(5):424-433.)

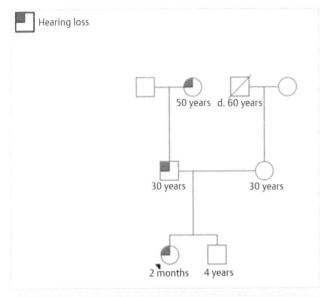

Hearing loss

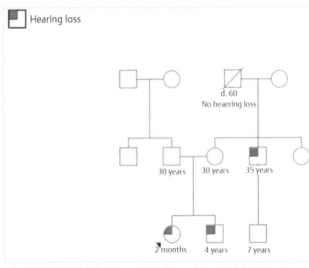

Hearing loss

Fig. 49.6 A pedigree demonstrating possible autosomal dominant hearing loss.

Fig. 49.7 A pedigree demonstrating possible autosomal recessive hearing loss.

age of onset, sensorineural versus conductive, and if the affected individuals have any other features, such as vision concerns, heart defects, or dysmorphic features, which may indicate a syndromic versus a non-syndromic type of inherited hearing loss. Once these questions have been asked and the answers added to the pedigree, the entire picture will help direct the differential diagnoses and possible genetic testing for each family.

49.4 Genome Analysis Can Reveal Mendelian Inheritance Patterns

The vast majority of clinical genetic testing available today is aimed at diagnosing genetic disorders of **Mendelian inheritance**. The term Mendelian inheritance comes from laws proposed by Gregor Mendel, a nineteenth-century Austrian monk who founded the modern science of genetics. A Mendelian trait is a genetic condition that is controlled by a single pathogenic variant. This is in contrast to multifactorial conditions, such as diabetes or high blood pressure, which may be due not only to multiple pathogenic variants that interact with each other, but environmental factors as well.

Although pedigree analysis can be used to determine if a genetic condition appears to be Mendelian, it must be followed up with genetic testing for a definitive diagnosis. Originally, genetic testing meant cytogenetics and chromosome analysis, which will be covered in Chapter 48), but molecular methods have now advanced genetic testing to allow for identification of pathogenic variants in individual genes. And when each patient is considered individually, medical practitioners can use genetic testing as a tool for the diagnostic odyssey each patient faces.

For example, if a patient has clinical features that align with a known genetic syndrome, the first step may be to pursue sequence analysis for the gene associated with that syndrome. A completed pedigree intake for the family presented in ▶ Fig. 49.7 would reveal that the hearing loss of those affected is known to be congenital, sensorineural, and non-progressive.

We would also learn that none of the affected family members have any other syndromic features. Armed with this information, the next step for this family would be to pursue genetic testing for the *GJB2* gene (also known as connexin 26), as pathogenic variants in the *GJB2* gene are the most common cause of autosomal recessive non-syndromic hearing loss. In European populations, there is a very common *GJB2* pathogenic variant, 35delG, which can be targeted with variant-specific analysis as a first step. Given the pedigree and the suggestion of autosomal recessive inheritance, the detection of two pathogenic variants would be expected. Therefore, if the variant-specific analysis detects the 35delG variant in the homozygous state, this is a diagnosis for the family. A diagnosis of hearing loss due to *GJB2* gene variants will not only provide this family with a recurrence risk, but may also provide reassurance that their affected family members will not be at risk for features associated with syndromic hearing loss. However, if the 35delG variant is found in the heterozygous state, or not detected at all, the clinician would continue to include not only full sequencing of the *GJB2* gene, but possibly a full panel of genes associated with autosomal recessive hearing loss.

We can also consider a patient who presents with a more generalized situation. Consider a non-dysmorphic 4-year old girl with ataxia and delayed gross motor development with a non-contributory family history. Since ataxia is not known to be associated with a single gene, the first step in her case would be to pursue a panel of genes associated with ataxia. However, should that panel return with negative or non-diagnostic results, the next step may be exome sequencing.

Exome sequencing is a technique which sequences the exons, or protein-coding regions, of the genome, which account for only 1% of the entire human genome. Exome sequencing is a fairly new diagnostic tool that has the ability to not only identify variants in known genes, but can also identify new candidate genes as well as clarify atypical presentations of well-known syndromes.

In the example of the current ataxia patient, exome sequencing identified that this little girl was homozygous for a pathogenic variant in the *SACS* gene, which is associated with an autosomal recessive form of ataxia. Since this gene is not commonly included on ataxia panels traditional testing would likely have been negative. Exome sequencing not only provided her with a diagnosis which can aid in her medical management, but also provided her family with future recurrence risk.

Exome sequencing may also identify variants in novel candidate genes, or genes that have not yet been implicated previously in human genetic disease. For instance, exome sequencing in three cohorts of individuals with unexplained intellectual disability revealed 39 females with *de novo* variants in the *DDX3X* gene. Exome sequencing and the results of this study changed the status of *DDX3X*, a gene previously unknown to be associated with human disease, to a disease-causing gene now known to be one of the more common causes of intellectual disability.

Exome sequencing has also provided a way to expand on previously well-known phenotypes. For example, Marfan syndrome, which is an autosomal dominant connective tissue disorder that mainly affects the cardiovascular, skeletal, and ocular systems, is known to be associated with pathogenic variants in the *FBN1* gene. An individual who is referred in for dilated aortic valve, ectopia lentis, and common skeletal findings such as tall stature and pectus excavatum will routinely undergo *FBN1* gene sequencing in order to obtain a diagnosis of Marfan syndrome. However, *FBN1* gene sequencing is typically not the first step in a patient who does not present with any of the classical features. In the case of a father and son who were both diagnosed with a diaphragmatic hernia, exome sequencing was pursued for a diagnosis and revealed a *FBN1* variant in both affected family members, as well as an additional unaffected, younger son.

Although a diagnosis of Marfan syndrome had not been previously considered for this family, the exome result prompted the clinicians to look closely at the family's additional clinical features. On clinical re-evaluation, the father and his affected son exhibited no physical features of Marfan syndrome, although the younger, unaffected son had a few minor clinical features. The clinicians also pursued echocardiograms for these three individuals, which would not have been part of their medical management before this exome sequencing finding, and found that while the father and his unaffected son had normal results, the affected son was found to have an aortic root diameter in the 95th centile. Additional ophthalmological exams also revealed findings associated with ectopia lentis in both the father and his affected son. The combination of the *FBN1* variant and closer clinical evaluation led to a diagnosis of Marfan syndrome for this family. Exome sequencing provided not only a diagnosis which changes this family's medical management, but it also expanded the phenotype of Marfan syndrome to include diaphragmatic hernia.

Suggested Readings

Beck TF, Campeau PM, Jhangiani SN, et al. FBN1 contributing to familial congenital diaphragmatic hernia. Am J Med Genet A. 2015; 167A(4):831–836

Bennett RL, French KS, Resta RG, Doyle DL. Standardized human pedigree nomenclature: update and assessment of the recommendations of the National Society of Genetic Counselors. J Genet Couns. 2008; 17(5):424–433

Liew WK, Ben-Omran T, Darras BT, et al. Clinical application of whole-exome sequencing: a novel autosomal recessive spastic ataxia of Charlevoix-Saguenay sequence variation in a child with ataxia. JAMA Neurol. 2013; 70(6):788–791

Snijders Blok L, Madsen E, Juusola J, et al. DDD Study. Mutations in DDX3X Are a Common Cause of Unexplained Intellectual Disability with Gender-Specific Effects on Wnt Signaling. Am J Hum Genet. 2015; 97(2):343–352

Review Questions

1. Which of the following is a characteristic feature of a pedigree for autosomal recessive disease?
   A) Disease is observed in every generation
   B) Females are affected more often than males
   C) Males are affected more often than females
   D) Males and females are equally affected

2. What is the probability that a father who is affected by an autosomal dominant disease will transmit this disorder to his daughter? Assume that the mother is phenotypically normal.
   A) 0%
   B) 25%
   C) 50%
   D) 75%
   E) 100%

3. A genetic analysis reveals that a patient carries the following variants in the beta-globin gene: the Glu6Val mutation on one chromosome and the Glu6Lys mutation on the other chromosome. Which of the following terms best describes this situation?
   A) Affected carrier
   B) Compound heterozygous
   C) Heterozygous
   D) Homozygous

4. Which of the following is the probability that a couple who are both carriers of the Glu6Val mutation in the beta globin gene will have a child with sickle cell disease?
   A) 0%
   B) 25%
   C) 50%
   D) 75%
   E) 100%

5. A phenotypically normal 30-year-old male has just found out that his parents and his sister are carriers of PKU. What is the probability that he is also a carrier of this disease?
   A) ¼
   B) ½
   C) 2/3
   D) ¾
   E) 1

Answers

1. **The correct answer is D**. Since the mutation is in the somatic genes, males and females are affected equally. Thus, answer choices B and C are incorrect. Typically with autosomal recessive diseases, two unaffected heterozygote parents (carriers) have affected child, who have inherited two copies of the disease causing allele, one from the mother and one from the father. Therefore, recessive inheritance generally skips generations. Autosomal dominant diseases are seen in every generation (answer choice A).

2. **The correct answer is C**. Punnett square can be used for determining recurrent risk of a known genetic disease. In this case the disease is autosomal dominant and, therefore, the most likely genotype of the father is Dd, where the capital letter "D" represents the allele associated with the autosomal dominant disease, and the lower case letter "d" represents the wild type allele. The mother's genotype is dd, since she is not affected (phenotypically normal). Punnett Square shows that the father with the autosomal dominant genetic disease has a 50% chance to pass on the allele associated with disease and a 50% chance to pass on the wild type allele. Therefore, for each pregnancy, there is a 50% chance to have a child with the same autosomal dominant condition and a 50% chance to have a child who is unaffected (dd).

3. **The correct answer is B**. Homozygous means that both alleles of the gene have the exact same DNA sequence. If an individual has one copy of the sickle cell variant (Glu6Val) on one chromosome and a wild type (6th position is Glu) on the homologous chromosome, then this would be a heterozygous situation and the genotype of such an individual is known as Hb-AS (sickle cell trait). The term **compound heterozygous** means that the individual has two different pathogenic variants, as is the case for this patient since one allele is Glu6Val and the other allele is Glu6Lys. The genotype of such an individual is known as Hb-SC.

4. **The correct answer is B**. Both father and mother are carriers for the Glu6Val mutation in the beta globin gene, which is an autosomal recessive condition, and their genotype is therefore AS. The possible outcomes for a single pregnancy as obtained on the Punnett square are as follows: a 25% chance of a child who inherits two wild type alleles (AA) and is therefore homozygous wild type, a 50% chance of a child who inherits both a wild type allele and a pathogenic allele (AS) and will be an unaffected carrier like the parents, and a 25% chance of a child who inherits a pathogenic allele from each parent (SS) and therefore has sickle cell disease.

5. **The correct answer is C**. Since his sister and his parents are carriers of PKU, their genotype must be Pp (heterozygous for the PKU mutation). From the Punnett square of the mating between his two heterozygous parents, there are three different possible genotypes of the offspring: PP (homozygous normal), Pp (heterozygous carrier), and pp (homozygous PKU patient). In the Punnett square only 2 out of 4 represents a carrier genotype. However, we know that he is phenotypically normal, so we know that he cannot be homozygous for the mutant allele (pp) and we can eliminate this as a possibility. Therefore, the probability that his genotype is Pp, the carrier genotype, is 2/3.

VII

50 Sex-linked and Non-traditional Modes of Inheritance

At the conclusion of this theme, students should be able to:
- Distinguish between the characteristic features of X-linked recessive and dominant inheritance patterns
- Analyze the inheritance patterns and pedigrees of sex-linked diseases
- Describe the molecular mechanisms of X-inactivation
- Analyze inheritance patterns and pedigrees for mitochondrial traits
- Explain the non-Mendelian modes of inheritance
- Explain factors that influence phenotypic expression of single-gene disorders

In this theme, we will review the patterns of sex-linked inheritance as well as non-Mendelian modes of inheritance. Like autosomal dominant and recessive inheritance, there are specific patterns one can look for in a family history or pedigree that can give clues about the mode of inheritance. In this theme not only will we explore sex linked and mitochondrial inheritance, but we will explore other factors influencing phenotypic expression in single gene disorders as well.

50.1 Sex Linked Inheritance

Sex linked inheritance is observed in disorders in which the disease-causing variant is in a gene located on one of the sex chromosomes. When considering sex linked inheritance, it is important to discuss the genetic differences between the sexes and how this may affect the phenotypic expression of the genotype in a family. The sex chromosomes are the major genetic difference between males and females with genetic females harboring two X chromosomes and genetic males harboring one X chromosome and one Y chromosome.

As demonstrated by a typical male cell, only one active X chromosome is necessary to produce a functional cell. Therefore, typical female cells with two X chromosomes have a method to silence one of the X chromosomes in each cell. This process is called **X-inactivation**, which is also referred to as lyonization (▶ Fig. 50.1). **Lyonization** is a permanent occurrence, which means all cells that descend from that particular cell line will feature the same inactive X.

Typically, the specific X chromosome inactivated in a cell is a random occurrence and is thought to be like the flip of a coin: a 50/50 chance. However, as illustrated by ▶ Fig. 50.1, X-inactivation can also be skewed. Skewed X-inactivation occurs when one X chromosome is preferentially inactivated. If a female inherits an abnormality on one of her X chromosomes and that is the X chromosome that is preferentially inactivated, then the remaining normal X is the only active form in these cells and will produce the expected functional proteins. On the other hand, if the abnormal X is active while the normal X is silenced by X-inactivation in the majority of cells, the female may display features of the disease which may have otherwise been hidden. Hence, X-inactivation is an important process to consider in the context of X-linked disorders.

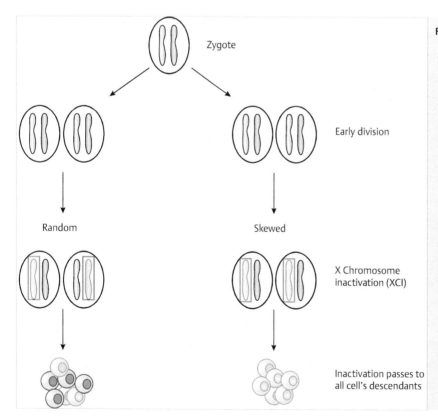

Fig. 50.1 The process of X-inactivation.

50.1.1 Link to Physiology: Barr Body is an Inactive X Chromosome

The inactive X chromosome can be observed under the microscope as a condensed dark spot in the nuclei of cells of females with a 46-XX genotype. This inactive chromosome is called a Barr body. The heterochromatic condensation of the X chromosome is accomplished by methylation of genes on the chromosome and by the coating of the chromosome by non-coding RNA molecules that are expressed from the *XIST* gene on the X chromosome. As we reviewed in Chapter 48, males with Klinefelter syndrome have a genotype of 47-XXY, and this extra X chromosome appears as a Barr body in these patients' cell nuclei.

X-linked disorders are those caused by pathogenic variants on the X chromosome. When a male has a pathogenic variant on their X chromosome it is referred to as **hemizygous**, because only one allele is present. X-linked disorders can be inherited in two different patterns: X-linked recessive and X-linked dominant.

50.2 X-linked Dominant Inheritance

In X-linked dominant inheritance, one pathogenic variant on one X chromosome is sufficient to cause disease. Though X-linked dominant disorders affect both males and females, they are often seen more frequently in females as they can be lethal in males. Typically, males are more severely affected than females with the same variant since their X chromosome is expressed in every cell and, therefore, the altered gene is expressed in every cell. As shown in ▶ Fig. 50.2a, when a female harbors a pathogenic variant that causes an X-linked dominant condition, she will pass the X chromosome with the variant to, on average, half of her daughters and half of her sons. A male harboring the same variant will pass his X chromosome to all of his daughters and none of his sons (▶ Fig. 50.2b).

50.2.1 Link to Pathology: Rett Syndrome is a MECP2-Related Disorder in Females

One example of a genetic disorder with X-linked dominant inheritance is the *MECP2*-related disorders (▶ Fig. 50.3). The

MECP2 gene is located at chromosome Xq28. Hence, females have two copies of the *MECP2* gene, though one is silenced in each cell via X-inactivation, while males only have one copy. The most common form of *MECP2*-related disorder in females is Rett syndrome, which is a progressive neurodevelopmental disorder. In males, pathogenic variants of *MECP2* cause severe neonatal encephalopathy resulting in death before the age of two, or, if they survive, syndromic or non-syndromic severe intellectual disability. This example illustrates the common findings in a pedigree with X-linked dominant inheritance: both males and females are affected, and males are phenotypically more severely affected, which can result in lethality. Considering the fact that some X-linked dominant disorders cause male lethality *in utero*, it is important to note that some family histories may not display any affected males since those that inherited the defect did not produce viable embryos.

50.2.2 Link to Pathology: X-Linked Hypophosphatemia

X-linked hypophosphatemia (XLH), also known as hypophosphotemic- or vitamin D resistant-rickets, is an X-linked dominant disease is caused by mutations in the *PHEX* gene. The wild type PHEX protein is a zinc-metallopeptidase which normally regulates fibroblast growth factor-23 (FGF-23). The loss-of-function mutations in the PHEX protein result in an overactive FGF-23. Overactive FGF-23 decreases both the hydroxylation of vitamin D and the reabsorption of phosphate from the kidneys. Due to insufficient calcitriol (1,25 dihydroxycholecalciferol) formation and loss of phosphate, individuals suffer from impaired bone growth, rickets and bone deformities. Patients present with low blood phosphate and calcium levels, but slightly elevated alkaline phosphatase levels. Although this is an X-linked dominant disease, heterozygote females are less severely affected compared to hemizygous males.

VII

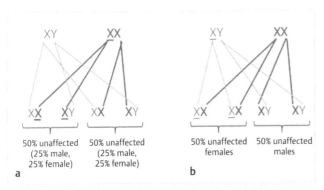

Fig. 50.2 X-linked dominant inheritance where the underlined X possesses the pathogenic variant (a) transmission through an affected female (b) transmission through an affected male.

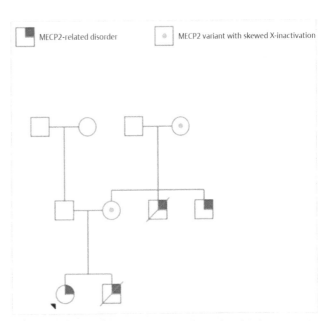

Fig. 50.3 A pedigree illustrating X-linked dominant inheritance in *MECP2*-related disorders.

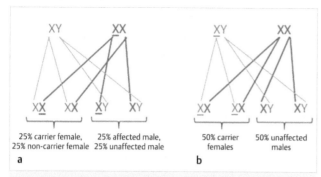

Fig. 50.4 X-linked recessive inheritance where the underlined X harbors the pathogenic variant (a) transmission through a carrier female (b) transmission through an affected male.

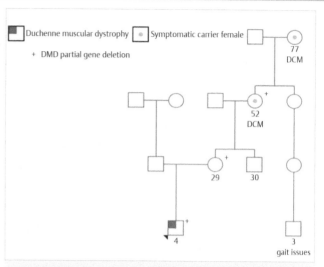

Fig. 50.5 A pedigree illustrating X-linked recessive inheritance in dystrophinopathies.

50.3 X-Linked Recessive Inheritance

Pathogenic variants on the X chromosome are more commonly recessive than dominant. In the past, it was felt that females were always asymptomatic carriers of X-linked recessive disorders. However, there is now documentation for multiple X-linked recessive disorders proving that not all female carriers are asymptomatic, although females tend to have mild or fewer symptoms of the genetic condition than their male counterparts.

As shown in ▶ Fig. 50.4a, when a female harbors a pathogenic variant that causes an X-linked recessive condition, there are four possible outcomes for each pregnancy: (1) a male inherits the X chromosome with the pathogenic variant and will develop the associated genetic disorder, (2) a male inherits the X chromosome with the functioning gene, and is thus unaffected (3) a female inherits the pathogenic variant and is a carrier who may or may not develop symptoms of the genetic disorder, and (4) a female inherits the functioning gene and, therefore, will not pass on the disorder to her offspring. Thus, an X-linked recessive disorder should be suspected when males are the primary individuals in the family affected with the disorder and/or when there is no male-to-male transmission of the condition's phenotype However, when a male harbors an X-linked recessive pathogenic variant, all of his daughters will inherit his X chromosome and will therefore be carriers of the genetic disorder and may develop the associated symptoms (▶ Fig. 50.4b).

50.3.1 Link to Pathology: Duchenne Muscular Dystrophy is an X-Linked Recessive Disorder

A common example of an X-linked recessive condition is the family of diseases known as dystrophinopathies, more commonly referred to as Duchenne and Becker muscular dystrophy, which are caused by pathogenic variants of the *DMD* gene located at chromosome Xp21.2p21.1. Pathogenic variants of the *DMD* gene can prevent the *DMD*-encoded protein, dystrophin, from being produced or can cause an abnormal, minimally

functional protein. Because genetic males only have one *DMD* gene, penetrance of dystrophinopathies is complete. The specific phenotype seen in males is typically dependent on the specific inherited variant and how that variant affects the protein product. ▶ Fig. 50.5 illustrates a pedigree analysis of the X-linked recessive inheritance of dystrophinopathies. The phenotypes can include markedly elevated serum creatine phosphokinase (CPK), progressive muscle weakness, calf hypertrophy, and eventual wheelchair dependency to dilated cardiomyopathy. Female carriers are also at risk to develop mild dilated cardiomyopathy in the fourth or fifth decade, and some even demonstrate additional features typical of muscular dystrophy. Current thought is that penetrance in females may in part be dependent on X-chromosome inactivation. Nonetheless, cardiac screening is recommended for all female carriers.

50.3.2 Link to Pathology

X-linked recessive diseases affect males more often than females. As we reviewed above, Duchenne and Becker muscular dystrophy most often affects males, although females can be affected if the X-inactivation is skewed unfavorably. The other X-linked recessive diseases are deficiencies of glucose 6-phosphate dehydrogenase (G6PD) and ornithine transcarbamoylase (OTC), Lesch-Nyhan syndrome, red-green color blindness, hemophilia A and B, Menkes syndrome, and severe combined immunodeficiency syndrome (SCID) due to IL-receptor ɣ-chain deficiency.

50.4 Mitochondrial Inheritance

Mitochondria are the powerhouses of eukaryotic cells, converting oxygen and nutrients into energy. Given the imperative roles of these organelles, it is no wonder that they have their own DNA. Each mitochondrion has between two and 10 copies of mitochondrial DNA, which exists as double stranded circular DNA made up of an H-strand and an L-strand which together

VII

code for 37 genes, most of which are ETC enzymes and components. Pathogenic alterations in any of these 37 genes can lead to mitochondrial disorders. Of note, most mitochondrial disorders are actually caused by pathogenic variants in nuclear genes encoding mitochondrial proteins.

50.4.1 Link to Pathology: Mitochondrial Diseases Affect Neurons and Muscles

Leber hereditary optic neuropathy (LHON) is caused by gene mutations in cytochrome reductase of the mitochondrial electron transport chain. The gene is located within mitochondrial DNA (mtDNA), and therefore LHON follows a mitochondrial inheritance pattern.

MELAS syndrome is also a mitochondrial disorder characterized by encephalopathy, myopathy, lactic acidosis and stroke symptoms, particularly in children and adolescents. The classic clinical presentation is an adolescent complaining of one or more of the following: muscle pain, weakness, twitching, and seizures or stroke-like symptoms (ie. vision loss, hemiparesis). It is caused by gene mutations in NADH-dehydrogenase (*ND4*), or complex I, of the ETC.

Leigh Syndrome is a mitochondrial disorder characterized by severe neurological degeneration, particularly in the first year of life. The classic clinical presentation is a young child with a history of difficulty swallowing, vomiting, failure to thrive, and weak muscle tone with gross movement or balance problems. Weakness may also occur in muscles of eye movement and respiration. Consequently, affected individuals usually die by age 3 due to respiratory failure. Leigh syndrome is caused by gene mutations in NADH dehydrogenase (complex I) or cytochrome-c oxidase (complex IV) of the electron transport chain. The genes for complex I and IV are located within mitochondrial DNA (mtDNA), and therefore Leigh syndrome follows a mitochondrial inheritance pattern.

Mitochondrial DNA (mtDNA) is maternally inherited since sperm has very little mitochondria and, therefore, the majority of an embryo's mitochondria is contributed by the egg. Given this inheritance pattern, mitochondrial disorders can only be passed from mother to child and not from father to child. If a male has a mitochondrial disorder caused by a pathogenic variant in his mtDNA, none of his children will inherit it since they will always inherit their mother's mtDNA. And, conversely, if a female has a mitochondrial disorder caused by a pathogenic mtDNA variant, she will pass her mtDNA to all of her children, regardless of their gender.

However, due to a phenomenon called **heteroplasmy**, which is a case in which some of the mtDNA within a cell harbor a pathogenic variant while others do not, the specific recurrence risk to each child is unknown because the number of mtDNA with the pathogenic variant can vary from child to child. **Homoplasmy**, on the other hand, refers to a cell that has a uniform collection of mtDNA, either all without pathogenic alterations or all harboring a pathogenic variant. Heteroplasmy and homoplasmy can be seen in families with mitochondrial disease due to what is called a genetic bottleneck. When primary oocytes are produced, a select number of mtDNA molecules are transferred to each oocyte. As illustrated in ▶ Fig. 50.6, during this process a heteroplasmic primordial germ cell can produce oocytes with an intermediate number of altered mtDNA molecules, a small percentage of mtDNA harboring a pathogenic variant, or even a large percentage of mtDNA harboring a pathogenic variant.

Because of this bottleneck, as shown in ▶ Fig. 50.7, an unaffected woman with a small percentage of heteroplasmy can have affected homoplasmic children. If enough of the mtDNA in a mitochondrion have a pathogenic variant, the mitochondrion will not function properly, and when too many mitochondria within the cells are non-functional, the features of mitochondrial disease occur. Therefore, the proportion of mtDNA molecules with a pathogenic variant determines both the penetrance and severity of expression of some mitochondrial diseases.

50.5 Non-Mendelian Inheritance

As has been previously discussed, genes are typically expressed from both alleles in diploids, one from the mother and one from the father. However, there are some instances that deviate from the Mendelian inheritance patterns. One of these is X-inactivation, which is described earlier in this chapter. Mitochondrial inheritance is also considered a non-Mendelian type

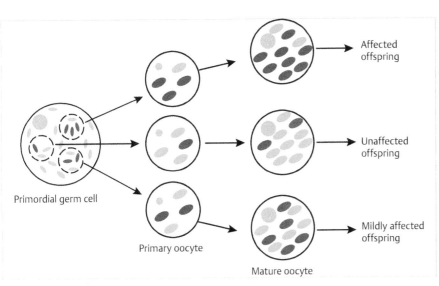

Fig. 50.6 The mitochondrial genetic bottleneck. Mitochondria with abnormal mtDNA are red, those with normal mtDNA are blue. (Adapted from Nature Publishing Group, Nature Reviews Genetics, 6, R. W. Taylor and D. M. Turnbull, Mitochondrial DNA mutations in human disease, 389–402)

Affected offspring

Unaffected offspring

Mildly affected offspring

Primordial germ cell

Primary oocyte

Mature oocyte

VII

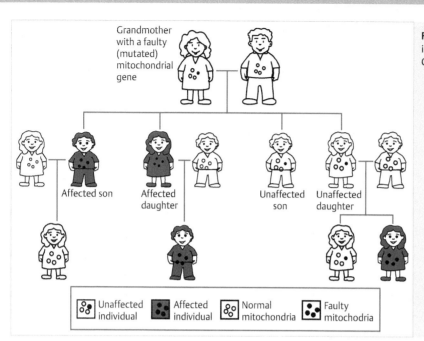

Grandmother with a faulty (mutated) mitochondrial gene

Affected son

Affected daughter

Unaffected son

Unaffected daughter

Unaffected individual

Affected individual

Normal mitochondria

Faulty mitochodria

Fig. 50.7 Pedigree illustrating mitochondrial inheritance. (Adapted from: Greenwood Genetic Center [1995]: Counseling Aids for Geneticists.)

VII

of inheritance. In addition to X-inactivation and mitochondrial inheritance, other non-Mendelian inheritance patterns include epigenetics, genomic imprinting, and uniparental disomy, which are reviewed in Chapter 52.

There are also environmental and some genetic factors that affect the phenotypic expression and recurrence risk of genetic disorders. These factors are discussed below and include penetrance, expressivity, multiple alleles, incomplete dominance, codominance, epistasis, mosaicism, allelic heterogeneity, locus heterogeneity, pleiotropy, and anticipation.

Penetrance is the likelihood someone with a variant in a particular gene will have the characteristic phenotype of the associated disorder, while **expressivity** describes phenotypic variation between individuals with the same genotype. Some genetic disorders have complete penetrance, meaning every individual with a variant in the associated gene will have symptoms of the disorder. However, there are a large number of genetic conditions in which penetrance is reduced (or incomplete). Hence, some individuals with a disease-causing variant do not show signs of the disorder. Numerous factors can affect penetrance of a disorder such as sex, age, environment, and modifier genes.

The penetrance of mitochondrial diseases caused by pathogenic variants in the mtDNA is mainly dependent on the percentage of heteroplasmy in each individual. Incomplete penetrance can sometimes explain why an affected individual inherits a disease-causing variant from an unaffected parent. Incomplete penetrance is also seen in hereditary breast and ovarian cancer syndrome since individuals with a pathogenic variant in the *BRCA1* or *BRCA2* genes have an increased risk for specific types of cancer, but not every individual with a variant in one of these genes develops cancer. Fragile X syndrome penetrance varies between the number of repeats and between males and females.

Due to **variable expressivity**, individuals in the same family with the same pathogenic mtDNA variant can display different phenotypic features of the condition. The variable expressivity can also be seen in other genetic disorders such as neurofibromatosis type 1 (NF1). Though the same pathogenic variant is being passed down through a family, severity of symptoms varies between parent and their affected child. For example, the parent may have 12 café-au-lait spots and axillary freckling while the child has numerous subcutaneous neurofibromas, café-au-lait spots, and an optic glioma. The cause of variable expressivity is currently unknown. However, it is likely a combination of genetic and environmental factors. Another example of variable expressivity is hemochromatosis. Males are affected at an early age, around 30-year-old, as compared to females, who are generally not affected until after menopause.

When a particular gene may exist in three or more allelic forms it is said to have **multiple alleles.** Several genetic disorders are known to have over 100 defective alleles, including cystic fibrosis and Marfan syndrome. Typically, when multiple alleles affect a trait, there are different of types of dominance patterns that occur.

Incomplete dominance, sometimes called partial dominance or semi-dominance, refers to instances when the dominant allele is not completely dominant over the other allele, resulting in a blending of traits. As such, a heterozygote's phenotype is distinct and often less severe than a homozygote's phenotype (▶ Fig. 50.8). This is seen in familial hypercholesterolemia, a genetic disorder that prevents tissues from removing low-density lipoproteins from the blood. Homozygotes have six times the amount of cholesterol in their blood and are at increased risk for a myocardial infarction before the age of two years, while heterozygotes have about twice as much cholesterol in their blood and may have a myocardial infarction before the age of 35 years.

Rather than a blending of traits, **codominance** is seen when both alleles are expressed and, hence, both traits are visible. We see codominance in our blood group system, which has three possible alleles: A, B, and O. Each of these letters

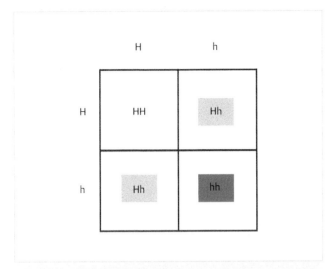

Fig. 50.8 Punnett square illustrating incomplete dominance in familial hypercholesterolemia. The H allele is the normal allele while the h allele is the variant allele. HH's phenotype is low cholesterol, Hh is mild cholesterol, and hh is very high cholesterol.

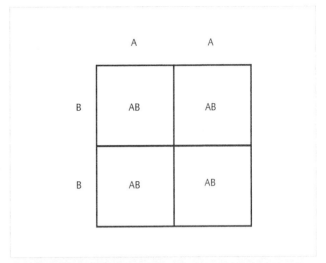

Fig. 50.9 Punnett square illustrating codominance with the combination of A and B blood types resulting in blood type AB.

corresponds to antigens found on the red blood cells. There are four phenotypes (blood types) in total: A, B, AB, and O. Blood type AB is an example of codominance as both the A and B alleles are expressed, causing both A and B antigens on the red blood cells.

There are also instances in which one gene depends on another gene for it to be expressed, which is called **epistasis**. This can be seen in common traits such as hair color or skin color. Though an individual's DNA may code for darker skin, if their *TYR* gene harbors a pathogenic variant resulting in a non-functioning tyrosinase protein, no melanin will be produced in the body, which will lead to a lack of pigment in the skin that is termed albinism. Thus, the DNA coding for skin color requires a functioning *TYR* gene to be expressed in the body.

Another phenomenon which can result in an altered or milder phenotype than is expected for the disorder is mosaicism. **Mosaicism** is the presence of two or more cell lines in an organism. There can be chromosomal mosaicism, discussed further in Chapter 48, but individuals can also be mosaic for a pathogenic variant. This type of mosaicism is typically caused by a spontaneous new mutation in one cell early in embryonic development. Each of that cell's daughter cells will go on to have the variant while the daughter cells of the unaltered cells do not have the variant, creating a mixture of two cell lines in the body. Depending on what percentage of, and what types of cells harbor the pathogenic variant, a mosaic individual may only exhibit some of the features associated with the genetic condition, which can result in a milder phenotype.

Genetic heterogeneity is when the same or similar phenotypes may be caused by different genetic mechanisms. There are two types of genetic heterogeneity: allelic heterogeneity and locus heterogeneity. **Allelic heterogeneity** refers to different alleles, or variants, in the same gene leading to the same phenotype. An example of this is Marfan syndrome, an autosomal dominant connective tissue disorder characterized by symptoms involving the skeletal, ocular, and cardiovascular systems, and caused by pathogenic variants in the *FBN1* gene.

According to the Human Gene Mutation Database, there have been nearly 2000 different disease-causing variants identified in the *FBN1* gene, which is allelic heterogeneity.

Locus heterogeneity is the term used to describe a specific phenotype that can be caused by alterations in more than one gene (more than one locus). Osteogenesis imperfecta (OI), sometimes referred to as brittle bone disease, is one example of locus heterogeneity, since pathogenic variants in either the *COL1A1* or *COL1A2* genes can cause the classic features of OI: fractures with minimal or no trauma, dentinogenesis imperfecta, and adult-onset hearing loss.

The opposite of genetic heterogeneity is **pleiotropy**, the phenomenon of variants in a single gene causing more than one phenotypic trait or disorder. Pleiotropy is seen with a number of different genes including one we already discussed, *FBN1*. Pathogenic variants in *FBN1* can cause the common symptoms of Marfan syndrome but may also cause isolated ectopia lentis (displacement of the lens in the eye) in some families. Another example of pleiotropy is porphyria variegate, an autosomal dominant disorder characterized by chronic blistering skin lesions and episodic neurovisceral symptoms, which is due to a heterozygous pathogenic variant in the *PPOX* gene. These symptoms may develop individually or in combination, and in some of these patients, there may be no symptoms at all.

Anticipation in genetic disease is the phenomenon by which symptoms of a disorder may appear not only at a younger age compared to the previous generation, but also with more severe symptoms. In other words, the age of onset can change from generation to generation. This phenomenon is typically seen in disorders caused by trinucleotide (triplet) repeat expansions. A trinucleotide repeat expansion can occur in areas of the DNA in which trinucleotide repeats such as CGG, CGG, CGG are found. There is typically an expected number of trinucleotide repeats for each of these regions, and if the number of repeats is outside the expected range, the features of the disease will appear in the individual. Trinucleotide repeat expansions, or an increase in the number of trinucleotide repeats, occur due to slipping of the polymerase during DNA replication.

The polymerase slip creates more copies of the triplet repeat, which is then passed on to the next generation.

Paternal anticipation is seen in Huntington disease, meaning that the CAG trinucleotide repeats in a male's *HD* gene can expand when passed on to their offspring. A significant expansion in CAG repeats in the *HD* gene can cause an earlier age of onset in this disease, referred to as juvenile Huntington disease. Individuals with juvenile Huntington disease typically have symptoms before the age of 21, which is much younger than the mean age of onset in Huntington disease of 35 to 44 years.

Fragile X syndrome is an example of maternal anticipation. Women who are carriers of a fragile X syndrome premutation (a number of CGG repeats larger than normal but below the pathogenic variant range) in their *FMR1* gene can conceive children who inherit a larger number of CGG repeats due to an expansion. Therefore, a seemingly unaffected parent can have an affected child due to trinucleotide repeat expansion.

Suggested Reading

Stenson, et al. Human genetics. 2014; 133(1):1–9 (PMID: 24077912)

VII

Review Questions

1. Which of the following is a characteristic feature of a pedigree for X-linked recessive diseases?
   A) Disease is observed in every generation
   B) Females are affected more often than males
   C) Males are affected more often than females
   D) Males and females are equally affected
   E) Disease is transmitted from father to sons

2. Which of the following diseases is X-linked dominant?
   A) Duchenne muscular dystrophy
   B) Glucose 6-phosphate dehydrogenase deficiency
   C) Lesch-Nyhan syndrome
   D) Menkes syndrome
   E) Rett syndrome
   F) Severe combined immunodeficiency syndrome

3. What is the probability that a couple will have an affected child if the woman is a carrier of an X-linked recessive disease?
   A) ½
   B) ¼
   C) 2/3
   D) ¾
   E) 1

4. There are many mutations within the *CFTR* gene that cause cystic fibrosis. Which of the following is the best terminology for this phenomenon?
   A) Allelic heterogeneity
   B) Anticipation
   C) Locus heterogeneity
   D) Pleiotropy

5. A couple who are both heterozygous for an autosomal dominant disease with an 80% penetrance are planning to have a child. What is the probability that their child will have the disease?
   A) 20%
   B) 40%
   C) 60%
   D) 80%
   E) 100%

Answers

1. **The correct answer is C**. Since the mutation is in the X-chromosome, hemizygous males are affected more often than females. Thus, answer choices B and D are incorrect. Typically, carrier females transmit the disease to sons. Since a father will give a Y chromosome to his sons, and not an X chromosome, a father to son transmission of an X-linked disease is never observed. Therefore, answer choice E is incorrect. Furthermore, recessive diseases generally skip generations. Thus, answer choice A is incorrect.

2. **The correct answer is E**. Rett syndrome is caused by mutations in the methyl CpG binding protein 2 (*MECP2*)

gene, which is located at chromosome Xq28. This X-linked dominant neurodevelopmental disease affects only girls because the pathogenic variants of *MECP2* cause male lethality *in utero*. Answer choices A, B, C, D, and F are incorrect because these diseases are X-linked recessive, not dominant. Although some females may display mild symptoms with X-linked recessive mutations, males are mainly affected.

3. **The correct answer is B**. Recall that in the case of X-linked recessive diseases, boys will be affected but not girls. We can therefore eliminate any girls from the equation since the question is asking the probability of having an affected child. We know that there is a 50% chance that they will have boy, and a boy's X chromosome is always inherited from the mother. Therefore, there is a 50% chance that any boy that they have will inherit the affected X chromosome from the mother. So, a 50% chance of having a boy times a 50% chance of the boy having the affected chromosome equals 25%.

4. **The correct answer is A**. Allelic heterogeneity refers to different alleles, or variants, in the same gene leading to the same phenotype. Many different mutations in the *CFTR* gene cause the same phenotype cystic fibrosis. Besides cystic fibrosis, Marfan syndrome is another example of allelic heterogeneity. Mutations in the *FBN1* gene cause Marfan syndrome. Marfan syndrome is also an example of pleiotropy, which is the phenomenon of variants in a single gene causing more than one phenotypic trait or disorder. Pathogenic variants in *FBN1* can cause the common symptoms of Marfan syndrome but may also cause isolated ectopia lentis (displacement of the lens in the eye) in some families. Therefore, answer choice D is incorrect. Locus heterogeneity is the term used to describe a specific phenotype that can be caused by alterations in more than one gene (more than one locus). Osteogenesis imperfecta (OI) is one example of locus heterogeneity, since pathogenic variants in either the *COL1A1* or *COL1A2* genes can cause the classic features of OI. Therefore, answer choice C is incorrect. Answer choice B is incorrect because anticipation in genetic disease is the phenomenon by which symptoms of a disorder may appear not only at a younger age compared to the previous generation, but also with more severe symptoms.

5. **The correct answer is C**. Both parents are heterozygous (Aa) for an autosomal dominant disease. From the Punnett square, we can determine the offspring genotypes and phenotypes: one incidence of AA genotype (disease), two incidences of Aa genotype (disease), and one incidence of aa genotype (normal). Therefore, ¾ or 75% of the offspring will have the disease. However, the disease has 80% penetrance, thus we need to multiply 75% x 80%, which equals 60%. Thus, 60% of the offspring will have the disease.

VII

51 Population Genetics

At the conclusion of this chapter, students should be able to:
- Define gene pool, allele, and phenotype frequency
- Apply the Hardy-Weinberg equation to calculate carrier, gene, and phenotype frequency
- Explain the assumptions/conditions necessary for the Hardy-Weinberg equation
- Define and explain conditions/events for changing allele frequencies including non-random mating, migration, genetic drift, gene flow, new pathogenic variants, and natural selection

This chapter explains the distribution of alleles in populations and how the frequencies of alleles, genotypes, and their subsequent phenotypes may or may not be maintained through generations. This study is referred to as **population genetics**, which relies on genetic factors, environment, reproduction, and social factors to determine the frequency and distribution of genetic disorders in families and their communities. We will review applications of the Hardy-Weinberg equilibrium to calculate allele and phenotype frequency in a population. We will also discuss the factors that may cause deviation from this equilibrium.

In thinking of population genetics, we must first consider a population's **gene pool**, which consists of all the alleles possible at a particular locus. As a simple example, let's focus on a gene that has two alleles, a wild type (W) and a pathogenic (P). If we were interested in determining the proportion of the W allele at that locus, we would be determining the **allele frequency**.

In order to discuss population genetics, it is important to understand the **Hardy-Weinberg equilibrium**, which states that allele and genotype frequencies in a population will remain constant from one generation to the next in the absence of other evolutionary influences (▶ Fig. 51.1). This equilibrium relies on several assumptions about a population, including that the population itself is large, matings are random in regards to the locus in question, and allele frequencies remain constant over time.

This principle relies on the fact that when two alleles (W = wild type, P = pathogenic) are segregating in a large, randomly mating population, the proportions of the three possible genotypes (WW, WP (or PW), and PP) remain constant and can be obtained from the binomial expression $p^2 + 2pq + q^2 = 1$, where p and q represent the frequencies of the wild type and pathogenic alleles respectively, and $p + q = 1$. Therefore, if a population is in Hardy-Weinberg equilibrium, the frequency of the wild type, homozygote genotype (WW) will be p^2, carriers (WP and PW) will be 2pq, and affected, pathogenic homozygotes (PP) will be q^2 (▶ Table 51.1).

51.1 Example 1: Determining Carrier Frequency in a Population

As an example, the incidence of a certain autosomal recessive disorder is approximately 1 in 3200. Given this information, we can determine the number of heterozygous carrier individuals in this population using the Hardy-Weinberg equilibrium.

q^2 (homozygous recessive allele) is 1/3200 = 0.03%

Therefore, q (recessive allele) = (the square root of 0.03%) = 0.02

And since p = 1 – q

p = 1 – 0.02

p (dominant allele) = 0.98

Then, we use 2pq to determine carriers = 2(0.98)(0.02) = 0.04.

Therefore, the number of heterozygous carriers in this population is 4%. This can also be demonstrated by doubling the square root of the disease incidence.

Table 51.1 Genotype frequencies when a population is in Hardy-Weinberg equilibrium

Genotype	Frequency
WW	p^2
WP or PW	2pq
PP	Q^2

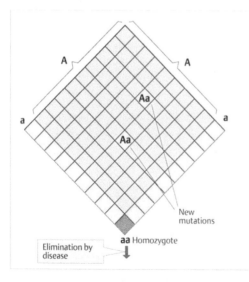

Alleles **a** are eliminated from the population owing to severe illness in homozygotes **aa**, but are replaced by new mutation. Equilibrium results

Genotypes		Frequency
Homozygote	**AA**	p^2
Heterozygote	**Aa**	2pq
Homozygote	**aa**	q^2

Fig. 51.1 Constant allele frequency (Hardy-Weinberg equilibrium). (Source: Passarge E, ed. Color Atlas of Genetics. 4th Edition. New York, NY. Thieme; 2012.)

51.2 Example 2: Determining Genotype Frequencies

The incidence of an autosomal recessive condition is known to be 36%. Given this information, we can determine the frequency of the homozygous, wild type genotype, as well as the frequency of the carrier genotype.

q^2 (homozygous recessive allele) = 0.36

Therefore, q (recessive allele) = (the square root of 0.36) = 0.6

And since p = 1 – q

p = 1 – 0.6

p (dominant allele) = 0.4

As a reminder, the incidence of homozygous, dominant alleles is represented by p^2

Therefore, $p^2 = (0.4)^2 = 0.16$

Again, carriers are represented by 2pq = 2(0.4)(0.6) = 0.48

Therefore, the incidence of individuals with homozygous, dominant alleles in this population is 16%, and the incidence of carriers is 48%.

This work can be double checked by plugging our new incidence values into the Hardy-Weinberg equilibrium as well.

$p^2 + 2pq + q^2 = 1$

0.16 + 0.48 + 0.36 = 1

51.3 Factors in Nature that Disrupt the Hardy-Weinberg Equilibrium

There are natural factors that disturb the Hardy-Weinberg equilibrium. These include nonrandom mating, genetic drift, gene flow, novel pathogenic variants, and natural selection (▶ Fig. 51.2).

Nonrandom mating occurs when the probability that two individuals in a population will mate is not the same for all possible pairs of individuals, as cultural values and social rules primarily guide mate selection. This can be due to a few factors including stratification, assortive mating, and consanguinity. **Stratification** refers to subgroups within a large population that mainly remain genetically separate and often have common alleles due to a shared ancestor, such as different ethnicities in a particular country. This can also explain why many autosomal recessive genetic disorders have carrier frequencies that differ between ethnic groups (▶ Table 51.2).

However, when mate selection is restricted to a particular subgroup, this can lead to an increase in homozygotes and a decrease of heterozygotes for the population as a whole, and as such, stratification has an effect on autosomal recessive conditions, and very little effect on autosomal dominant conditions or X-linked conditions.

Assortive mating refers to choosing a mate due to a particular trait expressed by that individual. Positive assortive mating refers to choosing a mate with similar features in regards to language, intelligence, stature, abilities, and so on, whereas negative assortive mating refers to the selection of a mate with very different features from those of oneself. However, when individuals choose partners with the same autosomal dominant genetic condition, such as achondroplasia or neurofibro-

Table 51.2 Carrier frequency for cystic fibrosis between three different ethnic populations [1]

Population	Approximate Carrier Frequency
Ashkenazi Jewish	1/29
North American of northern European heritage	1/28
African American	1/61

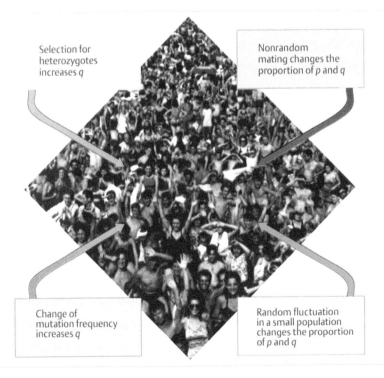

Fig. 51.2 Some factors influencing the allele frequency. (Source: Passarge E, ed. Color Atlas of Genetics. 4th Edition. New York, NY. Thieme; 2012.)

Selection for heterozygotes increases q

Nonrandom mating changes the proportion of p and q

Change of mutation frequency increases q

Random fluctuation in a small population changes the proportion of p and q

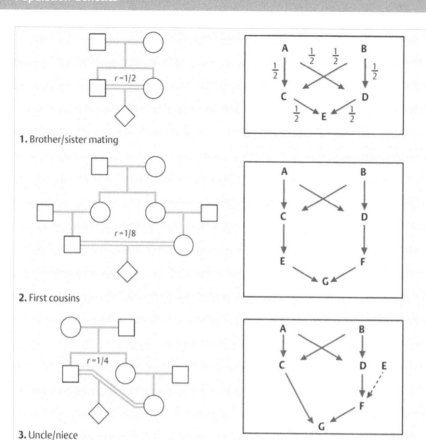

Fig. 51.3 Simple types of consanguinity. (Source: Passarge E, ed. Color Atlas of Genetics. 4th Edition. New York, NY. Thieme; 2012.)

1. Brother/sister mating

2. First cousins

3. Uncle/niece

matosis, the rules of the Hardy-Weinberg equilibrium do not apply, since the genotype of the mate at the genetic disorder's allele is not determined by the allele frequency of the general population. Additionally, in regards to a single family, if both individuals are affected with an autosomal recessive condition, with our without genetic heterogeneity, the fact that both carry genetic disorder alleles at the same locus increases their children's risk for the condition above what it would be if true random mating occurred.

Consanguinity is mating between two individuals who descended from a common ancestor. This type of nonrandom mating can increase the frequency of autosomal recessive disorders as it has the possibility of allowing very rare alleles to become homozygous (▶ Fig. 51.3).

Genetic drift relies on *chance* that an allele associated with a genetic disorder increases in frequency due to the increased fertility or survival of that carrier, both of which occur for reasons that have nothing to do with that disorder's allele. While this may not have an effect in a large population, this can lead to random fluctuation of the allele's frequency in a small population. One type of genetic drift is known as a **founder effect**, which is when an individual who carries a relatively rare allele is included in a small subpopulation that breaks off from their original group. That rare allele will have a much higher frequency in this smaller group than it had in the larger group.

While genetic drift relies on small populations and random allele fluctuations, **gene flow** relies on large populations and the gradual change in gene frequencies. Generally, this involves one population with their own particular allele frequencies gradually merging or **migrating** into the gene pool of a new population by crossing a reproductive barrier. This barrier can be geographic, but it can also be a social barrier based on ethnicity or cultural views.

Novel pathogenic variants, which can occur randomly in a population, can alter allele frequencies in a population by introducing new alleles. Additionally, **natural selection**, which is a non-intentional choice in response to an environment, leads to survival and reproduction of individuals due to a particular phenotype. This, in turn, can lead to a change in a population's allele frequency over time as there is a selection for a particular allele.

Reference

[1] Moskowitz SM, Chmiel JF, Sternen DL, et al. CFTR-Related Disorders. 2001 Mar 26 [Updated 2008 Feb 19]. In: Pagon RA, Adam MP, Ardinger HH, et al., editors. GeneReviews® [Internet]. Seattle (WA): University of Washington, Seattle; 1993–2016. Available from: https://www.ncbi.nlm.nih.gov/books/NBK1250/

Review Questions

1. The frequency of Gaucher disease in a population is 1/40,000. What is the frequency of heterozygotes in this population?
   A) 0.1%
   B) 0.5%
   C) 1.0%
   D) 2.0%

2. The incidence of red-green color blindness in a population is 1/2,000. What is the frequency of the allele that causes red-green color blindness in this population?
   A) 1/500
   B) 1/1,000
   C) 1/1,500
   D) 1/2,000
   E) 1/2,500

3. The incidence of red-green color blindness in a population is 1/2,000. What is the carrier frequency of the red-green color blindness in this population?
   A) 1/500
   B) 1/1,000
   C) 1/1,500
   D) 1/2,000
   E) 1/2,500

4. The carrier frequency of Tay-Sachs disease is more common in the Ashkenazi Jew population than in the general population. Which of the following terms best describes this occurrence?
   A) Consanguinity
   B) Founder effect
   C) New mutations
   D) Natural selection
   E) Stratification

Answers

1. **The correct answer is C.** For an autosomal recessive condition, disease frequency is q^2, which is given in the question as 1/40,000. If $q^2 = 1/40,000$ then $q = 0.005$. Based on the Hardy-Weinberg equation, the frequency of carriers or heterozygotes is 2pq. We can calculate the p from $p + q = 1$, then p= 1 − 0.005
   p= 0.995
   $2pq = 2(0.995)(0.005) = 0.01$ or 1.0%

2. **The correct answer is D.** For X-linked recessive conditions (such as red-green color blindness), the frequency of the disease is equal to the frequency of the allele. The frequency of the condition (q) is 1/2,000. Since males have only one X chromosome, the incidence of the disease in males is the same as the allele frequency.

3. **The correct answer is B.** For X-linked recessive conditions, the frequency of female carriers is 2q. The frequency of the X-linked recessive disease is equal to the frequency of the allele. Thus, q is 1/2000. This means that the carrier frequency is 2q = 2 (1/2000) = 1/1000.

4. **The correct answer is E.** Stratification refers to subgroups within a large population that mainly remain genetically separate than the general population. Consanguinity (answer choice A) is marriage between blood-related individuals, such as cousins. Founder effect (answer choice B) is when an individual who carries a relatively rare allele is included in a small subpopulation that breaks off from their original group. That rare allele will have a much higher frequency in this smaller group than it had in the larger group. New mutations (answer choice C) can occur randomly in a population, and can alter allele frequencies in a population by introducing new alleles. Natural selection (answer choice D) is a non-intentional choice in response to an environment, leading to survival and reproduction of individuals due to a particular phenotype. This is true for sickle cell trait, beta-thalassemia, and G6PD deficiency, but not the case for Tay-Sachs disease.

VII

52 Genomic Imprinting and Epigenetics

At the conclusion of this chapter, students should be able to:
- Define epigenetics
- Explain the cellular processes that are classified as epigenetic modifications, including genomic imprinting, DNA methylation, chromatin remodeling, and X chromosome inactivation
- Explain the epigenetic mechanisms that play a role in the regulation of gene expression
- Describe the roles of epigenetics in medicine

The term **epigenetics**, which was coined by Conrad Waddington, literally means "above genetics" and refers to heritable alterations that are not due to changes in the DNA sequence and, therefore, do not disrupt the genetic code. Epigenetic regulation is thought to work through three types of processes, including DNA methylation, post-translational modification of histone proteins, and ATP-dependent chromatin remodeling. These modifications, which alter DNA accessibility and chromatin structure, are responsive to cellular and environmental cues and regulate gene expression by controlling the potential of a genomic region to be transcribed. Additionally, epigenetic changes are thought to be triggered by some environmental or behavioral circumstances.

Epigenetic modifications are thought to be heritable, not only from a parent cell to a daughter cell when cells divide, but possibly also from one generation to the next. Transgenerational epigenetic inheritance is the idea that these epigenetic alterations can be acquired by the DNA of one generation and stably passed on through the gametes to the next generation. Ongoing transgenerational research in regards to the effects of the Dutch Hunger Winter, a period of famine for individuals in the western part of The Netherlands that was caused by the German-imposed food embargo during the Second World War, has revealed that not only did the children of women who were pregnant during the famine experience higher rates of morbidity, including obesity, diabetes, and mental illness, but the grandchildren of the men and women exposed to the famine also had higher rates of illness. This finding from the Dutch Hunger Winter suggests that an environmental epigenetic effect can be transferred through the generations. However, much about the transmission of epigenetic information across generations is unknown and further research is required.

Genetic imprinting is a process which leads to the monoallelic expression of a gene that is dependent on that gene's parent of origin. In order to discuss imprinting, **methylation** must first be reviewed. DNA methylation is a mechanism used by cells to control gene expression by adding a methyl group to the cytosine residues of CpG dinucleotides within the gene's promoter region. This methylation of cytosine prevents the gene from being transcribed, or expressed. Imprinting occurs due to methylation of the gene that is to be permanently inactivated in that individual. This type of alteration is known as an **epigenetic** change since it modifies the gene's expression without disrupting the genetic code, and in fact, is reversible.

For example, when a female inherits a paternally imprinted gene, this means she inherited an imprinted, or inactivated, gene from her father, while the homologous gene she inherited from her mother will be active, or not imprinted. However, as a female, all paternal imprinting must be erased and converted so that each of these genes capable of imprinting can be passed on as maternally-imprinted genes to her own children. This conversion of imprints occurs in each individual's gamete cells prior to fertilization.

Imprinting does not occur on every chromosome, and while there are multiple chromosomes with minor imprinted regions, there are also two major clusters of imprinted genes in humans, one on the short arm of chromosome 11 (11p15) and the second on the long arm of chromosome 15 (15q11 to 15q13).

As you might expect, it is also possible for genetic disorders to occur due to improper imprinting, which can give rise to an individual having two active copies of a gene or two inactive copies of a gene when only one gene is supposed to be active. Deletions or pathogenic variants can also affect imprinted genes.

52.1 Link to Pathology

Prader-Willi and Angelman syndromes are both linked to the imprinted region of chromosome 15. The classic example of genetic disorders related to errors in the imprinting of specific genes and chromosomal regions include Prader-Willi syndrome (PWS) and Angelman syndrome (AS), which are both linked to the imprinted region of chromosome 15 (▶ Fig. 52.1). Abnormalities of this one region lead to two very different phenotypes due to the fact that there are both maternally and paternally imprinted genes in this region. In the region where

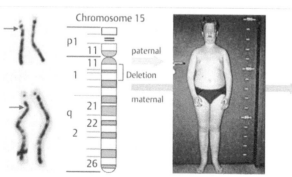

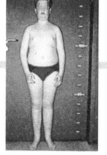

Fig. 52.1 Two syndromes associated with the same chromosomal region. (Source: Passarge E, ed. Color Atlas of Genetics. 4th Edition. New York, NY. Thieme; 2012.)

1. Interstitial deletion 15q11-13   **2.** Prader–Willi syndrome   **3.** Angelman syndrome

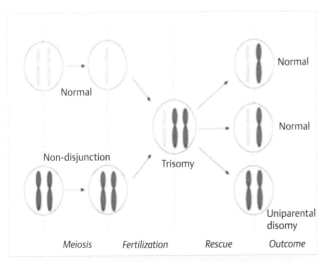

Fig. 52.2 Trisomy rescue can cause uniparental disomy (UPD), inheritance of two copies of the same chromosome.

imprinted genes on the maternal chromosome 15 are inactive, PWS occurs when the homologous region on the paternal chromosome has been deleted. When there are no active paternal genes present, the individual will have features of PWS, which include infantile hypotonia resulting in poor feeding, followed by compulsive eating beginning during the toddler years, leading to obesity. These individuals also tend to have short stature and learning difficulties. On the other hand, in a region where imprinted paternal genes are inactive, AS occurs when the genes inherited from the mother have been deleted. Without active maternal genes, an individual will have features of AS, which include learning difficulties, speech problems, seizures, movement issues, and an unusually happy disposition.

Another cause of genetic disorders related to imprinted genes is a phenomenon referred to as **uniparental disomy** (UPD), or the inheritance of two copies of the same chromosome, or even part of a chromosome, from one parent and no copy from the other parent. UPD can occur in many cases without having an effect on an individual's health, since that individual still inherits two copies of a gene. However, genetic disorders can occur if an individual inherits two copies of an imprinted gene, and no copy of an active gene. One mechanism for this occurrence can be nondisjunction, which results in three copies of a chromosome present in one cell, followed by trisomy rescue, where the cell

removes one copy so that the correct diploid number of chromosomes is then present (▶ Fig. 52.2). However, if the active, non-imprinted gene is located on the chromosome that was removed from the cell, leaving only inactive, imprinted genes on the two remaining chromosomes, an individual will be expected to have features of a genetic condition. UPD is another cause of both Prader-Willi syndrome and Angelman syndrome, since it can also result in the loss of either the active maternal or paternal contribution for this region.

Other types of uniparental disomy include hydatidiform moles, or molar pregnancy, and ovarian teratomas, which both result in non-viable pregnancies. Hydatidiform moles are products of abnormal conception where all of the DNA material is paternal in origin, known as paternal uniparental disomy, while ovarian teratomas are the result of maternal uniparental disomy.

Ongoing research concerning epigenetics in medicine includes its function in regards to certain syndromes as well as cancer. In fact, the disruption of epigenetic modifications is already closely associated with some well-known genetic syndromes, including the imprinted disorders Prader-Willi syndrome and Angelman syndrome, and the hypermethylated trinucleotide repeats of the *FMR1* gene associated with fragile X syndrome (reviewed in Chapter 50).

Additionally, researchers have discovered that tumor suppressor genes in cancer cells are hypermethylated in the promoter region, which essentially turns the gene off, while cancer cell oncogenes are hypomethylated and abnormally active. This has led to some insight in regards to cancer initiation, development, and drug resistance, to the point where researchers can focus efforts on cancer drugs that target DNA methyltransferases in order to attempt to correct the abnormal methylation pattern of tumor suppressors and oncogenes.

Suggested Readings

van Bokhoven H, Kramer JM. Disruption of the epigenetic code: an emerging mechanism in mental retardation. Neurobiol Dis. 2010; 39(1):3–12

Visual: By National Institutes of Health - http://commonfund.nih.gov/epigenomics/figure.aspx, Public Domain, https://commons.wikimedia.org/w/index.php?curid=9789221

Zhang J, Huang K. Pan-cancer analysis of frequent DNA co-methylation patterns reveals consistent epigenetic landscape changes in multiple cancers. BMC Genomics. 2017; 18 Suppl 1:1045

VII

Review Questions

1. Which of the following post-translational modifications is responsible for epigenetic regulation of gene expression?
   A) Methylation of cytosine
   B) Mutations of CpG islands
   C) Thymine dimers
   D) Trinucleotide expansions

2. Which of the following DNA regions is the most likely target for methylation?
   A) Acetylated histones
   B) GC dinucleotides
   C) Mutated DNA
   D) TATA box regions
   E) Trinucleotide expansions

3. Which of the following diseases is caused by the inefficient repression of transcription from methylated promoters?
   A) Angelman syndrome
   B) Fragile X syndrome
   C) Prader-Willi syndrome
   D) Rett syndrome

Answers

1. **The correct answer is A.** Epigenetic regulations are not due to changes in the DNA sequence and therefore, do not disrupt the genetic code. However, these epigenetic alterations are found to be inheritable. Epigenetic regulation is thought to work through three types of processes, including DNA methylation of cytosine residues, post-translational acetylation and deacetylation of histone proteins, and ATP-dependent chromatin remodeling.

Answer choices B, C, and D are incorrect because all of these alterations disrupt the DNA sequence.

2. **The correct answer is B.** One of the most common epigenetic alterations of DNA is methylation, which is catalyzed by the methyltransferase enzymes that add methyl group from SAM to the cytosine residues of DNA. These enzymes recognize the CpG dinucleotides and transfer methyl group onto the cytosine nucleotides.

3. **The correct answer is D.** Defects in the DNA methylation machinery or mutations in the gene *MeCP2* are involved in Rett syndrome, which is an X-linked dominant disease. Normally, the MeCP2 protein binds to methylated CpG regions of promoters and represses the transcription of the associated gene. Therefore, mutations in the *MeCP2* gene cause inefficient repression of transcription, and thus abnormal expression, of genes involved in early development. Fragile X syndrome (answer choice B) is an X-linked dominant disease which is caused by expansion of the trinucleotide repeat CGG. This expansion of CGG repeats causes hypermethylation of the *FMR1* gene promoter region and, therefore, turns off the transcription of the *FMR1* gene. Hence, answer choice B is incorrect. Answer choices A and C are incorrect because Prader-Willi syndrome is due to deletion of a specific set of genes on the paternal chromosome 15, while expression of the genes in the homologous region on the maternal chromosome is prevented through genetic imprinting The opposite is true for Angelman syndrome, where a deletion of a genomic region on the maternal chromosome 15 is paired with genetic imprinting of the homologous region on the paternal chromosome.

VII

53 Gene Interactions and Multifactorial Inheritance

At the conclusion of this chapter, students should be able to:
- Describe polygenic and multifactorial traits and diseases
- Explain the principles of polygenic inheritance
- Explain the multifactorial nature and inheritance of complex diseases
- Explain traditional ways to study multifactorial traits including empiric risk, adopted individuals, twins studies, and genome-wide association studies

While an alteration in a single gene can cause a specific trait or disease, there are diseases and traits that are influenced by the interaction of multiple genes, which is called multifactorial inheritance. In fact, many common diseases fall under this category. In this chapter, we will review how these types of diseases are studied in regard to their heredity.

53.1 Polygenic Inheritance

When one trait or disease is controlled by more than one gene, the inheritance is called **polygenic**. Traits that are purely polygenic without any environmental influence are rare. However, polygenic inheritance can be seen in traits such as hair color and eye color. Eye color is currently thought to be determined by a combination of 16 different genes (▶ Fig. 53.1). Two genes located on chromosome 15, *OCA2* and *HERC2*, play a major role in human eye color. The protein encoded by the *OCA2* gene, called the P protein, is involved in the maturation of melanosomes. Variants in *OCA2* affect the amount of functional P protein produced: the less P protein produced, the less melanin present in the iris, and the lighter the eye color. There is a well-known polymorphism located in intron 86 of *HERC2* which when present decreases the expression of *OCA2*, hence reducing the amount of melanin in the iris and leading to lighter-colored eyes.

53.2 Multifactorial Inheritance

While **polygenic traits** are those controlled by multiple genes, **multifactorial traits** are extremely common in humans and are determined by a combination of genetic as well as environmental factors such as diet, lifestyle, and environmental exposures and. Not only are height and weight examples of multifactorial traits, but many common diseases are also associated with multifactorial inheritance. In these diseases, alterations in the DNA alone may only serve as a predisposition for disease until combined with a threshold of environmental factors.

53.2.1 Link to Pathology

Some examples of multifactorial diseases and disorders include type1 and type2 diabetes mellitus, obesity, alcoholism, Alzheimer's disease, schizophrenia, and Crohn's disease. For example, type 2 diabetes mellitus is caused by a combination of multiple genetic variants associated with susceptibility to the disease, in addition to environmental factors such as obesity. This also explains why a single clinical genetic test for type 2 diabetes mellitus is not available, as multifactorial inheritance is much more complicated than a single gene disorder.

53.3 Determination of Recurrence Risk for Multifactorial Diseases

Because they are caused by interactions between genes and environment, multifactorial traits do not follow a Mendelian

	OH	Oh	OH	Oh
OH	OOHH	OOHh	OoHH	OoHh
Oh	OOHh	OOhh	OoHh	OOhh
OH	OoHH	OoHh	OOHH	OOHh
Oh	OoHh	Oohh	ooHH	oohh

Fig. 53.1 OCA2 and HERC2 genes play a major role in human eye color. O = full amount of P protein produced by OCA2 gene; o = decreased amount of P protein produced by OCA2 gene; H = normal HERC2 gene; h = HERC2 polymorphism decreasing expression of OCA2.

Table 53.1 Empiric risk data for siblings of an individual with cleft lip with or without cleft palate.

Phenotype of Proband	Recurrence for Siblings
Unilateral cleft lip	0.5%
Unilateral cleft lip and palate	2.3%
Bilateral cleft lip	4.2%
Bilateral cleft lip and palate	6.2%
Adapted from Grosen et al.[1]	

Table 53.2 Degree of relationship and percentage of alleles in common.

Relationship to proband	Percentage of alleles in common with the proband
Monozygotic twin	100%
First-degree relative (dizygotic twin, sibling, parent, child)	50%
Second-degree relative (aunt/uncle, grandparent)	25%
Third-degree relative (first cousin, great-aunt/uncle, great-grandparent)	12.5%

inheritance pattern. However, we know these traits have some amount of heritability, given that the cause is not completely environmental. Unfortunately, the combination of both genetic and non-genetic factors makes it difficult to provide a precise recurrence risk. Therefore, the recurrence risk is typically provided using empiric risk. **Empiric risk** is an observed population statistic used to predict the probability that a trait will occur.

For instance, because there is no known genetic cause for non-syndromic cleft lip and palate, the recurrence risk of this disease can be determined by using empiric data (▶ Table 53.1).

It is important to note that empiric risks are not an exact science nor are they static. For example, if data is collected on only one population it may not be transferable to other populations. This is because if the genetic and environmental components are different, the incidence will be different. In addition, since specific genetic variants and environmental factors are associated with multifactorial traits, the risk estimates will likely require modification.

Recurrence risk of multifactorial diseases or conditions depends not only on the diagnosis but also on the severity of the trait, the number of affected individuals in the family, increased relatedness to the affected individual, and sometimes the sex of the family members.

53.4 Traditional Ways to Study Multifactorial Traits

Multifactorial traits are typically studied through three means: **adoption studies, twin studies,** and **genome-wide association studies**. Individuals who are adopted by non-relatives can be useful study subjects because they share genetic variants, but not environmental factors, with their biological parents, and they share environmental factors, but likely not many genetic variants, with their adoptive parents. Hence, traits shared with the biological parents are likely to be influenced more by genetics while traits shared with the adoptive parents are felt to be environmental. These types of studies, however, have largely been replaced by twin studies.

Twin studies are performed by observing multifactorial traits in twins. The focus of twin studies is the concordance rates, which is the presence of the same trait in both members of a pair of twins. The only difference between monozygotic (MZ) and dizygotic (DZ) twins are the number of genes in common (▶ Table 53.2). Their environment is, for all intents and purposes, constant.

In the twin studies, the environment is constant but the difference is the number of genes. Therefore, the difference in concordance rates in MZ versus DZ twins is attributed to genetics. If there is significantly greater concordance between MZ twins, the evidence points toward a strong genetic component of the trait.

The ideal study would be performed on MZ twins separated at birth, as their DNA is the same but their environments are different. There are limitations, however, to this type of study as twins raised in the same household will likely have different environments as adults and may even have had different environments in utero (e.g., blood supply, intrauterine development).

The heritability of multifactorial traits can be calculated using data obtained through twin studies. Although the data from these studies is useful to estimate the heritability and the proportion of a multifactorial trait that is due to genotype, the heritability data does not provide information about which genes are involved or how many genes interact to influence the specific trait. Heritability (h2) is calculated using the following formula:

H2 = Variance in DZ pairs – Variance in MZ pairs / Variance in DZ pairs.

For example, the heritability of obesity using body mass index would be calculated as follows:

Variance in DZ pairs = 50
Variance in MZ pairs = 8
50 – 8 / 50 = 0.84

Hence, the estimated heritability of obesity is 0.84, indicating an individual's genotype has a strong influence on the trait. In other words, 84% of the phenotypic variance is due to genetic contribution and 16% of the phenotypic variance is due to environmental contribution.

53.5 Genome-Wide Association Studies

The latest technique used to study multifactorial traits is genome-wide association studies (GWAS). GWAS focus on **single nucleotide polymorphisms (SNPs)**, which are single nucleotide changes that vary in at least one percent of the population. Researchers compare the SNPs in populations with a specific trait to those that do not have the trait in a case-control design. GWAS also help to identify risk genes of complex diseases. Identifying specific SNPs and genes that lead to increased disease risk may help guide medical care in the form of preventative medicine.

The overall theme of multifactorial traits and diseases is that they remain difficult to not only study but also to provide accurate recurrence risk information given the lack of knowledge regarding the genetic basis and how this interacts with environmental factors.

Reference

[1] Grosen D, Chevrier C, Skytthe A, et al. A cohort study of recurrence patterns among more than 54,000 relatives of oral cleft cases in Denmark: support for the multifactorial threshold model of inheritance. J Med Genet. 2010; 47(3): 162–168

VII

Review Questions

1. Data from a study reveals that the heritability of problem drinking is 40%. Which of the following is the most likely inheritance pattern of this condition?
   A) Autosomal dominant
   B) Multifactorial
   C) Mitochondrial
   D) Polygenic inheritance
   E) X-linked dominant

2. Which of the following data is most useful for identification of genes that are involved in multifactorial complex diseases?
   A) Concordance rates
   B) DNA sequencing
   C) Empirical
   D) Twin studies
   E) Single nucleotide polymorphism

3. The concordance data for diseases D, H, P, and S is shown below:

   Disease MZ concordance DZ concordance
   D 0.80 0.30
   H 0.89 0.88
   P 0.95 0.47
   S 0.55 0.15

   Based on the data, which of the following diseases exhibit multifactorial inheritance?
   A) Diseases D and H
   B) Diseases D and P
   C) Diseases D and S
   D) Diseases H and P
   E) Diseases H and S

Answers

1. **The correct answer is B.** The heritability data, which is most often obtained by twin studies, indicates that genetic and environmental factors contribute to problem drinking. Therefore, it is most likely multifactorial inheritance. Answer choice D is incorrect because polygenic inheritance is controlled by multiple genes. Autosomal, X-linked, and mitochondrial inheritance patterns are controlled by variants of single genes, thus answer choices A, C, and E are incorrect.
   Reference: De Moor MHM, Vink JM, van Beek JHDA, et al. Heritability of Problem Drinking and the Genetic Overlap with Personality in a General Population Sample. *Frontiers in Genetics*. 2011;2:76. doi:10.3389/fgene.2011.00076.

2. **The correct answer is E.** Researchers design genome-wide association studies (GWAS) to compare the single nucleotide polymorphisms (SNPs) between the control group and the affected individuals. The findings of the SNPs can be useful for identification of genes that may contribute to risk of complex diseases.
   Twin studies are performed by observing multifactorial traits in twins. The focus of twin studies is the concordance rates, which is the presence of the same trait in both members of a pair of twins. This data does not provide information about specific genes but provides if the trait is heritable. For example, if the concordance rate in MZ is equal to the concordance rate in DZ then the trait is strictly environmental. On the other hand, if the concordance rate in MZ is higher than the concordance rate in DZ, then the trait/disease has a genetic component and it is heritable. However, if the concordance rate in MZ is ~100% and concordance rate in DZ is ~50%, then the trait/disease is strictly genetic. Therefore, answer choices A and D are incorrect. Recurrence risk of multifactorial inheritance is typically provided using empiric risk, which is obtained from family history, diagnosis, and medical records. Empiric risk is useful to predict the probability that a trait will occur. However, it does not provide information about specific genes, thus answer choice C is incorrect. Answer choice B is incorrect because complex diseases are not caused by single genes, thus sequencing the entire genome would not be beneficial to identify specific genes.

3. **The correct answer is C.** Comparing concordance rates in MZ and DZ pairs of twins provide information about whether the trait/disease is heritable. The higher concordance rate in MZ than in DZ suggests that the trait or disease has some genetic component and the trait/disease is heritable. Based on this, diseases D, P, and S are heritable. However, if the concordance rate in MZ is equal to the concordance rate in DZ then the trait is strictly environmental. In this case, the disease H is strictly environmental, or acquired, such as diseases caused by viral or bacterial infections. The disease P is most likely not multifactorial because the concordance rate in MZ is ~100% and the concordance rate in DZ is ~50%. This suggests that the disease P is most likely strictly genetic.

54 Personalized Medicine

At the conclusion of this chapter, students should be able to:
- Define pharmacokinetics and explain the two phases of drug metabolism involved
- Describe the cytochrome P450 family of enzymes and how polymorphic variations in their alleles contribute to drug response
- Give specific examples for three modes of phase II drug metabolism and describe how polymorphic variations in each can influence an individual's response to a particular drug
- Define pharmacodynamics and give an example of how allelic variations in a drug's downstream targets can contribute to its effectiveness

Personalized medicine refers to a healthcare approach that is tailored to an individual patient based on genetic factors. This can exist in many different forms. One is the screening of asymptomatic individuals for susceptibility to disease, which may involve family history, population screening, newborn screening, genetic susceptibility screening, and heterozygote screening. Another potential component of personalized medicine is the utilization of pharmacogenetic information for the purposes of determining the most appropriate treatment regimen. This allows for the identification of the best drug therapy option based on allelic variations in genes affecting drug metabolism, efficacy, and toxicity. Two factors that influence the effectiveness of drugs are their pharmacokinetic and pharmacodynamics characteristics. Finally, personalized medicine may include the prediction of the severity of a disease based on polymorphic variations in genes that are not directly responsible for the disease.

Pharmacokinetics is the rate at which the body absorbs, transports, metabolizes, or excretes drugs or their metabolites. The abbreviation ADME (**A**bsorption, **D**istribution, **M**etabolism, **E**xcretion) is often used to describe these pharmacokinetic properties. Absorption refers to the process by which the compound first enters the blood stream, usually through intestinal absorption. Distribution is the transport of the compound to the effector site. The breakdown of the compound into metabolites, resulting in either pharmacologically inert or pharmacologically active products, occurs in the metabolism stage. And finally, during excretion the compound and its metabolites are removed from the body.

The major site of metabolism for drugs, also known as xenobiotic metabolism, is the liver and requires the activity of the cytochrome p450 family of enzymes. Xenobiotics are toxic compounds containing aromatic rings or heterocyclic rings that the body is unable to degrade or recycle into useful components. Since these molecules are hydrophobic, they are either retained in adipose tissue or sequestered by the liver. In the liver, these lipophilic compounds are metabolized into water-soluble products, which can then be excreted in the urine or bile. This metabolism of xenobiotics occurs in two distinct phases.

In the first phase, called Phase I, hydroxylation occurs, adding a hydroxyl group to the compound. This serves as a site for the addition of a side group later on in Phase II (▶ Fig. 54.1). Through this hydroxylation, the lipophilic compound is converted into a water soluble product. The enzymes responsible for this hydroxylation belong to the cytochrome p450 family, which are induced by the presence of their substrate. Thus, the major role of the cytochrome p450 system is to oxidize substrates (▶ Fig. 54.2).

54.1 Cytochrome p450

Inside the cell, many cytochrome p450 molecules are embedded in the membrane of the endoplasmic reticulum, also known as the microsome. The activity of these cytochrome p450 enzymes requires another enzyme called cytochrome p450 reductase. The combined activity of these two enzymes ultimately results in the transfer of electrons from NADPH to O_2 and the addition of a hydroxyl group to the substrate. In the cytochrome p450 reductase, FAD and FMN relay the electrons from the donor NADPH molecule to the Fe-heme inside the cytochrome p450, which then transfers the electrons to oxygen, resulting in the hydroxylation of the substrate (▶ Fig. 54.3). Cytochrome p450s are also found embedded in the inner membrane of the mitochondria. In this case, adrenodoxin reductase and adrenodoxin are responsible for the transfer of electrons from NADPH to the cytochrome p450.

The cytochrome p450 family constitutes a large family of enzymes, with 56 different genes in humans, subsets of which are responsible for the breakdown of specific xenobiotics. Of these subsets, the most important for pharmacogenetics are CYP1A1, CYP1A2, CYP2C9, CYP2C19, CYP2D6, and CYP3A4 (▶ Fig. 54.4). Combined, these enzymes are responsible for the phase I metabolism of more than 90% of all commonly used drugs. In fact, CYP3A4 alone is involved in the metabolism of over 40% of all drugs, and CYP2D6 is responsible for the metabolism of many commonly prescribed antidepressants and antipsychotics. Polymorphisms in these genes can determine an individual's response to specific drugs.

A genetic polymorphism is a DNA sequence variation that is common in the population. There is no single allele that is regarded as the standard sequence, but rather there are two or more equally acceptable alternatives. This is in contrast to a mutation, which is commonly considered a change in the sequence that is found in less than 1% of the population. If this change is seen in more than 1%, it is typically referred to as a genetic polymorphism.

A majority of genetic polymorphisms are referred to as Single Nucleotide Polymorphisms (SNPs). There are two types of SNPs: synonymous and non-synonymous. Synonymous SNPs are those in which the nucleotide change does not result in a change in the amino acid sequence encoded by the gene. This is equivalent to a silent mutation. A non-synonymous SNP is one in which the nucleotide change does result in a change in the amino acid sequence encoded by the gene.

Polymorphic variations in the genes that encode the cytochrome p450 family of enzymes can influence the activity of these enzymes. Depending on the nucleotide variation, these changes can result in three potential phenotypes: normal

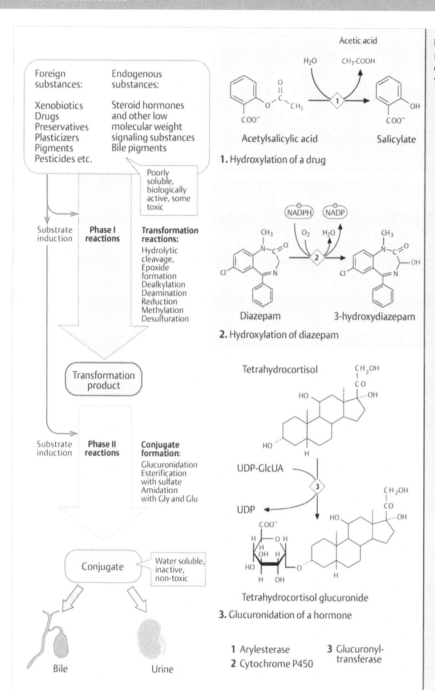

Acetic acid

1. Hydroxylation of a drug

Acetylsalicylic acid — Salicylate

2. Hydroxylation of diazepam

Diazepam — 3-hydroxydiazepam

3. Glucuronidation of a hormone

Tetrahydrocortisol — Tetrahydrocortisol glucuronide

1 Arylesterase
2 Cytochrome P450
3 Glucuronyl-transferase

Fig. 54.1 Phase I and Phase II reactions of drug metabolism. (Source: Koolman J, Röhm K, ed. Color Atlas of Biochemistry. 3rd Edition. New York, NY. Thieme; 2012.)

metabolizers, poor metabolizers, and ultrafast metabolizers. Poor metabolizers are at risk for accumulation of toxic levels of drugs, whereas ultrafast metabolizers are at risk for being undertreated with normal doses of drug.

An extreme case of polymorphic variation can be seen with the *CYP2D6* gene, which has 26 different variations. These variants are classified as reduced, absent, or increased activity. Reduced activity is caused by missense point mutations and, as the name suggests, results in decreased activity. The absent phenotype is due to splicing or frame shift mutations and results in the complete loss of the enzyme's activity. In the increased phenotype, variation in the copy number of the gene (either three, four, or more copies) leads to high levels of enzyme activity.

Some cytochrome p450 variants that affect the ability to metabolize specific drugs are known to exist at high frequency in certain ethnic groups. For example, individuals belonging to Asian or Pacific Islander ethnic groups are at higher risk of accumulating toxic levels of some antidepressant or antianxiety drugs if they possess the poor metabolizer allele of CYP2C19.

During the second phase of drug metabolism (Phase II) a sugar, acetyl, or methyl group is attached to the hydroxyl group created during phase I (▶ Fig. 54.1). This creates a water-soluble product, allowing it to be excreted more readily by the body and effectively detoxifying the drug. For each of these reactions, a specific enzyme is required, and genetic polymorphic variations can influence the efficiency of these reactions.

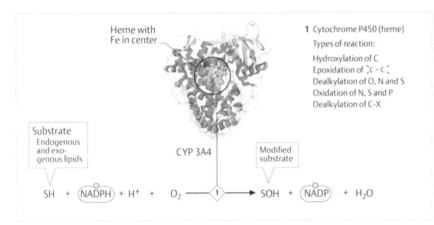

Fig. 54.2 Hydroxylation reactions of cytochrome p450 enzymes. (Source: Koolman J, Röhm K, ed. Color Atlas of Biochemistry. 3rd Edition. New York, NY. Thieme; 2012.)

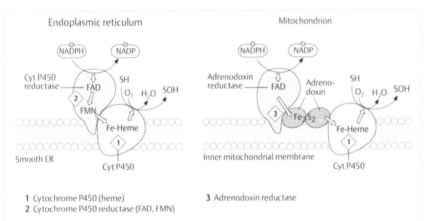

Fig. 54.3 The structure of cytochrome p450 enzymes. (Source: Koolman J, Röhm K, ed. Color Atlas of Biochemistry. 3rd Edition. New York, NY. Thieme; 2012.)

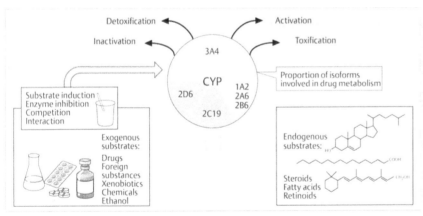

Fig. 54.4 4 functions of cytochrome p450 enzyme families. (Source: Koolman J, Röhm K, ed. Color Atlas of Biochemistry. 3rd Edition. New York, NY. Thieme; 2012.)

Glucoronidation, or the addition of the sugar glucuronic acid to the substrate, is one of these phase II events (▶ Fig. 54.1). This process is mediated by the enzyme glucoronate transferase, which is encoded by the gene *UGT1A1*. Allelic variations in this gene affect its level of expression, resulting in an increased risk of serious toxicity from some common chemotherapy regimens, such as camptothecin. The polymorphic variations in *UGT1A1* do not occur in the coding region of the gene, but rather the promoter region, and are due to a variable number of tandem repeats in the TATAA box. The variant designated as

*UGT1A1*1* is the normal allele and has six repeats. In contrast, *UGT1A1*28*, a common variant, has seven repeats. The presence of this extra repeat results in a reduction in the transcription of the gene, leading to decreased levels of the enzyme.

Another genetic polymorphism is found in the *NAT2* gene, which encodes the enzyme N-acetyltransferase. This enzyme is responsible for the transfer of an acetyl group to its substrate during phase II drug metabolism. Individuals with the slow inactivator variant of this gene, also known as slow acetylators, have decreased levels of N-acetyltransferase in the liver and, therefore, an increased risk of accumulating target drugs to toxic levels in the body. On the other hand, fast inactivators, or fast acetylators, may require higher doses of a drug in order to maintain therapeutic levels. Slow inactivators that are given isoniazid to treat tuberculosis will inactivate the drug more slowly, causing them to be at higher risk for toxic effects, including peripheral neuropathy and bone marrow suppression.

In addition to glucuronidation and acetylation, the addition of a methyl group can also increase the excretion of certain drugs. One of the enzymes responsible for this methylation is thiopurine methyltransferase, which is encoded by the gene *TPMT*. Polymorphic variations arising from missense point mutations in this gene lead to the production of an unstable enzyme, resulting in its rapid degradation and, therefore, loss of its activity. Partial deficiency, as in the case of an individual that is heterozygous for this variant, results in slower metabolism of the affected drug. This decreased metabolic rate could either increase drug effectiveness or increase toxicity, depending on the dose. Polymorphisms in thiopurine methyltransferase can be used to predict a patient's response to chemotherapeutic agents such as 6-mercaptopurine, which is used in the treatment of leukemias.

54.1.1 Link to Pharmacology: Role of CYP2E1 and Acetaminophen in Alcoholic Liver Injury

Besides genetic polymorphic variations, there are also external influences that can affect the ability of cytochrome p450 enzymes to metabolize their substrate. For example, chronic alcohol intake is known to induce the CYP2E1 enzyme, which is responsible for metabolism of many toxicological substrates including alcohol and acetaminophen. The metabolism of alcohol by CYP2E1 also known to produce reactive oxygen species, superoxide radicals and hydrogen peroxide. Metabolism of acetaminophen (Tylenol®) by CYP2E1 produces the very reactive intermediate N-acetyl p-benzoquinone imine (NAPQI), which immediately forms a conjugate with glutathione. Although the NAPQI-glutathione conjugate is nontoxic to the cells, it is excreted from the body. As a result, this depletes the glutathione stores of the hepatocytes. Therefore, a combination of alcohol and acetaminophen could cause significant liver damage due to accumulation of toxic levels of reactive oxygen species and depletion of reduced glutathione.

54.2 Pharmacodynamics

In addition to pharmacokinetics, pharmacodynamics must also be taken into consideration when a drug regimen is prescribed.

Pharmacodynamics refers to genetic causes of variability in the response to a drug due to allelic variation in the drug's downstream targets, such as receptors or enzymes, or in components of metabolic pathways. In other words, the response to the drug is not due to genetic variation in the enzymes that metabolize the drug, but rather in the genes that encode downstream targets of the drug. An example of this can be seen in individuals with G6PD deficiency.

G6PD deficiency is the most common disease-producing enzyme defect in humans. It is due to the loss of function of glucose-6-phosphate dehydrogenase (G6PD), an enzyme that generates NADPH and is responsible for protecting red blood cells from oxidative damage, reviewed in chapter 23. If a G6PD deficient individual is given an oxidant drug, such as the anti-malaria drug primaquine, their red blood cells will be overwhelmed with oxidative damage and hemolysis will occur. This hemolysis can ultimately result in drug-induced hemolytic anemia.

The therapeutic efficacy of some drugs can be influenced by variations in both phase I metabolism and downstream effector enzymes. One such drug is warfarin. Warfarin is an anticoagulant used for the prevention of thromboembolism. Its mechanism of action is the inhibition of the enzyme VKORC1 (vitamin K epoxide reductase complex I), which is involved in the recycling of vitamin K and is important for the activation of some blood coagulation proteins, such as factor VII and prothrombin. Since warfarin is metabolized by *CYP2C9*, polymorphisms in this gene interfere with the proper breakdown of this drug. In this case, lower doses of warfarin are needed since the drug is not eliminated as quickly. There are also polymorphic variations in the *VKORC1* gene that are associated with a reduced production of the VKORC1 protein. Warfarin would also be effective in lower doses in individuals with this polymorphism since there would be less of the VKORC1 enzyme for the warfarin to inhibit.

54.3 Pharmacogenomics

Pharmacogenomics is the assessment of common genetic variants in a population for their impact on the outcome of drug therapy. In pharmacogenomics, sets of alleles at a large number of polymorphic loci are identified that distinguish patients who have responded adversely to what was considered a beneficial drug from those who had no adverse response. The effectiveness or toxicity of a particular drug can then be predicted based on the genetic profile of the patient.

Review Questions

1. Which of the following is the major site of drug metabolism in the body?
   A) Kidneys
   B) Liver
   C) Pancreas
   D) Stomach

2. Genetic variations in which of the following genes/enzymes would be expected to influence the glucoronidation of xenobiotics?
   A) G6PD
   B) UGT1A1
   C) VKORC1
   D) NAT2

3. Allelic variations in which of the following cytochrome p450 enzymes has the most effect on the metabolism of alcohol?
   A) CYP2E1
   B) CYP3A4
   C) CYP2D6
   D) CYP2C19

4. Which of the following allelic variations affects the intracellular pool of the reduced form of vitamin K?
   A) CYP2C9
   B) VKORC1
   C) CYP2E1
   D) G6PD
   E) TPMT

Answers

1. **The correct answer is B.** The major site of metabolism for drugs is the liver and requires the activity of the cytochrome p450 family of enzymes. These potentially toxic compounds contain aromatic rings or heterocyclic rings that the body is unable to degrade or recycle into useful components. Since these molecules are hydrophobic, they are either retained in adipose tissue or sequestered by the liver. In the liver, these lipophilic compounds are metabolized into water-soluble products, which can then be excreted in the urine or bile. This metabolism of xenobiotics occurs in two distinct phases.

2. **The correct answer is B.** Glucoronidation is the process of covalently linking a glucoronic acid to a substrate. During Phase II of xenobiotic metabolism, glucoronidation is used to transform a drug from an insoluble form into a water soluble form so that they can be eliminated from the body. This is accomplished through the activity of the glucoronate transferase, which is encoded by the gene UGT1A1. Polymorphic variations that result in an abnormal number of tandem repeats within the promoter region of this gene lead to a lower than normal production of the enzyme. NAT2 (answer choice D) is the gene that encodes N-acetyltransferase, another enzyme involved in some of the Phase II reactions, although in this case, acetylation occurs rather than glucoronidation. G6PD (answer choice A) and VKORC1 (answer choice C) are examples of genes that encode proteins that are important for aspects of pharmacodynamics.

3. **The correct answer is A.** The cytochrome p450 enzyme CYP2E1 is responsible for metabolism of alcohol and is induced by chronic alcohol intake. This enzyme is also responsible for the metabolism of many toxicological substances, including acetaminophen. The metabolism of alcohol by CYP2E1 also produces reactive oxygen species, superoxide radicals and hydrogen peroxide, and the metabolism of acetaminophen produces the very reactive intermediate N-acetyl p-benzoquinone imine (NAPQI), which forms a conjugate with glutathione. Although the NAPQI-glutathione conjugate is nontoxic to the cells, it is excreted from the body. As a result, this depletes the glutathione stores of the hepatocytes and leads to liver damage. The other cytochrome p450 enzymes, CYP3A4, CYP2D6, and CYP2C19, are important metabolizers of many types of xenobiotics, but they are not responsible for the metabolism of alcohol. Therefore, answer choices B, C, and D are incorrect.

4. **The correct answer is B.** The enzyme VKORC1 (vitamin K epoxide reductase complex I) is involved in the reduction and recycling of vitamin K, which is important for the activation of some blood coagulation proteins, such as factor VII and prothrombin. Polymorphic variations in this gene, therefore, can affect the reduction of vitamin K. There are also polymorphic variations in the *VKORC1* gene that are associated with a reduced production of the VKORC1 protein. CYP2C9 (answer choice A) and CYP2E1 (answer choice C) are members of the cytochrome p450 enzyme family and are responsible for the metabolism of xenobiotics. G6PD (answer choice D) is an enzyme that generates NADPH and is responsible for protecting red blood cells from oxidative damage. TPMT (answer choice E) is an enzyme that covalently attaches a methyl groups to substrates in the Phase II of drug metabolism, which allows for the elimination of some types of drugs from the body.

VII

55 Developmental Genetics

At the conclusion of this chapter, students should be able to:
- Describe the molecular and cellular processes in embryonic and fetal development
- Explain the functions/roles of four major classes of paracrine signaling molecules involved in embryonic development
- Describe the roles of transcription factors and extracellular matrix proteins in developmental pathways
- Differentiate among the types of congenital abnormalities (malformations, deformations, disruptions) caused by teratogenic agents e.g. genetic pathogenic mutations and environmental factors

Developmental genetics is the study of the genes that control the growth and differentiation of an organism. The question at the heart of this field of study is "How do cells with the same genetic information develop their diverse structures and functions?" This chapter will focus on the genes that have functions in developmental pathways, as well as examine the genetic pathogenic mutations that give rise to birth defects.

Since most developmental pathways are highly conserved among species, a great deal of knowledge about human developmental events and the cause of birth defects have been garnered from studies in model organisms. Some of the most commonly used model organisms are *Caenorhabditis elegans* (roundworm), *Drosophila melanogaster* (fruit fly), *Danio rerio* (zebrafish), *Xenopus laevis* (frog), *Gallus gallus* (chicken), *Mus musculus* (mouse), and *Papio hamadryas* (baboon).

55.1 Overview of Embryonic Development

Zygote formation starts in the fallopian tube with the fusion of sperm (22 + Y or 22 + X) and egg (22 + X) pronuclei, also known as fertilization. Considered as the beginning of embryonic development, zygote formation triggers rapid mitotic divisions called cleavage.

Cleavage is a series of mitotic divisions that begins within 24 hours of zygote formation and occurs as the zygote passes down the uterine tube to the uterus (▶ Fig. 55.1). At this stage, the daughter cells of the zygote are called blastomeres. When the zygote reaches the eight-cell stage, the blastomeres flatten and connect to each other through tight junctions. This process is called compaction.

The compacted cells continue to divide to form **morula**. By the third or fourth day, the morula contains 16–32 cells. By the fifth day, the morula reaches the uterus and forms a central cavity known as a blastocele. The zygote, now called a **blastocyst,** contains two different groups of cells. The embryoblast, or inner cell mass, and trophoblast, or outer cell mass. The embryoblast will ultimately give rise to embryo, and the trophoblast will give rise to the placenta.

Implantation of the embryoblast begins by the end of first week. When the embryo implants into the endometrial epithelium of the uterus, the trophoblast produces human chorionic gonadotropin (hCG) hormone. Pregnancy testing detects the presence of this hCG in urine or in plasma.

55.1.1 Link to Pathology

Ectopic pregnancy is the implantation of a blastocyst in an abnormal site such as the uterine tube, surface of the ovary, or in the abdomen.

Around the second week of development, the embryoblast differentiates into two germ layers, the epiblast and the hypoblast, to form a bilaminar embryonic disk. In the third week, when **gastrulation** begins, the bilaminar disk becomes a trilaminar disk and the major structure known as the primitive streak forms from the epiblast of the blastocyst. Soon after, the three germ layers (ectoderm, mesoderm, and endoderm) are formed. This special arrangement of differentiated cells will later give rise to the different tissues and organs of the body. All of these events are facilitated by cell and tissue migration, a process called **pattern formation**.

The third week through the eighth week of development is called the embryonic period (embryogenesis). In embryogenesis, organs and systems of the body form from the three germ layers (organogenesis). The fetal period starts after the embryonic period has concluded. During fetal development, the organ systems mature and body growth takes place.

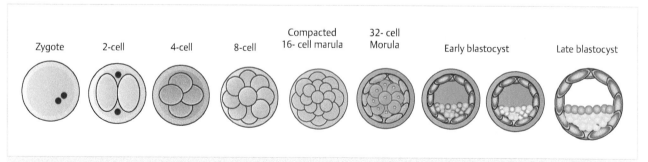

Fig. 55.1 Stages of human embryo development. Following fertilization, the zygote undergoes a series of mitotic divisions that ultimately result in the formation of a blastocyst containing an inner cell mass (embryoblast) and an outer cell mass (trophoblast).

Zygote   2-cell   4-cell   8-cell   Compacted 16- cell marula   32- cell Morula   Early blastocyst   Late blastocyst

55.2 Tissues and Organs that Derive from Ectoderm

The central nervous system (brain and spinal cord), epidermis of the skin (hair, nails, tooth enamel), sensory neurons (organs of hearing and equilibrium), lens, pituitary and mammary glands all develop from the ectoderm layer.

55.3 Tissues and Organs that Derive from Mesoderm

Connective tissue (skeletal bone and cartilage), skeletal muscle, muscle of the eyes and head, dermis, kidneys, urogenital system, vascular structures (heart, vessels, and lymphatics), adrenal cortex, spleen, and gall bladder are derived from the mesoderm.

55.4 Tissues and Organs that Derive from Endoderm

Gastrointestinal organs, thyroid and parathyroid glands, thymus, liver, pancreas, bladder, urethra, and lining of respiratory tract are all derived from endoderm.

55.5 Molecular Mediators of Development and Their Functions

During embryonic development, the differentiation of the three germ layers into specialized cells and tissues is a tightly controlled process that is influenced by three major factors: paracrine signaling, transcription factor activity, and extracellular matrix composition. Recall from chapter 3 that paracrine signaling is a form of cellular communication mediated by extracellular signaling molecules that occurs between neighboring cells. Four families of these paracrine signaling molecules have central roles in several aspects of embryonic development: FGF, Hh, Wnt, and TGF-β.

The fibroblast growth factors (FGFs) constitute a family of signaling molecules that are essential for several events necessary for proper development including cell migration, growth, and differentiation. The cellular effects of FGF are mediated by tyrosine kinase receptors found on the surface of target cells. As we saw in Chapter 10, binding of a signaling molecule to a tyrosine kinase receptor initiates a series of phosphorylation events in the cell that ultimately lead to cellular growth and division. In the context of embryonic development, FGF signaling is an important factor in normal bone development. Individuals with achondroplasia, a form of short-limbed dwarfism, have germline pathogenic mutations in the gene encoding one particular form of the FGF receptor, FGFR3. Instead of creating an inhibited form of the receptor, these particular mutations result in an over-active molecule. This results in an inhibition of chondrocyte proliferation and differentiation, giving rise to the skeletal defects that are characteristic of this disease.

The Hedgehog family of molecules was originally identified in *Drosophila melanogaster* as a mutation that resulted in the growth of bristles in an area of the fly that is normally devoid of these structures. The human genome includes several members of this family with sonic hedgehog (Shh) being the most important in development. Shh has key roles in both axis specification and the patterning of limbs. The receptor for Shh, patched (Ptch), is a transmembrane protein that inhibits the activity of another transmembrane protein, smoothened (Smo) (▶ Fig. 55.2). In the absence of Shh, Ptch keeps Smo in an inactive state. When Shh binds to Ptch, however, this inhibition is relieved and Smo can then activate downstream intracellular signaling molecules, leading to activation of the Gli1 transcription factor and induction of cell division (chapter 10). Pathogenic mutations in PTCH1, the gene that codes for the Ptch protein, lead to a condition called Gorlin syndrome, which is characterized by cysts of the jaw, rib abnormalities, and a form of skin cancer known as basal cell carcinoma.

The Wingless (Wnt) family of genes was also first identified in Drosophila, where mutations were noted to cause a loss of polarity in limb formation. In vertebrates, the Wnt signaling molecule binds to a cell surface receptor called frizzled (chapter 3) (▶ Fig. 55.3). This interaction is important for a variety of developmental processes, such as establishment of the dorsal/ventral axis and formation of the brain, muscle, gonads, and kidneys. Pathogenic mutations that result in the loss of Wnt signaling are responsible for a condition called tetra-amelia, which is characterized by the absence of all four limbs. Wnt signaling has also been shown to be disrupted in a number of different tumors.

The TGF-β superfamily includes the TGF-β gene family as well as the BMP, activin, and Vg1 families of genes. The BMP family plays an essential role in embryonic development, especially in the development of bone. Pathogenic mutations in members of this family result in a number of skeletal abnormalities, including shortening of the digits (brachydactyly) and shortening of the long bones of the limbs (acromesomelic dysplasia and Grebe chondrodysplasia).

Another way in which developmental processes are controlled is through the activity of proteins called transcription factors that act to either turn on or turn off specific genes. Most transcription factors are capable of regulating several different genes at the same time. Therefore, gene mutations in the genes that encode transcription factors typically have pleiotropic effects. There are several families of transcription factors including the homeobox-containing genes, the high-mobility group (HMG)-box-containing genes, and the T-box family (▶ Fig. 55.4). Within the HMG-box-containing group is the SOX family of genes. One member of this family, SRY, encodes the testis-determining factor. This transcription factor is responsible for the regulation of another SOX family member, SOX9, which turns on a series of genes during embryogenesis that are necessary for promoting the development of testes while also repressing ovarian development (▶ Fig. 55.5).

The extracellular matrix also serves an important function in embryonic development. As discussed in chapter 8, molecules such as collagen, fibrillins, laminin, tenascin, fibronectin, and proteoglycans not only provide a medium for holding cells and tissues together, but they also form a matrix on which cells can migrate. This migration of cells from one embryonic structure to another is an essential element in the proper formation of all of the structures of the body. The ability of the cells to migrate along the ECM is dependent on transient interactions that facilitate this movement. Integrins and glycosyl transferases found

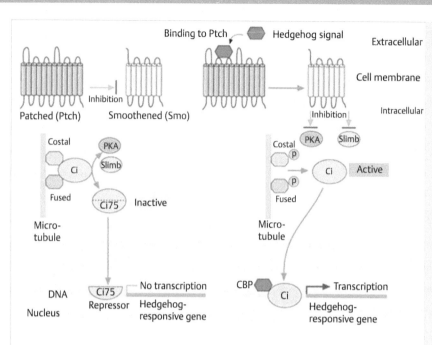

Fig. 55.2 The hedgehog (Hh) signaling pathway. (Source: Passarge E, ed. Color Atlas of Genetics. 4th Edition. New York, NY. Thieme; 2012.)

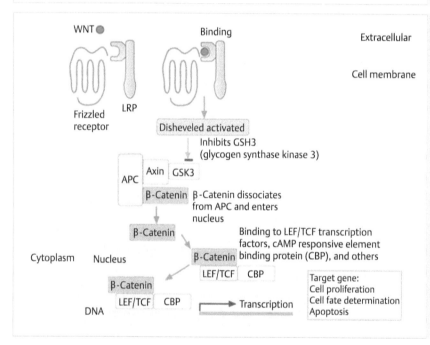

Fig. 55.3 The Wnt β-catenin signaling pathway. (Source: Passarge E, ed. Color Atlas of Genetics. 4th Edition. New York, NY. Thieme; 2012.)

on the surface of the cells allow for relatively weak binding to the ECM, which allows the cells to grab hold and release as they move along the ECM matrix. Once the cell has reached its destination, it can make more permanent interactions with laminin components of the ECM that keep the cell in place.

55.6 Axis Specification

Specification and formation of axes of symmetry are important events during development. In mammals, the anterior/posterior axis is derived from the primitive streak. Expression of the gene nodal in this structure is essential for later events that lead to the formation of the left/right axis. A family of transcription factors, encoded by the HOX genes, is responsible for establishing the anterior/posterior axis. In humans, 39 different HOX genes exist and are grouped into four gene clusters. One gene cluster is found on each of the chromosomes 2, 7, 12, and 17. The expression of the HOX genes is regulated in both a temporal and spatial manner, so that genes located in the 3' end of the genomic cluster are expressed earlier in development and in anterior sections of the body, whereas those in the 5' end are expressed later and in posterior portions of the body. This differential expression contributes to the patterning that is

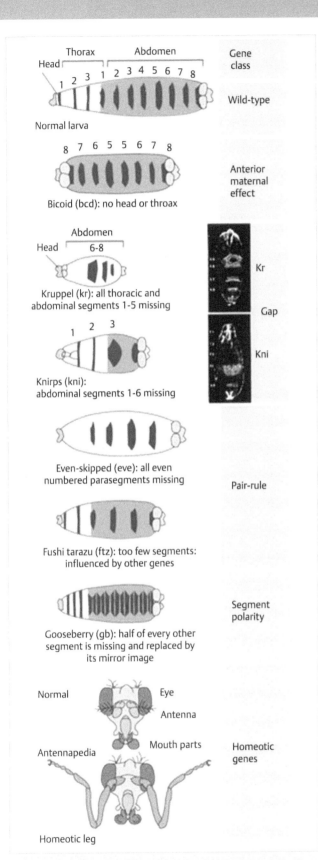

Fig. 55.4 Embryonic developmental genes. (Source: Passarge E, ed. Color Atlas of Genetics. 4th Edition. New York, NY. Thieme; 2012.)

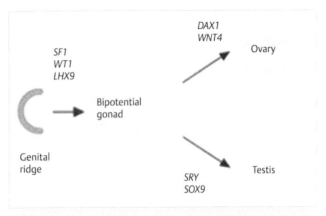

Fig. 55.5 Genes required for gonad development. (Source: Passarge E, ed. Color Atlas of Genetics. 4th Edition. New York, NY. Thieme; 2012.)

necessary to produce the symmetries and asymmetries of the fully developed body.

Formation of the dorsal/ventral axis involves regulation of the genes noggin, chordin, and Bmp-4. Noggin and chordin coordinate the formation of the dorsal anatomy, and Bmp-4 expression designates ventral structures. In the dorsal region, noggin and chordin prevent Bmp-4-mediated ventralization by binding to the receptors for Bmp-4, preventing the activation of these receptors by Bmp-4 and, thus, inhibiting the development of ventral structures.

55.7 Limb Development

Of all the events in embryogenesis, the process of limb development is the best-understood. Limb development begins with the formation of a mesenchymal outgrowth covered by a layer of ectoderm. The creation of the fully-formed limb from a region of the embryonic trunk involves the tightly regulated expression of multiple genes, both spatially and temporally. The transcription factors Tbx4 and Tbx5 are required for the initial formation of the limb bud by inducing the expression of growth-inducing target genes, such as the FGFs. In turn, gradients of FGF protein along the proximal/distal axis control the proliferation of cells in the mesoderm that are responsible for the limb's lengthwise growth. As mentioned earlier, the HOX genes also play a key role in axis specification, as their expression at specific times during development, as well as within specific anatomical locations, is crucial in establishing the anterior/posterior axis of the limb. Gradients of another signaling molecule, Shh, also help to establish the anterior/posterior axis. In addition, Shh also ensures that the digits are formed properly.

VII

Review Questions

1. Which of the following events is considered as the beginning of embryonic development?
   A) Fertilization
   B) Cleavage
   C) Formation of blastocyst
   D) Development of the morula
2. Which of the following molecules is a ligand for the Ptch receptor and has an essential role in the patterning of limbs?
   A) BMP
   B) FGF
   C) SHH
   D) SOX9
3. A gain-of-function mutation in which of the following proteins gives rise to the characteristic phenotype of achondroplasia?
   A) BMP
   B) FGFR3
   C) Frizzled
   D) TGF-β
4. Which of the following tissues is derived from the mesoderm?
   A) Brain
   B) Liver
   C) Skin
   D) Skeletal bone

Answers

1. **The correct answer is A.** Zygote formation, also known as fertilization, is the point at which the sperm pronucleus fuses with the egg pronucleus. This is considered to be the beginning of embryonic development, and is followed 24 hours later by a series of rapid mitotic divisions called cleavage (answer choice B). Once the embryo contains 16–32 cells, it is called the morula (answer choice D). Once the morula reaches the uterus, it forms a central cavity known as a blastocele. The zygote is now called a blastocyst (answer choice C).
2. **The correct answer is C.** The human genome includes several members of the Hedgehog family of signaling ligands. Of these, sonic hedgehog (Shh) is the most important in development. Shh has key roles in both axis specification and the patterning of limbs and relays their signal to the cell through the interaction with Ptch receptors on the cell's surface. The BMP family of extracellular signaling molecules (answer choice A) plays an essential role in embryonic development, especially in the development of bone. The receptors for these molecules are members of the BMP receptor family. The fibroblast growth factors (FGFs) (answer choice B) are a family of signaling molecules that are essential for several events necessary for proper development including cell migration, growth, and differentiation. These molecules transmit their signals through tyrosine kinase receptors, such as the FGFR3. SOX9 (answer choice D) is a member of the HMG-box-containing SOX family of genes. This gene is responsible for turning on and off a series of genes during embryogenesis and is regulated by another member of the SOX family, SRY.

3. **The correct answer is B.** Achondroplasia is a form of short-limbed dwarfism caused by germline pathogenic mutations in the gene encoding one particular form of the FGF receptor, FGFR3. Instead of creating an inhibited form of the receptor, these particular mutations result in an over-active molecule, causing an inhibition of chondrocyte proliferation and differentiation, and giving rise to the skeletal defects that are characteristic of this disease. The BMP family (answer choice A), as well as members of the TGF-β family (answer choice D) play an essential role in embryonic development, especially in the development of bone. Pathogenic mutations in members of these families result in brachydactyly, acromesomelic dyplasia, and Grebe chondrodysplasia). Frizzled (answer choice C) is a receptor for the signaling molecule Wnt, which is important for a variety of developmental processes, such as establishment of the dorsal/ventral axis and formation of the brain, muscle, gonads, and kidneys. Mutations that result in the loss of Wnt signaling are responsible for a condition called tetra-amelia, which is characterized by the absence of all four limbs.

4. **The correct answer is D.** The mesoderm gives rise to the following tissues and organs: connective tissue (skeletal bone and cartilage), skeletal muscle, muscle of the eyes and head, dermis, kidneys, urogenital system, vascular structures (heart, vessels, and lymphatics), adrenal cortex, spleen, and gall bladder. Ectoderm produces the central nervous system (brain and spinal cord), epidermis of the skin (hair, nails, tooth enamel), sensory neurons (organs of hearing and equilibrium), lens, pituitary and mammary glands. The endoderm gives rise to gastrointestinal organs, thyroid and parathyroid glands, thymus, liver, pancreas, bladder, urethra, and lining of respiratory tract.

56 Cancer Genetics

At the conclusion of this chapter, students should be able to:
- Describe the three classes of genes that contribute to oncogenesis
- Compare sporadic cancers and inherited cancer syndromes
- Associate the eight most prevalent inherited cancer syndromes with the type of tumor most likely to be present in each
- Identify specific genes that are mutated in these inherited cancers, describe the normal functions of these genes, and explain how disrupted function of these genes can lead to cancer

Malignant neoplasms are the second leading cause of death in the United States. It is estimated that approximately one out of every two men and one out of every three women will develop some form of cancer during their lifetime. Given its prevalence, cancer is a widely studied disease, especially in the field of genetics. This chapter will focus on the molecular causes of hereditary cancer and will explore the main features of the most common inherited cancer syndromes and the genes associated with each.

Genetics is a particularly important area of study when it comes to cancer because all malignancies are due to DNA mutations. These mutations cause cells to proliferate inappropriately, resulting in cell growth and division that is out of control. There are currently three classes of genes that are known to contribute to oncogenesis (▶ Fig. 56.1).

The first class of genes are the tumor suppressor genes, which are aptly named because they are involved in halting abnormal cell proliferation and preventing tumor growth. This is typically achieved through the induction of apoptosis and, ideally, occurs in the early stages of tumor formation. Humans possess two copies of each tumor suppressor gene, one on each chromosome in a pair. When one of the tumor suppressor genes is disrupted by a gene mutation, the individual's risk for cancer increases. However, oncogenesis typically does not occur until both tumor suppressor genes have been disabled by mutations (▶ Fig. 56.2).

A simple analogy can be used to explain the function of tumor suppressor genes. Much like brakes can prevent a car from going too fast, tumor suppressor genes can prevent cells from dividing too quickly. If only the back brakes malfunction, similar to having only one of the tumor suppressor genes disrupted in a human, the car is still able to stop. However, if both the front and back brakes fail, the car cannot be stopped, much

like oncogenesis cannot be stopped if both tumor suppressor gene alleles are not functioning.

The second class of genes known to contribute to oncogenesis are care-taker, or stability, genes. An example of this class is the mismatch repair (MMR) genes. The function of mismatch repair genes is to correct single base pair abnormalities which were not corrected during the proofreading process during DNA replication. These single base pair abnormalities can include mismatches in which one base is improperly paired with another base or those caused by small insertions or deletions. In addition, mismatch repair genes can aid in the repair of DNA damage. DNA damage or single base pair changes are a natural part of DNA replication and recombination. However, if not corrected, they can alter the DNA alignment, leading to cell

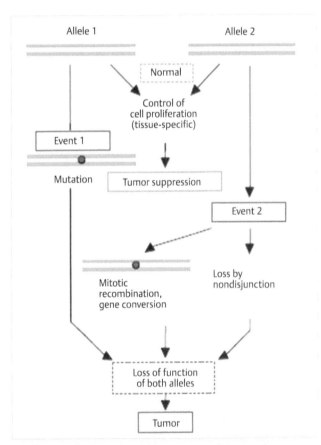

Fig. 56.2 Tumor suppressor gene. (Source: Passarge E, ed. Color Atlas of Genetics. 4th Edition. New York, NY. Thieme; 2012.)

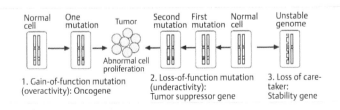

Fig. 56.1 Three categories of cancer genes. (Source: Passarge E, ed. Color Atlas of Genetics. 4th Edition. New York, NY. Thieme; 2012.)

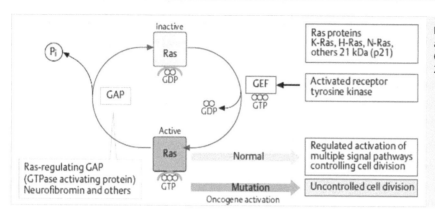

Fig. 56.3 Ras mutations leading to oncogene activation. (Source: Passarge E, ed. Color Atlas of Genetics. 4th Edition. New York, NY. Thieme; 2012.)

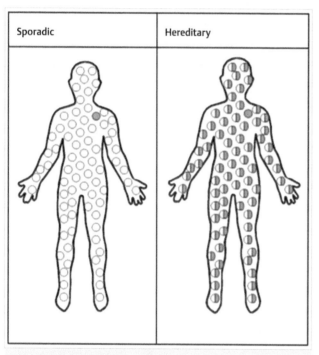

Fig. 56.4 Sporadic versus hereditary cancer.

instability, and, eventually, oncogenesis. Like tumor suppressor genes, oncogenesis typically does not occur until both MMR genes, one on each chromosome, have been disabled by mutation.

Finally, the third class of genes are the proto-oncogenes. Proto-oncogenes are involved in normal cell growth and differentiation. When altered in such a way that increases gene expression, a proto-oncogene becomes an oncogene. By increasing gene expression, oncogenes increase the protein product, which increases cell division, decreases cell differentiation, and inhibits cell death, which can result in oncogenesis (▶ Fig. 56.3).

A car analogy can also be used when discussing proto-oncogenes. Mutations that convert proto-oncogenes to oncogenes cause uncontrolled cell growth, much like an accelerator that is stuck to the floor in a car. A single oncogene is not typically sufficient to cause oncogenesis because tumor suppressor genes are still putting the brakes on to keep cell growth from getting out of control. Therefore, tumor suppressor genes, in addition

to oncogenes, must undergo mutations in order for oncogenesis to occur.

When an individual harbors a germline, or inherited, pathogenic mutation in one of these types of cancer predisposition genes, they have a hereditary cancer syndrome (▶ Fig. 56.4). Although cancer is a prevalent cause of disease, only 5–10% of cases are due to a hereditary predisposition. Hence, though all cancer is genetic, because it is due to mutations in our DNA, only a small portion is actually hereditary. More commonly, cancer occurs sporadically. However, before we discuss the differences between sporadic and hereditary cancer, we must first discuss Knudson's Two Hit Hypothesis.

In 1971, Alfred Knudson (1922–2016) published the formulation of what is now referred to as the Knudson hypothesis, or the two-hit hypothesis. The two-hit hypothesis states that oncogenesis is caused by an accumulation of pathogenic mutations. This hypothesis was originally proposed to explain why some tumors are seen in both sporadic and hereditary forms. It is suggested that the hereditary form of a tumor is caused by a heterozygous germline pathogenic mutation along with a second somatic pathogenic mutation in the other allele of the gene, while the sporadic form is caused by two independent somatic events, as shown in ▶ Fig. 56.4. In either case, pathogenic mutations in both alleles are necessary for the cancer phenotype to appear.

In individuals with a heterozygous germline pathogenic mutation in an inherited cancer syndrome-associated gene, clinical features develop when the remaining wild-type allele is inactivated by a somatic mutation. The average onset of hereditary cancers is typically younger than cancers that occur sporadically. Another common characteristic of hereditary cancer is the development of the same or related type of cancer in multiple individuals within the same family. For instance, sporadic cancers may be common in a family but include numerous cancers that do not fit in an inherited cancer syndrome, and they tend to be diagnosed over the age of 50 years. On the other hand, if multiple individuals in a family have the same type of cancer or have cancers that are associated with a specific hereditary cancer syndrome, it is more likely they are hereditary, especially if diagnosed under the age of 50. Additional characteristics of hereditary cancer include multiple primary cancers in one individual, as well as bilateral or multifocal cancers. Finally, most inherited cancer syndromes (including all of those discussed later in this chapter) are inherited in an autosomal dominant manner and hence this pattern of cancer

Table 56.1 Characteristics of hereditary cancer

Early age of cancer onset

Multiple individuals in the family with the same or related cancers

Multiple primary cancers

Bilateral or multifocal cancers

Typically autosomal dominant transmission of related cancers

incidences in a family should be suspicious for hereditary cancer (▶ Table 56.1).

Most inherited cancer syndromes have reduced penetrance and variable expressivity. Reduced penetrance in regards to hereditary cancer means not every individual with a pathogenic mutation will develop cancer associated with that pathogenic mutation. In other words, carrying the pathogenic mutation does not guarantee cancer will develop. Variable expressivity in inherited cancer syndromes refers to the fact that oncogenesis occurs in different organs and at different ages, even among individuals with the same pathogenic mutation. Hence, obtaining a complete and accurate family history is crucial in individuals with a personal and/or family history of cancer.

Occasionally, even a well-constructed, three-generation pedigree can be misleading or uninformative. For example, there are individuals who have little to no information about their family's health, perhaps due to adoption or their family's unwillingness to share health information. In addition, even if a patient can provide a detailed family health history, there may not be enough individuals affected with cancer or enough individuals in the family as a whole for the inherited cancer syndrome to be recognized. These situations and others like them are termed limited family histories or limited family structures. In the case of the pedigree shown (▶ Fig. 56.5), both sides of the family are considered limited because the proband does not have at least two first- or second-degree female relatives who lived past the age of 45 years.

A limited family structure can be particularly deceiving in the case of hereditary cancer syndromes in which there is an increased risk of malignancy that is far greater in one sex than the other. For example, currently described hereditary cancer syndromes that include an increased risk for breast cancer include a much higher risk of female versus male breast cancer. Hence, if the germline mutation associated with hereditary breast cancer is being passed through males in the family, the hereditary cancer phenotype is likely to be masked. This is sometimes referred to as sex-limited expression (▶ Fig. 56.6).

It is important to keep in mind that even individuals with a family history of an inherited cancer syndrome who did not inherit the disease-causing mutation may develop sporadic cancer. This is illustrated in the pedigree above (▶ Fig. 56.6) in the paternal grandmother. Though she does not carry the pathogenic *BRCA1* mutation (denoted by the "+"), she developed sporadic breast cancer at the age of 60. The grandmother's cancer in this case is referred to as a phenocopy. A phenocopy is when the abnormal phenotype is due to factors that are not inherited, while consistent with a phenotype due to hereditary factors.

There are multiple reasons to consider genetic testing for hereditary cancer syndromes. One is to insure appropriate management is pursued in the form of screening, risk reduction, and/or treatment. In addition, someone with a current

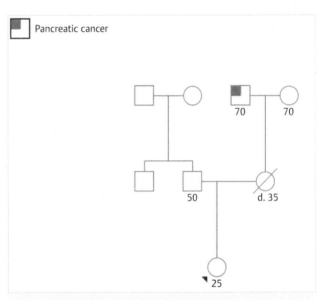

Fig. 56.5 An example of a limited family history, fewer than 2 first- or second-degree female relatives surviving beyond age 45 years in either lineage.

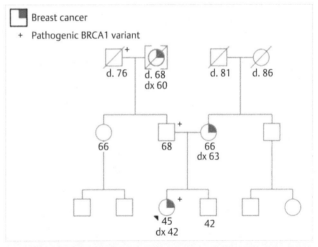

Fig. 56.6 An example of a limited family history, sex-limited expression.

diagnosis or personal history of cancer may benefit from oncology genetic testing as it may inform them of additional cancer risks (e.g., the risk for a second breast cancer or a risk for ovarian cancer). Once a pathogenic mutation is detected in an individual, their family members may pursue site specific testing for the known familial mutation to determine whether they also have an increased risk of cancer due to the mutation. Finally, some individuals choose to use this information for family planning, with one option being to screen for the pathogenic mutation using prenatal genetic diagnosis.

It is also important to take into consideration who is the most appropriate person in the family to initially pursue oncology genetic testing to determine whether the tumors in the family are indeed hereditary. For example, consider the proband in the pedigree drawn in ▶ Fig. 56.7 who has concern about Lynch syndrome being transmitted through her maternal

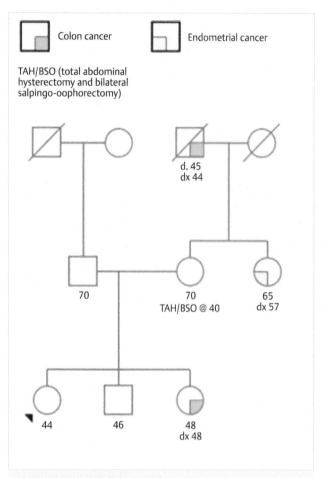

Colon cancer

Endometrial cancer

TAH/BSO (total abdominal
hysterectomy and bilateral
salpingo-oophorectomy)

d. 45
dx 44

70

70
TAH/BSO @ 40

65
dx 57

44

46

48
dx 48

Fig. 56.7 Lynch syndrome pedigree.

Table 56.2 Common inherited cancer syndromes

Inherited Cancer Syndrome	Associated Genes	Common Tumor Type (s)
Hereditary breast and ovarian cancer (HBOC)	BRCA1, BRCA2	Breast, ovary
Lynch syndrome	MLH1, MSH2, MSH6, PMS2, EPCAM	Colorectal, uterine, gastric, ovary
Familial adenomatous polyposis (FAP)	APC	Colorectal, colorectal polyps
Li-Fraumeni syndrome	TP53	Sarcoma, breast, brain, adrenocortical carcinoma
Von Hippel-Lindau syndrome	VHL	Hemangioblastoma, clear cell renal cell carcinoma, pheochromocytoma
Retinoblastoma	RB1	Retinoblastoma
Multiple endocrine neoplasia, type 1	MEN1	Parathyroid, pituitary, gastro-entro-pancreatic
Multiple endocrine neoplasia type 2	RET	Medullary thyroid carcinoma, pheochromocytoma, parathyroid adenoma
Neurofibromatosis type 1	NF1	Neurofibroma (cutaneous or plexiform), optic nerve glioma

VII

family. The proband pursues genetic testing, which turns out negative. She will likely feel relieved that she does not have a mutation that causes an increased risk for cancer. However, because none of the affected relatives have pursued genetic testing, we cannot be sure she was tested for the pathogenic mutation causing an increased risk in this family. If her sister then pursues testing and is found to have a pathogenic mutation in a Lynch syndrome-associated gene that the proband was already tested for, you could confidently tell the proband they do not have Lynch syndrome and their risk for Lynch syndrome-related cancers is that of the general population. However, should her affected sister's testing reveal no pathogenic mutation, the cause of the cancer in the family remains unknown. Thus, the patient remains at an increased risk for cancer, as it is not known whether she harbors the DNA mutation being passed through the family.

This example illustrates the importance of testing the appropriate person in the family initially in order for other family members to pursue appropriate testing. In order to select the most informative person in the family, a good rule of thumb is to use an affected individual who was clinically diagnosed at the youngest age. The most appropriate person in this family to pursue testing would be the sister diagnosed with colon cancer because the individual diagnosed at the youngest age in the family (the maternal grandfather) is deceased. If the sister pursued testing first and a pathogenic germline mutation was

detected, this patient would have a 50% chance of also harboring the mutation. She and other at risk relatives are then able to pursue targeted testing in order to determine whether or not the mutation responsible for the cancers in this family was inherited.

As genetic technology advances and genes that are specifically related to the development of cancer are identified, hereditary cancer syndromes have begun to be recognized and described in detail in regards to the types of cancer associated with each syndrome, common ages of diagnosis, and additional syndromic features that may not be related to cancer. And we will now review the most common inherited cancer syndromes listed in ▶ Table 56.2.

56.1 Li-Fraumeni Syndrome

TP53 is a tumor suppressor gene located on chromosome 17p13.1 which codes for tumor suppressor protein p53, a protein that reacts to cell stressors by regulating other genes responsible for inducing apoptosis, senescence, cell cycle arrest, and cell cycle repair (▶ Fig. 56.8). Tumor protein p53 is kept inactive in unstressed cells via ubiquitination which promotes its degradation. However, if p53 is inactive in stressed cells either due to germline or somatic *TP53* gene mutations or loss of cell signaling upstream or downstream from *TP53*, oncogenesis occurs.

Individuals with a germline heterozygous mutation in the *TP53* gene are said to have Li-Fraumeni syndrome. Hence, this syndrome is inherited in an autosomal dominant

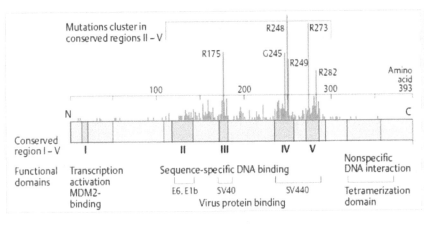

Fig. 56.8 Functional domains and common mutations in the p53 gene. (Source: Passarge E, ed. Color Atlas of Genetics. 4th Edition. New York, NY. Thieme; 2012.)

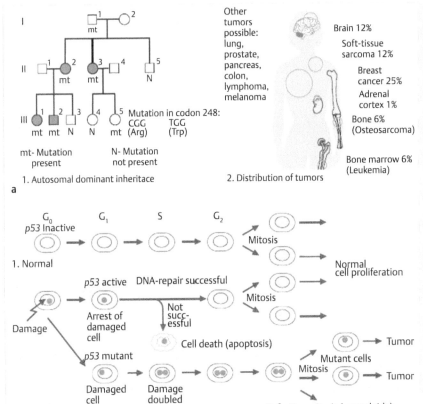

Fig. 56.9 (a) The human TP53 protein. (b) Mutations of the *p53* gene in familial multiple tumors (Li_Fraumeni syndrome). (Source: Passarge E, ed. Color Atlas of Genetics. 4th Edition. New York, NY. Thieme; 2012.)

manner (▶ Fig. 56.9). Approximately 80% of individuals with a clinical diagnosis of Li-Fraumeni syndrome (LFS) have a detectable alteration in the *TP53* gene, 95% of which are sequence mutations. Though most individuals with LFS have a family history of related cancers, it is estimated that 7–20% of individuals with LFS have *de novo* mutations. There has also been some debate about whether heterozygous mutations in the *CHEK2* gene, which is involved in DNA damage responses, also cause Li-Fraumeni syndrome. However, these studies have been conflicting.

Individuals with Li-Fraumeni syndrome are at an increased risk to develop soft tissue sarcomas, osteosarcomas, breast cancer, brain cancer, leukemia, and adrenocortical carcinomas. Additional cancers that have been reported in families with LFS include neuroblastoma, gastric cancer, lymphoma, and genitourinary cancer, as well as skin, lung, and non-medullary thyroid cancers. To date, breast cancer has rarely been reported in males with Li-Fraumeni syndrome. As with any inherited cancer syndrome, the lifetime cancer risk for individuals with LFS is increased. LFS is a highly penetrant cancer syndrome with an estimated cancer risk of 60–90% by age 30 years and 60 years, respectively.

Clinically, the definition of classic LFS is the presence of all of the following in a family:

1. A proband with a sarcoma diagnosed before 45 years of age
2. A first-degree relative with any cancer diagnosed before 45 years of age

3. A first- or second-degree relative with any cancer diagnosed before 45 years of age or a sarcoma at any age

However, LFS should be suspected in the following situations as well:

1. When an individual meets the Chompret criteria for *TP53* genetic testing
2. When a woman has a personal history of early onset breast cancer and neither a *BRCA1* nor a *BRCA2* mutation has been detected
3. When an individual has a personal history of adrenocortical carcinoma
4. When an individual has a personal history of choroid plexus carcinoma

56.2 Hereditary Breast and Ovarian Cancer syndrome

In regards to hereditary breast cancer, the most common syndrome is Hereditary Breast and Ovarian Cancer syndrome (HBOC), which affects approximately 1 in every 400 to 800 people of the general population in the United States and approximately 1 in 40 individuals in the Ashkenazi Jewish population. HBOC is caused by heterozygous mutations in the *BRCA1* and/ or *BRCA2* genes. Discovered in 1994 and 1995 respectively, the *BRCA1* and *BRCA2* genes were among the first cancer susceptibility genes in mainstream media with television commercials urging women to pursue testing in order to "be ready against cancer." In addition, *BRCA1* and *BRCA2* received media attention in June of 2013 when the Supreme Court ruled human genes could not be patented.

BRCA1 and *BRCA2*, located at chromosome 17q21.31 and 13q13.1, respectively, are both tumor suppressor genes (▶ Fig. 56.10). The *BRCA1* gene encodes the tumor suppressor breast cancer type 1 susceptibility protein, BRCA1. Typically expressed in the cells of breast and other tissues, this protein forms multiple distinct complexes responsible for DNA repair, cell cycle checkpoint control, and maintenance of genomic stability. The *BRCA2* gene encodes a tumor suppressing protein which plays a critical role in DNA double-stranded break repair as well as regulation of *RAD51*. Both of these genes are part of the Fanconi anemia pathway (▶ Fig. 56.11) along with numerous other genes known to be vital for DNA repair in breast cells, including the already reviewed *TP53* gene.

Due to overlapping features with other hereditary breast cancer syndromes, gene panel testing has become standard of care, replacing single gene testing. Many hereditary breast cancer gene panels include a number of the genes in the Fanconi anemia pathway, given their effect on breast cancer risk.

Germline pathogenic mutations in *BRCA1* and *BRCA2* cause an increased risk for breast, ovarian, prostate, and pancreatic cancer. Female breast cancer risk in HBOC is between 40–80% and ovarian cancer risk is 11–40%. Of note, the risk of ovarian cancer in women with a *BRCA1* germline pathogenic mutation

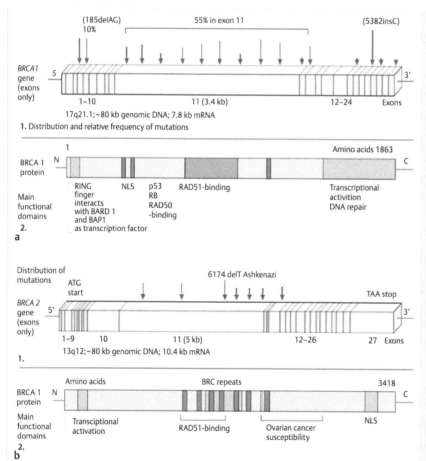

Fig. 56.10 (a) the breast cancer susceptibility gene *BRCA1*. **(b)** The breast and ovarian cancer susceptibility gene *BRCA2*. (Source: Passarge E, ed. Color Atlas of Genetics. 4th Edition. New York, NY. Thieme; 2012.)

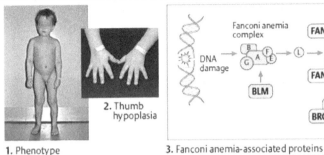

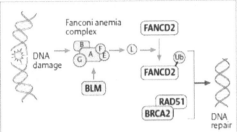

2. Thumb hypoplasia

1. Phenotype

3. Fanconi anemia-associated proteins

Fig. 56.11 Fanconi anemia (FA). (Source: Passarge E, ed. Color Atlas of Genetics. 4th Edition. New York, NY. Thieme; 2012.)

Table 56.3 BRCA mutations

BRCA1	c.68_69delAG
BRCA1	c.5266dupC
BRCA2	c.5946delT

is considerably higher compared to women with a germline *BRCA2* mutation. Male breast cancer risk is as high as 10% and the risk of prostate cancer is up to 39%. Pancreatic cancer risk for males and females is 1–7%.

Hereditary breast and ovarian cancer syndrome should be suspected if an individual has a personal or family history (first-, second-, or third-degree relative) of any of the following:

- Breast cancer diagnosed at age 50 or younger
- Ovarian cancer
- Multiple primary breast cancers either in the same breast or opposite breast
- Both breast and ovarian cancer
- Male breast cancer
- Triple-negative (estrogen receptor negative, progesterone receptor negative, and HER2/neu [human epidermal growth factor receptor 2] negative) breast cancer
- Pancreatic cancer with breast or ovarian cancer in the same individual or on the same side of the family
- Ashkenazi Jewish ancestry
- Two or more relatives with breast cancer, one under age 50
- Three or more relatives with breast cancer at any age
- A previously identified *BRCA1* or *BRCA2* pathogenic mutation in the family

Of note, biallelic germline mutations of *BRCA2* are a rare cause of Fanconi anemia, a severe type of anemia that is also associated with short stature, failure to thrive, microcephaly, cafe-au-lait-spots, and bone marrow failure with an increased susceptibility to malignancy. Hence, individuals with a germline *BRCA2* mutation not only have a 50% chance to pass the mutation on to each of their offspring, if their partner also harbors a *BRCA2* mutation their children have a 25% chance to inherit both mutations and develop Fanconi anemia (▶ Table 56.3).

56.3 von Hippel-Lindau Syndrome

Another example of an inherited cancer syndrome caused by mutations in a tumor suppressor gene is von Hippel-Lindau syndrome (VHL). VHL is caused by heterozygous mutations in the *VHL* gene. The *VHL* gene is located at chromosome 3p25.3 and codes for the von Hippel-Lindau disease tumor suppressor

protein. This protein is a component of the E3 ubiquitin ligase complex which targets multiple proteins for degradation when they are no longer needed in the cell.

One of the proteins the E3 ubiquitin ligase complex targets is hypoxia-inducible factor (HIF), a protein which responds to decreased oxygen in cells by controlling genes involved in cell division, the formation of new blood vessels, and the production of red blood cells. Since the VHL tumor suppressor protein is responsible for recruiting HIF for degradation, when the VHL protein is non-functional, HIF is not degraded and cell division and angiogenesis continue to occur, causing the highly vascularized tumors seen in von Hippel-Lindau syndrome.

Approximately 20% of individuals with VHL have a de novo heterozygous mutation in the *VHL* gene while the other 80% inherited their germline mutation from a parent. In other words, 1 in 5 individuals with VHL syndrome do not have a family history of the disease. Therefore, von Hippel-Lindau syndrome should be at the top of the differential list in any individual diagnosed with one of the rare tumors associated with this condition, regardless of family history.

Clear cell renal cell carcinoma is the most common tumor diagnosed in individuals with VHL, with an estimated 70% of individuals affected by the age of 60 years. This condition is also the leading cause of mortality in these individuals. Because of this risk, annual abdominal ultrasound is recommended for individuals with VHL syndrome, individuals with a germline mutation in the VHL gene, and first-degree relatives who have not pursued genetic testing.

Hemangioblastomas are also commonly seen in association with von Hippel-Lindau syndrome. These can occur in the retina or in the central nervous system (CNS). Retinal hemangioblastomas, also referred to as retinal angiomas, are detected in approximately 70% of individuals with VHL and are diagnosed at an average of 25 years of age. It is estimated that 80% of CNS hemangioblastomas develop in the brain while the other 20% occur in the spinal cord. Rarely, peripheral nerve hemangiomas have been reported.

Von Hippel-Lindau syndrome is characterized not only by renal cell carcinoma and hemangioblastomas of the brain, spinal cord, and retina but also additional visceral tumors and cysts. Multiple renal cysts are quite common in individuals with VHL syndrome. Typically benign, pheochromocytomas may be present in one or both adrenal glands. Paragangliomas can also develop in the abdomen or thorax. Most pancreatic lesions in individuals with VHL syndrome are cysts. However, anywhere from 5–17% of people develop neuroendocrine tumors of the pancreas, which tend to be benign and slow growing. Endolymphatic sac tumors, seen in 10–16% of

VII

individuals, may cause hearing loss. Finally, epididymal and broad ligament cystadenomas are fairly common in males and if bilateral can cause infertility. Rarely, females with VHL syndrome develop a papillary cystadenoma of the broad ligament.

56.4 APC-Associated Polyposis Conditions

Another tumor suppressor gene associated with hereditary cancer is the *APC* gene, located at chromosome 5q22.2. The *APC* gene product is adenomatous polyposis coli (APC), a multidomain tumor suppressor protein which antagonizes the WNT signaling pathway (▶ Fig. 56.12a). The APC protein is part of a complex which promotes phosphorylation of β-catenin, triggering ubiquitination, and subsequent degradation. When APC is dysfunctional, β-catenin is not degraded. Instead, it is translocated into the nucleus where it acts as a transcription factor for genes associated with cell proliferation. The majority of germline pathogenic mutations in *APC* are sequence mutations, with an estimated 10% of individuals having a partial or whole gene deletion or duplication.

The most striking APC-associated polyposis condition is familial adenomatous polyposis (FAP). This inherited cancer syndrome is characterized by hundreds to thousands of precancerous polyps carpeting the colonic mucosa (▶ Fig. 56.12b). These polyps begin developing at a mean age of 16, with a range between 7 and 36 years of age. Approximately 95% of individuals with FAP have colon polyps by the age of 35. Given the sheer number of colon polyps, the risk of colon cancer in untreated individuals with FAP is virtually 100%, with an average age of diagnosis around 40 years.

Individuals with FAP are also at increased risk to develop small bowel polyps (50–90%) and cancer (4–12%). They can develop polyps in the gastric fundus as well. Additional cancer risks include papillary thyroid carcinoma, pancreatic cancer, medulloblastoma, stomach cancer, hepatoblastoma, and bile duct cancer. Extra-colonic manifestations also include osteomas, soft tissue tumors, desmoid tumors, congenital hypertrophy of the retinal pigment epithelium (CHRPE), and dental anomalies.

Individuals with the classic colonic manifestations of FAP along with osteomas and soft tissue tumors were once said to have Gardner syndrome. Similarly, individuals with the central nervous system tumors and the colonic manifestations of FAP were once thought to have a distinct clinical entity called Turcot syndrome. However, it is now understood that these combinations of clinical features are not distinct syndromes, but are due to variable expressivity in APC-associated polyposis. Of note, the combination of CNS tumors and colon cancer can also be seen in individuals with pathogenic mutations in one of the genes associated with Lynch syndrome (discussed later in this chapter).

Given the early onset of clinical manifestations and cancer in individuals with germline pathogenic APC mutations, this is one of the few inherited cancer syndromes in which molecular testing of minors is recommended. Screening for the colonic manifestations of FAP is recommended to begin as early as age 10 years, which certainly warrants presymptomatic testing in children prior to this age.

56.5 Lynch syndrome

Though the disease burden on those with FAP is high, the most common cause of hereditary colorectal cancer is actually Lynch

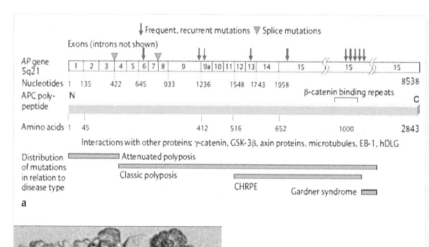

Fig. 56.12 (a) Structure and function of the *APC* gene. **(b)** Polyposis coli. (Source: Passarge E, ed. Color Atlas of Genetics. 4th Edition. New York, NY. Thieme; 2012.)

Table 56.4 Lynch syndrome-related cancer risks by gene

Cancer type	MLH1[2]	MSH2[2]	MSH6[2]	PMS2[3]
Colorectal	85%	52%	30%	19% (male) 11% (female)
Endometrial	82%	21%	47%	12%
Stomach	17%	0.2%	low	unknown
Ovarian	66%	3.8%	3%	Increased
Kidney/urinary tract	2.6%	2.2%	2.1%	Increased
Small intestine	3%	1.1%	low	increased
Brain	N/A	N/A	N/A	unknown
Biliary tract	15%	0.02%	low	unknown

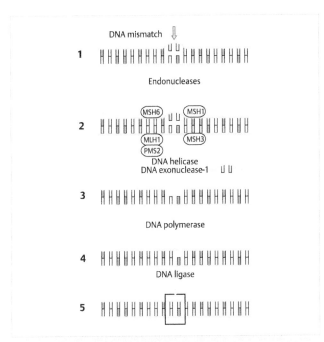

Fig. 56.13 Mismatch repair process.

syndrome (previously called hereditary nonpolyposis colon cancer, or HNPCC). Individuals with Lynch syndrome have an increased risk of many types of cancer including colorectal, endometrial, ovarian, small intestine, liver, gallbladder duct, upper urinary tract, and brain. This syndrome is caused by heterozygous mutations in mismatch repair genes including *MSH1*, *MSH2*, *MSH6*, and *PMS2*. The *de novo* mutation rate of MMR genes is felt to be very low, so the vast majority of Lynch syndrome-associated mutations are inherited from a parent. *MLH1*, located at chromosome 3p22.2, and *MSH2*, located at chromosome 2p21-p16, code for the DNA mismatch repair proteins Mlh1 and Msh2, respectively. *MSH6*, located at chromosome 2p16.3, and *PMS2*, located at chromosome 7p22.1, code for the DNA mismatch repair proteins Msh6 and Pms2, respectively.

When a mismatch occurs, a heterodimer between the Msh2 and Msh6 proteins binds to the mismatch and a second heterodimer between Mlh1 and Pms2 is recruited (▶ Fig. 56.13). The mismatch repair endonuclease Pms2 causes single strand breaks near the mismatch creating an entry point for the exonuclease, which degrades the strand containing the mismatch. The strand is then resynthesized and ligated. When a pathogenic mutation in one of the Lynch syndrome genes alters the protein product, the mismatch repair function is destroyed, leading to the increased cancer risks associated with Lynch syndrome.

The cancer risks associated with Lynch syndrome vary and are conditional upon the gene in which the disease-causing mutation presides, giving us a good example of a genotype-phenotype correlation, as certain mutations result in specific phenotypic clinical manifestations of the disorder. In Lynch syndrome, individuals with mutations in *MLH1* and *MSH2* have a higher risk of cancer than individuals with mutations in *MSH6* or *PMS2* as seen in ▶ Table 56.4.

While germline genetic testing is available for Lynch syndrome, testing can also be performed on tumor tissue, typically colon or endometrial tumors, in order to estimate the probability of Lynch syndrome and to identify which Lynch syndrome-associated gene is most likely to harbor a pathogenic germline mutation. Microsatellite instability testing interrogates short, repetitive sequences of DNA within the tumor called microsatellites, which are susceptible to accumulating mutations when a mismatch repair (MMR) gene is altered. Therefore, tumors with microsatellite instability are much more likely to occur in individuals with a germline pathogenic mutation in an MMR gene. Hence, a diagnosis of Lynch syndrome should be considered in all individuals with tumor microsatellite instability (MSI-H). Immunohistochemistry is another type of tumor testing which detects the presence of proteins expressed by the MMR genes in the tumor. If a protein is not detected in the tumor, the gene that expresses it may not be functioning properly due to a germline mutation. For example, loss of expression of the Mlh1 protein may indicate a mutation in *MLH1*. Furthermore, Pms2 and Msh6 are unstable when not in a heterodimer. Therefore, a germline pathogenic mutation in *MLH1* results in loss of Mlh1 and Pms2 protein and a germline pathogenic mutation in *MSH2* typically results in loss of Msh2 and Msh6 protein.

One mutation of Lynch syndrome typically caused by germline pathogenic mutations in *MSH2* is Muir-Torre syndrome. Muir-Torre syndrome is used to describe the combination of one or more of the internal malignancies seen in Lynch syndrome and sebaceous neoplasms of the skin. These skin neoplasms can include sebaceous adenomas, sebaceous epitheliomas, sebaceous carcinomas, and keratoacanthomas. In addition, as discussed in the familial adenomatous polyposis section, the combination of CNS tumors and colon cancer is sometimes referred to as Turcot syndrome and can be caused by germline mutations in *APC* or one of the MMR genes associated with Lynch syndrome.

56.6 Retinoblastoma

Retinoblastoma was the tumor most studied by Alfred Knudson during his work on the two-hit hypothesis. He noted that there were two forms of retinoblastoma. The first was a bilateral, familial form where individuals were more likely to develop other malignancies later in life. The second was a unilateral, sporadic form with no increased risk of later-onset malignancies. Knudson studied retinoblastomas in order to explain the difference between sporadic tumors and those associated with an inherited cancer syndrome.

Both hereditary and sporadic retinoblastoma are caused by biallelic pathogenic mutations in the *RB1* gene located at chromosome 13q14.2. The *RB1* gene encodes the tumor suppressor protein called retinoblastoma-associated protein (pRB). The normal function of pRB in the cell is to prevent excessive cell growth. It does this by interacting with the E2F transcription factors to form a repressor complex in order to arrest the cell cycle in G1 (▶ Fig. 56.14). The interruption of pRB function allows for unregulated cell cycle progression, leading to oncogenesis.

Individuals with heterozygous germline pathogenic *RB1* mutations have a greater than 90% risk to develop retinoblastoma, with most individuals being diagnosed with bilateral retinoblastoma in infancy. Additionally, these individuals are at increased risk for retinoma, pinealoblastoma, and extraocular

primary neoplasms such as melanoma, osteosarcoma, and soft tissue sarcomas. The extraocular primary neoplasms typically develop in adolescence or adulthood.

It is estimated that only 10% of germline pathogenic mutations in *RB1* are inherited from an affected parent. The other 90% of mutations are mostly *de novo*, with gonadal mosaicism reported in 6–10% of families. Hence, though a parent may not have a detectable pathogenic mutation via blood testing, recurrence risk for future pregnancies may be significant.

56.7 Neurofibromatosis

Neurofibromatosis type 1 (NF1) is a genetic condition commonly diagnosed based on nonmalignant clinical findings. The classic clinical phenotype seen in individuals with NF1 includes multiple cafe-au-lait macules, axillary and inguinal freckling, cutaneous neurofibromas, and Lisch nodules (▶ Fig. 56.15). Clinical diagnostic criteria currently exist for neurofibromatosis type 1 and are still used in the clinical setting when molecular testing is not used.

While neurofibromatosis type 1 shows variable expressivity between and within families, most affected individuals meet diagnostic criteria in childhood. Approximately half of germline *NF1* pathogenic mutations occur *de novo*. Therefore, there may be no family history of the disorder.

Malignancies associated with NF1 can include malignant peripheral nerve sheath tumors (MPNST), optic gliomas, other gliomas, leukemias, and breast cancers. The vast majority of individuals with NF1 never develop malignancy. However, given the function of the gene associated with this disorder, the risk of oncogenesis is greater than average.

Neurofibromatosis is caused by dysfunction of the tumor suppressor protein neurofibromin 1. This protein is predomi-

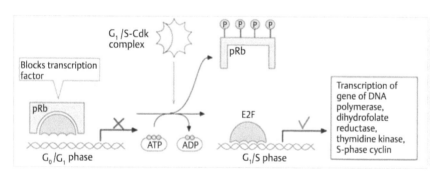

Fig. 56.14 Retinoblastoma protein. (Source: Koolman J, Röhm K, ed. Color Atlas of Biochemistry. 3rd Edition. New York, NY. Thieme; 2012.)

Neurofibromatosis 1 (NF1)
(von Recklinghausen disease)

Autosomal dominant
Frequency 1 in 3000
Gene locus on 17q11.2
Café-au-lait spots
Lisch nodules in the iris
Multiple neurofibromas
Skeletal anomalies
Predisposition to tumors
of the nervous system
50% new mutations

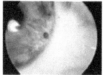

1. Lisch nodule

2. Café-au-lait spot

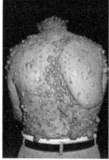

3. Neurofibromas

Fig. 56.15 Main manifestations of neurofibromatosis 1. (Source: Koolman J, Röhm K, ed. Color Atlas of Biochemistry. 3rd Edition. New York, NY. Thieme; 2012.)

nantly expressed in cells of the nervous system such as neurons, Schwann cells, and oligodendrocytes, as well as in leukocytes. Neurofibromin 1 is known to interact with multiple proteins and appears to be a negative regulator of the RAS pathway. Consequently, loss of neurofibromin 1 is associated with the activation of RAS, which fuels cell growth and division. The combination of a germline *NF1* mutation and a somatic *NF1* mutation is responsible for the clinical manifestations of neurofibromatosis type 1.

56.8 Multiple Endocrine Neoplasia type 2

Another inherited cancer syndrome that presents with additional clinical features consistent with a syndromic appearance is multiple endocrine neoplasia type 2B (MEN2B). Individuals with MEN2B often have a distinctive facial appearance including an elongated, sometimes coarse-appearing face with large lips. Mucosal neuromas of the lips and tongue are also common, the former of which can result in thick vermilion of the upper and lower lip. Additionally, a Marfanoid body habitus is also common which is typically tall and lean and often includes joint laxity and kyphoscoliosis or lordosis.

Of greater concern is the increased risk of malignancy associated with MEN2B, as individuals with MEN2B have a nearly 100% chance of developing early-onset medullary thyroid carcinoma, which typically occurs in early childhood. In addition, approximately 50% of individuals with MEN2B develop pheochromocytomas, which are often bilateral.

MEN2B accounts for approximately 5% of cases of multiple endocrine neoplasia type 2, while the more common type is multiple endocrine neoplasia type 2A (MEN2A). MEN2A is an inherited cancer syndrome which confers an increased risk for medullary thyroid carcinoma, pheochromocytoma, and parathyroid adenoma or hyperplasia. There are neither specific facial features nor body habitus associated with MEN2A.

While MEN2B and MEN2A are both due to pathogenic mutations in the *RET* gene, the vast majority of individuals with MEN2B are found to have germline pathogenic mutations in exon 16, while the pathogenic mutations associated with MEN2A are typically located in exons 10, 11, and 13–16. Additionally, only 50% of *RET* pathogenic mutations are inherited in individuals with MEN2B, while approximately 95% of individuals with MEN2A have an affected parent. *RET* is a proto-oncogene located at chromosome 10q11.2 which codes a tyrosine kinase transmembrane receptor involved in a plethora of cellular mechanisms including cell proliferation and differentiation. It is thought that this gene is expressed in neural crest cells including thyroid C cells and adrenal medulla. When *RET* is converted to an oncogene via a gain of function mutation, the receptor is constitutively activate, causing uncontrolled cell proliferation.

Of note, germline gain of function mutations in the *RET* gene can also cause familial medullary thyroid carcinoma, an increased risk of medullary thyroid carcinoma in families, while loss of function mutations are associated with the development of Hirschsprung disease.

56.9 Multiple Endocrine Neoplasia type 1

Multiple endocrine neoplasia type 1 (MEN1) is caused by a germline pathogenic mutation in the tumor suppressor gene *MEN1* located at chromosome 11q13.1, which encodes the protein menin. Though menin has been found to localize to the nucleus, its specific function in the cell remains unknown. Individuals with germline pathogenic *MEN1* mutations are at risk for more than 20 endocrine and non-endocrine tumors, which can be benign or malignant.

The most common endocrine tumors seen in MEN1 can be remembered with the "**3 Ps**": **parathyroid** tumors, **pituitary** tumors, and well-differentiated endocrine tumors of the gastro-entero-**pancreatic** tract. Additional endocrine tumors include carcinoid tumors and adrenocortex tumors, while the non-endocrine tumors include benign skin lesions, lipomas, leiomyomas, and CNS tumors. An estimated 50% of individuals with MEN1 are symptomatic by age 20, while a staggering 95% have clinical features by age 40. Multiple endocrine neoplasia type 1 should be suspected in any individual with any of the associated endocrine or non-endocrine tumors.

Approximately 90% of MEN1 cases are inherited from a parent and 80–90% of inherited MEN1 are due to germline pathogenic sequence mutations in *MEN1*. Partial or whole gene deletion or duplication has been reported in 1–4% of individuals with MEN1.

Tumor testing vs Germline testing: Mutations in many of the genes discussed in this chapter can also be seen in sporadic cancers as well. For example, it is well known that many clear cell renal cell tumors have mutations in the *VHL* gene that were acquired rather than inherited. Hence, the pathogenic mutation (s) found in tumor testing do not typically provide insight into a possible inherited cancer syndrome but rather may provide insight into disease treatment targeting specific pathways or genes.

Suggested Readings

Bonadona V, Bonaïti B, Olschwang S, et al. French Cancer Genetics Network. Cancer risks associated with germline mutations in MLH1, MSH2, and MSH6 genes in Lynch syndrome. JAMA. 2011; 305(22):2304–2310

Broeke SW, et al. Lynch syndrome caused by germline PMS2 mutations: delineating the cancer risk. J Clin Oncol. 2014; •••:32

Hersh JH, American Academy of Pediatrics Committee on Genetics. Health supervision for children with neurofibromatosis. Pediatrics. 2008; 121(3):633–642

King MC, Levy-Lahad E, Lahad A. Population-based screening for BRCA1 and BRCA2: 2014 Lasker Award. JAMA. 2014; 312(11):1091–1092

Knudson AG, Jr. Mutation and cancer: statistical study of retinoblastoma. Proc Natl Acad Sci U S A. 1971; 68(4):820–823

Ohh M. Ubiquitin pathway in VHL cancer syndrome. Neoplasia. 2006; 8(8):623–629

Vasen HF, Watson P, Mecklin JP, Lynch HT. New clinical criteria for hereditary non-polyposis colorectal cancer (HNPCC, Lynch syndrome) proposed by the International Collaborative group on HNPCC. Gastroenterology. 1999; 116(6):1453–1456

Weitzel JN, Lagos VI, Cullinane CA, et al. Limited family structure and BRCA gene mutation status in single cases of breast cancer. JAMA. 2007; 297(23):2587–2595

VII

Review Questions

1. Oncogenes have which of the following features?
   A) Their mutations result in a gain-of-function
   B) Both alleles must be mutated to have an effect
   C) They are commonly inherited in an autosomal recessive fashion
   D) They are responsible for mismatch repair

2. Which of the following is characteristic of sporadic cancers?
   A) Late age of onset
   B) Similar cancer in multiple family members
   C) Bilateral cancers
   D) Autosomal dominant inheritance pattern

3. Which of the following types of cancers is associated with von Hippel Lindau syndrome?
   A) Endometrial cancer
   B) Clear cell renal cell carcinoma
   C) Breast cancer
   D) Colon cancer

4. Li Fraumeni syndrome is caused by an inherited mutation in which of the following genes?
   A) APC
   B) BRCA1
   C) RB1
   D) TP53

Answers

1. **The correct answer is A**. The normal version of an oncogene, also known as a proto-oncogene, is typically involved in normal cell growth and differentiation. When altered in such a way that increases gene expression, a proto-oncogene becomes an oncogene. By increasing gene expression, oncogenes increase the protein product, which increases cell division, decreases cell differentiation, and inhibits cell death, which can result in oncogenesis. Therefore, the main feature of an oncogene is that the mutation present results in a gain-of-function. Answer choices B and C describe features of tumor suppressors, which follow the Knudson Two Hit Hypothesis. This hypothesis states that both alleles of a tumor suppressor must be disrupted in order to have an effect on the cell. Answer choice D describes a characteristic of care taker genes, many of which are responsible for the maintenance of DNA integrity in the cell.

2. **The correct answer is A**. Sporadic cancers typically originate at only one site in the body, do not appear in multiple members of a family, and have a late age of onset. Therefore, answer choice A is correct. In contrast, answer choices B, C, and D all describe characteristics of inherited cancers, including similar cancers in multiple family members, bilateral cancers, and an autosomal dominant inheritance pattern. In addition to these characteristics, most inherited cancer syndromes have reduced penetrance and variable expressivity

3. **The correct answer is B**. von Hippel Lindau syndrome is an inherited cancer syndrome caused by mutations in the *VHL* gene. The *VHL* gene codes for the von Hippel-Lindau disease tumor suppressor protein. This protein is a component of the E3 ubiquitin ligase complex which targets multiple proteins for degradation when they are no longer needed in the cell. Although this disease is associated with hemangioblastomas of the brain, spinal cord, and retina, the most common cancer in these patients is clear cell renal cell carcinoma. Endometrial cancer (answer choice A) is a feature of Lynch syndrome, while breast cancer (answer choice C) is the hallmark of hereditary breast and ovarian cancer (HBOC) syndrome. Individuals with familial adenomatous polyposis (FAP) develop colon cancer (answer choice D) early in life.

4. **The correct answer is D**. Li Fraumeni syndrome is a hereditary cancer syndrome in which individuals are at increased risk of developing a variety of different cancers including soft tissue sarcomas, osteosarcomas, breast cancer, brain cancer, leukemia, and adrenocortical carcinomas. This disease is caused by an inherited mutation in the *TP53* gene, which encodes for the tumor suppressor protein p53. *APC* (answer choice A) is mutated in familial adenomatous polyposis, and *BRCA1* (answer choice B) is the gene that is mutated in hereditary breast and ovarian cancer syndrome. Mutations in *RB1* (answer choice C) are responsible for the pediatric tumor retinoblastoma.

VII

Review Questions

1. An apparently healthy 25-year-old woman remarks during her history and physical that one of her siblings died from infantile Krabbe disease. What is the risk that this woman is a heterozygous carrier for a mutation in the galactosylceramidasegene (*GALC*)?
   A) 25%
   B) 50%
   C) 67%
   D) 75%

2. Analysis of a pedigree from a large family afflicted with an unidentified genetic disease reveals that affected males never produce affected children, but affected females produce affected children of both sexes when they mate with unaffected males. What is the most likely inheritance mode?
   A) Autosomal Dominant
   B) Autosomal Recessive
   C) X-Linked Dominant
   D) X-linked Recessive
   E) Mitochondrial

3. A woman with Leber hereditary optic neuropathy (LHON) and her unaffected husband seek genetic counseling. What is the likelihood that her child will be afflicted with the same disease?
   A) <1%
   B) 25%
   C) 50%
   D) 75%
   E) 100%

4. An 8-year-old girl presents with ectopic lentis. Physical examination reveals she has arachnodactyly with joint hypermobility. It is noted that she is very tall, with a height above the 90th percentile for her age. An extensive pedigree analysis of her family indicates no family history of this disorder. What is the most likely diagnosis?
   A) Fragile-X syndrome
   B) Huntington disease
   C) LHON (Leber hereditary optic neuropathy)
   D) Marfan syndrome
   E) Osteogenesis Imperfecta

5. Based on the pedigree above, what is the most likely inheritance pattern of thisdisease?
   A) Autosomal dominant
   B) Autosomal recessive
   C) X-linked dominant
   D) X-linked recessive
   E) Mitochondrial

6. A 6 year-old girl presents with xanthomas on her elbows and her knees. Blood tests indicate very high levels of serum LDL and total cholesterol. What is the most likely inheritance pattern of her disease?
   A) Autosomal dominant
   B) Autosomal recessive
   C) X-linked dominant
   D) X-linked recessive
   E) Mitochondrial

7. Cystic fibrosis patients may present with many unrelated symptoms such salty skin, recurrent lung infections, and fatty stools. Which genetic term best describes the presentation of multiple unrelated symptoms in an individual with a genetic disease, such as the case of salty skin, recurrent lung infections, and fatty stools in cystic fibrosis patients?
   A) Age of on-set
   B) Anticipation
   C) Incomplete penetrance
   D) Pleiotropy

8. Although Duchenne and Becker muscular dystrophies are more common in males than females, due to the *DMD* genebeing located on the X-chromosome, occasionally females who carry *DMD* gene mutation may have muscle weakness and cramping. What is the most likely explanation for the presence of the syndrome in females?
   A) Manifesting heterozygote
   B) Homozygous for the *DMD* mutation
   C) Incomplete penetrance
   D) Variable expression

9. A two-year-old boy presents with learning difficulties, developmental delay, and cleft palate. His pediatrician suspects DiGeorge, or 22q11.2 deletion, syndrome. Which of the following cytogenetic techniques would best confirm his diagnosis?
   A) Chromosomal microarray
   B) FISH
   C) G-banding
   D) SNP microarray

10. Genetic analysis reveals a mutation in a component of the synaptomenmal complex is present that prevents formation of the complex. These cells are expected to have a deficiency in a process in which of the following cell division phases?
   A) Prophase I
   B) Metaphase I
   C) Prophase II
   D) Metaphase II

VII

Answers

1. **The correct answer is C.** Krabbe disease is a lysosomal storage disease (reviewed in Theme 41) that is inherited in an autosomal recessive manner. Since she is phenotypically normal but her sister died from this disease, both of her parents must be carriers (Aa) for this mutation. Choice A is not correct because this disease is lethal in infancy and therefore it is clear that she does not have the disease. and she is not homozygous recessive (aa) for the mutation. This means there are only three possibilities for her genotype: (1) she inherited a wild type copy from both her mother and her father, (2) she inherited one normal copy from her father and one mutant copy from mother, (3) or vice versa. Since there is a two out of three possibility that she is heterozygous, the risk that she is a carrier is 2/3 or 67%.

2. **The correct answer is E.** Since sperm cells do not contribute mitochondria when it fertilizes an egg, the only mitochondria in the zygote originate from the egg cell. Therefore, the DNA inside the mitochondria (mitochondrial

DNA) is inherited exclusively through females. Pedigrees for mitochondrial diseases thus display a distinct mode of inheritance. The disease is transmitted only from affected females to their offspring, and affected males never produce affected children.

3. **The correct answer is E**. Leber hereditary optic neuropathy (LHON) is an inherited form of vision loss. The gene *ND4*, which encodes NADH dehydrogenase, is located in the mitochondrial genome. Mutations in the *ND4* gene disrupt the production of ATP in the electron transport chain. This results in cell death, especially in the neurons. Since mitochondrial DNA is inherited exclusively from mother, if the mother has a mitochondrial disease, it is transmitted to all of her offspring.

4. **The correct answer is D**. This child most likely has Marfan syndrome, which is an autosomal dominant connective tissue disorder that mainly affects the cardiovascular, skeletal, and ocular systems, and is associated with pathogenic variants in the *FBN1* gene. An individual with a dilated aortic valve, ectopic lentis, and common skeletal findings such as tall stature and pectus excavatum will routinely undergo *FBN1* gene sequencing in order to obtain a diagnosis of Marfan syndrome. Because Marfan syndrome can affect multiple organs and clinical symptoms do not always manifest in individuals with the mutation, this disease is said to be pleotropic.

Fragile-X syndrome (answer choice A) is an X-linked dominant disease and therefore occurs at a much higher frequency in males than females. Signs and symptoms include intellectual disability, large ears and jaw, and post-pubertal macro orchidism. Although rare, females can be affected and generally present with attention deficits and intellectual disability.

Huntington disease (answer choice B) is inherited in an autosomal dominant manner and is characterized by movement abnormalities, emotional disturbances, and cognitive and motor function impairments.

Osteogenesis imperfecta (answer choice E) is a bone disease resulting from defective collagen protein. It generally presents with blue sclera, skeletal deformities, and frequent bone fractures.

Leber hereditary optic neuropathy (LHON) (answer choice C) is a mitochondrial inherited disease and is due to the mutation in NADH dehydrogenase (*ND4*) gene. Most mitochondrial gene mutations present with neuropathies and/or myopathies.

5. **The correct answer is B**. Diseases that are inherited in an autosomal recessive manner are expressed when both alleles are defective. This inheritance pattern typically skips generations, but can be seen in multiple people in one generation. Males and females are equally affected. It cannot be answer choice A, because autosomal dominant diseases are observed in every generation. It cannot be answer choice C either, since X-linked dominant diseases affect males more often than females and is seen in every generation. X-linked recessive diseases (answer choice D) are also more commonly seen in males. It is most likely not a mitochondrial disease (answer choice E) either since her mother does not have the disease and these conditions are inherited

from mother and affect both males and females at equal frequency.

6. **The correct answer is A**. This child is most likely suffering from an inherited form of hypercholesterolemia, which can cause the buildup of excess cholesterol in tendons. This accumulation of excess cholesterol in the tendons is called tendon xanthomas. Familial hypercholesterolemia, due to genetic mutations in the LDL receptor (*LDLR*) gene, is inherited in an autosomal dominant manner. Heterozygote individuals generally develop cardiovascular problems at an early age of around 30 - 40 years. Homozygote individuals, however, present with severely high LDL cholesterol and cardiovascular problems in childhood. This patient is most likely homozygous.

7. **The correct answer is D**. Pleiotropy is the phenomenon of variants in a single gene causing more than one phenotypic trait or disorder. Pleiotropy is seen with a number of different genes, one of which is CFTR, which causes cystic fibrosis. Another one is the fibrillin gene, *FBN1*. Pathogenic variants in *FBN1* can cause the common symptoms of Marfan syndrome but may also cause isolated ectopic lentis (displacement of the lens in the eye) in some families. The symptoms of cystic fibrosis can be seen in multiple organ systems. Therefore the best explanation for these different symptoms is pleiotropy.

Anticipation (answer choice B) in genetic disease is the phenomenon by which symptoms of a disorder may appear not only at a younger age compared to the previous generation, but also with more severe symptoms. In other words, the age of onset can change from generation to generation. This phenomenon is typically seen in disorders caused by trinucleotide (triplet) repeat expansions, e.g. Huntington disease, Fragile-X syndrome.

If individuals have the disease genotype but they do not display the disease, this phenomenon is called incomplete penetrance (answer choice C), which is not the situation described in the question.

8. **The correct answer is A**. Duchenne and Becker muscular dystrophies are X-linked recessive diseases, which are seen at a much higher rate in males than females since males are hemizygous for the X chromosome. Females are not often affected with the X-linked recessive diseases because of compensation by the second normal X chromosome. However, in rare cases, inactivation of the normal X-chromosome can occur in an unusually high percentage of her cells. If this happens, most cells will have the mutated X-chromosome. So even though the female is heterozygous, she may manifest symptoms (manifesting heterozygote). In some situations, answer choice B may be correct, however since this is an X-linked recessive disorder, both of her parents have to have the mutated gene, which means her father would have to be affected, and no family history of the disease is indicated.

Incomplete penetrance (answer choice C) occurs when an individual has a disease genotype but does *not* display the disease phenotype.

Variable expression (answer choice D) can be a result of several factors. Mostly these factors are environmental effects, allelic heterogeneity or heteroplasmy. With variable expression, individuals with the disease producing genotype display varying degrees of phenotypes. Although in some cases, this answer choice may be correct, the best explanation for the X-linked recessive diseases is manifesting heterozygote, in which a heterozygote female occasionally expresses an X-linked mutation because her X chromosome inactivation is skewed unfavorably.

9. **The correct answer is B**. FISH is an established cytogenetic technique used to identify some genetic diseases such as 22q11.2 deletion syndrome (DiGeorge syndrome) and cri du chat syndrome. FISH uses a fluorescently labeled probe to visualize duplications and deletions of specific regions of DNA on chromosomes. Chromosomal microarray (answer choice A) is a genetic test that can be used to identify chromosomal abnormalities of less than 5 Mb in size. G-banding (answer choice C) is used to visualize gross chromosomal abnormalities and chromosome number changes, but does not have the resolution to reveal small deletions or insertions. SNP microarray (answer choice D) can be used to detect copy number variations and allow for the detection of regions of homozygosity, which may result from uniparental disomy.

10. **The correct answer is A**. The synaptonemal complex is a structure composed of three parallel dense elements that mediates the recombination, or crossing over, of genetic information between homologous chromosomes. This process occurs when pairs of homologous chromosomes are aligned in a tetrad, which is a defining event during prophase I of meiosis. In metaphase I (answer choice B), homologous pairs formed during prophase I line up, with one chromosome from each pair on either side of the equator. In this case, paternal and maternal chromosome pairs align randomly, leading to additional genetic diversity in a process called independent assortment. Prophase II (answer choice C) marks the beginning of meiosis II but does not include recombination, and the subsequent cell division following metaphase II (answer choice D) results in the formation of four individual haploid cells.

VII

Review Questions

1. If the frequency of sickle cell disease in the general population is 0.1%, what is the probability of a 27-year-old male with the genotype Hb-AS will marry a sickle cell carrier?
   A) Between 5-10%
   B) Between 10-20%
   C) Between 20-30%
   D) Less than 5%
   E) More than 50%

2. The frequency of familial hypercholesterolemia in a population is found to be 1 in 250 individuals. What is the frequency of homozygous affected individuals in this population?
   A) 1 in 25
   B) 1 in 250
   C) 1 in 2,500
   D) 1 in 25,000
   E) 1 in 250,000

3. The frequency of G6PD deficiency among males is found to be 10% in the Mediterranean population. What is the frequency of this disease among females in this population?
   A) 0.1%
   B) 1%
   C) 10%
   D) 50%
   E) Cannot be observed

4. The frequency of maple syrup urine disease(MSUD) in the Mennonites population is 1 in 400 births. A heterozygous male carrier marries a woman from this population. Which of the following is the probability that they will have a child with MSUD?
   A) 0.5%
   B) 1%
   C) 1.5%
   D) 2%
   E) 2.5%

5. Which of the following findings would support the hypothesis that epigenetic modifications are transferred through generations?
   A) A disease shows early age of onset compared to unexposed family members
   B) Only males develop a disease when compared to the general population
   C) There are different patterns of DNA methylation compared to unexposed family members
   D) There is an increased rate of DNA mutation compared to general population

6. A 4-year-old girl presents to her pediatrician's office for a well-child examination. Her mother states that when the child was younger, it was difficult to feed her but now she has an uncontrollable appetite. History reveals that the child had neonatal hypotonia and delayed motor development during her toddler years. Physical examination shows that her height is less than 5th percentile. Cytogenetic analysis shows a normal karyotype of 46,XX. Her physician orders FISH analysis. Which of the following is the most likely cause of her condition?
   A) Genomic imprinting of the entire chromosome 15

B) Imprinted genes in thepaternal chromosome 15
C) Microdeletion inthe maternal chromosome 15
D) Microdeletion in the paternal chromosome 15

7. A 6-year-old girl recently diagnosed with Acute Lymphoblastic Leukemia (ALL) begins treatment with the chemotherapeutic agent 6-mercaptopurine. What should be considered before administering this drug?
   A) The amount of acetylation in her cells
   B) Identification of any polymorphic variations in her*NAT2* gene
   C) The amount of thiopurine methyltransferase activity in her cells
   D) Ensure that there are a normal number of repeats in the promoter region of her*UGT1A1* gene

8. A 32-year-old male with recurrent migraine headaches is suggested by his physician to take an over-the-counter headache medication that contains caffeine. However, this treatment does not relieve his symptoms. Genetic testing reveals a polymorphic variation in his cytochrome P450 gene *CYP1A2*, which creates an ultrafast metabolizer phenotype for caffeine. This variant most likely results in a loss of which type of chemical modification?
   A) Hydroxylation
   B.Methylation
   C.Acetylation
   D.Glucuronidation
   B) Phosphorylation

Answers

1. **The correct answer is A.** The frequency of sickle cell disease is given as $q^2 = 0.1\%$ or 0.001
   We can solve for q= 0.03
   From the allele frequency equation p+q=1, we can solve for p= 1-0.03, thus, p=0.97
   From the Hardy-Weinberg equation 2pq is the carrier frequency;
   2pq= 2(0.03)(0.97)= 0.06 or 6%

2. **The correct answer is E.** Since familial hypercholesterolemia is an autosomal dominant disease, most affected individuals will be heterozygotes, which is also the disease frequency 2pq. We can determine 2q by making an assumption that p will be close to 1. Thus, 2q=1/250
   We can solve q= 0.002 or 0.2%
   From the Hardy-Weinberg equation, the homozygous affected individuals are
   q^2= (0.002)(0.002)= 0.000004 or 1 in 250,000

3. **The correct answer is B.** Since the G6PD deficiency is a X-linked recessive disease, the frequency of disease in males is the same as the allele frequency (q). Therefore, q=10% or 0.1
   Affected females must be homozygous. Therefore, the frequency of disease in females is q^2= (0.1)(0.1) = 0.01 or 1%

4. **The correct answer is E.** From the Hardy-Weinberg equation, homozygous affected individuals is q^2, which is given in the question as 1/400 or 0.0025 or 0.25% Therefore, q= 0.05

However, we know that the female is not homozygous affected since these individuals die early in childhood from the effects of the disease.

From p+q=1 we calculate the p=0.95

The frequency of the heterozygous individuals 2pq= 2(0.95)(0.05) = 0.095 or 9.5%

If two heterozygous carriers marry, the probability that they will produce an affected child is 25%, which is found from the Punnett square.

We need multiply 2 probabilities: 0.25 x 0.095 = 0.02375 or 2.4%. Answer choice E is closest to this calculated probability and is therefore the best answer choice.

5. **The correct answer is C.** Epigenetic alterations are heritable. However, these alterations are not due to changes in the DNA sequence and, therefore, do not disrupt the genetic code. Epigenetic regulation is thought to work through three types of processes, including DNA methylation, post-translational modification of histone proteins, and ATP-dependent chromatin remodeling.

Answer choices A and D are incorrect because these involve changes in the DNA sequence. Answer choice B is incorrect because epigenetic alterations would most likely affect the entire genome, not only the X-chromosome.

6. **The correct answer is D.** The patient most likely has Prader-Willi syndrome, which presents with neonatal hypotonia and delayed motor developments. Although infants exhibit feeding problems, after the age of 2 this switches to hyperphagia and obesity. Prader-Willi syndrome and Angelman syndrome are both linked to the imprinted region of chromosome 15. While Angelman syndrome is caused by deletion of a band in the maternal chromosome, Prader-Willi syndrome is caused by deletion of a band in the paternal chromosome. Thus, answer choice C is incorrect. Answer choice B is incorrect because microdeletion, not imprinting, on paternal chromosome 15 causes this disease. In unaffected individuals, the genes in this region on the paternal chromosome 15 would be expressed. This region, however, is imprinted on the maternal chromosome 15, thus it is silenced.

7. **The correct answer is C.** 6-mercaptopurine is a chemotherapeutic agent commonly used to treat leukemias. In the liver, this enzyme is metabolized by the enzyme thiopurine methyltransferase (TPMT), which transfers a methyl group to its substrate, aiding in its metabolism. In the population, there are polymorphic variations in the gene that encodes this enzyme. These variations result in enzymes with different levels of activity. If an individual has the variation that makes an enzyme of lower activity, then there is a risk for accumulation of toxic levels of the drug. Therefore, it is important to determine the activity of this patient's thiopurine methyltransferase before administering 6-mercaptopurine. Acetylation (answer choice A) is also a process that occurs in the Phase II metabolism of some drugs and involves the activity of N-acetyltransferase (NAT) (answer choice B). An abnormal number of repeats in the TATAA box of the promoter region of the *UGT1A1* gene (answer choice D) results in a lower production of the enzyme glucoronate transferase, another enzyme involved with some Phase II processes of drug metabolism.

8. **The correct answer is A.** The first phase of drug metabolism, known as Phase I, involves the activity of a family of enzymes called the cytochrome P450 family. These enzymes add a hydroxyl group (hydroxylation) to their substrates. This serves as a site for the addition of a side group later on in Phase II. Through this hydroxylation, the lipophilic compound is converted into a water soluble product. Answer choices B, C, and D are post-translational modifications associated with the Phase II processes of drug metabolism. Methylation is through activity of the enzyme thiopurine methyltransferase, acetylation through N-acetyltransferase activity, and glucuronidation via glucoronate transferase. Phosphorylation (answer choice E) is a common post-translational modification, but is not directly involved with Phase I or Phase II of drug metabolism.

VII

Review Questions

1. A researcher is studying the effects of a compound on the proliferation of neuronal cells. The results of the experiment indicate that the compound is a potential carcinogen, as it has mutagenic effects that promote cellular proliferation. Further sequencing of the DNA in these cells reveals point mutations in one allele of a particular gene, while the other allele is unaffected. Which of the following genesis most likely mutated by this compound?
   A) RAS
   B) RB
   C) BRCA1
   D) MLH1

2. A 31-year-old male presents to his primary physician with complaints of fatigue. He states that he has been experiencing constipation and has noticed blood in his stool several times. His family history reveals that his mother had endometrial cancer before the age of 50. Which of the following is the most likely diagnosis?
   A) Lynch syndrome
   B) von Hippel Lindau syndrome
   C) Familial adenomatous polyposis
   D) Hereditary breast and ovarian cancer syndrome

3. A 17-year-old female presents with unexplained weight loss and chronic fatigue. She also complains of frequent gas pains, bloating, and cramps. Colonoscopy reveals hundreds of adenomas in her colon. What molecular alteration has most likely occurred in these adenomas?
   A) Inactivation of BRCA1 that inhibits the repair of double strand DNA breaks
   B) Loss of function of APC, resulting in accumulation of ?-catenin
   C) Loss of heterozygosity in the gene that encodes Rb
   D) Inactivation of an E3 ligase, resulting in loss of HIF1-? ubiquitination

4. A 38-year-old male with a history of kidney stones presents with nausea, persistent headache, and bone pain. Blood tests indicate hypercalcemia. What is a likely characteristic of this man's tumor?
   A) It has undergone loss of function of both RB alleles
   B) The RET gene has been inactivated
   C) There is a loss of function of the menin protein
   D) APC is highly active

5. A 56-year-old male presents with muscle weakness and intense pain in his left arm. CT scan reveals a highly vascularized mass on the posterior side of the spinal cord. A defect in what type of protein is most likely responsible for the tumor's vascularization?
   A) Transcription factor
   B) Cdk inhibitor
   C) DNA mismatch repair protein
   D) E3 ligase

6. A 17-year-old female gives birth to a girl signs of holoprosencephaly. The infant's facial features include cleft palate and closely placed eyes. Genetic analysis indicates a mutation in the gene encoding Shh. Which of the following receptors is most likely affected by this mutation?
   A) FGF receptor
   B) Frizzled

C) PTCH
D) BMP receptor

7. An adopted 3-year-old male is brought to a pediatrician for his first well-child visit. Examination reveals dwarfism with short knob-like fingers and toes. Which paracrine signaling pathway is most likely affected?
   A) FGF
   B) Hedgehog
   C) WNT
   D) TGF-?

8. A developmental biologist is studying embryonic development in a newly derived mutant strain of mice. It is found that the formation of the dorsal anatomy is normal, but ventral structures are disrupted. Disruption of which of the following molecules is most likely responsible for the loss of ventral structures?
   A) Bmp-4
   B) B: Chordin
   C) C: Gli1
   D) D: Noggin

Answers

1. **The correct answer is A**. Mutations that promote cellular proliferation and affect only one allele of the gene are called oncogenes. Of the answer choices given, only RAS (answer choice A) is a oncogene. RAS is a component of the RAS-MAPK pathway and is activated in response to growth factor signaling. Activated RAS then activates the downstream component RAF, which in turn activates MEK. MEK induces enzymes and transcription factors in the cell that promote cellular proliferation. The other answer choices are all tumor suppressors. RB (answer choice B) is a tumor suppressor protein also known as retinoblastoma-associated protein, or pRB. The normal function of pRB in the cell is to prevent excessive cell growth by inhibiting the E2F transcription factor in order to arrest the cell cycle in G1. Since both alleles of RB must be interrupted to allow for unregulated cell cycle progression, RB is an example of a tumor suppressor. BRCA1 (answer choice C), also known as the breast cancer type 1 susceptibility protein, is expressed in the cells of breast and other tissues. This protein forms multiple distinct complexes responsible for DNA repair, cell cycle checkpoint control, and maintenance of genomic stability. BRCA1 is a tumor suppressor, as both alleles of BRCA1 must be disrupted in order to affect the cell. MLH1 (answer choice D) is a component of the mismatch repair machinery that is responsible for repairing errors that are introduced during DNA replication. As with the other tumor suppressor genes, both alleles of MLH1 must be disrupted to have an effect on the cell.

2. **The correct answer is A**. Lynch syndrome, also known as hereditary non-polyposis colorectal cancer is caused by an inherited mutation in one of the mismatch DNA repair enzymes (MLH1 or MSH2), which are classified as tumor suppressor genes. Individuals with Lynch syndrome have an increased risk of many types of cancer including colorectal, endometrial (as seen in this patient's family history), ovarian,

stomach, small intestine, liver, gallbladder duct, upper urinary tract, and brain. Von Hippel Lindau syndrome (answer choice B) is an inherited cancer syndrome characterized by highly vascularized tumors such as hemangioblastomas. However, clear cell renal cell carcinoma is the most common tumor diagnosed in individuals with VHL, with an estimated 70% of individuals affected by the age of 60 years. . von Hipel Lindau is due to an inheritance of a mutated allele of the VHL gene, which encodes for an E3 ubiquitin ligase that normally targets a protein called HIF for degradation. Although familial adenomatous polyposis (answer choice C) is also an inherited colorectal cancer, it is caused by mutations in APC and is not associated with a family history of endometrial cancer. Hereditary breast and ovarian cancer syndrome (answer choice D) is associated with breast cancer and ovarian cancer, but not colorectal cancer or endometrial cancer.

3. **The correct answer is B.** The early age of onset for this patient's disease and the location of the tumors indicate that she has an inherited type of colorectal cancer. The two major types of hereditary colorectal cancer are Lynch syndrome and familial adenomatous polyposis. However, only FAP is associated with the development of hundreds of adenomas in the colon at a young age. FAP is caused by the inheritance of a mutation in APC, which results in the loss of ?-catenin degradation, causing an increase in proliferation of the cells and ultimate formation of adenomas. Inactivation of BRCA1 (answer choice A) causes hereditary breast and ovarian cancer syndrome, whereas loss of heterozygosity of RB (answer choice C) results in retinoblastoma. Inactivation of VHL, an E3 ligase (answer choice D), results in von Hippel Lindau syndrome, which is characterized by highly vascular hemangioblastomas and clear cell renal cell carcinoma.

4. **The correct answer is C.** The hypercalcemia in this patient suggests a defect in calcium regulation by the endocrine system. Therefore, this individual most likely has multiple endocrine neoplasia (MEN), specifically MEN1. The most common endocrine tumors seen in MEN1 can be remembered with the "**3 Ps**": **parathyroid** tumors, **pituitary** tumors, and well-differentiated endocrine tumors of the gastro-entero-**pancreatic** tract. MEN1 is caused by mutations in menin and, since this is a tumor suppressor, loss of function of this protein is required in order to produce a tumor. Loss of function of RB (answer choice A) gives rise to retinoblastoma, a tumor of the eye. Inactivation of RET (answer choice B) would not be expected to result in cancer since this is an oncogene, which must undergo activating mutations to promote cellular proliferation. On the other hand, since APC (answer choice D) is a tumor

suppressor, its activity must be eliminated, not highly active, in order to produce a cancer.

5. **The correct answer is D.** Highly vascularized tumors are characteristic of von Hippel Lindau syndrome. This disease is caused by mutation in VHL, an E3 ligase that is normally responsible for the degradation of the angiogenesis-promoting HIF protein. Therefore, when VHL is mutated, HIF cannot be degraded properly, and new blood vessel formation is enhanced. Li-Fraumeni syndrome is caused by mutation in TP53, a transcription factor (answer choice A). This syndrome is associated with a wide variety of tumors, but is not specific for highly vascularized cancers. Mutations in Cdk inhibitors (answer choice B) are common in many types of cancers. DNA mismatch repair mutations (answer choice C), such as MLH1, are found in individuals with Lynch syndrome, a type of hereditary colorectal cancer.

6. **The correct answer is C.** As indicated in this case, holoprosencephaly is caused by mutations in the Shh gene. Shh is an extracellular signaling molecule that must bind to the PTCH receptor on the surface of its target cells in order to transmit developmental cues. The FGF receptors (answer choice A), Frizzled (answer choice B), and BMP receptors (answer choice D) are all important signaling components in embryonic development. However, the ligand for FGF receptors is FGF, for Frizzled is Wnt, and for BMP receptors is BMP.

7. **The correct answer is D.** The BMP family, which includes TGF-?, plays an essential role in embryonic development, especially in the development of bone. Pathogenic mutations in members of this family result in a number of skeletal abnormalities, including shortening of the digits (brachydactyly), as seen in this patient, as well as shortening of the long bones of the limbs. Mutations in FGF signaling (answer choice A) result in a condition known as achondroplasia, which is a form of short-limbed dwarfism. Hedgehog (answer choice B) mutations produce holoprosencephaly, whereas WNT (answer choice C) mutations cause a condition known as tetra-amelia, which is characterized by the absence of all four limbs.

8. **The correct answer is A.** Formation of the dorsal/ventral axis involves regulation of the genes noggin, chordin, and Bmp-4. Noggin (answer choice D) and chordin (answer choice B) coordinate the formation of the dorsal anatomy, and Bmp-4 (answer choice A) expression designates ventral structures. Gli1 (answer choice C) is activated inside the cell in response to Shh binding to its receptor PTCH. Activated Gli1 induces the expression of genes necessary for cellular proliferation.

Index